Unsteady Combustion

NATO ASI Series

Advanced Science Institutes Series

A Series presenting the results of activities sponsored by the NATO Science Committee, which aims at the dissemination of advanced scientific and technological knowledge, with a view to strengthening links between scientific communities.

The Series is published by an international board of publishers in conjunction with the NATO Scientific Affairs Division

A Life Sciences **B Physics**	Plenum Publishing Corporation London and New York
C Mathematical and Physical Sciences **D Behavioural and Social Sciences** **E Applied Sciences**	Kluwer Academic Publishers Dordrecht, Boston and London
F Computer and Systems Sciences **G Ecological Sciences** **H Cell Biology** **I Global Environmental Change**	Springer-Verlag Berlin, Heidelberg, New York, London, Paris and Tokyo

PARTNERSHIP SUB-SERIES

1. Disarmament Technologies	Kluwer Academic Publishers
2. Environment	Springer-Verlag / Kluwer Academic Publishers
3. High Technology	Kluwer Academic Publishers
4. Science and Technology Policy	Kluwer Academic Publishers
5. Computer Networking	Kluwer Academic Publishers

The Partnership Sub-Series incorporates activities undertaken in collaboration with NATO's Cooperation Partners, the countries of the CIS and Central and Eastern Europe, in Priority Areas of concern to those countries.

NATO-PCO-DATA BASE

The electronic index to the NATO ASI Series provides full bibliographical references (with keywords and/or abstracts) to more than 50000 contributions from international scientists published in all sections of the NATO ASI Series.
Access to the NATO-PCO-DATA BASE is possible in two ways:

– via online FILE 128 (NATO-PCO-DATA BASE) hosted by ESRIN, Via Galileo Galilei, I-00044 Frascati, Italy.

– via CD-ROM "NATO-PCO-DATA BASE" with user-friendly retrieval software in English, French and German (© WTV GmbH and DATAWARE Technologies Inc. 1989).

The CD-ROM can be ordered through any member of the Board of Publishers or through NATO-PCO, Overijse, Belgium.

Series E: Applied Sciences - Vol. 306

Unsteady Combustion

edited by

F. Culick

California Institute of Technology,
201 Karman Laboratory,
Pasadena, California, U.S.A.

M. V. Heitor

Department of Mechanical Engineering,
Instituto Superior Técnico,
Technical University of Lisbon,
Lisbon, Portugal

and

J. H. Whitelaw

Department of Mechanical Engineering,
Imperial College of Science, Medicine and Technology,
University of London,
London, U.K.

Kluwer Academic Publishers

Dordrecht / Boston / London

Published in cooperation with NATO Scientific Affairs Division

Proceedings of the NATO Advanced Study Institute on
Unsteady Combustion
Praia da Granja, Portugal
September 6–17, 1993

A C.I.P. Catalogue record for this book is available from the Library of Congress.

DOI 10.1007/978-94-009-1620-3

Published by Kluwer Academic Publishers,
P.O. Box 17, 3300 AA Dordrecht, The Netherlands.

Kluwer Academic Publishers incorporates the publishing programmes of
D. Reidel, Martinus Nijhoff, Dr W. Junk and MTP Press.

Printed on acid-free paper

TABLE OF CONTENTS

PREFACE

This book contains selected papers prepared for the NATO Advanced Study Institute on "Unsteady Combustion", which was held in Praia da Granja, Portugal, 6-17 September 1993. Approximately 100 delegates from 14 countries attended. The Institute was the most recent in a series beginning with "Instrumentation for Combustion and Flow in Engines", held in Vimeiro, Portugal 1987 and followed by "Combusting Flow Diagnostics" conducted in Montechoro, Portugal in 1990. Together, these three Institutes have covered a wide range of experimental and theoretical topics arising in the research and development of combustion systems with particular emphasis on gas-turbine combustors and internal combustion engines. The emphasis has evolved roughly from instrumentation and experimental techniques to the mixture of experiment, theory and computational work covered in the present volume.

As the title of this book implies, the chief aim of this Institute was to provide a broad sampling of problems arising with time-dependent behaviour in combustors. In fact, of course, that intention encompasses practically all possibilities, for "steady" combustion hardly exists if one looks sufficiently closely at the processes in a combustion chamber. The point really is that, apart from the excellent paper by Bahr (Chapter 10) discussing the technology of combustors for aircraft gas turbines, little attention is directed to matters of steady performance.

The volume is divided into three parts devoted to the subjects of combustion-induced oscillations; combustion in internal combustion engines; and experimental techniques and modelling. Although historically and intellectually there has long been considerable common ground occupied by the first and third subjects, it seems that too little interaction has occurred between the researchers concerned with internal combustion engines and those working in the other two areas. That apparently arbitrary separation may be largely a consequence of the distinctions in practice between aerospace and non-aerospace applications. Whatever the reason, the 1993 Institute and its proceedings published in this volume, offer an unusual opportunity to observe the differences and to evaluate understanding of the common ground.

Part 1 of this book comprises three major subjects: combustion instabilities (papers 1 to 5); pulsed combustors (paper 6); and active control of unsteady motions in combustors (papers 7 and 8). The context for unsteady motions in propulsion systems and in gas turbine combustors is admirably set by the material covered in papers 9 and 10, respectively. Primarily because of the high power density in a volume possessing low losses, excitation of oscillations, or combustion instabilities, is a likely event in combustion chambers. Hence the subject has long been object of research and of concern in practical systems. The papers included in this part of the book constitute a good survey of the range of problems central to combustion instabilities.

Although the idea of active control of combustion instabilities was invented nearly forty years ago, it has been possible only within the past decade to begin to realize the possibilities. The potential practical applications are considerable and papers 7 and 8 provide hints of what may be possible.

Internal combustion engines will undoubtedly be the prime movers in ground transportation for many years, if not decades, to come. The major problems in practice are fuel efficiency and emission of pollutants. In some respect or other, all the papers in Part II address those two problems. The coverage of this subject in this book is particularly timely in view of the growing emphasis on regulations imposed by many states around the world. In this context, the severe conditions in internal combustion engines set by the wide extremes of conditions during a cycle of operation and the rapid time variations, have forced development of special experimental techniques. Many of the foremost contemporary methods are discussed in papers 11-17. The reports in this book provide a broad view of current work and give a good idea of what must be accomplished in order to meet increasingly strict controls of emissions.

The third part of this book deals with quite general laser diagnostics (paper 18) and modelling aspects (papers 19-22). Contemporary modelling means also computer simulations, particularly in this field to predict the production of pollutant products of combustion. The subject is still young with an important and long future. The treatments in this book represent some of the leading current works.

All associated with this, enthusiastically thank the authors for their contributions. We are especially indebted to Graça Pereira, Carlos Carvalho and Anabela Almeida who handled efficiently all the administrative details of the Institute, including the difficult task of ensuring that the papers in this book were completed in timely fashion.

We wish to acknowledge the financial support of NATO and the other organizations who have made possible both a highly productive Institute and publication of this book.

Fred Culick
Manuel Heitor
Jim Whitelaw

Part I Combustion-induced Oscillations: Principles, Practice and Control

1. UNSTEADY FLAMES AND THE RAYLEIGH CRITERION

E.C. FERNANDES and M.V. HEITOR
Instituto Superior Técnico
Mechanical Engineering Department
Av. Rovisco Pais
1096 Lisboa Codex
Portugal

ABSTRACT. The basic principles associated with unsteady flames coupled by pressure waves are reviewed in this paper. The various modes of oscillations and sources of instability which may arise in a combustion system are briefly described, together with the Rayleigh criterion for heat-driven oscillations. The applicability of this criterion to explain situations of practical interest, such as a Rijke pipe and a pulse combustor, is discussed and its experimental verification is analysed. This is achieved on the basis of simultaneous measurements of the fluctuations of sound pressure level and of signatures of the rate of change of heat release, which confirm that the stability of the pulsating operation of practical combustors is obtained by releasing and removing energy during each cycle of operation.

1. Introduction

Turbulent reacting flows are normally associated with "stable" processes, in the sense that the fluid dynamic and thermodynamic properties of the flames are characterised by random fluctuations with an amplitude which is small when compared with their average values (e.g. Libby and Williams, 1994). In this case, there exists no periodic correlation between the fluctuations at a given point of the combustion chamber and the fluctuations at another point, exception made to the turbulent structure.

On the other hand, when correlations in time or in space exists between fluctuations, the oscillatory motions become organised with high amplitudes, leading to unsteady flames, as recently reviewed by Candel (1992) and Zinn (1992). In general, the unsteadiness results from coupling between combustion and gas dynamic processes, with an oscillatory energy supplied to sustain the oscillations higher than a critical value, which is related with viscous dissipation, heat transfer and acoustic radiation, Putnam (1971).

A continued, but renewed, attention has recently been given to unsteady flames because of their attractive features recognised in the search of raising the efficiency and reducing the pollutant emissions from industrial combustion devices. Examples include the various domestic and industrial burners reviewed by Putnam et al. (1986) and Mansour et al. (1991), the drying applications of Ozer (1993), the cement calciners of Rabhan et al. (1991) or the incineration applications of Stewart et al. (1991). In addition, low-quality liquid and solid fuels and waste materials have been burned under pulsating conditions as shown, for example, by Torres et al. (1992).

These developments have been reported in parallel with continued research efforts in improving the understanding of unsteady and pulsating flames and, for example, the experiments of Heitor et al. (1984), Keller and Hongo (1990), Gutmark et al. (1991), Matsumoto et al. (1992), Sivasegaram and Whitelaw (1991, 1993), Keller et al. (1994) and Barr and Keller (1994) provide insight into the phenomenon. It is clear that much has been learnt from previous work applied to rocket motors, jet engine afterburners and ramjets (see, for example, Crocco, 1965; Harije and Readon, 1972; Culick, 1988), but

F. Culick et al., (eds.), Unsteady Combustion, 1–16.

whether the unsteadiness is undesirable (as in the aeronautic and aerospace applications) or desirable, the phenomenon is similar and derive from the acoustic interactions between the flow field and the combustion process itself.

Recent theoretical and numerical methods have also been derived during the last years (e.g., Bloxsidge et al., 1988; Candel, 1992; Margolis, 1993) but, in general, the Rayleigh criterion stating that oscillations are sustained if heat release takes place in phase with the pressure fluctuations, has become standard in the analysis of unsteady flames.

In this context, this paper is aimed to briefly review how combustion instabilities couple with pressure waves on the basis of the Rayleigh criterion. The theoretical background is given, together with examples of practical interest. The next section reviews the modes of oscillations which may arise in a combustion system and section 3 describes the related sources of instabilities. Section 4 presents the theory associated with the Rayleigh criterion and two typical cases of its applicability. An experimental verification of this criterion is described in section 5, based on a set of experiments reported elsewhere. The main conclusions are presented in the final section.

2. Modes of Oscillations

Since the work of Barrère and Williams (1969) that a basic classification of instabilities was proposed. In general, the presence of combustion instabilities is normally verified through the typical sinusoidal characteristics of the flow properties and their frequency of oscillation has been used as a means to classify the instabilities (e.g. Keller et al., 1982). Low, intermediate and high frequency instabilities have been identified and this is important in the sense that it determines, for example, the characteristics of the acquisition and signal processing systems for a complete and detailed experimental analysis. These three modes of instabilities can be characterised as follows:

- The *low frequency* type of oscillations, also called "chugging", refers to a coupling of the combustion chamber with the feeding system and the frequency encountered is then several hundred of Hz. It occurs during transient phases of either starting or shutdown of practical engines, with life time of only tenths of seconds and without any wave motion in the combustion chamber. Their presence has been correlated with the relative difference between the pressure drop in the fuel injectors and the combustion chamber mean pressure.
- The *intermediate-frequency*, also called "buzz" or "system instability", refers to the type of "chugging" instabilities where the entire combustion system is involved, which includes the supply line and the exhaust system. There is a wave motion inside the chamber, which is essentially longitudinal and may be described in terms of plane modes. The phase and the frequency of this wave does not correspond to any acoustic mode.
- The *higher frequency* instability, referred as "screeching" or "screaming", refers to a acoustic instability of resonant combustion, with the frequency and phase of the wave oscillations corresponding to those for the acoustic resonance of the chamber. Any type of waves may be present as the longitudinal, radial, tangential, standing or spinning modes.

The related implications on the combustion performance of any system depends upon the type of instability, as well as on the specific geometry under analysis, and this increases the complexity of the phenomenon. In general, analysis has shown that the high frequency instabilities tend to increase combustion efficiency and the heat transfer coefficient, but may reduce the pollutant emissions, while the low frequency oscillations tend to decrease the performance of practical combustion systems. For example, Perry and Culick (1974) verified the monotonically dependence of the heat transfer coefficient on pressure amplitude and frequency. Keller et al. (1992), through experimental analysis, concluded about the importance of strong reverse flow on the augmentation of convective heat transfer in oscillating turbulent flows and Dec et al. (1992) presented a correlation for the Nusselt number for this case. Nevertheless, besides the potential benefits associated with the intrinsic nature of an oscillatory flow field, combustion instabilities normally result in extensive damage to the combustors structure through the destructive vibration caused by the pressure fluctuations and the high heat transfer coefficient to the combustor walls. It is clear that the amount of damage is dependent on the amplitude and time duration of the oscillation and, again, this has led to much of the recent work on the control of combustion instabilities (e.g., Gutmark et al. 1989, Hendricks et al., 1992; Schadow and Gutmark, 1992; Candel, 1992; McManus et al., 1993).

3. Sources of Instabilities

Combustion instabilities may either develop spontaneously within the reacting gas flow, or be initiated by any natural or artificial disturbance external to the flow field. In the first case, the pressure fluctuations inherent to the turbulent flow field can trigger shear layer instabilities, forcing a closed loop to occur, which gives rise to the oscillating process. In the second case, the pressure signal is generated externally, but leads to a similar effect. The question of whether the pressure disturbance will lead to a "stationary" instability, depends upon the excitation mechanism and the nature of the coupling and damping processes between the variables, which will reduce, amplify or maintain them.

Since the chemical reaction is the main source of energy driving the pressure field, the related response of the heat release rate is of essential importance, being largely controlled by the local availability of the air-to-fuel mixture ratio, together with the status of instantaneous pressure and temperature variables. Specific coupling mechanisms which have been described in the literature include:

i) Changes in the turbulent mixing rate (Heitor et al., 1984);
ii) Flame area variation (Willis et al., 1991);
iii) Periodic supply of the upstream flow, induced by the pressure field (Reuter et al., 1986; Zinn, 1986) and
iv) Vortex shedding due to hydrodynamic instabilities (Toong et al., 1965; Schadow et al., 1989).

Nevertheless, recent work on acoustic driven-oscillation (Keller et al., 1989) has shown that the coupling mechanisms depend on the characteristic time scales, which include those of the acoustic field, turbulence and chemical reaction. For example, if the time scale associated with the acoustic field is much higher than that of the turbulent fluctuations, both fields are decoupled. In general, the coupling between the three main time scales mentioned above is not easily identified and Oran and Gardner (1985) and McIntosh (1991), among others, have shown that the chemical kinetics dependence on these time scales may be responsible for significant changes of the reacting flow field.

An additional difficulty in the interpretation of combustion-induced oscillations is the need to account for the dependence of the pulsations frequency upon the phase difference between the pressure and the heat release oscillations. This has unabled many past investigators to correlate measured pulse combustor frequencies with predicted acoustic modes of the systems under analysis, but has been clearly analysed by Zinn (1992), as described below.

4. The Rayleigh Criterion

4.1 THEORETICAL BACKGROUND

Whatever the mechanism is responsible for generating and/or maintaining the oscillations, the energy must be supplied to the oscillating flow field at a specific location in space and time, although, the thermoacoustic mechanisms may differ in the manner in which heat is produced and transferred to an acoustic medium. Regardless of the heat source, the so-called Rayleigh criterion must be satisfied to drive an heat-driven oscillation and to predict, at least qualitatively, whether the system is actually stable or not. Proposed by Lord Rayleigh (1878), this criterion expresses that:

" *If heat be periodically communicated to, and abstracted from, a mass of air vibration in a cylinder bounded by a piston, the effect produced will depend upon the phase of the vibrating at which the transfer of heat takes place. If heat be given to the air at the moment of greatest condensation, or be taken from it at the moment of greatest rarefaction, the vibration is encouraged. On the other hand, if heat be given at the moment of greatest rarefaction, or abstracted at the moment of greatest condensation, the vibration is encouraged* "

The mathematical representation of the Rayleigh criterion was first proposed by Putnam and Dennis (1953) as:

$$\int P'(t)\, Q'(t)\, dt > 0 \qquad [1]$$

where P' is the instantaneous pressure fluctuation and Q' is the instantaneous heat release rate. This can be physically explained based on the thermal expansion that arises from a suddenly heated gas layer that forces back the surrounding. If a pressure pulse arrives, or is present at that moment at the interface of the layer while the gas is expanding, the pulse pressure is reflected with greater amplitude. Furthermore, it is also clear that the real criterion of amplification of a disturbance is that the net mechanical work done by the gas, per cycle, must be at least greater than the acoustic loss through viscous dissipation (Chu, 1956).

In addition, although exact phase correspondence is not required between pressure and heat, the temporal energy release distribution within the combustor is a major factor of excitation. This nonsteady release of heat in the system, coupled with the acoustic waves, is often also convected hydrodinamically through the flow field. Therefore, the heat release is not only a function of time, but also of space, which creates a more difficult analysis due to the spatial dependence on the turbulence structure of the velocity flow field. To take into account this behaviour, a more appropriate form for the Rayleigh criterion is given as follows (Zinn, 1992):

$$\int_V \int_T P'(x,t)\, Q'(x,t)\, dt\, dv \geq \int_V \int_T \sum Li(x,t)\, dt\, dv, \qquad [2]$$

where the integral is explicitly extended to include the volume of the chamber and where $Li(x,t)$ represents the i-th damping process (i.e., viscous dissipation, heat release, acoustic radiation).

Among others, a complete analysis of the criterion was performed by Culick (1987), including linear and non-linear thermoacoustic oscillations for chambers of any shape. The work led to the understanding about the influence of Q' and dQ'/dt on creating an acoustic wave and concluded about the similarity of the Rayleigh criterion and the principle of linear stability in approximate analysis. Also, the analysis has shown the similarity between the effects of an oscillating heat source and the effect of an oscillating piston in generating waves.

Carvalho et al. (1989) used the criterion to derive an equation to predict the maximum amplification of the nth harmonic order of oscillations in a Rijke tube, as a function of flame location, of the phase between the pressure and the heat release, and of the ratio between the mean velocity and the amplitude of the acoustic velocity. Here, a simple analysis will be described to emphasise the dependence of the Rayleigh criterion on various important parameters that need to be considered in the evaluation of the driving process.

For example, in the presence of an acoustic standing-wave, either quarter- or half-wave, the natural resonant frequency of the system for a non-uniform temperature distribution along the tube, is given by (Culick et al., 1987):

$$F(x) = \frac{N(n)}{k \int_0^L \frac{dx}{\sqrt{\gamma R_o T(x)}}} \qquad [3]$$

where:

$T(x)$	-	temperature spatial function, K
L	-	length of the combustor, m
$N(n) = (2n-1)$	-	for a quarter-wave acoustic oscillation
$N(n) = n$	-	for a half-wave acoustic oscillation
n	-	harmonic order
$k=2$	-	for a half-wave acoustic oscillation
$k=4$	-	for a quarter- wave acoustic oscillation
γ	-	specific heat ratio
R_o	-	perfect gas constant, kJ/kgK

and the instantaneous pressure distribution for quarter- and half-wave acoustic oscillations is given respectively by

$$P'(x, t) = P_o \cos\left(\frac{n\pi x}{2L}\right) \sin(\omega t) \qquad [4]$$

and

$$P'(x, t) = P_o \sin\left(\frac{n\pi x}{L}\right) \sin(\omega t) \qquad [5]$$

where P_o and ω represent respectively the amplitude and the frequency of the pressure fluctuations.

The instantaneous acoustic velocity for these two cases is obtained, respectively, through the following linearized Euler equations:

$$U'(x, t) = - U_o \sin\left(\frac{n\pi x}{2L}\right) \cos(\omega t) \qquad [6]$$

$$U'(x, t) = + U_o \cos\left(\frac{n\pi x}{L}\right) \cos(\omega t) \qquad [7]$$

with U_o being the amplitude of the velocity fluctuations.

The heat release rate is a function of space and time and, in general, can be written as

$$Q'(x, t) = Q_o (1 + H(x) \sin (\omega t + \phi)) \quad [8]$$

where

Q_o	-	mean heat release
$H(x)$	-	location of heat release
ϕ	-	phase between heat release and pressure oscillations

It should be noted that the heat release can be located in a specific point in space, therefore H(x) should be proportional, for example, to a Delta or a Gaussian function. Alternatively, if one considers reactant pockets of gas convected hidrodynamically downstream, H(x) should depend on the mean and periodic velocity and, consequently, on space.

For the left side of equation [2] to be maximised, with P'(x, t) and Q'(x, t) given by equations [4] and [8] respectively, the phase between the heat release rate and the pressure ocillations must be less than 90°, with energy being released at the pressure antinode. If the phase is zero, the integral can be said to be partially maximised.

The phase affects the resonant frequency of the combustor, as noted before and described by Keller et al. (1989) and Zinn (1992). When $\phi=0$ the heat released by the combustion system does not affect the natural frequency of the system, and the peak of heat release coincide with that of the pressure wake, amplifying the pressure oscillation. When $0<\phi<90°$, the heat release rate occurs after the pressure peak, and the frequency emitted is lower than the natural frequency, while the pressure amplitude remains unchanged. For $-90<\phi<0$, the frequency emitted is higher than the natural frequency of the system and the pressure amplitude is not affected. For values of ϕ out of these ranges, the oscillations are damped.

It should be noted that the phase ϕ is essentially dependent on the type of fuel (solid, liquid or gaseous), on the way the fuel is injected, and on the burner type. For example, when the fuel is injected in the chamber with the admission controlled by pressure or velocity fluctuations, the reaction does not occur instantaneously and a *time-lag* arises between the injection and the burning instants.

This is important because the admission of fuel can be controlled by two main process, namely: i) the chamber acoustic pressure fluctuations, inducing a velocity fluctuation on the liquid or gas fuel tube supply, which is the mechanism typical of in pulse combustors (Zinn, 1992); or by ii) the chamber acoustic velocity fluctuations, which is the case of a Rijke tube with an electric heater or burning solid fuels (Torres et al., 1992).

If only gaseous combustion is considered, the mixing and reaction are conventionally characterised by the Damkholer number, which represents a ratio between the characteristic chemical and mixing times (Libby and Williams, 1994). Depending on the process involved, the mixing time can be desintegrated and, for example, Keller et al. (1989) identified three characteristic times for a Helmhotz combustor, as follows: i) the characteristic time required to mix the fuel with the oxidiser; ii) the time required to mix the reactants with hot products; and iii) the chemical characteristic time. The sum of all these times is the total ignition delay time for the release of energy to occur. Consideration of this time delay is crucial to the attainment of pulsations and the required developing capabilities for controlling it.

4.2 PRACTICAL APPLICABILITY: SAMPLE EXAMPLES

To help understand the Rayleigh criterion, two typical cases reported in literature are briefly analysed here, namely the Rijke tube and the pulse combustor (e.g. Zinn, 1992).

4.2.1. *The Rijke Tube.* In the case of a Rijke tube, figure 1, where the pressure field is characterised by a standing half-wave, Carvalho et al. (1989) obtained an equation which predicts the maximum pressure amplification as a function of flame location, of the ratio between the mean velocity and the amplitude of the acoustic velocity, and of the phase between the energy release and pressure.

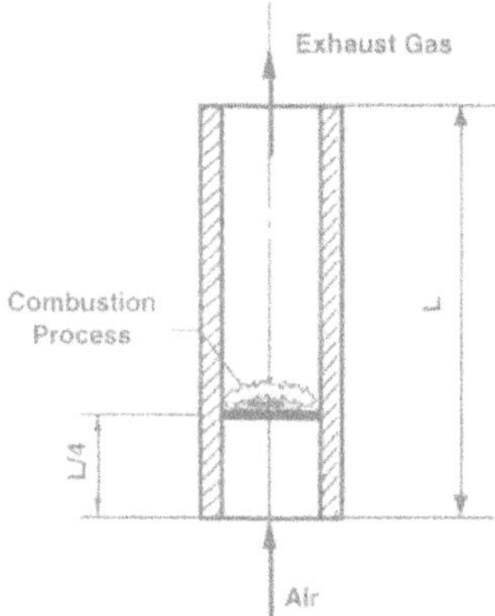

Figure 1. Schematic diagram of a Rijke tube

The heat release model for the planar heat bed was modelled as

$$Q'(x, t) = a + \text{const.}\left|\overline{U}(x, t) + U_o(x, t)\right| \qquad [9]$$

where 'a' is a constant, $\overline{U}$ is the mean flow velocity and U_0 is the amplitude of the acoustic velocity fluctuations. If 'a' is taken as zero for the case of a combustion process, the result of the integration expressed in [3] is given by:

$$\int P'(x, t)\, Q'(x, t)\, dt \equiv \text{const.}\sin\left(2n\,\pi\frac{x}{L}\right)\left(\sin 2\theta + 2\theta\right)\cos\phi \qquad [10]$$

where

$$\theta = \frac{\overline{U}}{U_o}\frac{1}{\cos n\,\pi\frac{x}{L}} \qquad [11]$$

It should be pointed out that this equation and the following analysis is only valid for first-order acoustics, when U_0 is much lower than $\overline{U}$ and when the gas temperature difference across the burner is small. For the purposes of the present analysis, it is also important to understand the three terms of equation [10], as follows:

i) $\sin\left(2n\,\pi\frac{x}{L}\right)$: this term gives the location of the heat release, where the maximum for any acoustic mode, is expressed by

$$\frac{x}{L} = \frac{1+4k}{4n} \quad \text{with } k = 0, 1, 2, 3\ldots \text{ and } n = 1, 2, 3\ldots \qquad [12]$$

For example, the burner position for the amplification of the fundamental (n=1, k=0) and second (n=2, k=0,1) harmonics is presented in figure 2.

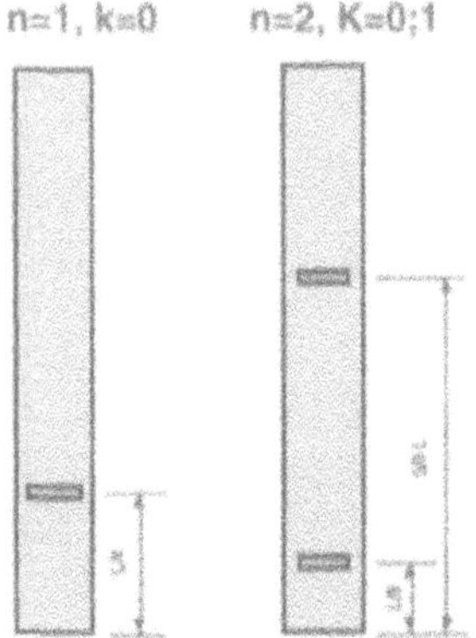

Figure 2. Schematic presentation of the burner position for the amplification of fundamental and second harmonics.

ii) $\left(\sin 2\theta + 2\theta\right)$: this term quantifies the influence of $\overline{U}/U_0$ on the location of the heat release. The location of the heat release to maximise this term is given by

$$\frac{X}{L} = \frac{\cos^{-1}\left(\frac{\overline{U}}{U_0}\right)}{n\,\pi} \qquad [13]$$

and is presented in figure 3 as a function of $\overline{U}/U_0$.

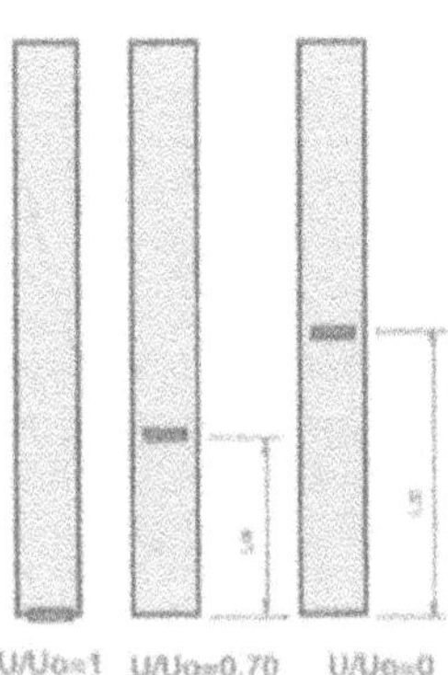

Figure 3. Schematic presentation of the burner position for the amplification of fundamental mode of oscillation as a function $\overline{U}/U_0$

Since the first two terms of the right hand side of equation [10] are functions of the longitudinal coordinate x, their product is given by

$$\sin\left(2\,n\,\pi\,\frac{x}{L}\right)\left[2\,\sin^{-1}\left(\frac{\overline{U}}{U_o}\frac{1}{\cos n\,\pi\,\frac{x}{L}}\right)+2\left(\frac{\overline{U}}{U_o}\frac{1}{\cos n\,\pi\,\frac{x}{L}}\right)\sqrt{1-\left(\frac{\overline{U}}{U_o}\frac{1}{\cos n\,\pi\,\frac{x}{L}}\right)}\right] \qquad [14]$$

For the fundamental mode, n=1, this equation shows that the optimum location for the heat release to occur moves towards L/2, when $\overline{U}/U_o$ decreases, and goes to a limit of L/4 when $\overline{U}/U_o$ tends to infinity.

iii) cos(ϕ): the angle ϕ represents the phase between the P'(x, t) and Q'(x, t). Carvalho et al. (1989) did not report any value for this variable, and the dispersion of values presented in the literature reveals that it should depend strongly on the type of the solid fuel and burner (e.g., particle size, bed deposition). While Mandarame (1981) concluded that there must exist a phase lag between heat release and pressure, the analytical resuts of Carrier (1955) applied to the above formulation leads ϕ to become of the order of $\pi/8$.

4.2.2. Pulse combustor. Keller and Westbrook (1986) study the effect of changes of fuel composition on the response of a pulse combustor of the Helmotz type, figure 4. The task was accomplished using a form of the Rayleigh criterion based on the application of the first law of thermodynamics to an open system consisting of a mass layer inside the combustor.

Assuming the gas to be perfect, the work done by the gas as it expands is given by:

$$dw = \frac{R_o}{\widehat{M}\,\overline{Cp}}\left[\overline{Q}\,(t)\,\overline{P}\,(t)\,dt + Q'\,P'\,dt\right] \qquad [15]$$

where $\overline{Cp}$ is the average value of the gas specific heat.

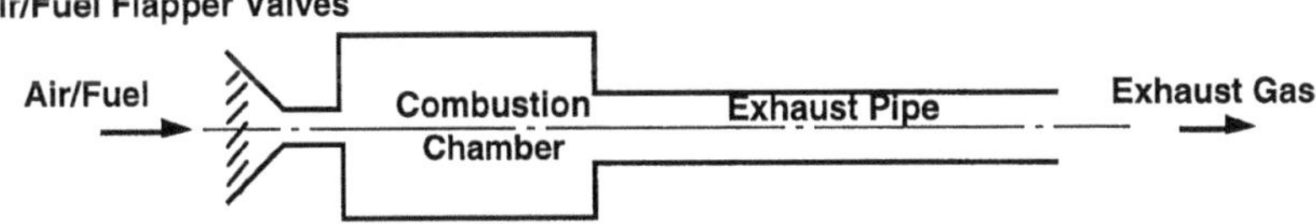

Figure 4. Schematc diagram of an Helmotz pulse combustor (see Keller and Westbrook, 1986 for details).

The first term on the right hand side of equation [15] represents the mean work to exhaust the gases out of the pipe, while the second term represents the amount of work necessary to sustain the pressure wave.

Assuming that Q'(t) is linearly proportional to the signal of an excited radical existing in the flame (see, for example, Price et al., 1969), the amount of energy that goes into the pressure wave, or the efficiency of the combustor in creating the oscillation, is obtained by dividing equation [15] by the integral of Q'(t) over one cycle. The derivative of this expression is given by:

$$\frac{d\eta}{dt} = \frac{\dfrac{R}{\widehat{M}\,\overline{Cp}} \langle OH^* \rangle' \; P'}{\displaystyle\int_0^t \langle OH^* \rangle \, dt} \qquad [16]$$

where <OH*> represents a signature of the rate of change of heat release, as measured by the time-averaged emissions of OH radicals.

This equation has been used as an alternative formula of the Rayleigh criterion, since it gives the instantaneous Rayleigh criterion along a cycle of oscillation. The analysis performed with this expression for two types of fuels, namely methane and methane-ethane, allowed Keller and Westbrook (1986) to conclude about the better performance of pure methane in creating the resonant wave.

5. An Experimental Verification of the Rayleigh Criterion

To verify experimentally the Rayleigh criterion and to obtain conditionally phase-averaged signals in an unsteady flame, a non-intrusive technique was used based on the simultaneous detection of the light emitted by the flame and the fluctuations of sound pressure level (see Fernandes, 1991, for details).

Figure 5 shows a schematic diagram of the experimental setup and data acquisition system used, which is comparable to those reported by other investigators to correlate pressure and heat release fluctuations (e.g., Price et al., 1969; Ramachandra and Strahle, 1983; Katsuki et al., 1986; Sankar et al., 1990; and Kotake and Takamoto, 1990). The instrumentation includes pressure transducers, microphones or piezoelectric transducers, and a monochromator in order to obtain a spectral analysis of the flame light.

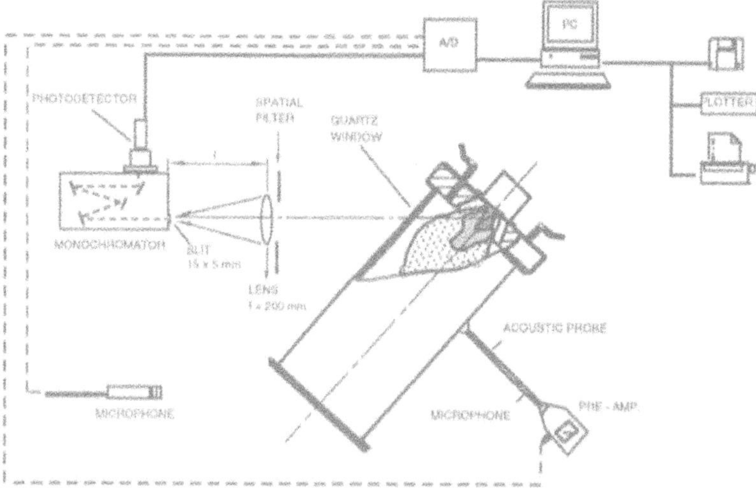

Figure 5. Schematic diagram of the experimental arrangement and of the instrumentation used for simultaneous analysis of the sound pressure levels and cheluminescence signals emited by the combustor.

The microphone used to quantify the sound pressure levels is typically a free-field condenser microphone with a flat response over a frequency band from 20Hz to 20KHz.

The microphone is either positioned away from the combustor, to characterise the free-field, or inserted in the holes provided along the combustor wall through a purpose-built probe in order to quantify the instantaneous static pressure distribution. The probe used was carefully design to avoid resonance frequencies close to the values measured, although the time delay of the wave propagation inside the tube was taken into account. Alternatively, a piezoelectric sensor could have been used as a reference signal transducer, which could have been located on the surface of the combustor due to its high mechanically robustness.

The light emitted from the flame was used as a signature of the rate of change of heat release (e.g. Willis et al., 1991; Keller and Saito, 1987) in terms of the chemiluminescence emission due to the radiative decay of electronically excited radicals existing in the reaction zone, such as <OH^*> , <CH^*> or <$C_2{}^*$> (Gaydon and Wolfhard, 1979). The light from the flame is guided to the entrance slit of a monochromator, where is dispersed in a rainbow by a 1200 lines/mm grating (WDG 30) covering the wavelength range from 380-760nm, and detected by a photomultiplier (EMI-9658A) operating at room temperature. The optical collection system comprised a 200mm focal distance lens placed in front of the monochromator slit with a field of view of 15x5mm. The signal output of the photomultiplier was digitally sampled at a rate of at least 10kHz. The spatial resolution of the system and wavelength readout were evaluated and calibrated using the blue and green beams of an Argon-Ion laser and uncertainties in the measurements are estimated to be less than 10%, with a resolution of 7nm at half peak transmission. Figure 6 shows a wavelength scanning for a typical non-luminous flame and identifies the <CH^*> (i.e. at 431.5nm) and the <$C_2{}^*$> (469nm, 516.5nm) bands which occurs in the visible range. The results quantify the resolution of the present analysis, and have made possible the selection band of 516.5nm for the phase average measurements of light and pressure presented below.

The technique is limited to non-luminous flames because the radical emission can be contaminated with the continuous radiation from soot, as well as background radiation, as discussed and analysed by Beyler and Gouldin (1981).

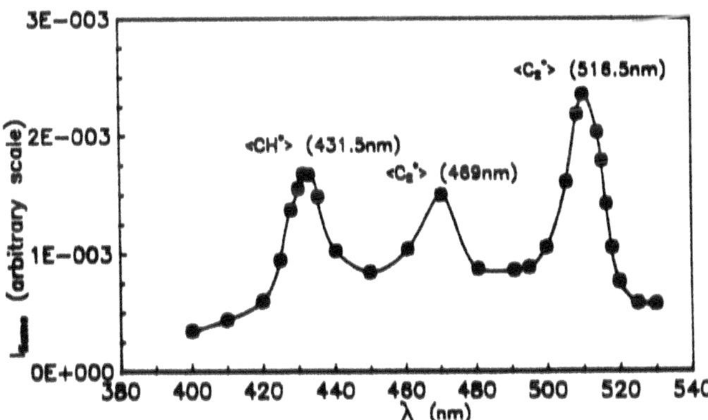

Figure 6. Wavelength scanning of the light emitted by a typical non-luminous propane-air flame

The signal output of the microphone together with that of the photodetector was digitally sampled at a Nyquist sample rate and post-processed. The statistical analysis was post-processed and included joint-statistic moments, spectral analysis determined by digital Fast Fourier Transforms and cross-correlations in order to quantify the time delay between signals.

Figure 6 shows the ensemble average over hundreds of cycles of P '(t) and Q'(t), for a complete cycle of unsteady operation of the combustor of figure 5, together with the result of the expansion work calculated from equation [15]. While the pressure signal behaves as a sine wave, the time series of <C_2*> reflects an evolution close to a squared-triangle wave type, with the beginning of the rapid combustion occurring earlier than the maximum pressure. The phase between the two variables, as quantified by a temporal cross-correlation techniques, is around 52°, which is a value similar to those reported by Keller and Saito (1987) in a valved pulse combustor and Langhorne (1988) in a disc-stabilised oscillating flame. The results verify the Rayleigh criterion mentioned before and confirm that the stability of the pulsating operation of the combustor is obtained by releasing and removing energy during each cycle of operation.

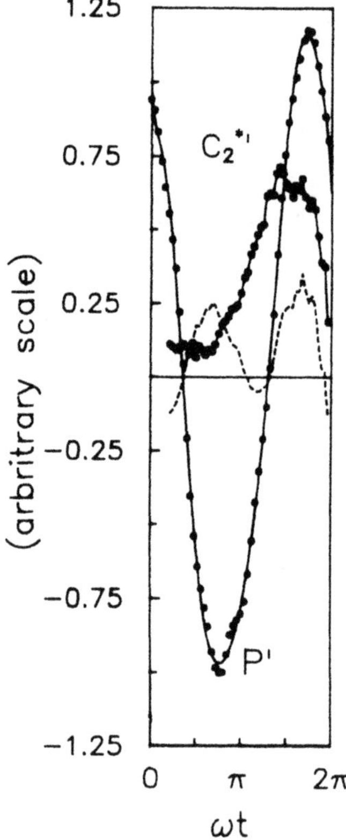

Figure 7. Sample results of the simultaneous measurements of pressure fluctuations, P'(t), and heat release, measured as <C_2>* signals, over a cycle of pulsating operation of the combustor.
NOTE: the dotted line represents the work of gas expansion, as calculated from eq. [15].

It is clear that the type of results plotted in figure 7 do not convey any information on the driving mechanisms responsible for the on-set of the oscillations, but they are important to improve knowledge of the characteristic times and phase-lag associated with the pulsating process.

6. Conclusions

Pulsating combustion describes a combustion process that occurs under oscillatory conditions, with the physical variables varying periodically with time. Under certain conditions, when the frequency of the pulsation is close to the frequency of the fundamental acoustic mode of the combustor, this technology appears to offer enormous possibilities and recent studies have shown that the unsteady gas motion associated with acoustic waves can enhance transport processes of mass and increases the rate of thermal energy transfer, as well as the mean residence time of the flow.

The ability to predict and design practical systems taking a full advantage of the enhanced processes, has been a challenging compromise because of the complex coupling of the various controlling parameters. In addition, due to lack of a well-founded theory for the pulsating phenomenon, which is reflected in the absence of a simple and direct explanation for the heat and mass transfer enhancement within the classical theories, the analysis based on the Rayleigh criterion supports the explanation of the unsteady behaviour of typical systems. This has been confirmed experimentally based on the simultaneous acquisition of sound pressure-level fluctuations and signatures of light emission from a research laboratory combustor.

7. Acknowledgements

The authors are pleased to thank many useful discussions with colleagues in the Department of Mechanical Engineering. The assistance of Messrs Carlos Carvalho and Jorge Coelho in the preparation of the final form of the manuscript is appreciated.

8. References

Barr, P.K. and J.O. Keller (1994) "Premixed combustion in an oscillating/resonant flow field. Part III: the importance of flame extinction by fluid dynamic strain", Proceedings of the International Symposium on Pulsating Combustion, August.

Barrère, M. and Williams, F.A. (1969) "Comparison of combustion instabilities found in various types of combustion chambers". Twelfth Symp. (Intl.) on Combustion, The Combustion Institute, pp. 169-181.

Beyler, C.L. and Gouldin, F.C. (1981) "Flame structure in a swirl stabilised combustor inferred by radiant emission measurements". Eighth Symp. (Intl.) on Combustion, The Combustion Institute, pp. 1011-1019.

Bloxsidge, G.J., Dowling, A.P. and Langhorne, P.J. (1988) "Reheat buzz: an acoustically coupled combustion instability. Part 2: theory". J. Fluid Mech., 193, pp. 445-473.

Candel, S.M. (1992) "Combustion instabilities coupled by pressure waves an their active control". Twenty Fourth Symp. (Intl.) on Combustion, The Combustion Institute, pp. 1277-1296.

Carrier, G.F. (1955) "The mechanics of a Rijke tube". Quart. Appl. Math., 12, pp. 383-395.

Carvalho, J.A. Jr., Ferreira, M.A., Bressan, C. and Ferreira, J.L.G. (1989). "Definition of heater location to drive maximum amplitude acoustic oscillations in a Rijke tube". Combust. and Flame, 76, pp. 17-27.

Chu, B.T. (1956) "Stability of systems containing a heat source - The Rayleigh Criterion". NACA RM 56D27.

Crocco, L. (1965) "Theoretical studies on liquid-propellant rocket instability". Tenth Symp. (Intl) on Combustion, The Combustion Institute, pp. 1101-1128.

Culick, F.E.C. (1987) "A note on Rayleigh's criterion". Combust. Sci and Tech., 56, pp. 159-166.

Culick, F.E.C. (1988) "Combustion instabilities in liquid-fueled propulsion systems: an overview". AGARD Conference on Combustion Instabilities in Liquid-fuelled Propulsion Systems.

Dec, J.E., Keller, J.O. and Arpaci, V.S. (1992) "Heat transfer enhancements in the oscillating turbulent flow of a pulse combustor tail pipe". Int. J. Heat Mass Transfer, 35, (9), pp. 2311-2325.

Fernandes, E.C. (1991) "Stability and structure of turbulent flames confined in axisymmetric geometries". Ms.C. Thesis, Mech. Engng. Department, Instituto Superior Tecnico, Technical University of Lisbon. (In Portuguese)

Gaydon, A.G. and Wolfhard, H.G. (1979) Flames - their structure radiation and temperature, Ed. Chapman and Hall, Fourth Edition, London.

Gutmark, E., Parr, T.P., Hanson-Parr, D.M. and Schadow, K.C. (1989) "On the role of large and small-scale structures in combustion control". Combust. Sci. and Tech., 66, pp. 107-126.

Gutmark, E., Schadow, K.C., Sivasegaram, S. and Whitelaw, J.H. (1991) "Interaction between fluid-dynamics and acoustic instabilities in combusting flows". Combust. Sci. and Tech., 79, pp. 161-170.

Harije, D.J. and Reardon, F.H. (1972) "Liquid propellant rocket instability". NASA SP-194.

Heitor, M.V., Taylor, A.M.P.K. and Whitelaw, J.H. (1984) "Influence of confinement on combustion instabilities of premixed flames stabilised on axisymmetric baffles". Combust. and Flame, 57, pp. 109-121.

Hendricks, E., Sivasegaram, S. and Whitelaw, J.H. (1992) "Control of oscillation in ducted premixed flames". Proceedings of IUTAM Symposium on the Aerothermodynamics of Combustors. Edited by R.S. Lee, J.H. Whitelaw and T.S. Wung, Springer Verlag.

Katsuky, M., Mizutani, Y., Chikami, M. and Kittaka, T. (1986) "Sound emission from a turbulent flame". Twenty-first. Symp. (Intl.) on Combustion, The Combustion Institute, pp. 1543-1550.

Keller, J.O. and Hongo, I. (1990) "Pulse combustion: the mechanism of NO_X production". Combust. and Flame, 80, pp. 219-237

Keller, J.O. Vaneveld J., Korschelt D., Ghonem A.F, J., Daily J. W and Oppenheim, A.K. (1982) "Mechanism of instabilitites in turbulent combustion leading to flashback". AIAA Journal, 20, pp. 254-262.

Keller, J.O., Bramlette, T., Dec, J.E. and Westbrook, C.K. (1989) "Pulse combustion: the importance of characteristic times". Combust. and Flame, 75, pp. 33-44.

Keller, J.O., P.K. Barr, R.S. Gemmen (1994) "Premixed combustion in a periodic flow field. Part I: experimental investigations". Combust. and Flame, in press.

Keller, J.O., Saito, K. (1987) "Measurements of the combusting flow in a pulse combustor". Combust. Sci. and Tech., 53, pp. 137-163.

Keller, J.O., Smith, L, Dibble, R.W. (1992) "Phase resolved PLIF measurements of CH in a strongly oscillating flow field". Poster presentation at the Twenty-Fourth Symp. (Intl.) on Combustion, Sydney, Australia.

Keller, J.O. and Westbrook, C.K. (1986) "Response of a pulse combustor to changes in fuel composition". Twenty-first Symp. (Intl.) on Combustion, The Combustion Institute, pp. 547-555.

Kotake, S. and Takamoto, K. (1990) "Combustion noise: effects of the velocity turbulence of unburned mixture". Journal of Sound and Vibration, 139, (1). pp. 9-20.

Langhorne, P.J. (1988) "Reheat buzz: an acoustically coupled combustion instability. Part 1 - Experiment". J. Fluid Mech, 193, pp. 417-443.

Libby, P.A. and Williams, F.A. (1994) Turbulent Reacting Flows. Academic Press, New York.

Mandarame, H. (1981) "Thermally induced acoustic oscillations in a pipe". Bull. JSME, 24, pp. 1626- 1633.

Mansour, M.N., Durai-Swamy, K., Chandran, R.R. and Duqum, J.N. (1991) "Pulse combustion systems for commercial, industrial and gas turbine applications". Proceedings of Symp. (Intl.) on Pulsating Combustion, Sponsored by Sandia National Laboratories and the Gas Research Institute, Monterey, California.

Margolis, S.B. (1993) "Nonlinear stability of combustion-driven acoustic oscillations in resonance tubes". J. Fluid Mech., 253, pp. 67-103.

Matsumoto, R., Nakajima, T., Kimoto, K., Noda, S. and Maeda, S. (1982) "An experimental study on low frequency oscillation and flame-generated turbulence in premixed/diffusion flames". Combust. Sci. and Tech., 27, pp. 103-111.

McIntosh, A.C. (1991) "Pressure disturbances of different length scales interacting with conventional flames". Combust. Sci. and Tech., 75, pp. 416-424.

McManus, K., Poinsot, T. and Candel, S. (1993) "A review of active control of combustion instability". Prog. Energy Combust., Sci, 19, pp. 1-29.

Oran, E.S., Gardner, J.H. (1985) "Chemical-acoustic interactions in combustion systems". Prog. Energy Combust., Sci, 11, pp. 253-276.

Ozer, R.W. (1993) "Pulse combustion drying", Proceedings of the Workshop on Pulsating Combustion and its Applications. Lund University, Lund, Sweden.

Perry, R.B., and Culick, F.E.C. (1974) "Measurements of wall heat transfer in the presence of large amplitude combustion-driven oscillations". Combust. Sci. and Tech., 9, pp. 49-53.

Price, R.B., Hurle, I.R. and Sulden, I.M. (1969) "Optical studies of the generation of noise in turbulent flames". Twelfth Symp. (Intl.) on Combustion, The Combustion Institute, pp. 1093-1102.

Putnam, A.A. (1971) Combustion driven oscillations in industry. Elsevier, New York.

Putnam, A.A. and Dennis, W.R. (1953) "Organ-pipe oscillations in a flame-filled tube". Fourth Symp. (Intl.) on Combustion, The Combustion Institute, pp. 566-574.

Putnam, A.A., Belles, F.E. and Kentfield, J.A.C. (1986) "Pulse combustion". Prog. Energy and Combust. Sci., 12, pp. 43-79.

Rabhan, A.B., Dubrov, E., Alvey, D.A., Daniel, B.R. and Zinn, B.T. (1991) "Industrial pulse combustor development and its applications in spray dryers and cement calciners". GRI-90/0203, Final Rept. May 1986-Sept. 1990, Gas Research Inst.

Ramachandra, M.K. and Strahle, W.C. (1983). "Acoustic signature from flames as a combustion diagnostic tool". AIAA Journal, 21, (8), pp. 1118-1125.

Rayleigh, L.J.W.S. (1878) "The explanation of certain acoustic phenomena". Doctoral Thesis, Ecole Centrale Paris, Chatenay-Malabry.

Reuter, D., Daniel, B.R., Jagoda, J. and Zinn, B.T. (1986) "Periodic mixing and combustion processes in gas fired pulsating combustors". Combust. and Flame, 65, pp. 281-290,

Sankar, S.V., Jagoda, J.I. and Zinn, B.T. (1990) "Oscillatory velocity response of premixed flat flames stabilized in axial acoustic fields". Combust. and Flame, 80, pp. 371-384.

Schadow, K.C. and Gutmark, F. (1992) "Combustion instabilities related to vortex shedding in dump combustors ad their passive control". Prog. Energy Combust. Sci., 18, pp. 117-132.

Schadow, K.C., Gutmark, E., Parr, T.P. and Wilson, K.J. (1989) "Large-scale coherent structures as drivers of combustion Instability". Combust. Sci. and Tech., 64, pp. 167-186.

Sivasegaram, S. and Whitelaw, J.H. (1991). "The influence of swirl on oscillations in ducted premixed flames". Combust. and Flame, 85, pp. 195-205.

Sivasegaram, S. and Whitelaw, J.H. (1993) "Active control of oscillations in combustors with several frequency modes". ASME, DSC-38, pp. 69-76.

Stewart, C.R., Lemineux, P.M. and Zinn, B.T. (1991) "Application of pulse combustion to solid and hazardous waste incineration". Proc. Symp. (Int.) on Pulse Combustion, Aug. 6-8, Monterey, California.

Toong, T., Salant, R.F., Stopford, J.M. and Anderson, G.Y. (1965) "Mechanisms of combustion instability". Twentieth Symp. (Intl.) on Combustion, The Combustion Institute, pp. 1301-1313.

Torres, E.A., Victório, J.R.S., Ferreira, M.A. and Carvalho, J.A. (1992) "Pulsating combustion of palm oil fruit bark". Fuel, 71, pp. 257-261.

Willis, J., Cadou, C., Karagozian, A. and Smith, O. (1991) "Diagnostic methods for visualization of heat release in unsteady combustion". Paper WSS/CI 91-68, presented at the 1991 Fall Meeting of the Western States Section/The Combustion Institute. October, 14-15, 1991, Los Angeles-CA, USA.

Zinn, B.T. (1986) "Pulsating combustion". In: Advanced combustion methods, ed.. Weinberg, F.J., Academic Press.

Zinn, B.T. (1992) "Pulse combustion: recent applications and research issues". Twenty-Fourth Symp. (Intl.) on Combustion, The Combustion Institute, pp. 1297-1305.

2. PREMIXED COMBUSTION IN A PERIODIC FLOW FIELD

J. O. Keller and P. K. Barr
Combustion Research Facility
Sandia National Laboratories
Livermore, CA 94551-0969 USA

ABSTRACT

The detailed mixing, combustion and ignition processes occurring in a flow field under the influence of resonant acoustic perturbations are discussed. These acoustic perturbations were created by a pulse combustor of the "Helmholtz" type.

During the first two-thirds of the resonant cycle the inlet jet forms a well-defined toroidal vortex that is phase-locked with the combustor cycle. This toroidal vortex is responsible for the convection and mixing of the reactants with the residual hot products. Although chemiluminescence is always present in quiescent regions of the combustor, none exists in the region of the reactants during injection. Moreover, the combustion of the fresh reactants begins late in the cycle along the outer edges of the rolled up toroidal vortex, and not in the quiescent regions of the flow. The combustion of the reactants moves into the center of this toroidal vortex where rapid uniform ignition and subsequent combustion of the fresh reactants occurs. Vortex dynamic modeling of this flow using a flamelet formulation for combustion shows that a principle ignition delay mechanism is that of fluid dynamic stretch. This highly strained flow field suppresses ignition during the injection part of the cycle providing a mechanism to thoroughly mix the fresh with residual products preparing them for a rapid, almost volumetric, combustion process initiated by thermal and/or radical ignition.

---

† Most of the work presented in this document is reprinted with permission from the articles by J.O. Keller, P.K. Barr, R.S. Gemmen, "Premixed Combustion in a Periodic Flow Field. Part I: Experimental Investigation" and P.K. Barr, J.O. Keller, "Premixed Combustion in a Periodic Flow Field. Part II: The Importance of Flame Extinction by Fluid Dynamic Strain" both are in press in *Combust. Flame*, 1994. The purpose for this manuscript is to provide a brief review of the relevant findings from our own work and not to report on new research results. The interested reader is referred to these publications for more detailed discussions of the findings reported on here.

‡ This work was performed at the Combustion Research Facility Sandia National Laboratories and supported by the U.S. Department of Energy, Office of Industrial Technologies, Advanced Industrial Concepts Division.

F. Culick et al., (eds.), Unsteady Combustion, 17–32.

INTRODUCTION

Acoustic interactions with combustion processes are the source of combustion instabilities in many systems (e.g. ramjets and rocket engines), and are the source of stable oscillations in devices such as pulse combustors. In systems such as ramjets these acoustic interactions can cause catastrophic failure, and in systems such as pulse combustors these interactions are desirable in that they cause significant increases in the rates of heat and mass transfer. Contemporary efforts in this area range from the study of acoustic interactions leading to flash back [1], the active control of instabilities by acoustic feed back [2], the study of acoustic fluctuations in ramjet-type combustors [3], the study of acoustic interactions in a low-speed dump combustor for potential use as hazardous waste incinerators [4, 5], a study of chemical-acoustic interactions [6], and the study acoustic interactions in pulse combustors [7-23]. All of these problems are similar in that they all have combustion chambers with like geometries (sudden expansions, stagnation plate flame holders, etc.) and the fluid dynamics in the combustion zone and the combustion of the reactants are affected by the presence of acoustic vibrations. These oscillations may be in resonance depending on the interactions between the combustion process, the acoustics and the system geometry. Whether these interactions are viewed as undesirable (ramjets or rocket engines) or desirable (pulse combustors or some dump combustors) the acoustic interactions with the flow field and the subsequent combustion process are the same. The flow field in a pulse combustor is one in acoustic resonance and is very repeatable. Hence, a pulse combustor provides a convenient method to study the effects of resonant pressure fields on fluid dynamics and combustion processes.

In the 1986 review article of Putnam *et al.* [13], a brief discussion of the fundamental operation of a pulse combustor is presented. They discuss a model of pulse combustor operation involving injection, ignition, and wave propagation. Ignition in their model occurred by isentropic heating when two shock waves intersected. Although these authors pointed out that the assumptions used in their ignition model were questionable, their model represented the state-of-the-art at that time. Being aware of this issue investigations were initiated by two separate investigators to answer some of these questions [14-16]. Relying on spatially integrated data, these later works found that the measured chemiluminescence never went to zero, and thus concluded that combustion never ceased. Based on these results the combustion process has been described as one of flame propagation growing and shrinking in size during each cycle [17] or that the new reactants are ignited by remnants of reacting pockets from the previous cycle [16]. Much information is lost by the spatially integrated data presented in Refs. 14, 15, and 16; it is distinctly possible that this continuous combustion source is not a discrete flame front. In support of this last argument, work by N. Ohiwa *et al.* [18] deduced from flame ionization measurements that combustion did not occur during the period of injection and in the region of the injected reactants. Based on these ionization measurements and schlieren images these authors concluded that a pair of large-scale eddy motions promoted mixing of hot products with the cold reactants, resulting in a distributed reaction zone.

The combustion process has also been modelled as one in which the reactants first are mixed with residual products and then undergo thermal ignition [10, 11]. Although this combustion model was successfully used by Barr *et al.* [19], these authors recognize that its validity relied on the success of the application. More recent work of Keller *et al.* [7] and Barr and Keller [8] shows that a principle delay mechanism is that of fluid dynamic stretch followed by ignition of a well stirred mixture of reactants and hot products. In this manuscript an overview of this work [7 and 8] will be presented.

FACILITY

Combustor

Figure 1 presents a schematic of the combustion system described in this work. A one-way valve (flapper-valve) is located in the reactant supply line just upstream of the square combustion chamber (80 mm square by 100 mm long). This valve opens and closes depending on the flow direction of the reactants from the supply line into the combustion chamber. For this work, the combustor was operated in a premixed mode where the fuel (99.99% methane) and air were mixed upstream of the combustion chamber. The injection system was configured so that the reactants were injected axially into the combustion chamber (with an inlet pipe diameter of 20.7mm) creating an axisymmetric velocity and temperature field (see Refs. 7-9 and 20). The interface between the inlet pipe and the combustion chamber is defined to be $x = 0$. The reactants impinge on a stagnation plate, located on the centerline at a nominal streamwise position of $x = 12$ mm and with a diameter of 19 mm. The square cross-section combustion chamber was connected to a square contraction section that joins the combustion chamber to the variable length tail pipe. For this work a tail pipe length of 880 mm was chosen. At these conditions the combustor frequency is nominally 100 Hz.

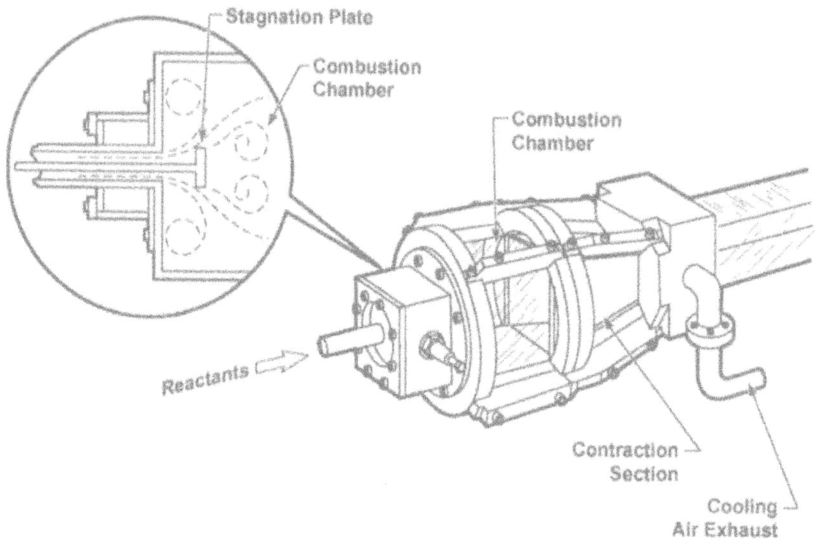

Figure 1. Schematic of the valved pulse combustor showing the injection geometry.

Diagnostics, Numerical Model and Operating Conditions

All measurements were cycle-resolved by phase-locking on the combustion chamber pressure. The downward-going zero-crossing of the oscillating component of pressure represents time zero. The phase-locked data for this work are presented in terms of the normalized cycle time τ which is defined to be t/t_{cycle}, where t is the time in a given cycle and t_{cycle} represents the total cycle time. Shown in Fig. 2 are typical data representing the phase relationships between the pressure in the combustion chamber, period of reactant injection, and the temporal location of energy release.

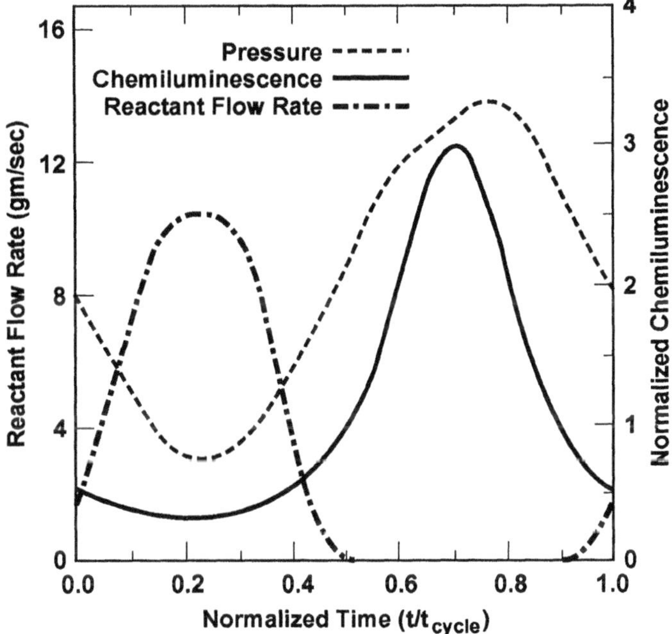

Figure 2. Shown are typical combustion chamber pressure, reactant injection profiles, and energy release rates for a well running pulse combustor.

During our investigations of this flow we have made phase-resolved, spatially detailed measurements of gas phase velocity, temperature, fluid dynamic mixedness and combustion intermediates (CH radicals). In addition to the measured quantities we have developed a detailed vortex dynamics numerical model of this periodic reacting flow that reproduces much of the relevant physics as determined by the experimental data. See Refs. 7-9, 19 and 20 for detailed descriptions of the facility, diagnostics and modelling activities reported on here.

DISCUSSION

The Flow Field

Figure 3 encapsulates the main features of this flow field. (See Refs. 7-9 for detailed presentations of these data and relevant discussions). Shown are four rows of images at four different key times throughout the cycle. The first row of images present particle path data, obtained from laser doppler velocimetry (LDV), overlayed on schlieren images of the flow.[†] The second row of images present the same particle path data as in the first row, but are now overlayed on phase resolved chemiluminescence images of the reacting flow.[‡] The third and fourth rows present the magnitude of the fine scale fluctuations, measured from LDV, for the streamwise and transverse directions respectively. The peak velocity fluctuation for the streamwise direction and the transverse direction was 19 m/s and 13 m/s, respectively; the peak of injection corresponding to an injection velocity of about 50 m/s. These values reach their minimum value during the peak of combustion of about 3.0 m/s for the streamwise and transverse directions.

The strong injection velocity quickly establishes a well-defined large toroidal vortex. This toroidal vortex is best seen at $\tau = 0.222$, which is the time corresponding to the Peak of Injection.

---

† Symmetry about the centerline was used to create the LDV data shown on the top half of the flow field.

‡ Chemiluminescence has been used as a powerful tool to qualitatively mark when and where energy release is occurring in pulse combustors, see Refs. 7-11, and 16-19.

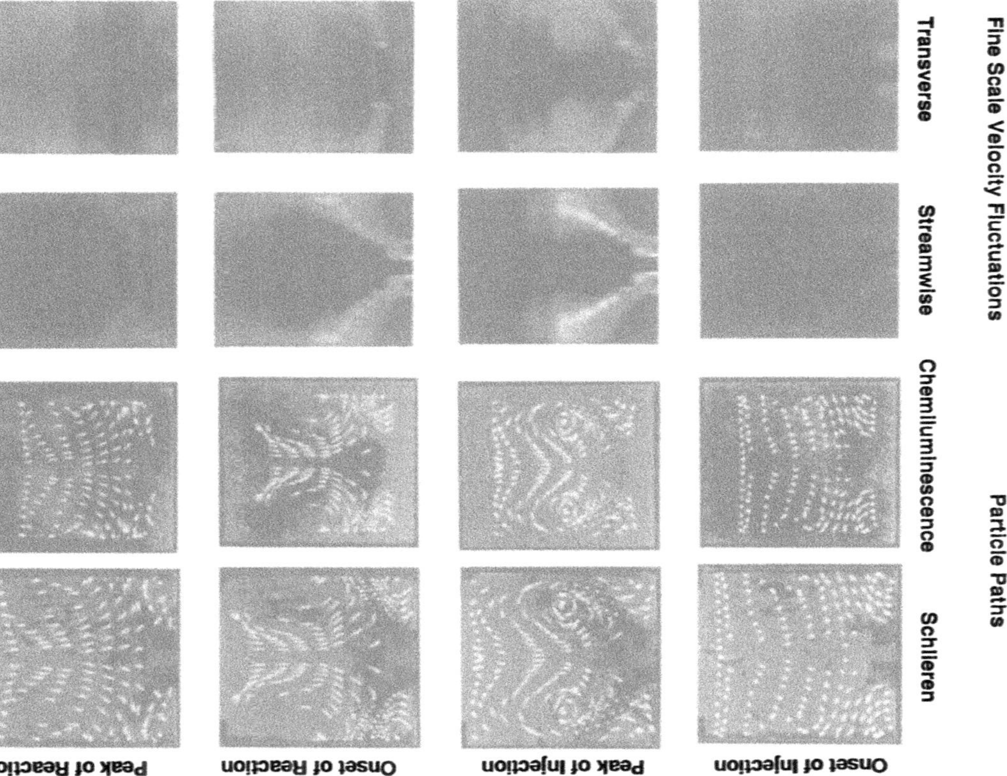

Figure 3. Shown are selected images of schlieren and chemiluminescence overlayed on particle path data obtained from phase resolved LDV measurements. Also shown are ensemble-averaged velocity fluctuations for the streamwise direction and transverse direction. These data are shown for four key times in the cycle: the onset of injection ($\tau \simeq 0.94$), the peak of injection ($\tau \simeq 0.22$), the onset of reaction ($\tau \simeq 0.44$), and the peak of reaction ($\tau \simeq 0.67$). This figure is reproduced from Fig. 8 of reference 7 with permission from Combustion and Flame.

Also evident is the effect this toroidal vortex has on the creation of the large recirculation zone. In the early stages of development the recirculation zone is weak, but as the toroidal vortex gains strength the influence of this increased circulation causes a strong negative velocity in the center of the combustion chamber. The mechanism for the convection of reactants to the downstream side stagnation point of the stagnation plate is attributed to the strong circulation of this toroidal vortex. This toroidal vortex diverts a major portion of the flow to the walls outside of the LDV measurement domain. The particle paths at the Peak of Injection and far downstream are almost radial providing a mechanism to convect fluid from the wall region into the centerline. These results show that this toroidal vortex creates a feature in the flow capable of convecting reactants to both the near and far fields where ignition of these reactants is observed to begin (see the chemiluminescence image shown for the Onset of Reaction.)

As the reactant's injection rate increases, the circulation in this toroidal vortex increases. This results in a buildup of momentum in the center of the flow that overwhelms the inlet jet and pushes it back toward the inlet, as indicated by the almost radial particle paths shown near the stagnation plate at the Onset of Reaction.

Another feature of this flow is a smaller toroidal vortex located in the corners near the entrance. The circulation of this smaller toroidal vortex is of the opposite sign from the toroidal vortex located in the center of the combustion chamber. This small toroidal vortex grows and decays in strength throughout the cycle, but is always present.

Early in the combustion phase the primary toroidal vortex has grown in size and moved out toward the walls (Onset of Reaction). The volumetric expansion due to combustion pushes this toroidal vortex completely out of the center of the flow and makes the flow pattern more uniform throughout the measurement domain (see the image shown for the Peak of Reaction). Recall from Fig. 2 that during the period of maximum injection the reaction goes through a minimum. Note also from the Peak of Injection frame of Fig. 3 that there is no evidence of reaction occurring along the interface between the inlet jet and the hot residual products. Indeed, the residual reaction occurring during the Peak of Injection occurs in the quiescent regions of the reactor and not near the reactant jet. In addition, the times associated with this delay are longer than those required for a chemical kinetic ignition delay by about a factor of three. *Ignition of the fresh charge of premixed reactants has been suppressed during the injection process.*

The first indication of ignition (Onset of Reaction) occurs simultaneously near the downstream side of the stagnation plate, and farther downstream near the centerline of the combustion chamber on the inside of the large recirculation zone. As the reaction proceeds, the intensity rapidly increases in the stagnation region behind the plate and along the edges of the recirculation zone, until the Peak of Reaction where it appears as though reaction is filling the combustion chamber. Ignition has occurred along the edges of this vortical structure near the centerline and at large streamwise locations, (Onset of Reaction). Reaction along the injection jet's boundary has been suppressed, allowing the toroidal vortex to effectively entrain and mix the cold reactants with the residual hot products sufficiently to remove the schlieren signal from the images shown depicting the Onset of Reaction and the Peak of Reaction. The chemiluminescence appears to be reaction initiated along the outer edges of the vortical structure, then the ignition of the reactants rapidly proceeds to the center of the vortical core, and occurs in an almost volumetric fashion. This rapid movement of the combustion zone is an ignition of reactants rather than a propagation of the reaction region seen in the Onset of Reaction chemiluminescence image.

The reactants that were not entrained into this large vortical structure react more slowly and are reacting on the outside of this primary reaction zone, along the walls and far downstream. This residual reaction is the source of the weak signal found at the Peak of Injection (see Fig. 4). These areas are not as strongly affected by the injection velocity, yet still have combustion-stabilizing mechanisms (for example, the toroidal vortex found in the corner at the leading edge of the combustion chamber, this point is better shown in the data presented in Ref. 7).

A principle mechanism for the suppression of ignition along the interface between the jet and the hot products is that of stretch [26-27]. From the LDV data the estimated strain rate between the injected reactants and that of the residual products during injection is 10 times greater (10,000 / sec) than that required for extinction of laminar flames for the same fuel and equivalence ratio used here

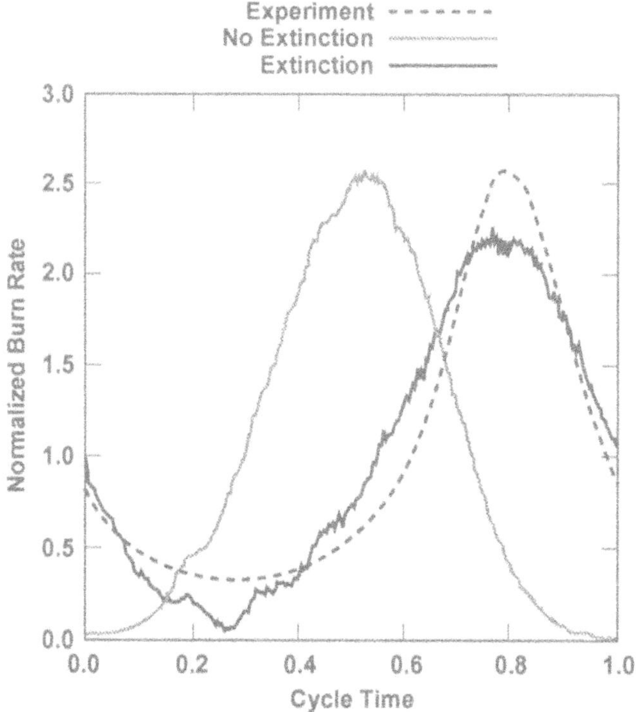

Figure 4. Integrated normalized burn rates for two different calculations compared with the experimental results. This figure is reproduced from Ref. 8, with permission from Combustion and Flame.

(1000 / sec) [28].

Axisymmetric vortex dynamics using a simple line interface flamelet combustion model was developed that reproduced the essential features of this flow well (see Ref. 8). Shown in Fig. 4 is the normalized spatially integrated calculated flame surface area (representing the calculated burn rate) as a function of cycle time shown for two different combustion models. These calculations are compared against spatially integrated measured chemiluminescence also representing the burn rate. In the calculation marked no-extinction, reaction is allowed to occur instantaneously when cold reactants come in contact with hot products supplying an ignition source. This formulation predicts that the reactants are consumed by combustion too early in the cycle. Even if one were to include a chemical kinetic ignition delay in the calculation the peak would shift by 8% to 12% still significantly short of the needed 35% shift. On the other hand, the calculation marked extinction compares well with the experimentally determined burning rate. In this second calculation combustion was allowed to proceed only if the local instantaneous fluid dynamic strain rate was below that required to extinguish laminar flames under the same operating conditions as this combustor. The strain rates in this flow are orders of magnitude larger than those required for extinction, thus, this unit step function representation for the flame speed versus stretch proved to be a perfectly good simplification.

The existence of fine-scale velocity fluctuations during injection enhances the molecular mixing of the reactants with the residual hot products. This fine-scale motion peaks during injection, and has dissipated to near its minimum value by the time combustion of the reactants occurs. Macroscopic and microscopic mixing occurs during injection while ignition is suppressed. A principle mechanism suppressing the ignition of fresh reactants is that of extinction by flame stretch caused by the high fluid dynamic strain rates. By assuming unity for the turbulent Schmidt number and knowing that the fine-scale fluctuations have dissipated by the end of the injection period, the reactants can be assumed to be well mixed by the end of the injection period. Thus, the velocity oscillations affect the system *before* combustion occurs and not *during* the combustion portion of the cycle.

The ignition process in this periodic flow field is comprised of two primary steps. The onset of ignition is suppressed during the injection process by elevated fluid dynamic strain rates. The reactants and hot products are then mixed on a fine scale. As the combustion chamber pressure increases the injection velocity decreases, decreasing the strain rate, thereby, allowing ignition of the mixed reactants and products to proceed. The reactants ignite by thermal and/or radical ignition in an almost volumetric fashion. Combustion in this system occurs as an ensemble of well-stirred reactors, undergoing temporally *staged ignition*, rather than a rapidly *propagating flame sheet*.

Note that both the schlieren and the chemiluminescence data discussed here are line-of-sight measurements that will mask the existence of flamelet-like structures. One can imagine an ensemble of well-stirred reactors each undergoing temporally staged ignition, where a flamelet-like structure traverses the small well-stirred reactor. The preliminary results presented in Ref. 9 do show the existence of flamelet-like structures traversing an ensemble of small well-stirred reactors.

Recall that Rayleigh's criterion for stable operation of flows in acoustic resonance requires that

$$\oint (\dot{Q}(t) - \bar{Q})(P(t) - \bar{P})dt \geq Acoustic\ Losses$$

where $\dot{Q}(t)$ and $P(t)$ represent the energy release rate and the pressure as a function of cycle time respectively. $\bar{Q}$ and $\bar{P}$ represents the time averaged quantities of the energy release rate and pressure. Simply stated, in order to sustain stable oscillations sufficient energy must go into the resonant pressure field to over come all the acoustic losses of the system. This requires that the energy release be in phase with the resonant pressure field. This series of events automatically phases the energy release with the resonant pressure field needed to satisfy Rayleigh's criterion.

The fluid dynamic flow field in the combustion chamber for an isolated cycle is a direct result of the resonant pressure field. The primary elements of this flow field are created early in the cycle during the injection of the reactants as a direct result of the favorable pressure gradient between the combustion chamber and the inlet. It is the injection of these reactants that establishes the toroidal vortex and the resultant flow field. As a result of the flow in acoustic resonance, the combustion chamber pressure begins to rise *before* ignition of the reactants occurs. Indeed, it is this rising resonant pressure field that creates the adverse pressure gradient during the second half of the cycle. This adverse pressure gradient decelerates the inlet jet and eventually stops the injection process decreasing the strain rate, thus allowing the ignition of the reactants. This fluid dynamic behavior is a direct result of the oscillating combustion chamber pressure and the resulting pressure gradient between the inlet and the combustion chamber, and hence, being a flow in acoustic resonance it exists whether or not a combustion event occurs during the current cycle. Considering only an isolated cycle the rising pressure in the combustion chamber is a result of the flow in acoustic resonance and not the combustion process. However, the energy release, if properly phased, does reinforce the resonant pressure field in preparation for the next cycle. The resonant pressure field creates a fluid dynamic flow field that controls combustion so that the energy released is in phase with the resonant pressure field to satisfy Rayleigh's criterion, the requirement necessary to maintain the stable limit cycle.

The previous discussion is not to imply that the fluid dynamics and the combustion processes are decoupled, this clearly is not the case. However, the coupling occurs as a result of Rayleigh's criterion over multiple cycles. The events that occur during the current cycle are a result of the resonant pressure field that was reinforced by the combustion event from the previous cycle. If the pressure field was not reinforced adequately then the current fluid dynamic flow field would be modified, affecting the current combustion event which in turns modifies the pressure signature for the succeeding cycle. For improperly phased events the volumetric expansion due to combustion will cause the pressure signature to differ from its typical sinusoidal shape, resulting in a modification of the operating frequency (see Refs. 5, 10, 11, 19 and 22). However, this occurs late in the cycle after the favorable pressure gradient part of the cycle during which time the dominate features of the flow field were established. The events for the current cycle are a result of the resonant pressure field driven by the combustion process from previous cycles. The primary role for the combustion process is to reinforce the resonant pressure field, to overcome the associated acoustic losses.

SUMMARY

A discussion of the detailed mixing, combustion and ignition processes occurring in a flow field under the influence of resonant acoustic perturbations created by a "Helmholtz" type pulse combustor was presented.

During the first two-thirds of the cycle the inlet jet forms a well-defined large toroidal vortex that is phase-locked with the combustor cycle. This vortical structure is responsible for the convection and mixing of the reactants with the residual hot products. Although the chemiluminescence is always present in quiescent regions of combustor, no luminosity exists in the region of the reactants during injection. Moreover, ignition of the fresh reactants occurs first along the outer edges of the rolled-up toroidal vortex, and not in the small corner recirculation regions of the flow. The combustion of these reactants develops rapidly in the center of this rolled up toroidal vortex where nearly uniform combustion of the fresh reactants occurs.

Local extinction by flame stretch has been identified as a principle mechanism delaying the onset of ignition during injection and allowing time for the reactants to mix with the hot products. As a result of the flow in acoustic resonance the pressure in the combustion chamber raises slowing the injection of reactants decreasing the strain rate. This allows an ensemble of well-stirred reactants and hot residual products to undergo thermal and/or radical ignition. Although combustion in this reactor occurs in a well-mixed mode, reaction proceeds flamelet-like manner.

ACKNOWLEDGEMENTS

The work reported on here is the culmination of the efforts of many people. Special recognition goes to T.T. Bramlette, R.S. Gemmen, L. Smith, R.W. Dibble, and J.R. Alvarez for their significant contributions to this overall effort. This work was performed at the Combustion Research Facility Sandia National Laboratories and supported by the U.S. Department of Energy, Office of Industrial Technologies, Advanced Industrial Concepts Division.

REFERENCES

1. Keller, J.O., L. Vaneveld, D. Korschelt, G.L. Hubbard, A.F. Ghonem, J.W. Daily, and A.K. Oppenheim, "Mechanism of Instabilities in Turbulent Combustion Leading to Flashback," *AIAA Journal,* Vol. 20, 254, 1982.

2. Poinsot, T., D. Veynante, F. Bourienne, S. Candel, E. Esposito and J. Surget, "Initiation and Suppression of Combustion Instabilities by Active Control," Twenty-Second Symposium (International) on Combustion, 1363-1370, The Combustion Institute, Pittsburgh, 1988.

3. Bhafouriana, A., J. Frost, and J. W. Daily, "Instability Peak Frequency Fluctuations in a Model Ramjet," Presented at the joint spring technical meeting of the Western States/Canadian Sections of the Combustion Institute, Baniff, Alberta Canada, 1990.

4. Smith, O.L., R., Marchant, J. Willis, L. M. Lee, P. Logan, and A. R. Karagozian, "Incineration of Surrogate Wastes in a Low-Speed Dump Combustor," *Combust. Sci. Tech.*, 74, 199, (1990).

5. Logan, P., J. W. Lee, L. M. Lee, A. R. Karagozian, and O. L. Smith, "Acoustics of a Low-Speed Dump Combustor," *Combust. Flame*, 84, 93, (1991).

6. Oran, E.S., J.H. Gardner, "Chemical-Acoustic Interactions in Combustion Systems," *Prog. Energy Combust., Sci.* 11, 253-276, 1985.

7. Keller, J.O., P.K. Barr, R.S. Gemmen, "Premixed Combustion in a Periodic Flow Field. Part I: Experimental Investigation" In press in *Combust. Flame*, 1994.

8. Barr, P.K., J.O. Keller, "Premixed Combustion in a Periodic Flow Field. Part II: The Importance of Flame Extinction by Fluid Dynamic Strain", In press in *Combust. Flame*, 1994.

9. Keller, J.O., L. Smith, R.W. Dibble, "Phase Resolved PLIF Measurements of CH in a Strongly Oscillating Flow Field," Poster presentation Twenty-Fourth Symposium (International) on Combustion , Sydney Australia, 1992, in preparation for submission to *Combust. Flame*, 1994.

10. Keller, J.O., T.T. Bramlette, J.E. Dec, and C.K. Westbrook, "Pulse Combustion: The Importance of Characteristic Times," *Combust. Flame,* 75, 33-44, 1989.

11. Keller, J.O., T.T. Bramlette, J.E. Dec, and C.K. Westbrook, "Pulse Combustion: The Quantification of Characteristic Times," *Combust. Flame,* 79, 151-161, 1990.

12. Dec, J.E., J.O. Keller, and I. Hongo, "Time-Resolved Velocities and Turbulence in the Oscillating Flow of a Pulse Combustor Tail Pipe," *Combust. Flame,* 83, 271-292 (1991).

13. Putnam, A.A., F.E. Belles and J.A.C. Kentfield, "Pulse Combustion," *Prog. Energy Combust., Sci.* 12, 43-79, 1986.

14. Keller, J.O., K. Saito, K. Kishimoto, "An Experimental Investigation of a Pulse Combustor – Flow Visulization by Schlieren Photography," Presented at the 1984 Fall Meeting of the Western States Section of the Combustion Institute, Stanford, California

15. Keller, J.O., and K. Saito, "Measurements of the Combusting Flow in a Pulse Combustor," *Combust. Sci. Tech.*, 53, 137-163, 1987.

16. Reuter, D., B.R. Daniel, J.Jagoda, and B.T. Zinn, "Periodic Mixing and Combustion Processes in Gas Fired Pulsating Combustors," *Combust. Flame*, 65, 281-290, 1986.

17. Belles, F. E., "Final Report on Pulse Combustion Design Research," The Gas Research Institute, March 15. 1990, GRI-89/0106.

18. Ohiwa, N., S. Yamaguchi, T. Hasegawa, Q.B. Kui, and A. Okada, "An Experimental Study of the Ignition and Burning Mechanisms in a Pulse Combustor," *Japan Society of Mechanical Engineers,* Series B, No. 85-0524, 1913-1922 (1985).

19. Barr, P.K., J.O. Keller, T.T. Bramlette, C.K. Westbrook, and J.E. Dec, "Pulse Combustor Modeling: Demonstration of the Importance of Characteristic Times," *Combust. Flame,* 82: 252-269 (1990).

20. Keller, J.O., and I. Hongo, "Pulse Combustion: The Mechanisms of NO_x Production," *Combust. Flame,* 80: 219-237 (1990).

21. Dec, J.E., and J.O. Keller, "Time-Resolved Gas Temperatures in the Oscillating Turbulent Flow of a Pulse Combustor Tail Pipe," *Combust. Flame,* 80: 358-371 (1990).

22. Keller, J.O. and C.K. Westbrook, "Response of a Pulse Combustor to Changes in Fuel Composition," Twenty-First Symposium (International) on Combustion, 547-555, The Combustion Institute, Pittsburgh, 1987.

23. "Special Issue on Pulsating Combustion," Special Eds. P. K. Barr, M. E. Gunn, Jr. J. A. Kezerle, T. T. Bramlette, J. O. Keller, T. R. Roose, *Combust. Sci. Tech.*, 94, (1993).

24. Warnatz, J., "The Structure of Laminar Alkane-, Alkene-, and Acetylene Flames," Eighteenth Symposium (International) on Combustion, 369, The Combustion Institute, Pittsburgh, 1981.

25. Kanury, A. Murty, Introduction to Combustion Phenomena, Gordon and Breach Science Publishers, 1975, Pp 290.

26. Peters N., "Laminar Flamelet Concepts in Turbulent Combustion," Twenty-First Symposium (International) on Combustion, 1231-1250, The Combustion Institute, Pittsburgh, 1987.

27. Law, C.K., D.L. Zhu, and G. Yu, "Propagation and Extinction of Stretched Premixed Flames," Twenty-First Symposium (International) on Combustion, 1419-1426, The Combustion Institute, Pittsburgh, 1987.

28. Kee, R.J., J.A. Miller, and G.H. Evans, "A Computational Model of the Structure and Extinction of Strained, Opposed Flow, Premixed Methane-Air Flames," Twenty-Second Symposium (International) on Combustion, 1479-1494, The Combustion Institute, Pittsburgh, 1988.

3. AN EXPERIMENTAL AND NUMERICAL INVESTIGATION OF PREMIXED COMBUSTION IN A VORTEX IN A LABORATORY DUMP COMBUSTOR

D. W. Kendrick, T. W. Zsak and **E. E. Zukoski**
California Institute of Technology
Jet Propulsion Center, 301-46
Pasadena, CA 91125

ABSTRACT. A one-dimensional acoustic model was used to predict the resonant modes of the Caltech pulsed combustion facility. The model accurately predicted pressure amplitude spectra found through experiments for the 2.5 and 7.6 cm duct height configurations. Heat addition locations were found to have only marginal effects on shifting the location of the facility's acoustic modes. A detailed experimental analysis of the reacting vortex structures shed from a rearward-facing step was also performed using high speed shadowgraph and CCD cinematography. Premixed vortical combustion was found to have two ignition mechanisms depending on the prior status within the combustor. In the first, burning was initiated at the surface and proceeded toward the center while in the second ignition was initiated near the center and the flame propagated outward. Time delays measured from the start of vortex shedding to subsequent ignition or to the corresponding maximum burning intensity were found to vary inversely with combustor pressure during injection (shedding) and with combustor pressure during burning. Reducing the height of the combustor increased interactions between the burning vortex and the wall, inhibited vortex growth, and produced longer axial burning regions and higher overall straining throughout the structure's cycle. Vortex straining was defined in two ways: first, based on the growth rate of the core diameter of the structure and second, based on the effective length of the streamline separating hot combustion products and cold reactants. Straining provided a sufficient delay mechanism to shift vortex shedding from 237 to 188 Hz for the 5.1 cm duct height case.

1. Introduction

Previous studies at the Caltech JPC combustion facility have focused on the growth and sustenance of pressure oscillations associated with the facility's acoustic modes for a 2.5 cm deep combustion chamber [1,2,3]. Little attempt was made to investigate the reacting vortex structures which dominate the flow field while operating in the pulsed mode. Studying such coherent structures may provide insight as to why a particular flow condition favours one mode of oscillation as opposed to another. It should be noted that several modes can be simultaneously excited with the proper selection of flow rate, fuel stoichiometry, and geometry, resulting in severe beating of the associated velocity and pressure traces and permitting temporal variations in growth to be investigated.

F. Culick et al., (eds.), Unsteady Combustion, 33–69.

This investigation provides an experimental and numerical look at reacting vortex structure shed from a rearward facing step by detailing their growth, ignition, and the location and timing of heat release. In the first part of the investigation, a one-dimensional acoustic model was devised to characterize the system's acoustic modes for different geometries. The model describes the effect that the location of heat addition, geometry, etc. have on the values of these resonant frequencies. Striking similarity between numerically and experimentally derived pressure FFTs has been achieved for both the 2.5 and 7.6 cm combustor height configurations.

Fluid dynamics and combustion processes occurring within the dump combustor were examined by using high-speed Hycam and Kodak CCD cinematography. The aim of this work was to develop an explanation for the different shedding frequencies that have been observed with various values of the operating parameters. The temporal variations in growth, ignition, and burning were successfully captured by the high-speed techniques and the results illustrate the effects of geometry, stoichiometry, etc. on shedding. From the Hycam work, it is shown that straining along the interface separating hot combustion products and cold reactants has no observable influence on delaying the onset of combustion, however, it most probably facilitates observed axial shifts in peak burning location by delaying complete combustion. The lower shedding frequency case is shown to have its maximum burning location further downstream relative to the higher frequency case. Since continuation of the instability is linked to the phase between the acoustic oscillations and the heat release oscillations and not the onset of combustion, i.e. ignition, vortex straining appears to have a fundamental roll in determining shedding frequencies. These results are consistent with previous numerical work [4].

2. Experimental Facility

Figure 1 is a schematic of the experimental facility used for the present research. It has been modified from previous investigations [1,2,3] to allow a broader unstable envelope. Combustion chamber height can be altered from 2.5 to 15.2 cm and the premixed CH4-air flow can be artificially pulsed to initiate unstable flow for larger duct heights. This study will focus on the 2.5, 5.1, and 7.6 cm geometries, 21 and 35 m/s dump plane speeds (flow velocity at the entrance of the chamber), and 1.2 and 1.4 fuel equivalence ratios (ϕ).

A typical data trace for the 7.6 cm, 21 m/s, and ϕ=1.4 configuration is shown in Figure 2. The pressure was measured 1.27 cm downstream of the step while the velocity fluctuations were obtained from a hot wire placed 3.18 cm upstream of the step to prevent it from burning due to flashback. For all the configurations described in the paper, there are three dominant shedding frequencies which may occur individually or simultaneously: 188, 237, or 290 Hz. The 2.5 cm. duct height promotes purely 188 Hz shedding while the 5.1 and 7.6 cm. chambers can support either the 188 or 237 Hz modes as the dominant frequency. The 290 Hz mode plays a more discrete role, appearing for richer fuel runs or with hydrogen-methane fuel mixtures. All shedding frequencies have been substantiated by pressure, velocity, and photomultiplier FFTs and further reinforced by numerical modeling.

Dynamic pressure signals were measured at access ports located along the top of the combustor by PCB model 106b piezoelectric pressure transducers, while dump plane velocity measurements were

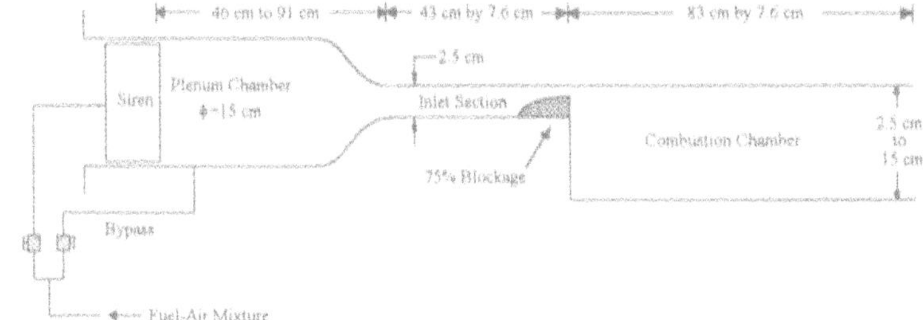

Figure 1: Diagram of pulsed combustion facility

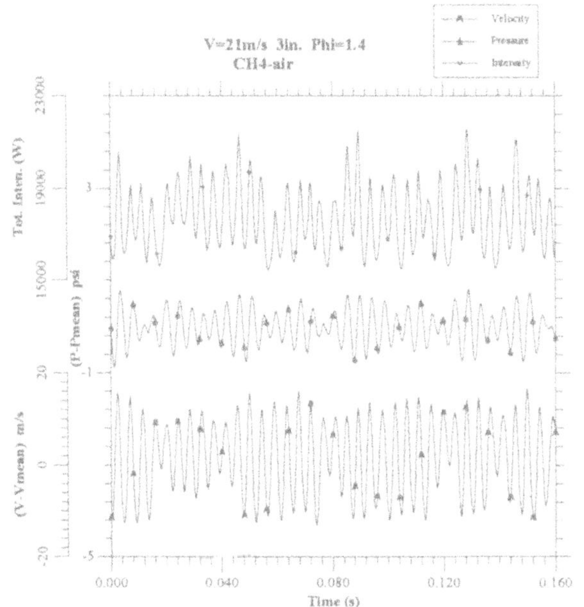

Figure 2: Sample data trace for the 7.6 cm duct height

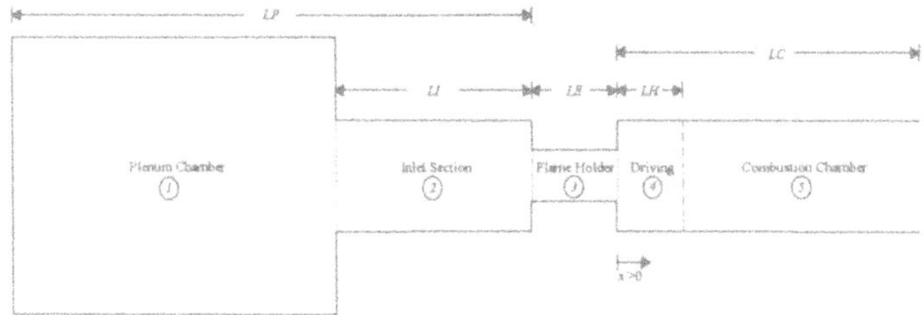

Figure 3: Schematic of experimental apparatus as used for the one-dimensional model

obtained by a constant temperature anemometry circuit. Flow visualization was carried out with both a Hycam camera operating in the shadowgraph mode framing at around 5100 Hz and a Kodak CCD TR1006 Intensified Imager framing at 3000, 4000, or 6000 Hz, depending on the duct height. The Kodak camera had a spectral response from 440 to 700 nm and full image resolution of 239 by 192 pixels (8 bits per pixel). No interference filters were used. Operating the CCD camera at such high speeds necessitated split framing the image though this decrease in vertical resolution provided no problem due to the duct's small aspect ratio. The flowfield was sufficiently frozen by gating the intensifier at 15 μs or approximately one degree of the pressure cycle. A typical test entailed running the facility, triggering the camera, storing the images on Kodak's own Ektapro Processor, and then downloading the images after a run to a 486-33MHz PC via an AT-GPIB board.

3. Results and Observation

3.1 NUMERICAL RESULTS

A one-dimensional acoustic model has been developed to predict the resonant acoustic modes and corresponding mode shapes of the laboratory dump combustor. This information is useful in characterizing the behavior of the apparatus and in analyzing the experimental results, for phenomena which are acoustic in nature can be separated from those emanating from other sources.

For planar acoustic waves, the conservation equations of mass and momentum can be written as

$$\frac{1}{\bar{\rho}}\frac{\partial \bar{\rho}}{\partial t}+\bar{u}\frac{1}{\bar{\rho}}\frac{\partial \bar{\rho}}{\partial x}+\frac{\partial \bar{u}}{\partial x}=0 \tag{1}$$

$$\frac{\partial \bar{u}}{\partial t}+\bar{u}\frac{\partial \bar{u}}{\partial x}+\frac{1}{\bar{\rho}}\frac{\partial \bar{p}}{\partial x}=0, \tag{2}$$

where $\bar{\rho}$, $\bar{p}$, and $\bar{u}$ are the density, pressure, and velocity, respectively, of the gas, and x denotes position and t denotes time.

Making use of the isentropic relation

$$\bar{p}(\bar{\rho})^{-\gamma}= constant,$$

where γ is the ratio of the specific heats, yields the following forms of the above equations

$$\frac{1}{\gamma\,\bar{p}}\frac{\partial \bar{p}}{\partial t}+\bar{u}\frac{1}{\gamma\,\bar{p}}\frac{\partial \bar{p}}{\partial x}+\frac{\partial \bar{u}}{\partial x}=0 \tag{3}$$

$$\frac{\partial \bar{u}}{\partial t}+\bar{u}\frac{\partial \bar{u}}{\partial x}+\bar{a}^2\left(\frac{1}{\gamma\,\bar{p}}\frac{\partial \bar{p}}{\partial x}\right)=0, \tag{4}$$

where $\bar{a}$ is the sonic speed.

Along with being isentropic, acoustic waves are also of small amplitude. The pressure and velocity variations, therefore, can be linearized by describing each property as the sum of an undisturbed, steady value, denoted by the subscript 'o', and a perturbation value as follows

$$\bar{p} = p_o + p$$
$$\bar{u} = u_o + u$$
$$\bar{a} = a_o.$$

The perturbation of the sonic speed is omitted because it has no bearing on the derivation of the acoustic equations.

Neglecting second-order terms in the fluctuating components, substitution of the linearized variables into equations 3 and 4, while making use of equations 1 and 2, gives

$$\left\{\frac{\partial}{\partial t} + a_o(M_o + 1)\frac{\partial}{\partial x}\right\}\left(\frac{u}{a_o} + \frac{p}{\gamma\, p_o}\right) = 0 \qquad (5)$$

$$\left\{\frac{\partial}{\partial t} + a_o(M_o - 1)\frac{\partial}{\partial x}\right\}\left(\frac{u}{a_o} - \frac{p}{\gamma\, p_o}\right) = 0, \qquad (6)$$

where $M_o = \frac{u_o}{a_o}$, the local Mach number. If a periodic time disturbance of the form

$$\frac{p}{\gamma\, p_o} = P(x)e^{-i\,\omega t}$$
$$\frac{u}{a_o} = U(x)e^{-i\,\omega t},$$

where ω is the frequency, is assumed, substitution into equations 5 and 6 and solving for $P(x)$ and $U(x)$ gives the following expressions for the fluctuating pressure and velocity

$$\frac{p}{\gamma\, p_o} = e^{-i\,\omega\, t}\left\{ P^{+} e^{\frac{i\,\omega\, x}{a_o\,(M_o+1)}} + P^{-} e^{\frac{i\,\omega\, x}{a_o\,(M_o-1)}} \right\} \tag{7}$$

$$\frac{u}{a_o} = e^{-i\,\omega\, t}\left\{ P^{+} e^{\frac{i\,\omega\, x}{a_o\,(M_o+1)}} - P^{-} e^{\frac{i\,\omega\, x}{a_o\,(M_o-1)}} \right\}, \tag{8}$$

where P^+ and P^- are complex coefficients representing the downstream- and upstream-traveling waves, respectively, computed from the boundary and matching conditions. Equations 7 and 8 are applicable in each of the regions of the experimental apparatus, and each quantity is in reference to the particular segment in question.

The laboratory dump combustor is modeled as a series of five constant area segments joined by abrupt area changes as shown in Figure 3. Because the acoustic wavelengths of the relevant frequencies are long compared to the lengths of the matching regions over which the areas change, the assumption of abrupt area changes is valid. For example, the plenum chamber can be assumed to be an acoustic body with solid boundaries at both ends because its cross-sectional area is much larger than either the sum of the areas of the fuel supply lines leading into the plenum or the area of the inlet section which connects to the downstream end of the plenum chamber. At resonance, therefore, the solid boundaries act as pressure antinodes so that the wavelength of the natural mode of the plenum is equal to twice the length of the plenum chamber, 182 cm. This is over seven times greater than the length of the transition region from the axisymmetric geometry of the plenum chamber to the rectangular cross-section of the inlet section.

The large volume plenum chamber is represented as section *1*, $-(LP+LE)\le x\le -(LI+LE)$, and section *2*, $-(LI+LE)\le x\le -LE$, is the inlet section. The contraction caused by the flame holder at the downstream end of the inlet section is denoted as section *3*, $-LE\le x\le 0$. In all cases, the temperature is taken as $T=T_{room}$ in the area upstream of the dump plane, covering all $x\le 0$. Within this region, $a=a_{room}=335$ m/sec and $\gamma=\gamma_{room}=1.4$ as well.

Two segments comprise the combustion chamber. Their boundary is denoted as the position $x=LH$, the location at which an oscillatory volumetric mass source is placed. This driving term, equal to $D(\omega)e^{-i\,\omega\, t}$, where $D(\omega)$ is the magnitude of the volumetric mass source introduced at frequency ω, is used to represent the heat addition process. The temperature within section *4*, $0\le x\le LH$, can be varied; however, it generally has the same value as that in section *5*, $LH\le x\le LC$, where $T=T_{combustion}$, $a=a_{combustion}=750$ m/sec, and $\gamma=\gamma_{combustion}=1.3$.

It is desired to use the model to predict the pressure and velocity responses of the apparatus at the dump plane to unit driving over the frequency range $0 \leq \omega \leq 600$ Hz for different combustion chamber configurations. Using the matching conditions that the pressure is continuous and that the velocity satisfies volumetric continuity at each segment boundary, one can obtain expressions for the coefficients $P^{+}(\omega)$ and $P^{-}(\omega)$ in any section in terms of known quantities. Equations 7 and 8 can then be used to obtain the amplitude spectra of the pressure and velocity within that section. Because pressure measurements are made within the combustion chamber, the coefficients $P_4^{+}(\omega)$ and $P_4^{-}(\omega)$ or $P_5^{+}(\omega)$ and $P_5^{-}(\omega)$ are solved for depending on the placement of the volume source. The coefficients $P_3^{+}(\omega)$ and $P_3^{-}(\omega)$ are solved for to determine the velocity response above the flame holder because the hot-wire anemometer is placed in the cold flow upstream of the dump plane.

In addition, the boundary conditions at the ends of the apparatus must be defined. The complex reflectance β relating the downstream- and upstream-traveling acoustic waves is given as

$$\beta = \frac{\Psi - 1}{\Psi + 1},$$

where Ψ is the acoustic impedance as defined in reference [5].

$$\Psi = \frac{p}{\gamma\, p_o} \frac{a}{\vec{u} \cdot \vec{n}}$$

with $\vec{n}$ being the outward-going unit normal vector.

For the plenum section, the complex reflectance at $x = -(LP + LE)$ is

$$\beta_{plenum} = \frac{P_1^{+} e^{-\frac{i\omega(LP+LE)}{a_1(M_1+1)}}}{P_1^{-} e^{-\frac{i\omega(LP+LE)}{a_1(M_1-1)}}}.$$

The relationship between the coefficients P_1^{+} and P_1^{-} is, therefore

$$P_1^{+} = \beta_{plenum} P_1^{-} e^{-\frac{2i\omega(LP+LE)}{a_1\left(M_1^2-1\right)}}.$$

At the combustor exit, the reflectance at $x = LC$ is

$$\beta_{exit} = \frac{P_5^- e^{\frac{i\omega LC}{a_5(M_5-1)}}}{P_5^+ e^{\frac{i\omega LC}{a_5(M_5+1)}}}.$$

The coefficients are related as follows

$$P_5^- = \beta_{exit} P_5^+ e^{-\frac{2i\omega LC}{a_5\left(M_5^2-1\right)}}.$$

The reflection coefficient at the upstream end of the apparatus is taken to be ideal because the area of the plenum section is much larger than the sum of the areas of the fuel supply lines leading into the plenum. At the downstream end, experimental correlations found in [6] for the exit impedance of a duct exhausting a hot gas are used. This impedance is dependent upon the temperature of the gas, the acoustic frequency, and the hydraulic radius of the rectangular duct.

3.1.1. Effects of Varying the Combustion Chamber Temperature

The validity of the model can be demonstrated by comparing the predicted acoustic modes with those determined experimentally for cold-flow conditions. For the model, this amounts to setting $T = T_{room}$ in each of the five segments so that $a = a_{room} = 335$ m/sec and $\gamma = \gamma_{room} = 1.4$. Experimentally, the cold-flow resonant modes were determined for the 2.5-cm tall combustor as they were excited by the sound produced as air passed through the siren located in the upstream end of the plenum chamber. The frequency of the siren was altered between 100 Hz and 600 Hz. Four modes were found within this range of frequencies and the results are tabulated in Table 1, where a comparison is made of the predicted and measured resonant mode frequencies.

As one can see, the agreement is very satisfactory, and it is felt that the one-dimensional model will predict the acoustic resonant modes of the experimental apparatus under operating conditions quite well.

Table 1. A comparison of predicted and measured cold-flow resonant modes for the 2.5-cm deep chamber and 100 Hz$\leq \omega \leq$600 Hz.

$FREQ_{model}$ (Hz)	$FREQ_{experimental}$ (Hz)	% DIFFERENCE
123	133	8.1
249	274	10.0
334	351	5.1
507	527	3.9

For the case of combustion with unit driving at the dump plane, LH is set equal to zero, and the amplitude spectrum of pressure as determined by the model shows three primary modes at frequencies of 185 Hz, 222 Hz, and 294 Hz. This is quite a different result from the cold-flow case, and the two spectra are shown graphically in Figure 4. Have the cold flow modes shifted to other frequencies or have they been damped or amplified by the heat release and temperature increase of combustion?

To investigate this, the temperature within the combustion chamber was varied from T_{room} to $T_{combustion}$ by varying the sonic speed in the combustor from 335 m/sec to 750 m/sec with $\gamma_{room} = 1.4$ and $\gamma_{combustion} = 1.3$. This amounts to a factor of 5.4 change in temperature, and it is interesting to note the movement of the cold-flow resonant modes to their hot-flow counterparts as the combustion chamber temperature increases.

Figure 5*a* shows the effect of increasing the combustor sonic speed from 335 m/sec to 488 m/sec for the 2.5-cm tall combustor over the entire frequency range of interest. The combustor sonic speed is further increased from 549 m/sec to 750 m/sec, and the resulting pressure amplitude spectrum is shown in Figure 5*b*. The most interesting activity occurs within the frequency range $100\ \text{Hz} \leq \omega \leq 300$ Hz, and the results given in Figures 5*a* and *b* are presented with an emphasis on this range.

The lowest cold-flow resonant mode, which is related to the entire length of the experimental apparatus, is at a frequency of 15 Hz. This value changes to 17 Hz as the combustor sonic speed is increased to 488 m/sec. At a combustor sonic speed of 549 m/sec, the lowest frequency mode is 18 Hz and becomes 19 Hz at the final combustor sonic speed of 750 m/sec.

The cold-flow mode at 123 Hz, which eventually becomes the hot-flow mode at 185 Hz, moves upward to 129 Hz, 142 Hz, and 154 Hz for the sonic speeds of 366 m/sec, 427 m/sec, and 488 m/sec, respectively, as shown in Figure 5*a*. Figure 5*b* shows that further increasing the combustor sonic speed to 549 m/sec, 610 m/sec, and 671 m/sec shifts this mode to 164 Hz, 173 Hz, and 180 Hz, respectively. Finally, at a sonic speed of 750 m/sec, the resonant acoustic mode which initially had the value of 123 Hz settles into its hot-flow value of 185 Hz. The amplitudes of the peaks are shown to decrease as the combustor sonic speed, *i.e.*, temperature, is increased. The damping in the system is evidently increasing with increasing combustor temperature and, in fact, the reflection coefficient at the exit of the combustion chamber moves further away from an ideal value of -1.0 as the temperature of the exhaust gas increases.

The hot-flow acoustic mode at 222 Hz begins as a very small response at 194 Hz under cold conditions. The response increases in amplitude as the peak shifts up the frequency scale to 195 Hz, 198 Hz, and 200 Hz as the temperature in the chamber is increased as given in Figure 5*a*. This mode, as shown in Figure 5*b*, moves from 202 Hz at a sonic speed of 549 m/sec to 206 Hz and 212 Hz at sonic speeds of 610 m/sec and 671 m/sec, respectively. At the final sonic speed of 750 m/sec, the resonant mode reaches a value of 222 Hz. One notes that the response of this mode increases in amplitude as the sonic speed is increased to 488 m/sec and then decreases in amplitude as the temperature within the combustion chamber is increased further.

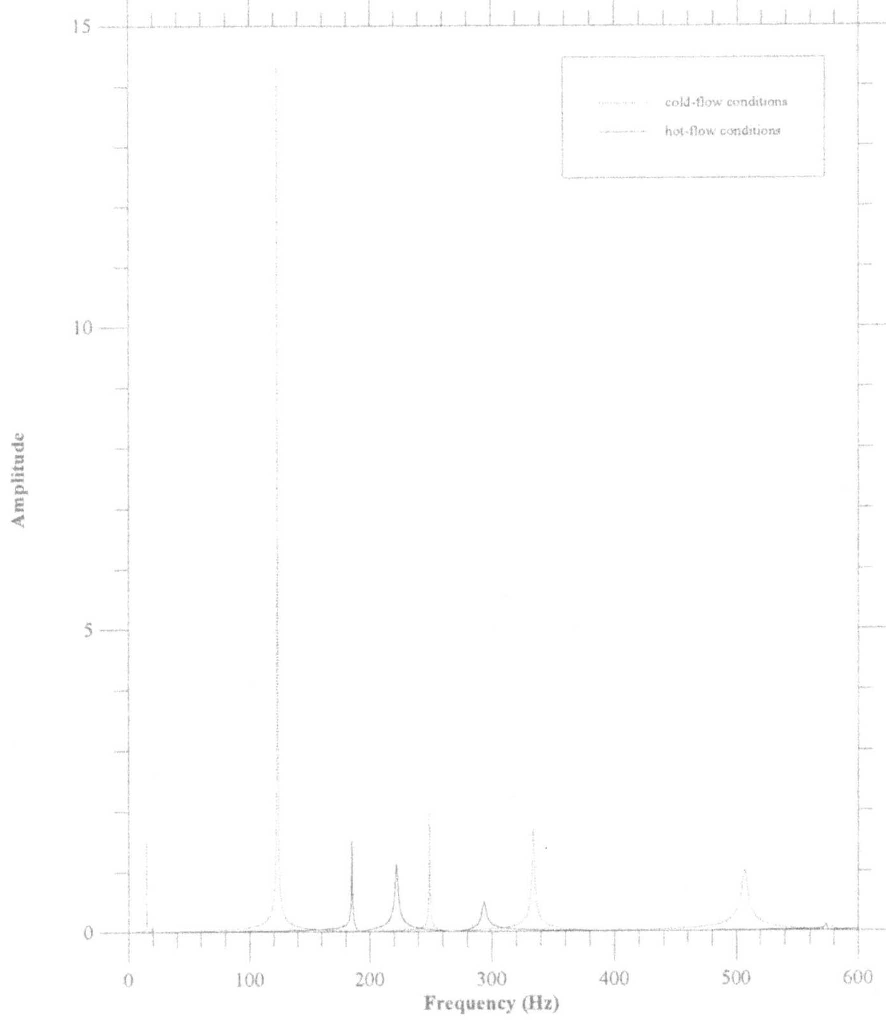

Figure 4: Pressure amplitude spectra for 2.5-cm deep chamber under cold- and hot-flow conditions with unit driving at dump plane.

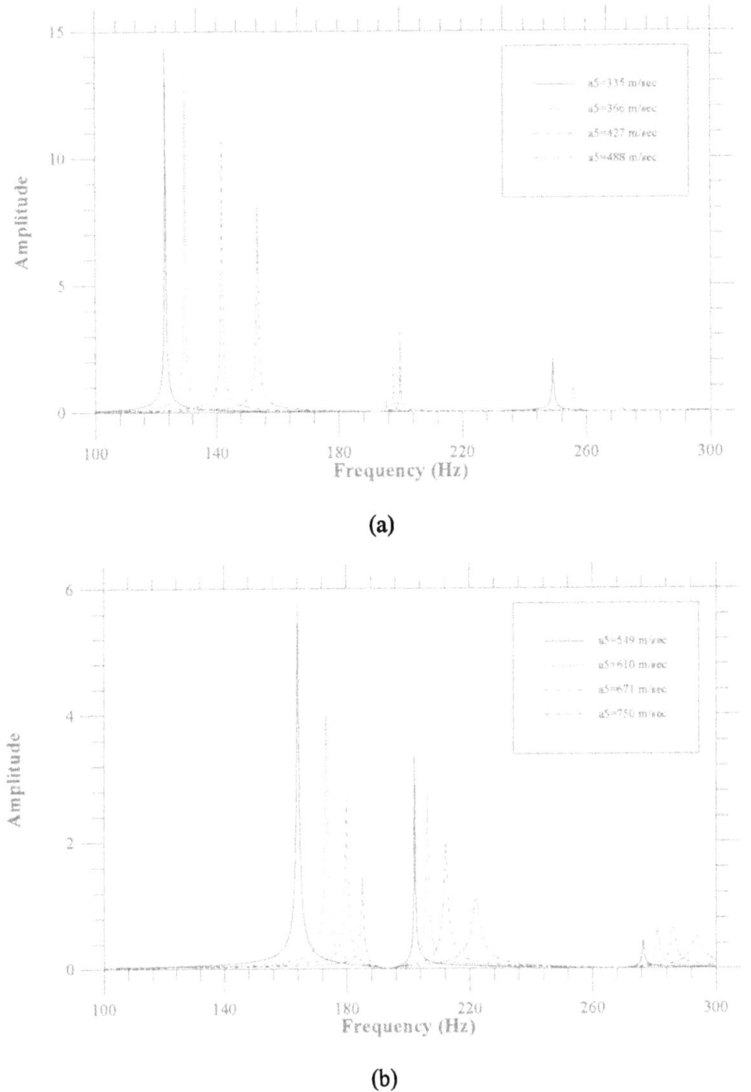

Figure 5: Pressure amplitude spectra for 2.5-cm deep chamber with unit driving at dump plane and 100 Hz $\leq \omega \leq$ 300 Hz for various combustor sonic speeds (*a*) 335 m/sec $\leq a_5 \leq$ 488 m/sec, γ_5=1.4 and (*b*) 549 m/sec $\leq a_5 \leq$ 750 m/sec, γ_5= 1.3.

The primary acoustic mode having the highest frequency under hot-flow conditions is predicted to be at 294 Hz for the 2.5-cm deep chamber. The origin of this mode is the 249 Hz cold-flow mode. As the temperature in the combustion chamber is increased, Figure 5*a* shows that this mode shifts to 256 Hz and to 265 Hz as the sonic speed changes from 366 m/sec to 427 m/sec, respectively. The amplitude of the response is decreasing rapidly so that at a frequency of 272 Hz, corresponding to a combustor sonic speed of 488 m/sec, the response is but a fraction of that under cold-flow conditions. Figure 5*b* shows that the natural mode occurs at frequencies of 277 Hz, 281 Hz, and 286 Hz for combustor sonic speeds of 549 m/sec, 610 m/sec, and 671 m/sec, respectively, before reaching the value of 294 Hz at a sonic speed of 750 m/sec.

The primary resonant modes of the larger, 7.6-cm deep combustion chamber having an expansion ratio of 12.0 are determined by the acoustic model to occur at frequencies of 110 Hz, 258 Hz, 316 Hz, and 501 Hz under cold-flow conditions. The cold-flow resonant modes were not determined experimentally for this chamber configuration. Under hot-flow conditions, the acoustic modes are predicted to occur at frequencies of 190 Hz, 222 Hz, and 278 Hz. The pressure amplitude spectra for the two flow conditions are plotted in Figure 6.

For hot-flow conditions, increasing the cross-sectional area of the combustion chamber by a factor of 3.0 has lead to a 2.6% increase in the frequency of the lowest primary acoustic mode and a 5.4% reduction in the frequency of the highest resonant mode as compared to the values of 185 Hz and 294 Hz, respectively, for the chamber with an expansion ratio of 4.0. Interestingly, the mode at 222 Hz has remained the same.

3.1.2. Shapes of Primary Resonant Modes

The one-dimensional model is also able to give the shape of a particular pressure and velocity mode. The pressure mode shapes of the three primary acoustic modes of 185 Hz, 222 Hz, and 294 Hz are given in Figures 7 for the 2.5-cm deep chamber where the dump plane is located at $x = 0$. The plenum chamber comprises the region -134 cm$\leq x \leq$-51 cm. The inlet section is between -51 cm$\leq x \leq$-7.6 cm, and the flame holder is at -7.6 cm$\leq x \leq$0. The combustion chamber occupies $0\leq x \leq 83$ cm. It can be seen that the mode at 185 Hz has a wavelength which is approximately twice as long as the length of the plenum chamber. Because the cross-sectional area of the plenum is large compared to that of the inlet section, it can be thought of having closed conditions at either end. The ends, therefore, are pressure antinodes and velocity nodes as depicted in Figure 7. In fact, dividing the sonic speed of 335 m/sec by twice the plenum length gives a frequency of 201 Hz, a difference of 8.6% from the 185-Hz value given by the acoustic model. This mode will be referred to as the 'plenum mode' at times in the following text.

The combustion chamber, whose cross-sectional area is much larger than that of the flame holder segment, behaves as if it had closed-open end conditions as does an organ pipe. Its natural frequency, therefore, has a wavelength which is four times the length of the combustion chamber and is equal to 226 Hz. This value is merely 1.8% different from the acoustic mode of 222 Hz predicted by the one-dimensional model. Henceforth, this mode will be referred to as the 'organ pipe mode'. One is unable to assign a particular segment of the apparatus to the 294 Hz acoustic mode.

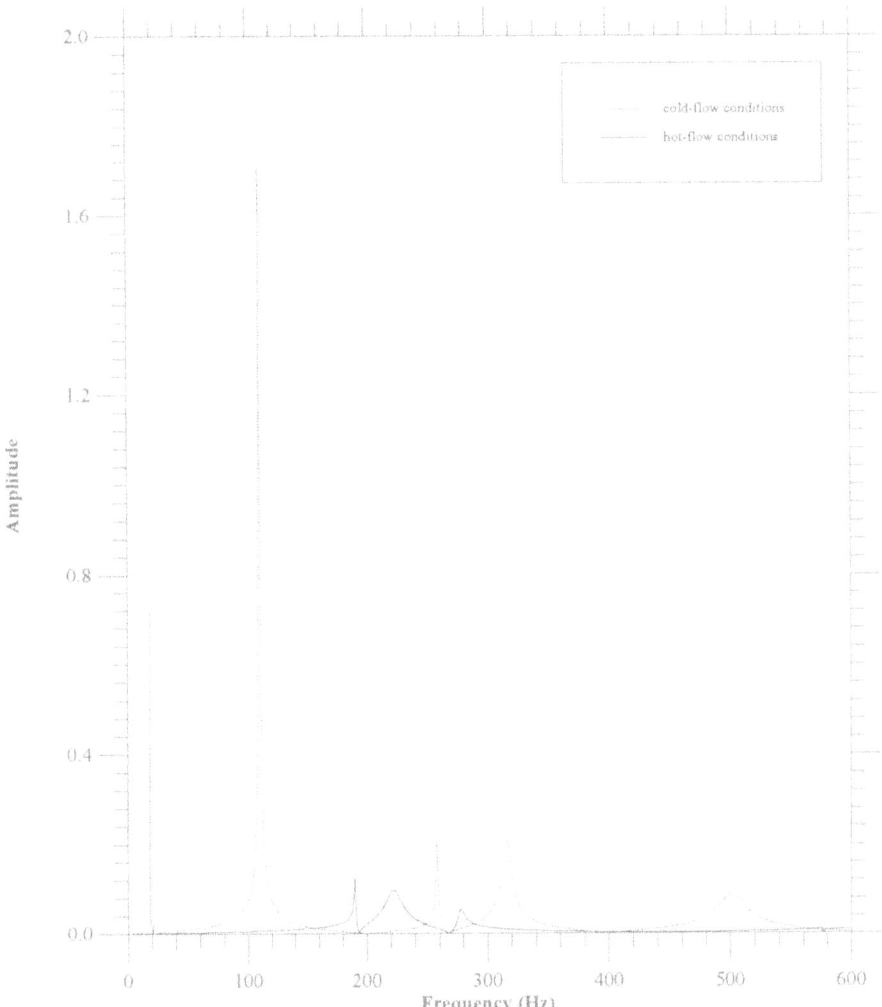

Figure 6: Pressure amplitude spectra for 7.6-cm deep chamber under cold- and hot-flow conditions with unit driving at dump plane.

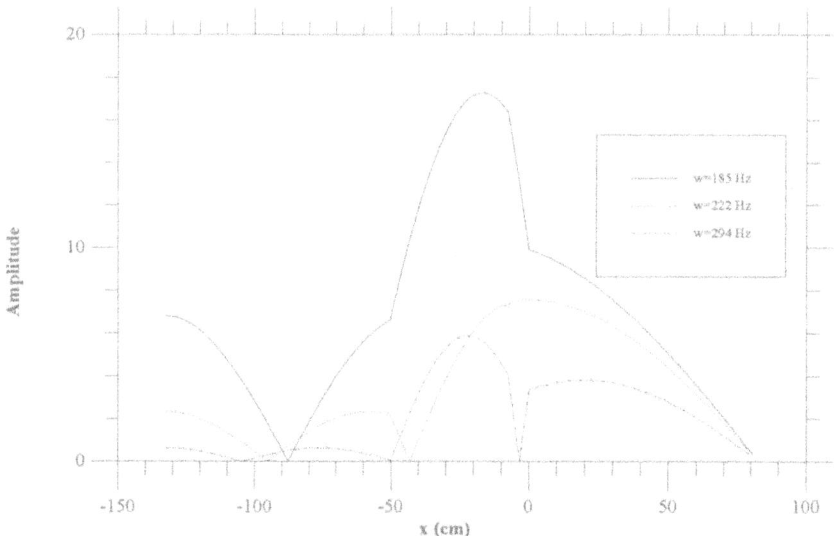

Figure 7: Pressure mode shapes of primary acoustic modes of 2.5-cm deep chamber.

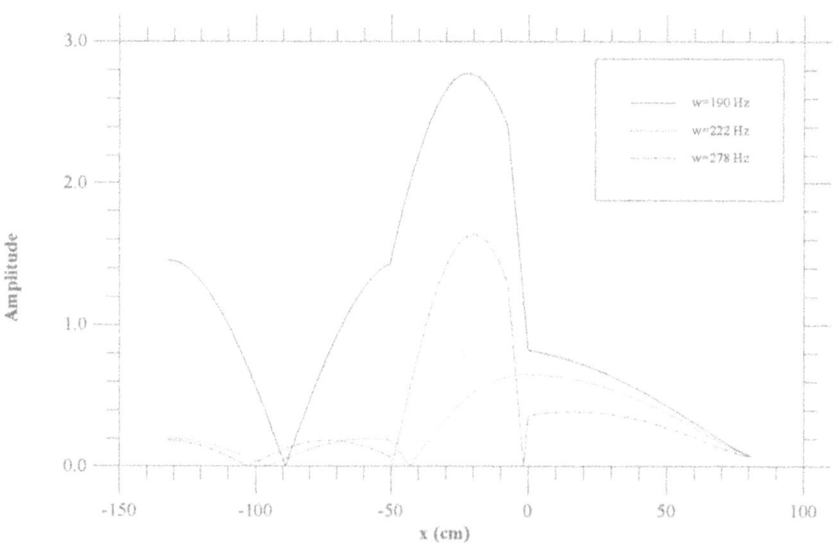

Figure 8: Pressure mode shapes of primary acoustic modes of 7.6-cm deep chamber.

Pressure mode shapes of the three primary acoustic modes at frequencies of 190 Hz, 222 Hz, and 278 Hz for the 7.6-cm deep chamber are given in Figure 8. Because the lengths of the components of the apparatus have not changed, the coordinates given above for the boundaries of each segment remain the same. Again, it is noted that the 'plenum mode' at 190 Hz has a wavelength which is nearly twice as long as the plenum chamber, and the wavelength of the 'organ pipe mode' of the combustion chamber at 222 Hz is approximately four times as long as the chamber.

3.1.3. Effects of Varying the Location of Heat Addition

The location of the driving or mass source within the combustor was found to have a negligible effect on the values of the resonant frequencies for either the 2.5-cm or 7.6-cm tall combustion chamber configurations under hot-flow conditions. To obtain a better understanding of how the amplitudes of the resonant modes change as the location of the driving varies, Figures 9*a* and *b* show the variation of pressure amplitudes with location *LH/LC* for the 2.5-cm and 7.6-cm tall combustors respectively.

3.1.4. Examination of Pressure and Velocity Data for the Two Expansion Ratios

Typically, the pressure oscillations within the combustor near the dump plane and the velocity fluctuations just upstream of the lip of the rearward-facing step are recorded during each experimental run. The pressure oscillations are much stronger for the smaller depth duct, where

$$\left|\frac{p}{p_o}\right|_{2.5\ cm} \leq 0.15$$

$$\left|\frac{p}{p_o}\right|_{7.6\ cm} \leq 0.05.$$

Flow visualization shows that in either case, the surging flow into the chamber is periodically halted and at times reverses direction over the step so that

$$\left|\frac{u}{u_o}\right|_{2.5\ cm\ and\ 7.6\ cm} \geq 1.0.$$

The pressure amplitude spectrum for the 2.5-cm deep chamber is shown in Figure 10 and Figure 11 is the pressure amplitude spectrum of the 7.6-cm deep combustor. As one can see, the spectra are strikingly different. The 2.5-cm deep chamber is excited almost exclusively at the frequency of 186 Hz while the 7.6-cm deep chamber responds with excitations of the 187-Hz and 235-Hz modes. It is apparent that the different geometries have produced different acoustic behaviors signifying a change in the coupling mechanism between the pressure and the heat release within the

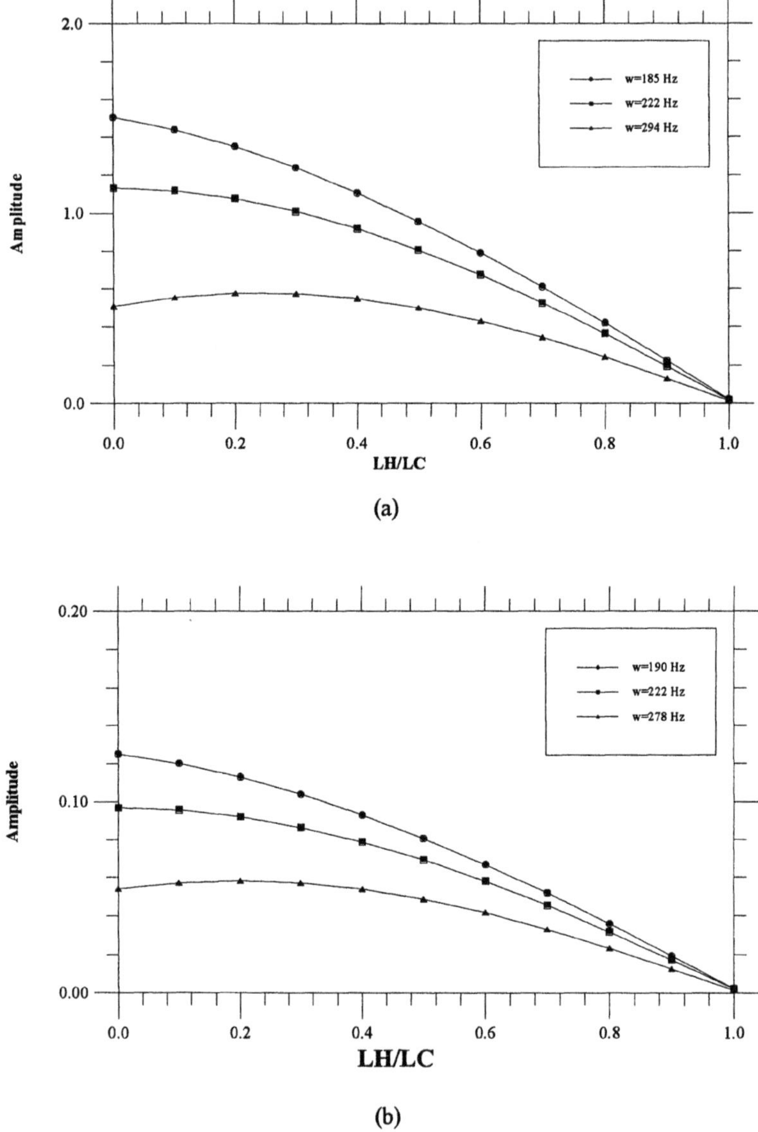

Figure 9: Pressure amplitudes of primary acoustic modes of (*a*) 2.5-cm deep chamber and (*b*) 7.6-cm deep chamber as a function of unit driving location.

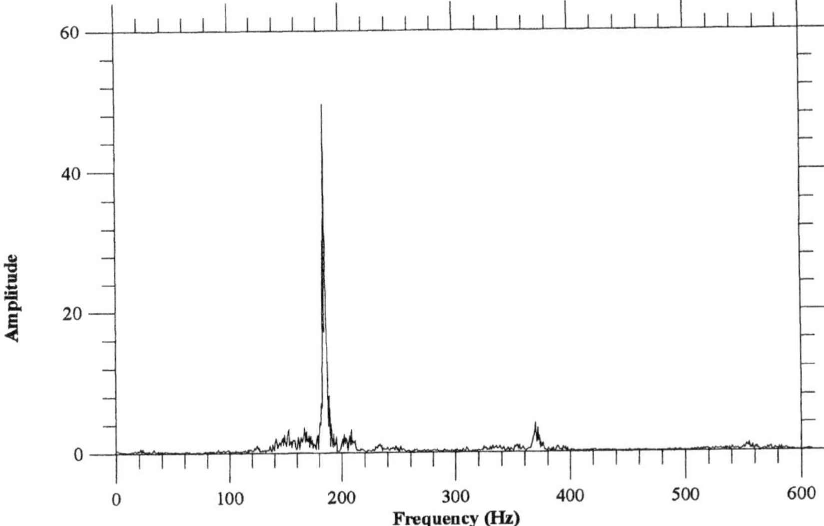

Figure 10: Experimental pressure amplitude spectrum for 2.5-cm deep chamber.

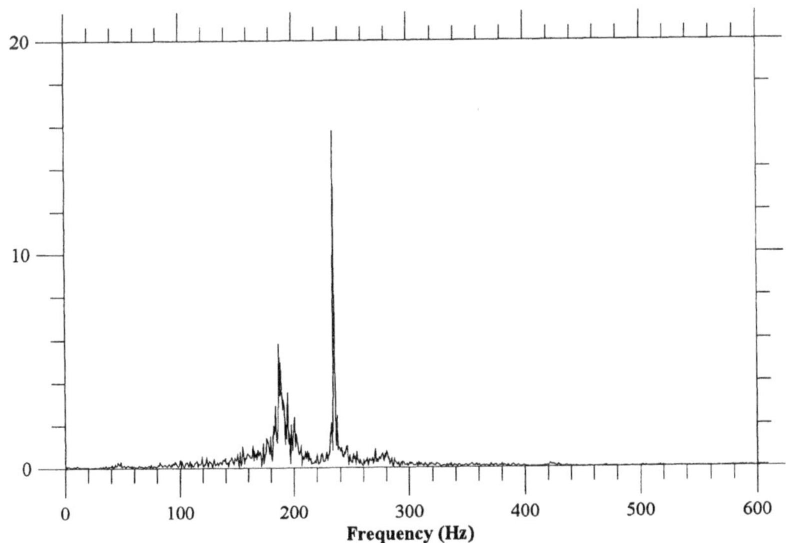

Figure 11: Experimental pressure amplitude spectrum for 7.6-cm deep chamber.

combustion chamber due to the configuration change. A further analysis of the experimental data and the one-dimensional model is required. The larger chamber appears to be the most interesting with its multiple mode excitation; therefore, this case will be examined first.

3.1.5. Comparison of Experimental Spectra with Those Predicted by Acoustic Model

It is evident that the acoustic model predicts the resonant frequencies of the system quite well. For example, there are significant peaks associated with the frequencies of 187 Hz and 235 Hz as well as a much smaller peak at 281 Hz for the 7.6-cm deep combustion chamber. Remember that the model gave the natural modes as 190 Hz, 222 Hz, and 278 Hz for this configuration. The model as derived, however, is unable to correctly predict the relative magnitudes of the three peaks in the $p/\gamma p_o$ amplitude spectrum. The reason for this is simply because the acoustic model assumes a unit driving for all frequencies. Clearly, the heat addition, or driving, is not of uniform magnitude over the frequency range of interest. Rather, it is coupled with the vortex shedding from the lip of the rearward-facing step flame holder. Figure 12*a* shows the velocity spectrum of the signal from the hot-wire anemometer which is placed just upstream of the dump plane in the cold flow for the 7.6-cm deep chamber. Once again the resonant modes are clearly represented in the spectrum with the mode at 235 Hz dominating the others. It is expected that the heat release follows the same pattern with frequency as the velocity; however, it is not possible to correctly assume what the characteristics of the driving amplitude spectrum are. It is possible, though, to experimentally determine the amplitude spectrum of the driving by taking advantage of the linear relationship between the light intensity of the chemiluminescence and the heat release of the combustion process. Once obtained, the driving spectrum can be inserted into the acoustic model, and the resulting $p/\gamma p_o$ spectrum can then be more realistically compared to the experimentally obtained pressure amplitude spectrum.

Light emitted from the combustion chamber over the first 30 cm downstream of the dump plane was focused into the photomultiplier tube for both combustion chamber configurations. In each case, visible combustion appeared to terminate at a distance of less than 30 cm downstream of the step, so it is felt that the entire region over which heat release occurs has been accounted for. The resulting light intensity measurement will contain the desired frequency content of the total heat release.

The amplitude spectrum of the heat release for the 7.6-cm deep chamber is shown in Figure 12*b*. Note the similarities with the velocity spectrum of Figure 12*a*, namely, a dominating peak at 235 Hz with smaller peaks at the other resonant modes on either side. This figure gives the frequency content of the total heat release for the 7.6-cm deep chamber under the given flow conditions. Clearly, the driving is not uniform over the frequency range of interest. A calculation of the $p/\gamma p_o$ amplitude spectrum as determined from the acoustic model using the experimentally derived driving spectrum gives the result shown in Figure 13*b*. Here, the resonant frequencies of the experimental driving spectrum have been shifted by a few hertz so that the experimental modes coincide with those of the one-dimensional model. A driving magnitude of 1.5 m^3/sec, such that

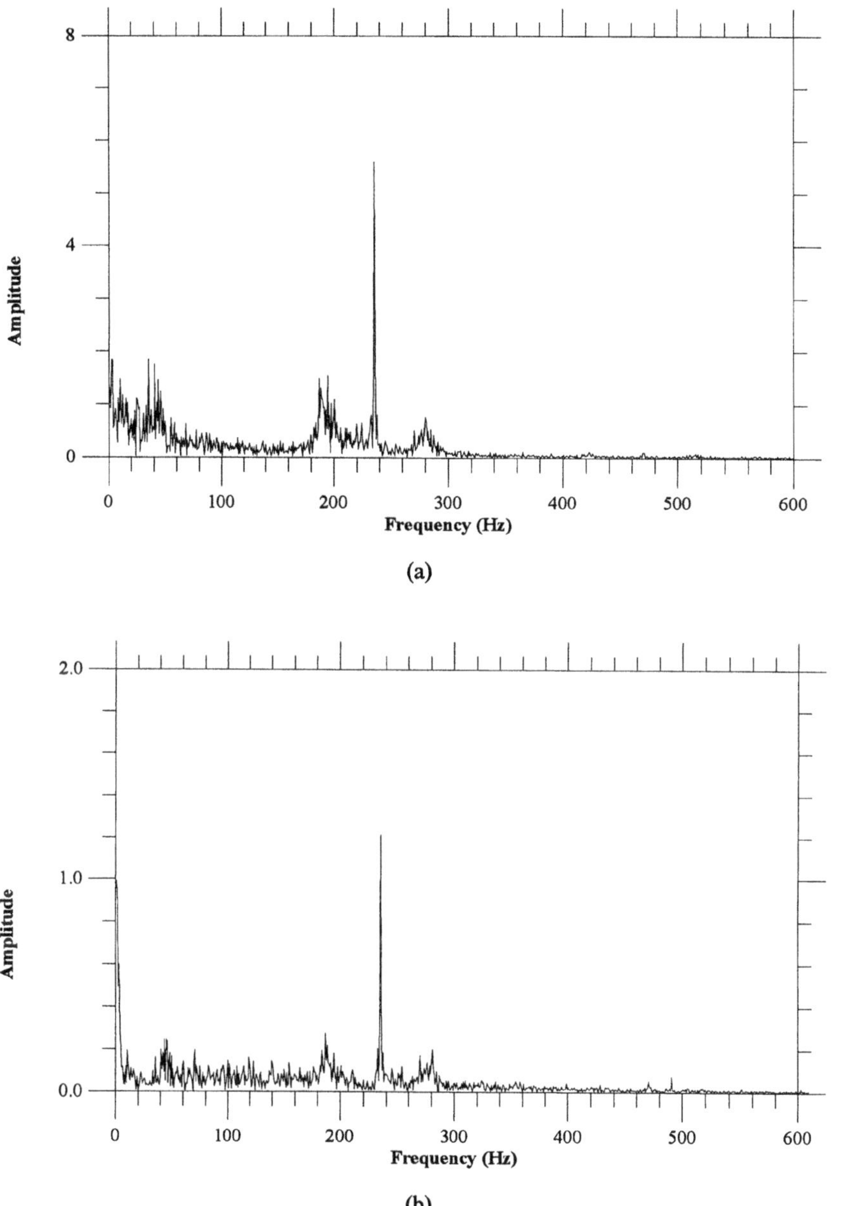

Figure 12: Experimental (*a*) velocity and (*b*) total light intensity amplitude spectra for 7.6-cm deep chamber.

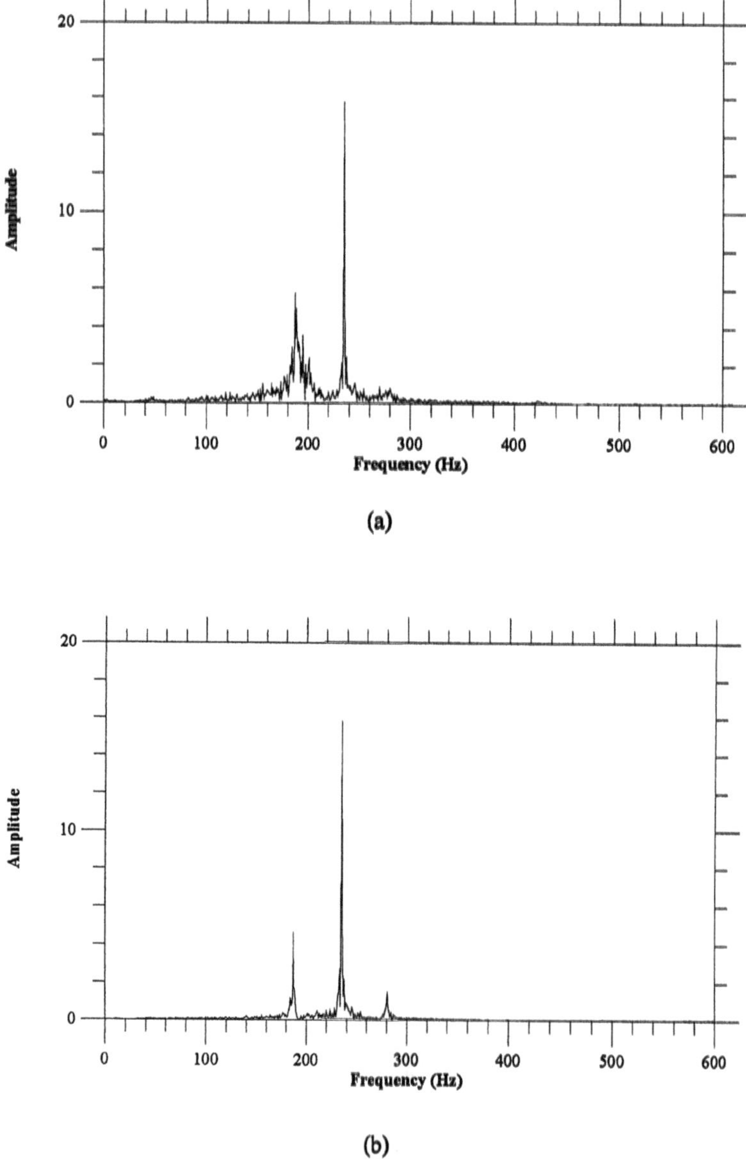

Figure 13: Comparison of (*a*) experimental and (*b*) predicted pressure amplitude spectra for 7.6-cm deep chamber.

$$D_{7.6\,cm}(\omega) = \frac{I_{7.6\,cm}(\omega)}{\sum_{\omega} I_{7.6\,cm}(\omega)} \times 1.5 \text{ m}^3/\text{sec},$$

has been used so that the peak at 235 Hz scales with that in the experimental pressure amplitude spectrum shown again in Figure 13*a* allowing the two spectra to be shown on the same scale. In the above equation, $D_{7.6\,cm}(\omega)$ is the driving at frequency ω and $I_{7.6\,cm}(\omega)$ is the experimentally determined value of the radiation intensity amplitude at a particular frequency ω for the 7.6-cm deep chamber. The two spectra of Figure 13 agree quite well. There is a slight underprediction by a factor of 0.96 in the amplitude of the 187-Hz mode and an overprediction by a factor of 2.42 for the 281-Hz mode. Because the mode at 281 Hz is of little acoustic consequence compared to the other modes, the overprediction of its amplitude by nearly a factor of 2.50 is not very disturbing.

As noted previously, the pressure within the 2.5-cm deep combustor responds almost exclusively to excitations occurring at a frequency of 185 Hz. There is a much smaller response near the harmonic of this frequency at 369 Hz. One would expect the dominant vortex shedding and heat release mode to be equal to 185 Hz as well, and the velocity and light intensity spectra of Figure 14*a* and 14*b*, respectively, indicate that indeed vortices are being shed and heat release is occurring primarily at a frequency of 185 Hz. Both plots also indicate a slight response at 208 Hz, a non-acoustic mode as determined by the one-dimensional model which gave the frequencies of 185 Hz, 222 Hz, and 294 Hz as the resonant modes for this configuration. Because the length scales have remained the same for this expansion ratio, it is not plausible that the 'organ pipe mode' of the combustion chamber has dropped 6.3% from its value of 222 Hz to a value of 208 Hz; therefore, the small response at this frequency must originate from the combustion process itself rather than from any acoustic means. This non-resonant mode in the velocity and heat release spectra does not appreciably appear in the pressure response.

When the heat release spectrum of Figure 14*b* is used as the driving spectrum for the one-dimensional model, the resulting $p/\gamma p_o$ spectrum is shown in Figure 15*b*. Here, a driving magnitude of 0.5 m^3/sec, such that

$$D_{2.5\,cm}(\omega) = \frac{I_{2.5\,cm}(\omega)}{\sum_{\omega} I_{2.5\,cm}(\omega)} \times 0.5 \text{ m}^3/\text{sec},$$

has been used so that the peak at 185 Hz scales with that at the same frequency in the experimental pressure amplitude spectrum shown again in Figure 15*a*. The quantity $D_{2.5\,cm}(\omega)$ in the above equation is the driving at frequency ω and $I_{2.5\,cm}(\omega)$ is the value of the radiation intensity amplitude spectrum at a particular frequency ω as determined experimentally for the 2.5-cm deep combustion chamber. One can see that the model predicted a much stronger response at the

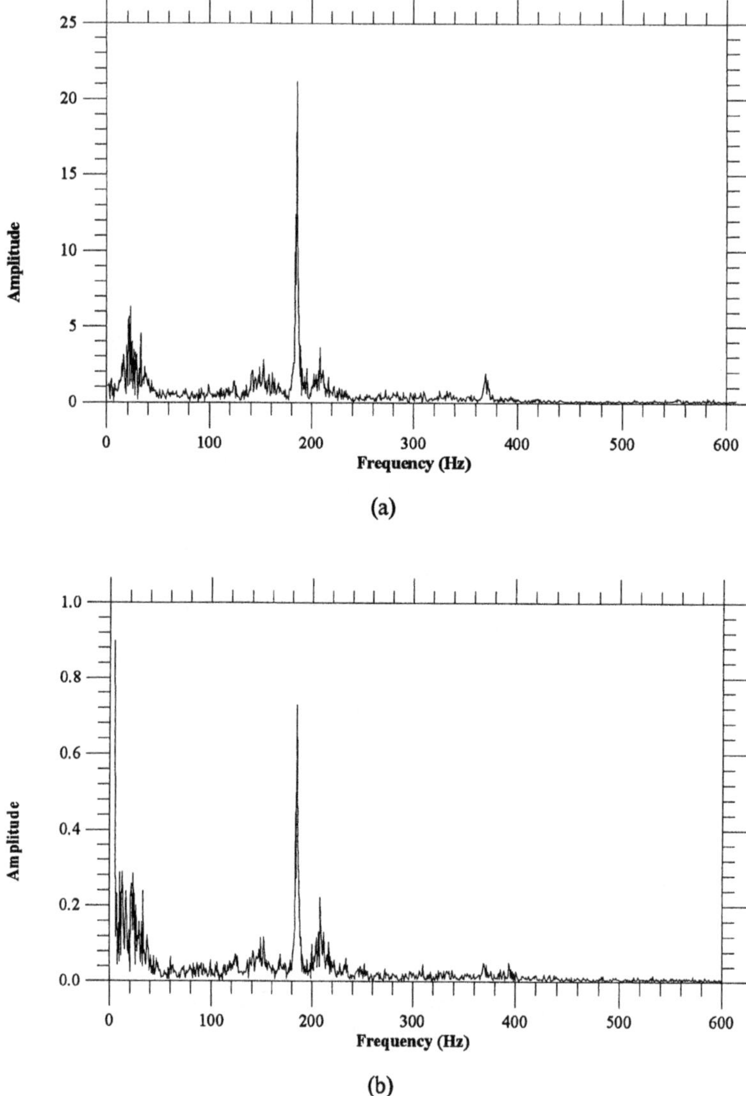

Figure 14: Experimental (*a*) velocity and (*b*) total light intensity amplitude spectra for 2.5-cm deep chamber.

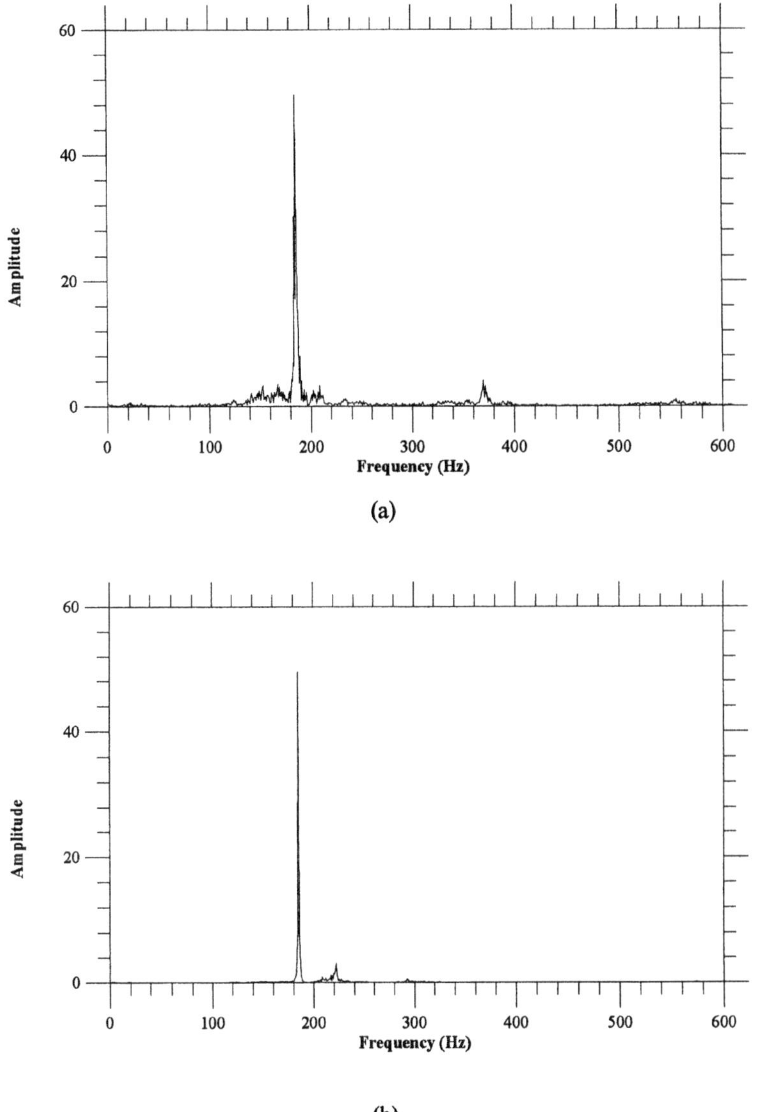

Figure 15: Comparison of (*a*) experimental and (*b*) predicted pressure amplitude spectra for 2.5-cm deep chamber.

resonant mode of 222 Hz, stronger by a factor of 17.5, and failed to predict the responses at the non-acoustic mode of 208 Hz and the harmonic at 369 Hz. Because these responses are still negligible when compared to the dominant response at 185 Hz, these shortcomings are not of great importance.

The investigation now shifts to a more focused examination of the reacting vortex structures shed from the lip of the rearward-facing step flame holder in an effort to explain the different behaviors of the dump combustor for the different chamber configurations.

3.2 EXPERIMENTAL RESULTS

3.2.1. Vortex Roll-up

Some 33 unstable cases have been investigated but only the four listed in Table 1 will be described. The 7.6 cm case will be discussed in detail to depict the overall injection, mixing, and burning of the vortex structures. In Table 2, "Frequency" refers to the dominant vortex shedding frequency, associated with the largest peak in the pressure and velocity FFTs.

Table 2. Experimental Cases

Case	Duct Ht.	Vdump	ϕ	Frequency	Fuel
1	7.6 cm	21 m/s	1.4	237/188 Hz	CH_4
2	2.5 cm	21 m/s	1.4	188 Hz	CH_4
3	5.1 cm	35 m/s	1.2	237 Hz	CH_4
4	5.1 cm	35 m/s	1.4	188 Hz	CH_4

A complete cycle of chemiluminescence images obtained with the Kodak CCD camera for Case 1 is presented in Figure 16. The images have been displayed using the graphics package PV-WAVE and artificially coloured using the displayed colour bar. From this sequence, we see there is a delay from the injection of fresh reactants into the chamber - coincident with a pressure maximum - to ignition. Ignition commences along the leading edge of the structure and proceeds circumferencially around it, continuing inward with time to zones of higher vorticity as evidenced from the finer scale, turbulent fluctuations existing in the structure's core [7]. This ignition sequence is denoted as Mechanism I and is depicted in Figures 17*a* and *b*. Such vortical burning patterns parallel previous experimental work as in reference [8] which details single shot PLIF images of vortex structures in a pulsed diffusion burner. Despite the fact that the burner in the previous reference is of the diffusion type, initial circumferential burning is observed, with quenching along the structure's braids. Quenching along the structure's tail is similarly experienced here, probably due to locally high strain rates or lack of hot combustion products in this area. Building a cycle of events from single shots as done here and in prior work at Caltech, however, prohibits investigating temporal variations which no doubt occur. Cycles with correspondingly

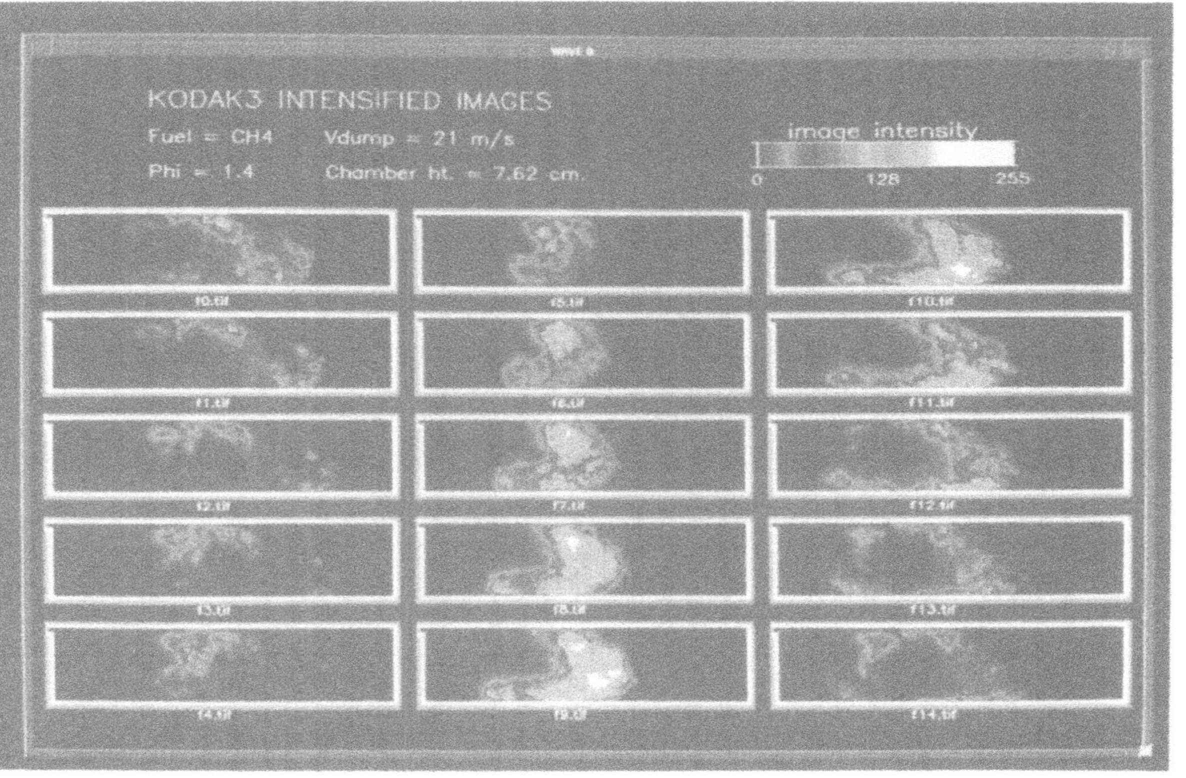

Figure 16: Complete cycle of chemiluminescent images for Case 1.

different pressure and velocity magnitudes reveal other interesting burning patterns as detailed below.

The previous cycle should be compared with others in which ignition occurs closer to the center of the vortex. In the latter case, combustion proceeds from the center to the edges of the structure and this behavior has been observed in previous numerical work [7,9]. This is denoted as Mechanism II type ignition and is also shown in Figures 17*a* and *b*. This difference can be attributed to the characteristics of the preceding structure because it typically provides the hot combustion products necessary for ignition of the subsequent structure. As will be shown later, cycles preceded by larger relative pressure amplitudes and consequently physically larger structures typically take longer time to ignite and burn fully. Consequently, the structure has more time to grow so that when it burns, it is much closer to the lower wall. The interaction between the vortex and the wall reduces the recirculation of hot products upstream toward the next vortex and also cools the recirculating flow by increasing heat transfer to the wall. Thus, the effective ignition mass has been reduced.

Conversely, smaller structures develop and burn earlier and have sufficient vorticity to force more hot products upstream because the structure is higher above the chamber floor at the point of maximum burning. The net result is the following structure is confronted with more hot products near its upstream edge and center causing a more outward burning pattern.

The relative frequency with which these two burning mechanism dominate the flow dictates the burning pattern. Mechanism II promotes a more even distribution of chemiluminescence as evidenced from later images compared with Mechanism I. When both mechanisms operate simultaneously, the greatest heat release is observed with an even more uniform burning pattern.

3.2.2. Ignition Delay and Delay to Intensity Maxima

In order to verify the conjecture that prior structures influence newly shed vortices, delays to ignition and to peak burning were evaluated for Case 1. Consult Figures 18 through 19. Integrated intensity profiles were constructed by averaging an entire image from the CCD camera for each time increment. Delays were plotted against combustor pressures amplitudes during injection (approximately two cycles prior to peak burning) and during maximum burning of the structure. Both delays were taken from the point of maximum pressure as this proved to be the most repeatable shedding location as evidenced from the Hycam work. The exact timing of vortex shedding changed somewhat from cycle to cycle, but typically occurred slightly after maximum pressure. As such, measurements given in the aforementioned figures may have overestimated ignition values by up to 0.3 ms and probably accounted for the observed scatter. In any event, the trends remain intact. Ignition delays and delays to maximum intensities varied inversely with combustor pressures during shedding and varied with combustor pressure during burning. What is seen is that a structure shed nearly two cycles ago influences the present structure's development in the form of burning delays, and a typical vortex lasts almost two complete pressure cycles in the duct. Obviously, the large chamber height promotes the lengthy development of the structure in contrast to other pulsed combustors which usually have only one injected mass burning each cycle. As will be shown, such development is hampered for smaller duct heights and causes ignition and peak burning to occur within one cycle after shedding.

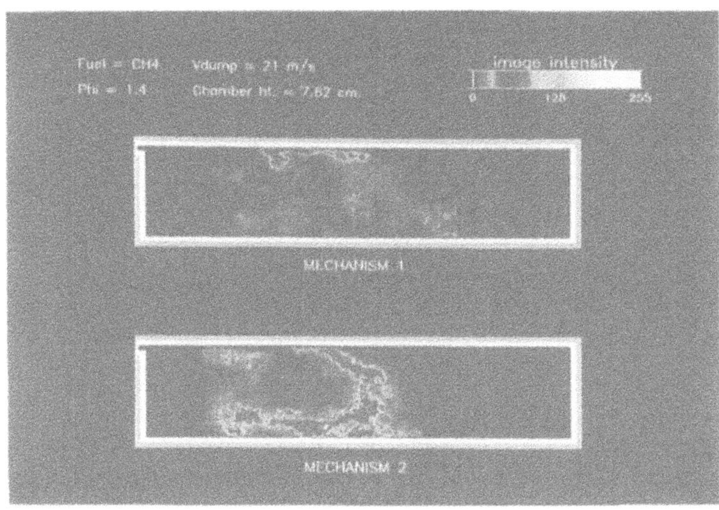

Figure 17*a*: Actual Ignition Mechanisms for Case 1.

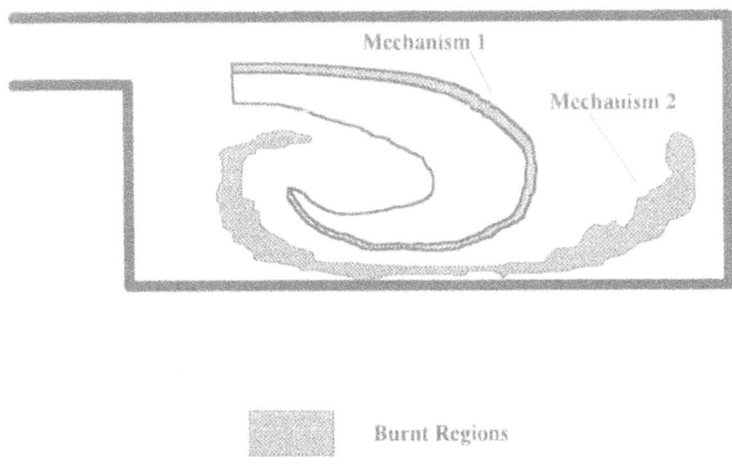

Figure 17*b*: Sketch of Ignition Mechanisms for Case 1.

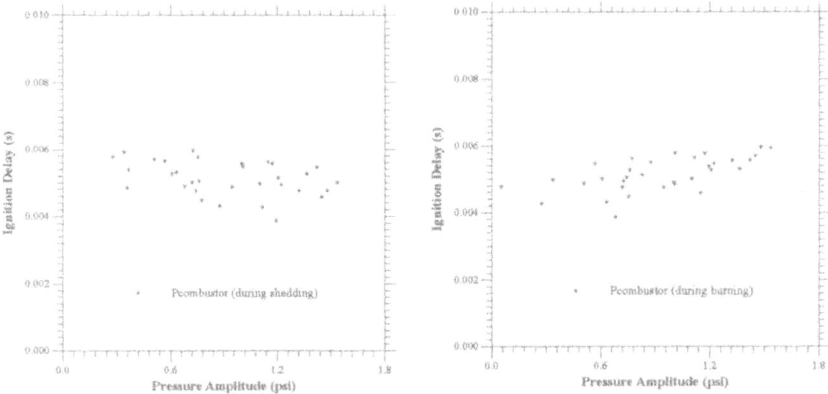

Figure 18: Ignition Delays for Case 1 plotted against combustor pressure during shedding and during burning.

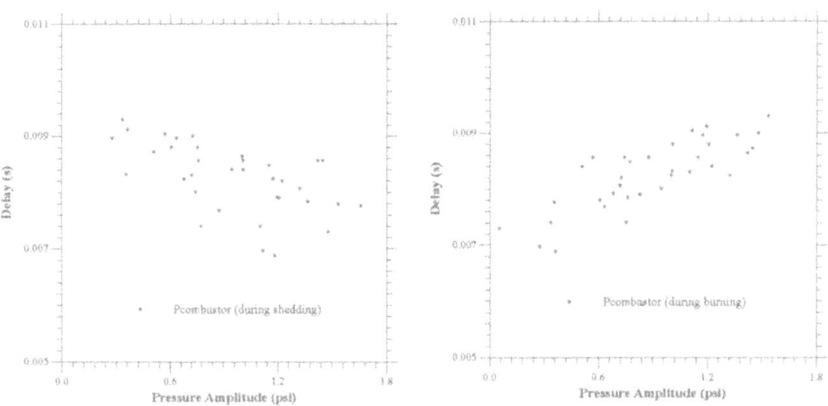

Figure 19: Delays from vortex shedding to their maximum burning intensities plotted against combustor pressure during shedding and during burning.

For Case I, there are typically three structures in the chamber at once: one that has just burnt (and consequently has produced a maximum in chamber pressure), one that was shed just prior to the previous pressure peak, and the latest structure just being shed. The first and second structures being early in their development have experienced no burning yet due to excessive strain rates (discussed later) and insufficient mixing with combustion products. The structure that has just burnt provides the ignition material to ignite the second structure by either of the two aforementioned mechanisms but also assists the latest vortex. As the older structure moves down the duct, combustion products from it move upstream into the recirculation zone beneath the impending or second structure aiding it to burn but also assisting the newest vortex. The arrival of hot gases typically coincides with the location of this newest structure. Hence it is this structure shed nearly two cycles ago that first influences the latest one. The larger the older structure, the larger the amount of products that are recirculated upstream and hence the shorter the ignition delay and delay to maximum intensity. This large quantity of hot products is useless for facilitating Mechanism 2 burning on the second structure since it occupies a thin layer along the combustor floor. As such, Mechanism I ignition is typically encouraged for the second structure when preceded by a larger vortex.

3.2.3. *Comparison between the Four Cases*

Perhaps the most efficient way to compare the four cases above is to show their average chemiluminescence images as done in Figure 20. The shift in burning region towards the dump plane with decreasing chamber height should provide no surprise. Evidently, the chamber floor retards the lengthy development of the structure in contrast for the 7.6 cm case and provides a more rapid and vigorous mixing environment for the structure. Consequently, a lengthening in the burning region is observed. Examination of the Hycam shadowgraph movies for this case clearly details the fine scale, turbulent mixing at work. As such, burning initiates closer to the dump plane due to a more abundant supply of hot combustion products in this region and typically permits only two structures in the chamber at once: the one burning and the one just being shed. A closer examination of the CCD movies reveals that new structures typically ignite from contact with a reacting shear layer starting a short distance from the step. As such, determination of an ignition time for this case seems unnecessary as the structure is typically burning immediately after injection into the chamber. On the other hand, the 5.1 cm cases seem to ignite by a combination of both the 2.5 and 7.6 cm mechanisms. In any event, larger duct heights encourage narrower combustion zones.

There exists, however, a slight variation in the locations of peak combustion for the two 5.1 cm cases: Case 4 (ϕ=1.4) appears to have its maximum intensity region located further downstream than Case 3 (ϕ=1.2). This axial shift in peak burning zones encourages speculation of a mechanism such as straining occurring throughout the structure's development to account for this observation.

3.2.4. *Vortex Straining*

It has been reported that straining can provide the necessary delay mechanism to ensure the correct phasing between the pressure and heat release distributions [4]. This part of the investigation is to

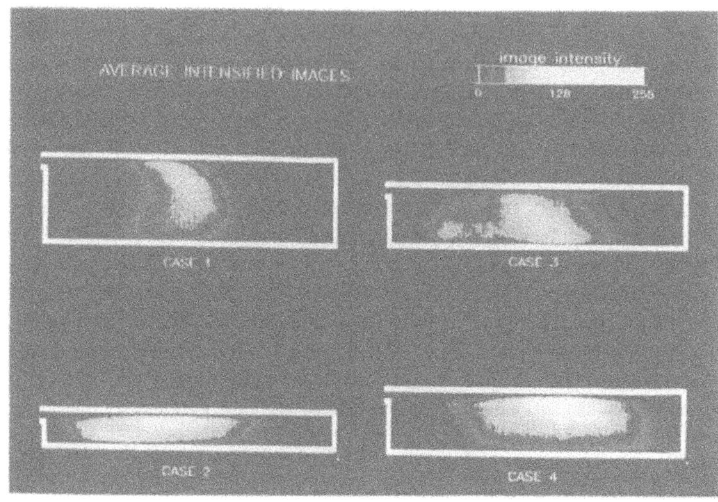

Figure 20: Averaged chemiluminescent images for all experimental cases.

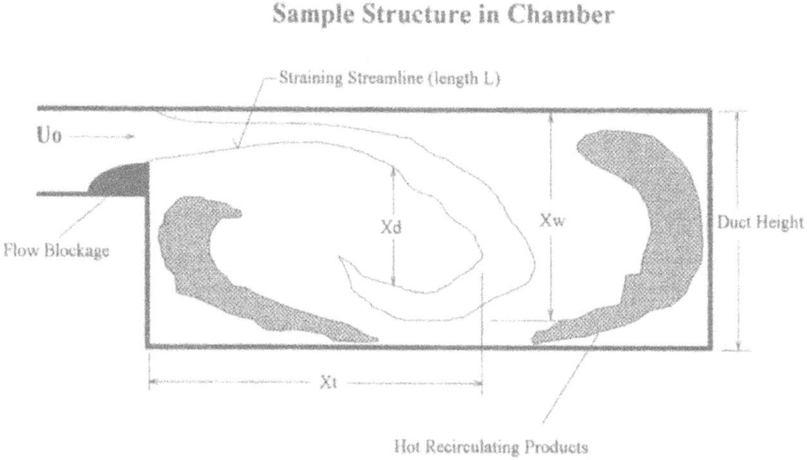

Sample Structure in Chamber

Straining Streamline (length L)

Uo

Flow Blockage

Xd

Xw

Duct Height

Xt

Hot Recirculating Products

Figure 21: Definition of vortex parameters.

experimentally validate whether the strain rate experienced on the structure can facilitate a sufficient mechanism to shift shedding from 188 Hz (Case 4) to 237 Hz (Case 3). The theory stems from the observation that leaner laminar flames can support more straining relative to their richer counterparts [10,11].

Two methods will be employed to study the straining experienced by the structure. Since combustion commences along the circular portion of the structure, one logical definition of the strain rate, $\dot{\varepsilon}$ is $(\dot{x}_d / x_d.)$. Here, X_d is the diameter of the structure, illustrated in Figure 21. A second and perhaps more appropriate method is to base the straining on the length:

$$L - \tfrac{1}{2} U_o\, t \qquad (9)$$

where L refers to the total length of the separation streamline between hot combustion products comprising the recirculation zone and cold reactants. The resulting strain would then be expressed as:

$$\dot{\varepsilon} = \frac{\left[\dot{L} - \tfrac{1}{2} U_o\right]}{\left[L - \tfrac{1}{2} U_o\, t\right]} \qquad (10)$$

The length $\tfrac{1}{2} U_o\, t$ (a simple estimate for the distance traveled by a particle at the center of the shear layer) has been subtracted so the resulting length represents the net stretching of fluid along the separation streamline. Both methods give similar results with the second providing smoother curves since the denominator in equation 10 is larger.

Figure 22*a* and *b* depicts the results for Case 1 for nine various cycles. The laminar extinction limit taken from [10] is included on the plot. From the plot, it is seen that straining can account for nearly half of the ignition delay. Of course other mechanisms (physical mixing of products with reactants, heat transfer to water-cooled wall, etc.) supplement the rest of the ignition delay [12,13]. Evidently, high initial strain rates prohibit early combustion of the structure which could facilitate the correct phasing between heat release and pressure fields to ensure continuation of the limit cycle. Also shown in Figure 23 is a comparison between the straining along the inner and outer radii represented by their respective dimensions X_d and X_w. The observation that the interior strain is generally higher throughout the cycle relative to its outer counterpart suggests a reason why burning tends to typically start along the outer edge of the structure. Of course, the outer edge of the structure has slightly more area of contact with hot combustion products than the interior, though both regions have comparable chances to ignite. The slight increase in the X_w based strain rate at around 3.7 ms is due to the volumetric increase in the vortex's size during combustion. In order to explore the straining theory more rigorously, an attempt will be made to use it to explain the shift in shedding from 188 Hz (Case 2) to 237 Hz (Case 3). Figures 24*a* and *b* depict their respective straining curves, together with their laminar extinction limits. Curves for Cases 3 and 4 have been average over five cycles. It can be seen that Case 3 (ϕ=1.2) should have a shorter ignition delay since it can support more highly strained flames relative to Case 2 (ϕ=1.4).

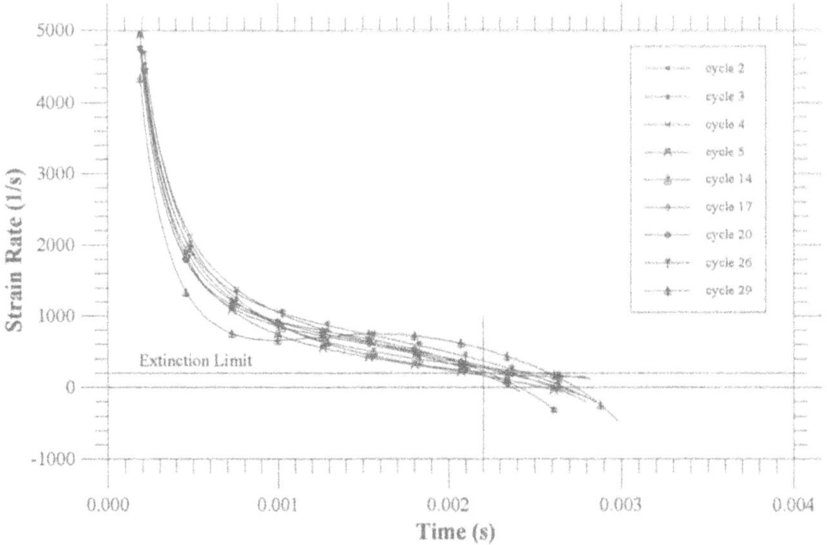

Figure 22*a*: Strain rate curves for Case 1 based on X_d.

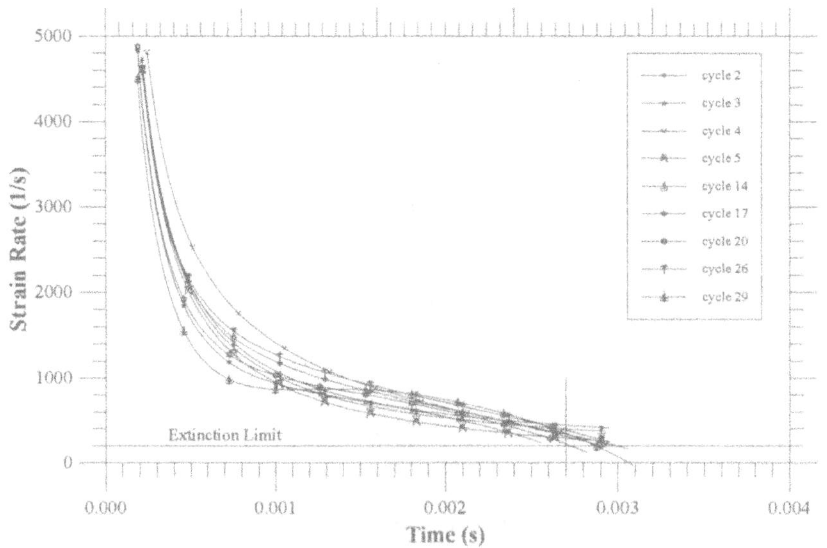

Figure 22*b*: Strain rate curves for Case 1 based on $L - \frac{1}{2}U_o t$

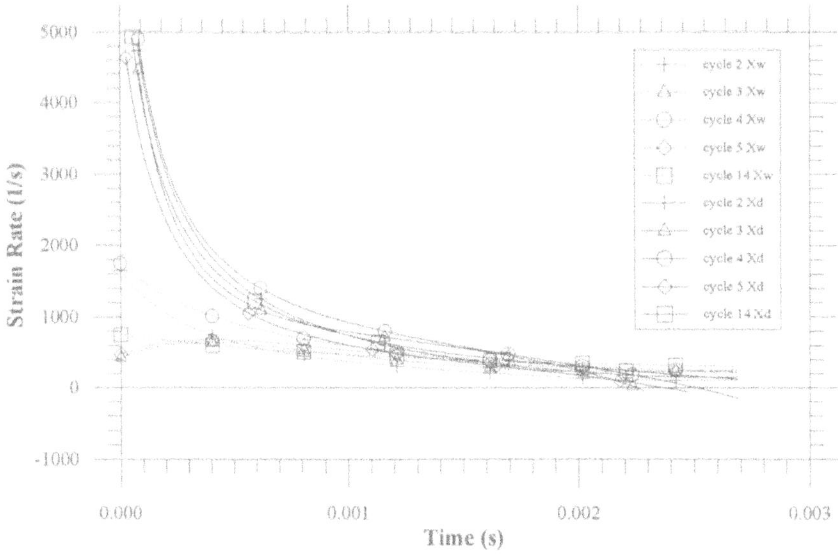

Figure 23: Comparison between strain rates calculated from the parameters X_d and X_w for Case1.

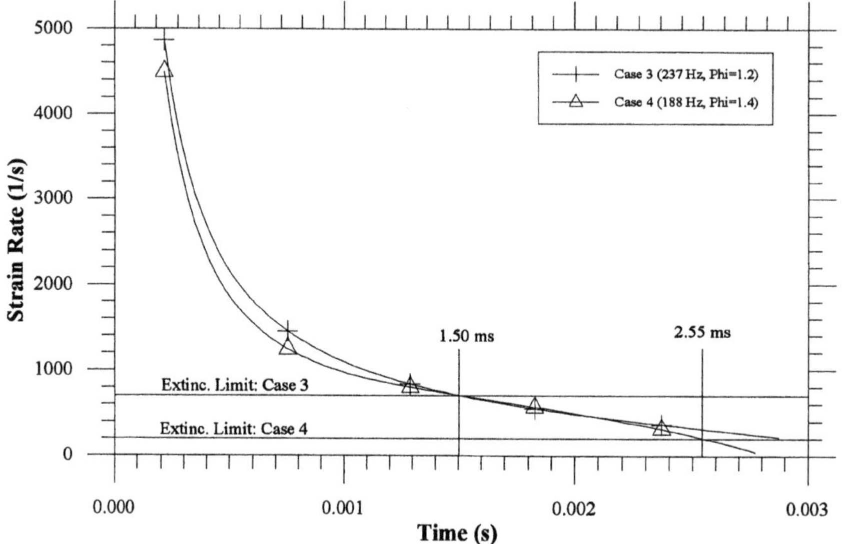

Figure 24*a*: Strain rate curves based on X_d for Cases 3 and 4.

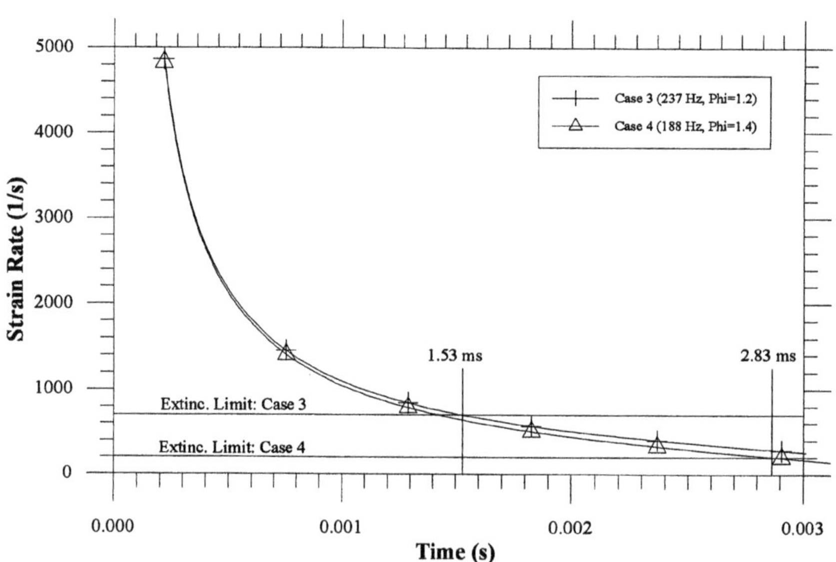

Figure 24*b*: Strain rate curves based on $L - \frac{1}{2}U_o t$ for Cases 3 and 4.

Assuming all other effects are equal between Cases 3 and 4, the shift in shedding should be attributed to the time difference when each case has reached appropriate straining values to support ignition. That is, the richer and consequently lower shedding case must allow more time for straining rates to attain acceptable levels before ignition. From the graphs we see the time difference between when the two cases have their straining rates at acceptable levels corresponds closely to the difference in shedding periods: 2.83ms-1.53ms ≅ 1/(188 Hz)-1/(237.Hz). Even closer agreement is achieved using straining based on X_d. Hence, if it can be shown that the ignition delays between the two cases typically differ by this amount, the variation in shedding frequency would most likely be attributable to straining.

Unfortunately, ignition delay data for the two cases yields very similar results. Although not shown, ignition may result from the burning shear layer commencing a short distance from the step as in the 2.5 cm case so calculation of an ignition delay may seem suspect since the structures may ignite immediately after injection into the chamber. There is, however, a definite shift in peak burning location between the two cases as previously noted and displayed in Figure 20. The 188 Hz case's maximum burning location is approximately 0.04 m further downstream than the 237 Hz case. This corresponds to roughly a 1.1 ms delay between the two cases and is remarkably close to the difference in shedding periods. What is typically observed is although combustion may commence at approximately the same axial location and hence time (assuming identical convection speeds as shown from the Hycam work), it proceeds more fully closer to the step and consequently earlier for the 237 Hz case. As such, peak burning locations are moved further upstream. We attribute this shift to variations in allowable straining values between the two case from changes in their fuel equivalence ratios. Further reinforcement of the roll straining plays in determining shedding frequencies is that when V_{dump} (velocity at entrance to combustor) is increased, straining rates will consequently increase and lower shedding frequencies encouraged. This is precisely what occurs for the majority of cases investigated.

4. Conclusions

One dimensional acoustics have proven successful in predicting the resonant frequencies for the combustion facility used in the investigation. Further, the model successfully predicted pressure FFTs for the 2.5 and 7.6 cm combustor configuration when compared to their experimental counterparts. Location of heat addition represented by a temperature discontinuity was found to have only a slight effect on shifting the frequency of its resonant peaks.

Premixed vortical combustion was found to have two general ignition mechanisms depending on prior conditions within the chamber. Mechanism I involved initial circumferential burning and was due to larger structures previously in the duct. Mechanism II resulted from smaller previous structures which allowed more hot products to recirculate upstream and cause ignition to occur near the center of the vortex.

Reducing the chamber height resulted in burning that occupied a longer axial distance. The smaller duct height inhibited vortex growth and forced relatively more hot products upstream and downstream.

Straining along the separation streamline between hot combustion products and cold reactants proved to account for nearly half of the ignition delay for the 7.6 cm case. It also accounted for the downstream shift in peak burning locations between the 188 Hz case (Case 4) and the 237 Hz case (Case 3). Delaying complete combustion through straining resulted in offsetting the coupling between the heat release and acoustic field and encouraged lower shedding frequencies.

5. References

1. Sterling, J. D. (1987), *Longitudinal Mode Combustion Instabilities in Air Breathing Engines.* Ph.D. thesis, California Institute of Technology, Pasadena, California.

2. Smith, D. A. (1985), *An Experimental Study of Acoustically Excited, Vortex Driven, Combustion Instability within a Rearward Facing Step Combustor.* Ph.D. thesis, California Institute of Technology, Pasadena, California.

3. Sterling, J. D. (1991), "Nonlinear Dynamics of Laboratory Combustor Pressure Oscillations", *Combustion Science and Technology*, Vol. 77, pp 225.

4. Barr, P. K. and J. O. Keller (1991), "Pulsed Combustion: The Importance of Flame Extinction by Fluid Dynamic Strain", Proceedings of the International Symposium on Pulsating Combustion, Vol. 1, paper D-5.

5. Currie, I. G. (1974), *Fundamental Mechanics of Fluids*, McGraw-Hill, New York.

6. Mahan, J. R. et al. (1984), 'A Temperature Correlation for the Radiation Resistance of a thick-walled circular duct exhausting a Hot Gas", *Journal of the Acoustical Society of America*, Vol. 75(1), pp 63.

7. Gohniem, A. F. (1991), "Vortex simulations of Reacting Shear Flow", *Progress in Astronautics and Aeronautics*, Vol. 135, pp 305.

8. Schadow, K. C. et al. (1989), "Large-Scale Coherent Structures as Drivers of Combustion Instabilities", *Combustion Science and Technology*, Vol. 64, pp 167.

9. Peters, N. and F. A. Williams (1988), "Premixed Combustion in a Vortex", *Twenty-Second Symposium (International) on Combustion*, The Combustion Institute, pp 495.

10. Law, C. K. et al. (1986), "Propagation and Extinction of Stretched Premixed Flames", *Twenty-First Symposium (International) on Combustion*, The Combustion Institute, pp 1419.

11. Hertzberg, M. (1984), "Flame Stretch Extinction in Flow Gradients", Flammability Limits Under Natural Convection", *Twentieth Symposium (International) on Combustion,* The Combustion Institute, pp 1967.

12. Keller, J. O. et al. (1990), "Pulse Combustion: The Quantification of Characteristic Times", *Combustion and Flame*, Vol. 79, pp 151.

13. Keller, J. O. et al. (1989), "Pulse Combustion: The Importance of Characteristic Times", *Combustion and Flame*, Vol. 75, pp 33.

6. Acknowledgments

This research was funded by the US Air Force Office of Scientific Research, Grant No. 89-0413, supervised by Dr. Julian Tishkoff. Thanks should also be given to F. E. Marble for his helpful suggestions.

4. THE INFLUENCE OF FUEL/AIR MIXTURE OSCILLATIONS ON THE FORMATION OF SELF-SUSTAINED COMBUSTION INSTABILITIES IN PREMIXED COMBUSTION SYSTEMS

H. Büchner and W. Leuckel
Engler-Bunte Institut,
University of Karlsruhe
Combustion Research Section, P.O. Box 6980,
76128 Karlsruhe, Germany

ABSTRACT- Combustion instabilities are characterized by periodic oscillations of the static pressure in the combustion chamber, in the fuel/air supply lines or in downstream flue gas sections. The energy supply to cover friction and acoustic losses of the oscillating system is provided by the flame in form of periodic heat release rates, which have to fulfill Rayleigh's criterion for the phase shift between the periodic fluctuations of heat release and pressure. Especially at low frequency high amplitude pressure oscillations, the pollutant emission characteristics are being altered and the fuel burnout is often reduced. Because of the increasing use of premixing type burners, as being promoted by the demand for minimized thermal NO_x emissions in industrial combustion facilities, the following investigations have been carried out for a premixed system fired with natural gas. With this system, the influence of periodic pressure oscillations on the fuel gas/air mixture formation was studied, and the effects of periodic changes of the air/fuel ratio on the strength of the pressure excitation and the burnout characteristic of the flame were analysed and discussed.

1. Introduction and Theoretical Considerations

In the last few years the application of premixing type burners in industrial combustion systems like gas turbine combustors, dump combustors or special industrial furnaces has strongly increased, in order to minimize the thermal NO_x emissions. Thereby, the characteristics of premixed turbulent combustion, i.e. high volumetric reaction densities and strong dependence of ignition stability on the equivalence ratio of the fuel/air mixture, cause a distinct rise of combustion instabilities. These instabilities, which are characterized by periodic fluctuations of the heat release of the flame and of the static pressure in the combustion chamber, are formed and sustained by the interaction of the unsteady flow field of the fuel/air mixture in the combustion chamber and the reaction kinetics in the flame on one hand and the resonance of the propagating pressure waves, generated by the periodic combustion process in the combustion chamber, on the other.

In Fig.1 the feedback mechanism for the formation of self-sustained pressure oscillations for premixed combustion is being demonstrated.

F. Culick et al., (eds.), Unsteady Combustion, 71–82.

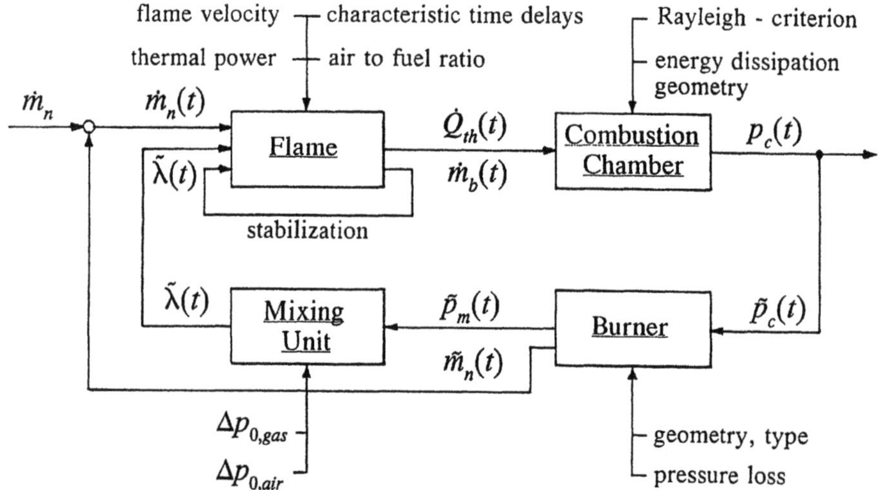

Figure 1: Feedback mechanism for the formation of self-sustained combustion instabilities in premixed combustion systems.

A disturbance of the static pressure in the combustion chamber may cause a fluctuation of the total flow rate of the mixture through the burner nozzle. For premixed combustion the characteristic time delay between the burner outflow of the mixture and its combustion in the main reaction zone of the turbulent flame system consists of contributions for the convective transport, for heating up to ignition temperature by mixing with hot products and the start-up time of the reaction kinetics [1,2,3]. This time delay inside the flame, which strongly depends on the thermal load as well as the equivalence ratio, affects the overall delay time or phase shift permitted by the Rayleigh criterion for the formation of self-sustained instabilities. A disturbance of the chamber pressure may as well cause a change in the mixture formation as shown in Fig.1. This mainly depends on the pressure transfer function of the flow line between the chamber and the mixing unit, i.e. a mixing section or a mixing chamber, and the pressure levels in the supply lines of the mixing unit.

In this study the unsteady mixing of the fuel (natural gas) with the air in the presence of periodic counterpressure oscillations using an often applied injector mixer was investigated and the effect of the oscillating mixture composition on the strength of the generated pressure waves and on the characteristic time delays inside the flame were analysed. To illustrate this intension, the following figure (Fig.2) shows a simple calculation which demonstrates qualitatively the effect of the phase shift between the sinusoidal oscillations of the mixture mass flow rate and the mixture composition, represented by the air/fuel or air equivalence ratio λ. This is the inverse of the fuel equivalence ratio Ø.
The input of this calculation are the sinusoidal time functions of the mixture mass flow rate and the air/fuel ratio λ with given amplitudes. For this estimation we used the kalorimetric flue gas temperatures calculated from the given air/fuel ratios for natural gas (97 % methane) and calculated the flue gas flow rates which are responsible for the pressure change in the combustion chamber, for two different phase shifts (case A and B).

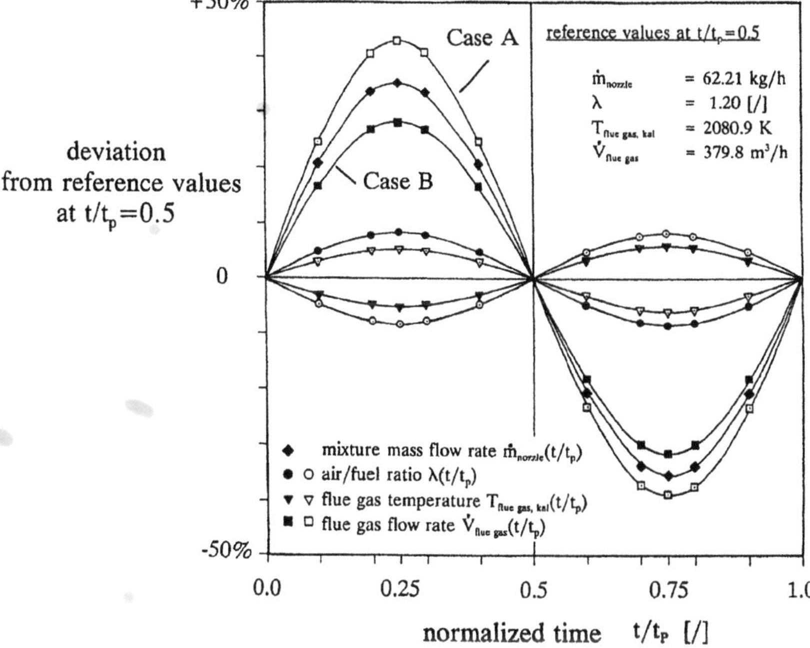

Figure 2: Influence of the phase shift between mixture mass flow rate and mixture composition on the flue gas flow rate.

In case B the oscillation of the air/fuel ratio λ is in phase with the mixture mass flow oscillation. When the mixture mass flow rate reaches its maximum, the mixture composition is fuel lean ($\lambda=1.3$) and therefore the flue gas temperatures are low. In case A there is a phase angle of 180 degrees between the two oscillations and the mixture mass flow rate reaches its maximum at higher flue gas temperatures ($\lambda=1.1$). The difference of the hot flue gas flow rates in the two cases is about 15 % referring to the mean value and therefore the sustainment of pressure oscillations will be favoured in case A.

In addition to this result the periodic change of the mixture composition also directly affects the characteristic time delays inside the flame, i.e. the convective time delay due to the axial position of the main reaction zone and the reaction kinetics. Therefore the relative timing of the reaction rate and the resonant pressure field is altered.

2. Experimental Setup and Measuring Techniques

The results presented in this paper were obtained with the experimental setup shown in Fig.3, which was developed to investigate the dynamic behavior of excited turbulent, premixed jet flames under enclosed and non-enclosed conditions as well as the unsteady natural gas/air mixture formation in the presence of well defined periodic pressure oscillations.

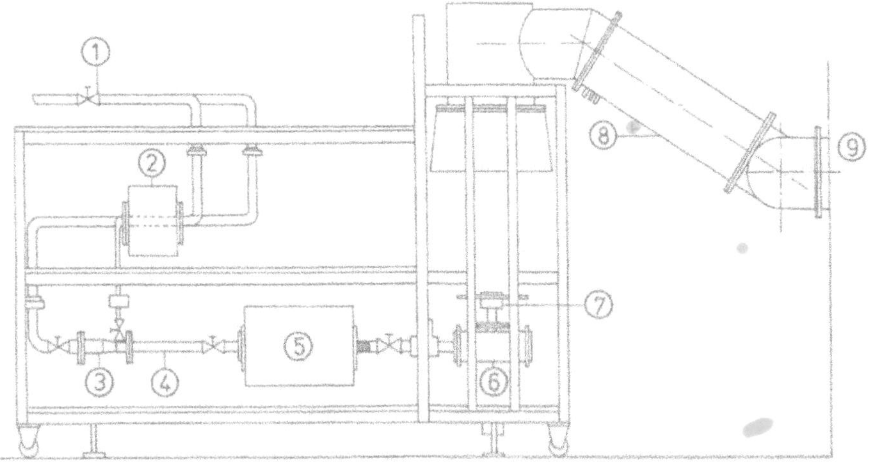

Figure 3: Experimental setup

The reduced schematic shows the main parts of the test rig: the injector mixer (3); the downstream mixing section (4); the pressure decoupling chamber (5) which was removed for the mixture investigations; the pulsating unit (6) and the burner (7).

Fig.4 displays the pulsating unit, which was applied to generate sinusoidal pressure oscillations with adjustable frequency and amplitude in the isothermal part of the test rig.

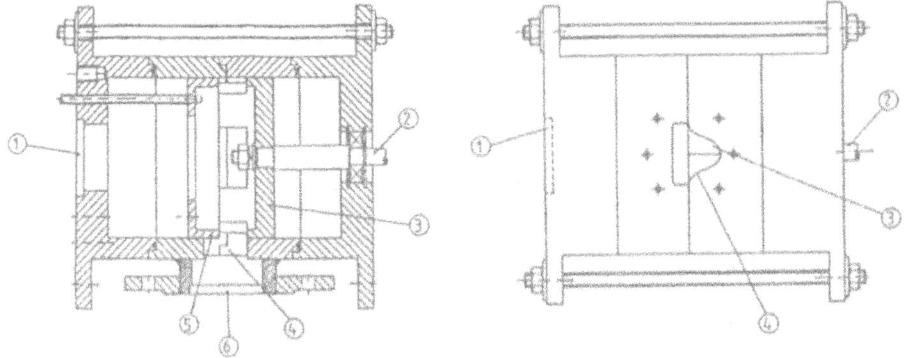

1. entry of fuel/air mixture
2. connection to stepping motor
3. rotating shutter
4. rectangular and sinusoidal area
5. shutterslide
6. outflow to burner

Figure 4: Pulsating unit for the generation of sinusoidal pressure oscillations

The pulsating unit consists of a cylinder (3) with four rectangular notches rotating in a cylindrical housing with a coaxial inflow area (1) and a radial outflow area (4), such that a sinusoidally modulated outflow is generated, which has been verified by phase-locked ensemble velocity measurements. When the sinusoidally shaped outflow area is closing, the static pressure upstream of the pulsating unit increases and the gas/air mixture formation in the injector becomes time dependent. Both frequency and amplitude of the generated pressure oscillations were measured using a microphone, which was installed at the end of the mixing section ((4) in Fig.3)

In order to obtain the time dependent mixture composition at the end of the mixing section where no radial concentration gradients inside the mixture could be detected, a measuring technique with a high time resolution was developed. It is based on the measurement of differences of the heat conductivity in isothermal mixture samples with a tungsten hot wire sensor (length $l_{hw}=2.0$ mm, diameter $d_{hw}=5\mu m$). The developed suction probe is shown in Fig.5.

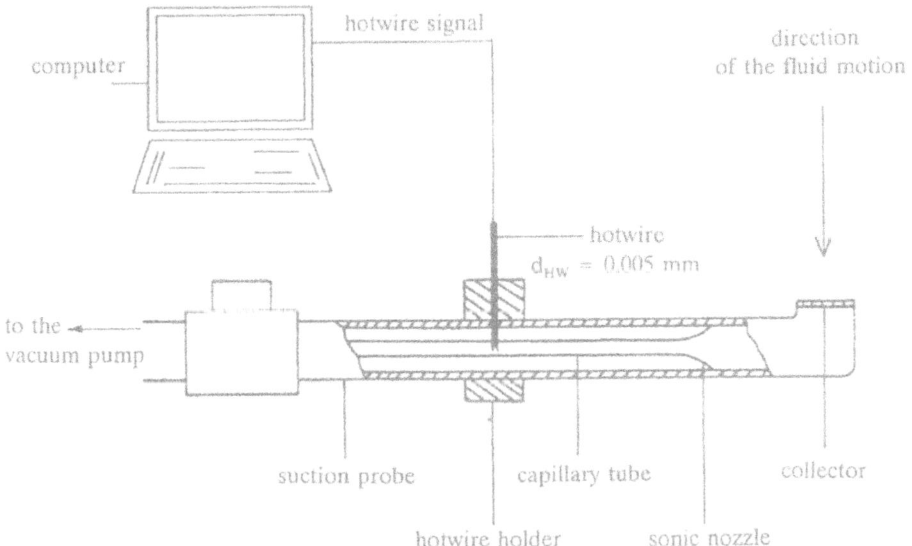

Figure 5: Suction probe for the measurement of the fluctuating composition of the fuel/air mixture

The main part of the suction probe is a sonic flow nozzle with an attached capillary tube, where the continuously sampled mixture is accelerated to sound velocity. As a result of the constant velocity and temperature of the mixture samples at the sensor, the measured changes of the heat transfer rate by forced convection can be correlated with the time dependent heat conductivity λ_{mix} of the samples. In Fig.6 the measured relative increase of the convective heat transfer rate with increasing methane concentration (represented by the air/fuel ratio λ) is plotted for various hot wire sensors and compared with an analytic calculation based on the theory of convective heat transfer in the laminar shear layer of a heated, two-dimensional cylinder in a cross flow (see equ.(1)).

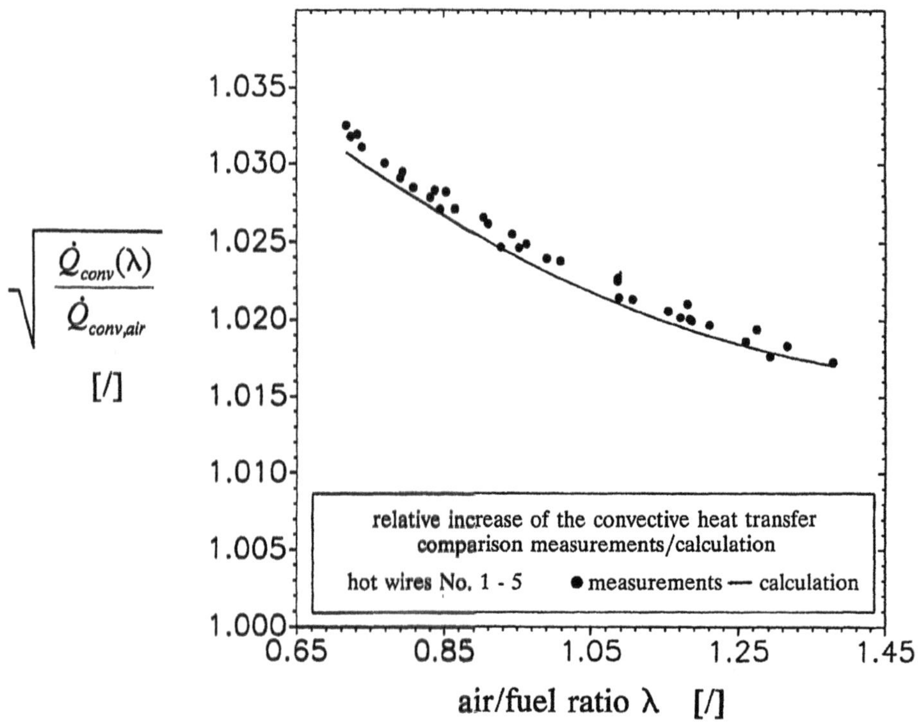

Figure 6: Measured and calculated calibration functions of the heat conductivity probe.

$$Nu = Nu_{min} + Nu_{lam} = 0.3 + 0.664 * Re^{1/2} * Pr^{1/3} \qquad (1)$$

$$Re = \frac{c_{sound} * l_{hw}}{\nu_{mix}} \quad with: \ l_{hw} \approx \frac{\pi}{2} * d_{hw} \qquad Pr = \frac{\rho_{mix} * c_{p,mix} * \nu_{mix}}{\lambda_{mix}}$$

In spite of high flow velocity in the capillary tube, the application of this laminar model is justified because the Reynolds numbers calculated with the flow (sound) velocity c_{sound}, the hot wire geometry l_{hw} and d_{hw} and the viscosity of the mixture ν_{mix}, are very low (Re_{max} = 120 - 150 << $Re_{crit} = 10^5$).

3. Experimental Results of the Steady and Unsteady Mixture Formation

At the beginning of the experimental work, the dependence of the mixture formation in the applied injector mixer was studied under steady flow conditions, starting from a nearly stoichiometric mixture composition. The static pressure levels in the natural gas and air supply lines were held constant for different values of the mean mixture counterpressure. The following diagram (Fig.7) shows the influence of a counterpressure variation on the air/fuel ratio λ for a given pressure difference in the fuel and air supply lines of 0.0 mbar.

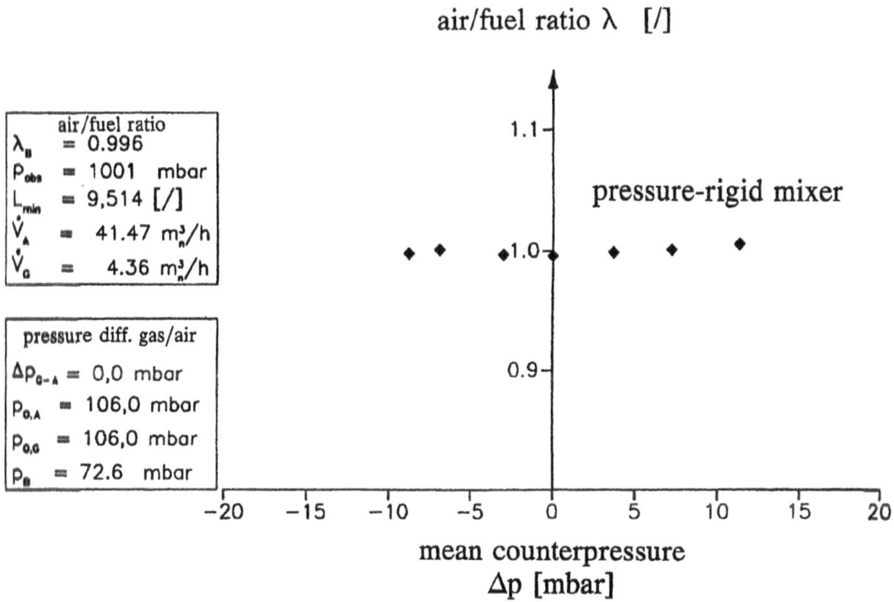

Figure 7: Influence of the mean counterpressure on the mixture formation under steady flow conditions (pressure difference gas/air: 0.0 mbar)

Due to equal values of static pressures in the mixer supply lines, and also equal pressure losses over the two slides, mixture composition remains constant independent of the adjusted value of the mixer counterpressure. The next figure displays the results for the same experiment, but now with a pressure difference in the mixer supply lines of 10.0 mbar (with the higher pressure level in the natural gas supply).

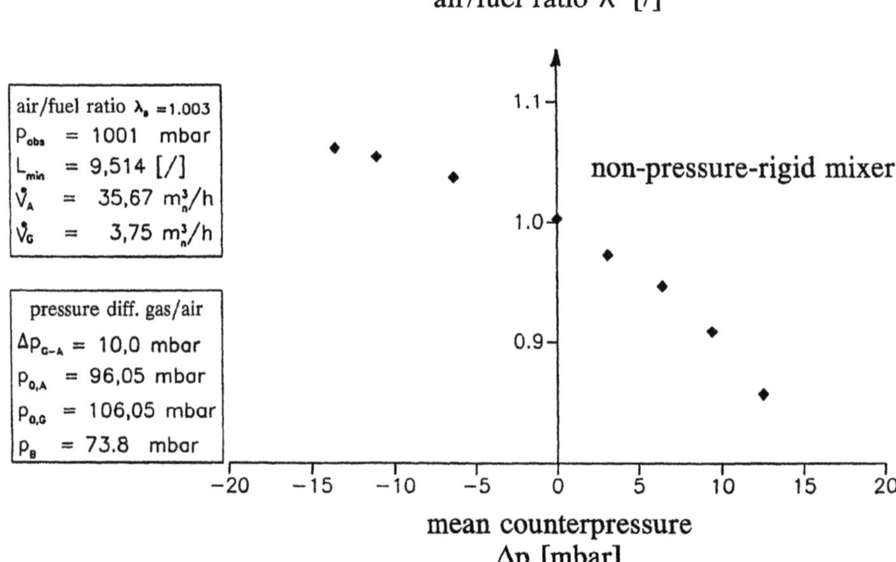

Figure 8: Influence of the mean counterpressure on the mixture formation under steady flow conditions (pressure difference gas/air: 10.0 mbar)

A rather strong dependence of the air/fuel ratio λ on the steady mean counterpressure has been found. An increase of the mixer counterpressure of only 10.0 mbar (at a mean counterpressure level of 1074.8 mbar) leads to a shift of the air/fuel ratio λ of 10 % in the fuel rich region. In addition the mixture formation was found to be non-linear and unsymmetric concerning its pressure dependence. One has to expect, therefore that a periodic and symmetric variation of the mixture counterpressure - caused by pressure oscillations in the downstream combustion chamber - may cause periodic but unsymmetric oscillations of the air equivalence ratio, and may therefore shift the mean mixture towards more fuel rich compositions. The following two diagrams (Fig.9 and 10) prove this in the case of unsteady sinusoidal counterpressure oscillations, generated with the pulsating unit at a frequency of 8 Hz for two different pressure characteristiscs of the natural gas/air mixer supply lines.

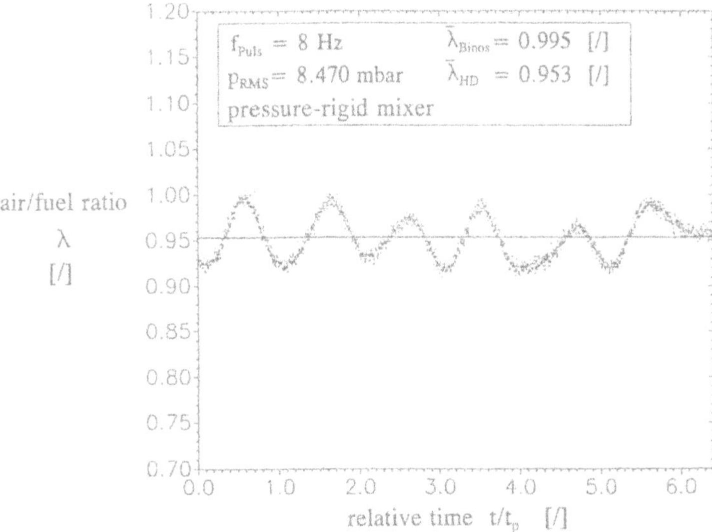

Figure 9: Time function of the mixture composition for pressure oscillations of 8.0 Hz (pressure-rigid mixer).

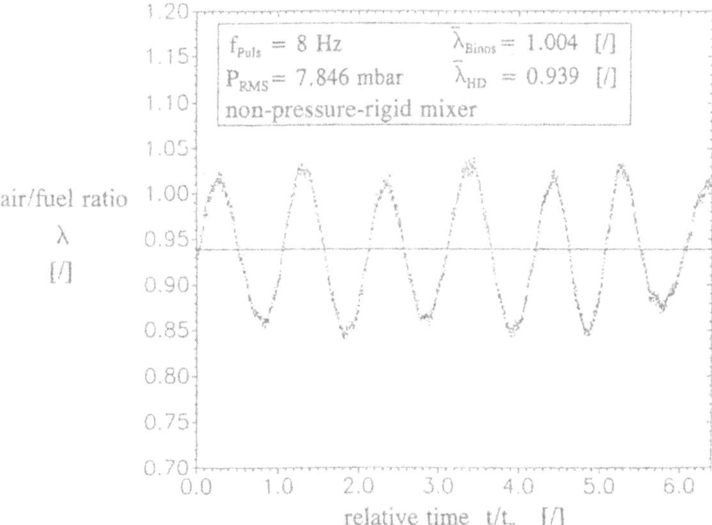

Figure 10: Time function of the mixture composition for pressure oscillations of 8.0 Hz (non-pressure-rigid mixer).

From Fig.9 we recognize the time function of the air equivalence ratio λ during unsteady operation of the mixer in the presence of forced pressure oscillations, with a frequency of 8.0 Hz and a RMS (root-mean-square) value of the counterpressure oscillation of 8.47 mbar. The excitation by the pressure signal was detected by using a microphone, the time dependent mixture composition by means of the described heat conductivity probe at the end of the mixing section. The amplitude of the air equivalence ratio oscillation is about 0.04, and a small shift of the mean value of λ under pulsated flow conditions towards fuel richer composition was found. In Fig.10 the same experiment was performed for the non-pressure-rigid mixer, when the pressure level in the air supply was lowered by approx. 10 mbar. At this case the amplitudes of the time dependent air equivalence ratio λ do reach values about 0.09 and the shift of the mean value towards fuel richer composition is increased, while the frequency and the amplitude of the exciting counterpressure remain nearly constant.

At the end of the experiments the dependence of the amplitude of the air equivalence ratio λ (represented by the RMS value in order to eliminate the superimposed turbulent fluctuations) on the frequency and amplitude of the exciting pressure oscillation were studied seperately. The next diagram (Fig.11) shows the amplitudes of the air equivalence oscillations for various pressure amplitudes at a fixed frequency of 8.0 Hz for two different mixer characteristics.

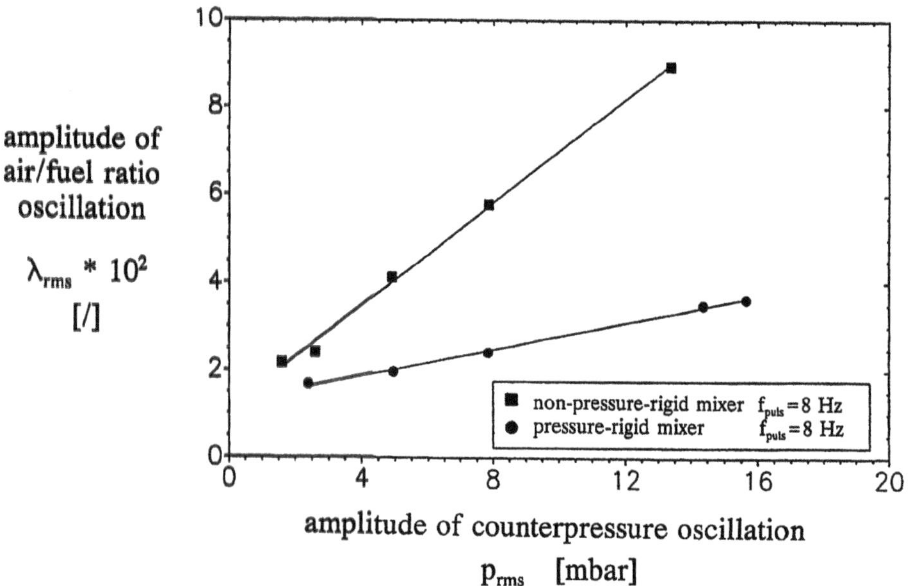

Figure 11: Influence of the counterpressure amplitude on the amplitude of the air equivalence ratio λ for a fixed frequency of excitation.

The amplitudes of λ increases linearly with increasing pressure amplitudes, but with different slopes for the two mixer characteristics. In the last figure the dependence on the pressure frequency (varied between 4.0 and 15.0 Hz) is plotted. In these experiments the amplitudes of the counterpressure oscillations were maintained constant at 10.5 mbar. In the range of frequencies studied in this investigation the mixture formation shows no dependence on the pressure frequency.

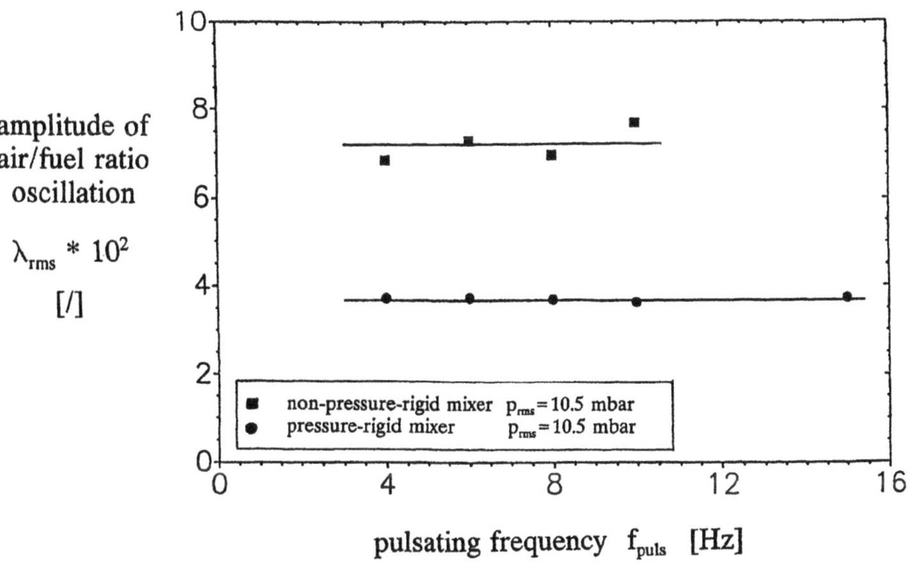

Figure 12: Influence of the counterpressure frequency on the amplitude of the air equivalence ratio λ for a fixed amplitude of excitation.

4. Summary

In the present paper the influence of mixer counterpressure oscillations, which can arise in the presence of self-sustained pressure oscillations in the combustion chamber, upon the fuel/air mixture formation is being demonstrated. Therefore, an exciting pressure signal, adjustable in frequency and amplitude, was generated with a pulsating unit. The time functions of the mixture compositions were detected by a concentration measuring technique, developed for this particular project and based on the measurements of heat conductivity differences in mixture samples. Due to the unsymmetric dependence of the mixture formation on the mixer counterpressure which was obtained from measurements under steady flow conditions, the expected shift of the mean air equivalence ratio λ under unsteady flow conditions (counterpressure modulated sinusoidally) could be proved. Especially under flame operation conditions near the stoichiometric mixture this shift towards fuel richer mixture compositions will cause a reduction of the fuel burnout. Furthermore, the amplitude of the air equivalence ratio can be predicted as function of the amplitude of the pressure fluctuation in the isothermal mixing section and the given pressure characteristics of the mixer gas/air supply system.

As demonstrated in a calculation (see Fig.2), the phase shift between mixture mass flow rate and mixture composition oscillations in the presence of self-sustained combustion instabilities will seriously affect the amplitude of the hot flue gas flow rate, which again causes pressure rise in the combustion chamber. In addition to that, the oscillating mixture composition alters the characteristric time delays inside the flame (time delay for convective transport and time delay for kinetics) and modifies by this, the relative timing of the heat release and the resonant pressure field and its strength inside the combustion chamber or downstream flue gas sections.

5. References

[1] Barr, P.K., Keller, J.O. and Bramlette, T.T. (1990): 'Pulse Combustor Modeling - Demonstration of the Importance of Characteristic Times'; Combustion and Flame 82 (1990), p. 252 - 269.

[2] Barr, P.K. and Keller, J.O. (1990): 'Premixed Combustion in an Oscillating /Resonant Flow Field - Part II: Modeling the Periodic Reacting Flow'; Conference Article, ASME, New York p. 23 - 39.

[3] Büchner, H., Hirsch, Chr. and Leuckel, W.: 'Experimental Investigations on the Dynamics of Pulsated Premixed Axial Jet Flames'; Proceedings of the 11. Intern. Symposium on Pulsating Combustion, Aug. 1991 in Monterey, Calif. USA.

5. SOME MODELING METHODS OF COMBUSTION INSTABILITIES

S. Candel*, C. Huynh* and T. Poinsot+

*EM2C Lab., CNSR
École Centrale Paris
92295 Chatenay-Malabry, France

+IMFT.
CNRS
31071 Toulouse, France

ABSTRACT. This article provides a review of modeling methods for combustion instability studies. After a presentation of classical material the paper focuses on modern methods. The survey concerns models of flame front dynamics, LES simulations, hydrodynamic instability studies of reactive flows, vortex ignition studies and nonlinear acoustic models for dynamical simulations.

1. Introduction

Combustion instabilities and oscillations constitute a major technical problem. Instabilities are often encountered in the high performance systems used in aerospace propulsion like rocket motors, jet engine afterburners, ramjets. They are also observed during anomalous operation of powerplants, industrial furnaces and process combustors. Under unstable operation the large amplitude oscillations of the flow induce many undesirable effects (large mechanical vibrations in the system, augmented heat transfer rates at the combustor walls, in extreme cases important damage and even a total loss of the system). In many circumstances the combustion process cannot sustain large oscillations and partial or total flame blow-off results.

In continuous combustion devices one generally wishes to avoid instabilities while oscillatory combustion constitutes the normal mode of operation of pulse combustors. The latter systems are designed to favor a resonant oscillation. Pulse mode operation has been found to enhance combustion efficiency and reduce the emission of pollutants like NO_x, CO and soot.

Because unsteady combustion effects arise in a range of engineering applications they have been a subject of continued research effort. The technical literature on this topic is in fact extensive. A large amount of experimental work has been carried out in many different configurations to identify the fundamental mechanisms.

While initial experiments carried out during the 1950ths where limited by the instrumentation, data acquisition and signal processing technology, detailed information has been gathered more recently with modern optical diagnostics and imaging

F. Culick et al., (eds.), Unsteady Combustion, 83–112.

techniques. The dynamics of instabilities and oscillatory combustion processes are now well described in a wide set of articles and review papers. New theoretical and numerical methods have also been devised during the last thirty years and used to test concepts and establish instability models. The modern developments are in fact a natural outcome of a long history. In an early analysis Rayleigh (1878) established a criterion stating that oscillations were sustained if the heat release took place in phase with the pressure fluctuation. This criterion has become a standard tool in the field of combustion instability. On the experimental side, there are many early observations of vibratory combustion. One notable study is that of Mallard and Le Chatelier (1883). During the initial development of rocket motors and jet propulsion engines many failures were caused by combustion instabilities and the topic became of central importance. Reviews of the early work on rocket instabilities were established by Crocco and Cheng (1956), Crocco, *et al.* (1960), Crocco (1965), Crocco, *et al.* (1968), a comprehensive reference on liquid rocket instabilities was assembled in the mid-seventies by Harrje and Reardon (1972). Barrère, *et al.* (1960) also provide an excellent chapter on the subject. A basic classification of instabilities was proposed by Barrère and Williams (1968). Modern reviews of combustion instabilities are due to Culick (1988) and Candel (1992). Many types of instabilities are described in Putnam (1971). The subject is examined in detail by Williams (1985) with special attention to studies of laminar flame instabilities. Pulse combustion is surveyed by Putnam, *et al.* (1986) and Zinn (1986). The special topic of active control is reviewed by McManus, *et al.* (1993).

We will here focus on some theoretical modeling methods and issues. We begin with a tutorial section on classical theory for combustion instability studies (Section 2). An acoustic wave equation is derived for reacting mixtures, the acoustic energy balance in combustor cavities is then established. The qualitative basis of the sensitive time lag concept is reviewed. We then deal with modeling issues and review some of the recent studies directed at a detailed description of the dynamics of instabilities (Section 3). An overview is provided on a limited number of topics (descriptions of flame front dynamics, direct simulations of instabilities, hydrodynamic instability analysis of reactive flows, vortex ignition dynamics, active control modeling). Other aspects of the instability and combustion oscillation problems are covered in companion papers included in this book.

2. Theoretical background

Combustion instabilities are a result of the resonant interaction between two or more physical mechanisms. A driving process generates the perturbations of the flow, a feedback process couples this pertubation to the driving mechanism and produces the resonant interaction which may lead to oscillatory combustion (Figure 1). In general, the feedback process relates the dowstream flow to the upstream region where the perturbations are initiated. As a consequence, acoustic wave propagation is most commonly responsible for the feedback path. While other processes like structural vibrations or convective modes constitute alternative feedback effects or play a role in the coupling process we will only concentrate on instabilities coupled by pressure waves.

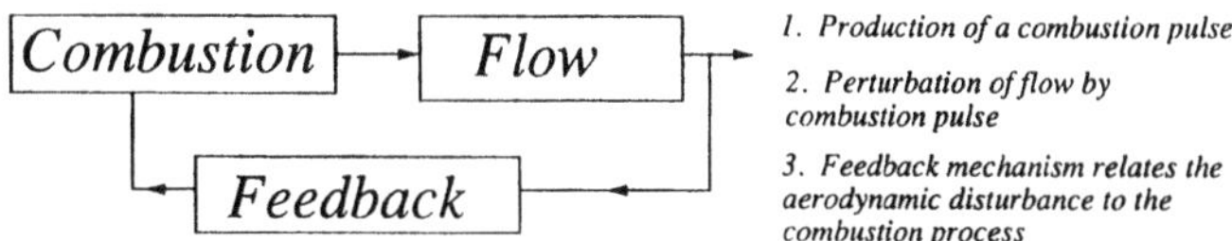

Instability grows when the combustion pulse is in phase with the feedback mechanism

Figure 1. Typical instability mechanism.

The accumulated experimental evidence indicates that instabilities belong to three main classes: *system instabilities*, *chamber instabilities*, *intrinsic instabilities*. The main features of these three classes are summarized in Fig.2. Instabilities of the first class involve the entire combustion system including the supply tanks, fuel lines, combustor and exhaust elements. The typical scales associated with such instabilities and the corresponding wavelengths are large compared to the transverse dimensions of each component so that the wave propagation is essentially longitudinal and may be described in terms of plane modes. The characteristic frequencies lie in the low frequency range (a few hundred Hz or less). When one component of the system is short with respect to the wavelength it may be treated as a compact element and described in terms of lumped parameter models. System instabilities appear in all types of situations. They are encountered during the ignition and shut-down sequences of liquid rocket engines when the pressure drop through the injectors has a reduced value and the supply system is closely coupled to the chamber. They also arise in certain ranges of the flight envelope of high performance afterburners and ramjets and they may perturb the operation of large scale powerplants or industrial burners and processes.

Instabilities of the second class are coupled by the chamber eigenmodes (resonant modes). In some cases the perturbations are indeed only detected in the combustor itself. In other configurations the oscillation is also sensed in other components. The size of the combustor fixes the instability wavelength and consequently the frequency lies in the high frequency range (typically above 1 kHz). In many circumstances chamber instabilities are coupled by higher order *transverse* modes of the combustor. The spatial structure of the pressure fluctuation then becomes an important aspect of the problem. When the combustor has an axis of symmetry (as in cylindrical or annular configurations) the transverse pressure pattern may be stationary or it may be spinning. Chamber instabilities observed in liquid or solid rocket motor or in ramjets and afterburners have destructive effects because they enhance the rates of heat transfer to the walls and injection head leading to hot spots and local melting of the structure. These instabilities require considerable attention in designing these devices and extensive testing is used to establish the domain of stable operation.

Instabilities of the third class only involve processes taking place in the flame itself. There is no external feedback path in this case.

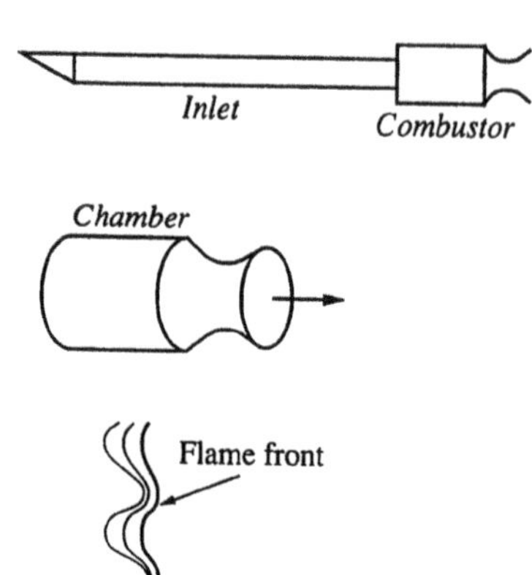

SYSTEM INSTABILITIES

They involve the entire combustion system

Wave propagation is essentially longitudinal

Characteristic frequency is low

CHAMBER INSTABILITIES

Coupled by the resonant modes of the chamber

Wavelengths of the order of a typical transverse dimension

High frequency range

INTRINSIC INSTABILITIES

Only depend on the combustion process itself

Figure 2. The three main classes of instabilities.

In what follows we will only consider the first two types of instabilities and we further restrict the review to the rather general situation where the instability is coupled by acoustic waves.

2.1. ACOUSTIC WAVE EQUATION FOR A REACTING MIXTURE

The description of acoustic wave generation and propagation in an inhomogeneous reactive medium may be based on a wave equation (see for example Strahle (1971), Kotake (1975)). This equation is briefly derived in this section to exhibit the source terms associated with the nonsteady heat release. The analysis also provides the proper background for studies of combustor eigenmodes (see Laverdant, *et al.* (1986)).

Equations obtained in the first sub-section are then used in section 2.2 to derive an acoustic energy balance for general combustor cavities.

2.1.1. Derivation of an acoustic wave equation for a chemically reacting mixture

We start from the general equations describing a chemically reactive mixture of N species. We use the following forms :

• *Conservation of mass*

$$\frac{\partial \rho}{\partial t} + \nabla \cdot \rho \boldsymbol{v} = 0 \tag{1}$$

• *Conservation of the $k-th$ chemical species*

$$\frac{\partial}{\partial t}(\rho Y_k) + \boldsymbol{\nabla} \cdot \rho Y_k(\boldsymbol{v} + \boldsymbol{v}_k^D) = \dot{w}_k \tag{2}$$

• *Conservation of momentum*

$$\frac{\partial}{\partial t}(\rho \boldsymbol{v}) + \boldsymbol{\nabla} \cdot \rho \boldsymbol{v}\boldsymbol{v} = -\boldsymbol{\nabla} p + \boldsymbol{\nabla} \cdot \boldsymbol{\tau} + \sum_{k=1}^{N} \rho \boldsymbol{f}_k Y_k \tag{3}$$

• *Conservation of energy (enthalpy form)*

$$\frac{\partial}{\partial t}\rho h + \boldsymbol{\nabla} \cdot \rho h \boldsymbol{v} = -\boldsymbol{\nabla} \cdot \boldsymbol{q} + \frac{dp}{dt} + \Phi + \sum_{k=1}^{N} \rho Y_k \boldsymbol{v}_k^D \cdot \boldsymbol{f}_k \tag{4}$$

• *Equation of state*

$$p = \rho R T \qquad \text{where} \quad R = R_0 \sum_{k=1}^{N} (Y_k / M_k) \tag{5}$$

• *Enthalpy of the mixture*

$$h = \sum_{k=1}^{N} h_k Y_k = \sum_{k=1}^{N} (h_k^{\circ} + \int_{T_0}^{T} c_{pk} dT) Y_k \tag{6}$$

In these equations ρ, p, T, v respectively designate the specific mass, pressure, temperature and velocity of the mixture. Y_k, $\boldsymbol{v}_k^D$, $\boldsymbol{f}_k$, M_k represent the $k-th$ species mass fraction, diffusion velocity, specific body force and molar mass. The mass rate of production of the $k-th$ species per unit volume $\dot{w}_k$ constitues the source term in the species equation. The global conservation of mass requires that $\sum_{k=1}^{N} \dot{w}_k = 0$ and $\sum_{k=1}^{N} \rho Y_k \boldsymbol{v}_k^D = 0$. The equation describing the balance of enthalpy involves the viscous dissipation function $\Phi = \boldsymbol{\tau} : \boldsymbol{\nabla} \boldsymbol{v}$ where $\boldsymbol{\tau}$ is the viscous stress tensor and Neglecting coupled effects and radiative heat transfer the heat flux $\boldsymbol{q}$ may be written in the form

$$\boldsymbol{q} = -\lambda \boldsymbol{\nabla} T + \sum_{k=1}^{N} \rho Y_k \boldsymbol{v}_k^D h_k \tag{7}$$

In general, the diffusion velocities are obtained by solving a system of multicomponent diffusion equations (see Williams (1985)).

Now the system of equations (1)-(6) may be simplified by neglecting the specific body forces $\boldsymbol{f}_k$. It is also convenient to introduce the material derivative $d/dt = \partial/\partial t + \boldsymbol{v} \cdot \boldsymbol{\nabla}$ and derive an equation for the temperature by combining the species and energy equations and introducing the heat flux expression (7). The following system is obtained :

$$\frac{d\rho}{dt} = -\rho \boldsymbol{\nabla} \cdot \boldsymbol{v} \tag{8}$$

$$\rho \frac{d\boldsymbol{v}}{dt} = -\boldsymbol{\nabla} p + \boldsymbol{\nabla} \cdot \boldsymbol{\tau} \tag{9}$$

$$\rho c_p \frac{dT}{dt} = \boldsymbol{\nabla} \cdot \lambda \boldsymbol{\nabla} T + \frac{dp}{dt} + \Phi - \sum_{k=1}^{N} h_k \dot{w}_k - \sum_{k=1}^{N} (\rho Y_k \boldsymbol{v}_k^D c_{pk}) \cdot \boldsymbol{\nabla} T \tag{10}$$

where $c_p = \sum_{k=1}^{N} c_{pk}$ designates the specific heat of the mixture. If it is assumed that all the specific heats of the species are equal $c_{pk} = c_p$ the last term of this equation disappears.

Dividing the energy equation (10) by $\rho c_p T$ using the equation of state (5) and the mass conservation (1) one obtains

$$\frac{1}{\gamma_p} \frac{d}{dt} \ln p + \boldsymbol{\nabla} \cdot \boldsymbol{v} = \frac{1}{\rho c_p T} \{ \boldsymbol{\nabla} \cdot \lambda \boldsymbol{\nabla} T + \Phi - \sum_{k=1}^{N} h_k \dot{w}_k - \sum_{k=1}^{N} (\rho Y_k c_{pk} \boldsymbol{v}_k^D \cdot \boldsymbol{\nabla} T) \} + \frac{1}{R} \frac{dR}{dt} \tag{11}$$

The momentum balance (9) may now be cast in the form

$$\frac{d\boldsymbol{v}}{dt} + \frac{c^2}{\gamma} \boldsymbol{\nabla} \ln p = \frac{1}{\rho} \boldsymbol{\nabla} \cdot \boldsymbol{\tau} \tag{12}$$

where $c^2 = \gamma p/\rho$ designates the local sound speed. The last two equations are now combined by taking the divergence of the second and substracting the material derivative of the first :

$$\boldsymbol{\nabla} \cdot \frac{c^2}{\gamma} \ln p - \frac{d}{dt} \left(\frac{1}{\gamma} \frac{d}{dt} \ln p \right) = \boldsymbol{\nabla} \cdot (\rho^{-1} \boldsymbol{\nabla} \cdot \boldsymbol{\tau}) - \frac{d}{dt} \left\{ \frac{1}{\rho c_p T} [\boldsymbol{\nabla} \cdot \lambda \boldsymbol{\nabla} T + \Phi - \sum_{k=1}^{N} h_k \dot{w}_k - \sum_{k=1}^{N} (\rho Y_k c_{pk} \boldsymbol{v}_k^D \cdot \boldsymbol{\nabla} T)] \right\} - \frac{d^2}{dt^2} (\ln R) - \boldsymbol{\nabla} \boldsymbol{v} : \boldsymbol{\nabla} \boldsymbol{v} \tag{13}$$

This is a wave equation for the logarithm of the pressure. An equation of this type was initially derived by Phillips (1960) in an analysis of the aerodynamic generation of sound in non reactive turbulent gas flows. The separation of terms between the right hand side and the left hand side is somewhat arbitrary and it has been argued that some terms appearing in the right hand side describe features of the propagation of sound in the medium and should be included in the left hand side. This aspect is also discussed by Kotake (1975) in a study of combustion noise. These objections are not too serious when the flow Mach number is low as in most subsonic combustion applications. One may the the right hand side of equation (15) as the source terms generating the pressure waves in the reactive mixture. An order of magnitude analysis of these terms (see Kotake (1975)) indicates that the dominant sources of pressure perturbations are associated with the chemical heat release fluctuations and to a lesser extent with the velocity perturbations. Neglecting all other terms one obtains

$$\nabla \cdot \frac{c^2}{\gamma} \nabla \ln p - \frac{d}{dt}\left(\frac{1}{\gamma}\frac{d}{dt} \ln p\right) = \frac{d}{dt}\left(\frac{1}{\rho c_p T} \sum_{k=1}^{N} h_k \dot{w}_k\right) - \nabla \boldsymbol{v} : \nabla \boldsymbol{v} \tag{14}$$

This equation is not linearized and hence it may be used to describe finite amplitude waves. If one considers low speed reactive flows, the convective term in the material derivative may be neglected $d/dt \sim \partial/\partial t$. One may assume in addition that the specific heat ratio γ is constant. Equation (14) then becomes

$$\nabla \cdot c^2 \nabla \ln p - \frac{\partial^2}{\partial t^2} \ln p = \frac{\partial}{\partial t}\left(\frac{1}{\rho c_v T} \sum_{k=1}^{N} h_k \dot{w}_k\right) - \gamma \nabla \boldsymbol{v} : \boldsymbol{v} \tag{15}$$

In many circumstances the pressure waves are relatively weak and linearization is appropriate. The pressure is then expressed as a sum of a mean and a fluctuating component

$$p = p_0 + p_1 \qquad \text{with} \quad p_1/p_0 \ll 1$$

Then $\ln p \simeq p_1/p_0$ and equation (15) becomes

$$\nabla \cdot c^2 \nabla \left(\frac{p_1}{p_0}\right) - \frac{\partial^2}{\partial t^2}\left(\frac{p_1}{p_0}\right) = \frac{\partial}{\partial t}\left(\frac{1}{\rho c_v T} \sum_{k=1}^{N} h_k \dot{w}_k\right) - \gamma \nabla \boldsymbol{v} : \nabla \boldsymbol{v} \tag{16}$$

In most practical combustors the mean pressure does not change by more than a few percent, the spatial derivatives of p_0 may be neglected and hence equation (16) may be written as :

$$\nabla \cdot c^2 \nabla p_1 - \frac{\partial^2}{\partial t^2} p_1 = \frac{\partial}{\partial t}\left[(\gamma - 1) \sum_{k=1}^{N} h_k \dot{w}_k\right] - \gamma p_0 \nabla \boldsymbol{v} : \nabla \boldsymbol{v} \tag{17}$$

In addition to this equation we also need an expression for the acoustic velocity. This may be obtained by linearizing the momentum equation (12) and neglecting the viscous stresses. This yields

$$\frac{\partial \boldsymbol{v}_1}{\partial t} = -\frac{1}{\rho_0} \nabla p_1 \tag{18}$$

Equations (17) and (18) describe the propagation and generation of small perturbations in nearly isobaric, low Mach number reactive mixtures.

To perform a modal analysis (ie to determine the resonant modes of a specific combustor) one drops the terms appearing on the right hand side of (17) and one assumes that the pressure field is harmonic and features a factor $\exp(-i\omega t)$:

$$p_1 = \psi_n(\boldsymbol{x}) \exp(-i\omega t)$$

The modes $\psi_n(\boldsymbol{x})$ are the eigensolutions of the Helmholtz equation

$$\nabla \cdot c^2 \nabla \psi_n + \omega^2 \psi_n = 0 \tag{19}$$

and the corresponding velocity field is obtained from

$$\boldsymbol{v}_1 = (1/\rho_0 i\omega) \nabla \psi_n$$

Equation (19) must be solved under some standard boundary conditions :

$$\begin{aligned} \boldsymbol{v}_1 \cdot \boldsymbol{n} &= 0 \quad \text{on a rigid wall} \\ \psi_n &= 0 \quad \text{at an open exit} \end{aligned} \tag{20}$$

Equation (19) and the related boundary conditions may be used to determine the resonant modes and the corresponding eigenfrequencies of combustor cavities. The effect of temperature non-uniformity is taken into account because the spatial variations of the sound speed are correctly represented in equation (19).

A modal analysis of this kind was performed by Laverdant (1986) in the case of a gaseous dump combustor (further details in this reference). It is now standard to use 3D finite element codes to solve the inhomogeneous Helmholtz equation (19) and determine the resonant modes of specific configurations.

2.1.2. The nonsteady heat release source term

Let us consider again the source term corresponding to the nonsteady heat release. This term has the form

$$\frac{\partial}{\partial t}[(\gamma - 1) \sum_{k=1}^{N} h_k \dot{w}_k]$$

It is often possible to assume that the chemical change occurs by a single reaction step and that all the c_{pk} are equal. Then if Δh_f° designates the change of formation enthalpy per unit mass of the mixture and if $\dot{w}$ represents the rate of reaction, the chemical source term becomes

$$-\frac{\partial}{\partial t}(\gamma - 1)(-\Delta h_f^\circ)\dot{w}$$

In most cases the only time dependence in this expression is due to the nonsteady rate of reaction term and as a consequence the acoustic source term associated with the chemical reaction may be written in the form

$$-(\gamma-1)(-\Delta h_f^\circ)\frac{\partial \dot{w}}{\partial t} \quad \text{or equivalently} \quad -(\gamma-1)\frac{\partial \dot{q}}{\partial t}$$

where $\dot{q}$ designates the rate of heat release per unit mass of fuel.

2.2. ACOUSTIC ENERGY BALANCE IN COMBUSTOR CAVITIES

2.2.1. Introduction

To examine the dynamical behavior of oscillations in a confined geometry, it is useful to consider the balance of the energy involved. Such a balance may be used to estimate the rate of growth or decay of these oscillations and it has been extensively used in the past.

Energy considerations are often put forward in solid rocket instability studies. The energy balance for acoustic perturbations is also closely related to the Rayleigh criterion which is used in many qualitative descriptions of flame instabilities (for example Putnam (1971)).

We begin by deriving an acoustic energy balance equation for a chemically reacting mixture. A global balance is then obtained for acoustic cavities. This energy balance is then related to the Rayleigh criterion.

2.2.2. An energy balance for acoustic perturbations in a reacting medium

The derivation of an energy balance equation for acoustic perturbations in a reacting medium may be based on the conservation relations given in Section 2.1. One only retains the heat release source term in equation (17) and replaces ∇p_1 in that equation by its expression in terms of $\boldsymbol{v}_1$ as given by (18). This yields :

$$\nabla \cdot \rho_0 c^2 \frac{\partial \boldsymbol{v}_1}{\partial t} + \frac{\partial^2 p_1}{\partial t^2} = -\frac{\partial}{\partial t}[(\gamma-1)\sum_{k=1}^{N} h_k \dot{w}_k] \tag{21}$$

Clearly only the nonsteady heat release source terms contribute to this equation and $\dot{w}_k$ may be replaced by $\dot{w}_k^{(1)}$ which represents the nonsteady rate of conversion of the $k-th$ chemical species.

The time derivative appearing in all three terms may be discarded and the following relation is obtained

$$\nabla \cdot \rho_0 c^2 \boldsymbol{v}_1 + \frac{\partial p_1}{\partial t} = -(\gamma-1)\sum_{k=1}^{N} h_k \dot{w}_k^{(1)} \tag{22}$$

Noting that $\rho_0 c^2 = \gamma p_0$ and that in many situations the pressure is nearly constant in the flow, equation (22) may be cast in the form

$$\nabla \cdot \boldsymbol{v}_1 + \frac{1}{\rho_0 c^2}\frac{\partial p_1}{\partial t} = -\frac{\gamma - 1}{\rho_0 c^2}\sum_{k=1}^{N} h_k \dot{w}_k^{(1)} \tag{23}$$

Multiplying this equation by p_1, taking the scalar product of equation (18) by $\boldsymbol{v}_1$ and adding the resulting expressions yields

$$\frac{\partial}{\partial t}\mathcal{E} + \nabla \cdot \mathcal{F} = -\frac{\gamma - 1}{\rho_0 c^2} p_1 \sum_{k=1}^{N} h_k \dot{w}_k^{(1)} \tag{24}$$

where $\mathcal{E} = \frac{1}{2}\frac{p_1^2}{\rho_0 c^2} + \frac{1}{2}\rho_0 v_1^2$ is the instantaneous acoustic energy density and $\mathcal{F} = p_1 \boldsymbol{v}_1$ stands for the acoustic energy flux.

The source term appearing on the right hand side corresponds to the nonsteady heat release in the reactive mixture. This term may be written in the form

$$\mathcal{S} = -\frac{\gamma - 1}{\gamma}\frac{p_1}{p_0}\sum_{k=1}^{N} h_k \dot{w}_k^{(1)} \tag{25}$$

The meaning of this term is best understood under the following assumptions. The set of reactions taking place in the mixture is replaced by a single global equation $F + sO \rightarrow P$. The rate of this reaction is $\dot{w}$ and the change of formation enthalpy per unit mass of mixture is Δh_f°. If the specific heat of the mixture does not depend on the composition it is possible to write

$$-\sum_{k=1}^{N} h_k \dot{w}_k^{(1)} = (-\Delta h_f^\circ)\dot{w}^{(1)} \tag{26}$$

and the source term in equation (6) becomes

$$\mathcal{S} = \frac{\gamma - 1}{\gamma}(-\Delta h_f^\circ)\frac{p_1}{p_0}\dot{w}^{(1)} \tag{27}$$

When this term is positive, energy is fed into the acoustic oscillation, when it is negative, energy is extracted from that oscillation.

A global energy balance is easily derived from expression (24) by integrating over the volume V of the combustor :

$$\frac{\partial}{\partial t}\int_V \mathcal{E} dV + \int_A \mathcal{F} \cdot \boldsymbol{n} dA = \int_V \mathcal{S} dV \tag{28}$$

One may now consider a situation where all the acoustic perturbations are quasi-periodic oscillations of the general form

$$p_1 = p_1(t)e^{-i\omega t}, \quad \boldsymbol{v}_1 = \boldsymbol{v}_1(t)e^{-i\omega t}, \quad \dot{w}^{(1)} = \dot{w}(t)e^{-i\omega t} \tag{29}$$

The complex amplitudes $p_1(t)$, $\boldsymbol{v}_1(t)$, $w(t)$ are slowly varying functions of time. An *average energy balance* may be obtained by integrating (29) over a period of oscillation $T = 2\pi/\omega$ and dividing the resulting expression by the period. The following expression is obtained

$$\frac{\partial}{\partial t}\int_V E dV + \int_A \boldsymbol{F}\cdot\boldsymbol{n} dA = \int_V S dV \tag{30}$$

where E, $\boldsymbol{F}$ and S are period averages of the instantaneous quantities $\mathcal{E}$, $\mathcal{F}$, $\mathcal{S}$

$$E = \frac{1}{4\rho_0 c^2} p_1 p_1^* + \frac{1}{4}\rho_0 \boldsymbol{v}_1\cdot\boldsymbol{v}_1^* \tag{31}$$

$$\boldsymbol{F} = \frac{1}{2} Re(p_1 \boldsymbol{v}_1^*) \tag{32}$$

$$S = \frac{\gamma - 1}{\rho_0 c^2}(-\Delta h_f^\circ)\frac{1}{2} Re[p_1 \dot{w}^{(1)*}] \tag{33}$$

To obtain these expressions we used a standard theorem to average the products of quasiperiodic quantities appearing in the instantaneous expressions.

Expression (30) together with the period averages E, $\boldsymbol{F}$ and S provide a global acoustic energy balance for the cavity of volume V surrounded by the surface A. According to this expression the energy contained in the acoustic perturbations inside the volume may be changed in two ways. One contribution is due to the volume integral of $Re[p_1\dot{w}^{(1)*}]$ while the other comes from the surface integral of the normal flux. The energy is increased if the surface integral of the flux is positive or if the integral of the volumetric source term is positive.

The growth rate of the acoustic perturbations may be deduced from the global energy balance.

For this one assumes that the slow variation of the perturbations is of the form

$$p_1 = p_1 e^{\alpha t},\ \boldsymbol{v}_1 = \boldsymbol{v}_1 e^{\alpha t},\ \dot{w}^{(1)} = \dot{w}^{(1)} e^{\alpha t} \tag{34}$$

where $\alpha T \ll 1$ (i.e. the growth of the perturbation over a single period is small).

The global energy balance then yields

$$2\alpha \int_V E dV = -\int_A \boldsymbol{F}\cdot\boldsymbol{n} dA + \int_V S dV \tag{35}$$

or

$$2\alpha = \left[-\int_A \boldsymbol{F}\cdot\boldsymbol{n} dA + \int_V S dV\right] / \int_V E dV \tag{36}$$

where factors of the form $e^{\alpha t}$ may be discarded from all the terms appearing in this expression.

To obtain a practical result one may characterize the boundary by an admittance Y and introduce a complex ratio D of the nonsteady volume rate of reaction to the pressure perturbation

$$Y = \boldsymbol{v}_1 \cdot \boldsymbol{n}/p_1 \tag{37}$$

$$D = \dot{w}^{(1)}/p_1 \tag{38}$$

Both Y and D are complex numbers. With these definitions the growth rate is given by

$$2\alpha = \left[-\frac{1}{2} \int_A p_1 p_1^* Re(Y^*) dA + \frac{1}{2} \int_V p_1 p_1^* Re(D^*) dV \right] / \int E dV \tag{39}$$

2.2.3. The acoustic energy balance and the Rayleigh criterion

It is now worth establishing the relation between the acoustic energy balance and the Rayleigh criterion. There are many possible derivations of this result, one of which relies on the acoustic energy balance relations presented in this section.

Consider once more the local energy balance

$$\frac{\partial \mathcal{E}}{\partial t} + \nabla \cdot \mathcal{F} = \frac{\gamma - 1}{\rho_0 c^2} (-\Delta h_f^\circ) p_1 \dot{w}^{(1)} \tag{40}$$

The source term on the right hand side will enhance the acoustic energy if p_1 and $\dot{w}^{(1)}$ have the same sign. If one takes the average of the previous equation over a period of oscillation it becomes

$$\frac{\partial E}{\partial t} + \nabla \cdot \boldsymbol{F} = S \tag{41}$$

where

$$S = \frac{\gamma - 1}{\rho_0 c^2} (-\Delta h_f^\circ) \frac{1}{2} Re(p_1 \dot{w}^{(1)*}) \tag{42}$$

Now let φ designate the phase between the pressure and the rate of reaction perturbations. The source term S becomes

$$S = \frac{\gamma - 1}{\rho_0 c^2} (-\Delta h_f^\circ) \frac{1}{2} P_1 \dot{W}^{(1)*} \cos\varphi \tag{43}$$

The source term will be positive and the amplitude of oscillation will increase if $-\pi/2 \leq \varphi \leq \pi/2$ (i.e. if the pressure and unsteady heat release are in phase). The source term will be negative and the amplitude of oscillation will decrease if

$\pi/2 \leq \varphi \leq 3\pi/2$ (i.e. if the pressure and heat release are out of phase). These results are summarized in Figure 3.

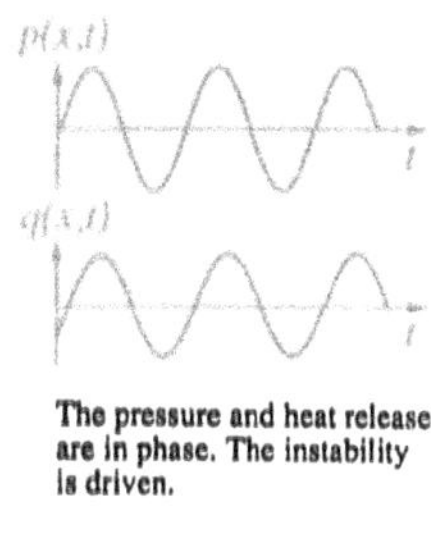

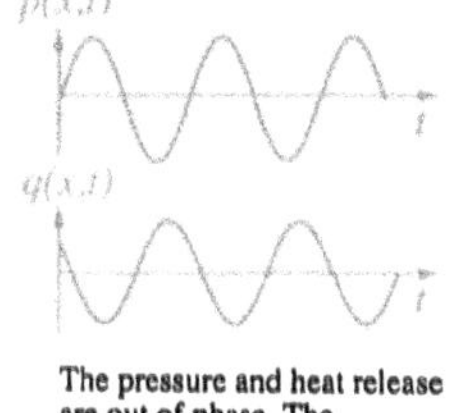

$$\int_{[T]} p(x,t)\, q(x,t)\, dt > 0 \qquad \int_{[T]} p(x,t)\, q(x,t)\, dt < 0$$

Figure 3. The Rayleigh criterion indicates that the oscillation is sutained if the heat release fluctuation is in phase with the pressure.

It is by now clear that the relationship which exists between the rate of heat release and the pressure perturbation is a central issue in combustion instability studies. The direct measurement of these two quantities is often attempted in experiments. The typical setup for such measurements is shown in Figure 4. The pressure is obtained from probes mounted on the combustor walls. It is not possible to measure the pressure in the medium itself but the field inside the combustor may be deduced by combining wall measurements with calculations. The heat release rate is deduced from free radical radiation imaging.

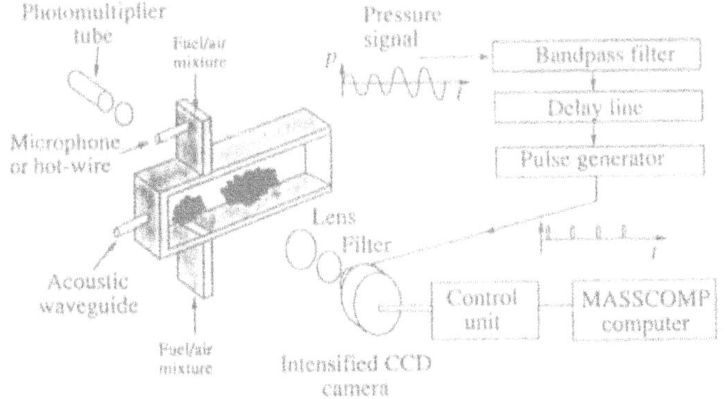

Figure 4. Typical experimental set-up for simulataneous emission imaging and pressure detection. The emission image provides the distribution of heat release in the combustor.

It is then possible to estimate the local Rayleigh index

$$R(\boldsymbol{x}) = \frac{1}{T} \int_{[T]} p_1 \dot{q}_1 dt$$

This index is directly proportional to the energy equation source term $R(\boldsymbol{x}) = (\rho c^2/(\gamma - 1))S(\boldsymbol{x})$ and it determines regions of driving and regions of damping as illustrated in Figure 5 (see Samaniego, *et al.* (1993) for an analysis of this type).

- The volumetric source of acoustic energy is proportional to the product of the pressure perturbation p by the heat release q
- The relation which exists between these two quantities constitutes a key issue in combustion instability problems

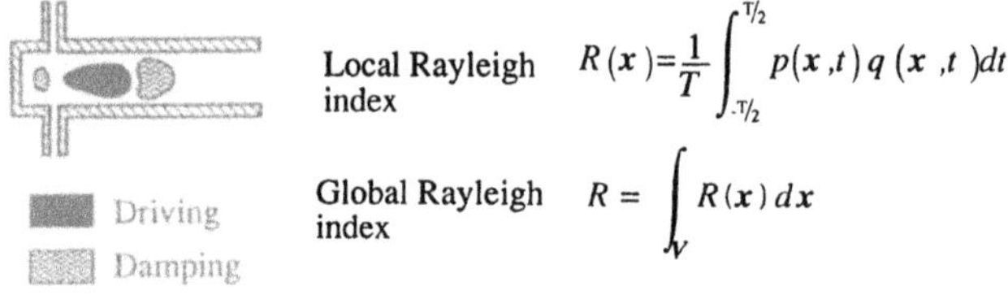

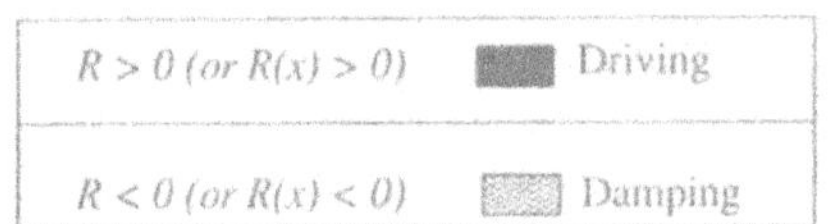

Figure 5. The local Rayleigh index may be used to determine regions of driving and regions of damping.

2.3. SENSITIVE TIME LAG CONCEPT

2.3.1. Introduction

The sensitive time lag is extensively used in the early literature dealing with combustion in rocket motors. This concept is still of interest because it allows the construction of simple instability models and because it provides a basic understanding of some important instability mechanisms. The time lag concept is briefly reviewed. Numerous applications may be found in the literature dealing with liquid rocket instabilities (for example Crocco and Cheng (1956)) and in a large number of propulsion related studies.

2.3.2. The time lag concept

The sensitive time lag concept was introduced in the early work on rocket instabilities (Crocco (1951), Summerfield (1951), Tsien (1952), Marble and Cox (1953), Marble (1955), Crocco and Cheng (1956)).

To introduce the STL concept we use the physical arguments employed by these various authors.

To fix ideas let us consider the case of monopropellant rocket chamber sketched in Figure 6.

Let $\dot{m}_i(t)$ denote the mass rate of injection of propellant into the chamber and let $\dot{m}_b(t)$ represent the mass rate of generation of hot gases by combustion of the propellant at time instant t. The conversion of propellant into hot products requires a certain time. This time lag is not constant and will depend on the physical and chemical processes that take place in the chamber. For example, atomization of the propellant jet will depend on the injector type, on the density of the ambiant gases and on many physical parameters. If the ambiant gas density increases due to a pressure increase, the droplet size in the propellant spray decreases. As a consequence the droplet vaporization will proceed faster and the chemical reaction of the propellant will take place sooner. One may conclude that an increase in pressure will produce a decrease of the time lag τ between injection and burning of the propellant.

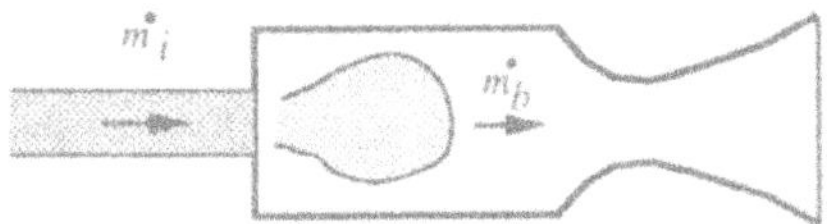

Figure 6. A monopropellant rocket motor.

Other factors such as the temperature in the chamber also influence the time lag and similar arguments may be used to treat this aspect.

On this basis one may derive a quantitative expression relating the propellant burning rate $\dot{m}_b$ to the injection rate $\dot{m}_i$. For this, consider a mass of propellant burning between time t and $t + dt$.

This mass $\dot{m}_b(t)dt$ must be equal to the mass injected from $t-\tau$ to $t-\tau+d(t-\tau)$: $\dot{m}_i(t-\tau)d(t-\tau)$. As a consequence

$$\dot{m}_b(t)dt = \dot{m}_i(t-\tau)d(t-\tau) \tag{44}$$

or

$$\dot{m}_b(t) = \dot{m}_i(t-\tau)(1-\frac{d\tau}{dt}) \tag{45}$$

In this expression, the term $d\tau/dt$ indicates that τ changes with time. For a constant delay, the derivative of the time lag $d\tau/dt$ vanishes and relation (45) becomes

$$\dot{m}_b(t) = \dot{m}_i(t-\tau) \tag{46}$$

It has been found convenient to write expression (45) in terms of relative variations of the burning and discharge rates which are defined with respect to the mean values of these quantities :

$$\mu_b(t) = \frac{\dot{m}_b - \overline{\dot{m}}}{\overline{\dot{m}}} \qquad \text{and} \quad \mu_i(t) = \frac{\dot{m}_i - \overline{\dot{m}}}{\overline{\dot{m}}}$$

where $\overline{\dot{m}}$ is the mean rate of injection and burning. Introducing these expressions into (45) and neglecting second order terms one obtains

$$\boxed{\mu_b(t) = \mu_i(t-\tau) - \frac{d\tau}{dt}} \tag{47}$$

The fractional rate of conversion of reactants μ_b is related to the fractional rate of injection μ_i and to the rate of change of the STL.

2.3.3. An equation for the rate of change of the time lag

The second step of the analysis consists of deriving an equation for the rate of change of the time lag. For this one considers again the process of conversion of propellant into combustion products. Let $f(p, T_g)$ designate a function which globally describes this process. As indicated above this function will depend on the pressure and gas temperature it may also depend on other parameters. If the time lag associated with vaporization dominates all other characteristic times the function f will represent for example the rate of heat transfer to the propellant. The liquid propellant will be vaporized when a certain amount of heat designated as Q_v will have been transferred. In other words, the time integral of the heat transfer f taken between the moment of injection and the moment of total vaporization should be equal to Q_v :

$$\int_{t-\tau}^{t} f(p, T_g)dt' = Q_v = \text{const.} \tag{48}$$

This equation governs the rate of change of the time lag τ. The dependence becomes more explicit if one differentiates the previous equation with respect to time :

$$f[p(t), T_g(t)] - f[p(t-\tau), T_g(t-\tau)](1 - \frac{d\tau}{dt}) = 0 \tag{49}$$

Now, assuming that $p(t)$, $T_g(t)$ remain close to their mean values $\overline{p}$ and $\overline{T}_g$ one may expand the function f in a Taylor series about $\overline{p}$ and $\overline{T}_g$. Substituting these expansions in (49) and neglecting second order terms one finds :

$$\frac{d\tau}{dt} = \frac{\partial \ln f}{\partial \ln p}\frac{p(t-\tau) - p(t)}{\overline{p}} + \frac{\partial \ln f}{\partial \ln T_g}\frac{T_g(t-\tau) - T_g(t)}{\overline{T}_g} \tag{50}$$

The dependence of τ on the pressure and gas temperature in the chamber is given explicitely in terms of two *interaction indices* :

$$n = \frac{\partial \ln f}{\partial \ln p}, \qquad q = \frac{\partial \ln f}{\partial \ln T_g} \tag{51}$$

It is then convenient to introduce the relative variations of pressure and gas temperature $\varphi = (p - \overline{p})/\overline{p}$ and $\Psi = (T_g - \overline{T}_g)/\overline{T}_g$ and use these definitions to transform expression (50) into

$$\frac{d\tau}{dt} = n[\varphi(t-\tau) - \varphi(t)] + q[\Psi(t-\tau) - \Psi(t)] \tag{52}$$

For monopropellant rockets, the gas temperature T_g remains essentially constant and the second term in (52) is usually neglected. The rate of change of the time lag is accordingly proportional to the difference of pressures at times t and $t - \tau$. Expressions (47) and (52) may be combined to relate the variation of the burning rate to the pressure and temperature perturbations. Specific instability mechanisms may then be examined by including these relations in dynamical models of the rocket chamber.

Despite its phenomenological nature the STL concept provides a comprehensive framework and is useful in the interpretation of experimental data. It may also help in design applications, system studies and the development of active control strategies.

3. Modern theoretical studies of combustion instabilities

Because combustion instabilities have so many facets they have been examined with a wide variety of theoretical models. Much of the recent theoretical work has focused on the response of laminar flames to pressure perturbations or pressure gradients (McIntosh (1986, McIntosh (1991, McIntosh and Wilce (1991), Clavin, *et al.* (1990), Clavin and Sun (1991), Leder and Kapila (1991), Marble and Hendricks (1986)). While a full analysis is provided in some cases, these studies cannot deal with the instability problems encountered in practical devices which are most often turbulent.

We will therefore survey some models which have more relevance in this respect. Among the many questions which have been investigated we have selected the following topics : models of flame front dynamics, direct simulations of instabilities, hydrodynamic instability modes of reactive flows, vortex ignition analysis. In addition we will also briefly describe the application of nonlinear instability models in active control simulations.

3.1 DESCRIPTIONS OF FLAME FRONT DYNAMICS

New experimental evidence and some of the earlier observations indicate that the instabilities involve large oscillations of the flame front, pulsations of the injected reactant streams, hydrodynamic instabilities, vortex roll-up, burning and interactions. The modeling has tried to describe these processes. In many cases the problem is analyzed in the absence of the feedback loop but in some instances the complete phenomenon is considered. Marble and Candel (1978) first indicated that thenonsteady response of the flame front could be a possible source of low frequency oscillations in afterburners or large utility powerplants. Their model describes the dynamics of fresh and burnt gases submitted to axial pressure waves as well as the motion of the flame sheet. The geometry is schematically that found in reheat devices where the flame is confined in a long duct and stabilized on a bluff body (Figure 7). A turbulent flame is stabilized in a duct of half width l. The flame is described as a sheet which is thin in comparison with the characteristic wavelength of the problem. The flame moves with respect to the fresh gases at a normal flame speed w_1. The flame location is described by its distance $\eta(x,t)$ with respect to the center line of the duct (the x-axis). In this thin flame sheet (TFS) model the flame front separates the incoming fresh gases (density ρ_1) from the hot combustion products (density ρ_2). The variables describing the upstream and downstream fields are indicated by subscripts 1 and 2. Ahead of the flame the gas is uniform, the flow is irrotational and the entropy is constant.

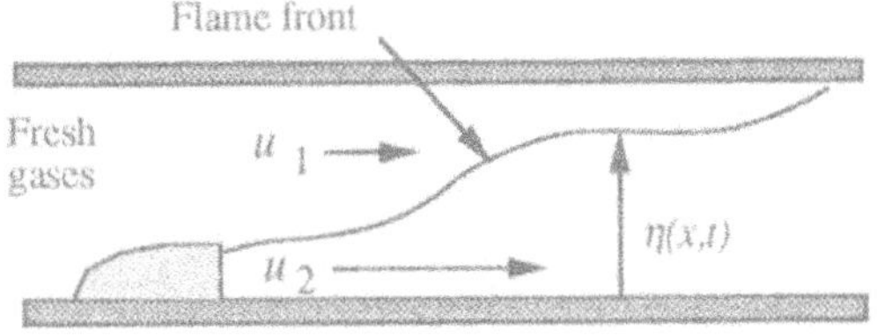

Figure 7. A turbulent flame stabilized in a rectangular duct.

The upstream and downstream flows are matched across the flame by writing mass, momentum and energy jump conditions and by stating that the kinematic deformation of the flame f is the same on both sides of the front. The kinematic conditions take the form :

$$\frac{\partial \eta}{\partial t} + u_1(x,\eta,t)\frac{\partial \eta}{\partial x} - v_1(x,\eta,t) = w_1 \sec\theta \tag{53}$$

$$\frac{\partial \eta}{\partial t} + u_2(x,\eta,t)\frac{\partial \eta}{\partial x} - v_2(x,\eta,t) = w_2 \sec\theta \tag{54}$$

In these expressions θ is the local angle of the flame, w_1 and w_2 are the normal flame velocities defined with respect to the upstream and downstream flows. Mass conservation across the flame takes the form :

$$\rho_1 w_1 = \rho_2 w_2 \tag{55}$$

The balance of normal momentum is

$$\rho_1 {w_1}^2 + p_1 = \rho_2 {w_2}^2 + p_2 \tag{56}$$

and the conservation of tangential momentum reduces to :

$$u_1 \cos\theta + v_1 \sin\theta = u_2 \cos\theta + v_2 \sin\theta \tag{57}$$

From the energy balance across the flame one finds that with good accuracy

$$\rho_1/\rho_2 \simeq 1 + Q/(c_p T_1) \tag{58}$$

where Q desigates the heat release in the flame. This expression together with the conservation of mass yields

$$w_2/w_1 = \nu \simeq 1 + Q/(c_p T_1) \tag{59}$$

Using the previous matching conditions and the local conservation equations of mass and momentum it is possible to derive a set of integral relations for the upstream and downstream flows. On the upstream side of the flame the local equations are integrated in the transverse direction from $y = \eta(x,t)$ to l. On the downstream side, this integration is carried out from $y = 0$ to $y = \eta$. The following set of equations is obtained :

• *Upstream (cold gases)*

$$\frac{\partial(l-\eta)}{\partial t} + \frac{\partial(l-\eta)\overline{u}_1}{\partial x} + w_1 \sec\theta = 0 \tag{60}$$

$$\frac{\partial(l-\eta)\overline{u}_1}{\partial t} + \frac{\partial(l-\eta)\overline{u}_1{}^2}{\partial x} + \frac{(l-\eta)}{\rho_1}\frac{\partial\overline{p}_1}{\partial x} + u_1(x,\eta,t)w_1 \sec\theta = 0 \tag{61}$$

• *Downstream (hot gases)*

$$\frac{\partial\eta}{\partial t} + \frac{\partial\eta\overline{u}_2}{\partial x} - w_2 \sec\theta = 0 \tag{62}$$

$$\frac{\partial\eta\overline{u}_2}{\partial t} + \frac{\partial\eta\overline{u}_2{}^2}{\partial x} + \frac{\eta}{\rho_2}\frac{\partial\overline{p}_2}{\partial x} - u_2(x,\eta,t)w_2 \sec\theta = 0 \tag{63}$$

The average velocities appearing in these expressions are defined as section averages :

$$\overline{u}_1 = \frac{1}{l-\eta}\int_\eta^l u_1 d\eta \quad \text{and} \quad \overline{u}_2 = \frac{1}{\eta}\int_0^\eta u_2 d\eta \tag{64}$$

The average pressures are defined by similar expressions. The system of equations (60) to (63) together with matching conditions across the flame front, closure assumptions on the local values of the variables at the flame and a set of boundary conditions define the dynamical response of the flow. In a given axial section, the flow is described by the transverse average velocities and pressures in the two regions and by the location of the front .

One advantage of this formulation is that it yields a one-dimensional time dependent model which includes a description of the flame front location. The set of unknowns is $\overline{u}_1(x,t), \overline{u}_2(x,t), \overline{p}_1(x,t), \overline{p}_2(x,t), \eta(x,t)$. As input, one has to specify the normal flame speed w_1. For turbulent flames stabilized in ducts, one may assign a constant value to the ratio of this speed to the incoming gas velocity. One may take for example $w_1/u_O = 0.1$. The value assigned to this ratio may chosen be to obtain the proper flame length. One may also include a more complicated dependence of w_1 on the flow parameters.

Using this formulation Marble and Candel (1978) were able to study the flame response to incoming acoustic perturbations. The work was later extended by Marble, *et al.* (1979), Subbaiah (1983) and Yang and Culick (1984). More recently, a nonlinear solution of this model was developed by Poinsot and Candel (1988). A typical result of calculation taken from this reference is shown in Figure 8.

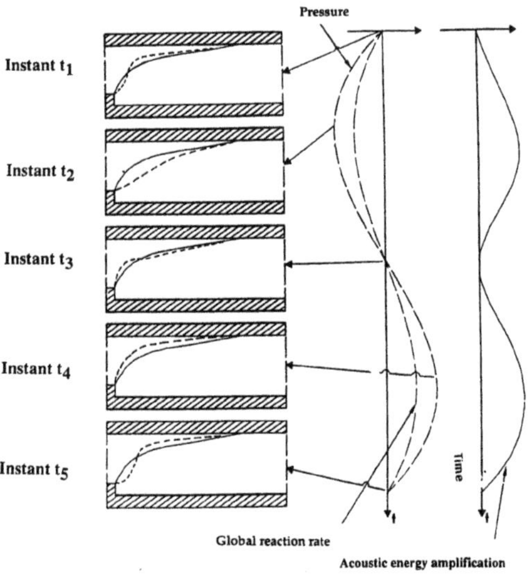

Figure 8. Quarter wave combustion instability of a ducted flame. Solid line gives the steady state position of the flame. The dotted line corresponds to the instantaneous position. (From Poinsot and Candel, 1988)

A second example of analysis which includes a description of the flame front dynamics is provided by Searby and Rochwerger[35]. These authors consider the instability of a laminar flame stabilized in a duct and submitted to an acoustic field or the oscillatory instability of a flame propagating downwards in a closed channel. The starting point of their analysis is an equation describing the response of a wrinkled flame to a weakly turbulent upstream flow :

$$A\ddot{\alpha}(\mathbf{k},t) + B\dot{\alpha}(\mathbf{k},t) + C\alpha(\mathbf{k},t) = u_e \tag{65}$$

where $\alpha(\mathbf{k},t)$ is the displacement of the flame front, A, B and C are coefficients depending on the flame characteristics and spatial wave number $\mathbf{k}$ of the wrinkling of the front and u_e is the external fluctuation imposed to the flame. In the absence of the external field u_e but in the presence of an acoustic field it is shown that the previous equation becomes :

$$A\ddot{\alpha}(\mathbf{k},t) + B\dot{\alpha}(\mathbf{k},t) + [C_0 - C_1\cos(\omega_a t)]\alpha(\mathbf{k},t) = O \tag{66}$$

This equation has the form of a parametrically driven harmonic oscillator. The last term represents the periodic acceleration of the flame front which separates regions of different density. Such an equation is already proposed on a phenomenological basis by Markstein (1951), Markstein and Somers (1953), Markstein and Squire (1955) but the exact forms of the coefficients A, B, C_0, C_1 are given by Searby and Rochwerger (1991). It is then shown that a Mathieu equation may be deduced from the dynamical equation for the front displacement. The domains of instability of the Mathieu equation are in agreement with those found in experiments on a stationary flame interacting with a standing acoustic mode.

3.2 DIRECT SIMULATIONS

While direct numerical calculations were initially limited to one-dimensional simulations of flow dynamics under unsteady combustion (for example Baum and Levine (1982), Baum, *et al.* (1986)) they are now carried out in multidimensional configurations. Much progress has been accomplished in the computational methodology and the related modeling. Multidimensional calculations reported by Kailasanath, *et al.* (1987),Jou and Menon (1987), Ghoniem, *et al.* (1987), Ghoniem and Najm (1989), Najm and Ghoniem (1991), Menon and Jou (1987), Menon and Jou (1990), Samaniego (1992) reveal many of the features found in combustion instability experiments such as vortex roll-up, vortex interactions with the dowstream boundary, pressure coupling of the initial shear layer.
One promising method devised by Menon and Jou (1991) combines a Large Eddy Simulation (LES) of the flow with a description of the flame front dynamics in terms of a kinematic equation for a progress variable G. The key idea here again is that one has to track the position of the flame front if one wishes to obain a suitable description of the reactive flow dynamics. This idea is related to those described in the previous subsection. A progress variable G is defined and satisfies a conservation equation of the form

$$\frac{\partial G}{\partial t} + \boldsymbol{v} \cdot \nabla G = u_F |\nabla G| \tag{67}$$

where $\boldsymbol{v}$ is the fluid velocity, $G = 1$ in the fresh mixture, $G = 0$ in the combustion products and u_F is a local turbulent flame speed depending on the subgrid-scale turbulence intensity u' and on the laminar flame speed u_L. The location of the flame front is given by a constant value contour of G. The heat release in the flame is described by writing the specific enthalpy in terms of G :

$$h = c_p T + (-\Delta h_0^f) G \tag{68}$$

The dynamical response of the front is infleunced by the local level of subgrid turbulence and by the stoichiometry through the dependence of u_F on u' and $u_L(\phi)$. Inital calculations carried out by assigning a constant value to u_F reproduce quite nicely experimental observations of the Smith and Zukoski (1985). More recently, Menon (1992) was able to simulate active control with this LES model by including an external loop describing an acoustic feedback.

3.3. HYDRODYNAMIC INSTABILITY MODES OF REACTIVE FLOWS

Problems of hydrodynamic instability are extensively studied in the case of nonreactive flows.

The stability of reactive shear flows has received much less consideration. The problem was considered by Blackshear (1956) who treated the stability of premixed flames stabilized in a duct using simplified velocity and temperature profiles. Growing experimental evidence of the presence of hydrodynamic instabilities in reactive shear flows has triggered a series of investigations of this topic. Various aspects of the problem are examined in Trouvè, *et al.* (1988), Koochesfahani and Frieler (1989), Mahalingam, *et al.* (1991), Shin and Ferziger (1990). In the more elaborate studies, the reaction terms are included. However, much can be learned by examining the spatial linear stability of a shear layer formed at the confluence of two streams having different velocities U_1 and U_2 and different temperatures T_1 and T_2. The analysis then relies on a Rayleigh equation including a temperature (or equivalently density) term :

$$(U-\omega/k)(\frac{d^2\varphi}{dy^2}-k^2\varphi)-\varphi\frac{d^2U}{dy^2}+\frac{1}{\overline{\rho}}\frac{d\overline{\rho}}{dy}[(u-\omega/k)\frac{d\varphi}{dy}-\varphi\frac{dU}{dy}]=0 \qquad (69)$$

In this equation U is the mean flow velocity, ω and k are the angular frequency and wavenumber of the flow perturbations and $\varphi(y)$ is the transverse structure of the perturbed streamfunction. When the temperatures are quite different as is the case in combustion applications, the stability characteristics are notably affected.

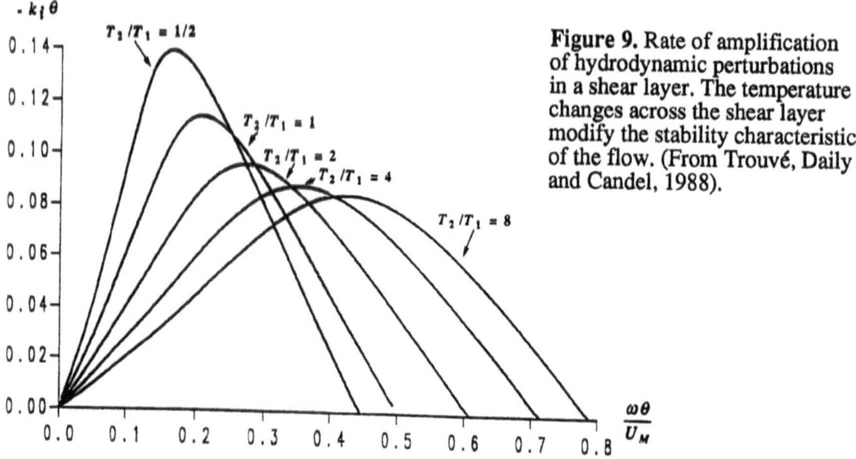

Figure 9. Rate of amplification of hydrodynamic perturbations in a shear layer. The temperature changes across the shear layer modify the stability characteristics of the flow. (From Trouvé, Daily and Candel, 1988).

Calculations of Trouvé *et al.* (1988) indicate that when $U_1 > U_2$ and when the temperature ratio $T_2/T_1 > 1$ the rate of amplification of the prefered mode of oscillation is diminished while the characteristic frequency of this mode is augmented. The inverse effect is observed when $T_2/T_1 < 1$. These features are displayed in Figure 9 which shows the amplification curves as a function of the reduced frequency $\omega\theta/U_M$ where θ is the shear layer thickness and $U_M = (U_1+U_2)/2$. If one considers a flame stabilized in a duct, there are two main regions to consider. In the first, close to the stabilizer lips the incoming flow velocity U_1 is much greater than the velocity behind the bluff-body while $T_1 < T_2$ and the shear layer responds to a wide band of frequencies as indicated in Figure 10.

Further downstream, the burnt gases have been accelerated and their velocity exceeds that of the fresh mixture. The shear profile is now inverted. If we now designate by 1 the stream of hot products in the center of the channel and by 2 the stream of fresh mixture flowing near the walls one has $U_1 > U_2$ but this time $T_1 > T2$ and the amplification curve is narrower and shifted towards the lower reduced frequencies.

Now the thickness θ increases slowly in the downstream direction while the mean velocity U_M increases by a larger amount of the order of the density ratio between fresh and burnt gases ($\simeq \rho_u/\rho_b$) As a consequence, the band of frequencies for which amplification takes place remains nearly constant and the most amplified frequency remains nearly unchanged. This may explain why ducted flames respond so well to external acoustic forcing. It also indicates why the maximum response will follow a Strouhal law as found by Yu *et al.* (1991).

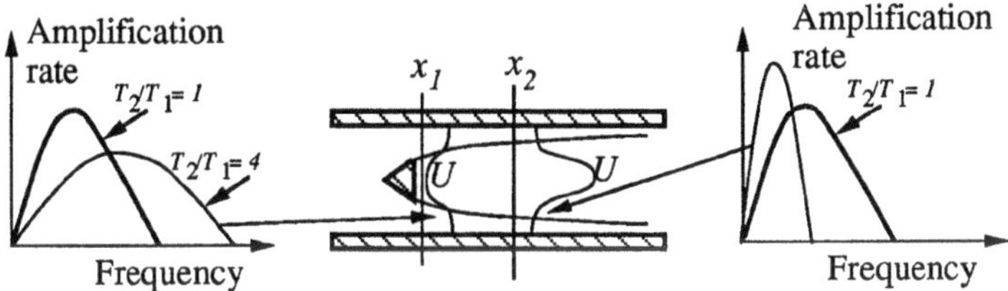

Figure 10. Hydrodynamic stability characteristics of a ducted flame.

3.4. VORTEX IGNITION STUDIES

Another problem that has received considerable attention recently is that of combustion in the field of vortex structures. Vortex burning is of special interest because it often drives the instability. Sketches of situations encountered in practice are shown in Figure . Much of the work concerns the process of roll-up of a flame sheet by a vortex and the subsequent combustion enhancementMarble (1985), Karagozian and Marble (1986). The process of extinction is studied by Poinsot, *et al.* (1991) while ignition in a vortex has been treated by Macaraeg, *et al.* (1991) and Thévenin and Candel (1991). The last analysis indicates that various modes of ignition are possible and that the time-delay is governed by a competition between processes of straining, mixing and reaction kinetics. The process of vortex ignition is examined by considering an initially plane interface separating fuel and oxidizer half planes. The interface is rolled up in a starting vortex. Typical results of calculations are shown in Figure 11.

In this case the temperature of the oxidizer is high and the reaction takes place in the braids and in the core. At a later time the core burns out. Consumption of the core thus takes place during a finite period of time. A heat pulse is produced in the process and energy may be fed into an acoustic mode of the combustor which in turn will trigger the next vortex shedding.

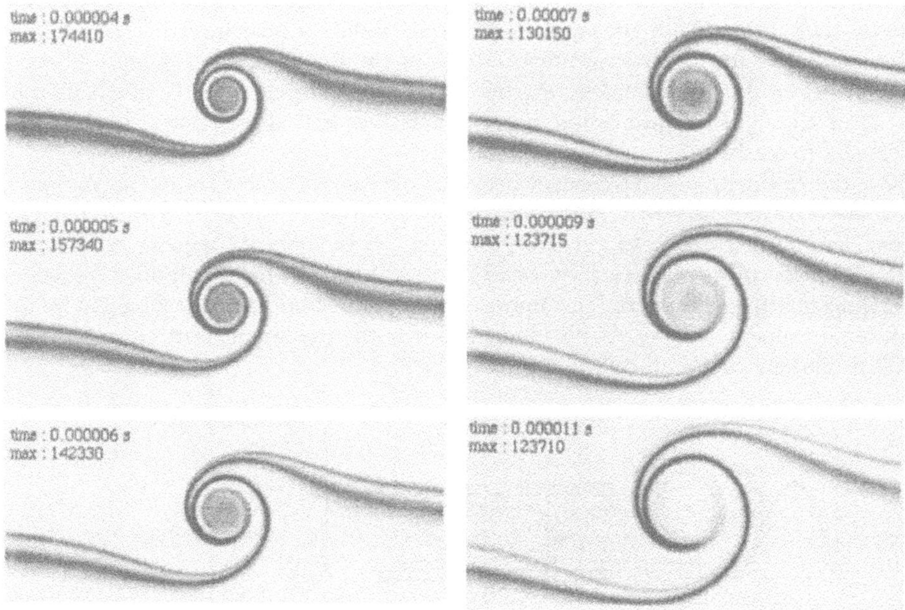

Figure 11. Heat release distribution during vortex ignition. Re = 700, T_{FO} = 300K, T_{OO} = 2000 K, H /Air reaction.(From Thévenin and Candel, 1992).

3.5. NONLINEAR ACOUSTIC INSTABILITY MODELS AND APPLICATIONS TO ACTIVE CONTROL SIMULATIONS

To conclude this section it is also worth citing the unified approach established by Culick (1976), Culick (1988) and coworkers Awad and Culick (1986), Yang, *et al.* (1987) to describe the nonlinear acoustics of combustion instability. A general framework is devised which yields a set of coupled ordinary differential equations representing the dynamics of the system. Limit cycles and nonlinear triggering features are conveniently examined within this framework. It is also possible to use this approach to simulate the dynamics of active control.

We will now briefly illustrate this last application. The configuration shown in Figure comprises a combustor and an external control loop. Modern applications of this principle use an adaptive filter in the control loop Billoud, *et al.* (1992). Experiments indicate that active control may be used to essentially suppress an instability. The controlled combustor operates over an extended domain and the flame stability is enhanced. The simulation of active control is of interest for many reasons. First, it is worth understanding why the control works and what determines its performance. Second, it is useful to develop a numerical model which may be used to design and test control schemes. The CI model should be realistic and in particular it should be able to describe the linear growth and the limit cycles. However if it is too complex, it will preclude time domain applications. The nonlinear acoustic model of Culick constitutes a good compromise between these requirements. It was used by Yang, *et al.* (1992) to test state space control methods and more recently by Huynh (1993) to study adaptive control algorithms.

In essence, the pressure field is expanded in terms of the eigenmodes of the combustion chamber :

$$p_1 = \overline{p} \sum_n \eta_n(t)\psi_n(\boldsymbol{x}) \tag{70}$$

where $\psi_n(\boldsymbol{x})$ designates the n-th mode of the chamber and $\eta(t)$ is the corresponding amplitude. The modal amplitudes satisfy a set of differential equations of the form :

$$\frac{d^2\eta_n}{dt^2} + \omega_n^2\eta_n = F_n + U_n \tag{71}$$

where

$$F_n = -\frac{c^2}{\overline{p}E_n^2}\int_V h\psi_n dV \quad U_n = -\frac{c^2}{\overline{p}E_n^2}\int_V h_c\psi_n dV \tag{72}$$

with $E_n^2 = \int_v \psi_n^2 dV$. In these expressions h and h_c respectively designate the nonsteady heat release and the nonsteady heat release associated with the modulation caused by the control actuator. Models for these two terms have to be suppliied in order to study the dynamics of the system (see Huynh for details). Figure 12 shows a typical result of simulation obtained with this model in combination with an adaptive control algorithm.

From top to bottom, the figure shows the pressure perturbation, the nonsteady heat release and the control signal. At the initial time the system is unstable and the pressure amplitude increases exponentially. The controller is started when the pressure amplitude reaches a predetermined level. After a certain time the controller has converged and the level of the pressure signal is brought back to a minimum. The simulation closely follows the typical response found in practical experiments indicating that the simulation model of the combustor retains the dynamical features of the real system.

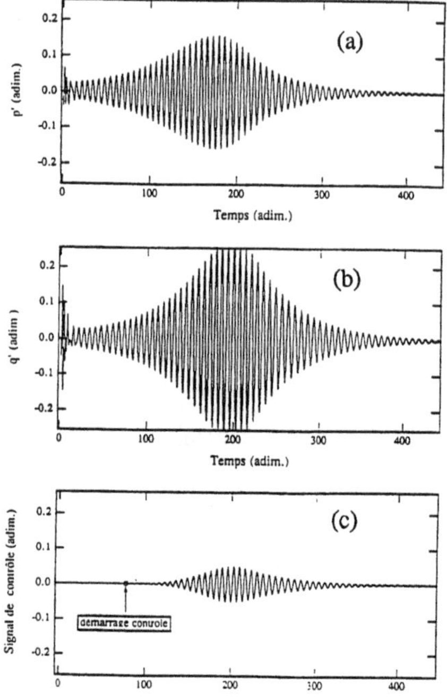

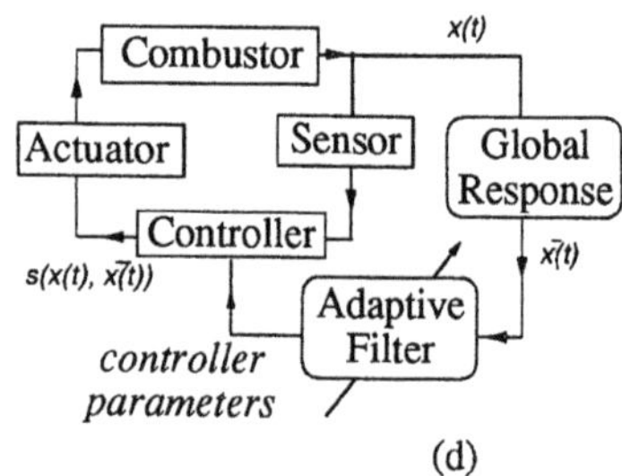

Figure 12. Time traces of (a) pressure, (b) heat release, (c) control signal during a simulated adaptive control sequence. The system is shown in (d). (From Huynh, 1993).

4. Conclusions

This article provides a review of modeling methods for combustion instability studies. After a presentation of classical material the paper focuses on modern methods. The survey concerns models of flame front dynamics, LES simulations, hydrodynamic instability studies of reactive flows and vortex ignition studies and nonlinear acoustic models for dynamical simulations. The new approaches together with recent experimental data have enhanced the current understanding of combustion instability. The prediction of instability of turbulent combustors is however not yet achievable and remains a problem for future research.

Acknowledgments

We wish to thank DRET, SEP, GdF for their support of our research in combustion instability and active control.

References

Awad, E. and Culick, F. E. C. (1986). Existence and stability of limit cycles for longitudinal acoustic waves in a combustion chamber. *Combustion Science and Technology* **46**, 169-181.

Barrère, M., Jaumotte, A., Fraeijis de Veubeke, B. and Vandenkerckhove, J. (1960). *Rocket propulsion.* Amsterdam : Elsevier.

Barrère, M. and Williams, F. A. (1968). Comparison of combustion instabilities found in various types of combustion chambers. In *Eleventh Symposium (International) on Combustion.* Pittsburgh : The Combustion Institute, 169-181.

Baum, H. R., Corley, D. M. and Rehm, R. G. (1986). Time dependent simulation of small scale turbulent mixing and reaction. In *Twenty first Symposium (International) on Combustion.* Pittsburgh : The Combustion Institute, 1263-1270.

Baum, J. D. and Levine, J. N. (1982). Numerical techniques for solving nonlinear instability problems in solid rocket motors. *AIAA Journal* **20**, 955-961.

Billoud, G., Huynh, C., Galland, M. and Candel, S. (1992). Adaptive active control of combustion instabilities. *Combustion Science and Technology* **81**, 257.

Blackshear, P. L. (1956). *Growth of disturbances in a flame generated shear region.* NACA Report 1360,

Candel, S. (1992). Combustion instabilities coupled by pressure waves and their active control. In *24th Symposium (International) on Combustion.* Pittsburgh : The Combustion Institute, 1277-1296.

Clavin, P., Pelc, P. and He, L. (1990). One-dimensional vibratory instability of planar flames propagating in tubes. *Journal of Fluid Mechanics* **216**, 299-322.

Clavin, P. and Sun, J. (1991). Theory of acoustic instabilities of planar flames propagating in sprays or particle-laden gases. *Combustion Science and Technology* **78**, 265-288.

Crocco, L. (1951). Aspects of combustion instability in liquid propellant rocket motors. *Journal of the American Rocket Society* **21**, 163.

Crocco, L. (1965). Theoretical studies on liquid-propellant rocket instability. In *Tenth Symposium (International) on Combustion.* Pittsburgh : The Combustion Institute, 1101-1128.

Crocco, L. and Cheng, S. L. (1956). *Theory of combustion instability in liquid propellant rocket motors.* AGARDOGRAPH N 8, Butterworths Science Publication.

Crocco, L., Grey, J. and Harrje, D. T. (1960). Theory of liquid propellant rocket combustion instability and its experimental verification. *J. Aeronautical Research Journal* **30**, 159-168.

Crocco, L., Harrje, D. T., Sirignano, W. A. and al., e. (1968). *Nonlinear aspects of combustion instability in liquid propellant rocket motors.* NASA CR 72426,

Culick, F. E. C. (1976). Nonlinear behavior of acoustic waves in combustion chambers. Parts I and II. *Acta Astronautica* **3**, 714-757.

Culick, F. E. C. (1988). Combustion instabilities in liquid-fueled propulsion systems-an overview. *AGARD Conference on Combustion instabilities in liquid-fuelled propulsion systems*

Ghoniem, A. F., Heidarinejad, G. and Krishan, A. (1987). Numerical simulation of a reacting shear layer using the transport element method.

Ghoniem, A. F. and Najm, H. N. (1989). *Numerical simulation of the coupling between vorticity and pressure oscillations in combustion instability.* AIAA Paper 89-2665,

Harrje, D. J. and Reardon, F. H. (1972). *Liquid propellant rocket instability.* NASA SP-194,

Huynh, C. (1993). *PhD thesis, Ecole Centrale Paris, Chatenay-Malabry.*

Jou, W. H. and Menon, S. (1987). Simulations of ramjet combustor flow fields. Part II. Origin of pressure oscillations.

Kailasanath, K., Gardner, J. H., Boris, J. P. and Oran, E. S. (1987). Numerical simulations of acoustic-vortex interactions in a central-dump combustor. *Journal of Propulsion* **3**, 525-533.

Karagozian, A. R. and Marble, F. E. (1986). Study of a diffusion flame in a stretched vortex. *Combustion Science and Technology* **46**, 65-84.

Koochesfahani, M. M. and Frieler, C. E. (1989). Instability of nonuniform density free shear layers with a wake profile. *AIAA Journal* **27**, 1735-1740.

Kotake, S. (1975). On combustion noise related to chemical reactions. *Journal of Sound and Vibration* **42**, 399-410.

Laverdant, A., Poinsot, T. and Candel, S. (1986). Influence of the mean temperature field on the acoustic mode structure in a dump combustor. *Journal of Propulsion and Power* **2**, 311-316.

Leder, G. and Kapila, A. K. (1991). The response of premixed flames to pressure perturbations. *Combustion Science and Technology* **76**, 21-44.

Macaraeg, M. G., Jacson, T. L. and Hussaini, M. Y. (1991). *Ignition and structure of a laminar diffusion flame in the field of a vortex.* ICASE 91-69, ICASE.

Mahalingam, S., Cantwell, B. J. and Ferziger, J. H. (1991). Stability of low-speed reacting flows. *Physics of Fluids* **A 3**, 1533-1543.

Mallard, E. E. and Le Chatelier, H. (1883). Recherches experimentales et thoriques sur la combustion de melanges gazeux explosifs. *Annales des Mines, Partie Scientifique et Technique* **Ser. 8, no. 4**, 274.

Marble, F. E. (1955). Servo-stabilization of low-frequency oscillations in liquid propellant rocket motors. *ZAMP J. Applied Mathematics and Physics* **VI/1**, 1-35.

Marble, F. E. (1985). Growth of a diffusion flame in the field of a vortex. In *Recent advances in the aerospace sciences.* Casci, E. New York : Plenum Press, 395-413.

Marble, F. E. and Candel, S. (1978). An analytical study of the nonsteady behavior of large combustors. In *17th Symposium(International) on Combustion.* Pittsburgh : The Combustion Institute, 761-769.

Marble, F. E. and Cox, D. W. (1953). Servo-stabilization of low-frequency oscillations in a liquid bipropellant rocket motor. *Journal of the American Rocket Society* **23**, 63-81.

Marble, F. E. and Hendricks, G. J. (1986). Structure and behavior of diffusion flames in a pressure gradient. In *Twenty first Symposium (International) on Combustion.* Pittsburgh : The Combustion Institute, 1321-1327.

Marble, F. E., Subbaiah, M. V. and Candel, S. (1979). Analysis of low frequency disturbances in afterburners. *Proceedings of the AGARD Specialists Meeting on Combustion Modelling*

Markstein, G. H. (1951). Experimental and theoretical studies of flame front stability. *Journal of the Aeronautical Sciences Journal of Aeronautical Science* **18**, 199-209.

Markstein, G. H. and Somers, L. M. (1953). Cellular flame structure and vibratory flame movement in N-butane-methane mixtures. 527-535

Markstein, G. H. and Squire, W. (1955). On the stability of a plane flame front in oscillating flow. *Journal of the Acoustical Society of America* **27**, 416-424.

McIntosh, A. C. (1986). The effect of upstream acoustic forcing and feedback on the stability and resonance of anchored flames. *Combustion Science and Technology* **49**, 143-167.

McIntosh, A. C. (1991). Pressure disturbances of different length scales interacting with conventional flames. . *Combustion Science and Technology* **75**, 287-309.

McIntosh, A. C. and Wilce, S. A. (1991). High frequency pressure wave interaction with premixed flames. *Combustion Science and Technology* **79**, 141-155.

McManus, K., Poinsot, T. and Candel, S. (1993). A review of active control of combustion instability. *Progress in Energy and Combustion Science* **19**, 1-29.

Menon, S. and Jou, W. H. (1987). *Simulations of ramjet combustor flow fields; PartI. Numerical model, large scale and mean motions.* AIAA Paper 87-1421,

Menon, S. and Jou, W. H. (1990). Numerical simulations of oscillatory cold flows in an axisymmetric ramjet. *Journal of Propulsion and Power* **6**,

Menon, S. and Jou, W. H. (1991). Large-Eddy simulations of combustion instability in an axisymmetric ramjet combustor. *Combustion Science and Technology* **75**, 53-72.

Najm, H. N. and Ghoniem, A. F. (1991). Modeling pulsating combustion due to flow-flame interactions in vortex-stabilized premixed flames. *International Symposium on Pulsating Combustion*

Phillips, O. M. (1960). On the generation of sound by supersonic turbulent shear layers. *Journal of Fluid Mechanics* **9**, 1-28.

Poinsot, T. and Candel, S. (1988). A nonlinear model for ducted flame combustion instabilities. *Combustion Science and Technology* **61**, 121-153.

Poinsot, T., Veynante, D. and Candel, S. (1991). Quenching processes and premixed turbulent combustion diagrams. *Journal of Fluid Mechanics* **228**, 561-606.

Putnam, A. A. (1971). *Combustion driven oscillations in industry.* New York : Elsevier.

Putnam, A. A., Belles, F. E. and Kentfield, J. A. C. (1986). Pulse combustion. *Progress in Energy and Combustion Science* **12**, 43-79.

Rayleigh, L. J. W. S. (1878). The explanation of certain acoustic phenomena. *Nature* **18**, 319-321.

Samaniego, J. M. (1992). *Etude des instabilites de combustion dans les statoreacteurs.* Doctoral Thesis, Ecole Centrale Paris, Chatenay-Malabry.

Samaniego, J. M., Yip, B., Poinsot, T. and Candel, S. (1993). Low frequency combustion instability in a side dump combustor. *Combustion and Flame* **94**, 363-380.

Searby, G. and Rochwerger, D. (1991). A parametric acoustic instability in premixed flames. *Journal of Fluid Mechanics* **231**, 529-543.

Shin, D. and Ferziger, J. H. (1990). *Linear stability of the reacting mixing layer.* AIAA Paper 90-0268,

Smith, D. A. and Zukoski, E. E. (1985). *Combustion instability sustained by unsteady vortex combustion.* AIAA Paper 85-1248,

Strahle, W. C. (1971). On combustion generated noise. *Journal of Fluid Mechanics* **49**, 399-414.

Subbaiah, M. V. (1983). Nonsteady flame spreading in two-dimensional ducts. *AIAA Journal* **11**, 1557-1564.

Summerfield, M. (1951). A theory of unstable propulsion in liquid propellant rocket systems. *American Rocket Society Journal* **21**, 108-114.

Thevenin, D. and Candel, S. (1991). Ignition dynamics of a diffusion flame rolled-up in a vortex. *SIAM Conference on Numerical Combustion*

Trouve, A., Candel, S. M. and Daily, J. W. (1988). *Linear stability of the inlet jet in a ramjet dump combustor.* AIAA Paper 88-0149,

Tsien, H. S. (1952). Servo-stabilization of combustion in rocket motors. *American Rocket Society Journal* **22**, 256-263.

Williams, F. A. (1985). *Combustion theory.* Menlo Park : Benjamin/Cummings.

Yang, V. and Culick, F. E. C. (1984). Analysis of low frequency oscillations in a laboratory ramjet combustor. *Combustion Science and Technology* **45**, 1-25.

Yang, V., Kim, S. I. and Culick, F. E. C. (1987). *Third-order nonlinear acoustic inbstabilities in combustion chambers, Part I : longitudinal modes.* AIAA Paper 87-1873,

Yang, V., Sinha, A. and Fung, Y. T. (1992). State-feedback control of longitudinal combustion instabilities. *Journal of Propulsion* **8**, 66-73.

Yu, K., Trouve, A. and Candel, S. (1991). *Combustion enhancement of a premixed flame by acoustic forcing with emphasis on role of large-scale vortical structures.* AIAA Paper 91-0367.

Zinn, B. T. (1986). Pulsating combustion. In *Advanced combustion methods.* Weinberg, F. J. New York : Academic Press,

6. PULSE COMBUSTION APPLICATIONS: PAST, PRESENT AND FUTURE

B.T. Zinn
Laboratory for Pulsating Combustion Processes
School of Aerospace Engineering
Georgia Institute of Technology
Atlanta, Georgia 30332-0150, USA

Abstract

This paper presents an introduction to pulse combustion and a discussion of its past, present and future applications. The types and principles of operation of pulse combustors are discussed. A linear theoretical treatment of the driving of pulsations by a periodic heat addition process is presented. This model shows that the combustion process drives or damps acoustic oscillations when the magnitude of the phase difference between the combustion process and flow oscillations is smaller or larger than 90°, respectively. Furthermore, the frequency of the pulse combustor oscillations is larger or smaller than the natural acoustic mode frequency of the pulse combustor when the magnitude of the phase difference between the combustion and pressure oscillations is smaller or larger than 180°, respectively. The theoretical discussion is followed by a brief historical review of pulse combustion applications. Next, the advantages of pulse combustion heating applications, which have been successfully commercialized in the US, Europe and Japan in recent years, are discussed. The article closes with a discussion of the recently developed resonant driving method that uses a tunable pulse combustor to drive large amplitude beneficial pulsations in energy intensive, combustion and incineration processes. It is shown that this approach can be used to produce fuel savings, increased productivity, reduced emissions and improved product quality in a wide range of industrial processes.

Introduction

Interest in pulse combustion has increased in recent years because of its potential to reduce fuel consumption and pollutant emissions and increase productivity in energy intensive processes. In addition, pulse combustion offers an attractive approach for combusting difficult-to-burn fuels such as heavy fuel oils, low grade coals and wastes. As is discussed later in this paper, the advantages of pulse combustion can be related to

F. Culick et al., (eds.), Unsteady Combustion, 113–137.

the excitation of large amplitude pressure and velocity oscillations by the pulse combustion process, and the challenge facing those who work in this field is to develop applications of the technology that will take advantage of the increases in the rates of mass, momentum (i.e., mixing) and heat transfer that accompany the excitation of these oscillations.

The mechanisms that control the operation of pulse combustors will be discussed by considering the operation of the mechanically valved, quarter-wave type, pulse combustor shown in Fig. 1. While practical pulse combustors are generally constructed from a number of tube-like sections having different diameters and lengths, the one shown in Fig. 1 consists of a constant diameter tube, for simplicity. This pulse combustor consists of fuel and air flapper valves, a combustor section and a tail pipe. Operation is started by injecting a combustible fuel-air mixture into the combustor section and igniting it by, for example, a spark plug. The ensuing combustion process is accompanied by a combustor pressure increase (i.e., period 0-1) that forces the flapper valves to close and the combustion products to move into the tail pipe. The outflow of combustion products is accompanied by a combustor pressure drop (period 1-3). Since the inertia of the moving gases prevents them from stopping when combustor pressure equals the pressure downstream of the tail pipe exit plane, the combustor pressure continues to decrease until it reaches a minimum value at point 3. When the combustor pressure decreases below the pressure levels upstream of the flapper valves, the valves open and admit fresh fuel and air into the combustor section. The mixing of the incoming reactants with one another and with combustion products and flamelets left over from the previous cycle leads to ignition and reaction of the incoming reactants and a "new" pressure rise (period 3-4). The initial cycle can now repeat itself indefinitely (without the use of a spark plug) and the pulse combustor will continue to operate until the fuel supply is stopped.

The combustor shown in Fig. 1 is referred to as a quarter-wave pulse combustor because its length equals $\lambda/4$, where λ is the wavelength of the oscillation. Maximum and minimum pressure amplitudes occur at the upstream and downstream ends of the pulse combustor, respectively. The operation of this pulse combustor is controlled by a feedback process involving a periodic combustion process, flapper valves and acoustic oscillations; the combustion process excites acoustic oscillations within the pulse combustor, these oscillations result in periodic opening and closing of flapper valves that produce a periodic reactants injection rates into the combustor, resulting in a periodic combustion process, which completes the feedback cycle.

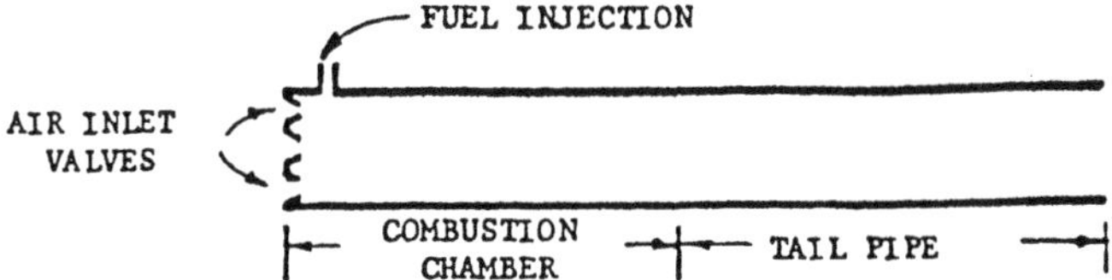

SCHEMATIC OF A QUARTER WAVE TYPE PULSATING COMBUSTOR

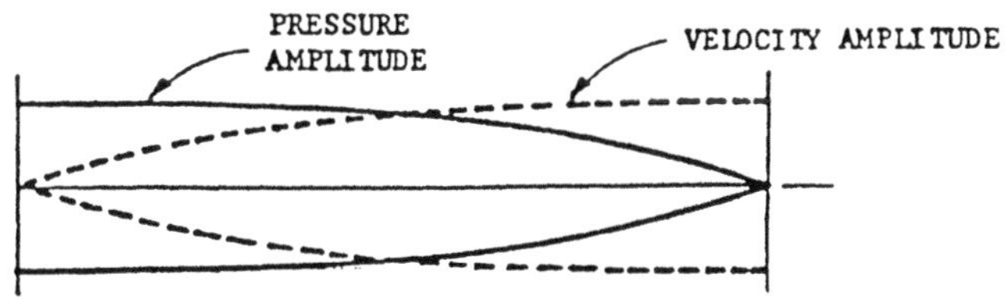

DISTRIBUTIONS OF THE ACOUSTIC PRESSURE AND VELOCITY AMPLITUDES

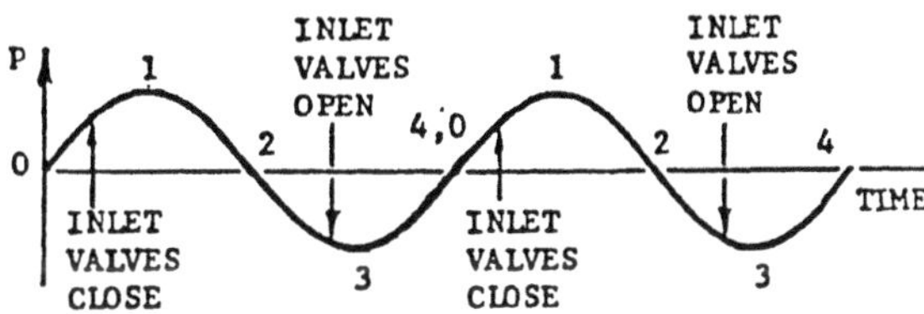

TIME DEPENDENCE OF COMBUSTOR PRESSURE

PERIOD 0-1: COMBUSTION OCCURS WITH PRESSURE RISE;
INLET VALVES CLOSE.

PERIOD 1-3: OUTFLOW OF COMBUSTION PRODUCTS ACCOMPANIED
BY PRESSURE DROP AND OPENING OF INLET VALVES.

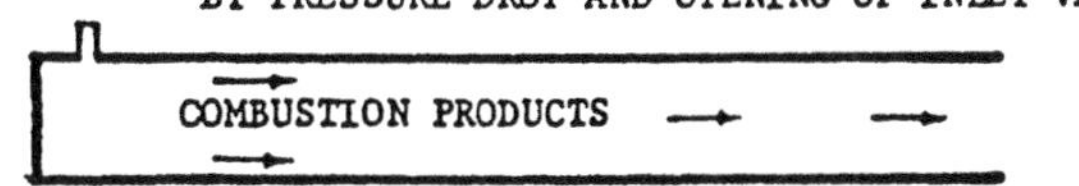

PERIOD 3-4: INFLOW OF FRESH AIR, FUEL, AND SOME COMBUSTION
PRODUCTS ACCOMPANIED BY PRESSURE RISE.

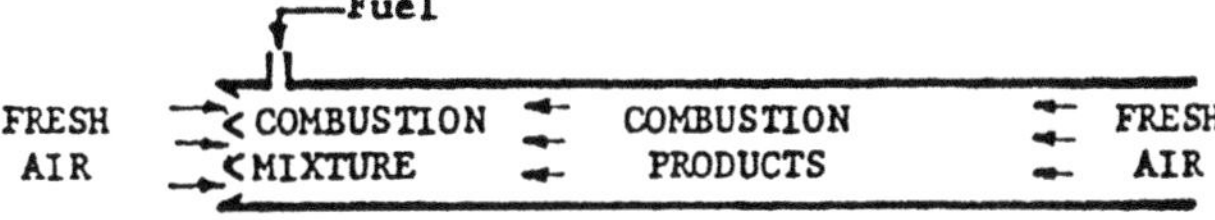

Fig. 1 Characteristics of a Quarter-Wave Type Pulsating Combustor.

To attain pulse combustion operation, the driving provided by the combustion process must be larger or equal to the damping experienced by the oscillations due to, for example, such processes as viscous dissipation, heat transfer, and acoustic radiation through the flapper valves and tail pipe opening. This condition is approximately described by the following inequality

$$\int_T\int_V p'(x,t)Q'(x,t)dvdt \geq \int_T\int_V\sum_i L_i(x,t)dVdt \qquad (1)$$

where L_i is the i-th damping process experienced by the pulsations and the integrations are performed over the volume, V, where driving and damping occur, and the period of the oscillations, T. The integrals on the left and right side of Eq. 1 describe the driving and damping experienced by the pulsations, respectively. Driving of the pulsations by the combustion process occurs when the integral on the right side of Eq. 1 is a positive quantity; a condition that is satisfied when the magnitude of the phase difference between the pressure, p', and heat addition, Q', oscillations is smaller than ninety degrees. This condition is widely known as Rayleigh's criterion[1,2]. It should be also pointed out that the inequality in Eq. 1 is generally applicable during the initial phase of the pulse combustor operation, after ignition, when the driving by the combustion process is larger than the damping experienced by the pulsations, while the equal sign is applicable when the pulse combustor has attained limit cycle operation where the amplitude of the pulsations no longer changes, indicating that the driving and damping experienced by the oscillations per cycle equals one another.

In addition to the Quarter Wave pulse combustor, the best known pulse combustors, which have been developed to date, are the Helmholtz and Rijke pulse combustors[2]. The operation of the Helmholtz pulse combustor is very similar to that of the Quarter Wave pulse combustor, and it involves "bulk" type pressure oscillations within a combustor section whose diameter is significantly larger than that of the tail pipe. To date, Helmholtz type pulse combustors have been successfully commercialized in heating applications, which are discussed later in this paper. The design of the Rijke pulse combustor is based upon the principles of operation of the acoustic Rijke tube[2,3] and is significantly different from the Quarter Wave and Helmholtz pulse combustors designs. It generally consists[2-4] of a straight tube that is open at both ends and is supplied with an oxidizer flow through one end. A combustion process is stabilized at a distance of $L/4$, where L is the tube length, from the tube entrance. Interaction between the heat released by the combustion process and the fundamental acoustic mode of the tube, consisting of half a wavelength with pressure nodes at both ends, results in a pulsating combustion process.

The conditions for the attainment of pulse combustion operation are similar for all pulse combustors. First, Rayleigh's criterion must be satisfied, requiring that the magnitude of the phase difference between the combustion process heat addition and pressure oscillations be smaller than 90°. Second, the driving provided by the combustion process must be larger than or equal to the damping experienced by the pulsations. Research conducted to date has shown that in all instances these conditions are satisfied when the driving mechanism is controlled by a feedback process involving interactions between combustion, fluid mechanical and acoustic processes.

It should be also pointed out that the principles of operation of pulse combustors are very similar to those that control the behavior of instabilities in various propulsion systems, and much of what has been learned to date in combustion instability research[5,6] is applicable to pulse combustors (and vice versa). This experience showed that combustion instabilities can occur in propulsion systems having drastically different geometries and operating conditions, and it may involve the excitation of longitudinal, transverse and three dimensional acoustic oscillations. These observations suggest that pulse combustors having different geometries, operating conditions and the excitation of different acoustic modes could be developed in response to various needs.

Since the details of the processes that control the operation of pulse combustors are discussed in an accompanying paper in this volume (i.e., the paper by Dr. Jay Keller), the remainder of this paper will attempt to complement Dr. Keller's paper by focusing on some theoretical aspects and applications of pulse combustion.

Theoretical Considerations

This section presents a simplified analysis of the driving or damping of pulsations by a periodic heat addition process similar to those observed in pulse combustors or unstable propulsion devices. Starting with the following acoustic axial momentum and energy equations[2],

Momentum: $$\bar{\rho}\frac{\partial u'}{\partial t}+\frac{\partial p'}{\partial x}=0 \qquad (2)$$

Energy: $$\frac{\partial p'}{\partial t}+\gamma\bar{p}\frac{\partial u'}{\partial x}=(\gamma-1)\bar{p}Q' \quad (3)$$

differentiating the momentum and energy equations with respect to x and t, respectively, and subtracting the resulting equations from one another yields the following inhomogenous wave equation

$$\frac{\partial^2 p'}{\partial t^2}-\frac{\gamma\bar{p}}{\bar{\rho}}\frac{\partial^2 p'}{\partial x^2}=(\gamma-1)\bar{p}\frac{\partial Q'}{\partial t} \quad (4)$$

where the term on the right side describes the driving or damping provided by the periodic heat addition process. Assuming that the heat addition and pressure are periodic functions of time; that is,

$$p'(x)=P(x)\bullet\exp(i\omega t) \quad (5\text{-a})$$

and

$$Q'(x)=Q(x)\bullet\exp(i\omega t)=R\bullet p'(x,t);\ R=R_r+iR_i \quad (5\text{-b})$$

and substituting Eqs 5-a,b into Eq. (4), reduces the latter to the following ordinary differential equation:

$$\frac{\partial^2 P}{\partial x^2}+(ik_nR+k_n{}^2)P=\frac{\partial^2 P}{\partial x^2}+\gamma_n{}^2P=0 \quad (6)$$

where

$$k_n=k_{n,r}+ik_{n,i}=\omega_n/\bar{c} \quad (7)$$

is a complex wave number, which, according to Eq. 6, is related to γ_n by the relationship

$$\gamma_n{}^2=ik_nR+k_n{}^2 \quad (8)$$

Expressing the solution of Eq. 6 as

$$P=A\bullet\cos(\gamma_n x) \quad (9)$$

where A is the amplitude of the pulse combustor oscillations, assuming that the acoustic velocity is approximately zero at both ends of the pulse combustor (i.e., at $x=0$ and

$x = L$), which requires that the solution P satisfy the boundary condition $\frac{dP'}{dx} = 0$ at these locatios, yields the following expression for the eigenfrequency γ_n:

$$\gamma_n = (n\pi / L) = \gamma_{n,r} + in_{n,i}; \quad n = 1,2,3,... \tag{10}$$

Using Eq. 8 and assuming that $|R| << 1$, it can be shown that:

$$k_n = (n\pi / L) - \frac{i}{2}(R_r + iR_i) = ((n\pi / L) + .5R_i) - \frac{i}{2}R_r \tag{11}$$

Using Eqs. (5-a) through (9), it can be shown that

$$p' = p(x) \cdot \exp[\underbrace{i\bar{c}((n\pi / L) + .5R_i)t}_{\substack{Effective \\ frequency}} + \underbrace{(.5R_r\bar{c}}_{\substack{amplification \\ coefficient}})t] \tag{12}$$

An examination of Eq. 12 shows that the heat addition process modifies the acoustic frequency by an amount that equals $\bar{c}R_i$, which is proportional to the imaginary part of the "Response Function" relating the pressure and heat addition oscillations, see Eq. 5-b. Furthermore, Eq. 12 indicates that pulsations will be driven (i.e., amplified) by the combustion process when $R_r < 0$, which, according to Eq. 5-b, is satisfied when the pressure and heat addition oscillations are in phase. These results are described in Fig. 2. It shows that the frequency of the excited oscillations will be higher and lower than the natural acoustic mode frequency of the device when $0° < \phi < 180°$ and $180° < \phi < 360°$, respectively. Figure 2 also shows that the heat addition oscillations drive or damp the pulsations when $270° < \phi < 90°$ and $90° < \phi < 270°$, respectively. Thus, it is important to keep in mind that heat addition oscillations can change the acoustic frequency of the pulse combustor as well as damp the pulsations; facts that have been neglected on occasion by previous investigators in the field.

Finally, it should be pointed out that the analysis presented in this section neglected important phenomena that should be included in any "reasonable" model of a pulse combustor; namely, nonlinear effects, the characteristics of the unsteady combustion process (e.g., mixing and chemical reactions), the effect of the flapper valves and tail pipe boundary conditions, and the effect of heat losses through the pulse combustor walls.

- In the diagram below, the vectors P and Q=P•exp[iΦ] and Φ are the pressure and heat addition oscillations and the phase between them, respectively.
- The diagram below indicates the effect of the oscillatory heat addition upon pulsations for different values of the phase Φ.

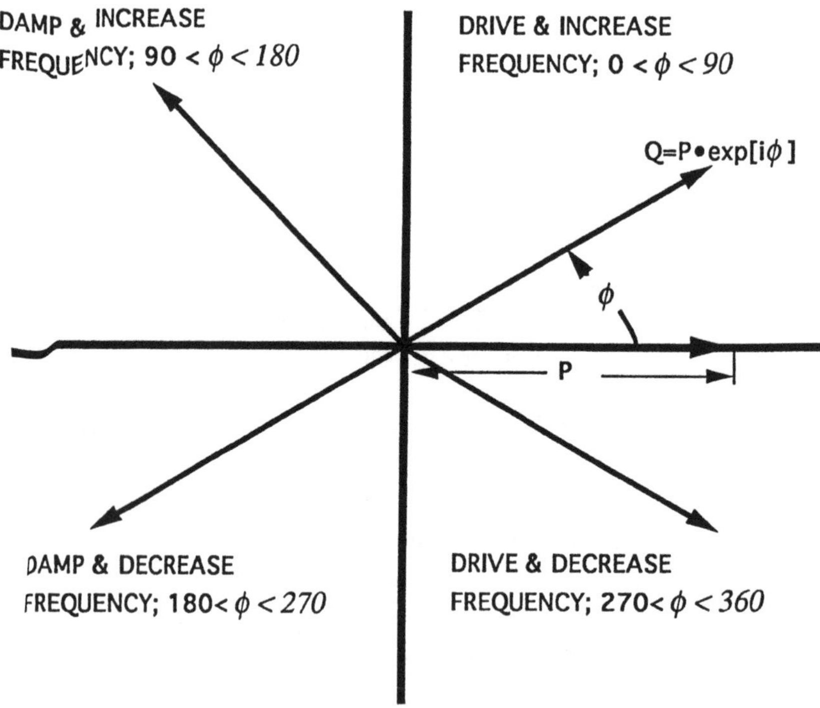

Driving/Damping by the Heat Addition Process = P•Q Cos.ϕ

Fig. 2 Effect of the phase Φ between the heat addition and pressure oscillations upon the driving and frequency of the pulsations (Rayleigh's Criterion).

Many of these effects have not been adequately modeled to date, and the development of capabilities for modeling these processes are some of the challenges that need to be addressed in future research efforts.

Applications

Historical Developments. References to pulse combustion and its applications date back to Dr. Higgin's "singing flame" experiments in 1777[1,7-9]. These references generally deal with the complex fluid-thermo-chemical processes that control the operation of various pulse combustors and oscillating flames, and pulse combustion applications. Development of pulse combustion applications[8,9] started around the turn of the century and led to the development of the Esnault-Peltrie, Holzworth and Karavodine gas turbines that utilized a pulse combustion process to burn the fuel. Subsequent efforts[9] in Germany, during the 1930s and 1940s, led to the development of the V-1 rocket (i.e., the "buzz" bomb), which was used by the Germans to propel destructive payloads from Europe to England near the end of the second world war. Following research efforts[9-11] studied the use of pulsejets (which was the name utilized to describe propulsion systems similar to the V-1 rocket), installed in the tips of helicopter blades, to supply the thrust needed for setting the blades into circular motion. Additional propulsion related applications, which have been investigated in the twenty years following the second world war, included the use of pulsejets to power gliders[12], airplanes, and drones.

While efforts to develop propulsion applications of pulse combustion have practically stopped in the 1960s due to their inability to compete with those that utilized gas turbines, the energy crisis of the 1970s had generated renewed interest in the development of energy applications of pulse combustion[13-16]. This interest was driven by the realization that such applications could produce significant fuel savings, increased productivity, reduced emissions, and systems that required lower capital expenditures. These efforts led to the successful commercialization of a number space and water heating systems that utilize gas burning pulse combustors[13-15] in the US, Europe and Japan.

The successful commercialization of heating applications of pulse combustion stimulated interest in extending the range of applications of pulse combustion to energy intensive and incineration processes. This interest led to the development of a pulse combustion based drying[16] and gasification[17] systems, and the resonant driving process[18,19] that utilizes a tunable pulse combustor to excite large amplitude, beneficial, pulsations in industrial scale energy intensive and incineration[20] processes. Because of the ongoing interest in improving the performance of energy intensive and incineration processes, past, present and future applications of pulse combustion in these areas are discussed in the remainder of this paper.

Heating Applications. Heating is one area in which the superior attributes of pulse combustion have been clearly demonstrated. In a typical heating application, see Fig. 3, the pulse combustor is immersed in the heated medium, which could be air, water, steam or an industrial liquid such as paint, asphalt or thermal fluid. Air and fuel are burned in an oscillatory combustion process within the pulse combustor and the released energy is transferred to the heated medium (i.e., the load) through the pulse combustor walls.

The advantages of pulse combustion heating applications are directly related to the presence of large amplitude pulsations within the combustor. These pulsations increase the rates of mass, momentum (i.e., mixing) and heat transfer within the combustor, resulting in a highly intense and efficient combustion process that does not require much excess air. Consequently, combustion is completed within a small volume, which reduces the size and, thus, the cost of the combustor. Furthermore, since the combustion process is highly efficient, the generated combustion products are practically free of products of incomplete combustion such as soot, CO and hydrocarbons. The high heat transfer rate from the flames and combustion products to the combustor walls significantly reduces the flame and combustion products temperature, resulting in low nitrogen oxides (i.e., NO_x) emissions. In addition, the high heat transfer rate permits the transfer of nearly all of the energy released by the combustion process to the heated medium (e.g., water or air) through a relatively small heat transfer area, resulting in a high thermal efficiency and, thus, fuel savings. Also, reducing the size of the required heat transfer area reduces the cost of the system. Finally, the combustor pressure oscillations set the air and fuel flapper valves, which have been used to date in all of the successfully commercialized pulse combustion heating systems, into motion, which provides a mechanism for pumping the fuel, air and combustion products through the system without the use of a fan or a chimney. The elimination of the fan or chimney further reduces capital investment costs of new installations and the costs of operating a fan or keeping a chimney clean and structurally sound.

The Lennox pulse furnace, see Fig. 4, which has been developed in the US,. is one of the most successful applications of pulse combustion to date. It consists of two

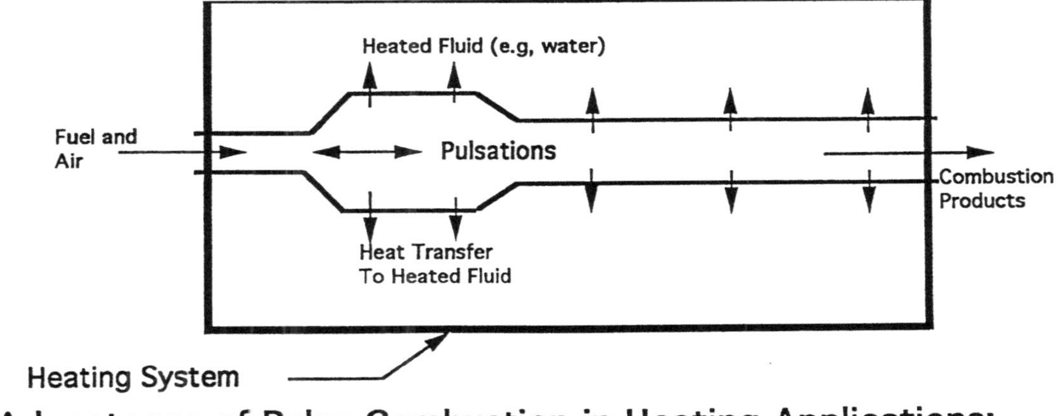

Advantages of Pulse Combustion in Heating Applications:

- High Heat Transfer Rate to the Walls (load)
  - Small heat tranfer area
  - "Cool" flame, resulting in low NOx
  - High thermal efficiency
- High rates of mixing within the pulse combustor
  - High combustion intensity - small combustor volume
  - Natural Exhaust Gas Recirculation (EGR)?
  - Low emissions of CO, soot and unburned hydrocarbons
  - Low excess air requirements
- Self aspiration
  - eliminates the need for a fan or a chimney

Fig. 3 Heating Applications of Pulse Combustion.

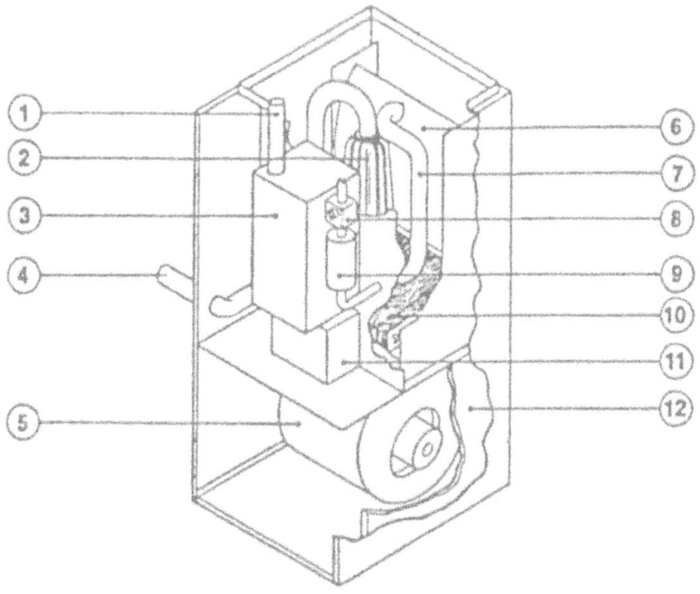

1. AIR INTAKE
2. COMBUSTION CHAMBER
3. AIR DECOUPLER
4. VENT PIPE
5. BLOWER
6. EXHAUST DECOUPLER
7. EXHAUST PIPE
8. GAS VALVE
9. GAS DECOUPLER
10. CONDENSING HEAT EXCHANGER
11. CONTROL BOX
12. FILTER

Fig. 4 A Schematic of a Pulsating Combustion Furnace.

compartments; the one on top houses the pulse combustor and the lower houses the fan that draws the house air and circulates it through the upper compartment where it is heated as it comes into contact with the combustor section, tail pipe, downstream decoupler and condensing heat exchanger. The inclusion of a condensing heat exchanger is optional; it is added to system if additional heat transfer area is required to obtain a higher thermal efficiency. It is of interest to note that the fuel and air flapper valves are enclosed within decoupling chambers, and that the tail pipe exit is connected to an exhaust decoupler to reduce sound emissions from the pulse combustor to the outside. It is also important to note that the pulse combustor exhaust pipe has been bent to reduce the space occupied by the pulse combustor. The bending of the exhaust pipe does not adversely affect the pulse combustor operation because the low frequency acoustic waves excited within the pulse combustor (e.g., in the 45-125 Hz. range) can readily propagate through the bends in the exhaust pipe. In fact, the introduction of such bends is beneficial because the rate of convective heat transfer from the hot gases to the walls in the bent sections of the tail pipe is higher than that in straight pipe sections, which reduces the required heat transfer area and, thus, the cost of the system.

While the benefits of pulse combustion applications are directly related to the presence of large amplitude acoustic oscillations within the combustor, successful commercialization of pulse combustion applications requires that sound emissions from these applications be acceptable to users and below government imposed limits. As indicated above, see Fig. 4, this problem has been successfully solved in heating applications by enclosing the air and fuel flapper valves within acoustic decouplers, attaching the tail pipe exit plane to an exhaust decoupler and lining the pulse combustor enclosure with sound absorbing materials.

__Large Scale Industrial Applications.__ The processes that control the performance of large scale energy intensive, industrial, processes are in many respects similar to those that control the performance of the heating systems that are discussed in the previous section. Consider, for example, a large scale boiler. In this case, energy released by the combustion process within the boiler is transferred by convection and radiation to the boiler walls that are constructed of tubes through which water and steam are flowing. The heating of the fluid in the tubes produces steam that is used for industrial processing and/or driving a turbine. It can be argued that the performance of such a boiler will improve if means for increasing the rates of transport processes within the boiler could be found. For example, increasing the rate of convective heat transfer will increase the fraction of the combustion process energy that will be transferred to the walls, which will increase the productivity and thermal efficiency of the process. Alternately, increasing the rate of convective heat transfer to the walls will decrease the size of the heat transfer area (and, thus, boiler) required to produce a given output, which will decrease the capital investment cost of the boiler.

Since pulsations generally increase the rates of transport processes, the above considerations of the operation of an industrial boiler suggest that exciting pulsations within such a boiler will produce benefits similar to those produced by the pulsations in the small scale heating applications that are discussed in the previous section. While the idea of developing a large scale pulsating boiler (or other energy intensive processes) is appealing, it is not clear whether such a pulsating system could be developed by a straight forward scale up of, for example, the small scale pulse combustors that are shown in Figs. 1, 3 and 4. In fact, it is likely that the the scaled up pulse combustor will stop pulsing when its size reaches a certain, as yet unknown, scale. This scale would most likely depend upon the fuel input rate, system configuration, combustion process characteristics, system acoustics, and heat transfer considerations. The development of such scale up capabilities (or large scale pulse combustors) will require development of understanding of such issues as: the stabilization of a pulsating combustion process within a large volume, the characteristics of the acoustic mode that will be excited in the larger system, the interaction of the excited acoustic mode with the pulsating combustion process, the relationship between the period of the excited acoustic mode and the characteristic times of the pulsating combustion process, and whether the needed fuel and air flow rates could be supplied with large scale flapper valves that will be structurally stable and respond sufficiently fast to the excited boiler pulsations.

Considerations of the complexity of the problems that will be encountered during any attempt to develop large scale pulse combustors stimulated interest in the development of an alternate approach for exciting large amplitude, beneficial, pulsations within large scale energy intensive systems without using large scale pulse combustors. These efforts led to the development of the resonant driving process[18-20] that utilizes a tunable pulse combustor to excite large amplitude pulsations within large scale systems. This approach is illustrated in Fig. 5 where the shown system is assumed to be a spray dryer. Typically, a spray dryer consists of a large drying volume, a main burner that supplies a stream of hot combustion products that supplies the energy required for drying, an injector that generates a spray of the slurry that is being dried, and means for removing the dry material and cool exhaust gases from the system. Drying occurs as the spray droplets are heated as they come into contact with the hot gases while they move through the dryer.

In the process, water is evaporated and absorbed by the surrounding gases. The performance of the dryer in Fig. 5 is improved by installing a tunable pulse combustor on the dryer wall. The tunable pulse combustor has two functions; it supplies a small fraction (e.g., 1-5%) of the total process energy and it excites large amplitude pulsations within the dryer. The remainder of the process energy is supplied by main burner at the top of the dryer. The tunable pulse combustor excites pulsations inside the dryer by supplying the dryer with a pulsating flow of hot combustion products. When the frequency of these pulsations equals that of one of the natural acoustic modes of the dryer, large amplitude, resonant, pulsations are excited within the dryer.

Since the natural acoustic mode frequencies of most practical systems are not known in advance, due to the complexity of the system geometry and the nonuniformity of the flow properties inside, and since these frequencies may vary during operation, due to changes in process operating conditions, tuning of the pulse combustor to operate at a resonant frequency of the process must be performed on site. This is accomplished by varying the frequency of the pulse combustor and monitoring the amplitude of the pulsations excited within the dryer with one or more pressure transducers that are installed on the dryer's walls. The output of these pressure transducers is at a maximum whenever the frequency of the pulse combustor equals that of one of the natural acoustic modes of the dryer volume. When this condition is satisfied, the oscillations within the pulse combustor and the dryer are in resonance, and the pulse combustor is set to operate at one of these frequencies.

In the process illustration in Fig. 5, the tunable pulse combustor excites a transverse acoustic mode within the dryer. In this case, the excited acoustic velocity oscillations are directed normal to the direction of the (axial) mean flow through the dryer. These oscillations increase the rates of heat and moisture transfer between the wet material, which is injected at the top of the dryer, and the hot combustion products that are supplied by the main burner and pulse combustor. The increases in the rates of these transport processes reduce the required residence time of the wet material within the dryer (i.e., the drying time), which permits increasing the velocity and, thus, the supply rate of wet material into the dryer. This, in turn, increases the dryer's productivity. Furthermore, the increase in the rate of heat transfer between the hot gases and wet material increases the fraction of input energy transferred to the load (i.e., wet material), which results in increased thermal efficiency and fuel savings. The pulsations also eliminate property nonuniformities within the dryer[21], resulting in more uniform drying and improved product quality. Finally, since in a system retrofitted with a tunable pulse

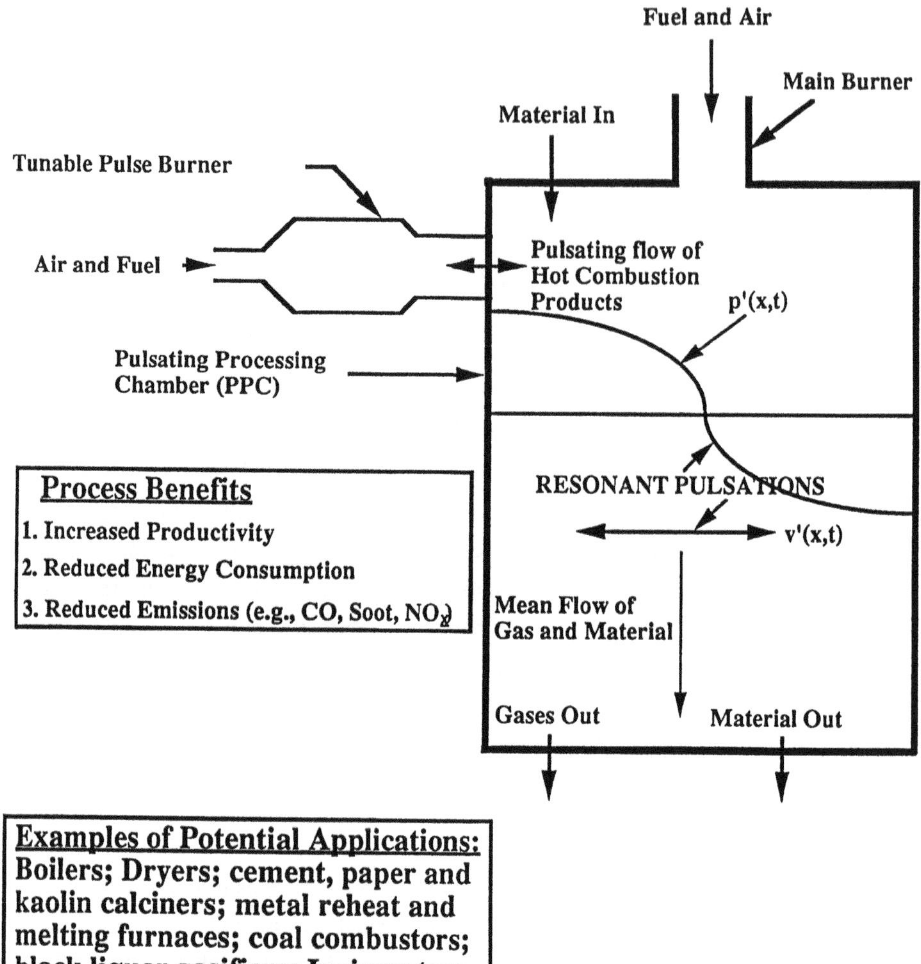

Fig. 5 A schematic describing the application of a tunable pulse combustor in the excitation of large amplitude resonant pulsations in energy intensive and incineration processes.

combustor the rates of transport processes are controlled by the pulsations and not the magnitude of the mean flow velocity, the performance of the process should not deteriorate when the load is reduced as in many conventional processes whose performance deteriorates due to the reduction in the mean flow velocity within the process (and, thus, the rates of transport processes) that accompanies a reduction in the process load.

The expected benefits would depend upon the characteristics of the process. For example, it is expected that additional benefits will be produced in a process such as boiler, where combustion occurs within the boiler volume, when pulsations are excited within the boiler by a tunable pulse combustor. The pulsations will increase mixing rates within the boiler's combustion region, resulting in decreased excess air requirements and, thus, lower exhaust stack losses and fuel savings. Furthermore, the higher mixing rate should increase the combustion process efficiency and reduce emissions of products of incomplete combustion such as soot, CO and unburned hydrocarbons. Finally, the increased rate of heat transfer from the main flame and combustion products to the load, should reduce the flame and gases temperature, which may reduce NO_x emissions.

To date, the resonant driving process has been investigated by the process developer[18,19], Sonotech from Atlanta, Georgia, the US Environmental Protection Agency[20] (EPA) and the Laboratory for Pulsating Combustion Processes (LPCP) at the Georgia Institute of Technology, in Atlanta, Georgia. Briefly, these studies demonstrated that commercially available Cello® tunable pulse combustors[22] can be tuned on site to excite large amplitude resonant pulsations within industrial scale volumes. For example, amplitude as large as 167 dB. were excited in a 9.5 feet diameter, 21 feet long tank with an industrial scale pulse combustor that operated over a 2-9.5 Mbtu/hr fuel input range and comparable amplitudes were excited at the LPCP in a 6 feet diameter, 18 feet high tank with a relatively small, 250,000 Btu/hr, Cello® pulse combustor. These studies clearly demonstrated that the resonant driving process can excite large amplitude pulsations in industrial scale processes and be used to convert both new and existing processes to pulsating operation.

Additional studies[19] investigated the effect of pulsations upon spray water evaporation, which simulates the initial phase of drying in which water films on the surface of the wet material is evaporated. Spray evaporation studies[19] conducted under various operating conditions showed that the excitation of acoustic resonances of various types (i.e., longitudinal, transverse and three dimensional) within the evaporator produced fuel savings and productivity increases of 8-16 and 5-17 percent, respectively. One of the utilized test configuration and test results are described in Fig. 6. The investigated evaporator consisted of a cylindrical section whose lower open end was welded to an inverted conical section. The vertex of the conical section was left open to permit unevaporated water to drip out of the evaporator. The evaporator was retrofitted with a tunable pulse combustor that was operated with a fixed fuel input of 500,000 Btu/hr in all tests. Different flow rates of dilution air were supplied to the evaporator through an annular section that was concentric with the pulse combustor, and used to vary the temperature and flow conditions within the evaporator. A water spray was injected into the evaporator through a two fluid injector that was attached to the top of the evaporator. The gases and the unevaporated water left the system through the exhaust duct and the opening at the bottom of the conical evaporator section, respectively.

The objectives of the tests that were performed in the setup shown in Fig. 6 were to determine the effect of resonant pulsations upon the total amount of water that could be completely evaporated within the evaporator and the fuel consumption per unit mass of evaporated water. A typical pulsating or nonpulsating experiment was started by letting the setup reach thermal equilibrium while injecting a given flow rate of water spray into the setup that produced an outflow of unevaporated water through the opening at the bottom of the conical section of the evaporator. Next, the flow rate of the injected water was gradually decreased until water stopped dripping through the opening at the bottom of the evaporator, indicating the flow rate of water that could be completely evaporated under the investigated test conditions. For each dilution air flow rate, tests were repeated with and without the excitation of resonant pulsations within the evaporator, and the water flow rates that could be completely evaporated under these test conditions are described in the table in Fig. 6. These data show that the pulsations always increased the amount of water flow rate that could be evaporated completely within the investigated evaporator setup by amounts that varied between 4.8 and 17.2 percent, depending upon the dilution air flow rate. Since the energy input remained fixed in all tests, these results also show that the pulsations produced energy savings. For example, when the dilution air flow rate was 11,000 SCFM, 1694.34 and 1940.99 Btu were required to evaporate a pound of water in the pulsating and nonpulsating tests, respectively, indicating that for these test conditions the pulsations also produced a fuel savings of 12.7 percent.

In an accompanying study[19], the effect of pulsations upon the energy required to completely evaporate a given flow rate of water sprays that were supplied into a

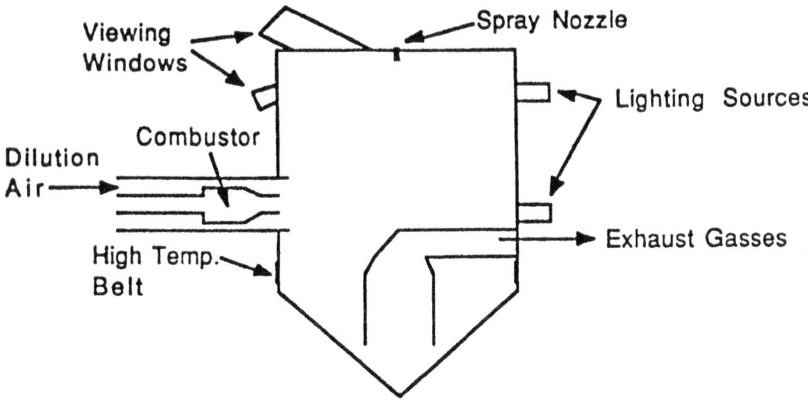

Table I: Maximum Water Flow Rates Completely Evaporated In Pulsating and Nonpulsating Tests With Different Dilution Air Flow Rates.

Fuel Flow Rate = 500,000 BTU/HR; Comb. Air Flow Rate = 6,250 SCFH
Resonant Amplitude in Evaporator During Pulsations = 149 dB.

Dilution Air Flow Rate (SCFH)	Maximum Water Flow Rate (Lbs./Hr.)	Pulsations (yes/no)	Percent Increase Due to Pulsations
6,000	290.4	yes	10.7
6,000	262.3	no	
11,000	295.1	yes	14.5
11,000	257.6	no	
13,000	318.5	yes	17.2
13,000	271.1	no	
18,000	304.5	yes	4.8
18,000	290.4	no	

Fig. 6 Effect of resonant pulsations excited with a tunable pulse combustor upon the rate of water spray evaporation.

horizontal tank by different types and arrangements of injectors was investigated. In this study, the flow rate of water supplied to the evaporator was kept fixed while the fuel supplied to the pulse combustor was increased incrementally until all of the injected water was evaporated. Such tests were conducted with and without the excitation of pulsations within the tank. Comparison of the results shows that the pulsations reduced the fuel consumption by amounts that varied between 8 and 16 percent, depending upon the water flow rate, injector types and injectors configurations.

Additional studies[19,22] investigated the effect of pulsations upon metal heating and limestone calcining. These studies revealed that the pulsations significantly increased the rates of the heating and calcining processes, suggesting that tunable pulse combustors can be used to improve the performance of furnaces that are used in metal production (e.g., reheat and melting furnaces), and calciners and rotary kilns that are used in cement making. These findings led to the initiation of two current field experiments, supported by GRI, that are investigating the effect of resonant pulsations, excited by Cello® burners, upon the performances of a cement calciner and steel ladle preheater. While the initial results of these studies appear promising, final conclusions will have to await the completion of these studies.

The benefits of the resonant driving process have been also demonstrated[20] in tests conducted on an EPA rotary kiln incinerator simulator that was retrofitted with a Cello® burner, see Fig. 7. These tests investigated the effect of pulsations upon the emission of "puffs" from incineration systems. Such "puffs" are emitted when fuel rich conditions are formed within the incinerator due to, for example, sudden release of waste or nonuniform waste and oxygen distributions within the incinerator. This study investigated the effect of resonant pulsations upon the emission of puffs formed during the incineration of toluene and polyethylene. The magnitude of a puff was determined by measuring the total emissions of soot, CO and unburned gas phase hydrocarbons during the combustion of a given amount of waste (e.g., toluene). Comparisons of the magnitudes of the puffs emitted in pulsating and nonpulsating tests showed that the pulsations reduced the magnitude of the puffs under all investigated conditions. Furthermore, the soot emissions were reduced by 50-75 percent, which is most significant as it is generally difficult to destroy soot emitted from a primary incinerator stage in the afterburner. Finally, no oxygen was found in the incinerator exhaust flow in pulsating tests, indicating that the pulsations increased the mixing rates within the incinerator, which led to complete consumption of the available oxygen. In a related

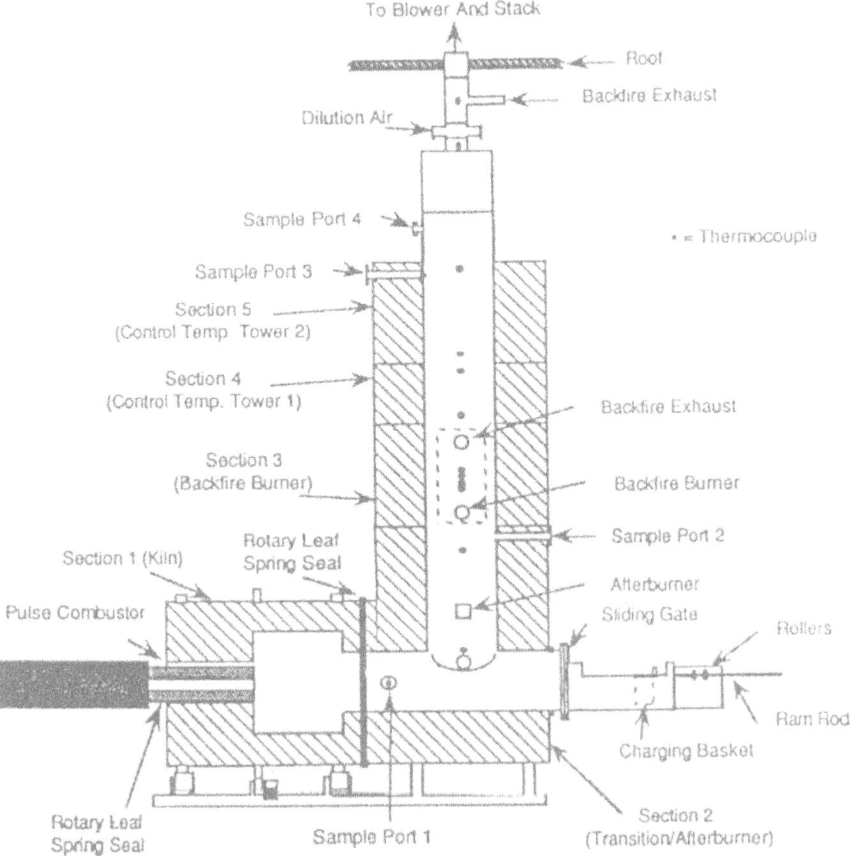

Fig. 7 EPA's Rotary Kiln Incinerator Simulator.

incineration study[23] the fuel supplied to a Cello® burner that was attached to EPA's Rainbow furnace (located in EPA's facilities in Research Triangle Park, North Carolina) was spiked with chlorinated hydrocarbons in an effort to determine the incineration of these compounds and their byproducts in a pulsating environments. Tests were conducted with and without the excitation of pulsations in the furnace. Analyses of the compositions of the exhaust products showed that in all cases the destruction efficiencies of the chlorinated hydrocarbons were higher than 99.9999 percent.

The improved performance of the EPA incinerators that were retrofitted with a tunable pulse combustor also suggests that resonant driving can be used to improve the performance of combustion systems that utilize difficult-to-burn fuels such as wood, heavy fuel oils and a variety of coals.

A different application of pulse combustion is demonstrated in the Bepex's Unison dryer[16]. It consists of a pulse combustor that is installed on top of large drying tank. The pulse combustor supplies a pulsating flow of hot gases that provides the energy required for drying. A slurry to be dried is injected into the pulse combustor exhaust flow. The interaction of the slurry and pulsating flow produces a spray that is subsequently dried as it moves through the tank below the pulse combustor. This tank, according to Bepex statements, also serves as an acoustic decoupler that minimizes sound emissions from the dryer.

A somewhat different application of pulse combustion is provided by the solid waste/black liquor fluidized bed gasifier that has been under development by MTCI in the US for a number of years[17]. The gasification occurs in a bed that is supplied with the gasified material (e.g., black liquor) and fluidized by, for example, steam. The heat required for the endothermic gasification reactions is supplied by tail pipes of one or more pulse combustors that are immersed in the bed. The pulse combustor(s) are fired with the generated gas. The objective of this system is to take advantage of the high convective heat transfer rates inside the pulse combustor tail pipes to reduce the heat transfer area required for transferring the required process energy. It has been claimed that small scale versions of this gasifier have produced gases with medium Btu values, and that these gases could be used to fire the pulse combustor. A pilot scale version of this gasifier is apparently under development. Final evaluation of this gasification process will have to await careful analyses of the pilot plant and full scale systems performance data.

Conclusions

Developments in the area of pulse combustion clearly show that this technology could be applied to improve the performance of energy intensive processes. To date, small scale heating (e.g., space and water) applications of pulse combustion have been successfully commercialized and current efforts focus on the developments of large scale applications of this technology. Based upon results obtained to date, it appears that pulse combustion applications could produce fuel savings, increased productivity, reduced emissions, and improved product quality; benefits that could translate into reduced capital investment and operating costs. It is also possible that as a larger fraction of the scientific community becomes familiar with pulse combustion technology, new applications of pulse combustion will emerge.

Nomenclature

English Symbols

c - speed of sound

i - imaginary unit; $\sqrt{-1}$

k - wave number, $\omega / \bar{c}$, see Eq. 7

L - length of the pulse combustor

n - denotes the order of the acoustic mode, see Eq. 10

p - pressure,

P - amplitude of the pressure oscillations

Q - heat addition by the combustion process

R - combustion process response function, see Eq. 5-b

t - time

T - period of the oscillations

x - axial location

V - volume where acoustic damping and driving occur

Greek Symbols

ρ - density

ϕ – phase difference between heat addition and pressure oscillations

γ - ratio of specific heats

γ_n - eigenvalue of the wave equation, see Eqs. 6 and 8.

ω – frequency of the oscillations, rad/sec.

Subscripts

i - imaginary part of a complex quantity
r - real part of a complex quantity

References

1. Rayleigh, L., "The Theory of Sound", Dover Pub., New York, 1945

2. Zinn, B. T., "Pulsating Combustion", Chapter II in "Advanced Combustion Methods", edited by F. Weinberg and published by Academic Press, 1986.

3. Dowling, A. D. and Pfowcs Williams, J. E., Sound and Sound Sources, John Wiley and Sons, New York, 1983.

4. Zinn, B. T.., Miller, N., Carvalho, J. A., and Daniel, B. R.., Pulsating Combustion of Coal in a Rijke Type Combustor, Proceed. 19th Int'l Symposium on Combustion, pp. 1197-1203, Haifa, Israel, August 8-13, 1982.

5. "Liquid Propellant Rocket Combustion Instability", Harjee, D. and Reardon, F., eds., NASA SP 194, 1972

6. Crocco, L. and Cheng, S. I., "Theory of Combustion Instability in Liquid Propellant Rocket Motors." Butterworths Scientific Publications. 1956.

7. Tyndall, J., "Sound." D. Appleton & Company. New York, 1897.

8. Stodola, A., "Steam and Gas Turbines", Vol. I, McGraw Hill Book Company, 1927.

9. Reynst, F. H., "Pulsating Combustion", Thring (ed.) 1952 Pergamon Press. New York.

10 Tharrat, C. E., "The Saunders-Roe Pulse Jet Engines", Transactions of the Society of Engineers, June 1958.

11 Tharratt, C. E., "The Propulsive Duct," Aircraft Engineering, Part I, Nov. 1965, pp. 327-337; Part II, December 1965, pp. 359-371; Part III, February 1966, pp. 23-25.

12. Bertin, J. and Le Foll, J., "The S.N.E.C.M.A Escopette Pulse Jet", Interavia, Vol. VIII, No. 6, 1953, p.343.

13 Brown, D. J. (ed.), "Proceedings of the 1st International Symposium on Pulsating Combustion", Sheffield University, England, 1971.

14. Proceedings of the 1st Symposium on Pulse Combustion Technology for Heating Applications, DOE Report, ANL/EES-TM-87, 1979.

13. Proceeding of the Symposium on Pulse Combustion Applications, Vol. 1, GRI-82/0009.2, GRI Sponsored, Atlanta, GA, March 2-3, 1982.

15. Proceed. of Int'l. Symposium on Pulsating Combustion, Sponsored by Sandia National Laboratories and the Gas Research Institute, Monterey, California, August 1991.

16. Ozer, R. W., "Pulse Combustion Drying", Proceedings of the Workshop on Pulsating Combustion and its Applications, Lund University, Lund, Sweden, 1993.

17. Mansour, M. N., Durai-Swamy, K., Chandran, R. R. and Duqum, J. N., "Pulse Combustion Systems for Commercial, Industrial and Gas Turbine Applications", Proceed. of Int'l. Symposium on Pulsating Combustion, Sponsored by Sandia National Laboratories and the Gas Research Institute, Monterey, California, August 1991.

18. Rabhan, A. B., Daniel, B. R. and Zinn, B. T., "Industrial Pulse Combustor Development", GRI Report No. GRI-87/0350, June 1987.

19. Rabhan, A. B., Dubrov, E., Alvey, D. A., Daniel, B. R. and Zinn, B. T., "Industrial Pulse Combustor Development and Its Application in Spray Dryers and Cement Calciners", GRI-90/0203, Final Rpt. May 1986-Sept. 1990, Gas Research Inst., July 1991.

20. Stewart, C. R., Lemieux, P. M. and Zinn, B. T., "Application of Pulse Combustion to Solid and Hazardous Waste Incineration", Proc. Int. Symp. on Pulse Combustion, Aug. 6-8, 1991, Monterey, California.

21. Vermeulen, P. J., Grabinski, P. and Ramesh, V., "Mixing of an Acoustically Excited Air Jet with a Confined Hot Crossflow", ASME 90-GT-28, 1990.

22. Zinn, B. T., Dubrov, E., Rabhan, A. B. and Daniel, B. R., "Application of Resonant Driving to Increase the Productivity and Thermal Efficiency of Industrial Processes", Proceed. of Int'l. Symposium on Pulsating Combustion, Sponsored by Sandia National Laboratories and the Gas Research Institute, Monterey, California, August 1991.

23. Plavnik, Z. and Zinn, B. T., "Application of Pulse Combustion in Solid and Hazardous Waste Incineration", Final Report Prepared by Sonotech, Inc. Under EPA Contract No. 68D10101, June 8, 1993.

7. RECENT PROGRESS IN THE IMPLEMENTATION OF ACTIVE COMBUSTION CONTROL

E. W. Hendricks *and K.C. Schadow+
*Naval Command, Control and Ocean Surveillance Center
San Diego, CA 92152-6040, USA
+Naval Air Warfare Centre
Weapons Division
China Lake, CA 93555-6001, USA

Abstract

The suppression of pressure oscillations and the extension of flammability limits in dump combustors and premixed combustors with flame-holders through the use of active feedback control is a relatively new technology with tremendous potential. To accelerate progress in this emerging area, a research program was initiated at the Office of Naval Research (ONR). The ONR program aims to explore new control strategies, develop new actuators, and explore the combustion dynamics of combustion systems with an eye towards improving their controllability. This paper summarizes progress made in the program. Through physical and numerical experiments the role of shear-flow dynamics in combustion control has been clarified. New, more effective, actuators that manipulate the shear layer by a periodic chemical energy release have been demonstrated. New control strategies based on neural networks, adaptive filters, and modern control synthesis procedures have been implemented.

I. Introduction

The suppression of combustion induced pressure oscillations and the extension of flammability limits are a major challenge in the design and development of high performance combustors. Passive techniques have historically been used to enhance combustor performance. However, in recent years active techniques for the control of combustion have been increasingly considered as an alternative to the less flexible passive techniques.

In the past (and present), the modification of fuel injection distribution patterns and combustor geometry were used to improve combustor performance. More recently, techniques for passive control of combustion characteristics have made use of an increased understanding of the shear-flow dynamics to improve combustor performance.[1] In the dump combustor, nonstandard inlet duct cross-sections have been used to control the generation and breakdown of large-scale structures, which play a critical role in driving pressure oscillations and determining the flammability limits.

In contrast to passive techniques, active control utilizes actuators which modify the pressure field of the combustor by modulating the air or fuel supply to suppress

F. Culick et al., (eds.), Unsteady Combustion, 139–160.

combustion oscillations. Typically, a feedback control loop is used to drive an actuator using the processed output from a sensor which monitors the flame characteristics or pressure oscillations. Several active control schemes have been used to suppress pressure oscillations and extend flammability limits in laboratory combustors (heat release rates up to 250 kW) at ambient pressure and with gaseous fuels. An extensive review of recent active control work has been done by McManus.[2]

In 1987, the US Office of Naval Research (ONR) initiated a research program in active combustion control. The program focusses on the application of novel control and actuation techniques to dump combustors and premixed combustors with flameholders (Figure 1). Specific goals of the program include; extend the demonstration of control for laboratory combustors with modest heat release rates to a combustor with higher energy release rates (greater than 1 MW); demonstrate the feasibility of control with liquid fuels; and determine the utility of control at higher combustor pressures.

The overall approach to achieving control at more practical operating conditions is two pronged : 1) continue to build upon the increasing physical understanding of the shear-layer and combustion dynamics to guide the development of new actuators and 2) examine new types of feedback control techniques. Specifically, active manipulation of a reacting shear layer downstream of a dump or behind bluff bodies is being examined in order to exploit its ability to control combustion dynamics. Experiments in flames and laboratory combustors using advanced diagnostics, combined with numerical large-eddy simulations (LES) are the tools being used to explore the shear layer. In the physical combustor experiments, novel actuators are being explored to actively control shear-flow development and allow operation at elevated pressures. Standard and advanced feedback control techniques including modern control synthesis procedures, adaptive filters, and neural networks are being studied. In the ONR program, standard sensors are used, including high-frequency response pressure transducers, microphones, and CH emission sensors.

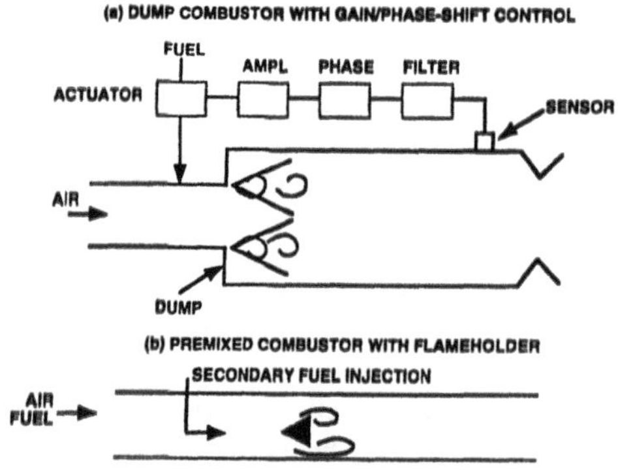

Figure 1. Dump Combustor and Premixed Combustor with Flameholder for Active Combustion Control Experiments.

This paper presents the progress to date made by several investigators in the ONR program. Participants in the Active Control Program include Barron Associates, Inc., Stanardsville, Virginia; California Institute of Technology, Pasadena, California; General Electric Corporate Research and Development, Schenectady, New York; Imperial College of Science, Technology and Medicine, London, England, Naval Air Warfare Center Weapons Division, China Lake, California; Naval Command Control and Ocean Surveillance Center, San Diego, California; Quest Integrated, Inc., Kent, Washington; Stanford University, Stanford, California. Progress in the following areas is reviewed: (1) physical understanding of the combustion dynamics, (2) development and testing of new types of actuators, (3) implementation of novel feedback control techniques, (4) demonstration of active combustion control in combustors up to 1 MW and (5) a demonstration of the feasibility of active control with liquid fuels.

II. Physical Understanding of Combustion Dynamics

Combustion characteristics are closely related to detailed fluid dynamic processes. In dump combustors, for example, combustion features such as flammability limits and combustion stability depend on the evolution of large-scale structures and their breakdown into fine-scale mixing which occurs in the shear layer developing downstream of the dump.[3,4] If the initial condition of the shear layer is changed by a non-circular dump geometry the downstream shear-flow dynamics are modified, flammability limits can be extended and improved combustion stability achieved and visualized.[1]

While the passive control results of the past are, in and of themselves impressive, active control of the shear layer is necessary when operational conditions change over a wide range. Figure 2 demonstrates the effect of a closed-loop feedback control system on a ducted flame during the transition from uncontrolled to closed-loop control operation.[5] During the uncontrolled (unstable) state, high amplitude pressure oscillations in the duct are associated with the development of large-scale vortices at the burner lip as visualized by a CH emission imaging system. To control the flame stability, the acoustic signal from the duct was time-delayed, filtered, amplified, and fed back into an acoustic actuator which modulated the air/fuel mixture. As seen in Figure 2, the suppression of the oscillations is associated with preventing the development of large coherent structures.[1,3]

This impressive closed-loop control was achieved by first examining the response of the flame to open-loop forcing. Using the acoustic actuator, the higher harmonics of the flame were excited to disrupt the development of large-scale structures and generate small-scale vortices which attenuated the pressure oscillations and enhanced flame stability and extended flammability limits. Based on the response of the shear layer to this open-loop forcing near the Kelvin-Helmholtz instability frequency, a closed-loop amplitude modulation controller was developed to modify the initiation of reaction in the flame.[6]

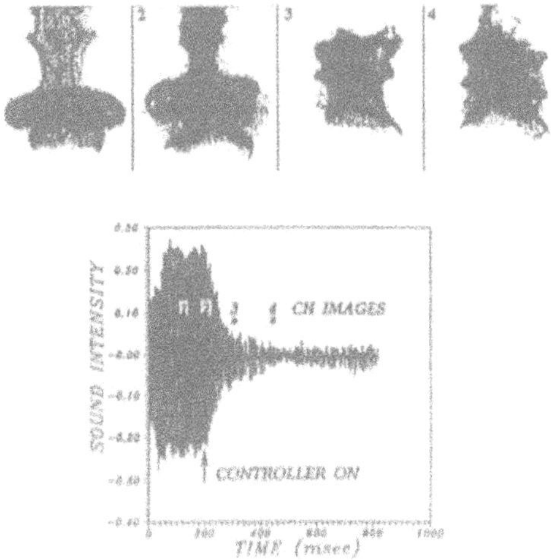

Figure 2. Flame Dynamics and Pressure Signal During Transition for Uncontrolled to Controlled Operation using Gain/Time-Delay Controller.

The flame experiments provided critical guidance in the subsequent application of active control to a 12.5-cm diameter dump combustor with a nominal heat release rate of 1 MW. The role of large-scale structures in driving the pressure oscillations in this combustor has been previously verified.[3] For this larger combustion system an acoustic actuator is not practical, so oscillating fuel injection was chosen as the control actuator. The fuel is injected through an orifice plate at the dump (Fig. 3).[7] The fuel injection is done in this manner to maximize the effectiveness of the actuator by introducing the fuel at the thin shear layer which separates from the orifice. The effectiveness of manipulating the shear layer with "fuel" modulations was demonstrated in non-reacting experiments.[8] In the non-reacting experiments, acoustic forcing of the inlet-duct flow was used to simulate combustion oscillations. Coherent vortices were shed from the dump as indicated by a peak in the velocity spectrum and flow visualization using Mie scattering. "Fuel" jets were modulated at the instability frequency, using acoustic forcing. When this control actuator was operated with the proper phase adjustment, the active control system was able to disrupt periodic vortex shedding at the dump, eliminating the peak in the velocity spectrum.

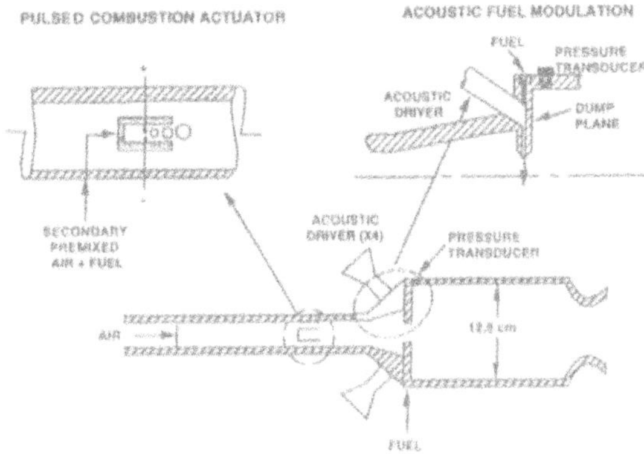

Figure 3. Schematic of Dump Combustor with Pulsed-Combustion Actuator and Acoustic Fuel Modulation.

An LES of a dump combustor provides insight into the combustion dynamics. The LES simulation contains the essential physics of the combustion instability, including unsteady, turbulent, compressible reacting flows and acoustics, vorticity, and entropy waves.[9,10] Figure 4a shows an example of flame and vorticity data for a simulation of an unsteady dump combustor in operation. Figure 4b shows a simulation of a combustor in which pulsed secondary fuel injection into the inlet duct is used to actively control the combustion.[11] When the control is effective, the pressure oscillations are significantly reduced, and large-scale structures are not present in the combustor. More recently the LES model has been extended to include a new subgrid model for kinetic energy which allows for coarser grids and consequently higher Reynolds numbers (essential to model more realistic combustors). In addition a more general expression of the turbulent flame speed (in terms of the laminar flame speed and subgrid turbulence intensity) is now being used [11]. Finally, in an effort to increase the Reynolds number range and perform a 3-D calculation, a new simulation has been implemented on a massively parallel processor (the iPSC/860). Figure 4c shows some initial results from these 3-D simulations. Validation to experimental data is ongoing, but initial qualitative comparisons look good.

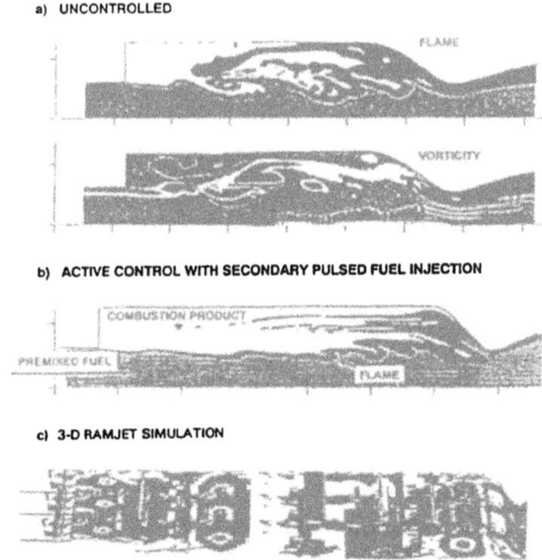

Figure 4. Large-Eddy Simulation of Uncontrolled and Actively Controlled Combustion Dynamics using Pulsed Secondary Fuel Injection.

Additional insight into the effectiveness of active control on shear-flow dynamics has been gained from experiments with a premixed, two-dimensional combustor.[12] Results of these experiments are discussed in the following section related to novel actuators. The physical understanding of the combustion dynamics provided by the previously described work is critical for the placement and choice of actuators to optimize control authority in the more practical combustors. The importance of the actuation scheme cannot be overemphasized. The limiting factor in almost all cases is not the control algorithm, but the lack of an effective actuator with sufficient control authority. Actuators explored in the present program are described in the following section.

III. Actuators

Several types of actuators were used in earlier active control experiments. These included (1) loudspeakers to modify the pressure field of the system [13] or to obtain gaseous fuel flow modulation,[14] (2) pulsed gas jets aligned across a rearward facing step,[15] (3) adjustable inlets for time-variant change of the inlet area of a combustor,[16] and (4) solenoid-type fuel injectors for controlled unsteady addition of secondary fuel into the main combustion zone.[17] Several new types of actuators were explored in the ONR program.

In a laboratory-scale premixed, two-dimensional dump combustor a combination of active and passive vortex generators (Figure 5) was studied. [12] Streamwise vorticity was introduced into the inlet flow with two jets, skewed at 45 degrees towards the side walls. These jets, with controllable momentum flux, produced one dominant pair of counter-rotating vortices, which increased the volumetric energy release but decreased combustion stability. To increase combustion stability a slot which spanned the inlet width was used to provide spanwise forcing. The spanwise slot excitation, which was forced by a loudspeaker produced reduced pressure oscillations by introducing a periodic cross-stream flow perturbation to the inlet boundary layer. Therefore the energy release and combustion stability of this combustor are a function of control jet momentum, speaker voltage and frequency making it an ideal test bed for an active control technique which is described later. The effect of this combined control forcing on the flame structures may be seen from Schlieren images in Figure 6. With spanwise and streamwise forcing (Figure 6d), the flame is highly three dimensional and contains many small-scale vortices. This type of flame structure is indicative of an efficient combustion with minimal pressure oscillations. In the experiments shown in Figure 6, the vortex generating jets were replaced by delta-wing vortex generators.

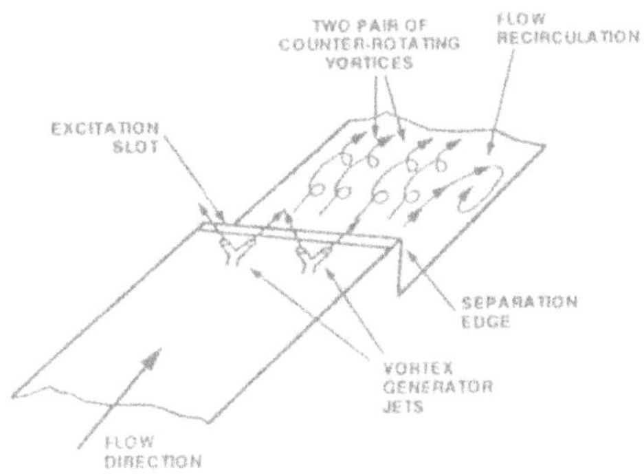

Figure 5. Combination of Streamwise and Spanwise Vortex Generators in Premixed, Two-Dimensional Dump Combustor.

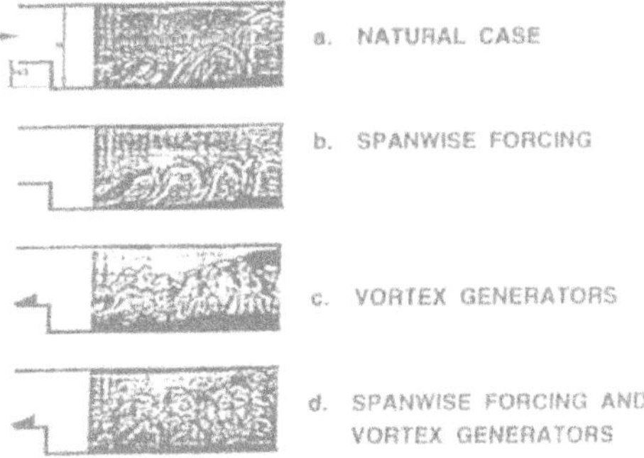

Figure 6. Effect of Spanwise and Streamwise Forcing on Flame Structure using Schlieren Photography.

Several techniques to provide a time-varying fuel injection forcing were explored. For gaseous fuels, a needle valve driven by a vibrator was used to oscillate a portion of the secondary fuel supply.[18] Tests with this actuator were made in a premixed duct combustor. Oscillating all of the secondary fuel and injecting it either radially or axially into the duct sometimes led to early blowoff. The amplitude of the oscillation of the secondary fuel was proportional to the amplitude of the needle-valve oscillation. However, the forcing amplitude was subject to damping in the feed lines and the phase and amplitude of the fuel oscillations were difficult to quantify for closed-loop active control for frequencies above 200 Hz. Nevertheless, successful suppression of combustion oscillations was obtained as presented later. Gaseous fuel modulation by loudspeaker was also explored in the 12.5-cm dump combustor.[7] Acoustic modulations were superimposed on the fuel stream through four tubes which were connected to 75 W acoustic drivers (Figure 3).

For liquid fuels, a Moog servo-valve was modified to obtain high-speed fuel-injector actuation up to 250 Hz.[19] To obtain the high frequency response, a process controller with dynamic signal analyzer was used for closed-loop servo-valve control (Figure 7). The valve was tested in a premixed combustor with the flame stabilized behind a standard V-gutter flameholder and a system with improved flameholding using a swirler. With the swirler, the reaction zone was shortened by a factor of about 4. For the V-gutter flameholder, an unsatisfactory correlation between the fuel oscillation pressure downstream of the valve and the chamber pressure downstream of the flameholder was obtained. With the swirl flameholder, the correlation was significantly improved (Figure 8).

In an attempt to achieve actuation at elevated combustor pressure, a disk was used as part of the combustor wall and oscillated by a vibrator.[20] However the acoustic power input of the oscillating disk diaphragm was insufficient at frequencies greater than 100 Hz.

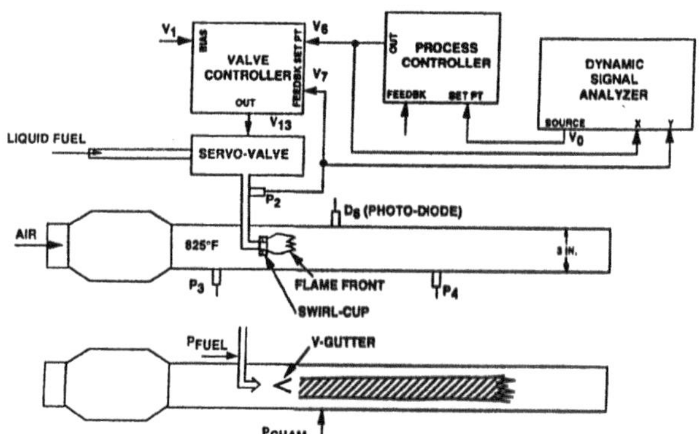

Figure 7. Relative Flame Zone Length for Swirl-Cup and V-Gutter Flameholder using Closed-Loop Control for High Frequency Fuel Modulation.

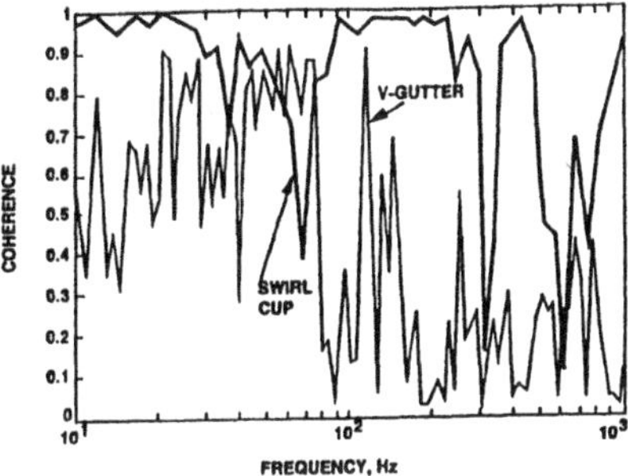

Figure 8. Coherence Between Fuel Pressure Modulation and Chamber Pressure Oscillation for Combustor Configurations Shown in Figure 7.

A most promising actuator is under development by NAWC. The NAWC actuator uses controlled periodic chemical heat release to produce high acoustic power levels while maintaining amplitude and phase control. This actuator is effective even at elevated operating pressures. The actuator uses controlled convected flame kernels in a duct of premixed fuel and air to produce pressure oscillations (Figure 3).[21] To avoid merging of the kernels, the flame speed is smaller than the gas velocity that is convecting the kernels. The actuator, with 70 W energy input for the spark ignition, can operate in a frequency range of 50-1000 Hz at up to 240 kPa operating pressure using gaseous ethylene fuel.

The novel actuators discussed above provided the potential for active control at higher energy levels. In concert with the actuator development, new stabilizing controllers were examined and progress made by the ONR program is described in the following section.

IV. Controller

Previous work in active control has utilized phase-shift/time delay type controllers (Figure 1). The utilization of active control concepts in practical combustion systems requires feedback control strategies which are adaptive and capable of multi-frequency instability control. Several novel approaches are being examined in the ONR program. As a part of the ONR effort, work which aimed to discover a reduced order approach to the problem of combustion control via an analytical approach to simulating the unsteady flow field related to combustion instabilities was studied. This work, which ultimately supports the design of controllers, is described first. Subsequently, different control methodologies are discussed and their application to combustion experiments is described in Section V (Active Control Experiments).

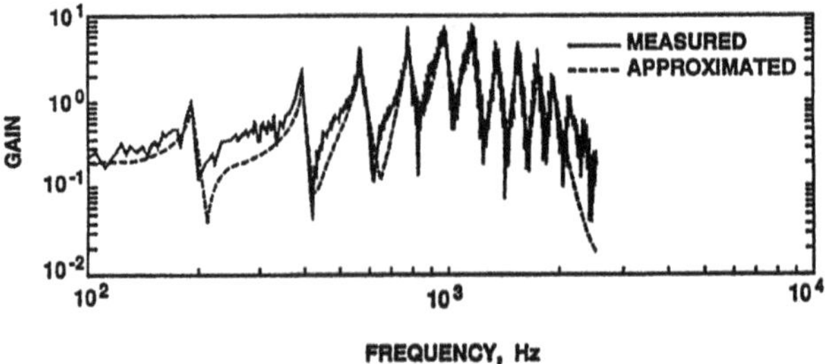

Figure 9. Measured Transfer Function and Analytical Approximation

Approximate Analysis

An approximate analysis is used as a theoretical framework to explicitly represent all gas dynamic processes including some nonlinear terms while accommodating other physical processes by modeling.[22] The analysis is considered general because when a specific case of combustion instabilities is considered, unsteady heat addition needs to be modeled in some fashion. In formulating the analysis, fluctuating value equations are obtained for the flow variables from the governing equations of fluid motion. Subsequently, a form of Galerkin's method is applied in which the acoustic modes of the combustion chamber are used as the basis functions. The governing partial differential equations are finally reduced to a set of second-order ordinary differential equations written for time-dependent amplitude of the corresponding acoustic mode. The approximate analysis is applied to a Rijke tube by incorporating a model for unsteady heat addition with two parameters. Results and comparison to experimental data are discussed later.

Controllers Using Loop Shaping Techniques

To extend the performance of simple phase-shift controllers, a classical approach using a frequency domain compensator was undertaken. This digital controller, which was designed for a ducted premixed flame, consisted of an 8th order Butterworth filter, a second-order notch filter, a first-order lead compensator, and a gain module. The design procedure is illustrated in Ref. 23, and its improved performance and stability robustness was experimentally demonstrated.

The ultimate implementation of active instability suppression in complex combustor geometries will require a multi-frequency controller with the ability to suppress closed spaced resonance modes. Controllers for this purpose have been designed and applied to the Rijke tube[24] using loop shaping H^{∞} techniques. By utilizing a simplified linear model for the acoustic dynamics and doing system identification measurements, a transfer function was approximated, which was in satisfactory agreement with the measured transfer function (Figure 9). The loop shaping controller design procedure, described in

Reference 25, was then used to design a controller for the Rijke tube experiments.

Knowledge Based Control

Knowledge based control was applied to a ducted premixed flame.[18] In a knowledge based controller, the signal from a pressure transducer in the combustor is fed back with constant gain through a bandpass filter to a phase shifting device to drive an actuator to obtain suppression. The phase shifter monitors the filtered feedback signal to give a synthesized sinusoid locked to the feedback with a switch-selectable phase shift at 256 equal intervals between 0 and 360°. The software-controlled output voltage (input voltage to the vibrator) is incrementable in the ratio 2n/4, where n is any integer between 0 and 63. The performance of the knowledge based control with variable input to the actuator is compared in the following section to a controller with constant input .

Adaptive Filter

An adaptive filter controller was also used on a ducted premixed flame. The adaptive controller was a modified adaptive filter, which utilized a digital implementation of the Wiener-Hopf least-mean-squares algorithm for adaptation.[26] The filter was modified to allow differencing of the performance and input signal externally. The bandpass filtered pressure transducer signal was amplified to produce the detector and error signals with voltage gains of around 10 (20 dB) over the filtered signal.

Dynamic Polynomial Neural Networks

The utility of neural networks for quasistatic problems such as pattern recognition is well established. Their suitability for the dynamics problems of system identification and control is less well known. In the present program, work has been supported to develop and test network architectures which are suited to the dynamic environment. They have concentrated on networks whose elements have internal time delays and/or feedback loops. Additionally, network synthesis methods have been developed which employ information theory to constrain the number of network nodes to the minimum required for accurate system identification. A discussion of this concept of polynomial neural networks (PNN) is given in [27].

The implementation of a controller based on PNN is being explored in a premixed combustor (Figure 5), utilizing combined streamwise and spanwise forcing and in a dump combustor, with acoustic fuel modulation (Figure 3).

For the premixed combustor,[28] control inputs are jet-to-cross flow ratio (the streamwise forcing), R, and speaker voltage, A, and frequency, f, for the spanwise forcing. These inputs are controlled for varying equivalence ratio, ϕ, and inlet velocity, U_o, to minimize RMS pressure fluctuation level, P, and maximize volumetric energy release rate, E, as indicated by CH emission. For these interrelated combustor parameters, a response surface is generated from static actuator inputs for varying combustor conditions. Subsequently a cost function is defined representative of the desired operating characteristics. Optimum control is obtained by minimizing the cost function

$$J = aP^2 - bE^2$$

where P and E are the respective response surfaces represented by the neural net:

$$P = f(R,A,f,\phi,U_o)$$

$$E = g(R,A,f,\phi,U_o)$$

The weighting coefficients a and b allow a balance to be prescribed between minimizing P and maximizing E. Figure 10 shows the effect of varying a, with b held constant, on the cost function. This plot is generated with actual combustor data and neural net predictions. Work is in progress to complete training of the neural net and allow on-line re-minimization of the cost function by active search in control parameter space.

For the dump combustor (Figure 11),[29] a PNN controller was used to determine the optimum time-delay, TD* between pressure oscillations and fuel modulation to minimize RMS pressure fluctuations (IFPAMP) and optimize heat release rate (EAVR2) at the combustion chamber center. The cost function for this controller was defined as follows:

$$J = 1FPAMP + 1/EAVR2$$

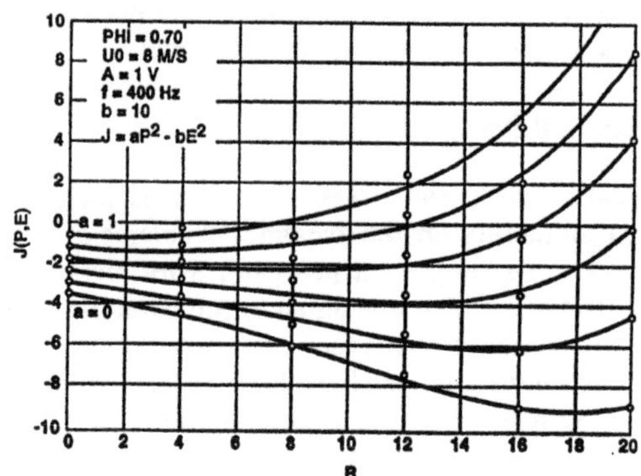

Figure 10. Effect of Pressure Response Surface Weighting Factor on Cost Function Determined from Static Experiments and Polynomial Neural Network.

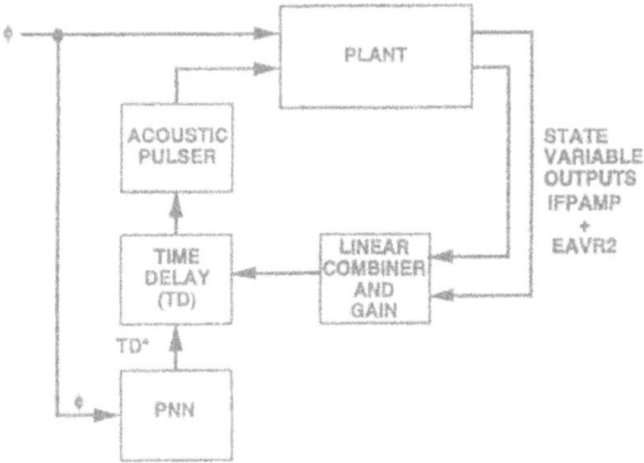

Figure 11. Feed Forward Neural Net Controller to Optimize Time Delay for Varying Equivalence Ratio.

Based on experiments in which IFPAMP and EAVR2 were systematically determined as a function of TD* and ϕ, the cost (or performance) function J in Figure 12 was calculated based on PNN. Depending on ϕ, this surface determines TD to minimize J (or maximize EAVR2 and minimize IFPAMP). For varying equivalence ratio, TD* was calculated using the performance function and manually adjusted in the control loop. At the present time, the PNN controller is being explored in off-line operation. Results are discussed in Section V.

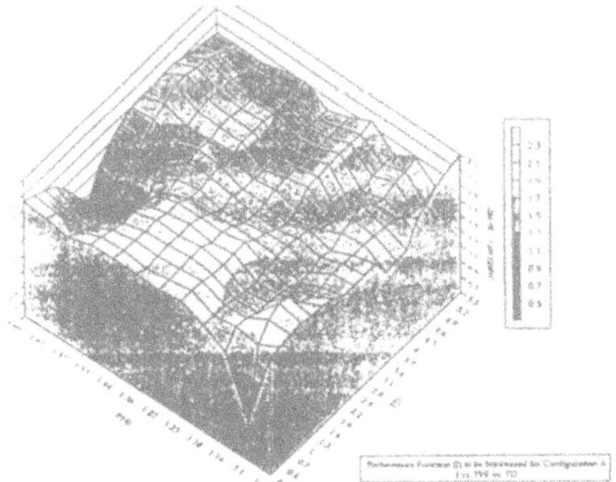

Figure 12. Performance (Cost) Function J to be Minimized for Optimum Performance as a Function of Equivalence Ratio.

V. Active Control Experiments

Closed-loop active control experiments were conducted in four combustion devices in the ONR research program. The devices were (1) Rijke tube with loudspeaker actuator

(multi-mode controller, evaluation of plant model), (2) 1-kW premixed combustor with loudspeaker actuator (improved gain/phase-shift controller with stabilizing compensator), (3) 50 and 100 kW premixed combustors, with loudspeakers and secondary fuel injection actuators (control with variable input amplitude and adaptive filter), and (4) 1 MW dump combustor with fuel modulation (loudspeaker) and pulsed-combustion actuator (gain/time-delay controller, neural net). Experiments performed in devices (1) to (3) were aimed at the suppression of pressure oscillations. The experiments done in (4) extended flammability limits in addition to suppressing combustion oscillations.

Rijke Tube

The acoustic mode predictions of the approximate analysis for the Rijke tube were compared to an experiment in which a thin electrically heated wire grid was used as the energy source. Comparison of the pressure spectra shown in Figure 13 reveals a good agreement of the instability frequencies, but the agreement between the relative amplitudes of each mode is still poor. Since the Rijke tube is one of the simplest examples of thermally driven oscillations, it is evident that much work is needed before an accurate theoretical prediction of combustion instability in a complex device can be made.

In the same device, the H^∞ controller designed with loop shaping techniques was able to control multiple unstable modes.[24] As shown in Figure 14, suppression of multi-frequency oscillations was obtained in closed loop operation.

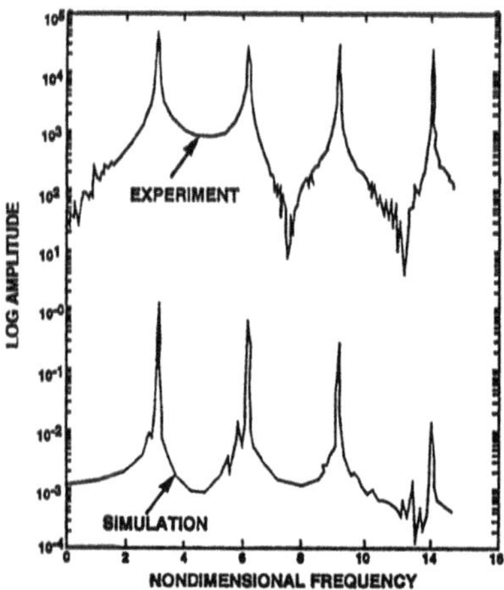

Figure 13. Experimental and Simulated Acoustic Modes in Rijke Tube.

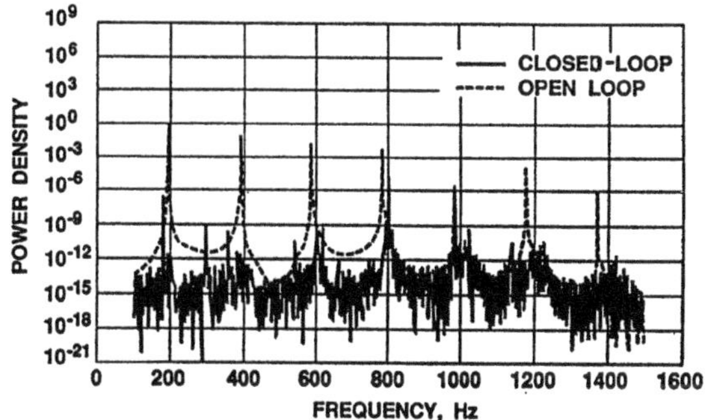

Figure 14. Closed and Open Loop Spectrum in Rijke Tube Using Controller With Loop Shaping H∞ Techniques.

Premixed Combustor (1 kW)

Experiments were conducted in a premixed methane-air combustor.[30] The flameholder consisted of a perforated plate with 80 holes and was located midway inside the combustor. The 1-kW combustor exhibited large self-sustained oscillations, predominantly at 380 Hz (three-quarter wave) and 760 Hz (second harmonic). A loudspeaker was mounted at the head end of the combustor. A simple gain/phase-shift controller suppressed the oscillations over a limited range of inlet flow conditions, in terms of total flow rate and equivalence ratio. Even in the effective range of suppression, the controller had a limited allowable gain and phase-shift margin, probably due a multiplicity of closely spaced instability modes. An improved controller with lead compensator and notch filter was designed using frequency-response data and standard design laws, resulting in a more robust controller with a wider gain and phase margin and improved range of effectiveness. An attenuation of 33 dB was achieved at the 1 kW energy level for the pressure oscillations with uncontrolled ΔP_{RMS} = 0.2 kPa.

Premixed Combustor (50 and 100 kW)

Several actuators were evaluated in this pipe combustor with premixed propane/air stabilized behind a bluff body. Only the experiments with secondary fuel injection, modulated by a needle valve, and the kerosene spray flame work are discussed here.[18] Combustor diameters of 40 mm, 50 mm and 80 mm, with heat release rates of between 50 and 100 kW, were used.

In the initial experiments a simple gain/phase-shift controller provided a constant input amplitude to the needle valve and resulted in a periodic loss of lock between actuator input and feedback. This loss of lock limited attenuation of the 120 Hz-quarter wave pressure oscillations to 10 dB (a factor of 3). A controller with variable input amplitude (knowledge-based controller) averted loss of lock between actuator input and feedback and improved attenuation of pressure oscillations by a factor of 6 up to 15 dB (Figure 15).

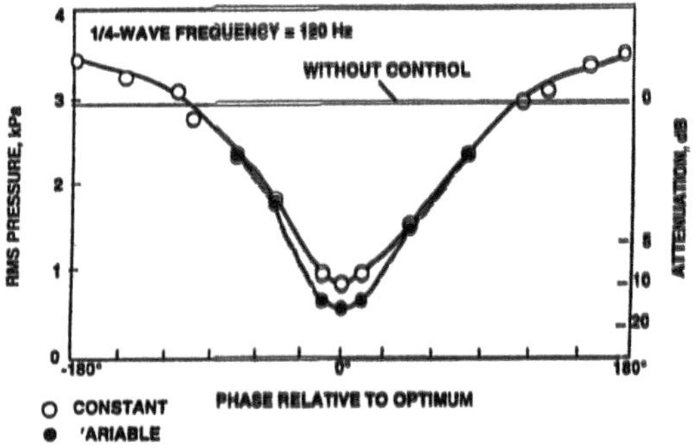

Figure 15. Suppression of Pressure Oscillations with Constant and Variable Input Amplitude to Secondary Fuel Actuator.

Tests with the adaptive filter showed that this control scheme was more effective than knowledge-based control because of its ability to track the instability frequency to lower levels of amplitude. A reduction of 20 dB (a factor of 10) of a 120 Hz pressure oscillation was obtained when amplitudes were around 4 kPa and the heat release was about 80 kW.[31]

The effectiveness of control in this type of combustor decreased as the heat release rate increased. Attenuation of oscillations up to 20 db was possible for heat release rates up to 100 kW. However, attenuation decreased to values below 5 dB for heat release rates greater than 160 kW in the 80-mm pipe (Figure 16). This result points out the need for actuators capable of providing a larger oscillatory input.

Combustion oscillation amplitudes of up to 3.5 kPa with a kerosene spray arrangement and heat release of around 70kW were reduced by up to 20 dB using a loudspeaker actuator and adaptive filter. Further reduction of up to 10 dB could be obtain using a pulsed injection of the fuel. The limiting factor was again the actuator in both cases.

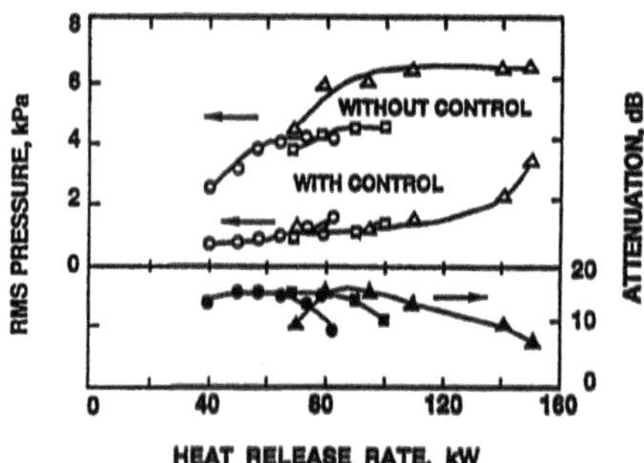

Figure 16. Influence of Combustor Heat Release on Attenuation.

Dump Combustor

Active combustion control was tested in the 12.5-cm diameter combustor for heat release rates up to 1 MW and combustor pressure up to 180 kPa. Three test conditions are reviewed. (1) For 250 and 500 kW operation at nearly ambient pressure, acoustic drivers modulated the ethylene flow rate at the combustion instability frequency with varying phase shift relative to the instability. (2) For 1 MW operation at 180 kPa, the pulsed-combustion actuator, using 2% of the total mass flow, was located in the inlet duct and operated with a simple gain/phase-shift controller. (3) Preliminary tests were made to evaluate the neural net controller at 33 kW and ambient pressure operation with fuel modulation using a loudspeaker.

For 250 kW and 500 kW operation at ambient pressure, peak amplitude and RMS of the 300 Hz instability was suppressed at 40 degrees phase angle. The maximum suppression was 47% of the peak amplitude (6.6 dB attenuation). The controlled fuel oscillations also extended the lean flammability limit.[7] For similar operational conditions, a dual-loop feedback control system was employed to suppress pressure oscillations during a bi-modal combustion instability. The sensor signal was fed back into the acoustic drivers via two separate channels, with different time delays (transfer functions). When the transfer functions for both channels were optimized for the suppression of its corresponding frequency, the combined dual-mode system was effective in suppressing the bi-modal frequency and delayed onset of high level instability oscillations at the lean flammability limit. The flame blow-out was reduced from an equivalence ratio of 0.72 to 0.54 (Figure 17).[32]

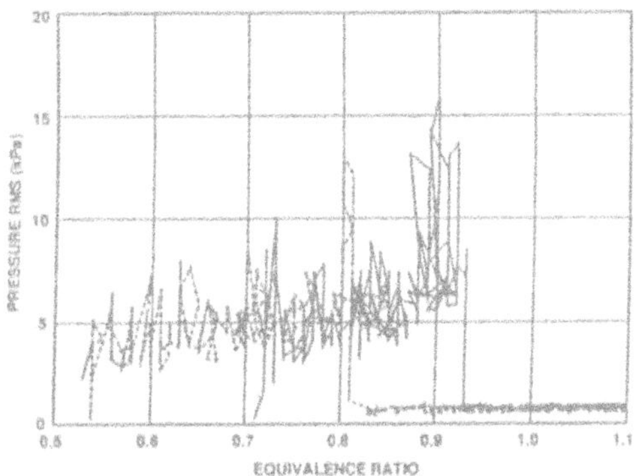

Figure 17. Extension of Lean Flammability Limit by Dual-Mode Active Control Using Gain/Time-Delay Controller and Acoustic Fuel Modulation, 500 kW Dump Combustor.

For 1 MW operation, with the pulsed combustion actuator, at 180 kPa, the peak value of a 140 Hz instability was reduced by 5 dB at a phase angle of about 0 degrees (Figure 18). The natural RMS pressure fluctuations were 25.4 kPa or 182 dB. For low pressure conditions in the dump combustor the instability could by suppressed 7 dB with closed loop control and 120 degree phase shift. This control was effective over a range of equivalence ratios from .04 to .53 despite the fact that it was optimized for $\Phi = .45$. Mie scattering pictures indicate that the major effect of the actuator is to modulate the shear layer at the dump, not to directly modify the heat release pattern and chamber acoustics. Higher attenuation is expected when the pulsed combustion actuator (flame kernel) concept is used for direct shear-layer excitation.

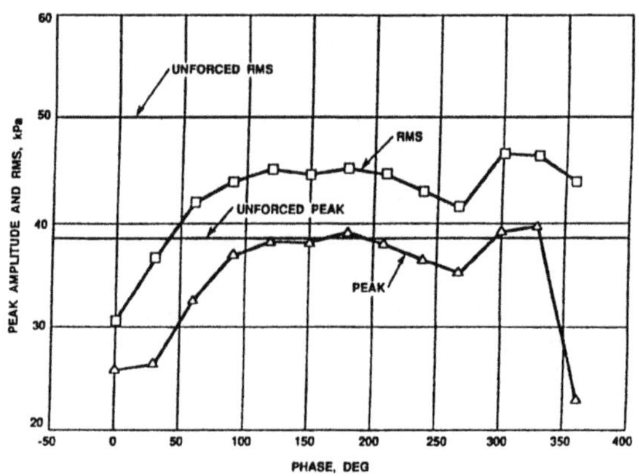

Figure 18. Suppression of Peak Amplitude Using Gain/Phase-Shift Controller and Pulsed Combustion Actuator, 1 MW Dump Combustor at 180 kPa Chamber Pressure

Active control of a 33 kW dump combustor using the PNN network was previously discussed (Figure 19). The pressure oscillations and average heat release, measured from the CH emission, was determined. A nonlinear cost function was defined to be directly proportional to the pressure fluctuation and inversely proportional to the energy release. A surface of the cost function was measured for a wide range of equivalence ratios, spanning from stoichiometric mixture to a fuel rich mixture of $\Phi = 1.7$ and a full cycle of time delays. The PNN was used to define an optimal time-delay which minimized the cost-function in the entire range of equivalence ratios. Reductions up to 14 dB in the pressure fluctuation level concurrent with an increase in the energy release were obtained. The lean flammability limit was also extended. Experiments are continuing to test the PNN controller in closed-loop operation, as shown in Figure 11.

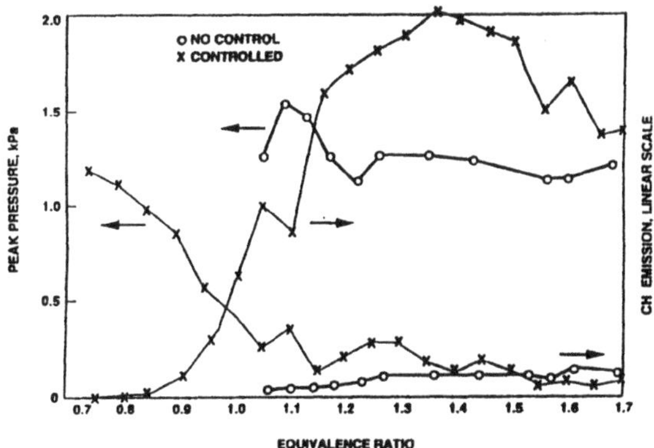

Figure 19. Peak Pressure and CH Emission Control with Acoustic Fuel Modulation Using Neural Net to Determine Time-Delay for Varying Equivalence Ratio, 33 kW Dump Combustor at Ambient Chamber Pressure

VI. Discussion and Conclusions

It is clear that active control is a viable approach to suppressing combustion instabilities and extending the flammability limits of a variety of combustion systems. A detailed physical understanding of the combustion/acoustic processes gained from small-scale laboratory experiments and simulations is an essential tool in the successful implementation of control. Several control methodologies have demonstrated potential.

Active control has been extended to test conditions which approach operational energy and pressure levels. The results of the ONR program point towards the need for a more effective actuator that could produce high acoustic power at elevated pressures. The development of the flame kernel actuator based on periodic chemical heat release is an important advance. Until this new actuator was developed, the trend in active combustion control experiments was towards decreasing performance with increasing energy and pressure levels.

Another encouraging result is some recently completed work at Imperial College. In a combustor with multiple unstable frequencies present the suppression of one mode using a loudspeaker actuator did not lead to the amplification of another. Experiments with an adaptive controller in the rough combustion regime of a disc stabilized duct combustor also yielded some interesting results. Where two unstable frequencies were present, the control system was able to prevent the onset of the alternative frequency mode when the principal mode was suppressed, by locking on to the subharmonic of the full-wave of the alternative mode and suppressing it.

A second area of continuing investigation should be the combined use of active and passive control methods. Experiments to date in laboratory scale combustion systems suggest that the two approaches can be used effectively in combination more satisfactorily than either one alone. It is plausible that the same will be true in other systems and processes, but the question has not yet been explored in detail.

VII. References

1. K. C. Schadow and E. Gutmark. "Combustion Instability Related to Vortex Shedding in Dump Combustors and Their Passive Control," Progress in Energy and Combustion Science, Vol. 18, pp. 117-132, 1992.

2. K. R. McManus, T. Poinsot, and S. M. Candel. "A Review of Active Control of Combustion Instabilities," International Symposium on Pulsating Combustion, Monterey, Calif., 1991.

3. K. C. Schadow, E. Gutmark, T. P. Parr, D. M. Parr, K. J. Wilson, and J. E. Crump. "Large-Scale Coherent Structures as Drivers of Combustion Instability," Combustion Science and Technology, Vol. 64, No. 4-6, 1989, pp. 167-186.

4. E. Gutmark, T. P. Parr, D. M. Hanson-Parr, and K. C. Schadow. "Stabilization of a Premixed Flame by Shear Flow Excitation," Combustion Science & Technology, Vol. 73, pp. 521-535, 1990.

5. E. Gutmark, T. P. Parr, D. M. Hanson-Parr, and K. C. Schadow. "Structure of a Controlled Ducted Flame," Western States Section/The Combustion Institute Spring Meeting, Boulder, Colorado, Paper 91-44, March 1991.

6. E. Gutmark, T. P. Parr, D. M. Hanson-Parr, and K. C. Schadow. "Closed-Loop Amplitude Modulation Control of Reacting Premixed Turbulent Jet," AIAA Journal, Vol. 29, No. 12, pp. 2155-2162, December 1991.

7. K. J. Wilson, E. Gutmark, K. C. Schadow, and R. Stalnaker. "Active Control of a Dump Combustor With Fuel Modulation," AIAA 29th Aerospace Sciences Meeting, Reno, Nevada, January 1991. (AIAA Paper No. 91-0368.)

8. K. Yu, E. Gutmark, K. J. Wilson, and K. C. Schadow. "Active Control of Organized Oscillations in a Dump Combustor Shear Layer," Proceedings of the International Symposium on Pulsating Combustion, Vol. II, Paper G-2, 16 pp., August 1991.

9. Suresh Menon and Wen-Huei Jou. "Large-Eddy Simulations of Combustion Instability in an Axisymmetric Ramjet Combustor," Combustion Science and Technology, Vol. 75, pp. 53-72, 1991.

10. S. Menon. "Numerical Simulation and Active Control of Combustion Instability in a Ramjet Combustor," AIAA 13th Aeroacoustics Conference, Tallahassee, Florida, October 1990. (AIAA Paper No. 90-3930.)

11. S. Menon. "A Numerical Study of Secondary Fuel Injection Techniques for Active

Control of Combustion Instability in a Ramjet," AIAA 30th Aerospace Sciences Meeting, Reno, Nevada, January 1992. (AIAA Paper No. 92-0777.)

12. K. R. McManus and C. T. Bowman. "Effects of Controlling Vortex Dynamics on the Performance of a Dump Combustor," 23rd Symposium (International) on Combustion/The Combustion Institute, pp. 1093-1099, 1990.

13. W. Lang, T. Poinsot, and S. Candel. "Active Control of Combustion Instability," Combustion and Flame, Vol. 70, pp. 281-289, 1987.

14. T. Poinsot, D. Veynante, F. Bourienne, S. Candel, and E. Esposito. " Suppression of Combustion Instabilities by Active Control," Journal of Propulsion and Power, Vol. 5, p. 14, 1989.

15. P. R. Choudhury, M. Gerstein, and R. Moharadi. "A Novel Feedback Concept for Combustion Instability in Ramjets, 22nd JANNAF Combustion Meeting, 1985.

16. G. J. Bloxsidge, A. P. Dowling, N. Hooper, and P. J. Langhorne. "Active Control of Reheat Buzz," AIAA Journal, Vol. 26, p. 783, 1988.

17. P. J. Langhorne, A. P. Dowling, and N. Hooper. "A Practical Active Control System for Combustion Oscillations," AGARD PEP 72nd-B Specialists' Meeting, 1988.

18. R. Hockey and S. Sivasegaram. Knowledge-Based Active Control of Oscillations in a Premixed Disk-Stabilized Flame," Department of Mechanical Engineering, Imperial College of Science, Technology and Medicine, London, Report FS/90/17, July 1990.

19. Private communication, Anil Gulati, GE Corporate Research and Development, 1992.

20. S. Sivasegaram. "Experiments in Active Control of Oscillations in a Premixed Disk-Stabilized Flame," Department of Mechanical Engineering, Imperial College of Science, Technology and Medicine, London, Report FS/89/309, 1989.

21. K. J. Wilson, E. Gutmark, R. A. Smith, and K. C. Schadow. "Development of a Pulse Actuator for Active Combustion Control," to be presented at the ASME 1992 Winter Annual Meeting Symposium on Active Control of Noise and Vibration., November 1992.

22. F. Culick, W. Lin, C. Jahnke, and J. Sterling. "Modeling for Active Control of Combustion and Thermally Driven Oscillations," Proceedings of the 1991 American Control Conference, 1991, pp. 2939-2947.

23. P. K. Houpt and G. C. Goodman. "Active Feedback Stabilization of Combustion for Gas Turbine Engines," Proceedings of the 1991 American Control Conference, Boston, Mass., June 1991.

24. J. E. Tierno and J. C. Doyle. "Experimental Active Stabilization of a Rijke Tube," Proceedings ASME Symposium on Active Control of Noise and Vibration," Anaheim,

California, November 1992.

25. D. C. McFarlane and K. Glover. "Robust Controller Design Using Normalized Coprime Factor Plant Descriptions," Lecture Notes in Control and Information Sciences, Vol. 138, Springer-Verlag, 1989.

26. B. Widrow and J. R. Glover, Jr. "Adaptive Noise Cancelling: Principles and Applications," Proc. IEEE 63, pp. 1692-1716, 1975.

27. A. R. Barron and R. L. Barron. "Statistical Learning Networks: A Unifying-View," Proceedings of the 20th Symposium on the Interfaces Computing Science and Statistics, pp. 192-203, April 1988.

28. Private communication, C. T. Bowman, Stanford University, 1992.

29. E. Gutmark, K. J. Wilson, K. C. Schadow, E. Parker, and R. Barron. "Dump Combustor Control Using Polynomial Neural Network (PNN)," submitted to AIAA 31st Aerospace Sciences Meeting, January 1993.

30. A. Gulati and R. Mani. "Active Control of Unsteady Combustion-Induced Oscillations," AIAA 28th Aerospace Sciences Meeting, Paper No. 90-0270, January 1990.

31. E. W. Hendricks, S. Sivasegaram, and J. H. Whitelaw. "Control of Oscillations in Ducted, Premixed Flames," Proceedings of IUTAM Symposium on Aerothermodynamics in Combustors, Taipei, Taiwan, June 1991.

32. E. Gutmark, K. J. Wilson, T. P. Parr, and K. C. Schadow. "Feedback Control of Multi-Mode Combustion Instability," AIAA 30th Aerospace Sciences Meeting, AIAA-92-0778, January 1992.

8. SUPPRESSION OF COMBUSTION OSCILLATION OF PREMIXED FLAMES BY ACTIVE CONTROL

M. Katsuki, Y. Mizutani, T. Miyauchi,*
T. Ochi**,
and Y. Morinishi***

*Osaka University
2-1 Yamada-oka, Suita,
Osaka, 565 Japan

**Osaka Prefectural College
of Technology
26-12 Saiwaicho, Neyagawa
Osaka, 572 Japan

***Kobe Steel
1-5-5- Takatsukadai
Nishiku, Kobe,
651-22 Japan

ABSTRACT. Combustion-induced oscillations in a duct were observed by changing the duct length and the mixture equivalence ratio. The wall pressure fluctuations in the combustion duct and the OH emission intensity of the flame were well-correlated with the free field sound pressure. A closed-loop active control using the above signals and a open-loop active control were studied to suppress combustion oscillations. Based on the results of the experiments, the possibility of suppression of combustion oscillation by a forced oscillation in the upstream feed tube of the facility was suggested.

1. Introduction

Recent demands for high combustion load aiming at minimizing the sizes of domestic and industrial combustion devices frequently provoke combustion instabilities or combustion oscillations. Since combustion oscillation is an undesirable source of noise and sometimes causes a disastrous destruction of the facility, the active control of combustion oscillation is recently a great interest in combustion technology. In order to develop an effective control method of combustion oscillation we are required to elucidate its mechanism and to understand the influential factors of the phenomena.

Combustion noise is often generated by the acoustic resonance of a flow system or by the combustion instability due to feedback phenomena between chemical reactions and fluid dynamics in the flow field. A number of researches concerning elucidation of noise generation mechanism are found in literature [1-4]. In course of those investigations passive controls by modifying operating and design factors of combustor to avoid or suppress combustion oscillations were studied in the past. Recent years, however, experiments on active controls have been intensively conducted by several groups [5-8]. A typical active control for low heat

F. Culick et al., (eds.), Unsteady Combustion, 161–171.

release rate combustors was the phase-shift type control with a loudspeaker to impose secondary sound waves into the feed flow. The fuel modulation by a phase-shifted primary or higher harmonic oscillation frequency was found more effective to control larger heat release rate oscillations.

In the present study we observed the combustion oscillation of disk-stabilized premixed flames in a duct with a sudden expansion not only to understand the interaction between combustion oscillation and flow conditions but also to obtain better understanding on the correlations between local heat release rate, represented by natural chemiluminescence of OH radicals, and pressure fluctuations. Another aim of this investigation was to develop an active control method for the suppression of combustion oscillation in small combustion devices.

2. Experimental set up and instrumentation

The vertical combustion rig, illustrated horizontally in Fig.1, was a 38-mm-wide square duct with a variable length between 300 mm and 700 mm. The walls of 64-mm-long observation section were plates of Vycor glass to permit optical observations. A mixture of propane and air was prepared in a swirl mixing chamber and was fed through an annular tube (28mm O.D. and 16mm I.D.) to the combustion duct. A flame was stabilized by a central disk placed at the entrance of the combustion duct. The combustion rate was 6.1 kW. A loudspeaker, attached on the bottom of the mixing chamber, was used as an oscillator of pressure field in the system. The open- and closed-loop active control systems used in the present experiment are shown in Fig.2.

The operating range of nominal flow velocity (air mass flow rate divided by the cross sectional area of the combustion duct), was 1.22 m/s to 5.9 m/s. The equivalence ratio of the mixture was varied between 0.6 and

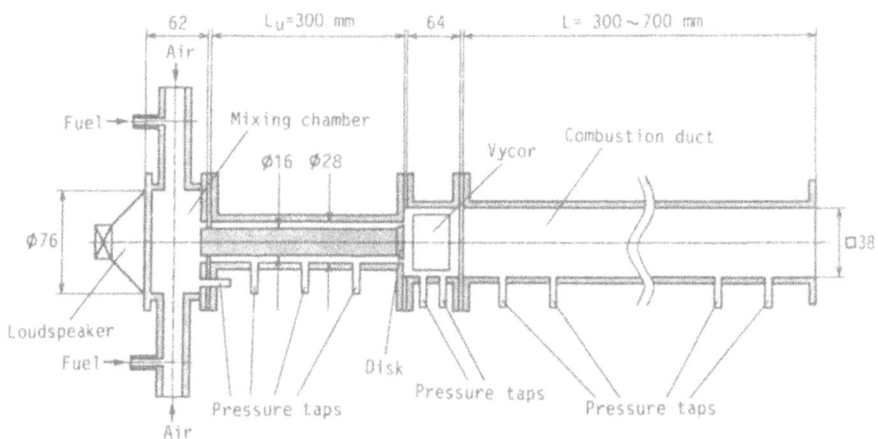

Fig,1 Combustion rig.

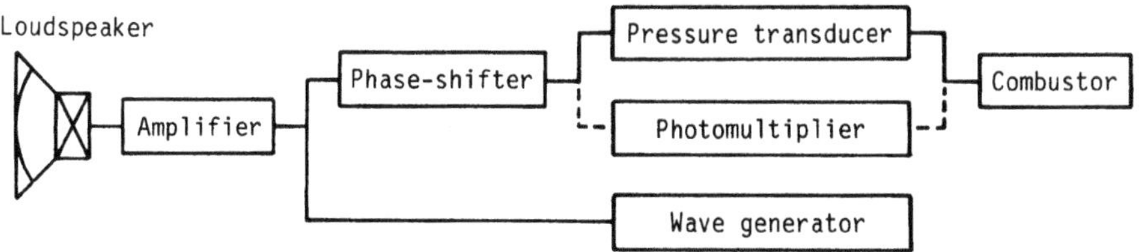

Fig.2 Closed- and open-loop control systems.

1.5 by changing the air flow rate. Field sound pressure was detected by a condenser microphone (B&K 5145) placed 1200 mm away from the combustion duct axis. The sound travel time from the flame to the microphone was estimated 3.5 ms. Local pressure fluctuations in the flow field were measured by a pressure transducer (KISTLER 7261) at each pressure tap shown in Fig.1.

A metal interference filter (central wave length of 308 nm and half-peak bandwidth of 14 nm) and a photomultiplier (HAMAMATSU R-106) were combined to detect natural chemiluminescence from OH radicals. The combination of a quartz lens (focal length 200 mm) and an aperture collected the global OH emission through the quartz window.

Simultaneous measurements of pressure fluctuation in the combustion duct, OH radical emission of the flame, and field sound pressure were carried out using the devices mentioned above. The signals of sound pressure, pressure fluctuation and OH emission were filtered by the same low-pass filters to prevent the resultant phase difference from spoiling one-to-one correspondence among the three signals. The maximum observed frequency was limited to 1.9 kHz, and the signals were recorded by a PCM (pulse cord modulation) recording system (NF RP-880 and PANASONIC NV-8730). These signals were digitized by an A/D converter with a sampling rate of 5kHz and analyzed statistically with a signal analyzer(IWATSU, SM2100) and a personal computer.

3. Results and discussion

Combustion oscillation, recognized by an abrupt increase in the free field sound level and in pressure fluctuation intensity, occurred for the mixtures around stoichiometric fuel-air ratio and for the duct length 500 mm and 700 mm as shown in Fig.3 and 4. Through the window flames filling the whole duct width were seen moving back and forth with a high frequency regardless the disk stabilizer. In case of 300 mm duct it was not like a harmonic oscillation although combustion was still rough and emitting considerable noise. The power spectrum of the sound pressure exhibited several peaks when combustion oscillation occurred as shown in Fig.5. The frequency corresponding to the highest peak were approximately 315 Hz and 190 Hz for the combustion duct of 500 mm and 700 mm, respectively, and varied slightly with the change in mixture equivalence ratio. These dominating frequencies were expected for a standing quarter wave in the combustion duct downstream of the sudden expansion [4], and this will be

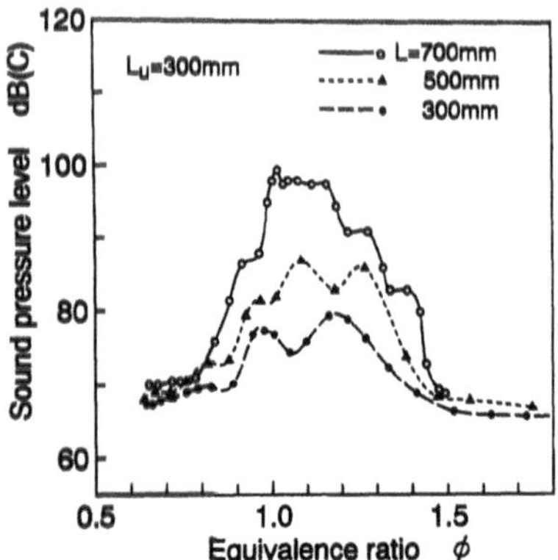

Fig.3 Effect of equivalence ratio on sound pressure level with the change of combustion duct length.

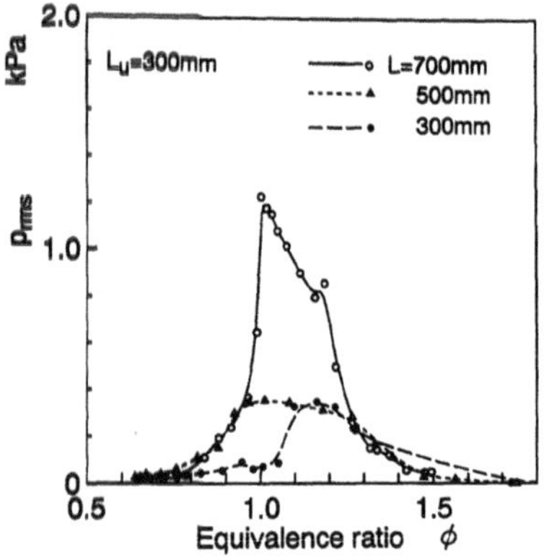

Fig.4 Effect of equivalence ratio on pressure fluctuation intensity in the combustion duct with the change of duct length.

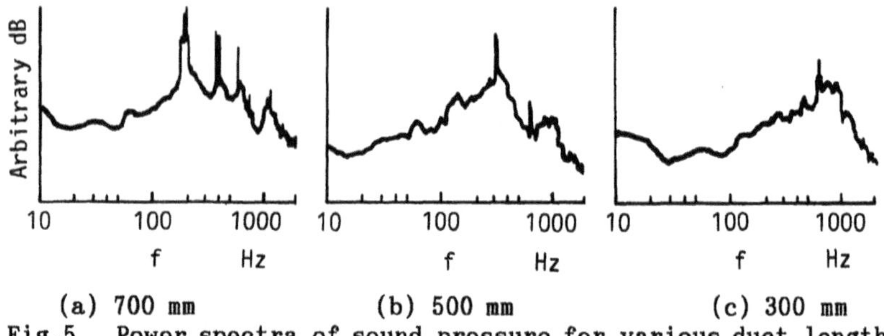

(a) 700 mm (b) 500 mm (c) 300 mm

Fig.5 Power spectra of sound pressure for various duct length.

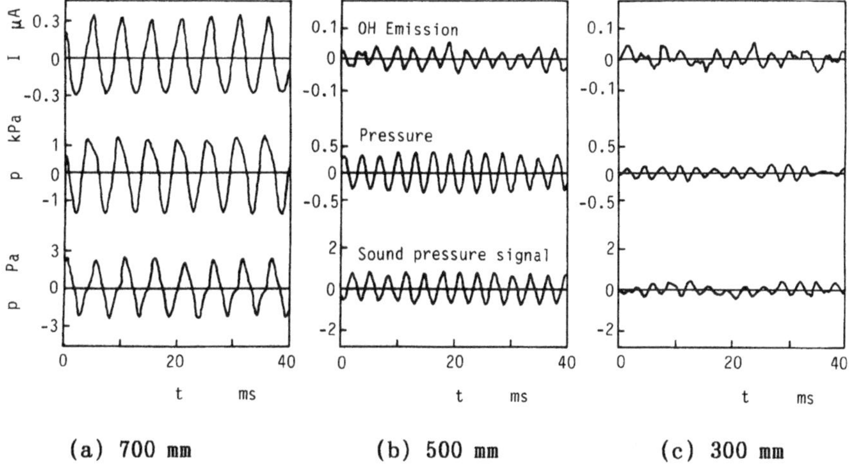

(a) 700 mm (b) 500 mm (c) 300 mm

Fig.6 Fluctuating signals of OH emission, pressure in the combustion duct and sound pressure for various duct length.

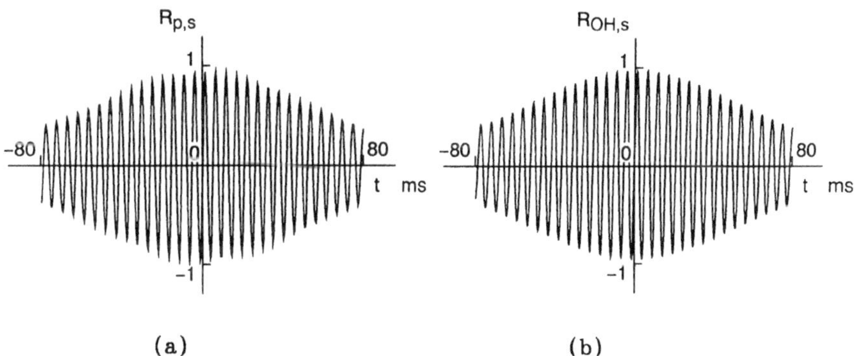

(a) (b)

Fig.7 Cross-correlation coefficients between fluctuating signals under combustion oscillation. (a) Sound pressure and pressure in the combustion duct. (b) Sound pressure and OH emission.

discussed later.

The fluctuation of chemiluminescence of OH radicals is held to correspond to the variation in heat release rate of a flame. Therefore, the OH emission intensity integrated over the combustion chamber corresponds to the instantaneous total combustion rate in the chamber, and the fluctuating OH emission from a local point may be an effective indicator of flame movement or fluctuation in local combustion rate.

Simultaneous measurements of pressure fluctuation in the combustion duct, global OH radical emission of flames in the observing section, and field sound pressure were carried out and it was found that fluctuating signals of three quantities were dominated by the same frequency depending on the duct length when a combustion oscillation occurred as shown in Fig.6. The beats in signals are recognized with the decrease in duct length, which indicates that the standing wave in the combustion duct interacts the travelling waves in the upstream feed tube.

Figure 7 shows cross-correlation coefficients between the fluctuations of three quantities. Intimate correlation was observed between the signals, which proved that OH emission of a flame can be a useful signal for detecting a combustion oscillation.

Figure 8 shows a typical example of the spatial distribution of the rms of wall pressure fluctuations measured by the pressure transducer at each pressure tap. The pressure fluctuations at the observing section of the combustion duct synchronize in phase with those measured at solid part of the line in the figure and out of phase with the dotted part. The strongest oscillation for 700 mm duct shows a half wave standing for the full-length of the system. For 500 mm duct a standing quarter wave between the disk stabilizer and the duct exit dominated the pressure field, but a weak travelling wave in the upstream feed tube was observed as indicated in Fig.6(b). Furthermore, pressure fluctuations at any point for the case of 300 mm duct did not coincide with each other in phase even within the combustion duct, which indicated non-existence of a harmonic oscillation.

The pressure in the observing section of combustion duct was used as the feedback signal for the closed-loop active control. Figure 9 shows the effect of phase-shift angle on noise reduction achieved with a closed-loop active control driven by the pressure signal. The phase-shift angle was

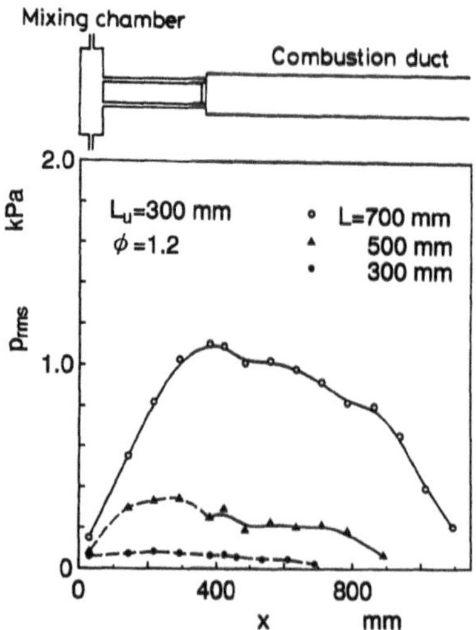

Fig.8 Spatial distribution of the rms of static pressure fluctuations.

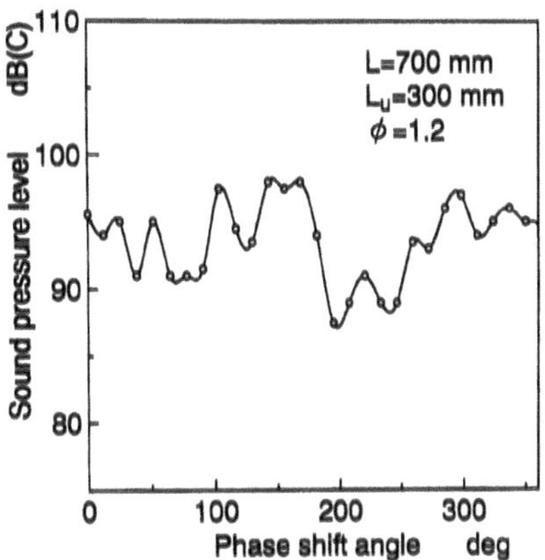

Fig.9 Reduction of sound pressure level achieved with the closed-loop active control driven by the phase-shifted pressure signal of combustion duct.

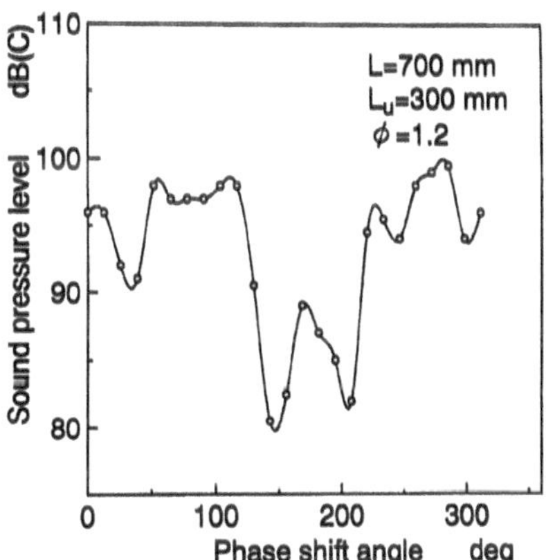

Fig.10 Reduction of sound pressure level achieved with the closed-loop active control driven by the phase-shifted pressure signal of supply tube.

calculated based on the first dominant frequency although all frequency components were contained in the signal. The driving input into the loudspeaker was 2.8W against the combustion rate of 6.1kW. Some reduction of sound pressure was achieved in this operating condition when the sift angle was set between 180 to 240 degrees. However, rough combustion still survived in spite of the decreased intensity of sound pressure because the input was not strong enough to suppress the oscillation thoroughly, and the stronger the oscillation was, the less effective the control worked.

Another trial of closed-loop control was carried out using pressure

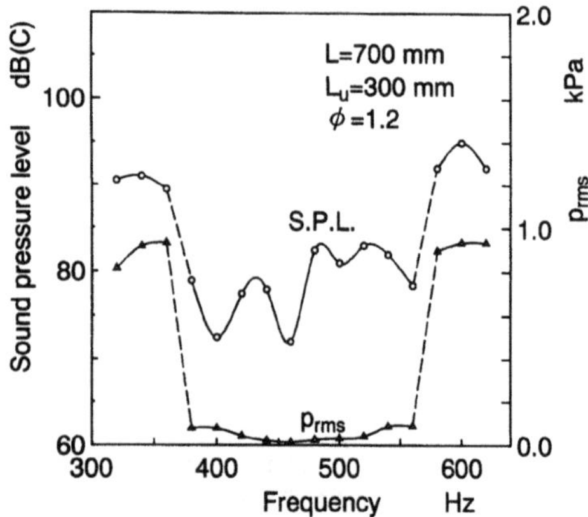

Fig.11 Reduction of sound pressure level and pressure fluctuation intensity achieved with the open-loop active control driven by single sinusoidal wave.

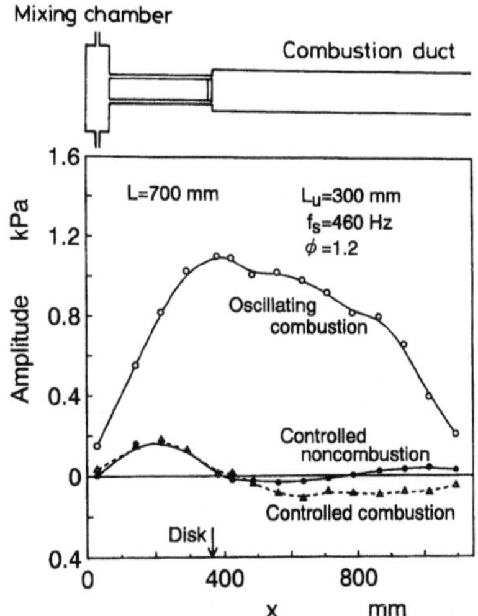

Fig.12 Effective amplitude and mode of oscillation with and without control.

fluctuations at the middle of the upstream feed tube as the feedback signal. Figure 10 shows considerable reduction of noise for the phase-shift angle between 140 to 200 degrees. In the feedback process of this control the loudspeaker possibly worked as an active absorber of pressure fluctuations in the feed tube which was exited by the standing quarter wave in the combustion duct. The controlled flame behaviour looked slightly quieter compared with the oscillating flame in the previous control.

In contrast, the open-loop active control with the input of a single sinusoidal wave worked more effectively, and the controlled flame looked perfectly stable on the central disk and the edge of sudden expansion at the combustion duct inlet. Figure 11 shows the reduction of sound pressure achieved with the open-loop active control. The sinusoidal wave between 380 Hz to 560 Hz suppressed the oscillation of 190 Hz effectively independent of the degree of phase-shift although the input was as small as 2.5W. The combustion noise level was decreased by almost 20 dB.

The mechanism of suppression of oscillation in this case was considered quite different from that of the phase-shift type control because the effect did not show any relationship with the degree of phase-shift angle and the dominating oscillation frequency. In order to understand the reason why the small input of a higher frequency independent of the occurring oscillation could suppress the oscillation, the rms of wall pressure fluctuations along the flow field and the mode of oscillation were measured imposing the optimum frequency of 460 Hz. The measured spatial distributions of the effective amplitude of pressure fluctuations are shown in Fig.12 including the one for noncontrolled oscillating combustion. The oscillating combustion was dominated by the standing half wave associated with the whole length of the flow system and the position of the disk stabilizer seemed to correspond to the antinode of pressure fluctuation. By imposing 460 Hz sinusoidal wave, on the other hand, a standing half wave in the upstream feed tube was formed in both combusting and noncombusting flows. In the downstream combustion duct another half wave was formed in combusting flow and one wave in isothermal noncombusting flow depending on the temperature level. The controlled flame anchored by the central disk and the edge of sudden expansion looked steadily propagating into the reactant flow. It is thus considered that the standing half wave

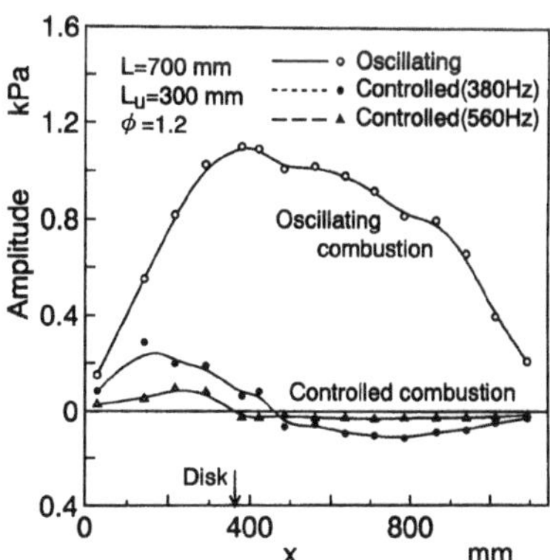

Fig.13 Influence of input frequency on the controlled mode of oscillations.

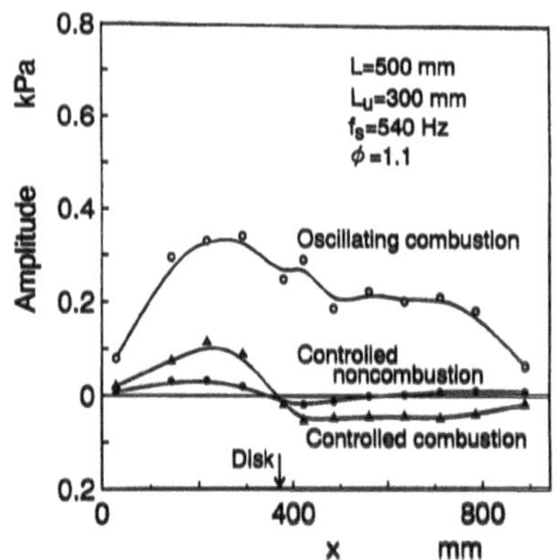

Fig.14 Effective amplitude and mode of oscillation.

oscillation in the upstream feed tube, which was triggered by the imposed wave, possibly produced the maximum velocity fluctuations at the entrance of the combustion duct which suppressed the onset of a quarter wave oscillation in the combustion duct.

Figure 13 shows the influence of the frequency of input signal on the effect of control and the forced oscillation modes. 380 Hz and 560 Hz were the limiting values of the effective frequency range of control which was suggested by Fig.11. The position of pressure node shifted slightly depending on the input frequency. Thus, combustion oscillation can be controlled by a forced oscillation in the upstream feed tube if the distance between the disk stabilizer and the pressure node of the forced oscillation is within a certain range, hence the pressure fluctuation intensity at the stabilizer is lower than a certain limit.

The change in combustion duct length showed the similar results as shown in Fig.14, although the optimum imput frequency moved to 540 Hz. Therefore, the possibility of suppression of combustion oscillation by a forced oscillation in the upstream feed tube was suggested although the strongest oscillation for the stoichiometric mixture appeared in Fig.4 could not be suppressed perfectly even with this type of control.

4. Conclusions

Combustion-induced oscillations in a duct were observed by changing the duct length and the mixture equivalence ratios. Both static pressure fluctuations in the combustion duct and OH radical emissions are suitable for the feedback signal of active control systems because they showed an intimate correlation with the free field sound pressure. Closed- and open-loop active control systems were tried to suppress combustion oscillations and an open-loop control using a frequency independent of combustion oscillation worked effectively. Experimental results suggested the possibility of the suppression of combustion oscillation by a forced oscillation forming a standing half wave in the upstream feed tube of the facility.

References

[1] Putnam, A.A. (1971) Combustion-Driven Oscillations in Industry, Elsevier, London.

[2] Kilham, J.K., Jackson, E.G. and Smith, T.J.B. (1961) 'An Investigation of Tunnel Burner Noise', Institution of Gas Eng. Journal 1, 251-266.

[3] Heitor, M.V., Taylor, A.M.K.P. and Whitelaw, J.H. (1984) 'Influence of Confinement on Combustion Instabilities of Premixed Flame Stabilized on Axisymmetric Baffles', Combustion and Flame 57, 109-121.

[4] Katsuki, M. and Whitelaw, J.H. (1986) 'The Influence of Duct Geometry on Unsteady Premixed Flames', Combustion and Flame 63, 87-94.

[5] Poisont, T., Bourienne, F., Candel, S. and Eposito, E. (1989) 'Suppression of Combustion Instabilities by Active Control', Journal of Propulsion and Power 5, 14-20.

[6] Langhorne, P.S., Dowling, A.P. and Hooper, N. (1990) 'Practical Active Control System for Combustion Oscillations', Journal of Propulsion and Power 6, 324-333.

[7] Hendricks, E.W., Sivasegaram, S. and Whitelaw, J.H. (1992) 'Control of Oscillation in Ducted Premixed Flames', in R.S.L. Lee, J.H. Whitelaw and T.S. Wung (eds.), Aerothermodynamics in Combustors, Springer-Verlag, Berlin Heiderberg.

[8] Schadow, K.C., Gutmark, E. and Wilson, K.J. (1992) 'Active Combustion Control in a Coaxial Dump Combustor', Combustion Science and Technology 81, 285-300.

9. COMBUSTION INSTABILITIES IN PROPULSION SYSTEMS

F. E. C. Culick
California Institute of Technology,
Daniel and Florence Guggenheim Jet Propulsion Center
MS 205-45 Pasadena,
91125 CA, USA

ABSTRACT. The purpose of this paper is to give a broad overview of the field of combustion instabilities in propulsion systems. Virtually all of the material included here has appeared elsewhere, either in primary research reports or in reviews. None of the propulsion systems are covered in great detail, but sufficiently to establish the fundamental point that while there are obvious practical differences among the systems, for understanding and treating combustion instabilities, much is to be gained by treating the various phenomena within a common framework. In that context, the systems are distinguished chiefly by geometry and the kinds of propellants used. On that basis, a general framework can be constructed to serve both practical and theoretical purposes.

1. Introduction

Chemical propulsion systems depend fundamentally on the conversion of energy stored in molecular bonds to mechanical energy of a vehicle in motion. Combustion of fuel and oxidizer is the first stage. Burning takes place in a vessel open only to admit reactants and to exhaust the hot products. Higher performance is achieved by increasing the energy release per unit volume. For example, probably the highest power density was reached in the F-1 engine for the Apollo vehicle (1960s), 146.4 gigawatts/m^3. The power densities in solid rockets are much less. For a cylindrical bore, the values are approximately $0.25(r/D)$ gigawatts/m^3, where r is the linear burning rate, typically a few centimeters per second, and D is the diameter. Thus the power densities rarely exceed one gigawatt/meter^3. An afterburner on a high-performance fighter may burn fuel at the rate of 75,000 pounds per hour, generating roughly 450 megawatts for a short period in a volume of less that two cubic meters.

These are indeed very large power densities. To appreciate *how* large, consider the fact that the average power consumption per person in a developed country is about 4 kilowatts (roughly also the same as that for astronauts on a mission). For the United States, with approximately 250 million people, the total power consumption is about 1,000 gigawatts. Hence, for a few minutes, the five F-1 engines in the first stage of the *Apollo* produced power equivalent to about 75% of the entire electrical power consumption of the U.S. — in a *very* small volume. We cannot be surprised that such enormous power densities should be accompanied by small, but annoying, fluctuations.

Combustion instabilities were discovered about the same time in solid and liquid propellant rocket engines, in the late 1930s. Since that time, unstable oscillations have occurred in most, if not in practically all, new development programs. Indeed, because of the high density of energy release

F. Culick et al., (eds.), Unsteady Combustion, 173–241.

in a volume having relatively low losses, conditions normally favor excitation and sustenance of oscillations in any combustion chamber intended for a propulsion system.

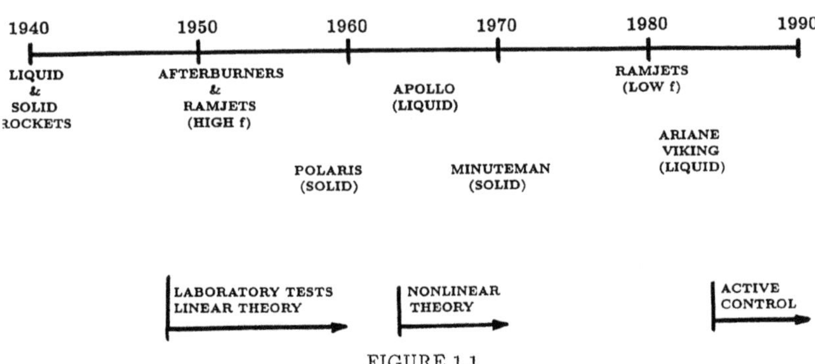

FIGURE 1.1

Figure 1.1 is an abbreviated chronology of some major events and features of the subject during the past 50 years, chiefly in the U.S. In one form or another, combustion instabilities have been under continuous study for that entire period. However, with time the emphasis naturally shifted, depending on what sort of full-scale system experienced difficulties. During World War II in the U.S., it seems that virtually all work in this subject was concerned with elimination of high-frequency 'resonant burning' (the term used at the time) in small tactical solid rocket motors. The common treatment was usually a form of passive control, involving installation of baffles, resonance rods, or some other modification of geometry. Since that time the need to solve problems of instabilities in solid rockets has continued for rockets of all sizes. Much of the basic understanding gained is applicable to liquid rockets despite the obvious differences in the systems.

Although work on combustion instabilities in liquid rockets began in the early 1940s, significant progress was not achieved, or required, until after World War II with the development of large intercontinental ballistic missiles. Later, during the 1970s, the needs of the Apollo program motivated a large amount of work on instabilities, rendered particularly important because of the astronauts on board the rocket. That experience formed part of the basis for developing the Space Shuttle main engine. Practically no progress was made in the U.S. from the early 1970s to the middle 1980s when interest rose again with the prospects for developing the Advanced Launch System (ALS), subsequently canceled. In France, a relatively small but significant program has been continuously supported since 1981 as part of the Ariane program. Three papers in a recent symposium (First International Symposium on Liquid Rocket Instability, *Proceedings*, 1993) have given most researchers in the West their first extensive acquaintance with the considerable body of work accomplished in the Soviet Union over many years.

Combustion instabilities in afterburners have been troublesome since the invention of the devices in the late 1940s. The sorts of problems encountered at at that time still persist: high-frequency transverse modes, similar to the oscillations often found in liquid rockets. In addition, however, with the development of high bypass engines, conditions became favorable for the entire length of the engine to participate in the motions, leading to the presence of longitudinal vibrations having lower frequencies which are much more difficult to treat simply by making changes in geometry.

Similarly, although earlier liquid-fueled ramjets had instabilities mainly in the high-frequency range, more compact designs in the late 1970s led to longitudinal oscillations having lower frequencies in the range of a few hundred Hz. Those problems received much attention during the 1980s and several concentrated research programs have produced quite good basic understanding of the causes and possible means of treatment.

Solid rockets also have exhibited instabilities over wide ranges of frequency, from the low frequencies arising in large ICBMs and strap-on booster rockets to very high frequencies in tactical solid rockets. Research on the problem in solid rockets has not ceased from the early 1950s to the present, at varying levels of intensity. Although in comparison with liquid rockets, details of the problem, notably the basic causes, are significantly different in solid rockets, much of what has been learned of the general aspects of the problem is directly applicable to liquid rockets.

Indeed, quite generally, progress made in any one of these systems is relevant, and often importantly so, to understanding and treating combustion instabilities in the other sorts of systems as well. This review of instabilities is brief, but we will discuss results achieved in all types of systems, especially to encourage workers in the field to be aware of, if not conversant with, combustion instabilities in propulsion systems not necessarily their immediate concern.

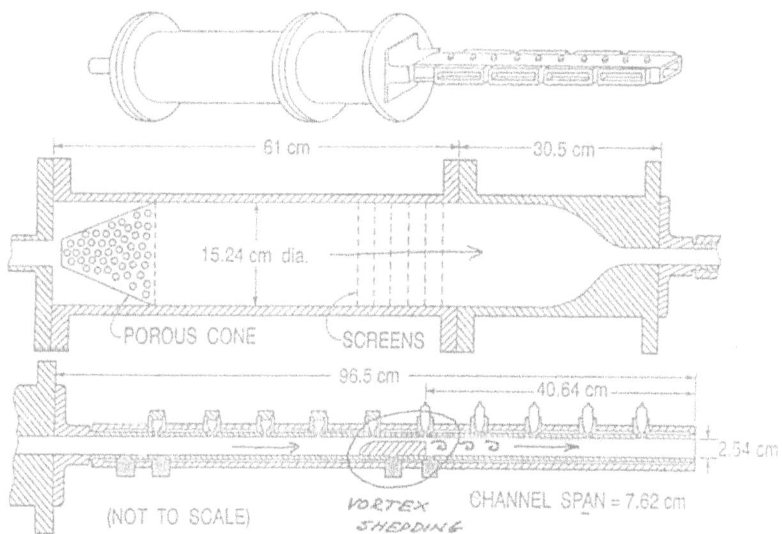

FIGURE 1.2

Given the experimental conclusion that combustion instabilities always share certain common features, we may illustrate basic properties with data taken in any system. A clear illustration of the phenomena may be given with data taken in the device sketched in Figure 1.2 (Smith and Zukoski 1985; Sterling and Zukoski 1987; Kendrick, Tsak, and Zukoski 1993), showing a laboratory dump combustor. The basic mechanism for combustion instabilities in this device is associated with periodic combustion in large vortices shed from the rearward facing step. Longitudinal modes are excited, having frequencies close to the values easily estimated by applying elementary methods of classical acoustics. Figure 1.3 reproduces a typical pressure record and Figure 1.4 is the spectral power density. The peaks correspond to the acoustic modes at 530 and 460 hertz, and their sub-harmonics at $\frac{1}{2}(530)$ and $\frac{4}{5}(460)$. That sub-harmonics are present suggest influences of nonlinear processes, a feature already established because the self-excited linearly unstable motions can be restricted to finite amplitudes only if some sort of nonlinear behavior is present.

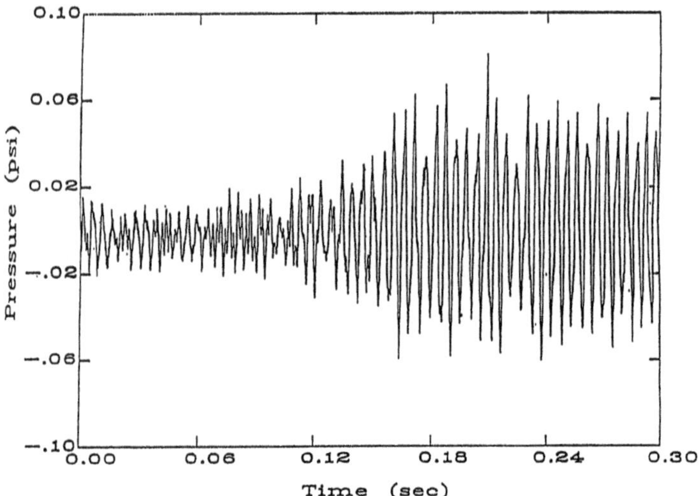

FIGURE 1.3

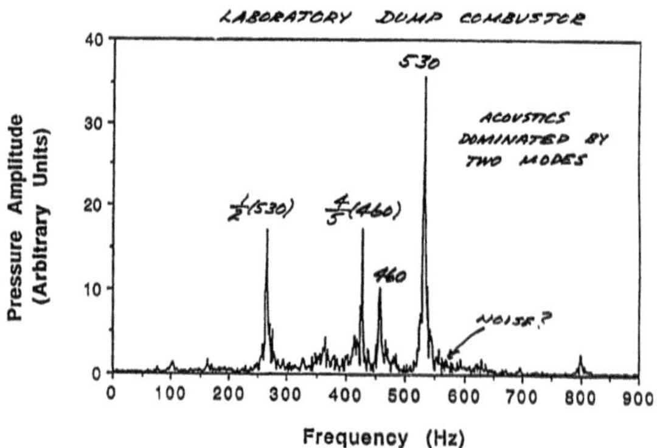

FIGURE 1.4

Also obvious in Figure 1.4 is the background of broad-band fluctuations or noise. That a combustion chamber is noisy is obvious during any firing. The details of the generating processes, and especially the possible connections between noise and combustion instabilities, are not understood. The latter may be part of nonlinear behavior as suggested in Figure 1.5, taken from Poinsot *et al.* (1989). Oscillations in a small, gas-filled laboratory combustor were actively affected ('controlled') by injecting acoustic oscillations having adjustable amplitude and phase. With control on, the amplitudes of the instabilities were reduced by several tens of decibels and in addition, the local noise was depressed over practically the entire range of frequencies. The experimental results are inadequate to reach unambiguous conclusions — e.g. whether the reduction of noise actually reaches levels below those occurring naturally in the absence of instabilities. However, these results suggest the intriguing prospect that organized oscillations and random fluctuations are connected by nonlinear processes.

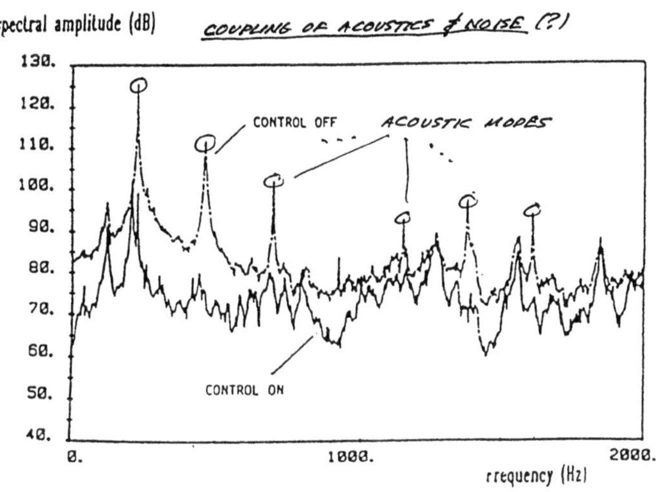

FIGURE 1.5

To begin understanding the essential characteristics of combustion instabilities, it is best first to distinguish linear and nonlinear behavior. Linear behavior presents only one general problem, linear instability, which received widespread attention during the 1950s and 1960s; see, c.g. the monograph by Crocco and Cheng (1956) and the lengthy compilation of works edited by Harrje and Reardon (1971). Any disturbance may be synthesized as an infinite series of harmonic motions. An approximate analysis developed over many years (Culick 1988, Culick and Yang 1992) allows one to use classical acoustic modes as the terms in the series and to compute the perturbations of the complex wavenumber for each mode due to the various contributing processes in a combustion chamber. The real part of the wavenumber gives the frequency shift and the imaginary part gives the growth (or decay) constant associated with each mode. Vanishing of the imaginary part then determines the formal condition for linear stability whose dependence on the parameters characterizing the system is then known.

If one is prepared to accept the approximations on which that analysis is founded, then for practical purposes the problem of linear stability is, in principle, solved for a wide range of applications. All actual processes are accommodated and the real difficulties in using the results consist in modeling the processes. This approach has been used successfully for analysis and prediction of instabilities in solid rockets; its validity has been confirmed both by application to experimental data and by comparison with numerical results specially obtained for the purpose. Application to liquid-fueled systems, in particular liquid rockets, has not been so widely accomplished. There are particularly difficult problems of modeling associated with the presence of significant amounts of liquid phase in the chamber, but there is presently no reason to doubt that the approximate linear analysis has a useful place in treating combustion instabilities in liquid propellant rocket engines. Owing to its generality and relative simplicity, we shall use the approximate analysis here when convenient to explain theoretical results and as a framework for interpreting observed behavior.

2. Three Examples of Combustion Instabilities in Operational Rocket Motors

In this paper we proceed from observation to theory and analysis. The analysis to be described in Section 6 is rooted in the data and observations accumulated chiefly from tests of liquid and solid rocket motors. As a prelude to a wider ranging discussion of characteristics of instabilities in various propulsion systems, we begin with brief examination of three examples, two of rockets using liquid propellants and one for a solid propellant rocket. Examples of instabilities in afterburners and ramjets are more difficult to find in the open literature, partly because less data has been taken than for rocket engines and, in the case of afterburners, partly due to commercial competition.

We have remarked that the distinction between linear and nonlinear behavior offers one means of classifying combustion instabilities. In some respects, that is an approach too broad; a more practical scheme is based on the kinds of propellants used and the geometry of the combustion chamber. There are many types of solid propellants and no purpose is served here by attempting to sort out the differences. There are three main combinations of liquid propellants: liquid oxygen (LOX) with a hydrocarbon fuel; LOX and liquid hydrogen (commonly referred to as cryogenic propellants); and storable propellants, the most common being nitrogen tetroxide as the oxidizer and monomethylhydrazine (MMH) or A-50 (a mixture of 50% hydrazine, N_2H_4, and 50% unsymmetrical dimethylhydrazine (UDMH) as the fuel. We discuss briefly here two cases of instabilities in liquid rocket engines, both taken from the Apollo program: the first-stage F-1 engine which used the propellants LOX/RP-1; and the lunar module descent engine (LMDE) which of course used storable propellants.

2.1 COMBUSTION INSTABILITIES IN THE F-1 ENGINE

The problem of instabilities in the F-1 engine was serious for more than seven years during the development of the engine. Recently, Oefelein and Yang (1992) have given a comprehensive review of the subject, the first such work based on a thorough reading of the original reports and memoranda. Here we shall give only a cursory discussion.

The F-1 engine was qualified for manned flight in 1966 but its origins can be traced back to several engines developed in the 1950s by Rocketdyne, principally the E-1; the Atlas MA-2; and a derivative of the engine used in the Thor/Jupiter vehicle, the H-1, itself later used in Saturn I. Initial development of the F-1 began as an Air Force project in 1955 but became a NASA program when Rocketdyne received a contract in 1959 to begin developing an engine to produce ultimately 1.5 million pounds thrust. Many components of the F-1 were modelled on those of the H-1 but serious problems associated at least partly with scaling arose in this period. The H-1 and E-1 produced 165–205 thousand pounds thrust and 360–400 thousand pounds respectively, significantly less than that required by the F-1.

In particular, of the 44 tests of the first F-1 design carried out from January 1959 to May 1960, twenty had combustion instabilities with peak amplitudes greater than or comparable to the average pressure. The first model of the engine used an injector based on that of the E-1 and produced one million pounds thrust. That significant oscillations were discovered was a result contrary to predictions based on experimental and theoretical work at the time and relatively poor instrumentation apparently provided little new useful information. The oscillations caused serious erosion and burning of the injector face, showing the presence of large radial and tangential motions.

A second design used an injector similar to that of the H-1, which was under concurrent development and had shown less problems with instabilities. From June 1960 to October 1962, 186 tests were performed; combustion instabilities were common, finally leading to total loss of an engine in June 1962. Various configurations of baffles were tried during this time, showing that without baffles the engine did not have 'dynamic stability,' i.e. stability to disturbances introduced with small explosive charges.

It was apparently during that period, 1960–62, that distinctions between linear and nonlinear instabilities were recognized and, at least implicitly, helped guide the development of tests. In the literature of that time, the term 'self-triggering' was often used, referring to spontaneous or linear instability; and 'dynamic instability' meant that a linearly stable system was nonlinearly unstable to a sufficiently large disturbance. In the F-1 program, the requirement for flight qualification was that the engine should be stable to the disturbance following ignition of a 13.5-grain explosive charge. An important conclusion established by the early tests (1959 – 1962) and confirmed throughout subsequent work was that baffles were necessary to give dynamic stability.

Several mechanisms for exciting instabilities were evidently active in the engine, of which the dominant ones seem to have been: coupling between oscillations in the chamber and unsteady motions within the injection elements (injection coupling); periodic pulsed combustion of excess liquid propellant accumulations on boundary surfaces accompanying film cooling; and transverse displacements of the injected fuel and oxidizer jets when exposed to oscillations of velocity parallel to the injector face. The first tangential mode was the troublesome oscillation, eventually subdued by a combination of modifications of the injector design and use of baffles extending downstream from the injector face.

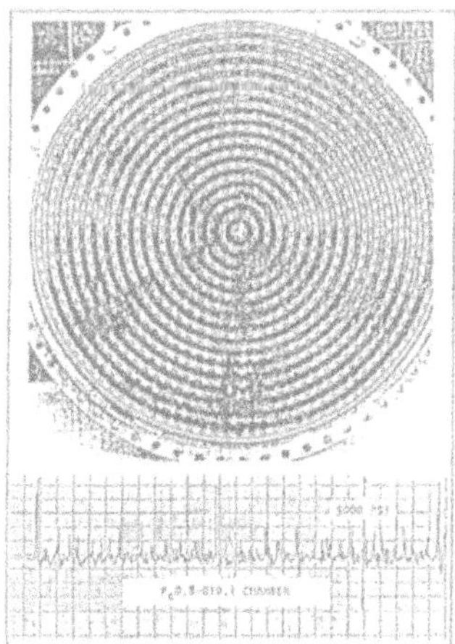

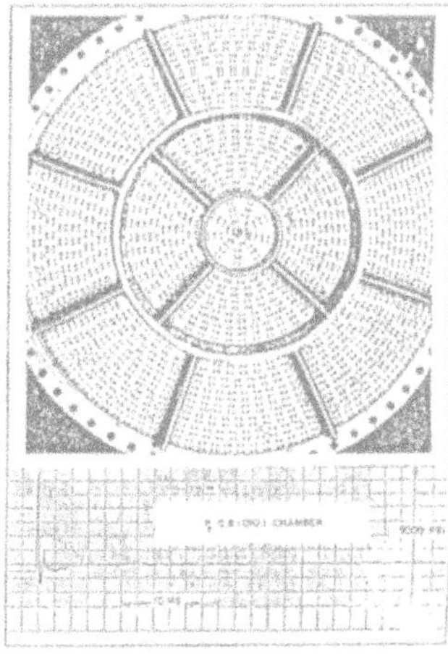

FIGURE 2.1

Figure 2.1 shows examples of the injector with and without baffles. Without baffles, the engine was prone to instabilities, a characteristic consistent with previous experience with engines using

LOX/RP-1 as propellants. With suitable combination of baffle configuration and injector design, the engine not only did not exhibit self-excited oscillations (linear instabilities), but was also stable to finite disturbances (nonlinear stability). Pressure records in the figure illustrate that behavior.

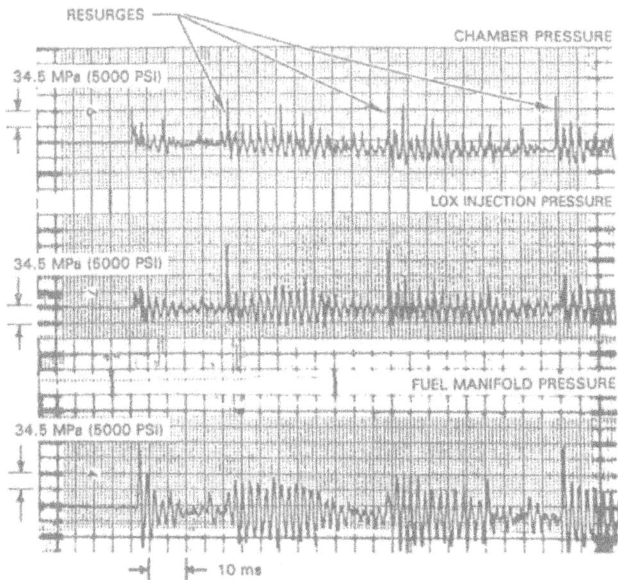

FIGURE 2.2

The mechanisms responsible for the instabilities in the F-1 engine were apparently located within three regions near the injector face and a fourth region associated with film cooling of surfaces. The last was identified with the dominant cause of a phenomenon called 'resurging' frequently observed following explosion of a 'bomb' for rating dynamic stability. Figure 2.2 reproduces pressure traces. Eventually, the problem was understood well enough to be solved. The dominant cause was pulsed combustion of liquid fuel detached from the liquid layer produced with film cooling of the baffles. A secondary cause may have been rapid combustion of droplets at supercritical pressures. In any event, resurging was essentially eliminated by determining the best conditions of film cooling, the thickness of the film being the main parameter. Film cooling of the injector and baffles was reduced by more than half and produced an increase in characteristic velocity of four percentage points.

While the mechanism associated with film cooling is largely specific to the injector design, other causes near the injector face have much more general significance and remain the subject of research. The three regions referred to above are: (1) that nearest the injector face, containing the spray fans and all processes generating liquid drops of fuel; (2) the region of fuel vaporization extending roughly 10 inches downstream of the face; and (3) the region further downstream in which both fuel and oxidizer are gaseous and where most of the combustion occurs.

In the region closest to the injector, the dominant processes involve the dynamics of the liquid jets and the spray fans formed by impinging jets. Within this region, extending roughly three inches from the injector face, formation of fuel droplets is nearly complete and LOX drops have almost entirely vaporized. The baffles finally used extend three inches downstream from the face, of which the diameter was 40 inches. Hence the baffles effectively 'shadowed' the region in which interactions between jets and the processes forming the liquid drops are most important. That was likely the chief benefit of the baffles which had very minor effects on the frequencies of oscillations and probably did not provide much attenuation. A secondary consequence, whose importance was not quantified, was conversion of tangential fluctuations to longitudinal motions which are then damped by the action of the exhaust nozzle (R. Levine, private communication).

Fuel drops suffer break-up due to aerodynamic forces and are vaporized in region 2. It is here that vaporization is encouraged by high relative velocity between gaseous oxygen and the liquid fuel drops. If the gaseous fuel and oxygen are not uniformly mixed as a result of the processes in regions 1 and 2, interactions with oscillations in the flow can cause fluctuations of the mixture ratio and hence of the burn rate.

The view that the flow downstream of the injector could be usefully considered as three regions evolved during the development program from 1962–1965. early work was directed primarily to identifying the causes of the linear instabilities and making the engine linearly stable. In fact, several of the dominant causes were discovered and motivated much of the subsequent work: initially, injection coupling was suspected as the root cause, so early tests and modifications of the injector design were intended to eliminate unsteady motions in the propellant-supply system. As a result, although the engine was not linearly stable, excitation of instabilities was less common. That statistical aspect of combustion instabilities has still not been satisfactorily treated and in fact has received little attention, a serious deficiency in the subject. However, some very interesting Russian work on that problem has been discussed in a recent symposium (First International Symposium on Liquid Rocket Instability, *Proceedings*, 1993).

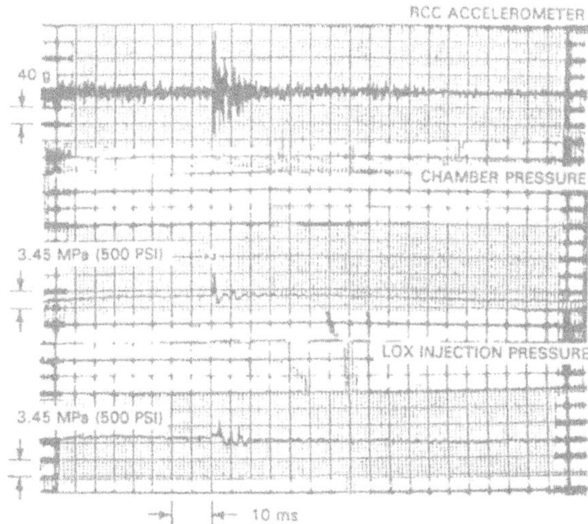

FIGURE 2.3

Attention then turned to the processes in the discharge region of the injector elements and to the flow downstream of the injector face. The sensitivity of impinging jets was recognized as a fundamental mechanism, a conclusion based on the work of several organizations, including the research group at Princeton University, led by Crocco. Also, the basic importance of the location of the combustion zone was established. Tests with design changes to produce larger fuel drops and higher relative velocity between the drops and the gaseous oxygen showed that the zone of vigorous burning moved downstream and improved stability. This part of the program also showed beyond question that baffles were necessary for dynamic (nonlinear) stability.

The main purpose of the program after June 1963 was to improve dynamic stability as indicated by the decay rate following the detonation of a 13.5-grain explosive charge. Attention was focused largely on influences of processes associated with LOX injection; and on the causes of resurging. Design changes were for the most part consistent with, or extensions of, understanding that had been gained earlier, i.e. the favorable influences of (1) large fuel drops and relative velocity between the fuel drops and gaseous oxidizer; (2) reduced sensitivity of impinging jets and spray fans; (3) uniformity of the distribution of vaporized fuel and oxidizers; and (4) combustion zone not too close to the injector face. It was during this time that the problem of resurging was practically (not completely) solved by reducing the rate of film cooling of the baffles.

Although the injector existing at the end of 1964 gave acceptable performance, development of the flight-qualified injector continued from January 1965 to September 1966 to obtain better stability. The goals were to improve and make consistent the decay of oscillations induced by 'bombing' and to improve performance. Buzzing was never completely eliminated, but the damping processes were sufficiently enhanced to give consistent attenuation of the oscillations to essentially noise level within 45 milliseconds (roughly 20–25 cycles). The final injector produced specific impulse of 265.4 seconds and characteristic velocity efficiency equal to 93.8%, confirmed with 703 full-scale engine tests. Figure 2.3 shows test results typical of the behavior following 'bombing.'

2.2 RESULTS FOR COMBUSTION STABILITY OF THE LUNAR MODULE DESCENT ENGINE (LMDE)

The Apollo LMDE operated with the storable propellants N_2O_4 oxidizer and A-50 fuel. Owing to the stringent requirements for descent to landing on the Moon, the engine had to be throttleable, ultimately over the range of 1,000 pounds to 9,850 pounds thrust, with maximum single firing time of 1,000 seconds, but with three starts required. Essentially no combustion instability could be tolerated in the entire range of operation. The TR 201 engine, conceived first by G. Elverum at the Jet Propulsion Laboratory and later developed and produced by TRW, successfully met all the requirements and remains one of the most remarkable liquid rocket engines ever used.

Figure 2.4 shows the general configuration of the engine and a sketch of the pintle injector, the crucial item for its success. Fuel is injected as a cylindrical tubular sheet along the axis and oxidizer is injected radially as a number of jets near the end of the moveable sleeve. At first acquaintance, one might expect this configuration to have poor performance, but that is not the case. The specific impulse exceeds 300 seconds above 3,000 pounds thrust. What is equally remarkable is that even during severe tests of dynamic stability by 'bombing,' the engine "has never experienced a case of sustained high-frequency instability," a statement referring to the LMDE operated within its design envelope. Although the engine has never been subject to a comprehensive analysis of its stability, it appears that the reason for its very stable behavior is that speculated during the development program and illustrated in Figure 2.5.

The argument rests on the relative distributions of pressure and combustion energy release. Owing to the way in which the liquid fuel and oxidizer are injected, the energy release tends to be most concentrated in an annular region between the axis and the chamber boundary, and roughly midway between the head end and the nozzle. The central axial location is unfavorable to exciting the fundamental longitudinal mode (and in fact all odd modes), which is additionally highly damped by the exhaust nozzle. Moreover, the intermediate radial position is best for least destabilizing influence on the first radial and first tangential modes.

In view of the remarks in the Introduction, one cannot expect one engine to be absolutely stable

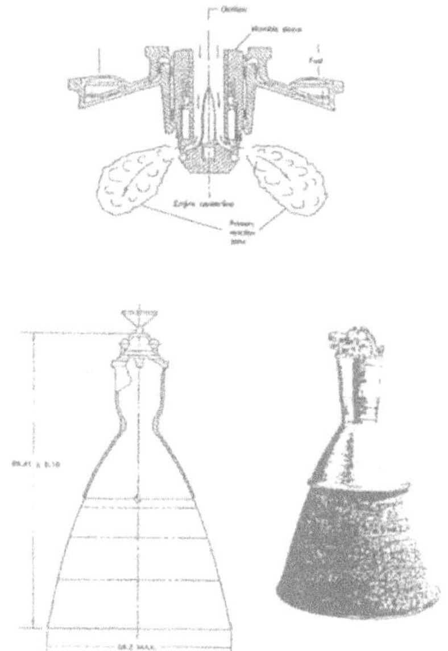

FIGURE 2.4

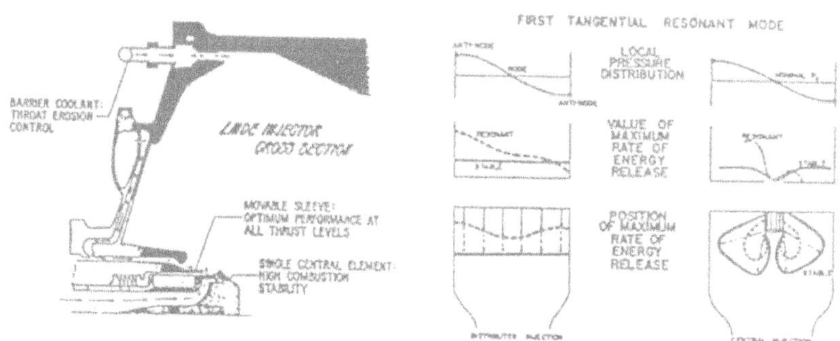

FIGURE 2.5

under all possible conditions. The TR 201 did show transient oscillations, particularly when rapidly throttled, or at full thrust with a fuel-rich mixture, or under some conditions of partial failure. Moreover, the envelope for stable operation changes when different propellants are used. However, it is fair to state that the engine is stable over a wide range of conditions for which it was qualified. The method of bombing to assess dynamic stability will be discussed briefly in Section 4.

2.3 COMBUSTION INSTABILITIES IN THE MINUTEMAN II STAGE 3 SOLID ROCKET MOTOR

This example has been documented in a group of papers presented in 1971 (Bergman and Jessen 1971; Browning and Krashin 1971; Fowler and Rosenthal 1971) although, owing to the obvious sensitivity of the matter, the problem occurred much earlier. The circumstances were briefly the following.

While production of the motors was still in progress, beginning with Lot 1-11 of the propellant manufacturing, three flight failures occurred, one because of an electronic failure, and two because a hydraulic valve in the thrust vector control system failed. The causes were traced to high-amplitude ascillatory accelerations having frequency around 500 hertz. Examination of flight and static test records showed the presence of oscillations of chamber pressure at the same frequency. Figure 2.6 shows a flight record of chamber pressure; the pressure oscillations are revealed as the broadening of the trace.

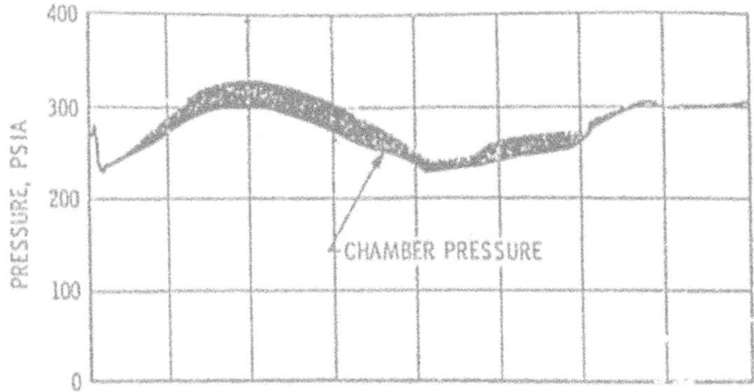

FIGURE 2.6

Investigation of the available data showed that the pressure oscillations in fact existed throughout the production program. However, prior to Lot 1-11, the amplitudes were lower than those often present after Lot 9. Figure 2.7 shows a few examples illustrating that conclusion.

The problems with structural vibrations occurred in the period 5–15 seconds of the firing when 0–peak pressure amplitudes were usually greater than 10 psi and occasionally much greater, evidently the cause for failures. Note that the time-history of the amplitudes tended to be a bit more reproducible for the motors fabricated with propellant from Lot 1-10 (and earlier). Figure 2.8 shows processed data for two motors, one successfully fired, and one which failed in flight.

The reason for this peculiar change of behavior during the manufacturing program was ultimately attributed to the fact that the source of aluminum used as an additive was changed between Lots 1-10 and 1-11. Apparently due to somewhat different initial content of oxide, and possibly other characteristics, the combustion properties were different. As a result, the size of oxidizer particles (Al_2O_3) produced in combustion changed. That in turn reduced the rate of dissipation of acoustic energy associated with gas/particle interactions. Consequently, as described in Section 9, the limiting values of the amplitudes increase, roughly as the square root of the dissipation rate.

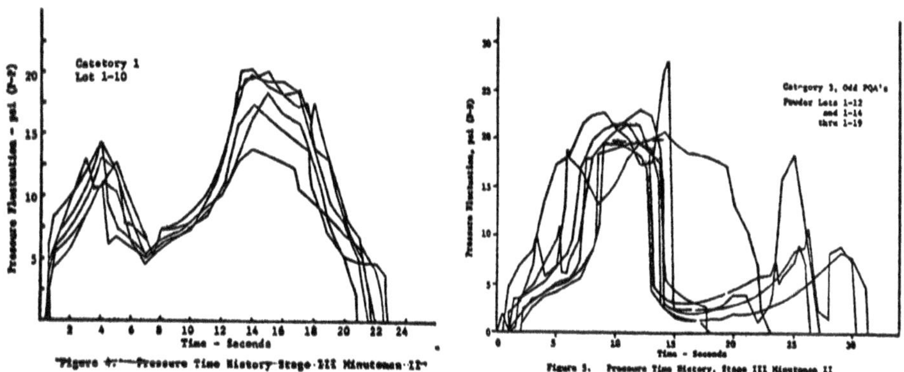

FIGURE 2.7

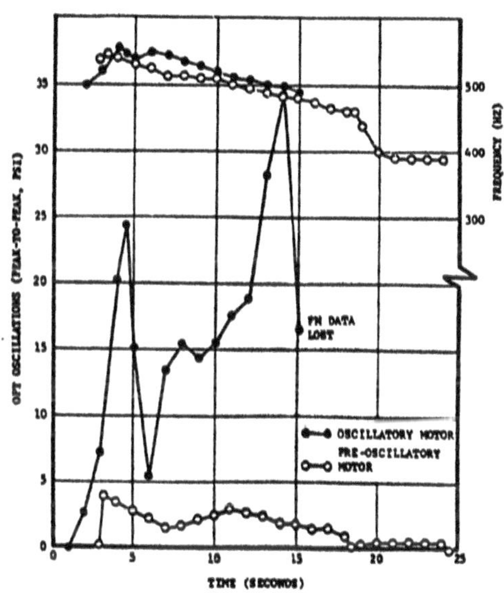

FIGURE 2.8

Perhaps the most important generic lesson to be learned from this example is that harmless oscillations may unexpectedly become harmful. Hence in general, even low-level oscillations should not be ignored. They should be attended to at least to the extent that one has confidence that their amplitudes will unlikely exceed an acceptable value for the system in question.

3. Stability Rating

As interesting and challenging as understanding combustion instabilities is, the eventual purpose in practice is to guarantee their nonexistence. Theory, analysis, and laboratory work should establish the basic laws governing the unsteady behavior: scaling laws, dependence on geometry, and other parameters defining the system, generally representing as thoroughly as possible linear and nonlinear behavior. If the task could be carried out to completion, then the basis would exist for minimum development costs of stable engines, with test programs planned mainly to confirm expected performance and stability. The actual case is far from the ideal and important testing procedures devised to establish stability characteristics of engines remain an important — and expensive — part of any development program. Those procedures define the process called 'stability rating.'

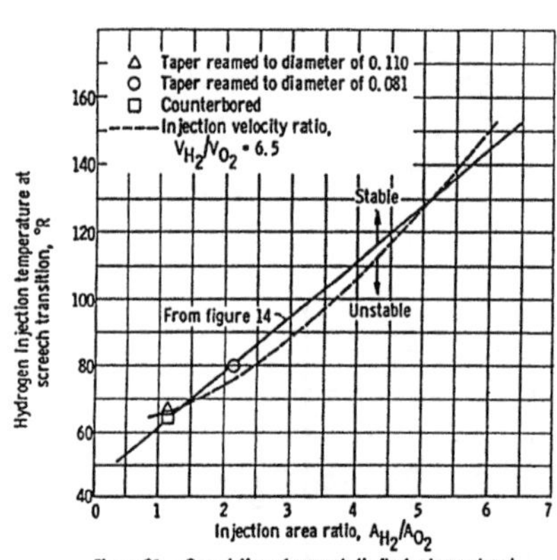

Figure 20. - Correlation of screech limits for tapered and counterbored oxidizer-tube configurations with area ratio and injection velocity ratio. Oxidant-fuel ratio, 5.0.

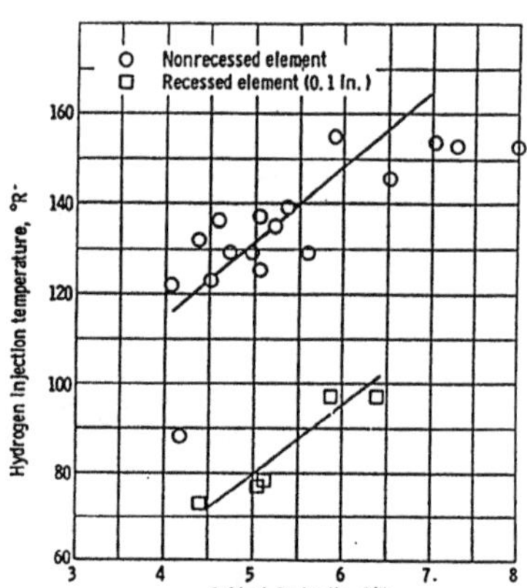

Figure 21. - Effect of oxidizer-tube recess on variation of screech transition temperature. Oxygen area, 0.89 square inch; hydrogen area, 4.84 square inches.

FIGURE 3.1

As in much of this subject, NASA report SP-194 (Harrje and Reardon 1972) remains the most complete reference for stability. Little seems to have changed in the past 20 years except for the improvements in instrumentation, data acquisition, and data processing. Although there are other methods used to investigate special aspects of stability (e.g. oscillations involving the propellant supply system), we shall discuss briefly here only the two basic methods used routinely to determine the stability of high-frequency wave modes in a chamber. 'Bombing' a chamber by initiating small explosive charges is the older of the two, having been introduced in the 1950s; the second method is called 'temperature ramping' in which the temperature of one of the propellants, usually the fuel, is increased more or less linearly in time until an unstable motion develops spontaneously, or, less commonly, the transient following a small explosion becomes a sustained oscillation. Because there is ample guidance available for applying these methods, nothing is added here by elaborating details. Rather, we shall only try to provide a general context with examples of observations.

In general, raising the inlet temperature of the injected propellant (e.g. hydrogen) promotes stability, for reasons not totally understood. Various proposals have been offered for the behavior, such as the change of density with temperature, change of pressure drop across the injector and 'injector coupling,' a term that normally refers to participation by the acoustical system in the injector elements and possibly as well the manifolds. Figure 3.1 shows some typical data, taken from Wanhainen *et al.* (1966).

Stability rating for 'bombing' has been more widely used than temperature ramping and in particular was the standard means for qualifying engines in the Apollo program. The arrangement of explosive charges and pressure transducers in the LMDE are shown in Figure 3.2. Some typical

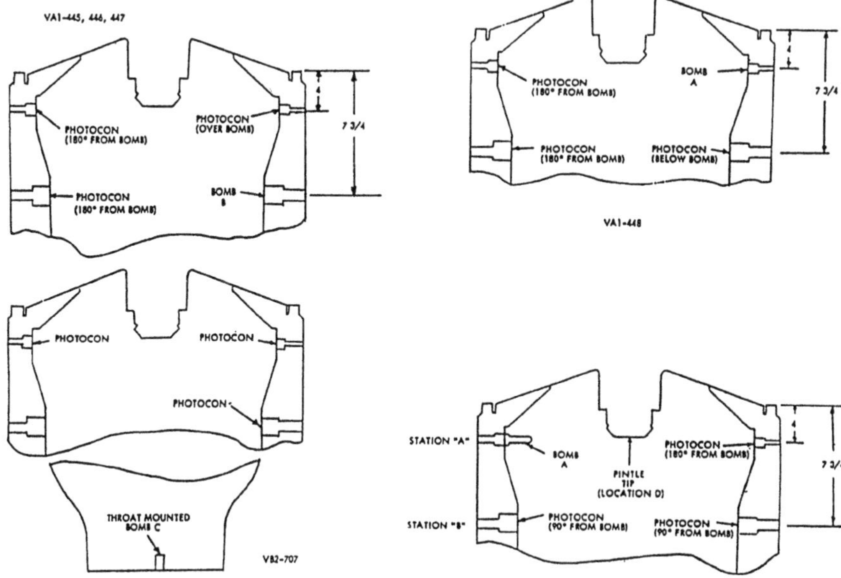

FIGURE 3.2

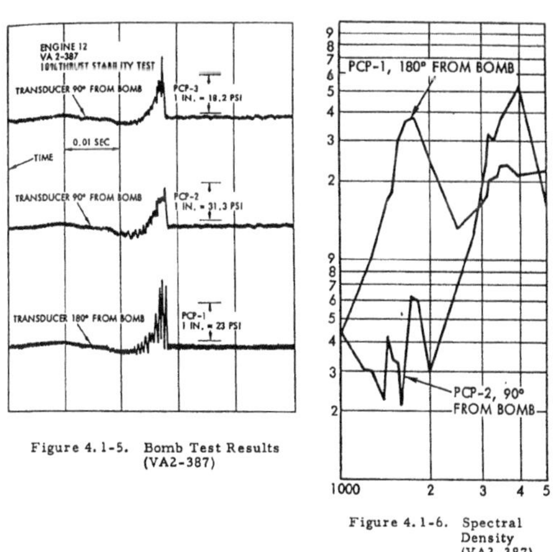

Figure 4.1-5. Bomb Test Results (VA2-387)

Figure 4.1-6. Spectral Density (VA2-387)

FIGURE 3.3

results are reproduced in Figure 3.3; both figures have been taken from an internal TRW document supplied by Dr. J. Miller to the author (Anonymous 1967).

4. Fundamental Processes and Mechanisms of Combustion Instabilities

As a practical matter, the most important aspect of treating a combustion instability is determining the mechanism. Once that has been accomplished, the matter of affecting the mechanism is such a fashion as to reduce or eliminate the instability can be addressed. Although the strategy is simply stated, it is often neither easy to identify the crucial mechanism nor easy or in some cases possible, to modify the mechanism in the desired fashion. That is essentially why practical means of control are attractive, as discussed in other papers in this program. It is generally true, whether or not a control scheme is to be used, that treatment of an instability is more effective the better understood the mechanism. Thus in that sense the subject of this section is truly fundamental to the entire subject of combustion instabilities.

4.1 FUNDAMENTAL PROCESSES IN LIQUID ROCKET ENGINES

The purpose of the injection system and combustion chamber in a liquid rocket engine is to accomplish controllably the conversion of liquid propellants to product gases at high temperature and pressure. Thrust is then produced by expansion through the exhaust nozzle, transforming potential and thermal energy into kinetic energy. Although the characteristics of the flow in the nozzle are significant in determining the overall efficiency of the engine, we are concerned here chiefly with unsteady processes upstream of the nozzle entrance. The only significant contribution of the nozzle to combustion instabilities arises from its position, setting the downstream boundary conditions on waves in the chamber; its most important influence is attenuating longitudinal oscillations.

Figure 4.1 is a broad summary of the various elementary processes. It is convenient to display them in serial form, roughly corresponding to the sequence of events from propellant supply to exhaust. However, it is important to realize that several different processes may take place simultaneously in a given region of space.

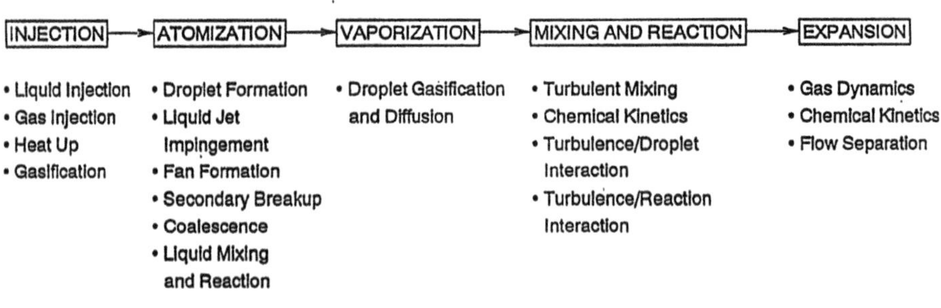

FIGURE 4.1

Several papers in a recent symposium (First International Symposium on Liquid Rocket Engine

Instability, *Proceedings*, 1993) and this program give good summaries of past work in this subject, the state of understanding presently, and reports of research in progress. There is therefore no need here to dwell on details. We shall rather try to clarify the context within which the elementary processes may act as mechanisms for exciting and sustaining combustion instabilities. The discussion, and in fact any consideration of fundamental processes, cannot be completely general in a practical sense; for a given engine, there are inevitably characteristics peculiar to the design. For example, we have seen in the case of the F-1 engine that accumulation of liquid fuel during film cooling of baffles was responsible for an unacceptable 'spiking' unsteady motion.

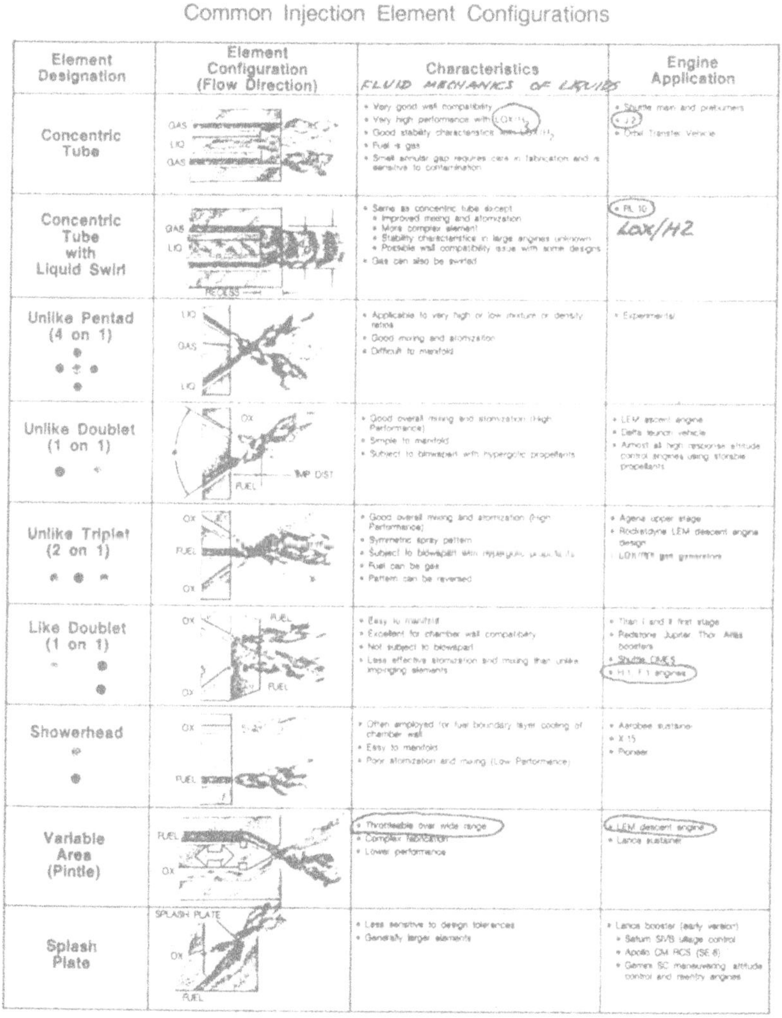

Common Injection Element Configurations

Element Designation	Element Configuration (Flow Direction)	Characteristics FLUID MECHANICS OF LIQUIDS	Engine Application
Concentric Tube	GAS, LIQ, GAS	• Very good wall compatibility • Very high performance with LOX/H$_2$ • Good stability characteristics with LOX/H$_2$ • Fuel is gas • Small annular gap requires care in fabrication and is sensitive to contamination	• Shuttle main and preburners • J 2 • Orbit Transfer Vehicle
Concentric Tube with Liquid Swirl	GAS, LIQ, RECESS	• Same as concentric tube except • Improved mixing and atomization • More complex element • Stability characteristics in large engines unknown • Possible wall compatibility issue with some designs • Gas can also be swirled	• RL 10 LOX/H2
Unlike Pentad (4 on 1)	LIQ, GAS, LIQ	• Applicable to very high or low mixture or density ratios • Good mixing and atomization • Difficult to manifold	• Experimental
Unlike Doublet (1 on 1)	OX, FUEL, IMP DIST	• Good overall mixing and atomization (High Performance) • Simple to manifold • Subject to blowapart with hypergolic propellants	• LEM ascent engine • Delta launch vehicle • Almost all high response attitude control engines using storable propellants
Unlike Triplet (2 on 1)	OX, FUEL, OX	• Good overall mixing and atomization (High Performance) • Symmetric spray pattern • Subject to blowapart with hypergolic propellants • Fuel can be gas • Pattern can be reversed	• Agena upper stage • Rocketdyne LEM descent engine design • LOX/RP gas generators
Like Doublet (1 on 1)	OX, FUEL, OX, FUEL	• Easy to manifold • Excellent for chamber wall compatibility • Not subject to blowapart • Less effective atomization and mixing than unlike impinging elements	• Titan I and II first stage • Redstone Jupiter Thor Atlas boosters • Shuttle OMES • H 1, F 1 engines
Showerhead	OX, FUEL	• Often employed for fuel boundary layer cooling of chamber wall • Easy to manifold • Poor atomization and mixing (Low Performance)	• Aerobee sustainer • X 15 • Pioneer
Variable Area (Pintle)	FUEL, OX	• Throttleable over wide range • Complex fabrication • Lower performance	• LEM descent engine • Lance sustainer
Splash Plate	SPLASH PLATE, OX, FUEL	• Less sensitive to design tolerances • Generally larger elements	• Lance booster (early version) • Saturn S IVB ullage control • Apollo CM RCS (SE 8) • Gemini SC maneuvering attitude control and reentry engines

FIGURE 4.2

To bring some order to the possible importance of the many processes shown in Figure 4.1, it is helpful to overlay a classification according to injector types. Figure 4.2, taken from Jaqua and Ferrenberg (1989), shows most of the commonly used injector configurations. For the purposes here, we lump all possibilities into three classes distinguished by the configurations of the liquids when they are discharged by the injection elements into the nozzle:

(1) impinging jets
(2) concentric jets
(3) jet/sheet (non-coaxial)

The last form arises notably in the TRW Lunar Module Descent Engine. Little is known about the fundamental processes involved in this injector although there is research in progress. Because the resistance of this engine to instability is well-established, but the reasons are not understood, this is an important subject for continuing work.

We have already discussed the F-1 engine using injectors based on impinging jets; as noted in Figure 4.2, other examples include the lunar ascent engine, several launch vehicles, and the Shuttle Orbital Manoeuvering Engine (Aerojet). Notable applications of concentric injectors include the RL-10 (Pratt and Whitney) and the Space Shuttle Main Engine (Rocketdyne). In this symposium, problems of impinging jets have recently been discussed by Santoro (1993); concentric or coaxial injection processes have been covered by Zaller (1993) and by Vingert *at al.* (1993), which includes an excellent review; all of these papers appear in the *Proceedings* of the First International Symposium on Liquid Rocket Instability, 1993. There seems to be no work providing a basis for assessing quantitatively the relative stability of the various configurations. In this regard, there has been a suggestion in work not available at the time this paper was prepared, that the Soviets regarded injectors with impinging jets as more prone, than those using coaxial jets, to causing combustion instabilities, even with hydrocarbon fuels. We have been unable to establish definitely that conclusion.

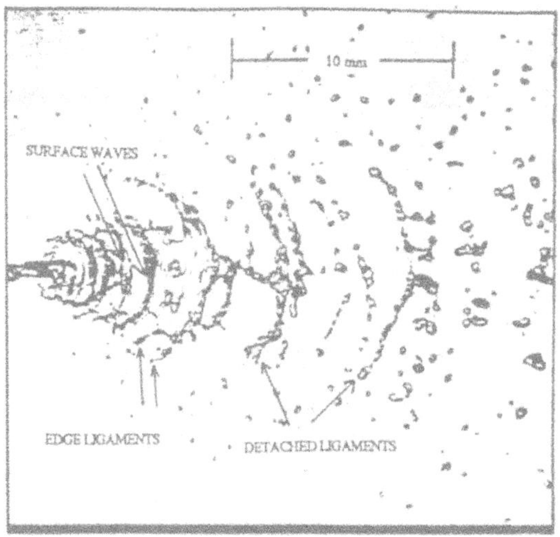

FIGURE 4.3

Figure 4.3 (Santoro *et al.* 1992) is a photograph showing one sort of situation for a case of impinging liquid jets. Immediately following impingement, a spray fan or sheet is formed. Either because of disturbances already growing in the jets, or because the sheet is itself unstable (possibly for both reasons), the sheet breaks up. There are several forms of the break-up, but in any event a cloud of liquid drops or droplets is formed, having a fairly broad distribution of size and velocities. Because there is a surrounding atmosphere of moving gases, the sheet and the drops are subject to shear forces that encourage break-up of both. In a combustion chamber, these processes take place in an atmosphere consisting of combustion product gases and reactant gases as well. In the F-1 engine, for example, the liquid oxygen vaporized much more rapidly than the RP-1. Under these circumstances, the picture presented by a test involving, say, two jets of water at room temperature, does not reproduce faithfully the situation in the chamber.

It seems accurate to state that the greater part of what is known about the events in which liquid jets are transformed to drops has been learned from tests with liquids at room temperature and with no combustion. While it is true that such tests are necessary, valuable, and do produce a great deal of useful information, one must never forget that the situation may be drastically different when combustion occurs. Heat transfer upstream and pressure disturbances generated by combustion of individual drops may have strong effects on the jets and spray sheets before and during break-up. For example, with hypergolic propellants, 'blow-a-part' of the reactive streams and 'popping' or intermittent explosion of liquid may be sufficiently vigorous to excite instabilities.

Whatever may be the environment, the first step in the transformation of liquid jets to drops must involve some sort of instability. For the case of impinging jets, the jets themselves may be unstable (a single jet breaks into drops) or the spray fan is unstable. Concentric jets are unstable at their interface. The basic processes responsible for the instabilities are characterized by various dimensionless groups that provide scaling laws (see Vingert *et al.* and Wu *et al.* in this symposium for particularly good discussions of scaling).

Determining the distributions of drop size and velocity theoretically is an extremely difficult problem, more so than the matter of incipient instability. In general, the influences of aerodynamic forces on the break-up of sprays and secondary break-up of drops is so complicated as to be beyond realistic analysis.

Combustion occurs in the gaseous phase and must be preceded by mixing. In a LOX/ hydrocarbon system, such as in the F-1, vaporization of the liquid oxygen occurs much more rapidly than that of the liquid fuel. Hence a significant part of the spray combustion involves vaporization and subsequent combustion of the fuel in an oxygen-rich atmosphere. It appears that some of the consequences of detailed design changes on the injector elements may be explained ultimately by the large difference in vaporization rates. That characteristic must also have much to do with differences in the stability of concentric injectors compared with impinging jets when the same propellant combination is used.

The situation for a cryogenic system is less clear, the vaporization rates of LOX and LH2 being closer than those of LOX and a hydrocarbon fuel. Available information is insufficient to allow general characterization.

All the processes discussed to this point have to do mainly with questions of fluid mechanics, or hydraulics, and depend primarily on the physical properties of density, viscosity, surface tension and, to a lesser extent, thermal conductivity. Most of the significant phenomena can be well understood by considering a single, pure liquid (thereby excluding, of course, combustion) and without wide variations of temperature and pressure. Thus, experiments at room temperature and pressure with various inert liquids provide a great part of the fundamental data required and in fact much work remains before the subjects of jet break-up, jet impingement and formation of sheets, disintegration of sheets, and the formation of drops from sheets can be considered well-understood.

While burning of single drops has been actively studied for many decades and is quite well understood, spray combustion, particularly for dense clouds of interacting drops under conditions found in a rocket chamber, is very poorly understood. Only ad hoc modeling is available for unsteady combustion of sprays.

In addition to the purely fluid-mechanical and chemical (i.e. combustion) problems, serious further complications arise in combustion chambers operated at high pressure, namely those arising from behavior near the critical point. This is not a new subject — see, for example, reviews by Faeth (1977), Law (1982), and Hsieh *et al.* (1991) — but thorough analysis for conditions appropriate to combustion instabilities in liquid rocket engines is recent and unfinished.

The fundamental origin of the special behavior accompanying combustion near the critical point is the rapid variation of thermodynamic properties, particularly the mass diffusivity and the latent heat of vaporization. For a given substance or mixture of substances, the pressure p_{cr} and the temperature τ_{cr} at the critical point are well-defined. However, the special characteristics of behavior at the critical point itself, extend, albeit attenuated, in some region about the critical point. Hence it is quite clearly unnecessary that the chamber pressure and temperature be precisely p_c and τ_c to cause unusual phenomena during combustion. Moreover, at a given pressure, the temperature near a vaporizing or burning liquid drop varies from the ambient value in the immediate environment to some value within the liquid.

To be specific, consider a liquid drop of oxygen injected as liquid, and hence, having temperature below the critical point, into a hot fuel-rich environment, the simplest case being hot hydrogen. Suppose that the chamber pressure is the critical value for some mixture of the fuel and oxidizer. As the drop follows its trajectory, its temperature rises, oxygen vaporizes, and gaseous hydrogen diffuses inward. Somewhere between the center of the drop and the environment far away, the temperature and pressure may, if conditions are suitable, assume the critical values for the local mixture ratio of oxygen and hydrogen. For a spherically symmetric flow, the critical point is reached on a spherical surface. Locally, the sharp variations of properties with temperature will produce drastic changes in the flow field at and near that surface. Globally, those effects have substantial influence on the steady burning rate.

If oscillations of pressure and temperature occur in the environment, then it is possible, even though the average chamber pressure may be far from critical, that critical conditions are reached in some point of the drop under unsteady conditions. Then transient fluctuations can be amplified by the large fluctuations of thermodynamic properties. The result is a new possibility for coupling between acoustical motions in the chamber and droplet burning.

Figure 4.4 shows an example of this behavior for transient vaporization of hydrocarbon fuel droplets into an atmosphere at supercritical pressure (80 bar) and at several subcritical pressures. Obvious is the substantial increase of vaporization rate when the pressure is supercritical. While this is a potential contributing mechanism of instabilities, there is no data supporting this result for full-scale engines or for laboratory tests.

4.2 FUNDAMENTAL PROCESSES IN SOLID ROCKET MOTORS

In some respects the simplest instabilities to understand are those arising in solid propellant rocket motors. Under almost all conditions in practical systems, the combustion processes are confined to a thin (less than one millimeter) region adjacent to the burning surface. For most geometeries flow separation is not a serious issue and it is has long been known that instabilities are excited and sustained by unsteady energy release in the combustion zone. Simple models of the process have provided crude approximations to observed behavior. Measurements in laboratory devices have supported certain important aspects of the predicted behavior, but two serious deficiencies remain: it is not possible to predict quantitatively with confidence the tendency for a given propellant to drive instabilities; and all known methods of measurement are both expensive and burdened by substantial uncertainties in the results.

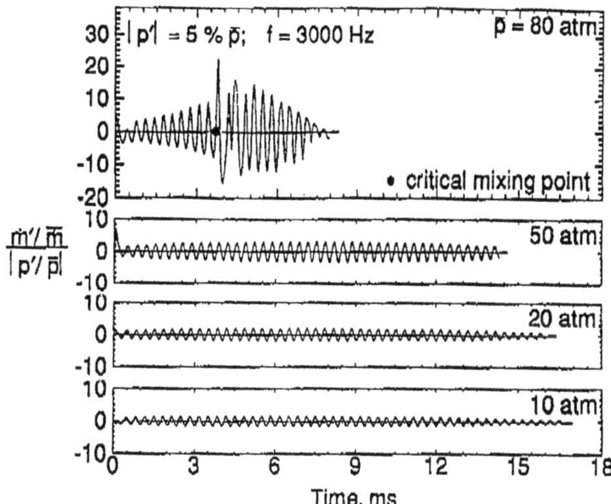

FIGURE 4.4

The problem comes down to determining the response or admittance functions. In physical terms, waves are driven because $p - v$ work done on the gas in the chamber by the material issuing from the surface is non-zero in one cycle of the motion. Averaged over the surface, that process expresses the content of the surface integral in equation (5.18). The pressure fluctuation is approximately proportional to ψ_n, and $A_b^{(r)}\psi_n$ is proportional to that part of the velocity fluctuation normal to the surface and in phase with the pressure; the contribution $\overline{M}_b\psi_n^2$ represents convection of acoustic energy into the chamber. Hence the theoretical problem consists in computing the velocity fluctuation. It happens to be more convenient and direct to determine the fluctuation of mass flux, proportional to $R_b\psi_n$.

A simple model of the burning solid suffices to capture much of the gross behavior. As illustrated in Figure 4.5, a one-dimensional approximation is adequate to explain the main features. A reference system fixed to the average position of the interface between the solid and gas phases is used. Solid material advances to the interface at the linear burning rate $\overline{r}$ and is converted to gas at $x = 0$. Within the one-dimensional approximation the interface is assumed to be flat. This is only a crude approximation for actual propellants which normally contain substantial amounts of oxidizer particles (as large as several hundred microns) and aluminum particles normally tens of microns or less in diameter. Photomicrographs and films have shown that burning surfaces are certainly not flat, so we are adopting here a spacially averaged picture. At some distance from the surface, the gases react and burn in a flame. Heat flowing back to the surface from the flame zone is the cause for raising the temperature of the solid to the value required for phase transition at the interface. There is therefore a fairly thin zone near the surface, in which the temperature decays, roughly exponentially, to its value in the cold solid.

A fairly straightforward calculation shows that the characteristic time is $\overline{r}^2/\kappa$ where κ is the thermal diffusivity of the solid material. Hence the proper dimensionless parameter characterizing oscillatory behavior is $\Omega = \omega\overline{r}^2/\kappa$.

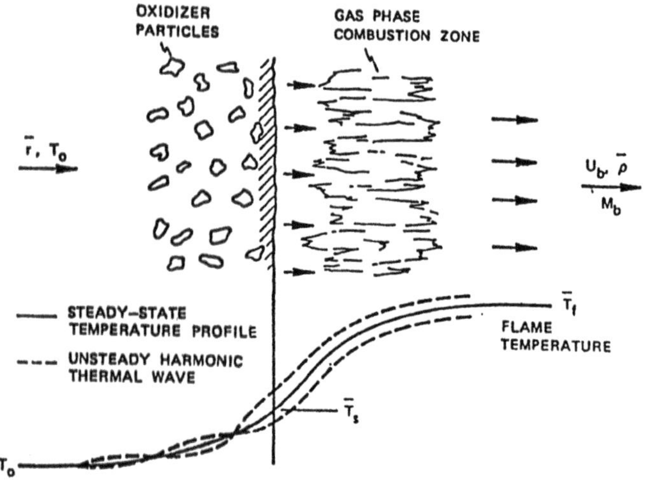

FIGURE 4.5

Culick (1968) reviewed various models based on this view, showing that with the approximations just described, all results produce the same two-parameters (A, B) form for the response function:

$$R_b = n \frac{AB}{\lambda + \frac{A}{\lambda} - (1 + A) + AB} \tag{4.1}$$

where n is the index in the burning rate law, $r \sim p^n$ and λ the complex function of frequency

$$\lambda = \frac{1}{2} + \frac{1}{2}\sqrt{1 + i4\Omega} \tag{4.2}$$

Figure 4.5 shows the real and imaginary parts of R_b for typical values of A and B. It is the real point that enters in the formula for the growth constant, $A_b^{(r)} + \overline{M}_b \approx \overline{M}_b R_b^{(r)}$ if the term $\Delta \hat{T}_s / \overline{T}_s$ is neglected. The value of $\Delta \hat{T}_s / \overline{T}_s$ is often small, but depends on the model of the flame; this matter has received little detailed attention.

The most important feature of Figure 4.6 is the broad peak in the real part of the response, generally falling in the range of a few hundred hertz to perhaps two kilohertz or so. That covers the frequencies of the lowest modes in virtually all solid rockets and is evidently a major reason for combustion instabilities. Thus a major part of the experimental work in this subject has been devoted to measuring the response function (Culick 1974; Brown, Culick, and Zinn 1978). Success with quantitative results has been limited and extremely difficult to achieve. Nevertheless, effective methods have been devised for qualitative comparison of different propellants and for detecting trends of behavior due to changes of composition. There are standard procedures followed in development of new propellants and motors. An extensive body of literature exists on this subject, the best source being the proceedings of the annual JANNAF Combustion Meetings.

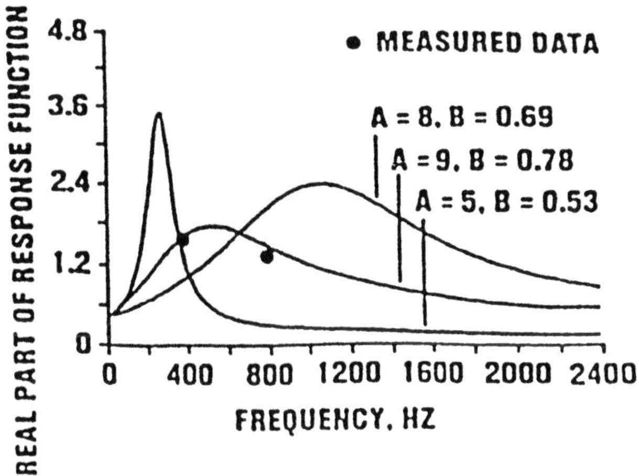

FIGURE 4.6

5. Formulation of an Approximate Analysis

The analysis described in this section has evolved from early work on linear combustion instabilities (Culick 1963, 1964) to extensions to accommodate nonlinear behavior (Culick 1971, 1975; Culick and Yang 1992; Culick 1993; Jahnke and Culick 1993). Thus the discussion here is quite parochial, although in Section 8 we give a brief review of the important earlier works on nonlinear combustion instabilities. The material provided here also appears in Chapter 4 of the proceedings of the ESA/ESTEC course (Schoeyer 1993).

5.1 CONSERVATION EQUATIONS FOR TWO-PHASE FLOW

Although the instabilities observed in combustion chambers involve primarily oscillations of the gas phase, it is essential to account for the presence of the liquid phase as well. To simplify the discussion, we lump the liquid fuel and oxidizer together as a single liquid phase and represent the multi-component gas mixture as a single average gas. Thus we treat a reacting two-phase mixture for which the equations of conservation are:

Conservation of Mass

$$\frac{\partial}{\partial t}(\rho_g + \rho_\ell) + \nabla \cdot (\rho_g \vec{u}_g + \rho_\ell \vec{u}_\ell) = w_{eg} + w_{e\ell} \tag{5.1}$$

Conservation of Momentum

$$\frac{\partial}{\partial t}(\rho_g \vec{u}_g + \rho_\ell \vec{u}_\ell) + \nabla \cdot (\rho_g \vec{u}_g \vec{u}_g + \rho_p \vec{u}_\ell \vec{u}_\ell) = \nabla \cdot \overleftrightarrow{\tau}_g + \vec{m}_{eg} + \vec{m}_{e\ell} \tag{5.2}$$

Conservation of Energy

$$\frac{\partial}{\partial t}(\rho_g e_{g_o} + \rho_\ell e_{\ell_o}) + \nabla\cdot(\rho_g \vec{u}_g \vec{u}_\ell e_{\ell_o}) = \nabla\cdot(\overleftrightarrow{\tau}_g \cdot \vec{u}_g) - \nabla\cdot\vec{q}_g(Q + Q_e) \tag{5.3}$$

Subscript $(\ )_g$ refers to the gas species and subscript $(\ )_\ell$ to the liquid; subscript $(\ )_o$ denotes the stagnation state. External sources are denoted by subscript $(\ )_e$. Those terms do not normally appear in the literature but are included here because they may be used to represent the presence of active control, a subject to be addressed briefly later. The stress tensor $\overleftrightarrow{\tau}_g$ is assumed to depend on properties of the gas phase and may be written as the sum of the isotropic pressure and the viscous stress tensor $\overleftrightarrow{\sigma}_v$:

$$\overleftrightarrow{\tau}_g = -p\overleftrightarrow{I} + \overleftrightarrow{\sigma}_v \tag{5.4}$$

All forces exerted in the flow by external influences are represented by the momentum sources m_{eg} and $m_{e\ell}$. Distributed forces (gravity, electromagnetic,...) are not an issue in the problems considered here; non-zero values of m_{eg} and $m_{e\ell}$ will arise due to momentum transfer associated with flow of material through the boundary of the combustion chamber and interactions associated with material injected for actuating control. Internal heat flow is represented by $\vec{q}_g$, sufficiently well approximated by Fick's law for the gas phase, and Q represents heat addition in the gas phase associated with combustion processes not accompanying conversion of condensed phase to gas. Hence the symbol e_{g_o} stands for stagnation thermal energy only, containing no part related to chemical processes. For a mixture of perfect gases, each species having constant specific heat,

$$e_{g_o} = C_v T_g + \frac{1}{2}u_g^2 \tag{5.5}$$

where C_v and $\vec{u}_g$ strictly stand for values mass-averaged over all gaseous species.

Let w_ℓ denote the rate at which the liquid phase is converted to gas and Eq. (4.28) can be written as the sum of the two equations:

Conservation of Mass (Gas Phase)

$$\frac{\partial \rho_g}{\partial t} + \nabla\cdot(\rho_g \vec{u}_g) = w_\ell + w_{e_g} \tag{5.6}$$

Conservation of Mass (Liquid Phase)

$$\frac{\partial \rho_\ell}{\partial t} + \nabla\cdot(\rho_\ell \vec{u}_\ell) = -w_\ell + w_{e\ell} \tag{5.7}$$

The following manipulations are directed to writing eventually a nonlinear wave equation governing the behavior of waves in the mixture. Elasticity (the 'spring constant') for wave propagation is provided by the compressibility of the gas only, while the mass of an elementary oscillator in the medium is the sum of the masses of gas and liquid in a unit volume.

Equation (4.28), the equation for the total density of the mixture, can be rewritten

$$\frac{\partial \rho}{\partial t} + \nabla\cdot(\rho\vec{u}) = \mathcal{W} \tag{5.8}$$

where $\mathcal{W} = -\nabla\cdot(\rho_e \delta\vec{u}_\ell) + w_e$, the total external source of mass, is

$$w_e = w_{eg} + w_{e\ell} \tag{5.9}$$

and the slip velocity between the condensed phase and the gas is

$$\delta\vec{u}_\ell = \vec{u}_\ell - \vec{u}_g \tag{5.10}$$

Now expand the momentum equation, Eq. (4.2), and substitute the definitions to give

Conservation of Momentum (Gas Phase)

$$\rho_g \frac{\partial \vec{u}_g}{\partial t} + \rho_g \vec{u}_g \cdot \nabla \vec{u}_g + \nabla p = \nabla \cdot \overleftrightarrow{\sigma}_r + \vec{F}_\ell - (\vec{\sigma} + \vec{\sigma}_e) + \vec{m}_e \tag{5.11}$$

where

$$\vec{F}_\ell = -\rho_\ell \left[\frac{\partial \vec{u}_\ell}{\partial t} + \vec{u}_\ell \cdot \nabla \vec{u}_\ell \right] \tag{5.12}$$

$$\vec{\sigma} = (\vec{u}_g - \vec{u}_\ell) w_\ell = -\delta \vec{u}_\ell w_\ell \tag{5.13}$$

$$\vec{\sigma}_e = \vec{u}_g w_{e\ell} + \vec{u}_\ell w_{e\ell} \tag{5.14}$$

$$\vec{m}_c = \vec{m}_{eg} + \vec{m}_{e\ell} \tag{5.15}$$

Equation (5.12) is the momentum equation for the condensed phase and σ represents the rate at which momentum is supplied *to* newly created gas phase *by* the gases already present. Hence $-\vec{\sigma}$ in (5.11) represents the force exerted on the gas phase by the vaporizing or evaporating condensed phase. To equation (5.11) add $\rho_\ell[\partial \vec{u}_g / \partial t + \vec{u}_g \cdot \nabla \vec{u}_g]$ to find

$$\rho \left[\frac{\partial \vec{u}_g}{\partial t} + \vec{u} \cdot \nabla \vec{u}_g \right] + \nabla p = \nabla \cdot \overleftrightarrow{\tau}_v + \delta \vec{F}_\ell - (\vec{\sigma} + \vec{\sigma}_e - \vec{m}_e) \tag{5.16}$$

where

$$\delta \vec{F}_\ell = -\rho_\ell \left[\frac{\partial \delta \vec{u}_\ell}{\partial t} + \delta \vec{u}_\ell \cdot \nabla \delta \vec{u}_\ell + \delta \vec{u}_\ell \cdot \nabla \vec{u}_g + \vec{u}_g \cdot \nabla \delta \vec{u}_\ell \right] \tag{5.17}$$

is the force of interaction between the condensed and gas phases.

More elaborate manipulations eventually lead to the form of the energy equation for the temperature of the gas phase:

$$\begin{aligned} \rho \bar{C}_v \left[\frac{\partial T_g}{\partial t} + \vec{u}_g \cdot \nabla T_g \right] + p \nabla \cdot \vec{u}_g &= (\overleftrightarrow{\sigma}_v \cdot \nabla) \vec{u}_g - \nabla \cdot \vec{q}_g + (Q + Q_e) + \vec{u}_g \cdot (\vec{\sigma} + \vec{\sigma}_e - \vec{m}_e) \\ &+ \delta Q_\ell + w_\ell \delta e_o + \delta \vec{u}_\ell \cdot \vec{F}_\ell (e_{g_o} w_{eg} + e_{\ell o} w_{e\ell}) \end{aligned} \tag{5.18}$$

where

$$\begin{aligned} Q_\ell &= -\rho_\ell \left[\frac{\partial e_\ell}{\partial t} + \vec{u}_\ell \cdot \nabla e_\ell \right] \\ \delta e_o &= e_{\ell_o} - e_{g_o} \end{aligned} \tag{5.19}$$

and, corresponding to (5.17), the heat exchange between the two phases is

$$\delta Q_\ell = -\rho_\ell C \left[\frac{\partial \delta T_\ell}{\partial t} + \delta \vec{u}_\ell \cdot \nabla \delta T_\ell + \delta \vec{u}_\ell \cdot \nabla T_g + \vec{u}_g \cdot \nabla \delta T_\ell \right] \tag{5.20}$$

The mass-averaged properties are

$$\begin{aligned} \bar{C}_v &= \frac{1}{\rho}(\rho_g C_v + \rho_\ell C) = \frac{C_v + C_m C}{1 + C_m} \\ \bar{C}_\ell &= \frac{1}{\rho}(\rho_g C_\ell + \rho_\ell C) = \frac{C_v + C_m C}{1 + C_m} \end{aligned} \tag{5.21)a,b}$$

and $C_m = \rho_\ell/\rho$ is the liquid phase loading, the fraction of mass in unit volume as liquid.

In summary, the three basic equations for unsteady motions in a two-phase mixture are:

$$\textbf{Conservation of Mass} \qquad \frac{D\rho}{Dt} = -\rho\nabla\cdot\vec{u} + \mathcal{W} + \mathcal{W}_e \tag{5.22}$$

$$\textbf{Conservation of Momentum} \qquad \rho\frac{D\vec{u}}{Dt} = -\nabla p + \vec{\mathcal{F}} + \mathcal{F}_e \tag{5.23}$$

$$\textbf{Conservation of Energy} \qquad \rho C_v\frac{DT}{Dt} = -\rho\nabla\cdot\vec{u} + \mathcal{Q} + \mathcal{Q}_e \tag{5.24}$$

where the sources are

$$\mathcal{W} + \mathcal{W}_e = -\nabla\cdot(\rho_\ell\delta\vec{u}_\ell) + w_e \tag{5.25}$$

$$\vec{\mathcal{F}} + \vec{\mathcal{F}}_e = \nabla\cdot\overleftrightarrow{\sigma}_v - \vec{\sigma} + \delta\vec{F}_\ell - (\sigma_e - \vec{m}_e) \tag{5.26}$$

$$\begin{aligned}\mathcal{Q} + \mathcal{Q}_e = {} & (\overleftrightarrow{\sigma}_v\cdot\nabla)\vec{u} - \nabla\cdot\vec{q}_g + Q + \vec{u}\cdot\vec{\sigma}_e + \delta Q_\ell + \delta e_o w_\ell + \delta\vec{u}_\ell\cdot\vec{F}_\ell \\ & + Q_e + \vec{u}\cdot(\vec{\sigma}_e - \vec{m}_e) - (e_{g_o}w_{e_g} + e_{\ell_o}w_{e_\ell})\end{aligned} \tag{5.27}$$

To simplify writing, we shall hereafter use the symbol $\vec{u}$ instead of $\vec{u}_g$ for the gas velocity; T instead of T_g for the gas temperature; and we drop the overbars on mass-averaged thermodynamic properties. The substantial derivative is defined in terms of the gas velocity,

$$\frac{D}{Dt} = \frac{\partial}{\partial t} + \vec{u}\cdot\nabla \tag{5.28}$$

The equation of state appropriate to this formulation is

$$p = \rho RT_g \tag{5.29}$$

where R is the mass-averaged gas constant $\bar{C}_p - \bar{C}_v$, equal to ρ_g/ρ times the gas constant for the gas mixture. From the preceding equations, the equation for the pressure can be derived:

$$\frac{Dp}{Dt} = -\gamma p\nabla\cdot\vec{u} + \mathcal{P} + \mathcal{P}_e \tag{5.30}$$

with

$$\mathcal{P} + \mathcal{P}_e = \frac{R}{C_v}Q - RT\nabla\cdot(\rho_p\delta\vec{u}_\ell) + p\frac{D\ln R}{Dt} + \left(\frac{R}{C_v}Q_eRTw_e\right) \tag{5.31}$$

Equations (5.22) – (5.30) and the equations of state (4.56) describe the motions of the two-phase mixture. Possibly with small modifications or extensions to accommodate many species, these are the basic equations for all analyses of combustion instabilities. To obtain solutions, in addition to setting boundary conditions, the condensed phase must be treated separately with Eqs. (5.7), (5.12), and (5.20), with suitable laws for the force and heat exchange between the two phases.

5.2 THE WAVE EQUATION FOR COMBUSTION INSTABILITIES

A wave equation may be constructed by analogy with a procedure followed in classical acoustics. Because we shall not deal with nonlinear behavior in the analysis described in this paper, we shall work from the beginning with the linearized forms of the governing equations. Write all variables as sums of mean and fluctuating parts, $p = \bar{p} + p'$, etc. and assume that the mean values are independent of time. Substitute in (5.23) and (5.30) and ignore terms of second order and higher in the fluctuations to find

$$\bar{\rho}\frac{\partial \vec{u}'}{\partial t} = -\nabla p' + \vec{\mathcal{F}}' + \vec{\mathcal{F}}'_e$$
$$\frac{\partial p'}{\partial t} = -\gamma\bar{p}\nabla\cdot\vec{u}' - \gamma p'\nabla\cdot\vec{\bar{u}} - \bar{u}\cdot\nabla p' + \mathcal{P}' + \mathcal{P}'_e \qquad (5.32)a,b$$

To simplify things, we have assumed also that the mean pressure and density are uniform. Now differentiate (5.32)b with respect to time, substitute (4.59)a, and re-arrange the result to find

$$\nabla^2 p' - \frac{1}{\bar{a}^2}\frac{\partial^2 p'}{\partial t^2} = h + h_e \qquad (5.33)$$

where

$$h + h_e = -\bar{\rho}\nabla\cdot(\vec{\bar{u}}\cdot\nabla\vec{u}' + \vec{u}'\cdot\nabla\vec{\bar{u}}) + \frac{1}{\bar{a}^2}\vec{\bar{u}}\cdot\nabla\frac{\partial p'}{\partial t} + \frac{\gamma}{\bar{a}^2}\frac{\partial p'}{\partial t}\nabla\cdot\vec{\bar{u}} + \nabla\cdot\vec{\mathcal{F}}' - \frac{1}{\bar{a}^2}\frac{\partial\mathcal{P}'}{\partial t} + \nabla\cdot\vec{\mathcal{F}}'_e - \frac{1}{\bar{a}^2}\frac{\partial\mathcal{P}'_e}{\partial t} \qquad (5.34)$$

Terms identified by subscript $(\ )_e$ always correspond, so h_e is defined as the last two terms on the right hand side.

The boundary condition is set on the normal gradient of the pressure by taking the scalar product of (5.32)a with the outward normal vector, giving

$$\hat{n}\cdot\nabla p' = -f - f_e \qquad (5.35)$$

where

$$f + f_e = \bar{\rho}\frac{\partial\vec{u}'}{\partial t}\cdot\hat{n} + \bar{\rho}(\vec{\bar{u}}\cdot\nabla\vec{u}' + \vec{u}'\cdot\nabla\vec{\bar{u}})\cdot\hat{n} - \vec{\mathcal{F}}'\cdot\hat{n} - \vec{\mathcal{F}}'_e\cdot\hat{n} \qquad (5.36)$$

The first term is commonly rewritten by introducing a response or admittance function characterizing the unsteady behavior of the boundary in response to imposed fluctuations. That it is the normal acceleration that matters (or frequency times velocity for sinusoidal disturbances) suggests correctly that the boundary appears as a fluctuating diaphragm (a 'speaker') to the unsteady field in the volume.

Otherwise, only gasdynamical interactions between the mean and fluctuating velocity fields are shown explicitly in (5.33) – (5.36). All other processes are hidden in $\mathcal{P}'$ and $\vec{\mathcal{F}}'$, those directly associated with external (control) actions being contained in $\vec{\mathcal{F}}'_e$ and $\mathcal{P}'_e$.

5.3 SPATIAL AND TIME AVERAGING OF UNSTEADY MOTIONS

Traditional calculations of linear behavior — meaning, almost always, linear stability — are usually based on the differential Eqs. (5.32)a, b, augmented by the linearized equation of state and possibly linearized forms of other equations given in Section 5.1. We shall use in the remainder of this paper the integrated form based on spatial averaging described, for example, by Culick (1975, 1988) and Culick and Yang (1992). The fundamental idea is that the actual problem defined by Eq. (4.60) with its boundary condition (5.35) differs in some sense by small amounts from the classical acoustics problem defined by setting $h = f = 0$. That is, the influences of all processes are small perturbations. This formulation, then, has the form naturally suited to an iteration procedure readily constructed by introducing a Green's function as first accomplished by Culick (1963). However, the identical first-order result (the only order legitimately considered with these equations) is obtained more directly in the following way.

5.3.1 *Spatial Averaging.* The unperturbed problem defines the normal modes, having mode shapes ψ_n and frequencies $\omega_n = \bar{a}k_n$ where k_n is the wavenumber:

$$\begin{aligned} \nabla^2\psi_n + k_n^2\psi_n &= 0 \\ \hat{n}\cdot\nabla\psi_n &= 0 \end{aligned} \tag{5.37a,b}$$

Then the idea is to compare the perturbed (actual) and unperturbed (classical) problems quite literally by comparing their difference in a spatially averaged sense. Multiply (5.33) by ψ_n, (5.37)a by p', subtract and integrate over the volume of the chamber to give

$$\int\left[\psi_n\nabla^2 p' - p'\nabla^2\psi_n\right]dV - \frac{1}{\bar{a}^2}\int\left[\psi_n\frac{\partial^2 p'}{\partial t^2} - \omega_n^2 p'\psi_n\right]dV = \int\psi_n(h+h_e)dV$$

Upon application of Green's theorem and substitution of the boundary conditions (5.31) and (5.37)b we find

$$\frac{1}{\bar{a}^2}\int\psi_n\left[\frac{\partial^2 p'}{\partial t^2} + \omega_n^2 p'\right]dV = -\left[\int\psi_n h dV + \oiint\psi_n f dS\right] = -\left[\int\psi_n h_e dV + \oiint\psi_n f_e dS\right] \tag{5.38}$$

The preceding calculations are formally correct and introduce no errors. Now we make the approximation that the pressure field in the actual chamber can be expanded in a series of normal modes for the classical unperturbed problem,

$$p' = \bar{p}\sum_{n=1}^{\infty}\eta_n(t)\psi_n(\vec{r}) \tag{5.39}$$

The error incurred arises because the $\psi_n(\vec{r})$, and therefore the right-hand side, do not satisfy the correct boundary condition (5.31) set on p' in the actual problem. When the series is used in the integrals above, the error is attenuated, and for other formal reasons not covered here (see, e.g. Morse and Feshback 1953, Chapter 10), this is not as serious an inaccuracy as one might think. We assume also, not a practical constraint, that the normal modes are orthogonal,

$$\int\psi_m\psi_n dV = E_n^2\delta_{mn} \tag{5.40}$$

Substitution in (5.37) leads finally to the second-order equations for the amplitudes $\eta_n(t)$:

$$\frac{d^2\eta_n}{dt^2} + \omega_n^2\eta_n = F_n + F_{ne} \tag{5.41}$$

and the force is

$$\begin{aligned} F_n &= -\frac{\bar{a}^2}{\bar{p}E_n^2}\left\{\int h\psi_n dV + \oiint\psi_n f dS\right\} \\ F_{ne} &= -\frac{\bar{a}^2}{\bar{p}E_n^2}\left\{\int h_e\psi_n dV + \oiint\psi_n f_e dS\right\} \end{aligned} \tag{5.42a,b}$$

With proper interpretation and suitable modeling of the processes represented in the functions h, f, h_e, and f_e, the representation of combustion instabilities by Eqs. (5.41) and (5.42)a,b has wide application to all propulsion systems. Relatively little has been done with this formulation applied specifically to liquid rocket engines, but as we shall see shortly, it is quite easy to obtain general results equivalent to those obtained with methods more familiar in this field.

An important point to emphasize is that the approach taken here provides a useful viewpoint as well as a framework useful for both theoretical and practical computations. According to (5.39) and (5.41), any unsteady motion may be regarded as a collection of oscillators, one associated with each normal mode of oscillation appearing in the synthesis (5.39). The problem then comes down to determining the time evolution of the amplitudes $\eta_n(t)$ of the oscillators (modes). Although we have assumed linear motions here, the same formal representation applies to nonlinear motions so long as the amplitudes do not become so large as to violate the approximations taken to give the governing differential equations.

5.3.2 *Time Averaging.* Combustion instabilities commonly appear as oscillations having slowly varying amplitudes and phases. That is, the fraction changes of amplitude and phase are small in one cycle of the oscillation. Under these circumstances, the method of averaging, or equivalently, expansion in two time scales, can be applied to reduce the set of N second-order Eqs. (5.41), when N modes are considered, to a set of $2N$ first-order equations. That is an enormously useful simplification, and while limits are placed on the range of validity of the results, this step in the method is extremely useful in practice. The limitations are discussed briefly in Section 10.

According to the behavior just mentioned, we assume that the amplitude $\eta_n(t)$ can be written as

$$\eta_n(t) = r_n(t)\sin[\omega_n t + \phi_n(t)] = A_n \sin\omega_n t + B_n \cos\omega_n t \tag{5.43}$$

where r_n, ϕ_n, A_n, and B_n are slowly varying functions. The simplest, and a physically appealing, way of applying time averaging is a modest variation of averaging developed by Krylov and Bogoliubov (1947). By analogy with a simple mass/spring system, the energy $\mathcal{E}_n$ can be associated with the oscillator governed by equation (5.41):

$$\mathcal{E}_n = \frac{1}{2}\omega_n^2\eta_n^2 + \frac{1}{2}\dot{\eta}_n^2 \tag{5.44}$$

The instantaneous velocity of the oscillator is $\dot{\eta}_n$, so that the rate at which work is done on the oscillator is $\dot{\eta}_n F_n$. Averaged over an interval τ at time t, the values are

$$< \mathcal{E}_n >= \frac{1}{\tau}\int_t^{t+\tau} \mathcal{E}_n(t')dt'$$

$$< \dot{\eta}_n F_n >= \frac{1}{\tau}\int_t^{t+\tau} \dot{\eta}_n \mathcal{E}_n F_n dt' \tag{5.45)a, b}$$

Conservation of energy for the averaged motion implies that the rate of change of time-averaged energy should equal the time-averaged rate of power in

$$\frac{d}{dt} < \mathcal{E}_n >=< \dot{\eta}_n F_n > \tag{5.46}$$

Because the single function η_n has been replaced by two functions, r_n and ϕ_n, we are free to place a restriction; following Krylov and Bogoliubov, we require

$$\frac{d\phi_n}{dt} r_n \cos(\omega_n t + \phi_n) + \frac{dr_n}{dt}\sin(\omega_n t + \phi_n) = 0 \tag{5.47}$$

Differentiating equation (5.43) and enforcing equation (5.47) gives the formula for the velocity in the same form as that for a classical conservative oscillator:

$$\dot{\eta}_n = \omega_n r_n \cos(\omega_n t + \phi_n) \tag{5.48}$$

and the energy is

$$\mathcal{E}_n = \frac{1}{2}\omega_n r_n \tag{5.49}$$

With Eq. (4.100), equation (5.47) gives

$$\frac{dr_n}{dt} = \frac{1}{\omega_n \tau} \int_t^{t+\tau} F_n \cos(\omega_n t' + \phi_n) dt' \tag{5.50}$$

The equation for $\phi_n(t)$ is found by substituting equations (5.44), (5.48), and (5.47) in the oscillator equation (5.46); multiplication by $\sin(\omega_n t + \phi_n)$ and time averaging gives

$$r_n \frac{d\phi_n}{dt} = \frac{-1}{\omega_n \tau} \int_t^{t+\tau} F_n \sin(\omega_n t' + \phi_n) dt' \tag{5.51}$$

It is often more convenient to use the equations for $A_n(t)$ and $B_n(t)$, found by solving equations (5.44), (5.47), (5.50), and (5.51) for $\dot{A}_n(t)$ and $\dot{B}_n(t)$, to give

$$\begin{aligned} \frac{dA_n}{dt} &= \frac{1}{\omega_n \tau} \int_t^{t+\tau} F_n \cos \omega_n t' dt' \\ \frac{dB_n}{dt} &= \frac{-1}{\omega_n \tau} \int_t^{t+\tau} F_n \sin \omega_n t' dt' \end{aligned} \tag{5.52}a, b$$

The assumption that the amplitudes and phases are slowly varying means

$$\begin{aligned} &\dot{r}_n \tau << 1 \qquad \dot{\phi}_n \tau << 2\pi \\ &\dot{A}_n \tau << 1 \qquad \dot{B}_n \tau << 1 \end{aligned} \tag{5.53}$$

These inequalities imply that the functions $A_n(t)$ and $B_n(t)$ will be taken as constant under the integrals in Eq. (5.52).

The original motivation for developing the method of averaging in the form expressed as equations (5.52) and (5.53) was to provide the basis for treating arbitrarily shaped chambers. Differences between geometries are reflected in the unperturbed mode shapes and frequencies. The mode shapes affect the values of parameters that arise in equations (5.52) and (5.53), but the frequency spectrum, as we shall see, influences the qualitative of the equations. It appears that the derivation of equations (5.52) and (5.53) is not restricted to particular geometries, but, to date, these results have been applied only to the simplest case of longitudinal modes for which the harmonic frequencies are integral multiples of the fundamental. Application to other cases requires further calculations, which we will not pursue here; preliminary examination suggests some difficulties that have not been resolved.

Culick (1975) has shown that F_n has the following form for second-order acoustics:

$$F_n = -\sum_{i=1}^{\infty}[D_{ni}\dot{\eta}_i + E_{ni}\eta_i] - \sum_{i=1}^{\infty}\sum_{j=1}^{\infty}[A_{nij}\dot{\eta}_i\dot{\eta}_j + B_{nij}\eta_i\eta_j] \tag{5.54}$$

The constants D_{ni}, E_{ni}, A_{nij}, and B_{nij} depend on the unperturbed mode shapes and frequencies; the D_{ni} and E_{ni} arise from linear processes and are usually proportional to the Mach number of the mean flow. (A notable exception arises with the presence of condensed material. The characteristic parameter then depends on the properties of the particles.) Consider the linear terms only. Substitution of equation (5.54) in equation (5.52), with τ equal to the period of the nth mode, leads to the following results:

$$\left(\frac{dA_n}{dt}\right)_{\text{linear}} = -\frac{1}{2}D_{nn}A_n + \frac{1}{2}\frac{E_{nn}}{\omega_n}B_n \tag{5.55}$$

$$\left(\frac{dB_n}{dt}\right)_{\text{linear}} = -\frac{1}{2}D_{nn}B_n + \frac{1}{2}\frac{E_{nn}}{\omega_n}A_n \tag{5.56}$$

Multiply the first of these by A_n and the second by B_n, and add the results to find the equation for the amplitude, $\dot{r}_n = (D_{nn}/2)r_n$, where $r_n^2 = A_n^2 + B_n^2$. Thus, $r_n \sim \exp(\alpha_n t)$, with $\alpha_n = -D_{nn}/2$, the growth constant for the nth mode.

Now, from the definition (5.43) of $\eta_n(t)$, we find

$$A_n \approx e^{\alpha_n t}\cos\phi_n \qquad B_n \approx e^{\alpha_n t}\sin\phi_n$$

Substitution in Eqs. (5.55) and (5.56) leads to the identification $E_{nn}/2\omega_n = \dot{\phi}_n$. But $\dot{\phi}_n$ can be interpreted as the frequency shift in the nth mode that is due to the perturbations because the perturbed frequency is

$$\omega = \frac{d}{dt}(\omega_n t + \phi_n) = \omega_n + \dot{\phi}_n = \omega_n + (\omega - \omega_n)$$

Hence we have established the two rules that the growth constant for the nth mode is 1/2 the coefficient of $\dot{\eta}_n$ in the form (5.55), and the frequency shift is $1/2\omega_n$ times the coefficient of η_n:

$$\alpha_n = -\frac{1}{2}D_{nn} \tag{5.57}$$

$$\theta_n = -(\omega - \omega_n) = -\frac{1}{2}\frac{E_{nn}}{\omega_n} \tag{5.58}$$

These rules are often conveniently applied after a representation of a particular process has been constructed.

6. Results for Linear Stability

We remarked earlier that there is really only one problem of linear stability and within the present analysis it is readily solved. According to the principles of Fourier analysis, with essentially no practical restrictions, any function can be synthesized as a finite or infinite series of orthogonal functions. Here, we apply the idea to a pressure record represented as a sum of the normal modes $\psi_n(\vec{r})$ of a chamber, with time-dependent amplitudes $\eta_n(t)$, equation (5.39). That representation presumes neither linear nor nonlinear behavior. If the behavior is linear, as we shall assume, and the modes are uncoupled, as we also assume, then it is a familiar result that to find the time-dependent properties of the normal modes, the amplitudes may be assumed to vary exponentially in time:

$$\eta_n(t) = \hat{\eta}_n e^{i\omega_n t} \qquad \text{(normal modes)} \tag{6.1}$$

Hence we assume that the normal modes are neither driven nor damped. Finding the normal modes requires solving the classical problem of the scalar wave equation in a closed volume having the same shape as the combustion chamber in question except that the exhaust nozzle is closed at its entrance. Both the mode shapes $\psi_n(\vec{r})$ and the natural frequencies are found in this calculation, the solution to (5.37)a, b.

For the actual problem with all relevant processes accounted for, the form (6.1) is again assumed, and for the problem of linear stability, the amplitudes are assumed to have the form

$$\eta_n(t) = \hat{\eta}_n e^{i\bar{a}K_n t} \qquad \text{(linear stability)} \tag{6.2}$$

where the actual complex wavenumber is

$$K_n = \frac{1}{\bar{a}}(\Omega_n - i\alpha_n) \tag{6.3}$$

Now the perturbed mode corresponding to the nth classical mode may either grow ($\alpha_n > 0$) or decay ($\alpha_n < 0$) in time. The actual frequency is

$$\Omega_n = \omega_n + \delta\omega_n \tag{6.4}$$

Hence the problem of linear stability comes down to determining for each perturbed mode the frequency shift $\delta\omega_n$ and the growth constant α_n. The basis for the calculation is the linearized form of the equations already treated.

Substitute (6.2) – (6.4) in equation (5.41), with $F_{ne} = 0$ to find

$$(\Omega_n - i\alpha_n)^i = \omega_n^2 + \frac{\hat{F}_n}{\hat{\eta}_n}$$

where, for linear behavior, we write $h = \hat{h}e^{i\bar{a}K_n}$, $f = \hat{f}e^{i\bar{a}K_n}$, so $F_n = \hat{F}_n e^{i\bar{a}K_n}$. For the usual case, $\alpha_n << \Omega_n$ and $\Omega_n - \omega_n << 1$, which must be satisfied to be consistent with approximations already made. Then separating the real and imaginary parts of the last equation, with the definition (5.44)a gives

$$\begin{aligned} \Omega_n &= \omega_n - \frac{\bar{a}^2}{2\omega_n \bar{p} E_n^2}\left\{\int \frac{\hat{h}^{(r)} + \hat{h}_e^{(r)}}{\hat{\eta}_n}\psi_n dV + \oiint \frac{\hat{f}^{(r)} + \hat{f}_e^{(r)}}{\hat{\eta}_n}\psi_n dS\right\} \\ \alpha_n &= \frac{\bar{a}^2}{2\omega_n \bar{p} E_n^2}\left\{\int \frac{\hat{h}^{(i)} + \hat{h}_e^{(i)}}{\hat{\eta}_n}\psi_n dV + \oiint \frac{\hat{f}^{(i)} + \hat{f}_e^{(i)}}{\hat{\eta}_n}\psi_n dS\right\} \end{aligned} \tag{6.5a,b}$$

in which $(\ )^{(r)}$ and $(\ )^{(i)}$ denote real and imaginary parts. These two formulas are the solution to the problem of linear stability. In particular, setting $\alpha_n = 0$ defines the stability boundary in terms of the various parameters arising in the functions h and f. That result is the basis for the Standard Stability Prediction program (Nickerson *et al.* 1983) written for the U.S. Air Force and widely used by industrial and governmental organizations to analyze the stability of motions in solid propellant rockets.

7. Rayleigh's Criterion

Probably the most widely quoted general principal in the field of combustion instabilities is Rayleigh's criterion formulated by Rayleigh in 1878:

"If heat be periodically communicated to, and abstracted from, a mass of air vibrating (for example) in a cylinder bounded by a piston, the effect produced will depend upon the phase of the vibration at which the transfer of heat takes place. If heat be given to the air at the moment of greatest condensation, or be taken from it at the moment of greatest rarefaction, the vibration is encouraged. On the other hand, if heat be given at the moment of greatest rarefaction, or abstracted at the moment of greatest condensation, the vibration is discouraged."

Roughly the gist of the principle is that heat addition tends most strongly to drive acoustic waves if the energy is added in the region of space where the oscillating pressure reaches greatest amplitude and is in phase. Suppose that the heat release associated with combustion is simply proportional to the pressure, so the fluctuation $Q' = \beta(p'/\bar{p})$ where β is a complex constant. Suppose further that only one mode is present, so we write $p' = \bar{p}\eta_n(t)\psi_n(\vec{r})$, a single term of the expansion (6.1). The simplest case is a steady oscillation, and $\eta_n(t) = \hat{\eta}_n e^{i\omega_n t}$. Write $\beta = |\beta|e^{i\phi_Q}$ and the oscillation of heat addition is

$$Q' = |\beta|Q_o\hat{\eta}_n\psi_n(\vec{r})e^{i(\omega_n t+\phi_Q)} \tag{7.1}$$

Rayleigh's criterion then states that heat addition is most destabilizing if the fluctuation Q' is in phase with the pressure oscillation, $\phi_Q = 0$. The assumed form already ensures that the magnitude of the oscillation of heat addition is maximum where the pressure oscillation has maximum amplitude. Although the principle seems fairly obviously true, these remarks do not serve as proof. The formulation described above can be used not only to establish the criterion but also to extend its application to all processes and nonlinear behavior (see Culick 1988, 1987). Zinn (1986) has also recently discussed Rayleigh's criterion, but only for linear heat addition.

We take advantage of the fact that each acoustic mode has an associated oscillator having unit mass and natural frequency ω_n, whose motion (i.e. time evolution of its amplitude η_n) is described by equation (5.41). The energy of a simple oscillator is the sum of its kinetic and potential energies,

$$\mathcal{E}_n = \frac{1}{2}\left(\dot{\eta}_n^2 + \omega_n^2\eta_n^2\right) \tag{7.2}$$

We interpret Rayleigh's criterion to be a statement concerning the change $\Delta\mathcal{E}_n$ of the energy in one cycle. The product of the force $(F_n + F_{ne})$ times the velocity $\dot{\eta}_n$ is the power input, so

$$\Delta\mathcal{E}_n(t) = \int\limits_t^{t+\tau_n} (F_n + F_{ne})\dot{\eta}_n dt' \tag{7.3}$$

In general $\Delta\mathcal{E}_n(t)$ varies with time, increasing from cycle to cycle for an unstable mode.

Indeed, for linear motions, if all processes are taken into account, $\Delta\mathcal{E}_n$ is proportional to the growth constant α_n and this extended form of Rayleigh's criterion is equivalent to the principle of linear stability

$$\Delta\mathcal{E}_n = 2\pi\omega_n\alpha_n \tag{7.4}$$

Thus the precise form of Rayleigh's criterion stated above is found by taking only the part of F_n representing heat addition and setting $F_{ne} = 0$. The result is

$$\Delta\mathcal{E}_n(t) = (\alpha - 1)\frac{\omega_n^2}{\bar{p}E_n^2}\int \psi_n dV \int\limits_t^{t+\tau_n} \eta_n Q' dt' \tag{7.5}$$

in which Q' must be expressed in terms of real quantities. Thus, instead of the complex form we write in the case Q' proportional to p',

$$Q' = Q_o\psi_n(\beta_1\eta_n + \beta_2\dot{\eta}_n) \tag{7.6}$$

where β_1, β_2 are real. The formula (4.120) leads to

$$\Delta\mathcal{E}_n = \pi(\alpha - 1)\beta_1\omega_n Q_o \tag{7.7}$$

Combination of (7.4) and (7.5) gives the convenient formula for the growth constant due to heat addition,

$$\alpha_n = \frac{(\alpha - 1)\omega_n}{2\pi\bar{p}E_n^2}\int \psi_n dV \int_0^{\tau_n} \eta_n Q' dt' \tag{7.8}$$

where we have used the fact that $\eta_n Q'$ has period $\tau_n = 2\pi/\omega_n$ to shift the limits of the integral. We should also note that whereas the time derivative $\partial Q'/\partial t$ appears as the source in the wave equation (5.33) and in the oscillator Equation (5.41), Q' itself occurs in (7.5) and (7.8), also a consequence of periodicity as explained in Culick (1987).

8. Review of Work in Nonlinear Instabilities

Before discussing the more recent results, we review briefly previous analyses other than numerical simulations. The following discussion is taken from the review by Culick (1993). Long before the work producing the approximate analysis reviewed here, much attention had been directed to nonlinear combustion instabilities. That the phenomena are intrinsically nonlinear was recognized practically from the earliest experimental results. In 1956, motivated partly by the common occurrence of transverse waves in combustors, Maslen and Moore (1956) treated finite amplitude waves, based on a power series expansion in the amplitude, but with no combustion and mean flow. Flandro (1967) extended that work to include combustion and flow, and succeeded in explaining the origin of roll torques observed occasionally in solid rockets.

Chu (1963) and Chu and Ying (1963) treated the problem of thermally driven nonlinear longitudinal oscillations with shock waves in a closed tube using the method of characteristics. Although the heat source was allowed to be sensitive to pressure in part of that work, no combustion and flow were accounted for.

Motivated partly by laboratory tests of gas-fueled rockets at Princeton, the first detailed analysis of the problem with combustion and flow was published by Sirignano (1964) and Sirignano and Crocco (1964), also using the method of characteristics, modified to include coordinate stretching following the PLK (Poincaré–Lighthill–Kuo) procedure. That work was followed by further analyses of longitudinal waves by Mitchell *et al.* (1969) and by Crocco and Mitchell (1969) using different methods allowing treatment of distributed combustion. At the same time, Zinn (1968) examined nonlinear behavior of transverse modes, based on expansion of the equation for the velocity potential in powers of the amplitude.

Those early works by the Princeton group were first to address quantitatively the broad subject of nonlinear motions in a combustion chamber. They were also first to expose some particular problems which remain unsolved in general, notably the possible existence of stable limit cycles and the conditions under which 'triggering' to stable or unstable limit cycles may occur. Apparently for at least three reasons, the approaches taken during the 1960s have not been pursued further by researchers outside the Princeton school: (1) particularly with the method of characteristics, the calculations become so detailed and divorced from physical behavior as to obscure understanding; (2) except in the analysis by Crocco and Mitchell (1969), the combustion zone was assumed to be thin and located at the end of the chamber, physically a serious limitation if one seeks general applications; and (3) perhaps more seriously, the 'time-lag model' of the combustion processes is used. It is true that any linear unsteady combustion process can be represented by a time-lag model, in practice amounting to definition of a transfer function. However, a serious difficulty arises with the usual assumption implied in all the works cited above, that the time lag is independent of frequency. That is rarely true. For example, it is known that the time lag is a critically strong function of frequency for unsteady burning of a solid propellant, being dominated by transient heat transfer in the condensed phase.

Nonlinear behavior of a dynamical system is both qualitatively and quantitatively dependent on the values of the parameters characterizing the linear behavior. Generalizations from particular results are almost never possible without actually performing the necessary analysis. Therefore, there is presently almost no basis for claiming generality for any of the results involving the time-lag model of linear behavior, however useful the model may be for other purposes.

The introduction by Zinn and Powell (1971) of the use of Galerkin's method to study instabilities in liquid rockets, and later applied by Zinn and Lores (1972) and Lores and Zinn (1973) marked an important shift of emphasis on the type of analytical methods used in this field. As in the earlier works, the time-lag model of combustion is used, so the results are similarly restricted. However, the general approach, pursued independently by Culick (1976) for application to solid rockets, will accommodate very general representations of all nonlinear processes.

9. Limit Cycles and the Two-Mode Approximation

Considerable effort has been spent, for more than four decades, on experimental and analytical programs devoted to solving problems of combustion instabilities. Much of the work has been required to measure quantities which, because of the complex processes involved, cannot be predicted accurately from first principles. Analytical work has been concerned largely with linear behavior, the chief purpose being to predict stability of small disturbances in combustion chambers. Many useful results have been obtained, serving in practice to help design experiments, correlate data, and estimate the stability of new systems.

In theory, the problem of linear stability is essentially solved and well-understood; in practice the situation is quite different. The reason is that the information required to predict stability is always imperfectly known. Predictions therefore carry with them substantial inaccuracies. Moreover, it is difficult to obtain reliable data for the growth or decay rates of small disturbances in operating engines. Few results are available.

The situation in respect to nonlinear behavior of liquid rocket engines is quite different. The theory is far more difficult, contains a broad array of important problems, and is only in the beginning stages of development. Yet it is the nature of combustion instabilities that almost all data show significant nonlinear behavior. Indeed, the most important means of assessing the stability of liquid rocket engines — observation of the physical behavior following initiation of finite explosive charges ('bombing') — involves intrinsically nonlinear behavior. The theory of bombing, really the theory of nonlinear instabilities (sometimes called 'triggering') hardly exists and has produced no useful results.

For application to combustion instabilities, there are two fundamental problems to be integrated:

(1) the conditions under which an *unstable* linear system will execute a stable limit cycle;

(2) the conditions under which a *stable* linear system can be caused, by suitable disturbance, to execute a stable limit cycle.

Both of these problems were discussed for simple cases in Section 4.1 where we emphasized that a linearly unstable motion in a combustion chamber must normally be viewed as a self-excited system:

an unstable motion therefore reaches finite amplitude only through the action of nonlinear processes. That is the substance of problem (1). Problem (2) is the problem of 'bombing' or 'triggering.' In the language of dynamical systems theory, problem (1) arises from a supercritical bifurcation and problem (2) from a sub-critical bifurcation.

We summarize briefly some of the recent results based on the analytical framework described in Section 5. Only nonlinear gasdynamics is accounted for, although it is well known that other nonlinear processes are significant — possibly essential — to explain some of the observed behavior. The reason for focusing on nonlinear gasdynamics is that, in a practical sense, this is the only nonlinear process that is known with full confidence. This the approach taken in recent work has been to understand as completely as possible the consequences and influences of the nonlinear gasdynamics. Differences from observed behavior should then, presumably, be ascribed to other processes. In liquid rocket engines, it seems probable that the next most important processes are those accomplishing the conversion of injected liquid to gases *prior* to combustion. Although ultimately the energy supplied to both the mean and unsteady flows must be generated by combustion, it appears likely that the chemical kinetics and burning processes are not normally the dominant tie-dependent processes in propulsion systems.

We emphasize that only gasdynamic nonlinearities will be considered here. Although related work is still in progress, much understanding has been gained. Probably the most important conclusion is that while stable limit cycles can be found to exist in a broad range of conditions with *only* nonlinear gasdynamics, it appears that if no other nonlinear processes are present, the motions in a combustion chamber are not nonlinearly unstable. That is, at least one mode must be linearly unstable in order for limit cycles to exist.

Most of the results for limit cycles have been obtained for longitudinal modes and second-order acoustics (i.e. quadratic nonlinearities in the fluctuations). For any geometry, F_n defined by (5.42)a has the form (5.54):

$$F_n = -\sum_{i=0}^{\infty}[D_{ni}\dot{\eta}_i + E_{ni}\eta_i] - \sum_{i=1}^{\infty}\sum_{j=1}^{\infty}[A_{nij}\dot{\eta}_i\dot{\eta}_j + B_{nij}\eta_i\eta_j] \tag{9.1}$$

Purely longitudinal modes have the important special characteristic that the unperturbed natural frequencies are integral multiples of the fundamental. As a result, the double series in F_n reduces to a single series, and the first-order equations obtained with time averaging become

$$\begin{aligned}\frac{dA_n}{dt} &= \alpha_n A_n + \theta_n B_n + \frac{n\beta}{2}\sum_{i=1}^{\infty}\Big[A_i\left(A_{n-i} - A_{i-n} - A_{i+n}\right) - B_i\left(B_{n-i} + B_{i-n} - B_{i+n}\right)\Big]\\ \frac{dB_n}{dt} &= -\theta_n A_n + \alpha_n B_n + \frac{n\beta}{2}\sum_{i=1}^{\infty}\Big[A_i\left(B_{n-i} + B_{i-n} - B_{i+n}\right) + B_i\left(A_{n-i} - A_{i-n} + A_{i+n}\right)\Big]\end{aligned} \tag{9.2a,b}$$

where $\alpha_n = -D_{nn}/2$, $\theta_n = -E_{nn}/2\omega_n$ and

$$\beta = \frac{\bar{\gamma}+1}{8\bar{\gamma}}\omega_1 \tag{9.3}$$

Although actual systems rarely show $\omega_n = n\omega_1$ exactly, there are many instances, particularly of solid and liquid rockets, when the spectrum of frequencies closely satisfies this condition. Hence this special case has received much attention since it was first studied by Culick (1976). Prior to that time, no formal theory existed to explain the observed occurrences of steady nonlinear periodic oscillations, i.e. limit cycles; calculations of limit cycles had been carried out in earlier works cited above, but analytical results had not been obtained owing to the nature of the methods used.

Most attention was therefore directed to discovering first whether limit cycles could be predicted, and second the conditions determining their existence and stability. Note that the constant β can be absorbed as a scaling parameter (β^{-1} is a time scale) so that all of the behavior — linear and nonlinear — is determined by the values of the linear parameters (α_n, θ_n), which contain completely the influences of all linear processes. In the earliest work, the existence of limit cycles was established by solving equations (9.2) numerically for ranges of the linear parameters.

Confidence in the approximate analysis was gained quite early with successful comparisons and numerical solutions to the partial differential equations (Levine and Culick, 1972, 1974). Several problems were examined, including development of a disturbance into a weak shock wave (no combustion or mean flow), decay of large amplitude waves due to the presence of condensed material, and the growth of a combustion instability to a limit cycle in a solid rocket. Using more efficient numerical routines developed by Levine and Baum (1983), Culick and Yang (1992) have reported the example shown in Figure 9.1 showing the waveform in a limit cycle computed with the numerical and approximate analysis (for five modes). The frequencies (shifted due to perturbations) and the amplitudes of the modes are predicted quite well by the approximate analysis as Table 9.1 shows.

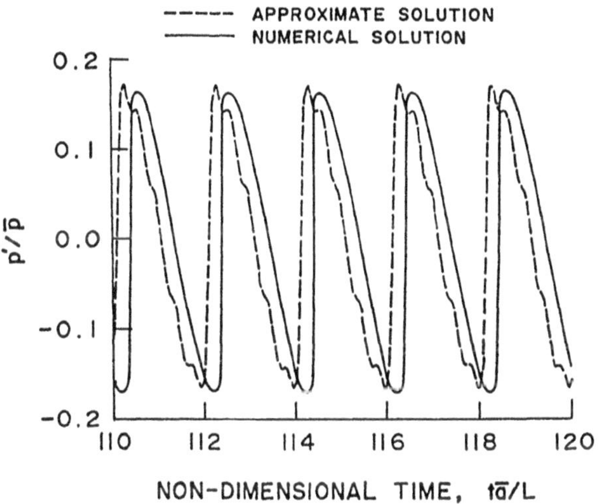

FIGURE 9.1

The frequencies agree within 3%, illustrating the familiar fact that accurate prediction of frequencies is not a demanding test of an approximate analysis — good approximations should (as here) be built into the formulation. More to the point is the good agreement of the amplitudes except for the highest ($n = 5$) mode. The reason for the higher approximate values is that truncation to five modes eliminates energy transfer to higher modes where it is ultimately dissipated. We shall see that the intrinsic tendency for energy transfer upwards is an important fundamental property of these problems, having a strong influence on the development of limit cycles from small initial disturbances.

While it is useful to be able to obtain results of this sort — inexpensive and accurate approximations to nonlinear behavior — this is only part of the story. It is perhaps even more significant both theoretically and practically, to establish the qualitative dependence of nonlinear behavior on

			Frequency		
Mode	1	2	3	4	5
Numerical	926	1824	2698	3595	4491
Approximate	895	1785	2683	3571	4449
			Amplitude $\lvert p'/\bar{p}\rvert$		
Mode	1	2	3	4	5
Numerical	0.151	0.042	0.234	0.0203	——
Approximate	0.151	0.0478	0.0280	0.0153	0.0188

TABLE 9.1

the linear parameters. The special structure of the nonlinear equations (9.1)a, b allows considerable progress in that respect. Awad and Culick (1986) reported the first formal results for existence and stability of limit cycles for purely longitudinal modes, using equations (9.1)a, b, although as noted above, the question had previously been primarily addressed with numerical solutions. Closed form results have been obtained only for the case of two modes; here we only describe the gist of the matter.

For two modes, equations (9.1)a, b become

$$\begin{aligned}
\frac{dA_1}{dt} &= \alpha_1 A_1 + \theta_1 B_1 - \beta(A_1 A_2 - B_1 B_2)\\
\frac{dB_1}{dt} &= \alpha_1 B_1 - \theta_1 A_1 + \beta(B_1 A_2 - A_1 B_2)\\
\frac{dA_2}{dt} &= \alpha_2 A_2 + \theta_2 B_2 + \beta(A_1^2 - \beta_1^2)\\
\frac{dB_2}{dt} &= \alpha_2 B_2 - \theta_2 A_2 + 2\beta B_1 A_1
\end{aligned} \qquad (9.4)a, b, c, d$$

Limit cycles correspond to equilibrium states for this system. There are two possible cases: $\dot{A}_n = \dot{B}_n = 0$, or $A_n = a_n \cos(\nu_n t + \psi_n)$, $B_n = a_n \sin(\nu_n t + \psi_n)$. In the first case, the functions A_n, B_n are constant in the limit cycle and in the second case they oscillate, with frequencies ν_n. The two possibilities were first discovered in the numerical calculations reported by Culick (1976) and confirmed theoretically by Awad and Culick (1986).

The analysis then proceeds as follows. First substitute in (9.4)a, b, c, d the assumed forms for the A_n, B_n in the limit cycle and solve the algebraic equations. It is a consequence of the special structure of those equations that formulas for the amplitudes can be derived, simultaneously with conditions for their existence. Let subscript $(\ )_0$ denote values in the limit cycle; then for the case when the amplitudes are constant in the limit cycle, the results may be written:

$$\begin{aligned}
A_{10} &= \frac{1}{\beta}\left[-\alpha_1\alpha_2\left(1 + \frac{\theta_1^2}{\alpha_1^2}\right)\right]^{\frac{1}{2}}\\
B_{10} &= 0\\
A_{20} &= \frac{1}{\beta}\alpha_1\\
B_{20} &= \frac{1}{\beta}\theta_2
\end{aligned} \qquad (9.5)a, b, c, d$$

These formulas are not unique. Because there is an arbitrary phase on the limit cycle, one constant is undetermined; here its value has been fixed by setting $B_{10} = 0$.

If the amplitudes are allowed to oscillate in the limit cycle, producing a small shift in the modal frequencies, then

$$A_n = a_n \cos(\nu_n t + \psi_n); \quad B_n = a_n \sin(\nu_n t + \psi_n) \tag{9.6}$$

The frequency shifts are found to be

$$\nu_1 + \frac{1}{2}\nu_2 = -\frac{\alpha_1\theta_2 + \alpha_2\theta_1}{2\alpha_2 - \alpha_2}$$

and the maximum amplitudes are

$$a_1 = \left\{-\frac{\alpha_1\alpha_2}{\beta_2}\left[1 + \left(\frac{2\theta_1 - \theta_2}{2\alpha_1 + \alpha_2}\right)^2\right]\right\}^{\frac{1}{2}}$$

$$a_2 = \frac{\alpha_1}{\beta}\left[1 + \left(\frac{2\theta_1 - \theta_2}{2\alpha_1 + \alpha_2}\right)^2\right]^{\frac{1}{2}} \tag{9.7)a, b}$$

Because A_{10}, equation (9.5)a, and a_1, equation (9.7)a, must be real, one necessary condition for existence is

$$\alpha_1\alpha_2 < 0 \tag{9.8}$$

The physical interpretation of this condition follows from the meaning of the α_n as growth or decay constants. If α_n is positive (negative) then in the linear approximation, energy is added (lost) to the nth mode. Hence (9.8) represents the requirement the one mode must gain energy and one must lose energy in order that the oscillations should reach a steady sustained amplitude. If both α_1 and α_2 are negative, the system is absolutely stable, and if α_1 and α_2 are both positive, the motions grow without limit. The nonlinear terms in (9.2)a, b, and hence in (9.4)a, b, c, cause transfer of energy from one mode to another.

Conditions for stability of the limit cycles are determined by examining the evolution of small disturbances from the stationary states. A set of inequalities is deduced from application of the Routh-Hurwitz or Lienard criteria to the characteristic polynomial. Figure 9.2 (taken from Paparizos and Culick 1989) shows one way of displaying the results in terms of the linear parameters. Ranges of the parameters for which stable limit cycles exist are indicated by the hatched lines.

It seems remarkable that simple conditions of existence and stability can be obtained for transverse modes, in a cylindrical chamber, by following the same approach used to treat longitudinal modes. The reason is that the time-averaged equations again have a special structure allowing the calculations to be carried out again for two modes. What makes this surprising is that because the natural frequencies do not satisfy the condition $\omega_n = n\omega_1$, the time-averaged equations contain modulation on the right-hand side.

Before averaging, use of (9.1) leads to the set of equations valid for second-order acoustics,

$$\frac{dA_n}{dt} = -\frac{1}{2\omega_n}\sum_{i=1}^{\infty}\left\{C_{ni}\left[\cos(\omega_n+\omega_i)t+\cos(\omega_n-\omega_i)t\right]+S_{ni}\left[\sin(\omega_n+\omega_i)t-\sin(\omega_n-\omega_i)t\right]\right\}$$

$$-\frac{1}{2\omega_n}\sum_{i=1}^{\infty}\sum_{j=1}^{\infty}\left\{F_{nij}a_{ij}\left[\cos(\omega_n+\omega_{ij+})t+\cos(\omega_n-\omega_{ij+})t\right]\right.$$

$$+G_{nij}b_{ij}\left[\cos(\omega_n+\omega_{ij+})t+\cos(\omega_n-\omega_{ij+})t\right]$$

$$-F_{nij}d_{ij}\left[\sin(\omega_n+\omega_{ij+})t-\sin(\omega_n-\omega_{ij+})t\right]$$

$$\left.+G_{nij}e_{ij}\left[\sin(\omega_n+\omega_{ij-})t-\sin(\omega_n-\omega_{ij-})t\right]\right\} \tag{9.9}$$

where

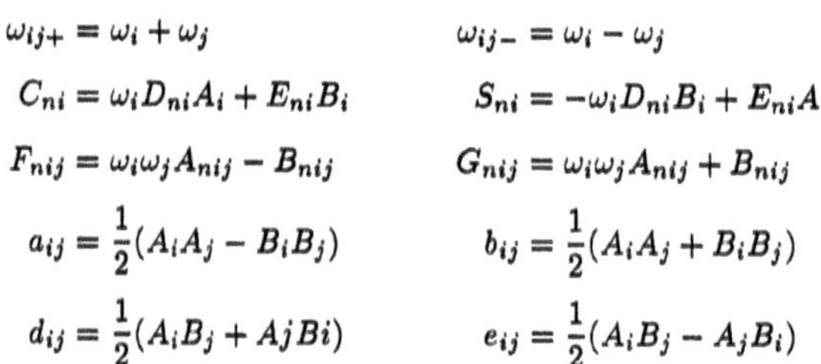

$$\omega_{ij+} = \omega_i + \omega_j \qquad \omega_{ij-} = \omega_i - \omega_j$$

$$C_{ni} = \omega_i D_{ni} A_i + E_{ni} B_i \qquad S_{ni} = -\omega_i D_{ni} B_i + E_{ni} A_i$$

$$F_{nij} = \omega_i \omega_j A_{nij} - B_{nij} \qquad G_{nij} = \omega_i \omega_j A_{nij} + B_{nij}$$

$$a_{ij} = \frac{1}{2}(A_i A_j - B_i B_j) \qquad b_{ij} = \frac{1}{2}(A_i A_j + B_i B_j)$$

$$d_{ij} = \frac{1}{2}(A_i B_j + AjBi) \qquad e_{ij} = \frac{1}{2}(A_i B_j - A_j B_i)$$

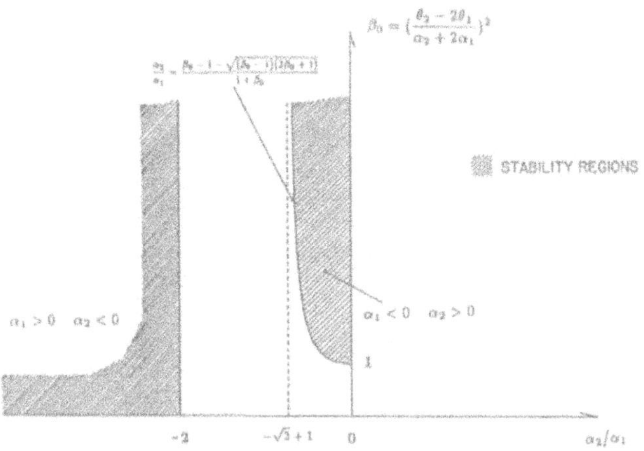

FIGURE 9.2

A similar equation defines $\dot{B}_n$; see Yang and Culick (1990) where further details of the analyses and the numerical values of the coefficients F_{nij} and G_{nij} are given. Three modes are considered — the first and second tangential and the first radial having the following wave numbers and mode shapes:

First Tangential Mode (1T)

$$\kappa_1 R = 1.8412 \qquad \psi_1 = \cos\theta J_1(\kappa_1 r) \qquad \psi_4 = \sin\theta J_1(\kappa_1 r) \tag{9.10}$$

First Radial Mode (1R)

$$\kappa_2 R = 3.8317 \qquad \psi_2 = J_o(\kappa_2 r) \tag{9.11}$$

Second Tangential Mode (2T)

$$\kappa_3 R = 3.0542 \qquad \psi_3 = \cos 2\theta J_2(\kappa_3 r) \qquad \psi_5 = \sin 2\theta J_2(\kappa_3 r) \tag{9.12}$$

Degeneracy of the tangential modes leads to the existence of standing or spinning waves, but here we consider only standing waves.

One interesting feature is that the coefficients F_{nnn} and G_{nnn} associated with nonlinear gasdynamical self-coupling are nonzero only for the first radial mode. However, these are eliminated in the averaging process so, as in the case of purely longitudinal modes, there is apparently no significant self-coupling. Absence of self-coupling is a qualitative property of the problem that has significant consequences in respect to nonlinear stability.

The averaged equations for A_n and B_n are:

First Tangential Mode (1T)

$$\frac{dA_1}{dt} = \alpha_1 A_1 + \theta_1 B_1 + 2\left[a_1 b_{12}\cos\Omega_1 t + a_2 b_{13}\cos\Omega_2 t\right] + 2\left[a_1 a_{12}\sin\Omega_1 t + a_2 a_{13}\sin\Omega_2 t\right] \tag{9.13}$$

$$\frac{dB_1}{dt} = -\theta_1 A_1 + \alpha_1 B_1 - 2\left[a_1 b_{12}\sin\Omega_1 t + a_2 b_{13}\sin\Omega_2 t\right] + 2\left[a_1 a_{12}\cos\Omega_1 t + a_2 a_{13}\cos\Omega_2 t\right] \tag{9.14}$$

First Radial Mode (1R)

$$\frac{dA_2}{dt} = \alpha_2 A_2 + \theta_2 \beta_2 + b_1\left[(A_1^2 - B_1^2)\cos\Omega_1 t - 2A_1 B_1 \sin\Omega_1 t\right] \tag{9.15}$$

$$\frac{dB_2}{dt} = -\theta_2 A_2 + \alpha_2 B_2 + b_1\left[(A_1^2 - B_1^2)\sin\Omega_1 t + 2A_1 B_1 \cos\Omega_1 t\right] \tag{9.16}$$

Second Tangential Mode (2T)

$$\frac{dA_3}{dt} = \alpha_3 A_3 + \theta_3 B_3 + b_2\left[(A_1^2 - B_1^2)\cos\Omega_2 t - 2A_1 B_1 \sin\Omega_2 t\right] \tag{9.17}$$

$$\frac{dB_3}{dt} = \alpha_3 B_3 - \theta_3 A_3 + b_2\left[A_1^2 - B_1^2)\sin\Omega_2 t + 2A_1 B_1 \cos\Omega_2 t\right] \tag{9.18}$$

where

$$a_1 = 0.1570\left(\frac{\bar{a}}{R}\right) \qquad a_2 = -0.0521\left(\frac{\bar{a}}{R}\right)$$

$$b_1 = -0.1504\left(\frac{\bar{a}}{R}\right) \qquad b_2 = 0.1873\left(\frac{\bar{a}}{R}\right)$$

and

$$\Omega_1 = 2\omega_1 - \omega_2 = -0.1493\left(\frac{\bar{a}}{R}\right) \qquad \Omega_2 = 2\omega_1 - \omega_3 = 0.6282\left(\frac{\bar{a}}{R}\right) \tag{9.19}$$

Note particularly that all of the nonlinear terms on the right-hand sides of equations (9.13) – (9.18) contain modulation factors oscillating at either Ω_1 or Ω_2.

Solutions for two combinations of two modes have been obtained: first tangential and first radial modes (1T/1R); and first and second tangential modes (1T/2T). For the case (1T/1R), write A_n and B_n in terms of an amplitude and phase:

$$A_n = r_n(t)\cos\Phi_n(t) \qquad B_n = r_n(t)\sin\Phi_n(t) \tag{9.20}$$

and substitute in (9.13) – (9.18). Dropping terms dependent on the second tangential mode leads to the system

$$\frac{dr_1}{dt} = \alpha_1 r_1 + a_1 r_1 r_2 \cos(2\Phi_1 - \Phi_2 + \Omega_1 t) \tag{9.21}$$

$$\frac{dr_2}{dt} = \alpha_2 r_2 + b_1 r_1^2 \cos(2\Phi_1 - \Phi_2 + \Omega_1 t) \tag{9.22}$$

$$\frac{d\Phi_1}{dt} = -\theta_1 - a_1 r_2 \sin(2\Phi_1 - \Phi_2 + \Omega_1 t) \tag{9.23}$$

$$\frac{d\Phi_2}{dt} = -\theta_2 + b_1 \frac{r_1^2}{r_2} \sin(2\Phi_1 - \Phi_2 + \Omega_1 t) \tag{9.24}$$

The fact that all right-hand sides contain the same time-varying argument, $2\Phi_1 - \Phi_2 + \Omega_1 t$ is crucial to producing simple results.

In the limit cycle, the amplitudes r_{no} are constant, and their values are found as the solutions to the algebraic equations given by setting to zero the left-hand sides of (9.21) – (9.24). The results are

$$r_{10} = \sqrt{\frac{\alpha_1\alpha_2}{a_1 b_1}}\left[1 + \left(\frac{2\theta_1 - \theta_2 - \Omega_1}{2\alpha_1 + \alpha_2}\right)^2\right]^{\frac{1}{2}} \tag{9.25}$$

$$r_{20} = -\frac{\alpha_1}{a_1}\left[1 + \left(\frac{2\theta_1 - \theta_2 - \Omega_1}{2\alpha_1 + \alpha_2}\right)^2\right]^{\frac{1}{2}} \tag{9.26}$$

Because $a_1 b_1$ is negative (see the values given above), $\alpha_1\alpha_2$ must be negative to make r_{10} real. The physical reason is the same as that explained in connection with the same result (9.8) for longitudinal modes.

Just as for longitudinal modes, we must allow for frequency shifts in the limit cycle. Eventually the results are found for the time-dependent amplitudes η_{no}:

$$\begin{aligned} \eta_{10} &= r_{10}\sin[(\omega_1 + \nu_1)t + \zeta_1] \\ \eta_{20} &= r_{20}\sin[2(\omega_1 + \nu_1)t + \zeta_2] \end{aligned} \tag{9.27a,b}$$

where

$$2\zeta_1 - \zeta_2 = \tan^{-1}\left[\frac{(\omega_2 - \theta_2) - 2(\omega_1 - \theta_1)}{2\alpha_1 + \alpha_2}\right] \tag{9.28}$$

We can set either ζ_1 or ζ_2 equal to zero because zero phase is arbitrary. In the limit cycle, the participating motions must have frequencies that are integral multiples of the fundamental in order that the motion be periodic. That (9.27)a, b satisfy this requirement is a partial confirmation that the approximations used in the averaging process are correct. The result was found also by Zinn and Powell (1971) in their numerical analysis of transverse modes. Figure 9.3 (taken from Yang and Culick 1990) shows two examples of limit cycles for this case; the values of the limiting amplitudes are independent of the initial conditions.

Establishing the conditions for stability of the limit cycles follows the procedure outlined in the preceding section. The necessary and sufficient conditions are

$$\begin{aligned} \alpha_1\alpha_2 &< 0 \\ \alpha_1 + \alpha_2 &< 0 \\ 2\alpha_1 + \alpha_2 &< 0 \end{aligned} \qquad (9.29)a,b,c$$

For $\alpha_1 < 0$ the additional condition must be met

$$\alpha_1 > \frac{-|K|}{2+\xi}\left[\frac{-\xi^2+2\xi+2}{\xi^2+2\xi+2}\right]^{\frac{1}{2}} \qquad (9.30)$$

where $K = 2\theta_1 - \theta_2 - \Omega_1$ and $\xi = \alpha_2/\alpha_1$ must lie in the range $1-\sqrt{3} < \xi < 0$.

Similar results have also been found for the case (1T/2T). Moreover, the case of three modes (1T, 2T, 1R) can be solved for the amplitudes in the limit cycle but simple formulas defining the stability of those limit cycles cannot be found.

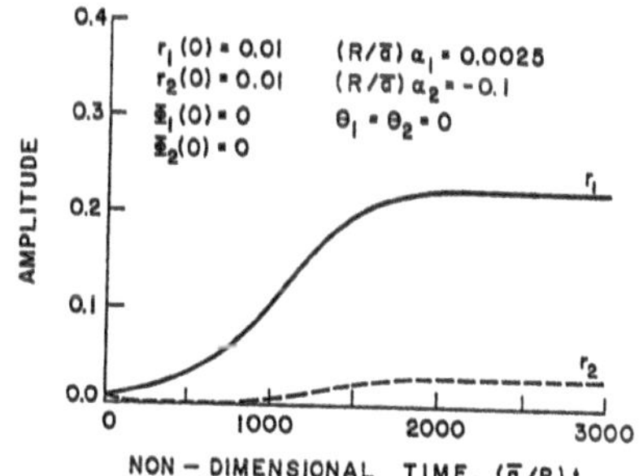

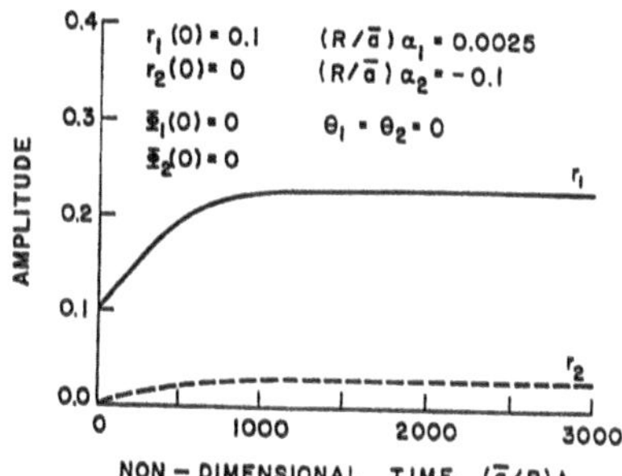

FIGURE 9.3

10. An Application of Dynamical Systems Theory: Continuation Theory of Limit Cycles and Bifurcation Theory

The approximate analysis described here is intended to serve two main purposes: to provide an efficient means of performing routine calculations for analysis and design; and to furnish a theoretical framework within which observed behavior of combustion instabilities may be understood. Even with the formal representation reduced to a set of coupled first-order equations, it is difficult to extract the sort of qualitative information necessary to satisfactory understanding. Accordingly, much effort has been expended in the past few years on the simplest possible case, the approximate model in which only two modes are accounted for. It is remarkable, a consequence of the special form of the gasdynamical nonlinearities, that simple explicit results can be obtained for a useful variety of special problems, as we have seen in the preceding section.

Because of its simplicity, the two-mode approximation based on time-averaging is an attractive basis for studying the nonlinear characteristics of combustion instabilities. This model can of course also accommodate other nonlinear processes, although presently results are available only for nonlinear gasdynamics. Within the model itself, it is impossible to assess the limits of validity, or the general consequences of, the two basic features defining the model: truncation of the modal expansion (5.39) to two terms; and time-averaging. One way of checking the accuracy of the results is to compare directly with results of numerical simulations, as in the example given for longitudinal modes in Section 9.

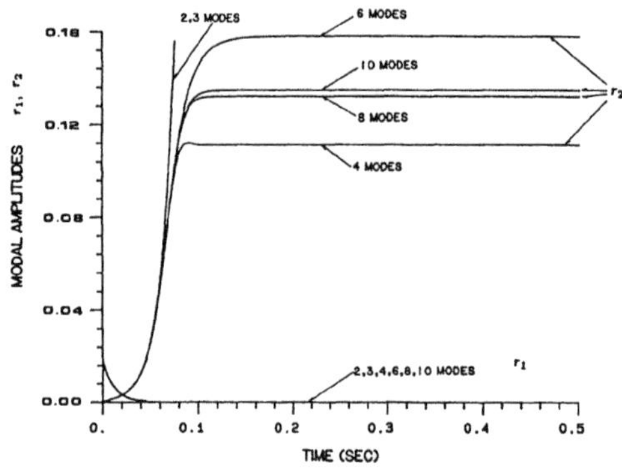

FIGURE 10.1

Some effects of truncation can be discovered simply by solving a succession of problems, increasing the number of modes with the same linear parameters α_n, θ_n. What one finds depends in the first instance where the values for α_1, θ_1, α_2, θ_2 lie on the stability diagram for limit cycles. Figure 10.1 (taken from Paparizos and Culick 1989) shows that, for example, a limit cycle unstable in the two-mode approximation can, under some circumstances, be made stable by accounting for more modes. One conceivable strategy would be a survey of special cases, solving the time-averaged equations and the second-order oscillator equations, with and without modal truncation.

It becomes an immense task to discover all possible sorts of behavior due to the large number of parameters: for a given geometry, there are two for each mode considered plus whatever quantities are introduced* by nonlinear processes. That is why exact solutions, such as those for the two-

* Note that nonlinear gas dynamics alone does not bring any new parameters, in addition to those set by the geometry, if the mean flow is not accounted for.

mode approximation, are so important. However, only a few exact solutions have been found, and only for problems restricted in three important aspects already noted: the only nonlinear process is gasdynamics; the modal expansion is truncated to a small number of modes; and the time-averaged equations have been used. For both theoretical purposes and for practical applications it is essential to avoid these restrictions.

In the absence of analytical solutions, it is necessary to resort to numerical methods. Numerical simulations for the system of coupled oscillators can be carried out without serious difficulties, but that is an expensive procedure. It is more efficient and productive of useful information to apply a continuation method for computing stationary states (e.g. limit cycles) of the dynamical system as one or more parameters are changed. Well known in other fields, the approach has been adapted by Jahnke and Culick (1993) to investigate nonlinear combustion instabilities. The procedure is broadly the following, for the case of longitudinal modes and only nonlinear second-order acoustics.

First the system (5.41) of oscillator equations is put in the form

$$\dot{x} = f(x, \mu(\epsilon)) \tag{10.1}$$

where x is a vector of dependent variables (the state vector), μ represents the parameters $(\alpha_n,\ \theta_n)$, and ϵ is a small parameter, here of the order of the average Mach numbers. After rescaling time, $t \to \omega_1 t$ and defining the variable $\delta_n = \dot{\eta}_n$, the equations (5.41) with F_n written for longitudinal modes, can be put in the form (10.1), with $x = (\eta_n,\ \delta_n)$:

$$\begin{aligned}
\dot{\eta}_n &= \delta_n \\
\dot{\delta}_n &= -n^2\eta_n + 2\alpha_n\delta_n + 2n\theta_n\eta_n \\
&= -\sum_{i=1}^{n-1}\left[C_{ni}^{(1)}\delta_i\delta_{n-i} + D_{ni}^{(1)}\eta_i\eta_{n-i}\right] \\
&= -2\sum_{i=1}^{\infty}\left[C_{ni}^{(2)}\delta_i\delta_{n-i} + D_{ni}^{(2)}\eta_i\eta_{n+i}\right]
\end{aligned} \tag{10.2}$$

To obtain numerical results, the modal expansion must be truncated at some number of modes $n = N$. Following the method outlined in Section 5.3.2, or similar procedures, the system (10.2) can also be time averaged. By carrying out calculations for both (10.2) and the time-averaged system, and for increasing $N \geq 2$, it is then possible to determine the consequences of modal truncation and time averaging.

The next step is to determine the stationary states defined by $\dot{x} = 0$, the solution to

$$f(x, u) = 0 \tag{10.3}$$

or the corresponding time-averaged equation. This is of course exactly what was done in Section 9 to determine the existence of limit cycles. In general, solving (10.3), which is a set of N coupled nonlinear equations, becomes difficult. A continuation method (we have used that worked out by Doedel and Kernevez 1984) is a recipe for finding the values of the dependent variables x_0 defining the stationary states as the parameters μ are varied ('continued') from some initial values. At each stage, after the values of μ have been changed, the stability of the new stationary state is determined by computing the eigenvalues of the Jacobian matrix arising in the linearized problem. That process also identifies bifurcations.

Use of a continuation method does not resolve the difficulty of determining the behavior over broad ranges of many parameters. However, the procedure does impose a certain systematic approach and yields much more information than one obtains from numerical simulations alone. One example makes the point.

The two-mode approximation with time averaging produces the result summarized in Figure 10.2 for longitudinal modes: stable limit cycles exist only if the values of the parameters α_n, θ_n $(n = 1,2)$ lie in certain ranges defined by a stability boundary. It is not possible within that restricted analysis to state whether the presence of the stability boundary is intrinsic to the physical mode, or is due to the approximations of modal truncation and/or time averaging. With the continuation method we have partly resolved the question. For a particular set of fixed parameters $(\theta_1, \alpha_2, \theta_2)$ and α_1 varied, the stability region for the original equations without time averaging is reduced and in fact a turning point bifurcation appears as α_1 is increased from zero.

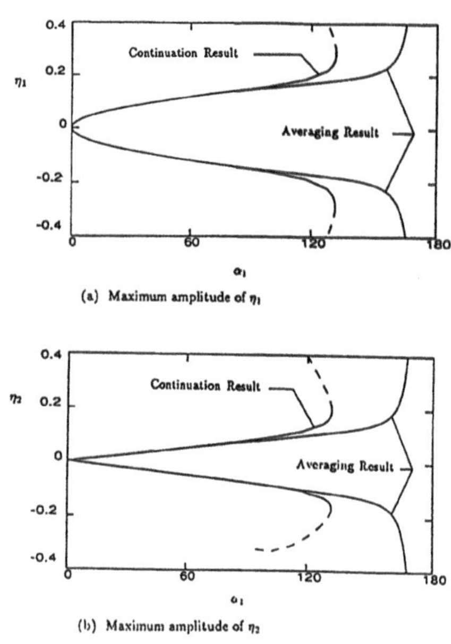

FIGURE 10.2

More interestingly and significant is the result, again for fixed values of the other parameters, the stability boundary apparently does not exist if enough modes are accounted for. Figure 10.1 already suggests the possible importance of truncation. With the continuation method, the effect of truncation can be found in straightforward fashion. The values of all parameters except α_1 are fixed ($\alpha_n < 0$ for $n \geq 2$) and α_1 is varied from zero so the first mode is linearly unstable. Figures 10.2 and 10.3 show the amplitudes for the cases of two modes and four modes. With the averaged equations and the two-mode approximation, the amplitudes become infinite when $\alpha_1 = 131$ (i.e. on the stability boundary). However, the second-order equations have a turning-point bifurcation at $\alpha_1 = 131$, Figure 10.2.

In contrast, when four modes are accounted for, as shown in Figure 10.3, there is neither a stability boundary nor a bifurcation for α_1 less than 300. No results have been computed in this case for the time-averaged equations although other examples suggest that the region of stability is at least expanded when more modes are included.

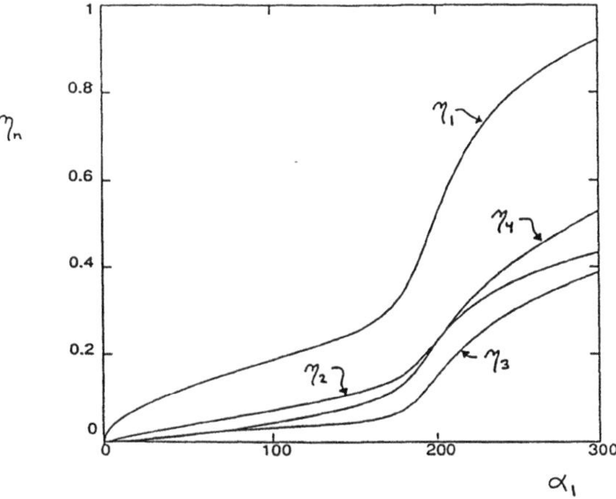

FIGURE 10.3

Further details and examples are given by Jahnke and Culick (1993). It seems at this time that the application of a continuation method offers the best means of understanding the global behavior of the dynamical system (5.41) representing combustion instabilities. Work is in progress to include other nonlinear processes and to obtain results for modes other than purely longitudinal.

11. Passive and Active Control of Combustion Instabilities

The ultimate practical aim of studying combustion instabilities is to develop sufficient understanding to be able to control the amplitudes to acceptable levels. Quite generally there are only two strategies to follow: change the basic design of the system (geometry, injector type, etc.), or introduce some sort of control. Because constraints set for other reasons often prevent sufficient changes in design, much effort has necessarily been exerted to devise means of control, always by passive means in operational propulsion systems.

11.1 PASSIVE CONTROL

'Passive control' means installation of rods, baffles, resonators, or acoustic liners that serve in the first instance to alter the natural modes and frequencies. The essential point is to force the resonance to occur in frequency ranges where the driving mechanisms are inadequate to sustain oscillations. A secondary, but potentially more significant reason for the effectiveness of baffles, remarked upon in connection with the F-1 engine in Section 2.1, is the possibility for shadowing regions of sensitive processes from disturbances. Figure 11.1 is a reproduction of a photograph showing the injector face of the Lunar Module Ascent Engine. The design included baffles and slots around the periphery that acted as acoustic resonators.

Figure 11.2 illustrates the idea that radial baffles extending axially into the chamber cause certain modes to be favored. A single baffle placed along a diameter, for example, will discourage spinning

FIGURE 11.1

modes. These symmetrically placed radial baffles act against the first and second tangential modes but do allow the third mode. The effectiveness of baffles is limited by the practical constraint that they cannot extend to far from the injector face because structural integrity may be sacrificed and flow losses may be unacceptable.

Whatever the device used, limitations always arise because of its frequency response. Resonators may be the most obvious example, typically showing a fairly narrow peaked response whose height is reduced and width is increased for larger amplitude motions. Figure 11.2 shows some results for the attenuation constant for three types of resonators (Nestlerode and Oberg 1968). Although the behavior of passive devices data taken in scale models at room temperature is inexpensive to acquire and usually serves as a useful guide in design. The performance of these control devices is most conveniently expressed directly as an attenuation constant.

11.2 ACTIVE CONTROL

Active control of combustion instabilities has received considerable attention in the past ten years although it is not a new idea. Tsien (1953) suggested that the chugging instability in a liquid rocket motor could be stabilized by controlling the supply of propellant. Figure 11.3 is a reproduction of his sketch explaining the proposal. The main item is a feedback control loop based on sensing the pressure and controlling the line capacitance. The system may be modeled approximated as a single mode in the approximate analysis described in Section 5. With a time-lag model of combustion, and using p in place of the amplitude η, the oscillator equation is

$$\ddot{p} + 2\alpha\dot{p} + \omega_0^2 p = \beta p(t - \tau) + u(t) \tag{11.1}$$

where $u(t)$ is the control. In the chugging mode the chamber pressure is practically uniform so p

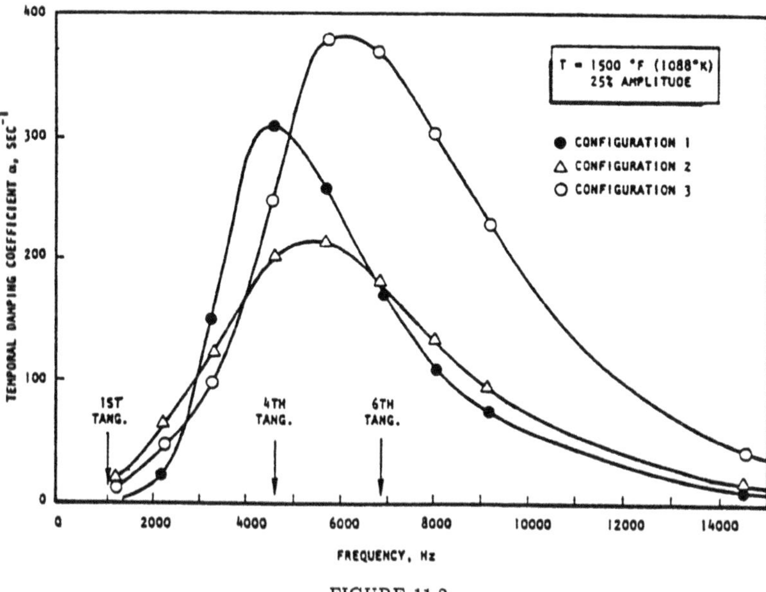

FIGURE 11.2

here indeed represents the pressure measured everywhere. Taking the Laplace transform of (11.1), we find the relation between the transform $P(s)$ of the pressure and that of the control force $V(s)$:

$$P(s) = \frac{G}{1 - GQ} V(s) \tag{11.2}$$

where G is the transfer function of the chamber (the "plant" in the terminology of control theory,

$$G(s) = \frac{1}{s^2 + 2\alpha s + \omega_0^2} \tag{11.3}$$

and

$$Q = \beta e^{-s\tau} \tag{11.4}$$

Well-known methods of classical control theory may be applied to investigate the performance and stability of this feedback system, although things are a bit complicated by the presence of the time lag.

Tsien's method of active control has never been successfully applied in practice. When it was first proposed, it is likely that the available instrumentation and transducers were inadequate. Subsequently the problem of chugging, most familiar in the form of the 'POGO' instability, has been satisfactorily solved by passive control in operational systems.

Contemporary ideas of active control began about ten years ago as an outgrowth of the field of noise control. A common and successful method of noise control is based on the idea of 'anti-sound' (Ffows-Williams 1984) in which destructive interference is caused by injecting an appropriate acoustic field to cancel the unwanted noise field. Almost all demonstrated applications of active control have been based on that idea. Culick (1988) has given a brief review of the work through

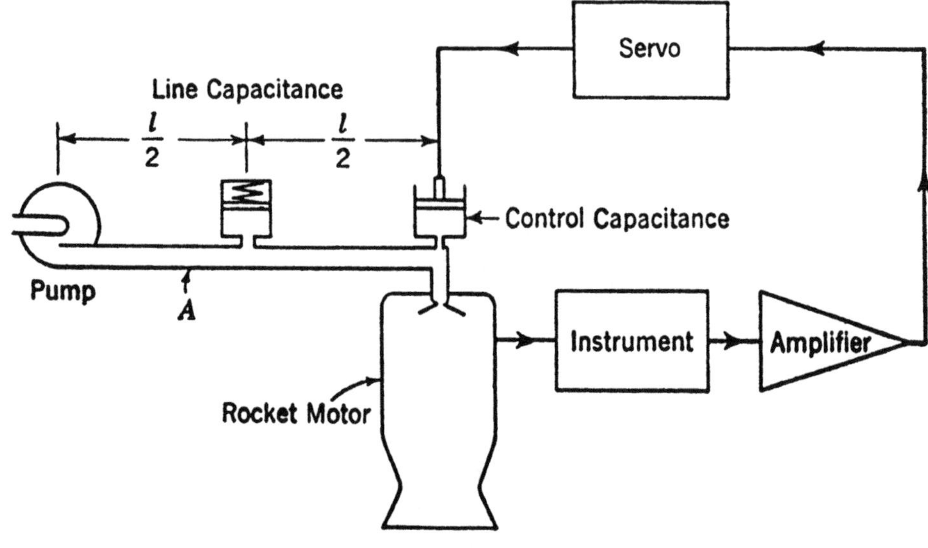

FIGURE 11.3

1988. Similar work has continued, largely in France (Candel 1992) and in a program sponsored by the office of Naval Research in the U.S. (see for example Gutmark *et al.* 1990 and Culick *et al.* 1991).

As emphasized earlier, the analysis described in Section 5 has a form that is naturally adapted to apply modern control theory. In the control field, the subject is called control of disturbed systems. The source terms denoted by subscript $(\ )_e$ in the analysis accommodate any practical means of control, including action at the boundary of the system. The only systematic work on control of combustion instabilities based on this approach appears in the Ph.D. thesis of Fung (1991) that has appeared in papers by Fung *et al.* (1991) and Yang *et al.* (1992).

In this context, any analysis of active control begins with the set of coupled oscillator equations (5.41). The theoretical work then divides into two main parts: modeling of the dominant physical processes, and designing a controller. The most difficult part of the problem is the modeling, which necessarily includes also modeling the actuators used. Combustion systems offer especially challenging problems in this respect because they are intrinsically nonlinear, contain substantial noise, and the activity of control changes the characteristics of the plant being controlled. Another way of expressing the last point is that there are inevitably uncertainties in the value of the parameters characterizing the plant.

If the linear behavior is assumed, and the system has been satisfactorily modeled, then the available methods of modern control theory can be applied. It is a much more difficult matter if nonlinear processes are accounted for. In his thesis, Fung has treated some aspects of nonlinear behavior, but with linear control theory.

For practical applications, the most serious obstacle is probably actuation. There seems to be general agreement that if active control is to succeed with operational systems, actuation will likely involve control of the fuel supply. There are only a few results available, and those have been obtained with gas-filled laboratory devices. Control with liquid fuels is evidently much more difficult and no successes have been reported.

A fundamental reason for difficulties of control with the fuel supply is understanding the internal processes: modeling is again a central problem. Moreover, it is likely that significant time lags are unavoidable — indeed the time lags are probably variable and non-uniform in space.

Hence, while active control of combustion instabilities seems to be a promising application, success with large-scale systems is by no means assumed. Much research remains to be accomplished.

12. Nonlinear Acoustics With Noise

Combustion noise in the absence of acoustic oscillations was a subject of research in the 1970s and early 1980s, but has received little attention since that time. Although Chu, Plett, and Summerfield (1975) carried out some calculations for noise in a duct sustaining linear resonances, the problem seems not to have been considered further. The first treatments of nonlinear problems of noise and acoustics are those by Culick *et al.* (1991) and Clavin, Kim, and Williams (1993). There are some similarities between the two works, but there are significant differences as well.Because (not surprisingly) the approach taken in the first of those references is based on the approximate analysis described here in Section 5, we shall reproduce here the relevant sections of that report.

The subject of nonlinear combustion instabilities in the presence of noise has practical importance. First, as any observations show immediately, combustion chambers contain substantial noise and although there seems to be little affect on instabilities, that conclusion has not been proved. Indeed, Clavin *et al.* suggest that if certain conditions are satisfied, there may be at least one quite interesting consequence. For a thorough understanding of combustion instabilities, the influences of noise must be comprehended. Second, if active control is to become a practical reality, then it is likely that the presence of substantial noise in the system being controlled must be accommodated. That is a phenomenon not readily treated with contemporary control theory and one of the characteristics that makes active control of a combustion system a challenging problem.

12.1 ACCOUNTING FOR NOISE SOURCES IN THE APPROXIMATE ANALYSIS

The basis for the analysis developed here is combination of the ideas discussed by Chu and Kovasznay (1958) with the approximate analysis developed previously for studying combustion instabilities. As we already noted in Section 5, equation (5.39), we assume that any unsteady field in a combustion chamber can be synthesized of normal modes of the chamber, with time-varying amplitudes, $\eta_n(t)$. After spatial averaging, the problem comes down to solving the system of nonlinear equations (5.41) for the $\eta_n(t)$. The right-hand sides will contain nonlinear terms of three sorts representing interactions among the acoustic modes themselves, among the non-acoustic motions. and coupling between the acoustic and non-acoustic fluctuations. Apart from use of spatial averaging, the analysis here differs from that of Chu and Kovasznay (1958) in the explicit consideration of nonlinear processes. Their treatment rested on a linearization of the equations of motion by successive or iterative approximation. Also, because eventually we divide the analysis into the two separate problems of finding the acoustic and non-acoustic amplitudes, we have the possibility of investigating stochastic processes. Therefore, unlike the case for combustion instabilities, in which only a few modes are normally present, here we must expect that a broad spectrum of modes must be present.

First we need to write the formula (5.42)a for the F_n explicitly showing dependence on the fluctuations; we drop the terms F_{ne} since we have no interest in control in this discussion. We assume that the average thermodynamic properties ($\bar{p}$, $\bar{\rho}$, $\bar{T}$) are constant in time and uniform in space. The surface integrals arising from the first, second, and last terms of f, equation (5.36), can be combined with corresponding volume integrals of h to give

$$-\frac{\bar{p}E_n^2}{\bar{a}^2}F_n = \bar{\rho}\int\left(\vec{\bar{u}}\cdot\nabla\vec{u}' + \vec{u}'\cdot\nabla\vec{\bar{u}}\right)\cdot\nabla\psi_n dV + \frac{1}{\bar{a}^2}\frac{\partial}{\partial t}\int\left(\vec{\bar{u}}\cdot\nabla p' + \gamma p'\nabla\cdot\vec{\bar{u}}\right)\psi_n dV$$
$$+\bar{\rho}\int\left(\vec{u}'\cdot\nabla\vec{u}' + \frac{\rho'}{\bar{\rho}}\frac{\partial\vec{u}'}{\partial t}\right)\cdot\nabla\psi_n dV + \frac{1}{\bar{a}^2}\frac{\partial}{\partial t}\int\left(\vec{u}'\cdot\nabla p' + \gamma p'\nabla\cdot\vec{u}'\right)\psi_n dV$$
$$+\bar{\rho}\iint\frac{\partial\vec{u}'}{\partial t}\cdot\hat{n}\psi_n dS + \int\left(\vec{\mathcal{F}}'\cdot\nabla\psi_n - \frac{\partial\mathcal{P}'}{\partial t}\psi_n\right)dV \tag{12.1}$$

Now use a vector identity to rewrite the integrand in the first integral

$$\vec{\bar{u}}\cdot\nabla\vec{u}' + \vec{u}'\cdot\nabla\vec{\bar{u}} = \nabla\left(\vec{\bar{u}}\cdot\vec{u}'\right) - \vec{u}'\times\nabla\times\vec{\bar{u}} - \vec{\bar{u}}\times\nabla\times\vec{u}'$$

A second identity gives

$$\nabla\left(\vec{\bar{u}}\cdot\vec{u}'\right)\cdot\nabla\psi_n = \nabla\cdot\left(\vec{\bar{u}}\cdot\vec{u}'\nabla\psi_n\right) - \left(\vec{\bar{u}}\cdot\vec{u}'\right)\nabla^2\psi_n$$
$$= \nabla\cdot\left(\vec{\bar{u}}\cdot\vec{u}'\nabla\psi_n\right) + k_n^2\vec{\bar{u}}\cdot\vec{u}'\psi_n$$

Then because we assume $\hat{n}\cdot\nabla\psi_n = 0$, we find

$$\int\left(\vec{\bar{u}}\cdot\nabla\vec{u}' + \vec{u}'\cdot\nabla\vec{\bar{u}}\right)\cdot\nabla\psi_n dV = k_n^2\int\left(\vec{\bar{u}}\cdot\vec{u}'\right)\psi_n dV - \int\left(\vec{\bar{u}}\cdot\nabla\times\vec{u}' + \vec{u}'\times\nabla\times\vec{\bar{u}}\right)\cdot\nabla\psi_n dV$$

and (12.1) becomes

$$-\frac{\bar{p}}{\bar{a}^2}E_n^2F_n = \bar{\rho}\int\left(\vec{\bar{u}}\cdot\vec{u}'\right)\psi_n dV - \bar{\rho}\int\left(\vec{\bar{u}}\times\nabla\times\vec{u}' + \vec{u}'\times\nabla\times\vec{\bar{u}}\right)\cdot\nabla\psi_n dV$$
$$+\frac{1}{\bar{a}^2}\frac{\partial}{\partial t}\int\left(\vec{\bar{u}}\cdot\nabla p' + \gamma p'\nabla\cdot\vec{\bar{u}}\right)\psi_n dV + \bar{\rho}\int\left(\vec{u}'\cdot\nabla\vec{u}' + \frac{\rho'}{\bar{\rho}}\frac{\partial\vec{u}'}{\partial t}\right)\cdot\nabla\psi_n dV$$
$$+\frac{1}{\bar{a}^2}\frac{\partial}{\partial t}\int\left(\vec{u}'\cdot\nabla p' + \gamma p'\nabla\cdot\vec{u}'\right)\psi_n dV + \bar{\rho}\iint\frac{\partial\vec{u}'}{\partial t}\cdot\hat{n}\psi_n dS \tag{12.2}$$
$$+\int\left(\vec{\mathcal{F}}'\cdot\nabla\psi_n - \frac{\partial\mathcal{P}'}{\partial t}\psi_n\right)dV$$

This formula for F_n shows explicitly only those processes associated purely with gasdynamics. In principle, with $\vec{\mathcal{F}}'$ and $\mathcal{P}'$, we can accommodate any physical process. Because the presence of the mean flow field is included to first order — a fundamental matter for problems of linear stability — the effects of convection and refraction are in some sense included here. However, because emphasis has been placed on standing or traveling acoustic waves, these effects have not been a matter of concern for studies of combustion instabilities, in contrast to the case of aerodynamic noise generation. That is, the effects of non-uniform flow to first order have been accounted for in analysis of combustion instabilities, but without attending to the detailed examination required to distinguish the particular consequences of convection and refraction. Some work in this direction has recently been done numerically (e.g. Baum 1988).

Although we shall not discuss extensively the functions $\vec{\mathcal{F}}'$ and $\mathcal{P}'$, they contain significant contributions. As we have discussed in the preceding section, the term $\partial Q'/\partial t$ in $\partial\mathcal{P}'/\partial t$ has been established as the dominant source for noise production in flames. Unsteady heat release is unquestionably the main reason for both acoustic instabilities and noise in combustion chambers. However, because they contain important sources of attenuation and noise production due to fluctuating flow variables, the remaining terms in (12.2) must be retained. The fluctuations $\vec{\mathcal{F}}'$ and $\mathcal{P}'$ are essential for analysis of control, for they contain the only means of actuation, through sources of mass, momentum, and energy.

The next step requires expressing the fluctuations p', $\vec{u}'$, and ρ' in terms of the time-dependent amplitudes $\eta_n(t)$. We assume that the flow variables can be split into acoustic and non-acoustic parts; write:

$$\begin{aligned} p' = p^a + \tilde{p} &= \bar{p} \sum_{j=1}^{\infty} \eta_j(t)\psi_j(\vec{r}) \\ &= \bar{p} \sum_{j=1}^{\infty} \left(\eta_j^a + \tilde{\eta}_j\right) \psi_j(\vec{r}) \end{aligned} \tag{12.3}$$

where the η_j^a are the amplitudes of acoustic oscillations (combustion instabilities). The $\tilde{\eta}_j$ are the amplitudes of the pressure fluctuations associated with the rest of the motions synthesized as a superposition of normal acoustic modes. To compute the amplitudes as solutions to the system (5.41) we need to devise an approximation to F_n, i.e. we must deal with p', ρ', and $\vec{u}'$ in the integrals. Following the tactic successfully used for studying combustion instabilities, we assume that the acoustic part of the motions satisfies the classical acoustic equations,

$$\begin{aligned} \frac{\partial \vec{u}^a}{\partial t} &= -\frac{1}{\bar{\rho}} \nabla p^a \\ \frac{\partial p^a}{\partial t} &= -\gamma \bar{p} \nabla \cdot \vec{u}^a \end{aligned} \tag{12.4a,b}$$

Because there is no entropy change associated with acoustic waves, $\bar{p} + p^a \sim (\bar{\rho} + \rho^a)^\gamma$ and we set $\rho^a = p^a/\bar{a}^2$. Hence the expansions of $\vec{u}^a$ and ρ^a are

$$\begin{aligned} \vec{u}^a &= \sum_{j-1}^{\infty} \frac{\dot{\eta}_j^a}{\gamma k_j^2} \nabla \psi_j \\ \rho^a &= \frac{\bar{p}}{\bar{a}^2} \sum_{j=1}^{\infty} \eta_j^a \psi_j \end{aligned} \tag{12.5a,b}$$

The formulas (5.39) and (12.5)a satisfy equations (12.4)a,b exactly because the eigenfunctions ψ_j satisfy (5.37)a and in zeroth order $\ddot{\eta}_j + \omega_j^2 \eta_j = 0$ with $\omega_j = \bar{a} k_j$.

Thus we approximate the acoustical variables in F_n by their values existing in zeroth order with no mean flow or sources. For the remaining parts, $\tilde{p}$, $\tilde{\rho}$, and $\tilde{u}$, we appeal to results established by Chu and Kovasznay (1958) who showed that the vortical and entropic modes of propagation carry velocity changes, but no pressure changes. We shall assume also that the density is constant; hence we *assume* in zeroth approximation

$$\tilde{p} = \tilde{\rho} = 0 \quad ; \quad \tilde{u} \neq 0 \tag{12.6}$$

It is not necessary at this stage to take $\tilde{\rho} = 0$; we do so to eliminate a few terms. However, if $\tilde{\rho}$ is taken to be non-zero, the problem arises later of relating $\tilde{\rho}$ to other thermodynamic properties. With these approximations, to second order in the fluctuations (12.2) becomes

$$F_n = F_n^a + \tilde{F}_n + \tilde{F}_n^a \tag{12.7}$$

where

$$F_n^a =$$
$$-\frac{\gamma}{\bar{p}E_n^2}\left[\bar{\rho}k_n^2\int(\vec{\bar{u}}\cdot\vec{u}^a)\,\psi_n dV - \bar{\rho}\int(\vec{u}^a\times\nabla\times\vec{\bar{u}})\cdot\nabla\psi_n dV + \frac{1}{\bar{a}^2}\frac{\partial}{\partial t}\int(\vec{\bar{u}}\cdot\nabla p^a + \gamma p^a\nabla\cdot\vec{\bar{u}})\,\psi_n dV\right]$$
$$-\frac{\gamma}{\bar{p}E_n^2}\left[\bar{\rho}\int\left(\vec{u}^a\cdot\nabla\vec{u}^a + \frac{1}{\bar{a}^2}p^a\frac{\partial\vec{u}^a}{\partial t}\right)\cdot\nabla\psi_n dV + \frac{1}{\bar{a}^2}\frac{\partial}{\partial t}\int(\vec{u}^a\cdot\nabla p^a + \gamma p^a\nabla\cdot\vec{u}^a)\,\psi_n dV\right]$$
$$-\frac{\gamma}{E_n^2}\iint\frac{\partial\vec{u}^a}{\partial t}\cdot\hat{n}dS - \frac{\gamma}{\bar{p}E_n^2}\int\left(\vec{\mathcal{F}}^a\cdot\nabla\psi_n - \frac{\partial\mathcal{P}^a}{\partial t}\psi_n\right)dV \quad (12.8)$$

$$\tilde{F}_n^a = -\frac{\gamma}{\bar{p}E_n^2}\left[\bar{\rho}\int\left(\vec{u}^a\cdot\nabla\tilde{u} + \tilde{u}\cdot\nabla\vec{u}^a + \frac{p^a}{\bar{a}^2}\frac{\partial\tilde{u}}{\partial t}\right)\cdot\nabla\psi_n dV + \frac{1}{\bar{a}^2}\frac{\partial}{\partial t}\int(\tilde{u}\cdot\nabla p^a + \gamma p^a\nabla\cdot\tilde{u})\,\psi_n dV\right]$$
$$-\frac{\gamma}{\bar{p}E_n^2}\int\left(\tilde{\mathcal{F}}^a\cdot\nabla\psi_n - \frac{\partial\tilde{\mathcal{P}}^a}{\partial t}\psi_n\right)dV \quad (12.9)$$

$$\tilde{F}_n = -\frac{\gamma}{\bar{p}E_n^2}\left[\bar{\rho}k_n^2\int(\vec{\bar{u}}\cdot\tilde{u})\,\psi_n dV - \bar{\rho}\int(\vec{\bar{u}}\times\nabla\times\tilde{u} + \tilde{u}\times\nabla\times\bar{u})\cdot\nabla\psi_n dV\right]$$
$$-\frac{\gamma}{E_n^2}\int(\tilde{u}\cdot\nabla\tilde{u})\cdot\nabla\psi_n dV - \frac{\gamma}{E_n^2}\iint\frac{\partial\tilde{u}}{\partial t}\cdot\hat{n}\psi_n dS - \frac{\gamma}{\bar{p}E_n^2}\int\left(\tilde{\mathcal{F}}\cdot\nabla\psi_n - \frac{\partial\tilde{\mathcal{P}}}{\partial t}\psi_n\right)dV \quad (12.10)$$

With the preceding definitions, the system (2.41) becomes

$$\frac{d^2\eta_n}{dt^2} + \omega_n^2\eta_n = F_n^a + \tilde{F}_n^a + \tilde{F}_n \quad (12.11)$$

The forcing functions F_n^a and $\tilde{F}_n$ contain the stochastic properties of the random field.

Previous work (Culick 1975, 1990) has shown that if only the gasdynamic nonlinear processes are accounted for to second order, F_n^a has the general form

$$F_n^a = \sum_{i=1}^{\infty}[D_{ni}\dot{\eta}_i^a + E_{ni}\eta_i^a] + \sum_{i=1}^{\infty}\sum_{j=1}^{\infty}[A_{nij}\eta_i^a\eta_j^a + B_{nij}\dot{\eta}_i^a\dot{\eta}_j^a] \quad (12.12)$$

where the coefficients $D_{ni}, \ldots$ are constants, depending on the linear processes (including those arising from $\vec{\mathcal{F}}^a$ and $\mathcal{P}^a$) and on the geometry of the chamber. If other nonlinear processes are accounted for, accommodated by the source functions $\vec{\mathcal{F}}^a$ and $\mathcal{P}^a$ (e.g. nonlinear interactions between combustion and the acoustic field), additional terms will appear, not necessarily having the form shown in (12.12).

According to its definition (12.9), the force $\tilde{F}^a$ contains terms linear in the acoustic amplitudes and in the random variables $\tilde{u}$ and $\partial\tilde{u}/\partial t$ plus terms arising from $\tilde{\mathcal{F}}^a$ and $\tilde{\mathcal{P}}^a$. If we ignore the possibility of higher order nonlinearities in those source terms, we may assume that $\tilde{\mathcal{F}}_n^a$ has the general form

$$\tilde{F}^a = \sum_{i=1}^{\infty}[\xi_i\eta_i^a + \xi_i^v\dot{\eta}_i^a] \quad (12.13)$$

where the ξ_i, ξ_i^v represent (stochastic) parametric excitation of the acoustic modes.

Finally, $\tilde{F}_n$ depends only on the stochastic fluctuations, spatially averaged over the node shapes. To make the notation more consistent, we replace $\tilde{F}_n$ by the symbol Ξ_n:

$$\tilde{F}_n \equiv \Xi_n \tag{12.14}$$

With these definitions, we write system (12.11) showing explicitly the dependence on stochastic processes:

$$\frac{d^2\eta_n}{dt^2} + \omega_n^2\eta_n = F_n^a + \sum_{i=1}^{\infty}[\xi_i\eta_i^a + \xi_i^v\dot{\eta}_i^a] + \Xi_n \tag{12.15}$$

Note that the ξ_i and the ξ_i^v depend linearly on the $\tilde{\eta}_i$ and $\dot{\tilde{\eta}}_i$ while Ξ_n is quadratic in those variables.

An important fundamental aspect of the systems treated here is the general problem of distinguishing deterministic and stochastic behavior. The natural splitting of the forces F_n into the components F_n^a, $\tilde{F}_n^a$, and $\tilde{F}_n$ defined by (12.8) – (12.10) is only an initial step. For unsteady motions in combustion chambers, we must pay particular attention to the source functions $\mathcal{P}^a$, $\tilde{\mathcal{P}}^a$, and $\tilde{\mathcal{P}}$, for they contain representations of energy addition. The acoustic field is purely deterministic, so $\mathcal{P}^a$ contains only deterministic energy sources. At the other extreme, $\tilde{\mathcal{P}}$ includes sources of energy addition that are purely stochastic, i.e., depend only on the turbulent motions and are insensitive to the deterministic acoustic field. Interactions between the deterministic and stochastic fields belong to $\tilde{\mathcal{P}}^a$. For example, turbulent fluctuations in a reacting shear layer may be sensitive to the local acoustic field, thereby affecting the energy release. A difficult part of the subject is construction of realistic models of these processes.

We have reduced the problem to finding the response of a system of nonlinear oscillators whose amplitudes $\eta_n = \eta_n^a + \tilde{\eta}_n$ must be determined when the system is subject to stochastic forces. The developments leading to the definitions of ξ_i, ξ_i^v, and Ξ_n have produced formulas (not given here) showing their explicit dependence on the random fluctuations of the velocity field. Before solutions can be found for the time-dependent amplitudes, those functions should evidently be expressed in terms of the $\tilde{\eta}_i$ and $\dot{\tilde{\eta}}_i$, analogous to the way in which the acoustic contributions in F_n were treated by using the approximations (12.4)a, b.

That is a large and difficult problem. There are at least four approaches to its solution: (1) construct an approximate representation of the flow variables $\tilde{\vec{u}}$, $\tilde{p}$, $\tilde{\rho}$ and the combustion processes, based on observations and guided by theoretical considerations, in the same spirit as we have treated the acoustic field; (2) extract a representation from numerical simulations using methods of computational fluid dynamics; (3) extract a representation by analysis of experimental data; and (4), the crudest approach, assume forms for the various contributions to $\tilde{F}_n^a$ and $\tilde{F}_n$ and determine the consequences by solving the system (12.11). None of the first three approaches have been investigated extensively for the class of problems considered here. The works by Chiu, Plett, and Summerfield (1975) and by Hegde, Reuter, and Zinn (1988) fall in the first category, but treat only linear behavior. A few elementary results have been reported by Menon and Jou (1990) following the second approach.

Methods of analyzing experimental data or the results of numerical simulations have become active subjects of research in the past few years. There is a growing body of literature dealing mainly with non-reacting flows. For example, the method called 'proper orthogonal decomposition,' originated in probability theory, has been applied to turbulent flows (Lumley 1967, Aubrey *et al.* 1988) and other problems of fluid mechanics (Sirovich 1987). That and related methods applied to a wide range of problems were central issues in a recent IUTAM/NATO Workshop (1991). Although those ideas are clearly applicable to the unsteady motions in combustion chambers, no results have been reported.

In order to gain some idea of the possible influences of stochastic sources on the behavior of acoustic modes, we have followed the last of the approaches listed above. We simply assume forms for the random functions of time, ξ_i, ξ_i^v, and Ξ_n, and treat the system of stochastic differential equations

$$\frac{d^2\eta_n}{dt^2} + \omega_n^2\eta_n = F_n^a + \sum_{i=1}^{\infty}[\xi_i\eta_i + \xi_i^v\dot{\eta}_i] + \Xi_n \tag{12.16}$$

This was the approach taken by Paparizos and Culick (1989)*a* for the simplest case of two longitudinal modes; the basic idea is discussed here in Section 9. Note that η_i^a, $\dot{\eta}_i^a$ have been replaced by $\eta_i = \eta_i^a + \tilde{\eta}_i$ and $\dot{\eta}_i = \dot{\eta}_i^a + \dot{\tilde{\eta}}_i$ on the right-hand side, implying the assumption that the random part of the amplitudes is relatively small.

In fact it appears that we may quite realistically assume that the ξ_i, ξ_i^v, and Ξ_i do *not* depend on the amplitudes η_n of the pressure fluctuations. This assumption has been made in all treatments of the aerodynamic noise problem. In other words, we assume that the random sources of noise do not themselves depend on the noise they generate. That does not exclude dependence of the noise field on the presence of coherent acoustic oscillations, nor does it imply that control of the noise field is impossible. The reason is that the acoustic and random pressure fluctuations are coupled by nonlinear processes, producing energy transfer between the two forms of motion. Practical matters of achieving successful control will rest partly on the strength of the energy exchange and on the possibility of discovering an effective means of control or 'actuation.' However, it may also be the case that if the random sources of noise are affected by the pressure fluctuations, then more effective means of control might be found.

Part of our justification for taking the stochastic sources independent of the pressure field rests on the results discussed by Chu and Kovasznay (1958) who established the independence of the three modes of propagation (acoustic, vortical, and entropic). The argument, however, is weak, ignoring some basic properties of processes present in combustion systems. For example, heat released in flames is indeed sensitive to pressure. Our assumption here is really made in the interests of obtaining some initial results.

12.2 COMBUSTION INSTABILITIES WITH STOCHASTIC SOURCES

The formulation developed in the preceding section accommodates a wide range of problems. Here we shall describe a few results for what is possibly the simplest case: the influence of stochastic sources on a combustion instability. Observations of instabilities in both laboratory and full-scale combustors normally show the presence of a small number of acoustic modes; the data cannot reveal how many modes must be accounted for in an analysis to explain satisfactorily the nonlinear behavior. In the theory used here, a minimum of two modes are required to produce a limit cycle. Energy is supplied to the acoustical system because one mode is unstable; the motions may then reach a steady state only if a stable second mode is available to dissipate energy, thereby providing the possibility for constant total energy in the acoustical motions. Nonlinear gasdynamical processes cause the flow of energy from the unstable mode to the stable mode.

Pressure records of instabilities rarely show purely harmonic motions. The amplitudes for unfiltered data usually fluctuate about some apparent average value. Those fluctuations are erased if the data are filtered. Otherwise, they will cause broadening of peaks in the spectra and as part of the broadband spectral background. The analysis described in this section is an initial effort to explain this behavior. Only limited results have been obtained.

For the acoustic system described by (12.16) with the stochastic sources equal to zero, and only gasdynamical nonlinear processes accounted for, exact solutions can be found for the time-averaged equations for two (and in some cases three) modes. Most of the known results have been reviewed by Culick (1990). In particular, the conditions for existence and stability of limit cycles can be written explicitly. Although the restriction to two modes carries limitations, serious under some conditions (Jahnke and Culick 1991), this representation is surprisingly accurate under wide realistic circumstances. To maintain simplicity, we therefore consider here only the two-mode approximation.

Moreover, we treat the case of longitudinal modes for which the frequencies are integral multiples of the fundamental, $\omega_n = n\omega_1$. With all linear processes and only the gasdynamic nonlinearities included, the first two of equations (12.16) are

$$\begin{aligned}\ddot{\eta}_1 + \omega_1^2\eta_1 &= 2(\alpha_1\dot{\eta}_1 + \theta_1\omega_1\eta_1) - (F_{11}\dot{\eta}_1\dot{\eta}_2 + F_{12}\eta_1\eta_2) + (\xi_1\eta_1^a + \xi_1^v\dot{\eta}_1^a + \Xi_1)\\ \ddot{\eta}_2 + \omega_2^2\eta_2 &= 2(\alpha_2\dot{\eta}_2 + \theta_2\omega_2\eta_2) - (F_{21}\dot{\eta}_1^2 + F_{22}\eta_1^2) + (\xi_2\eta_2^a + \xi_2^v\dot{\eta}_2^a + \Xi_2)\end{aligned} \qquad (12.17)a,b$$

where $\omega_2 = 2\omega_1$ and

$$\begin{aligned} F_{11} &= \frac{3-2\gamma}{2\gamma} \qquad & F_{12} &= \frac{5(\gamma-1)}{2\gamma}\omega_1^2\\ F_{21} &= -\frac{\gamma+3}{2\gamma} \qquad & F_{22} &= \frac{\gamma-1}{2\gamma}\omega_1^2\end{aligned} \qquad (12.18)$$

There are six stochastic functions in equation (12.17), in general mutually independent, to be specified. The functions bring with them a large number of parameters, too many at this stage for a sensible treatment of the problem. We shall therefore analyze a much simpler problem, motivated by the following reasoning based on previous results (Paparizos and Culick 1989*b*) obtained for the deterministic problem with no stochastic sources.

Combustion instabilities in practice commonly appear as oscillations having slowly varying maximum amplitudes and phases; the amplitudes η_i may then be assumed to have the form

$$\begin{aligned}\eta_i(t) &= r_i(t)\cos(\omega_i t - \phi_i(t))\\ &= A_i(t)\cos\omega_i t + B_n(t)\sin\omega_i t\end{aligned} \qquad (12.19)$$

where $\delta r_i/r_i$ and $\delta\phi_i/2\pi$, the changes over one cycle, are assumed small. The method of time-averaging is effective under these conditions, reducing the two second order equations for η_1 and η_2 to four first order equations for the $(r_1, r_2, \phi_1, \phi_2)$.

For technical reasons not covered here, the following transformation is an effective choice to simplify the structure of the first order equations:

$$\begin{aligned} y_1 &= r_1\\ y_2 &= r_2\sin(\phi_2 - 2\phi_1)\\ y_3 &= r_2\cos(\phi_2 - 2\phi_1)\end{aligned} \qquad (12.20)a,b,c$$

The deterministic equations for the y_i are:

$$\begin{aligned}\frac{dy_1}{dt} &= (\alpha_1 + \beta y_2)y_1\\ \frac{dy_2}{dt} &= \alpha_2 y_2 + \theta^* y_3 + 2\beta y_3^2 - \beta y_1^2\\ \frac{dy_3}{dt} &= -\theta^* y_2 + \alpha_2 y_3 - 2\beta y_2 y_3\end{aligned} \qquad (12.21)a,b,c$$

where

$$\beta = \frac{\gamma+1}{8\gamma}\omega_1$$
$$\theta^* = \theta_2 - 2\theta_1 \tag{12.22)a, b}$$

Equations (12.21) are unchanged if y_3 is replaced by $y_3|\theta_2-2\theta_1|/(\theta_2-2\theta_1)$, showing that no generality is lost by taking θ^* positive. Note that the original set of four equations (not given here) are reduced to three, essentially because there is an arbitrary phase available, the reason that the two phase angles appear only in the combination $\phi_2 - 2\phi_1$ in (12.20)a, b, c.

Fixed points of the system (12.21), defined by $\dot{y}_i = 0$, are the state of rest ($y_i = 0$) or limit cycles. It can be shown (Paparizos and Culick 1989b) that a unique limit cycle, whose properties are independent of initial conditions, exists if $\alpha_1\alpha_2 < 0$. As remarked above, if one mode (say the fundamental) is unstable, $\alpha_1 > 0$, the other must be stable, $\alpha_2 < 0$. In this case, the limit cycle itself is stable if $2\alpha_1 + \alpha_2 < 0$. Hence for existence and stability of the limit cycle, the conditions must be satisfied when $\alpha_1 > 0$ are

$$\alpha_2 < 0$$
$$2\alpha_1 + \alpha_2 < 0 \tag{12.23)a, b}$$

The remainder of this section is concerned with the influences of stochastic sources on the characteristics of this limit cycle.

12.4 STOCHASTIC AVERAGING

Stratonovich (1963, Vol. II, Chapter 4) introduced the method of stochastic averaging, an extension of the method of averaging outlined above for a deterministic system. See also Roberts and Spanos (1986) and Gardiner (1985) for reviews of the method. Here we simply quote the results of applying the method.

With stochastic sources in the second order equations (12.17)a, b, the averaging proceeds in two steps. First the equations are averaged over the period of the acoustic oscillations (τ_n for the nth mode) to give, eventually, equations (12.21)a, b, c in the variables y_i; that is, the oscillatory motions have been removed from the problem which has been reduced to one of finding the slowly varying amplitudes and phases (equivalent to the $y_i(t)$).

The second step is the 'stochastic averaging,' required to give the fluctuating (stochastic) sources in correct form for the averaged (i.e. first order) equations. We assume that the stochastic sources $\xi_i(t)$, $\xi_i^v(t)$, and $\Xi_i(t)$ have zero mean (eventually we assume that they have the form of white noise). However, the terms $\xi_i\eta_i$ and $\xi_i^v\dot{\eta}_i$, representing parametric excitations, have non-zero mean values. It is convenient to remove the mean values from the stochastic sources and place them with the deterministic part, accounting for the terms m_i appearing below. By definition, m_i is the average over one period of the oscillation of the expected value of the term in question.

It is also assumed that the auto-correlation times of the random processes are much smaller than the bandwidths (not the periods) of the oscillations. That means that the oscillations have non-zero decay (or growth) coefficients such that

$$\tau_{\text{corr}} \ll |\alpha_i| \tag{12.24}$$

a condition satisfied by the broad-band processes and acoustical motions expected in practice. In the limit considered here, we approximate the random processes by equivalent white noise.

Finally, in order to reduce the difficulty and complexity of finding solutions to equations (12.17)a, b we make two assumptions based partly on physical grounds:

(i) We replace η_i^a and $\dot{\eta}_i^a$ in the terms representing parametric excitation by η_i and $\dot{\eta}_i$. This implies that the stochastic perturbations of the acoustic modes should be small: $|\tilde{\eta}_i|/|\eta_i^a| \ll 1$. That happens to be true in the examples treated here, but not in general.

(ii) We ignore all stochastic sources in the equation (12.17)b for the amplitude of the second mode. This assumption produces considerable simplification, but is also motivated by the expectation that if the fundamental mode has a growth rate much larger than the decay rate of the second mode, then the first-order effects of stochastic sources will be primarily on the fundamental mode. Owing to nonlinear coupling between the modes, there will of course be a secondary random excitation of the second mode.

Omitting the lengthy details of stochastic averaging (which in fact begins with the second order equations) we eventually find the equations for the variables y_i:

$$\begin{aligned} dy_1 &= \left(\alpha_1^* y_1 + \beta y_1 y_2 + \frac{\pi}{2y_1\omega_1^2} S_3(\omega_1)\right)dt + \sqrt{Q_{11}}dw_1 \\ dy_2 &= (\alpha_2^* y_2 + \theta^* y_3 + 2\beta y_3^2 - \beta y_1^2)dt - 2y_3\sqrt{Q_0}dw_2 \\ dy_3 &= (\alpha_2^* y_3 - \theta^* y_2 - 2\beta y_2 y_3)dt + 2y_2\sqrt{Q_0}dw_3 \end{aligned} \qquad (12.25)a,b,c$$

The $w_i(t)$ are Gaussian (white noise) processes, the dw_i being defined in terms of the stochastic integral of a function $G(t)$ (see Gardiner 1985, p. 83),

$$\int_{t_0}^{t} G(t')dw_i(t')$$

The $S_i(\omega)$ are the spectral density functions of the w_i and the Q_{ij} are cross-variances of the stochastic sources, computed in the averaging process

$$\begin{aligned} Q_{11} &= \frac{\pi y_1^2}{4}\left(2S_1(0) + S_1(2\omega_1) + \frac{1}{\omega_1^2}S_2(2\omega_1)\right) + \frac{\pi}{\omega_1^2}S_3(\omega_1) \\ Q_0 &= \frac{\pi}{4}\left(\frac{2}{\omega_1^2}S_2(0) + S_1(2\omega_1) + \frac{1}{\omega_1^2}S_2(2\omega_1) + \frac{4}{y_1^2\omega_1^2}S_3(\omega_1)\right) \\ Q_{22} &= 4y_3^2 Q_0 \\ Q_{23} &= -4y_2 y_3 Q_0 \\ Q_{33} &= 4y_2^2 Q_0 \\ Q_{12} &= Q_{13} = 0 \end{aligned} \qquad (12.26)a-f$$

The mean values of the sources are

$$\begin{aligned} m_1 &= \frac{\pi}{4}\left[y_1 S_1(0) + \frac{3}{2}y_1\left(S_1(2\omega_1) + \frac{1}{\omega_1^2}S_2(2\omega_1)\right) + \frac{2}{\omega_1^2 y_1}S_3(\omega_1)\right] \\ m_2 &= 2y_2 Q_0 \\ m_3 &= -2y_3 Q_0 \end{aligned} \qquad (12.27)a,b,c$$

Finally, the growth and decay rates of the two modes are now

$$\begin{aligned}\alpha_1^* &= \alpha_1 + \frac{\pi}{4}\Big(S_1(0) + \frac{3}{2}S_1(2\omega_1) + \frac{3}{2\omega_1^2}S_2(2\omega_1)\Big)\\ \alpha_2^* &= \alpha_2 - \pi\Big(\frac{1}{\omega_1^2}S_2(0) + \frac{1}{2}S_1(2\omega_1) + \frac{1}{2\omega_1^2}S_2(2\omega_1)\frac{+2}{\omega_1^2 y_1}S_3(\omega_1)\Big)\end{aligned} \tag{12.28)a,b$$

12.5 THE CENTER MANIFOLD FOR EQUATIONS (12.25)

Although equations (12.25)a, b, c can be solved numerically in their present form, it is useful to simplify the matter further by taking advantage of some results obtained by Paparizos and Culick (1989b). It is relatively straightforward to find the equation for the center manifold, an approximation to the one-dimensional manifold (or locus) of limit cycles extending from the origin $y_1 = y_2 = 0$. We expect that a similar result should be a good approximation here when stochastic sources are present.

Again omitting details of the argument, the result is that in limit cycles of the deterministic part of the motions — i.e. $\dot{y}_1 = \dot{y}_2 = \dot{y}_3 = 0$ and dw_1, dw_2, dw_3 ignored in equations (4.9)a, b, c — the variable y_2 can be expressed as a function of y_1 only:

$$y_2(y_1) = \begin{cases} \frac{\alpha_2^*\beta}{(\alpha_2^*)^2+(\theta^*)^2} & (y_1 < y_1^*) \\ y_2^* & (y_1 > y_1^*) \end{cases} \tag{12.29}$$

where

$$y_2^* = \frac{\alpha_2^* - \alpha_1^* + 0.5\delta_1}{3\beta} \tag{12.30}$$

$$\delta_1 = \frac{\pi}{4}\Big(S_1(0) + S_1(2\omega_1) + \frac{1}{\omega_1^2}S_2(2\omega_1)\Big) \tag{12.31}$$

and y_1^* is the smallest positive root of

$$y_1^2\frac{\alpha_2^*\beta}{(\alpha_2^*)^2 + (\theta^*)^2} = y_2^* \tag{12.32}$$

Note that α_2^*, equation (12.28)b, is a function of y_1.

Equation (12.29) is the equation of the center manifold, essentially a relation between the amplitudes in the limit cycle; substitution of (12.29) in (12.25)c, with $\dot{y}_3 = 0$ and dw_3 dropped gives the values of y_3 in the center manifold.

12.6 THE FOKKER-PLANCK-KOLMOGOROV EQUATION FOR EQUATIONS (12.25)

We now confine attention to the behavior in the limit cycle executed in the presence of stochastic sources. Assuming that the variable y_2 is sufficiently well-approximated by its dependence on the center manifold, equation (12.25)a is now a stochastic differential equation for y_1 only:

$$dy_1 = \Big(\alpha_1^* y_1 + \beta y_1 y_2(y_1) + \frac{\pi}{2y_1\omega_1^2}S_3(\omega_1)\Big)dt + \sqrt{Q_{11}}dw_1 \tag{12.33}$$

where $y_2(y_1)$ is given by (12.29).

The amplitude y_1 is now a random variable characterized by a probability density $p(y_1, t)$. With known methods (e.g. Gardiner 1985, Chapter 5) we can construct the Fokker-Planck-Kolmogorov equation for p:

$$\frac{\partial p}{\partial t} = -\frac{\partial}{\partial y_1}\left[\left(\alpha_1^* y_1 + \beta y_1 y_2(y_1) + \frac{\pi}{2 y_1 \omega_1^2} S_3(\omega_1)\right) p\right] + \frac{1}{2}\frac{\partial^2}{\partial y_1^2}[Q_{11} p] \tag{12.34}$$

This equation possesses a stationary solution independent of time,

$$p^s(y_1) = C y_1 \exp\left\{\int_0^{y_1} \frac{x(\delta_2 + 2\beta y_2(x))}{Q_{11}(x)} dx\right\} \tag{12.35}$$

where

$$\delta_2 = 2\alpha_1^* - 3\delta_1 \tag{12.36}$$

and C is a constant fixed by requiring that $p^s(y_1)$ be normalized:

$$\int_0^\infty p^s(x) dx = 1 \tag{12.37}$$

The solution (12.35) is an approximation to the distribution of amplitude r_1, in the limit cycle when stochastic sources are assumed to affect directly only the fundamental mode. Just as for the deterministic behavior, the parameters arising in the problem must satisfy a stability condition in order that a stationary solution exist ("existence in mean square value"). Here, the deterministic conditions (12.23)b becomes

$$\alpha_2 + 2\alpha_1 + \pi\left[\frac{3}{4}S_1(0) - \frac{1}{\omega_1^2}S_2(0) + \frac{3}{8}S_1(2\omega_1) + \frac{3}{8\omega_1^2}S_2(2\omega_1)\right] < 0 \tag{12.38}$$

Because the noise field is coupled to the acoustic field, producing both parametric and 'external' excitation according to the general equations (12.15), a reasonable question is: can the presence of stochastic sources be responsible for observed combustion instabilities? That is, are the noise sources sufficiently strong to cause excitation and sustenance of acoustic modes that are deterministically stable ($\alpha_n < 0$)?

To obtain meaningful results, we must ensure that a stationary probability distribution exists. The conditions to be satisfied are

$$\alpha_n + \frac{\pi}{4}\left[S_1^{(n)}(2\omega_n) + \frac{1}{\omega_n^2}S_2^{(n)}(2\omega_n)\right] < 0 \tag{12.39}$$

where $S_1^{(n)}$ and $S_2^{(n)}$ are the power spectral density fluctuations of the parameter excitations ξ_n and ξ_n^v in the nth acoustic mode. For technical reasons of analysis, we are ignoring the 'external' excitations Ξ_n which cannot cause modes to be excited under the conditions treated here.

12.7 SOME NUMERICAL RESULTS

Because we have not attempted to work out physically realistic models of the non-acoustic fluctuations ($\tilde{u}$, $\tilde{p}$, $\tilde{\rho}$) we cannot claim or anticipate that the results faithfully represent observed behavior. The initial results discussed here indicate that the approach taken here, based on time and stochastic averaging, is an effective method for treating problems in which both instabilities and noise are present.

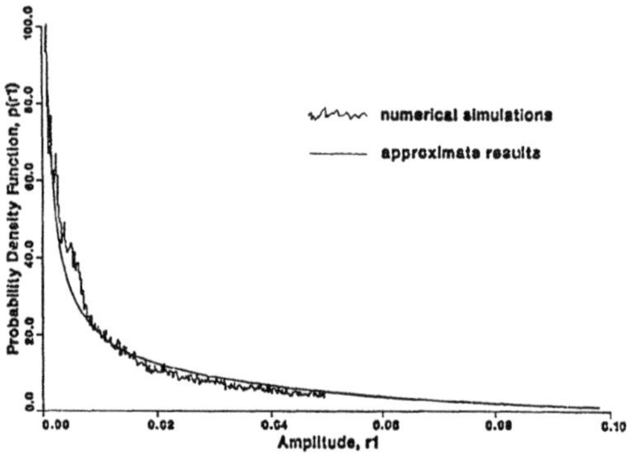

FIGURE 12.1

The analysis described above can be used to carry out three independent sorts of computations: Monte Carlo simulations with the second-order equations (12.17)a, b; Monte Carlo simulations with the first-order stochastic equations (12.25)a, b, c; and the probability distribution can be found directly as the solution to the Fokker-Planck-Kolmogorov equation (12.34).

Monte Carlo simulations with the second-order equations are very lengthy, requiring an estimated 100 CPU hours on a VAX 8800 for a simulation consisting of 100 samples, each covering 3,000 periods of the fundamental oscillation. Here we give some results of simulations with the first-order equations, requiring approximately 5 hours each. This work is continuing with other machines, but no results are available.

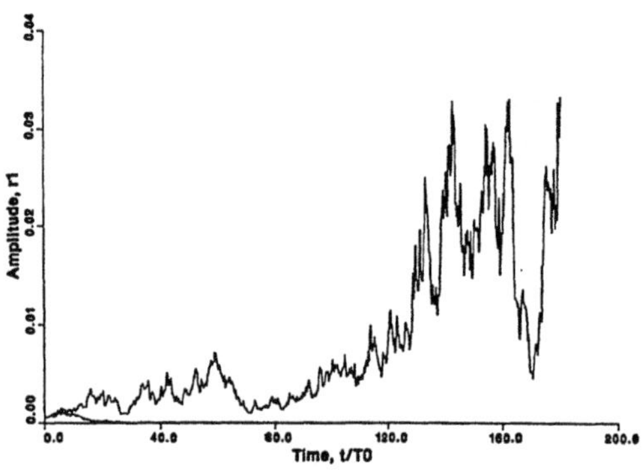

FIGURE 12.2

Attention here is confined to conditions under which stationary probability distributions should exist, with linear parameters ($\alpha_1 \ldots$) assigned realistic values. Figure 12.1 is a comparison of the distribution given by equation (12.25) with a Monte Carlo simulation. The frequency of the fundamental mode is 800 hertz, $\alpha_1 = 8\ \text{s}^{-1}$, and $\alpha_2 = -125\ \text{s}^{-1}$; both modes are damped. There is no external excitation ($\Xi_1 = 0$) and only the parametric excitation $\xi_1(t)$ is present. Good agreement is apparent. Figure 12.2 shows the time history of the amplitude of the first mode, computed for two samples chosen from the Monte Carlo simulations. Note that one is 'stable,' decaying within

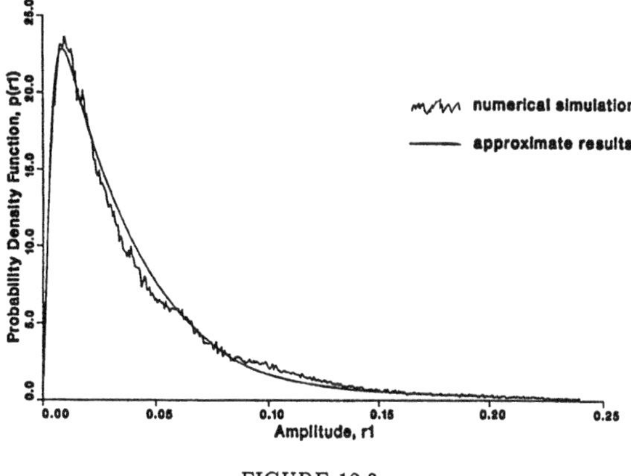

FIGURE 12.3

about 20 cycles, while the other exhibits 'asymptotic stability,' associated with the rather long tail of the distribution that is shown in Figure 12.1.

Figure 12.3 is a comparison of the distribution computed with equation (12.35) with a Monte Carlo simulation when both parametric and external excitations are present ($\xi_1(t)$, $\xi_2(t)$, and $\Xi_1(t)$ all non-zero). The parameters α_1, α_2 have the same values used to produce Figure 12.1. Again the approximation (12.35) for the probability distribution is satisfactory. Note that the external excitation is capable of causing the fundamental modes to be excited to an amplitude somewhat larger than that occasionally caused by parametric excitation alone: the most probable and average values of r_1 are larger than those predicted with the distribution given in Figure 12.1.

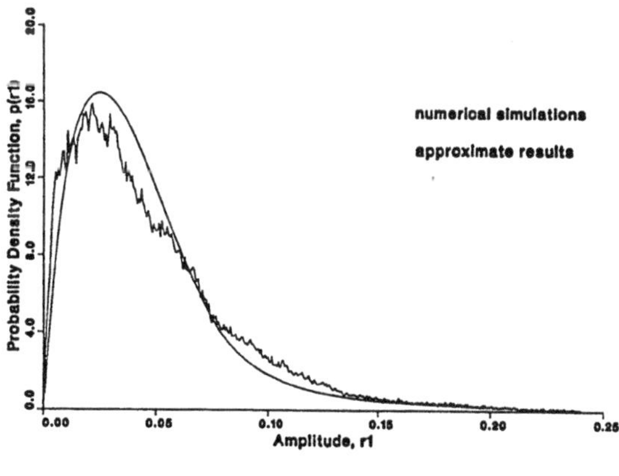

FIGURE 12.4

Figures 12.4 and 12.5 show cases in which the fundamental mode is unstable and the second is stable. The results therefore show the influence of stochastic sources on the amplitude of the first mode in a limit cycle. Now the most probable value of the amplitude naturally obtains a higher value, larger in Figure 9 for which α_1 has greater magnitude: in Figure 8, $\alpha_1 = 8\ \text{s}^{-1}$ and in Figure 12.5, $\alpha_1 = -25\ \text{s}^{-1}$.

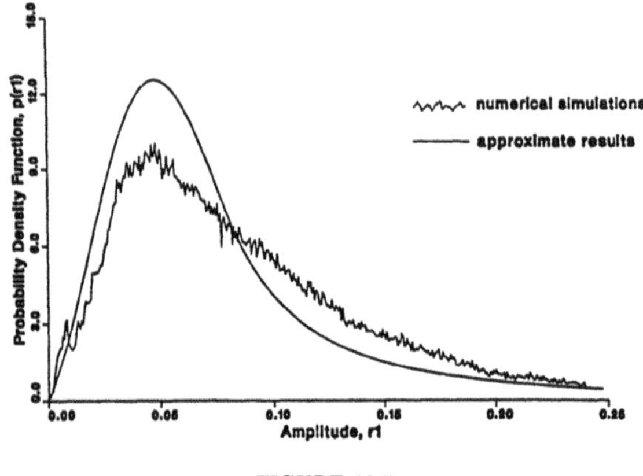

FIGURE 12.5

13. Concluding Remarks

Possibly the most significant obstacle to understanding combustion instabilities is the scarcity of quantitative data systematically acquired over broad ranges of the relevant parameters. That statement is particularly true for liquid-fueled systems, despite the significant accomplishments of extensive work carried out over four decades. It is not possible, given the geometry and the fuel/oxidizer combination, to predict the properties of linear or nonlinear motions in a combustion chamber.

The situation is considerably more favorable for solid rocket motors, chiefly because the dominant combustion processes are usually confined to thin regions adjacent to the burning solid surface. However it is also true that a great deal of effort was expended in the U.S. during the 1970s to develop both analytical and experimental capabilities for investigating combustion instabilities in solid rocket motors. As a result, if one has at hand the necessary input data, notably the response function for the propellant, it is possible in many cases to predict fairly well the stability of a motor. That conclusion is by no means universally true, of course, but it is true that what is possible to accomplish quantitatively for solid rockets substantially exceeds that possible for liquid-fueled systems.

It is important to realize that the subject of combustion instabilities will likely always be semi-empirical. Theory and analytical methods are essential to understand the problem, but it is not possible to produce usable results from first principles. Observational results and quantitative data are always required. The approximate analysis described in this paper displays clearly the sort of information required. As parts of the basic theory progress (e.g. transient combustion of sprays) the corresponding contributions to the analysis of combustion instabilities demand less experimental work.

A topic not covered in this paper is numerical analysis or, to use a standard term, computational fluid dynamics. Because the physical problems are so complicated, one may state with virtually complete confidence that a complete, formally exact mathematical theory of combustion instabilities will never exist. The best one might expect is that with accurate modeling of the various participating physical and chemical processes, precise numerical results might be produced.

Apart from the intrinsic limitation that each numerical result is a special case, so acquiring a deep understanding tends to be a tedious process, the practical fact is that now and for the foreseeable future, numerical analysis of these problems is extremely expensive. Indeed, numerical solution for time-dependent, three-dimensional flows of turbulent and chemically reacting two-phase mixtures is beyond current capabilities in any realistic sense. Nevertheless, computational fluid dynamics does, of course, have important applications discussed by others in this meeting.

In the first instance, because numerical methods continue to progress rapidly (particularly the accessibility and power of computing resources), it is important and useful to take advantage of the improvements. One significant application is to provide input information (e.g. transient spray combustion) for formal analysis of instabilities. A second use is to use numerical results as a means of assessing the accuracy of the approximate analysis by applying both methods to the same problem. This is generally difficult to do because intrinsic differences between the forms of the methods means that one cannot really solve the 'same' problem.

Up to the present time, numerical analysis has been used for time-dependent one-dimensional and axisymmetric two-dimensional problems. The approximate analysis can of course be used for any geometry. Hence the obvious strategy is to confirm the validity of analytical results for the limited cases that can be treated with numerical analysis. Then one assumes that the approximate analysis can be applied with some confidence to more complicated problems.

References

Anonymous (1967) "Engine Characteristics and Stability," Chapter 4 of TRW document No. 01827-6119-T000.

First International Symposium on Liquid Rocket Engine Instability (Proceedings) (1993) The Pennsylvania State University, January 18–20, 1993.

Aubrey, N., Holmes, P., Lumley, J. L., and Stone E. (1988) "The Dynamics of Coherent Structures in the Wall Region of Turbulent Boundary Layers," *J. Fl. Mech.*, Vol. 192, pp. 115–173.

Awad, E. and Culick, F E.C. (1986) "On the Existence and Stability of Limit Cycles for Longitudinal Acoustic Modes in a Combustion Chamber," *Combustion Science and Technology*, Vol. 46, pp. 195–222.

Bergman, G.H. and Jessen, E.C. (1971) "Evaluation of Conventional Rocket Motor Instrumentation for Analysis of Oscillatory Combustion," AIAA/SAE 7th Propulsion Joint Specialist Conference, AIAA Paper No. 71-755.

Brown, R.S., Culick, F.E.C., and Zinn, B.T. (1978) "Experimental Methods for Combustion Admittance Measurements," Chapter IV in *Experimental Diagnostics in Combustion of Solids*, Vol. 63 of *AIAA Progress in Astronautics and Aeronautics.*

Browning, S.C., Krushin, M., and Thacker, J.H. (1971) "Application of Combustion Instability Technology to Solid-Propellant Rocket Motor Problems," AIAA/SAE 7th Propulsion Joint Specialist Conference, AIAA Paper No. 71-754.

Candel, S. (1992) "Combustion Instabilities Coupled by Pressure Waves and Their Active Control," *24th Symposium (International) on Combustion*, The Combustion Institute, pp. 1277–1296.

Chiu, H. H., Plett, E., and Summerfield, M. (1975) "Noise Generated by Ducted Combustion," in *Aeroacoustics: Jet Combustion and Noise*, Vol. 37 of *Progress in Astronautics and Aeronautics*, AIAA, New York, pp. 249–276.

Chu, B.-T. (1963) "Analysis of a Self-Sustained Thermally Driven Nonlinear Vibration ," *The Physics of Fluids*, Vol. 6, No. 11, pp. 1638–1644.

Chu, B.-T. and Kovasznay, L.S.G. (1957) "Non-linear Interactions in a Viscous Heat-Conducting

Compressible Gas," *J. Fluid Mech.*, Vol. 3, No. 5, pp. 494-512.

Chu, B.-T. and Kovasznay, L.S.G. (1958) "Non-linear Interactions in a Viscous Heat-conducting Compressible Gas," *J. Fl. Mech.*, Vol. 3, No. 5, pp. 494–514.

Chu, B.-T. and Ying, S.J. (1963) "Thermally Driven Nonlinear Oscillations in a Pipe with Traveling Shock Waves," *The Physics of Fluids*, Vol. 6, No. 11, pp. 1625–1637.

Clavin, P., Kim, J.-S., and Williams, F.A. (1993) "Turbulence-Induced Noise Effects on High-Frequency Combustion Instabilities," submitted to *Combustion Science and Technology.*

Crocco, L. and Cheng, S.I. (1956) *Theory of Combustion Instability in Liquid Propellant Rocket Motors.* AGARDOGRAPH, No. 8, Butterworths Scientific Publications, London.

Crocco, L. and Mitchell, C.E. (1969) "Nonlinear Periodic Oscillations in Rocket Motors with Distributed Combustion," *Combustion Science and Technology*, Vol. 1, pp. 147–169.

Culick, F.E.C. (1963) "High-Frequency Pressure Oscillations in Rocket Chambers," *AIAA J.*, Vol. 1, No. 5, pp. 1097-1104.

Culick, F.E.C. (1974) "Acoustic Oscillations in Rocket Chambers," *Astronautica Acta*, Vol. 12, No. 2, pp. 113–126.

Culick, F.E.C. (1968) "A Review of Calculations for Unsteady Burning of a Solid Propellant," *AIAA J.*, Vol. 6, No. 6, pp. 2241-2255.

Culick, F.E.C. (1971) "Nonlinear Growth and Limiting Amplitude of Acoustic Oscillations in Combustion Chambers," *Comb. Sci. and Tech.*, Vol. 3, No. 1, pp. 1-16.

Culick, F.E.C. (1974) "T-Burner Testing of Metallized Solid Propellants," Air Force Rocket Propulsion Laboratory, Report AFRPL-TR-74-28.

Culick, F.E.C. (1976) "Nonlinear Behavior of Acoustic Waves in Combustion Chambers — Parts I and II," *Astronautica Acta*, Vol. 3, pp. 714–766.

Culick, F.E.C. (1987) "A Note on Rayleigh's Criterion," *Combustion Science and Technology*, Vol. 56, 1987, pp. 159–166.

Culick, F.E.C. (1988) "Combustion Instabilities in Liquid-Fueled Propulsion Systems — An Overview," AGARD 72B PEP Meeting, Bath, England.

Culick, F.E.C. (1993) "Some Recent Results for Nonlinear Acoustics in Combustion Chambers," to appear in *AIAA J.*

Culick, F.E.C., Paparizos, L., Sterling, J.D., and Burnley, V. (1991) "Combustion Noise and Combustion Instabilities in Propulsion Systems," *Proceedings of the AGARD Conference on Combat Aircraft Noise*, AGARD CP 512.

Culick, F.E.C., Lin, W.H., Jahnke, C.C., and Sterling, J.D. (1991) "Modeling for Active Control of Combustion and Thermally Driven Oscillations," American Controls Conference (June 1991).

Culick, F.E.C. and Yang, V. (1992) "Prediction of the Stability of Unsteady Motions in Solid Propellant Rocket Motors," Chapter 18 in *Nonsteady Burning and Combustion Stability of Solid Propellants*, Progress in Astronautics and Aeronautics (L. de Luca, E. W. Price, and M. Summerfield, Eds.).

Doedel, E.J. and Kernevez, J.P. (1984) *Software for Continuation Problems in Ordinary Differential Equations with Applications*, Applied Mathematics Publication 217–50, California Institute of Technology.

Faeth, G.M., Dominicus, D.P., Tulpinsky, J.F., and Olson, D.R. (1969) "Supercritical Bipropellant Droplet Combustion," *Proceedings of the Twelfth Symposium (International) on Combustion*, p. 9.

Ffowcs-Williams, J.E. (1974) "Sources of Sound," *Proceedings 8th Congress on Acoustics*, London, pp. 1–10.

Ffowcs-Williams, J.E. (1984) "Anti-Sound," *Proc. Roy. Soc. London,* A395, pp. 63–88.

Flandro, G.A. (1967) *Rotating Flows in Acoustically Unstable Rocket Motors*, Ph.D. Thesis, Daniel and Florence Guggenheim Jet Propulsion Center, California Institute of Technology.

Fowler, J.H. and Rosenthal J.S. (1971) "Missile Vibration Environment for Solid Propellant Oscillatory Burning," AIAA/SAE 7th Propulsion Joint Specialist Conference, AIAA Paper No. 71-756.

Fung, Y.-T. (1991) *Active Control of Linear and Nonlinear Pressure Oscillations in Combustion Chambers*, Ph.D. Thesis, Dept. of Mech. Eng., Propulsion Engineering Research Center, The Pennsylvania State University.

Fung, Y.-T., Yang, V., and Sinha, A. (1991) "Active Control of Combustion Instabilities With Distributed Actuators," *Combustion Science and Technology,* Vol. 78, pp.217–245.

Gutmark, E., Parr, T.P., Hanson-Parr, D.M., and Schadow, K.C. (1990) "Use of Chemiluminescence and Neural Networks in Active Combustion Control," *23rd Symposium (International) on Combustion,* The Combustion Institute.

Harrje, D.J. and Reardon, F.H. (Ed.) (1972) *Liquid Propellant Rocket Instability*, NASA SP-194.

Hegde, V.G., Rueter, D., and Zinn, B.T. (1988) "Sound Generation by Ducted Flames," *AIAA J.*, Vol. 26, No. 5, pp. 532–537.

Hsieh, K.C., Shuen, J.C., and Yang, V. (1991) "Droplet Vaporization in High Pressure Environments," *Combustion Science and Technology,* Vol. 76, pp. 111–132.

IUTAM & NATO Advanced Research Workshop (1991) "Interpretation of Time Series From Nonlinear Mechanical Systems," 19–23 August, University of Warwick, England.

Jagna, V.W. and Ferrenberg, A.J. (1989) "The Art of Injector Design," *Rocketdyne Corporation*, Spring Issue, pp. 3–11.

Jahnke, C.C. and Culick, F.E.C. (1993) "An Application of Dynamical Systems Theory to Nonlinear Combustion Instabilities," AIAA 31st Aerospace Sciences Meeting, AIAA Paper 93–0114.

Kendrick, D.W., Zsak, T.W., and Zukoski, E.E. (1993) "An Experimental and Numerical Investigation of Premixed Combustion in a Vortex in a Laboratory Dump Combustor," Proceedings of this conference.

Krylov, N. and Bogoliubov, N. (1947) *Introduction to Nonlinear Mechanics*, Princeton University Press.

Law, C.K. (1982) "Recent Advances in Droplet Vaporization and Combustion," *Progress in Energy and Combustion Sciences,* Vol. 8, p.171.

Levine, J.L. and Baum, J.D. (1983) "A Numerical Study of Nonlinear Instability Phenomena in Solid Rocket Motors," *AIAA Journal*, Vol. 21, No. 4 (April), pp. 557–564.

Levine, J.L. and Culick, F.E.C. (1972) "Numerical Analysis of Nonlinear Longitudinal Combustion Instability in Metalized Solid-Propellant Rocket Motors," Vol. 1, Analysis and Results, Ultrasystems, Inc., report prepared for the Air Force Rocket Propulsion Laboratory, AFRPL–72–88.

Levine, J.L. and Culick, F.E.C. (1974) "Nonlinear Analysis of Solid Rocket Combustion Instability," Ultrasystems, Inc., report prepared for the Air Force Rocket Propulsion Laboratory, AFRPL–74–74–45.

Lores, E.M. and Zinn, B.T. (1973) "Nonlinear Longitudinal Instability in Rocket Motors," *Combustion Science and Technology*, Vol. 7, pp. 245–256.

Lumley, J.L. (1967) "The Structure of Turbulent Flows," in *Atmospheric Turbulence and Radio Wave Propagation,* A. M. Yaglom and V. I. Tatarski, eds., Moscow: Nauka, pp. 166–178.

Maslen, S.H. and Moore, F K. (1956) "On Strong Transverse Waves Without Shocks in a Circular Cylinder," *J. Aero. Science,* Vol. 23, pp. 583–593.

Menon, S. and Jou, W.-H. (1990) "Modes of Oscillation in a Nonreacting Ramjet Combustor Flow," *J. Prop. and Power*, Vol. 6, No. 5, pp. 535–543.

Mitchell, C.E., Crocco, L., and Sirignano, W.A. (1969) "Nonlinear Longitudinal Instability in Rocket Motors with Concentrated Combustion," *Combustion Science and Technology*, Vol. 1, pp. 269–274.

Morse, P.M. and Feshback, H. (1953) *Methods of Theoretical Physics,* McGraw-Hill Book Company, New York.

Nestlerode, J.A. and Oberg, C.L. (1969) "Combustion Instability in an Annular Engine," Sixth ICRPG Combustion Conference.

Nickerson, G. R., Culick, F.E.C., and Dang, L.G. (1983) "Standard Stability Prediction Method for Solid Rocket Motors, Axial Mode Computer Program, User's Manual," Software and Engineering Associates, Inc., report prepared for Air Force Rocket Propulsion Laboratory, AFRPL–TR–83–017.

Oefelein, J.C. and Yang, V. (1992) "A Comprehensive Review of Liquid-Propellant Instabilities in F-1 Engines," Dept. of Mech. Eng., Propulsion Engineering Research Center, The Pennsylvania State University, July 1992.

Paparizos, L. and Culick, F.E.C. (1989*a*) "The Two-Mode Approximation to Nonlinear Acoustic in Combustion Chambers I. Exact Solution for Second-Order Acoustics," *Comb. Sci. Tech.*, Vol. 65, No. 1–3, pp. 39–65.

Paparizos, L. and Culick, F.E.C. (1989*b*) "The Two-Mode Approximation to Nonlinear Acoustic Waves in Combustion Chambers with Stochastic Sources," (unpublished).

Poinsot, T., Bourienne, F., Candel, S. H., and Esposito, E. (1989) "Suppression of Combustion Instabilities by Active Control," *J. Propulsion*, Vol. 5, No. 1, pp. 14–20.

Lord Rayleigh (1945) "The Explanation of Certain Acoustical Phenomena," *Royal Institution Proceedings*, Vol. VIII, pp. 536–542, 1878. See also *The Theory of Sound*, Dover Publications, Vol. II, p. 226.

Roberts, J.B. and Spanos, P.D. (1986) "Stochastic Averaging: An Approximate Method of Solving Random Vibration Problems," *Int. J. Non-Linear Mech.*, Vol. 21, No. 2, pp. 111–134.

Schoeyer, H. (Ed.) (1993) "Combustion Instability Course," ESA/ESTEC, Noordwijk Z-H, The Netherlands.

Sirignano, W.A. (1964) *A Theoretical Study of Nonlinear Combustion Instability: Longitudinal Mode*, Ph.D. Thesis, Department of Aerospace and Mechanical Sciences, Princeton University.

Sirignano, W.A. and Crocco, L. (1964) "A Shock Wave Model of Unstable Rocket Combustors," *AIAA Journal*, Vol. 2, No. 7, pp. 1285–1296.

Smith, D.A. and Zukoski, E.E. (1985) "Combustion Instability Sustained by Unsteady Vortex Combustion," AIAA/SAE/ASME/ASEE 21st Joint Propulsion Conference, AIAA Paper No. 85-1248.

Sterling, J. D. and Zukoski, E. E. (1987) "Longitudinal Mode Instabilities in a Dump Combustor," AIAA 25th Aerospace Sciences Meeting, AIAA Paper No. 87–0220.

Stratonovich, R.L. (1963) *Topics in the Theory of Random Noise*, Vols. I and II, Gordon and Breach, New York.

Wanhainen, J.P., Parish, H.C., and Conrad, E.W. (1966) "Effect of Propellant Injection Velocity on Screech in 20,000 Pound Hydrogen-Oxygen Rocket Engine," NASA TN D-3373.

Yang, V. and Culick, F.E.C. (1990) "On the Existence and Stability of Limit Cycles for Traverse Acoustic Oscillations in a Cylindrical Combustion Chamber, I. Standing Modes," *Combustion Science and Technology,* Vol. 72, pp. 37–65.

Yang V., Sinha, A. and Fung, Y. T. (1992) "State-Feedback Control of Longitudinal Combustion Instabilities," *J. Propulsion,* Vol. 8, No. 1, pp. 66–73.

Zinn, B. T. (1986) "Pulsating Combustion," Chapter 2 in *Advanced Combustion Methods,* F.J. Weinber (Ed.), Academic Publishers, London.

Zinn, B.T. (1968) "A Theoretical Study of Nonlinear Combustion Instability in Liquid-Propellant Rocket Engines," *AIAA Journal,* Vol. 6, No. 10, pp. 1966–1972.

Zinn, B.T. and Lores, E.M. (1972) "Application of the Galerkin Method in the Solution of Nonlinear Axial Combustion Instability Problems in Liquid Rockets," *Combustion Science and Technology,* Vol. 4, pp. 269–278.

Zinn, B.T. and Powell, F.A. (1971) "Nonlinear Instability in Liquid-Propellant Rocket Engines," *Thirteenth Symposium (International) on Combustion,* pp. 491–503.

10. AIRCRAFT TURBINE ENGINE NO_X EMISSION ABATEMENT

D.W. Bahr
GE Aircraft Engines
Cincinnati, Ohio
USA

ABSTRACT

Extensive efforts are currently underway to develop combustors with lower nitrogen oxides (NOx) emissions for use in both subsonic and supersonic civil aircraft engines. Basically, NOx abatement of any significance requires means of reducing the peak flame temperatures within the combustor. To obtain these flame temperature reductions, while also maintaining acceptable combustor performance and operability at low engine power conditions, combustion process staging methods are needed. For use in advanced subsonic aircraft engines, combustors with leaner primary combustion zone fuel/air mixtures, together with combustion process staging features, are being developed. Significant NOx level reductions have been demonstrated with these advanced combustor design concepts. Although the resulting combustors are more complex than current technology combustors, satisfactory performance and operability appear attainable with these configurations. The initial introduction of these combustors into operational engines during the latter part of this decade is a likely prospect. In the case of advanced supersonic transport aircraft engines, combustors with ultralow NOx levels will probably be needed to prevent adverse impacts on the stratospheric ozone layer. For these engine applications, the use of very advanced and complex combustor design concepts is being investigated. One important element of these efforts is the development of technology for the suppression of the high-amplitude pressure oscillations typically associated with lean fuel/air mixture combustion processes.

INTRODUCTION

Within recent years, the nitrogen oxides (NOx) emissions of aircraft engines have been the subject of growing attention, along with the NOx emissions of other sources.

Relative to other sources, aircraft engines are small contributors to

F. Culick et al., (eds.), Unsteady Combustion, 243–264.

global, national and local NOx emission burdens. On a global basis, aircraft-related NOx emissions constitute less than three percent of all manmade NOx emissions [1]. However, NOx emissions from any source are contributing factors in the formation of photochemical smog in urban regions and of acid rain. Some unique concerns are also associated with aircraft engine NOx emissions because the bulk of these emissions are released at high altitudes.

Environmental impact studies indicate that NOx emissions introduced into the upper troposphere and lower stratosphere may result in the formation of ozone at these altitudes and that the resulting ozone may, in turn, be a contributing factor in causing global warming [1 and 2]. Subsonic aircraft typically cruise at these altitudes. Other studies indicate that NOx emissions introduced at higher altitudes in the stratosphere can contribute to the depletion of the stratospheric ozone layer [3]. The optimum cruise altitudes of supersonic aircraft are generally at the altitudes where the ozone layer is located.

Because of these concerns, extensive efforts are underway to develop low NOx combustors for use in both subsonic and supersonic civil aircraft engines. The progress made to date in these abatement efforts and the longer-term prospects of these efforts are outlined herein.

BACKGROUND

The aircraft turbine engine exhaust emissions of key concern consist of smoke (soot), unburned hydrocarbons (HC), carbon monoxide (CO), NOx and carbon dioxide (CO_2). The engine operating conditions at which these emissions are primarily generated are shown in Table 1 and the most significant environmental impact issues associated with these emissions are indicated in Table 2.

TABLE 1. Aircraft turbine engine emissions of concern

Emission category	Primarily associated with engine operation at:
o Smoke (soot)	Takeoff and climbout
o Unburned hydrocarbons (HC)	Low power, especially ground idle
o Carbon monoxide (CO)	Low power, especially ground idle
o Nitrogen oxides (NOx)	High power, including cruise
o Carbon dioxide (CO_2)	All power settings

TABLE 2. Key environmental impact issues associated with aircraft engine emissions

Emission category	Impact
o Smoke	Visibility nuisance around airports
o HC	Contribution to urban smog burdens
o CO	Contribution to urban CO burdens
o NOx	
- Subsonic aircraft engines	Possible contribution to global warming
- Future supersonic aircraft engines	Stratospheric ozone depletion
o CO_2	Contribution to global warming

These emissions categories are typical of those of any heat or power plant operated with fossil fuels. CO_2 is a direct and major product of fossil fuel combustion. Abatement can only be achieved via the use of alternative fuels with lower carbon contents or reduced fuel consumption. Suitable alternatives to aviation kerosene fuels do not exist at this time. However, major emphasis has been focused for many years by the air transportation industry on reducing fuel consumption. As a result of both engine and aircraft technology advances, fuel usage per passenger seat-mile has been dramatically reduced during the past three decades.

Limits on the allowable levels of smoke, HC, CO and NOx emissions of civil turbojet and turbofan engines were promulgated by the International Civil Aviation Organization (ICAO) during the 1978 to 1982 time period. These standards, which are contained in ICAO Annex 16 - Volume II [4], are expressly designed to regulate the emission quantities that can be discharged in and around airports. Standards are specified for both subsonic and supersonic civil aircraft engines. In the case of the supersonic engine standards, the regulatory limit values are closely based on the emission characteristics of the Olympus engine, which powers the Concorde aircraft.

The smoke standard is expressed in terms of a Smoke Number (SN) which is determined by means of the specific test procedure defined in Reference 4. The gaseous emission limits are expressed in terms of a parameter which consists of the total mass, in grams, of a pollutant produced during a prescribed takeoff-landing operational cycle per kilonewton of rated takeoff thrust. The takeoff-landing cycle embodied in this parameter is expressly designed to regulate the quantities of gaseous emissions that can be discharged in and around airports.

Efforts to reduce the pollutant emission levels of aircraft engines were initiated well before the promulgation of regulatory limits. During the 1965 to 1975 time period, low smoke combustors were developed and incorporated into operational engines. With these combustors, smoke visibility is virtually eliminated in most cases. During the 1970 to 1980 time period, the introduction of more fuel-efficient engines resulted in significant reductions in HC and CO levels. Later, during the 1975 to 1985 time period, combustor design features for further reducing HC and CO emission levels were developed and incorporated into operational engines. A typical example of the significant HC and CO emission reductions that were realized via the incorporation of such design features is presented in Table 3. Accordingly, significant reductions in the pollutant emission contributions of aircraft engines to urban air pollution burdens have already been realized.

TABLE 3. Emission levels of CF6-50E2 engine with original production combustor and with subsequently developed low emission combustor

Emission category		Applicable ICAO regulatory limit	Status - original production engine	Status - low emission production engine
o Smoke (SN)		18.8	6.5	12.5
o HC	(g/kN)	19.6	57.8	3.4
o CO	(g/kN)	118.0	97.3	29.8
o NOx	(g/kN)	100.0	58.2	51.6

At cruise operating conditions, the pollutant emissions of modern aircraft engines consist primarily of NOx, as is illustrated in Table 4. In the case of subsonic aircraft engines, the cruise NOx emission indices (grams per kilogram of fuel) are always lower than the levels at takeoff conditions because of the lower combustor inlet air temperatures and pressures that prevail at cruise operating conditions. In the case of supersonic aircraft engines, the highest indices occur at the supersonic cruise conditions because the combustor inlet air temperatures and pressures are generally highest at these conditions.

TABLE 4. Typical engine emission levels at cruise and sea level static (SLS) takeoff operating conditions

Engine	Emission index (g/kg fuel)			
	NOx	HC	CO	Smoke
o CF6-80C2A3				
- at 10km/0.82 Mach Number	11.7	0.1	0.6	0.01
- at SLS takeoff	34.4	0.1	0.6	0.03
o CF6-50C2				
- at 10km/0.82 Mach Number	18.8	0.2	0.5	0.01
- at SLS takeoff	36.5	0.3	0.5	0.01
o CFM56-5A				
- at 10.7km/0.82 Mach Number	12.0	0.3	0.9	0.04
- at SLS takeoff	28.0	0.2	0.8	0.07

Because only NOx emissions are generated in any significant amounts during cruise and because the HC and CO levels of modern engines have been significantly reduced at idle and other low power conditions, the emissions typically generated in a flight of an aircraft with modern engines consist primarily of NOx. A typical example of the emission mass distribution associated with a flight of a modern engine-powered aircraft is presented in Table 5. As is shown, NOx emissions predominate during both the airport-associated operations and the cruise operations.

In view of these considerations, the development of engines with reduced NOx levels is currently being pursued on a high priority basis by several government and industry organizations.

NOx EMISSION CHARACTERISTICS

NOx emissions are generated in the primary combustion zone of the engine combustor. The primary combustion zone consists of the forward volume of the combustor. The fuel is injected directly into this zone. In most aircraft turbine engines, annular combustors are used. A typical configuration is shown in Figure 1.

Within the primary zone, localized regions containing stoichiometric and near-stoichiometric fuel/air mixtures generally exist at high engine power conditions and, as a result, localized high flame temperatures exist. These flame temperatures increase significantly as the combustor inlet air temperatures and pressures increase. Thus, the highest NOx levels of subsonic engines always occur at takeoff conditions.

TABLE 5. Typical distribution of total emission mass quantities generated during a flight of an aircraft equipped with modern engines

- o Aircraft: Twin-engine transport
- o Range: 500 nautical miles

Emission Category	% of total emission mass		
	During ICAO takeoff/landing cycle	During climbout/ cruise/descent	Overall
Smoke	---	0.1	0.1
HC	0.6	1.0	1.6
CO	5.4	7.0	12.4
NOx	7.8	78.1	85.9
Total	13.8 (56.5% NOx)	86.2 (90.6% NOx)	100.0

Because of these considerations, the suppression of NOx emissions primarily involves methods of reducing the peak flame temperatures within the combustor. Such reductions can be obtained by the use of alternative fuels such as methane which have lower stoichiometric flame temperatures, by the injection of inert fluids such as water or steam into the primary combustion zone or by the use of more advanced combustor designs with special features to reduce primary zone flame temperatures at high power conditions. All of these options can be considered in stationary turbine engines. However, in aircraft turbine engines, only the latter option is relevant.

The NOx reductions that can be realized by operating with reduced primary zone flame temperatures are illustrated in Figure 2. These results were obtained by adjustments of the combustor airflow distribution of a TF39 engine combustor so that the average primary zone equivalence ratio was varied from very lean to very rich values. As is shown, operation with lean average primary zone fuel/air mixtures resulted in significantly decreased NOx levels. These lean mixtures were obtained by large increases in the percentage of the total combustor airflow that was introduced into the primary zone. In current technology combustors, the use of such lean primary zone mixtures would, of course, be expected to result in unsatisfactory ignition characteristics and much increased HC and CO levels at idle, because of the very lean primary zone mixtures that would prevail at the low engine power operating conditions.

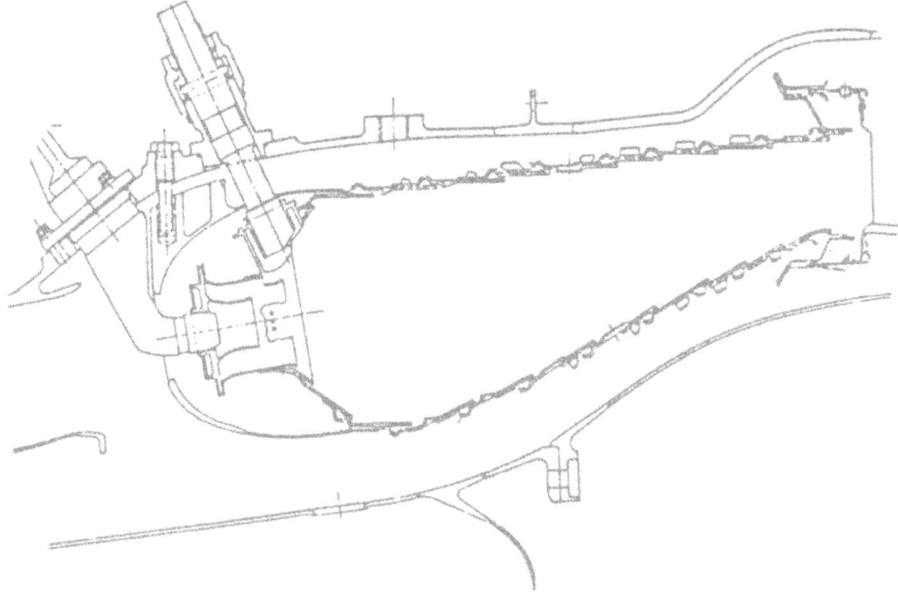

CF6-50 Engine Combustor

Figure 1. Cross-sectional drawing of typical annular combustor configuration

Smaller reductions were obtained in these tests with rich average primary zone equivalence ratios. The latter findings suggest that, to some extent, the NOx formation process shifted from the primary zone to the dilution zones immediately downstream. The use of such rich primary zone equivalence ratios in current technology combustors would be expected to result in increased smoke levels at high engine power operating conditions.

Overall, these findings suggest that a promising approach for obtaining low NOx levels is to operate with lean and uniform fuel/air primary zone mixtures, preferably homogeneous mixtures, at high engine power operating conditions. However, means of applying this general design approach are requred that do not result in unacceptable losses in ignition and lean blowout capabilities at low power conditions and in unacceptable increases in HC and CO levels at idle power. To obtain the requisite leaner primary zone mixtures at high power, increased proportions of the total combustor airflow must be introduced into the primary zone. Doing so in a conventional combustor results in unacceptable lean mixtures at low engine power conditions, because the overall engine fuel/air ratios are quite low

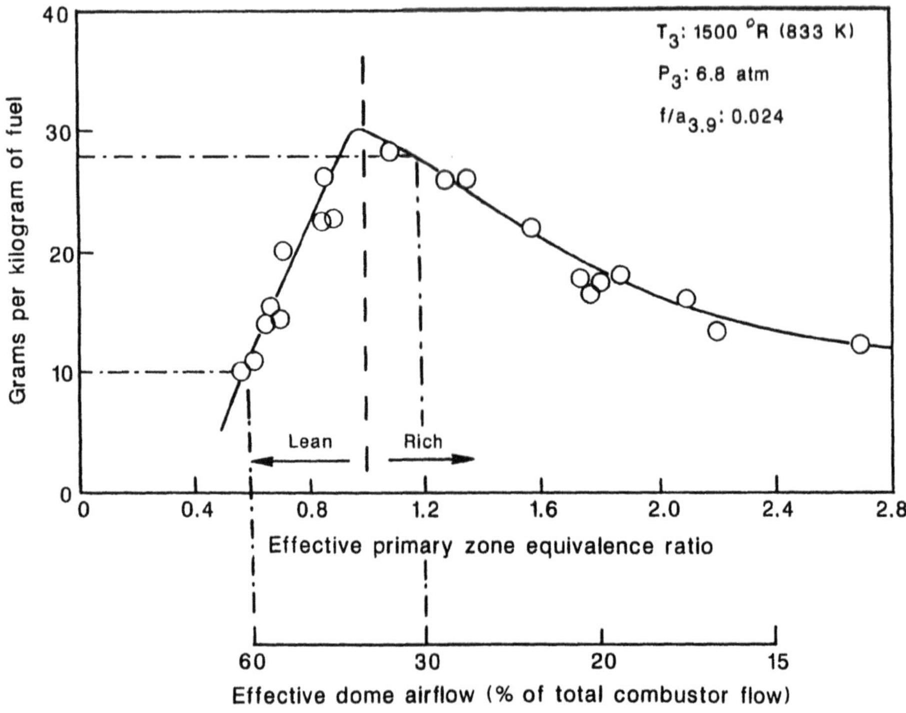

Figure 2. Typical effect of primary zone fuel/air equivalence ratio on combustor NOx levels

at these conditions. Therefore, methods are needed of obtaining both the required lean and uniform primary zone mixtures at high power conditions and of maintaining adequately rich primary zone mixtures at idle and other low power conditions. The attainment of these conflicting operating capabilities, in turn, necessitates consideration of combustors within which the combustion process can be staged via the use of two or more combustion zones or of combustors in which variable geometry features are incorporated to modulate the quantities of the total combustor airflow that are introduced into the primary combustion zone.

Another promising basic concept is the use of a rich/lean series-staged combustion process. In combustors of this kind, all of the fuel is injected into the first stage. This stage is operated with average fuel/air ratios above stoichiometric to suppress the formation of NOx. To obtain the fuel-rich mixtures, most of the combustor airflow is bypassed around the first stage and introduced further downstream. Very rapid mixing of this airflow with the first stage combustion products is needed to complete the combustion process, while at the time suppressing NOx formation as the rich gases are diluted. In the rich stage, the combustor liners must be cooled without the use of air film cooling, as is commonly used in current technology combustors. The use of film cooling is not acceptable because stoichiometric fuel/air mixtures would occur in regions close to the liner.

Some NOx abatement can also be realized by reductions in combustion gas residence time within the combustor. An example of the effect of residence time is presented in Figure 3. These data were obtained in component tests of several engine combustors. The tests of this family of annular combustors were conducted at the same inlet air conditions and with the same overall fuel/air ratio. The test configurations included the combustors of large-size engines, such as the CF6 models, and the combustors of small-size engines, such as the T700 and CF34 models. These results indicate that the NOx levels of small engine combustors are inherently lower than those of large engine combustors, when operated at the same conditions. This finding simply reflects the fact that combustor annular height and, thus, length and residence time increase directly as engine size, or thrust, is increased.

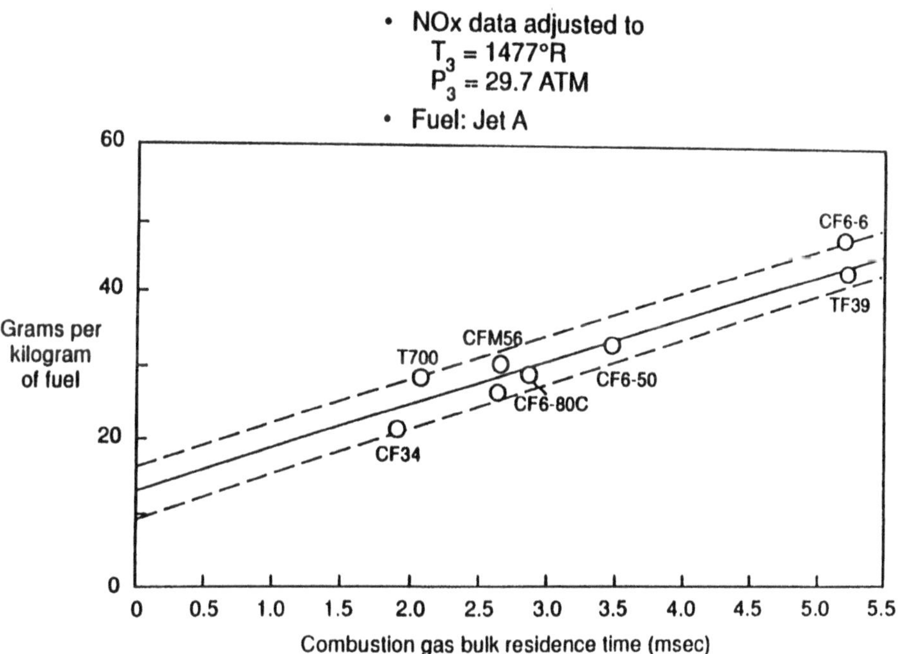

Figure 3. Effect of combustion gas residence time on combustor NOx levels

These results also provide an indication of the NOx level reductions obtainable with a combustor length, or residence time, reduction within an engine model of the same size class. The CF6-50 and CF6-80C engines have very similar thrust ratings. The CF6-80C engine is a later vintage design. Its combustor is shorter and its bulk combustion gas residence time is, therefore, shorter by about 20 percent, as is shown in Figure 3. As is also shown in Figure 3, this residence time reduction results in a NOx level reduction of only about 10 percent. Because of this consideration, together with limitations on how much more the conventional combustors for any given engine size class can be shortened without adverse performance and operability impacts, reductions in combustion gas residence time within current technology combustors appears to offer a somewhat limited NOx abatement potential.

Accordingly, the attainment of significantly reduced NOx levels in aircraft turbine engines requires the use of more advanced and complex combustor design concepts. As is summarized in Table 6, these advanced concepts involve the use of either much leaner or much richer primary zone fuel/air mixtures, at high engine power conditions, than is typical of current technology combustors. The use of concepts with lean primary zone mixtures must include features for combustion process staging to meet engine operability needs at low power conditions. The requisite lean primary zone mixtures can be obtained either without or with fuel/air mixing prior to combustion, as is discussed in the following sections.

LOW NOx COMBUSTORS FOR SUBSONIC ENGINES

In response to the need for lower NOx levels in the next generation of subsonic civil aircraft engines, efforts are currently in progress to

TABLE 6. Candidate combustor design approaches for reducing NOx emissions

Method	Required features
o Lean primary combustion zone fuel/air ratios - at high power - Without fuel/air premixing - With fuel/air premixing	Fuel or air staging for acceptable low power operability
o Series-staged rich/ lean combustion	Rapid air dilution/mixing of rich gas from first stage in second (lean) stage

develop combustors with leaner primary zones and staging provisions via the use of separate combustion zones. These efforts are primarily based on the results of previously conducted technology development efforts which were sponsored by the U.S. National Aeronautics and Space Administration (NASA) during the 1973 to 1985 time period [5 and 6]. These development efforts involved advanced combustor concepts with separately fueled combustion zones, with little or only partial fuel/air premixing ahead of the primary zone. Configurations with zones in parallel, as well as with zones in series, were investigated. Two promising concepts that were evolved are an axially-staged configuration, the Vorbix combustor, which was developed by Pratt and Whitney, and a radially-staged configuration, the dual annular combustor, which was developed by GE Aircraft Engines.

The operating characteristics of the dual annular combustor concept are generally typical of concepts with combustion process staging. This concept features the use of two primary combustion zones in parallel. Both annuli, or zones, are individually fueled. In this parallel-staged concept, one of the annuli, usually the outer annulus, is designed to operate with lower airflows than the other annulus and to serve as the pilot stage. The other annulus is designed with a higher airflow and serves as the main stage. Only the pilot stage is fueled at starting, altitude relight and idle conditions. In this manner, adequately rich fuel/air ratios and low air velocities are obtained in this annulus at low power conditions. At operating conditions above idle, both annuli are fueled. The fuel flow splits to the two annuli can be adjusted to provide lean fuel/air ratios in both annuli at high power conditions.

A version of this design concept was designed and developed in the NASA/General Electric Experimental Clean Combustor Program (ECCP) for use in the CF6-50 engine [7, 8 and 9]. A photograph of this dual annular combustor is presented in Figure 4. Subsequently, another version was designed and developed as a part of the NASA/General Electric Energy Efficient Engine (E^3) Program [10]. Cross-sectional drawings of these dual annular combustors are presented in Figure 5. Both of these dual annular combustors were evaluated and demonstrated in engine tests. In these tests, generally satisfactory combustion performance and operability chartacteristics were demonstrated.

A typical combustion process staging sequence carried out with one of these combustors is shown in Figure 6. In this sequence, staging from pilot only operation to operation with both stages occurred at a sea level static power setting of 30 percent of takeoff power. The fuel flow split following the staging sequence was 40 percent to the pilot stage and 60 percent to the main stage. In most dual annular combustors, the staging sequence can be executed over a range of power settings, including values just above idle. Also, a wide range of pilot/main stage fuel flow splits can usually be accommodated.

The degree of NOx abatement obtained in the CF6-50C engine tests of

Figure 4. Photograph of ECCP dual annular combustor

the ECCP configuration is about 35 percent, compared to the engine when equipped with its standard single annular combustor. With staged combustion concepts of this kind, leaner average primary zone fuel/air mixtures are obtained at high power operating conditions. However, simultaneous fuel/air mixing and combustion occur as in current technology combustors because of the direct injection of fuel into the primary zone. As such, local zones containing stoichiometric and near-stoichiometric mixtures are not eliminated, but only reduced in magnitude. For these reasons, the NOx abatement potential of concepts of this kind is contrained. This abatement potential in any given engine size class appears to be in the range of 30 to 40 percent, relative to current technology combustors.

In the prior and more recent evaluations of these low NOx combustors, generally acceptable low power operability has been demonstrated as a result of the combustion process staging features of the combustors. However, these staging features involve the use of new design technology and introduce added complexity into the combustor design. The currently underway development efforts are primarily structured to address these concerns. An overriding objective is to evolve combustor configurations that, in addition to providing lower NOx levels, have operability capabilities equivalent to the excellent capabilities of current technology combustors.

• Lean/lean dual annular configurations

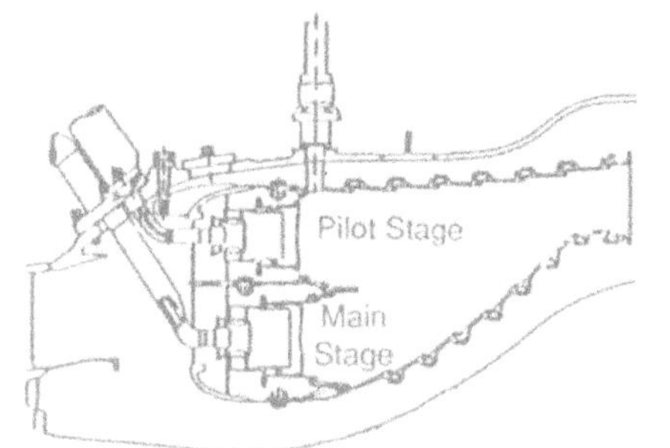

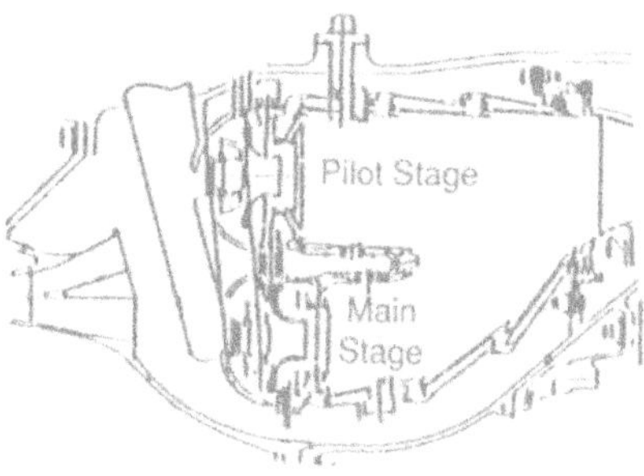

NASA/GEAE ECCP Configuration (for CF6-50 engine)

NASA/GEAE E^3 Configuration

Figure 5. Cross-sectional drawings of ECCP and E^3 dual annular combustors (both configurations drawn to the same scale)

ULTRALOW NOx COMBUSTORS FOR SUPERSONIC ENGINES

As an important part of the efforts to assess the environmental acceptability of a fleet of new supersonic transport aircraft, extensive studies are being conducted by NASA and others to determine the impacts of such aircraft fleets on the stratospheric ozone layer [3]. In these atmospheric modeling studies, the effects of aircraft fleet size and of combinations of flight Mach number and operating altitude are being examined. The preliminary predictions indicate that the degrees of ozone depletion, after several years of operation, are excessive if engines equipped with current technology combustors are used [3].

Based on these findings, a NOx emission index of 5 grams per kilogram of fuel at supersonic cruise conditions has been established by NASA as a goal for future supersonic transport engines [3]. To achieve this goal, combustor designs with NOx indices up to 90 percent lower than those typical of current technology combustors are needed. A major thrust of the overall NASA initiative to evolve technology for future supersonic transport engines, therefore, involves the development of ultralow NOx combustor technology. Two advanced combustor design concepts are being pursued - lean premixed/prevaporized combustion and rich/lean series-staged combustion.

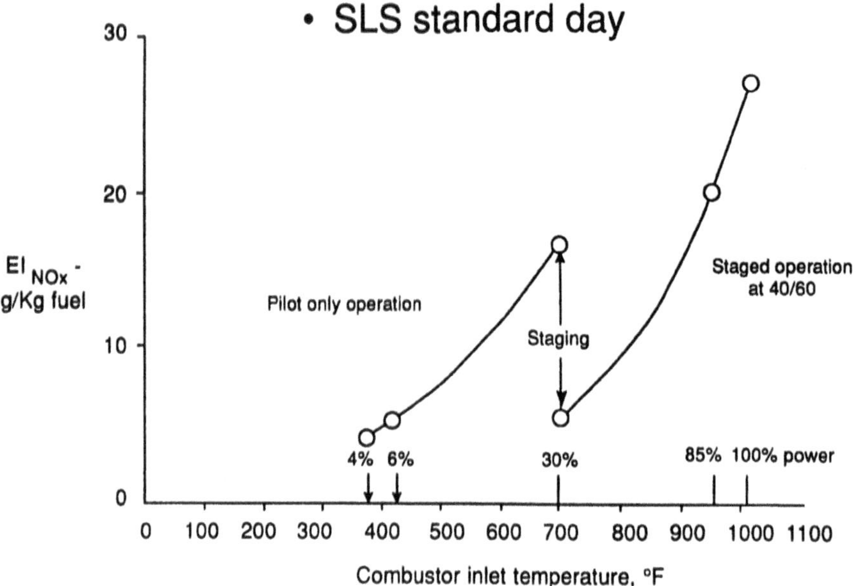

Figure 6. Example of combustion process staging sequence with E^3 dual annular combustor

Lean Premixed/Prevaporized Combustor Concepts

Lean premixed/prevaporized combustion involves the premixing of the fuel and its combustion air upstream of the combustion zone. By premixing and prevaporizing the fuel so that lean and homogeneous mixtures result, significant NOx reductions are possible. A demonstration of the feasibility of this approach is described in [11]. Some key results of this investigation are presented in Figure 7. As is shown, NOx emission index values below one were obtained, along with high combustion efficiency levels.

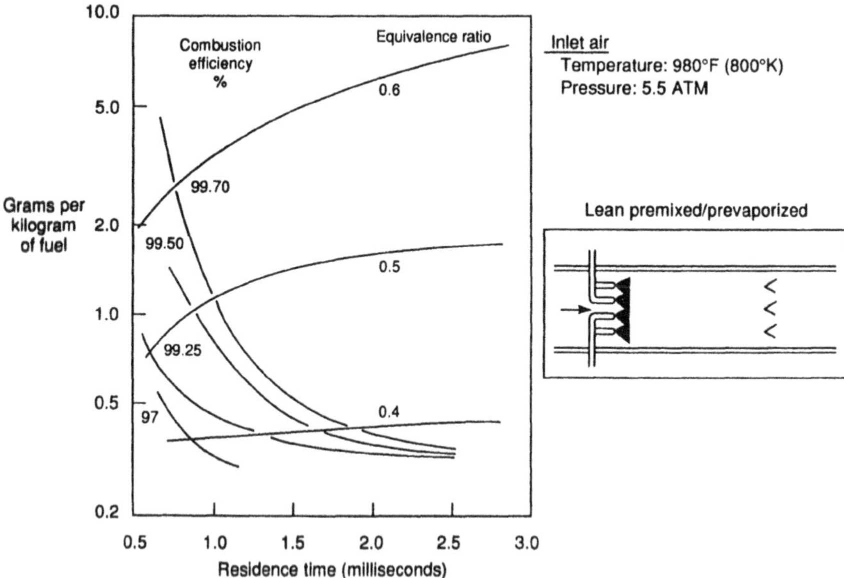

Figure 7. Example of NOx levels obtainable with lean and homogeneous hydrocarbon fuel/air mixtures

Some exploratory efforts to develop engine combustors of this kind have been conducted. One such effort is the NASA Advanced Low Emission Combustor (ALEC) Program [12]. In this program, combustors sized for the CF6-50 engine with features to provide lean premixed fuel/air mixtures were designed and developed The configurations that were designed and evaluated in full annular component tests are shown in Figure 8. Each design included a premixing duct to mix the main stage fuel with its combustion air. As in any combustor designed to operate with lean primary zone mixtures at high power, fuel and/or air staging features were included in the various design configurations to provide acceptable operability at low power operating conditions. These features added considerable complexity to the combustor designs.

A parallel effort, the NASA Clean Catalytic Combustor (C^3) Program, was conducted during the same time period [13]. In this effort, lean premixed combustor concepts, which included a catalytic reactor to enhance combustion of the lean mixtures, were investigated. The combustors were sized for the NASA/General Electric E^3. The configurations that were designed and evaluated in sector component tests are shown in Figure 9. An aft-looking-forward photograph of the parallel-staged configuration is presented in Figure 10. Fuel staging features, involving the use of a pilot combustion stage, were used to obtain acceptable low power operability.

Some of the key findings of these exploratory development investigations are summarized in Table 7. Significant NOx reductions, of more than 80 percent, were obtained at cruise operating conditions. However, at the inlet air temperatures associated with takeoff power, combustor operation was limited to a maximum inlet air pressure of less than 20 atmospheres because of the occurrence of fuel/air autoignition. Operation of the combustors with catalytic reactors was further limited because of the temperature limits and poor durability characteristics of the catalytic reactors.

Accordingly, lean premixed/prevaporized combustor concepts offer considerable promise as a means of significantly reducing NOx emissions. However, combustor designs of this kind are very complex because of the need for sophisticated combustion process staging features. Also, fuel/air autoignition is a barrier problem that must be overcome. At takeoff conditions in advanced subsonic engines and at supersonic cruise conditions in advanced supersonic engines, the autoignition dwell times are very short, typically one to four milliseconds. As such, the fuel/air premixing process must be accomplised within a very short time. Meeting this need with a liquid fuel is very challenging, because of the practical limitations on the number of fuel injection points that can be used.

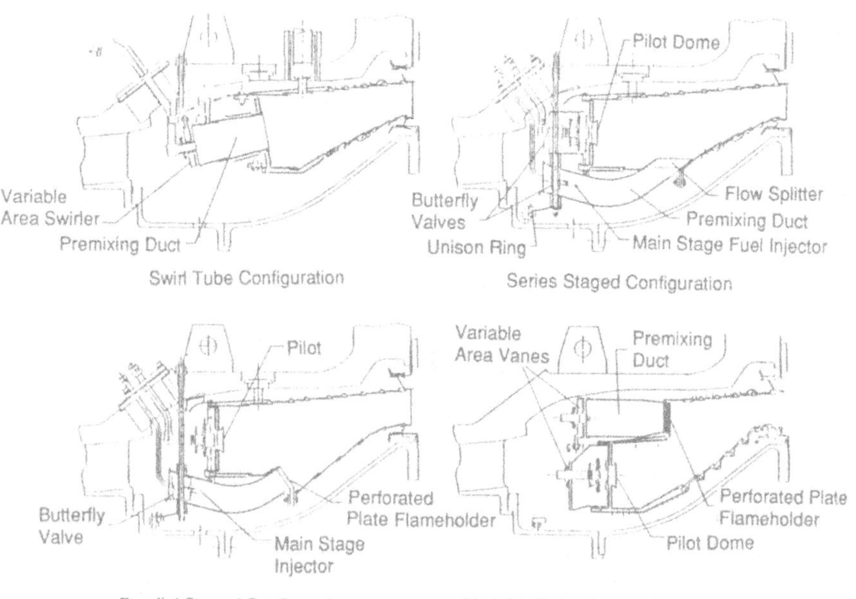

Figure 8. ALEC Program combustor test configurations

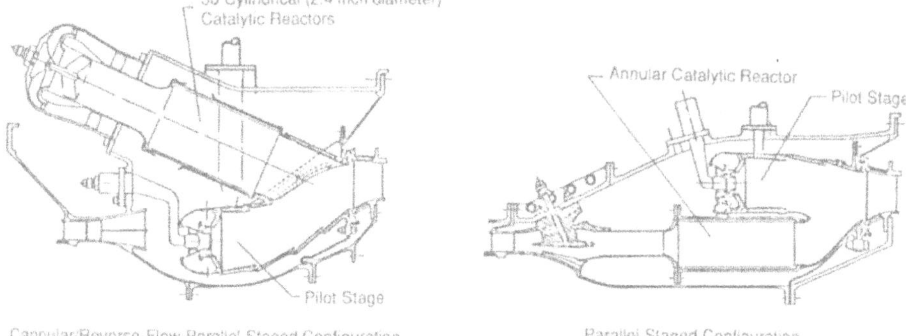

Figure 9. C^3 Program combustor test configurations

Figure 10. Photograph of parallel-staged combustor sector configuration tested in C^3 Program

Another operational concern is the liklihood of unsteady combustion occurrences at some operating conditions. Because these combustors typically have combustion zone fuel/air ratios only slightly above the lean blowout limits, to minimize NOx levels, dynamic combustion instabilities are an ever-present possibility. These dynamic instabilities can result in high amplitude pressure oscillations, which can cause combustor hardware damage. Passive and/or active control means of suppressing these instabilities are, therefore, an important technology need of lean premixed/prevaporized combustors.

Accordingly, considerable further research is needed to provide the technology needed to evolve suitable versions of these combustor concepts for use in aircraft turbine engines. The NOx abatement potential of these concepts is very considerable. Extensive efforts to provide this needed technology are, therefore, underway as a part of the NASA initiative related to advnced supersonic transport aircraft [3 and 6].

TABLE 7. Key findings of development efforts on lean premixed/prevaporized combustor design concepts

Combustor Concept	NOx Level at cruise (g/kg fuel)	Operability
o Lean premixed/ prevaporized combustion	~3	Maximum combustor inlet air pressure limited to 20 atmo-spheres due to fuel/ air autoignition
o Lean premixed/ prevaporized combustion with catalytic re-actor	~2	Maximum combustor inlet air pressure limited to 14 atmo-spheres due to fuel/ air autoignition Maximum combustor exit temperature limited to 1640K (2960°R) due to catalyst temperature limits and durability

Rich/Lean Series-Staged Concepts

A version of the rich/lean series-staged concept is shown in Figure 11. This specific design was developed as a part of the development effort described in Reference 14. Significant reductions in NOx levels were demonstrated in this investigation with both liquid and gaseous fuels.

In this low NOx combustor concept, all of the fuel is injected into the first stage and this stage is operated with average fuel/air ratios above stoichiometric to suppress the formation of NOx. In this stage, the combustor liners must be cooled without the use of air film cooling, as is commonly used in current technology combustors. The use of film cooling is not acceptable because stoichiometric fuel/air mixtures would occur in regions close to the liner. Most of the combustor airflow is bypassed around the rich first stage and is introduced further downstream. Very rapid mixing of this air with the

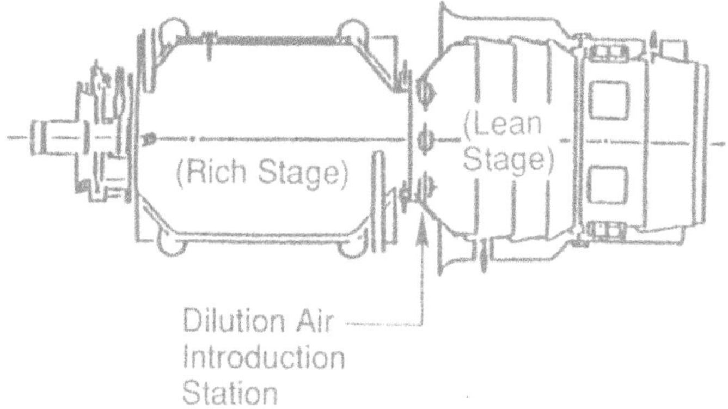

Figure 11. Typical series-staged rich/lean combustor configuration

rich first stage combustion products is needed to complete the combustion process, while at the time suppressing NOx formation as the rich gases are diluted.

Some progress has been made in the development of concepts of this kind for use in industrial gas turbine engines, with relatively long-length can type combustors, especially with gaseous fuels [15 and 16]. A key target of these efforts was to demonstrate a means of suppressing the conversion of the fuelbound nitrogen, contained in some liquid and gaseous fuels used in industrial gas turbine engines, to NOx emissions.

The problems associated with packaging series-staged combustors of this kind into the short available lengths of modern engine combustion sections are difficult. Also, cooling of the liners in the rich stage is difficult to manage without the use of air film cooling because of the high combustor inlet air temperatures and pressures of modern aircraft turbine engines. Because of these high inlet air conditions, suppression of NOx formation in the rapid mixing process between the rich and lean stage may be very difficult to achieve. In addition, prevention of excessive smoke levels in the rich stage with liquid fuels may prove to be difficult.

Accordingly, considerable further research is needed to resolve the several design and development problems associated with the use of concepts of this kind in aircraft turbine engines. However, the NOx abatement potential of this concept is substantial and its basic geometry and fuel injection provisions are relatively simpler than those of the lean premixed/prevaporized concepts. For these reasons, the use of combustor concepts of this kind in advanced supersonic transport engine applications is being explored, along with lean premixed/prevaporized concepts.

CONCLUDING REMARKS

Considerable progress has already been made in the development of aircraft turbine engine combustors with lower NOx levels than those of current technology combustors. One important finding of these efforts is that low NOx combustors are inherently more complex than current technology combustors. Additional and expanded technology development efforts are needed, especially with respect to combustor concepts with the potential for operating with ultralow NOx levels. The abatement potential of the key candidate low NOx combustor concepts is summarized in Table 8. Also included is a qualitative assessment of the development challenge associated with each of these concepts. The lean combustion concepts without fuel/air premixing involve less development challenge because considerable technology for the design of combustors of this kind is already available.

TABLE 8. Summary comparison of candidate combustor design approaches for reducing NOx emissions

Method	Abatement potential (%)	Development challenge	Prime emphasis - current technology development efforts for: Next generation subsonic aircraft engines	Advanced supersonic transport aircraft engines
Lean combustion zone fuel/air ratios - at high power				
- Without fuel/air premixing	30 to 40	Moderate	X	
- With fuel/air premixing	80 to 90	Major		X
Series-staged rich/ lean combustion	80 to 90	Major		X

Because of the latter consideration, the currently underway low NOx combustor development efforts targeted toward nearer-term subsonic lircraft engine applications primarily involve the lean combustion zone concepts without fuel/air premixing. With these advanced combustors, significant reductions in NOx levels are being obtained. Although the combustors are more complex than current technology combustors, satisfactory performance and operability appear attainable with these configurations.

The currently underway development efforts involving the other two concepts, which are more complex but offer more NOx abatement potential, are primarily targeted toward advanced supersonic aircraft engine applications. The latter engine applications are longer-term in nature. Although these development efforts are primarily targeted at this time to advanced supersonic transport engines, this technology is also relevant to future generations of subsonic civil engines.

The initial introduction of low NOx combustors during the latter part of this decade into the next generation of subsonic aircraft engines is a likely prospect. With these combustors, the resulting NOx levels of these advanced engines will be substantially lower than those of current technology engines. Accordingly, it is expected that aircraft engines will continue to be relatively small contributors to urban, nationwide and global air pollution burdens.

REFERENCES

1. Johnson, C., Henshaw, J., and McInnes, G. (January 1986) Impact of Aircraft And Surface Emissions Of Nitrogen Oxides On Tropospheric Ozone And Global Warming, Nature, Volume 335.

2. Haughton, J.T., Jenkins, G.J., and Exphraums, J.J. (1990) Climate Change - The IPCC Scientific Assessment, Intergovernmental Panel On Climate Change, Cambridge University Press.

3. Wesocky, H.L. and Prather, M.J. (September 1991) Atmospheric Effects of Stratospheric Aircraft: A Status Report From NASA'S High-Speed Research Program, Proceedings Of Tenth International Symposium On Air Breathing Engines.

4. Anon. (1981) Environmental Protection - Aircraft Engine Emissions, International Civil Aviation Organization, Annex 16 - Volume 2.

5. Anon. (1977) Aircraft Engine Emissions, U.S. National Aeronautics and Space Administration, NASA CP 2021.

6. Bahr, D.W. (September 1991) Aircraft Engine NOx Emissions - Abatement Progress And Prospects, Proceedings Of Tenth International Symposium On Air Breathing Engines.

7. Bahr, D.W. and Gleason, C.C. (June 1975) Experimental Clean Combustor Program - Phase I Final Report, NASA CR-134737.

8. Gleason, C.C., Rogers, D.W. and Bahr, D.W. (August 1976) Experimental Clean Combustor Program - Phase II Final Report, NASA CR-134971.

9. Gleason, C.C. and Bahr, D.W. (May 1979) Experimental Clean Combustor Program - Phase III Final Report, NASA CR-135384.

10. Sokolowski, D.E. and Rohda, J.E. (July 1981) The E^3 Combustors: Status and Challenges, NASA TM 82684.

11. Anderson, D. (March 1975) Effects Of Equivalence Ratio And Dwell Time On Exhaust Emissions From An Experimental Premixing-Prevaporizing Burner, NASA TM X-71592.

12. Ekstedt, E.E. and Fear, J.S. (June 1987) Advanced Low Emissions Combustor Program, AIAA Paper 87-2035.

13. Ekstedt, E.E., Lyon, T.F., Sabla, P.E. and Dodds, W.J. (November 1985) Clean Catalytic Combustor Program -Final Report, NASA CR-168323.

14. Novick, A.S. and Troth, D.L. (October 1981) Low NOx Heavy Fuel Combustor Concept Program, NASA CR-165367.

15. Cutrone, M.B. (October 1981) Low NOx Heavy Fuel Combustor Concept Program - Phase I Final Report, NASA CR-165449.

16. Cutrone, M.B., Symmonds, R.A. and Beebe, K.W. (April 1982) Low NOx Heavy Fuel Combustor Concept Program IA Gas Tests, NASA CR-167877.

Part II Combustion in Internal Combustion Engines

11. DIAGNOSTICS OF DIESEL COMBUSTION

O. Haahtela, W. Hentschel and K. P. S. Schindler
Volkswagen AG research and Development
38436 Wolfsburg
Germany

ABSTRACT. The demands in car development are largely contradictory. The most discussed items with regard to the environment are fuel consumption and the exhaust emissions of the combustion engine. Although the development of the engine during the last few decades has made good progress, further knowledge on combustion processes is required. In addition to exact measurements of the pollutants outside the engine modern diagnostic methods like the laser techniques enable a detailed insight in the engine.

The Diesel engine is the most efficient user of fossil fuels and causes only low pollutant levels. Within the European IDEA research programme a detailed experimental and theoretical study of the fundamental phenomena like injection, mixture formation, ignition, combustion and pollutant formation was carried out. Some optical measurement procedures are explained and exemplary results of experiments especially on an optically accessible engine are presented.

This research work is continued in the IDEA EFFECT research programme that includes the combustion in a petrol engine.

1. Introduction

Car development is a process, which necessitates many compromises because of the largely contradictory demands, Fig. 1. For example a higher level of comfort, safety and noise reduction lead to a higher car weight and thus to a higher fuel consumption. Depending on the special situation in each country the criteria have a different emphasis.

In recent years the ever increasing level of environmental protection has placed a high priority on the development of vehicles with high fuel economy and low exhaust emissions. Both demands are interdependant, principally a car with a low fuel consumption has also low emission levels. Especially the CO_2 emission, which is said to be a main contributory factor to the greenhouse effect, is proportional to the fuel consumption.

F. Culick et al., (eds.), Unsteady Combustion, 267–282.

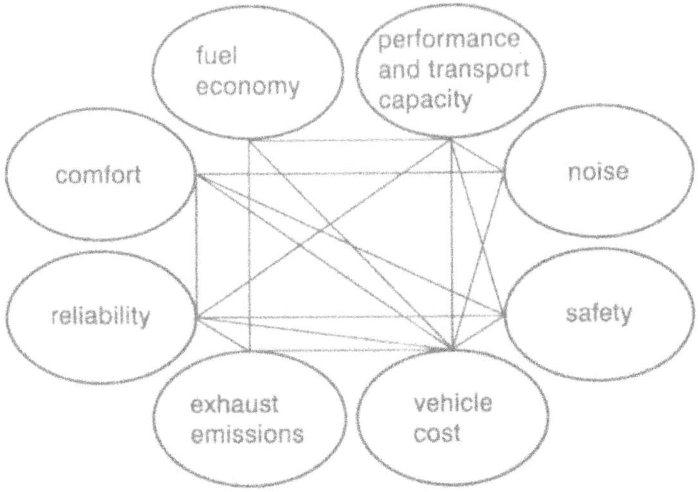

Figure 1. Conflicting demands in car development

Looking at the stages of a car's life, Fig. 2, research work does not show any immediate benefit. The cars planned today will be on the road for more than 20 years. Thus the high priority for the two demands is not a short term problem.

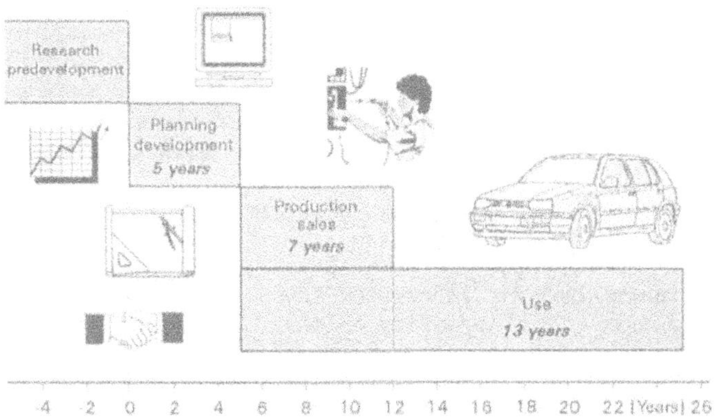

Figure 2. The stages of a car's life

The average fuel consumption of new cars is approximately 2 litres per 100 km less than that of the actual passenger car fleet, Fig. 3. But it will need more than 10 years to reduce the overall average to the level of the new cars. Even today the new TDI Diesel engines from Audi and since 1993 also from Volkswagen excel by a very high efficiency. The next goal is a car with a consumption of only 3 litres per 100 km.

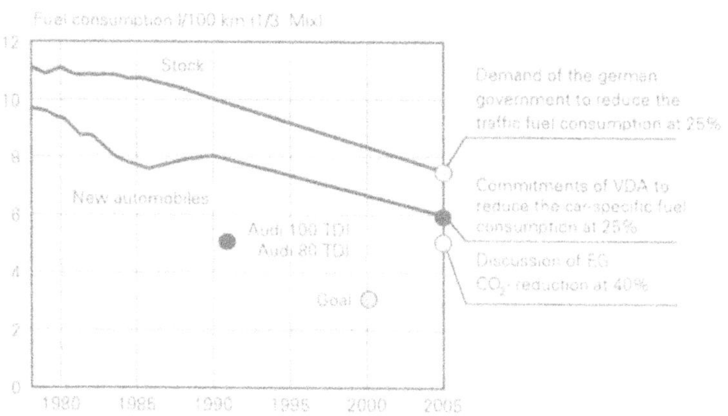

Figure 3. Average fuel consumption

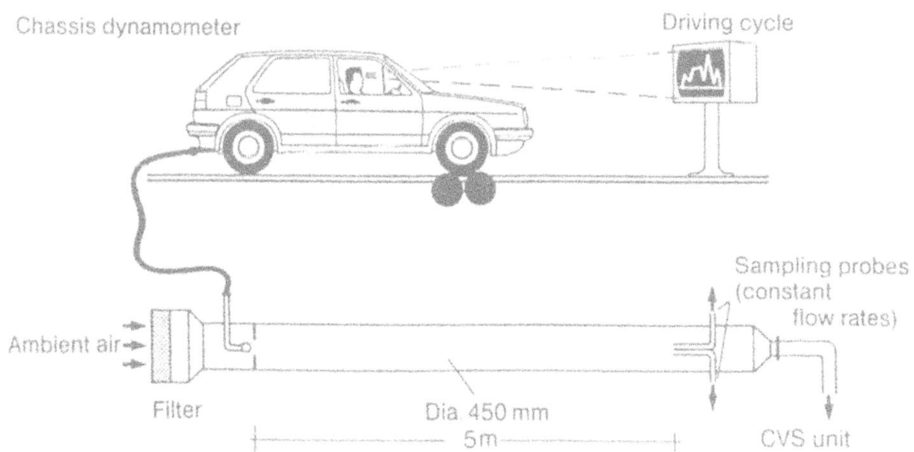

Figure 4. Set-up for measuring car exhaust emissions

2. Engine Emissions

The exhaust gases consist mainly of the non-toxic parts nitrogen, CO_2, water and oxygen. The rate of partly burned and unburned fuel is 2 % in petrol engines and only 1 % in Diesel engines because of their higher oxygen level. Exhaust emission standards exist for CO, HC, NOx and, for Diesel engines, soot level. The usual method for determining car exhaust emissions is to run vehicle on a dynamometer, Fig. 4.

The car speed is given by the test cycle and the exhaust gases are mixed in a dilution tunnel with the ambient air and gathered in the CVS unit. The different concentrations are measured and calculated to the result of grammes per distance unit of the test cycle (g/km or g/mile). The new European driving cycle "MVEG A", Fig. 5, consists of the earlier urban driving cycle (ECE 15) plus the new extra urban driving cycle (EUDC) with a higher speed range. At the start of the test the engine is cold (20° C ambient temperature).

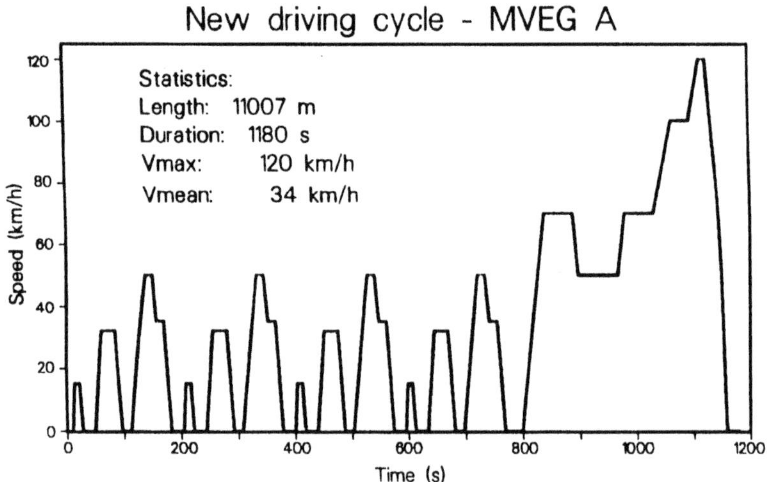

Figure 5. The new European driving cycle MVEG A

The CO_2 emission corresponds to the fuel consumption of the car and this is mainly dependent on the car weight in the driving cycle, Fig. 6.

In comparison with the petrol engine the prechamber IDI Diesel engine is almost 20% more efficient. An additional improvement of approximately 15% is obtained with the direct injection DI Diesel.

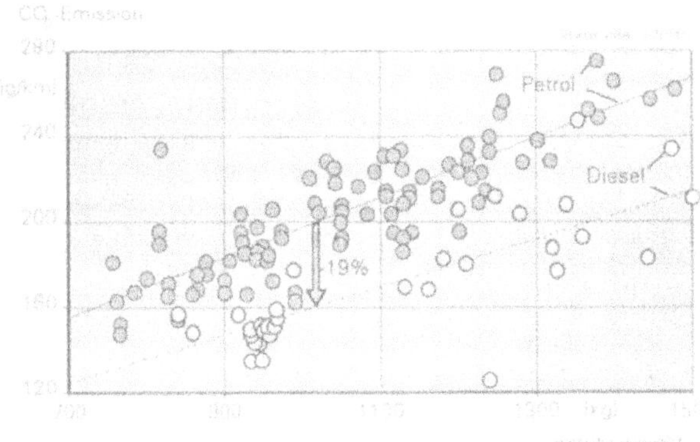

Figure 6. Vehicle CO_2-emissions in the FTP driving cycle

Since 1970 many changes in the passenger car emission standards have been enforced and they are now very stringent. For example, the maximum level of CO is only 7 % of that of 1970. Proposed EC directives for Diesel engines will only allow 0,08 g/km of particulates for the IDI engine and 0,10 g/km for the DI engine. As result of the improvements of the combustion the emissions of new progressive Diesel cars are already lower than those future directives.

3. Diesel Engine

The Diesel engine is the most efficient user of fossil fuels for vehicle propulsion. Its low CO_2 emissions contribute only marginally to the greenhouse effect. Improvements in the combustion process will lead to a greater reduction of the emissions, especially of particulates and NOx. The further reduction of soot formation by assuming a low NOx level is the key problem in Diesel engines.

Furthermore one of the advantages of the Diesel engine is that it works with oil fractions, which are otherwise used in furnaces. It is able to operate with alternative fuels like alcohol or vegetable oil.

Today there are two types of Diesel engine in production: the prechamber (IDI) and the direct injection (DI) engine. In Fig. 7 the two combustion chambers are compared. The DI engine has a combustion bowl in the piston crown and a helical port to generate swirl. One kind of prechamber is the swirl chamber, where the swirl is induced by the location of the port.

New DI engines use a distributor pump with electronic map control for the injection timing, the turbocharger pressure and the exhaust gas recirculation. Furthermore an oxidation catalytic converter reduces the amount of CO and HC.

In addition to the 15% higher fuel efficiency of the DI engine consumption when compared to the IDI engine it has even further advantages: the thermal stress is lower, a higher air charging is possible and this leads to an efficient high specific load and a better soot behaviour. This also permits a higher rate of exhaust gas recirculation to lower the NOx formation.

Although the Diesel engine is highly efficient and clean, there is a great potential for further improvements. For the DI engine it is expected that a 4-valve engine with a central injector and high pressure injection will bring benefits for the combustion. Regarding the after treatment of exhaust gases, researchers are working on a NOx-reduction catalytic converter and on a particle trap.

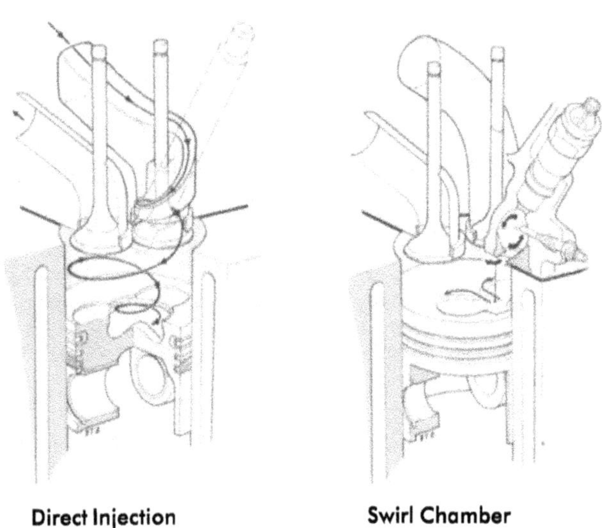

Figure 7. Comparison of Diesel combustion chambers

4. Diesel Combustion Research

4.1 COMBUSTION PROCESS

In order to meet the increasingly demanding requirements, it is necessary to understand the processes which take place throughout the combustion. Whilst a considerable amount of combustion research is continuing, it will certainly not be possible to eliminate the problems entirely. Therefore, it is necessary to carry out further pan-European research besides the research and development within the car companies and within national projects.

The association of European car manufactures JRC (Joint Research Committee), in co-operation with European research institutes, is carrying out fundamental research.

4.2 THE IDEA PROGRAMME

In the IDEA programme (Integrated Diesel European Action), the five JRC companies Fiat, Peugeot, Renault, Volkswagen and Volvo started joint research work on Diesel combustion. They decided to use the existing SPEED-SPRAY code as basic tool to develop a mathematical model of the combustion, which incorporates the total phenomena from the commencement of injection right through to pollutant formation.

Because this research is very extensive, fundamental and of common interest, it has been subsidized since 1990 by the CEC (Commission of the European Communities) within the JOULE research programme (Joint Opportunities for Unconventional or Long-term Energy Supply) and by the Swedish National Board for Industrial and Technical Development.

The programme consisted of 27 projects, which were grouped into five subprogrammes, Fig. 8.

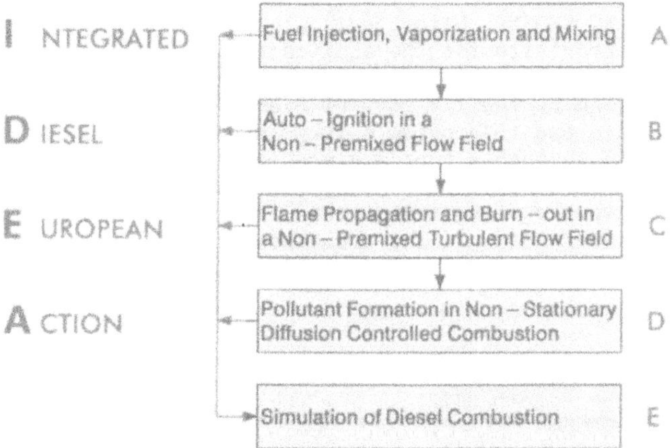

Figure 8. The IDEA research programme

The European universities and Research Institutes are located in France, Germany, Great Britain, Italy, Spain and Sweden.

The objectives of the IDEA Programme of the JRC were:

- Investigation of the spray behaviour, auto-ignition, flame propagation, and pollutant formation in real or simulated Diesel engine conditions
- Preparation of a data base for model development and validation
- Development of a documented and validated code (SPEED-DC) for Diesel combustion

The use of an engine model will support the design of more efficient engines and thus conserve energy resources and reduce harmful effects on the environment. It will also enable companies to reduce time and cost in developing new engines and enable the European industry to maintain its leadership of world Diesel technology.

4.3 DIAGNOSTICS OF THE DIESEL COMBUSTION

The experimental work within the IDEA project is carried out in the following facilities:

- in burners and similar facilities
- in a constant pressure chamber with/without swirl and with/without turbulence
- in a wind tunnel with different turbulence levels
- in model engines
- in a DI Diesel engine

The standard DI engine has four cylinders and a swept volume of 1.9 ltr., the combustion system is characterized by a swirl port, a bowl-in-piston combustion chamber and a five-hole injection nozzle.

4.3.1. TRANSPARENT ENGINE. Direct optical access to the combustion chamber is a prerequisite for the application of optical measuring techniques to internal combustion engines. Volkswagen has developed a transparent version of the 1.9 ltr. DI Diesel engine which is equipped with elongated pistons to provide a large scale visual access to the piston bowl, both through a quartz glass window in the bottom of the bowl and quartz glass plates in the upper bowl area. Fig. 9 shows the engine design with the elongated pistons in the intermediate housing between the cylinder head and the crankcase [1]. Measurements were performed at engine speeds of 1000 and 2000 rpm, motored and fired at different loads.

Figure 9. The Volkswagen transparent 1.9 ltr. DI Diesel engine

4.3.2. FLOW FIELD VISUALIZATION. To observe the development of the flow field inside the cylinder and the piston bowl, laser light-sheet studies were performed. The movement of the tracer particles (path lines)

were recorded with a high-speed camera through the window in the bottom of the piston bowl via a fixed mirror inserted from the side, Fig. 10. The measurements result in a qualitative visualization of the flow field in the illuminated plane, which is parallel to the cylinder head.

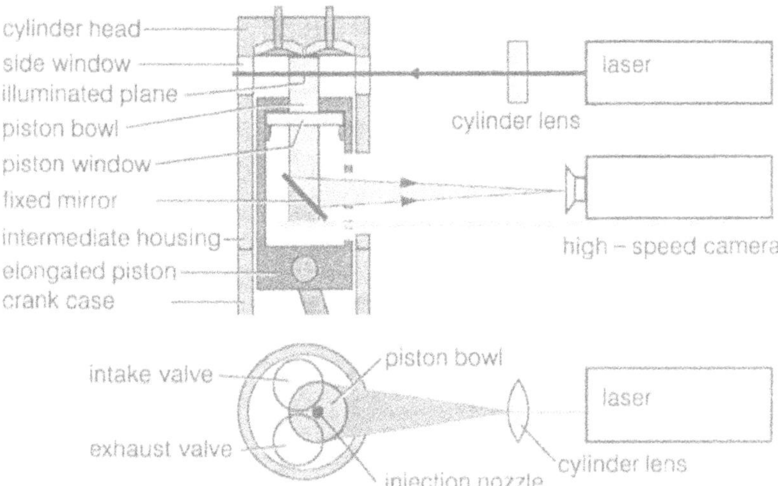

Figure 10. Set-up of the laser light-sheet technique on the transparent engine for flow analysis inside the cylinder

The development of the in-cylinder and in-bowl flow in a plane 5 mm below the cylinder head during intake and compression stroke within one cycle is shown in Fig. 11 and 12.

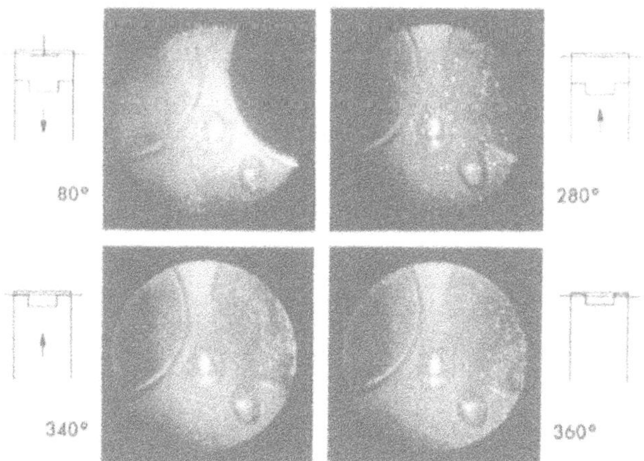

Figure 11. Four frames from a laser light-sheet film taken at different stages during intake and compression stroke (1000 rpm)

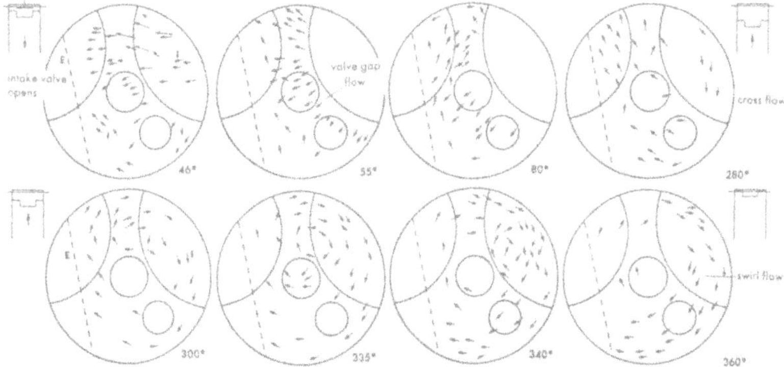

Figure 12. Sketches of the in-cylinder and in-bowl flow development during intake and compression stroke (1000 rpm)

The sketches in Fig. 12 give a clearer impression of the flow development than single high-speed film frames and show typical situations at selected crank angles. As long as the intake valve (I) is open, the flow passes through the valve gap with a tangential component - a consequence of the swirl port. Around bottom dead centre (BDC) no significant swirl is visible in this plane close to the cylinder head. In the second half of the compression stroke a well-defined swirl flow appears. During the last phase before the piston reaches TDC a strong interaction of the swirl flow and the squish flow is observed. This squish flow is formed by the air which is pressed into the bowl when the piston moves to TDC. But finally, at the time of injection (around TDC), the swirl flow dominates and is similar to a solid-body-rotation.

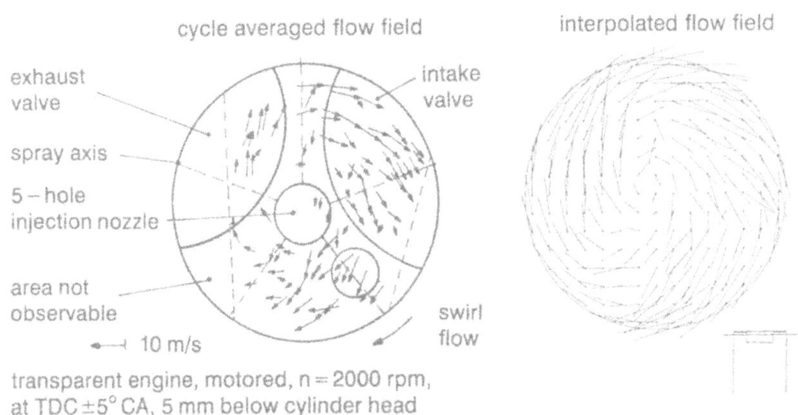

Figure 13. 2D cycle-averaged quantitative flow field at the time of injection; analysed flow vectors (left) and interpolated flow field

A frame-by-frame analysis of a large number of cycles leads to the determination of the quantitative average flow field at the time of injection. In Fig. 13 the evaluated flow vectors (left) are presented together with the interpolated flow field (right). The average value of the tangential component of the flow velocity approximately 10 mm from the bowl centre is about 9 m/s. This data was used to investigate the real flame velocities within the rotating flow field.

4.3.3. CRANK ANGLE RESOLVED FLOW VELOCITY MEASUREMENT. To measure the temporal development of the flow field (velocity and turbulence) LDV-measurements were taken through the piston bowl window. A two-component backscatter LDV system with frequency shifters and counters was used. A set of measuring locations inside the cylinder was scanned automatically by a tilted mirror traversing system. Ensemble-averaged mean velocities (bulk flow) and RMS-values (turbulence and cyclic variation) were calculated depending on the position of the moving piston (crank angle).

The LDV flow measurements were taken in a plane 30 mm below the cylinder head, Fig. 14. The swirl flow was initiated by the intake flow as explained above and remained during intake and compression stroke until TDC. The swirl centre moves around and finally centres in the piston bowl. For more details about LDV measurements in the transparent DI Diesel engine see [2].

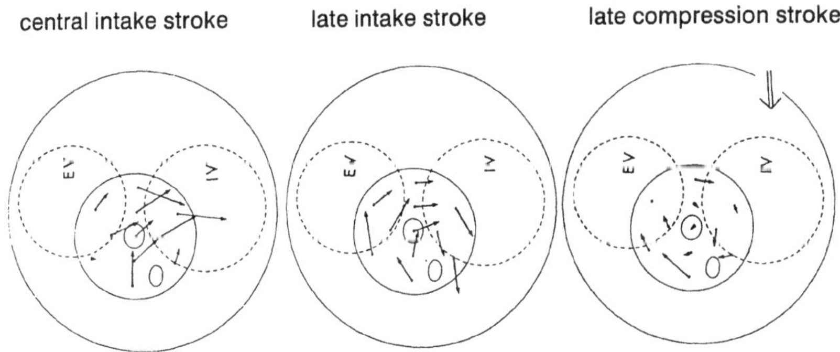

Figure 14. Swirl development in the cylinder by cycle-averaged time-resolved LDV flow measurements

4.3.4. SPRAY DEVELOPMENT MEASUREMENT. To visualize and measure the spray formation and penetration in the piston bowl and the interaction of the spray with the bowl wall, a modified laser shadowgraphy set-up is used in combination with high-speed film recording with up to 8000 frames/sec. The arrangement is shown in Fig. 15.

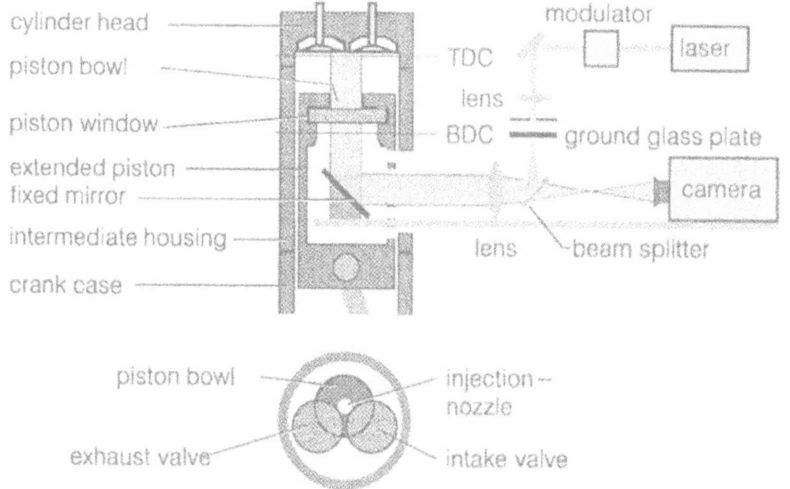

Figure 15. Set-up for spray development analysis by laser shadowgraphy and high-speed filming

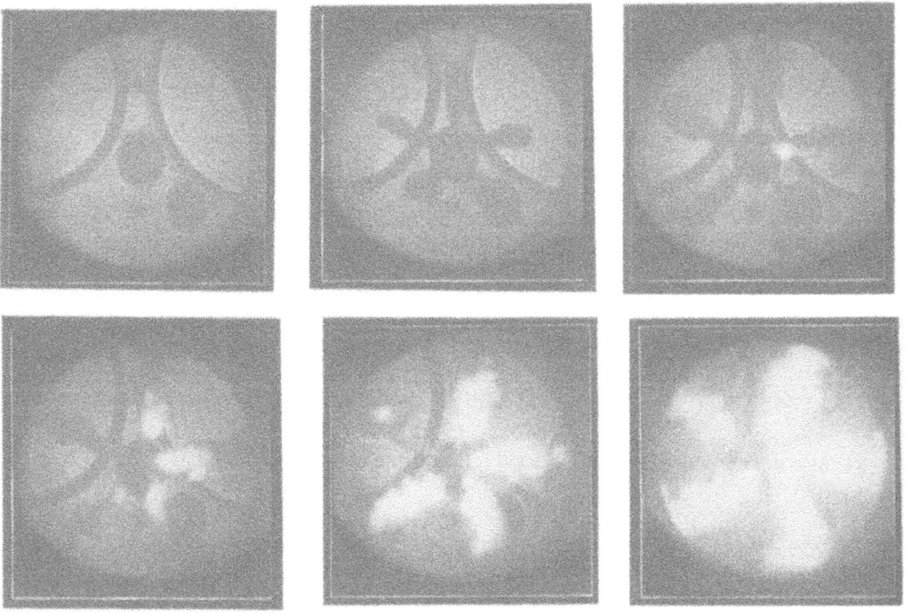

Figure 16. Spray development and combustion in the piston bowl; engine running at 2000 rpm and part load running

The laser illuminates the scene inside the combustion chamber from below through the piston window. If the blue line (488 nm) of the argon laser is used for illumination there is a good colour contrast to the bright yellow self luminescence of the soot. Detailed statistical analysis of many films with a great number of combustion cycles leads to quantitative data, e.g. about spray tip penetration and spray angles for the different sprays. This data is used for comparison with the 3D-simulation model of the in-cylinder processes.

An example for the spray formation and subsequent combustion in a single combustion cycle is shown in Fig. 16. There are slight differences in the shape of the sprays and in the penetration depth. At this engine speed and load - 2000 rpm and 8 mg - the sprays are hardly affected by the swirl flow. The droplets move straight on from the nozzle hole to the bowl wall.

Near the wall of the piston bowl the sprays are slightly shifted because the droplets have been slowed down by the high air density. The droplets hit the bowl wall and splash in a cloud of very tiny droplets which are then turned around by the swirl flow, Fig. 17. Along the traces of the droplets the fuel evaporates and an inflammable mixture is formed by turbulent mixing.

frame from a laser shadowgraphy high – speed film

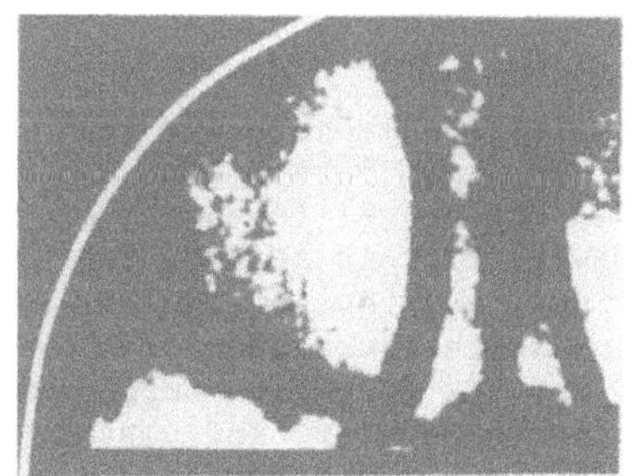

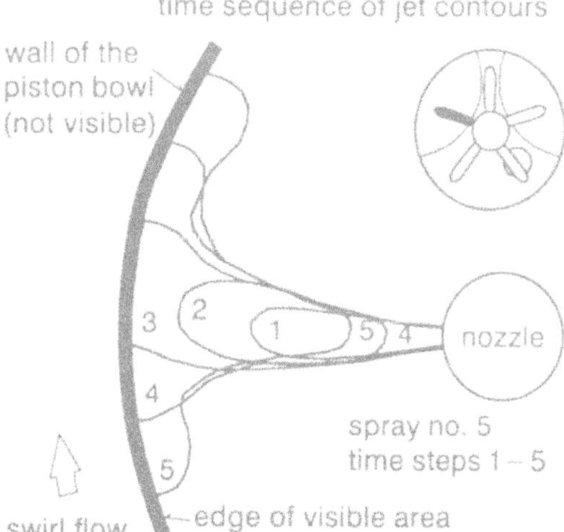

Figure 17. Spray-wall interaction at the piston bowl wall

4.3.5 AUTO-IGNITION. Auto-ignition is observed on the films by the bright emission of soot particles, as soot has been formed at the very beginning of the Diesel combustion. The spatial distribution of the auto-ignition locations averaged over many cycles is shown in Fig. 18. The fuel vapour is shifted by the swirl flow and ignites on the lee side of the sprays. The auto-ignition areas grow together and finally the sprays burn completely along the shifted spray axis.

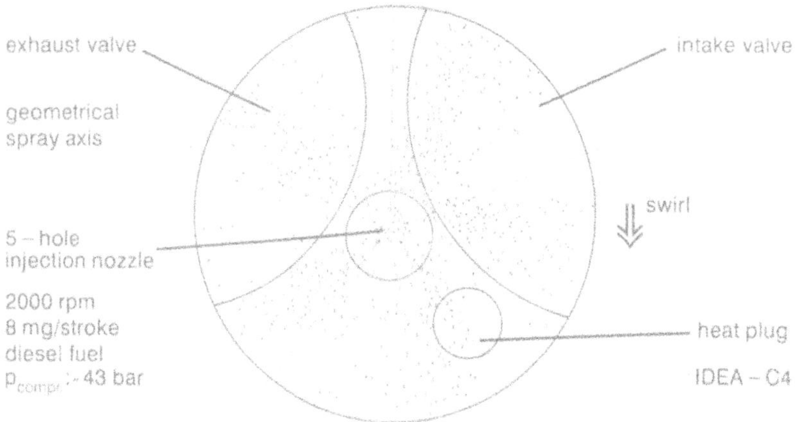

Figure 18. Spatial distribution of the auto-ignition locations in the piston bowl.

In summary, the results presented have clearly demonstrated the application of optical measuring techniques for flow, spray and combustion analysis in a production DI Diesel engine. At the point of injection a well-defined nearly solid-body swirl flow is formed. The fuel is injected into the swirl flow, vaporizes, and ignites on the lee side of the spray due to the high compression temperature and afterwards burns as a diffusion flame. Soot and NOx formation depend on the local air-fuel ratio and on the combustion temperature and can be controlled by the mixture preparation. In addition, laser optical techniques like Mie scattering and laser induced fluorescence have been applied by other research groups within the IDEA programme to gain more knowledge of the processes occuring during mixture preparation and combustion [3].

4.4. IDEA EFFECT PROGRAMME

The IDEA programme as a Diesel project was concluded in 1993, the research work is being continued in the new IDEA EFFECT programme (Integrated Development on Engine Assessment with Environment Friendly Fuel Efficient Combustion Technology) which started in July 1993 and has a duration of 3 years. This programme includes also the continuation of the JRC programme HOMOGENEOUS COMBUSTION for petrol engine research. The programme is supported by the Joint Research Committee members Fiat, Peugeot, Renault, Rover, Volkswagen and Volvo. It is also subsidized by the Commission of the European Communities within the JOULE programme, by the Swedish National Board for Industrial and Technical Development and the Austrian government.

The goal of the programme will be the application of a model for petrol and Diesel engines. The "numerical engine" will be applied to model and real engine geometries.

The main objectives of the IDEA EFFECT programme are:

- the experimental and theoretical study of the fundamental phenomena like injection, mixture formation, ignition, combustion and pollutant formation,
- the completion, improvement and extension (towards influence of turbulence, kinetics, and real fuels) of existing numerical submodels,
- the improvement of the existing 3D SPEED Code and the incorporation of the different submodels,
- the extension of the existing consistent data base for model development and validation,
- the application of the 3D SPEED Code to model and real engine geometries inside the JRC companies.

The programme endeavours to understand the fundamentals of the processes of internal combustion in detail and to model them. Totally, 34 projects have started in European universities, institutes and on the JRC companies.

Emphasis will be on the research and modelling of the fundamentals and on engine experiments. The main effort will be directed to the application to real engines by simulation and validation.

A well established collaboration between European car manufacturers and universities will ensure the success of this approach. This structure will benefit both sides and ensure results which can be applied in practice. The programme will support the design of more efficient engines and thus conserve energy resources and reduce emissions (NOx, unburned HC and particles).

Acknowledgement

Research funded by the Commission of the European Communities within the frame of the JOULE Programme, by the National Swedish Board for Industrial and Technical Development (NUTEK), and by the Joint Research Committee of European automobile manufacturers (Fiat, Peugeot S.A., Renault, Rover, Volkswagen and Volvo) within the IDEA and IDEA EFFECT Programme.

References:

[1] Schindler, K.-P., Hentschel, W.: Experimental characterization of Diesel experiments; in Durao et al. (eds): Combustion flow diagnostics, Kluwer Academic Publishers, Dordrecht (1992) 439-454

[2] Arcoumanis, C., Whitelaw, J.H., Hentschel, W., Schindler, K.-P.: Flow and combustion in a transparent 1.9 l direct-injection Diesel engine. Autotech '93 Seminar C462/22/041 I Mech E, 1993

[3] Arnold, A.; Dinkelacker, F.; Heitzmann, T.; Monkhouse, P.; Schäfer, M.; Sick, V., Wolfrum, J.; Hentschel, W.; Schindler, K.-P.: DI Diesel engine combustion visualized by combined laser techniques. 24th Symposium on Combustion / The Combustion Institute (1992) 1605-1612

12. Visualization and Quantitative Analysis on Fuel Vapour Concentration in a Diesel Spray

J. Senda, Y. Tanabe and **H. Fujimoto**
Doshisha University
Department of Mechanical Engineering
Karasuma - Imadegawa, Kamigyo
Kyoto 602, Japan

ABSTRACT. A single diesel spray of n - tridecane which was miscible with small dopants was injected from a hole nozzle into a quiescent high - temperature and high - pressure atmosphere , and was impinged upon a flat surface with high temperature. It is able to generate fluorescent emissions from vapor and liquid phases in this evaporating spray , when a laser light sheet from Nd : YAG laser is passing through the cross section of this spray. Then , clear 2 - dimensional images for vapor and liquid phases were obtained simultaneously , applying an exciplex fluorescence method. In particular, vapor concentration was assessed quantitatively by applying Lambert - Beer's law into measured fluorescence intensity in vapor phase and by considering the quenching process of the fluorescent emission due to the mixture temperature.

1. Introduction

A background of this study is shown in Fig.1. In small type high - speed diesel engines , the injected fuel spray impinges upon the piston cavity surface due to the short distance between the nozzle and cavity surface. The behavior of the impinging spray has a great influence on the dispersion of fuel , the evaporation and the mixture formation processes , and further on the combustion process.

The authors have presented spatial and temporal distributions of the fuel droplets density in an nonevaporating impinged spray upon a wall , by use of a laser extinction method and application of computed tomography [1]. With regard to real - time measurements which separately visualize and measure the fuel spray and its vapor , a novel technique , that is exciplex fluorescence method , has been devised by Melton [2]. In this method , spectrally separated fluorescence emissions from liquid and vapor can be obtained by adding an exciplex - forming dopants to the fuel , such as naphthalene / N,N,N',N' tetramethyl - p - phenylene diamine (TMPD). The authors also applied this exciplex technique to impinging diesel spray upon a flat wall at a normal angle , which is vaporizing under the quiescent high - temperature atmosphere of nitrogen [3]. Thus , we could obtained the concentration distributions of both phases qualitatively from the fluorescent emission intensities which is spectrally separated.

The purpose of this study is to quantitatively analyze the vapor concentration of an impinging diesel spray upon a flat wall surface. An outline of this study is summarized in Fig.2. This experiment was to serve as a simplified model of the actual state in a combustion chamber of diesel engines. As mentioned above , exciplex - forming dopants were mixed into the fuel to produce spectrally separated fluorescences by exciting those molecule with irradiation of a laser sheet . Two -

F. Culick et al., (eds.), Unsteady Combustion, 283–294.

dimensional visualized images for both phases were acquired simultaneously for the identical spray by applying an exciplex fluorescence method , using optical filters , an image - intensifier (I.I.) and a CCD camera system. Then , the vapor concentration was obtained quantitatively , applying Lambert - Beer's law and correcting fluorescence intensity for mixture temperature into measured its intensity in the vapor phase.

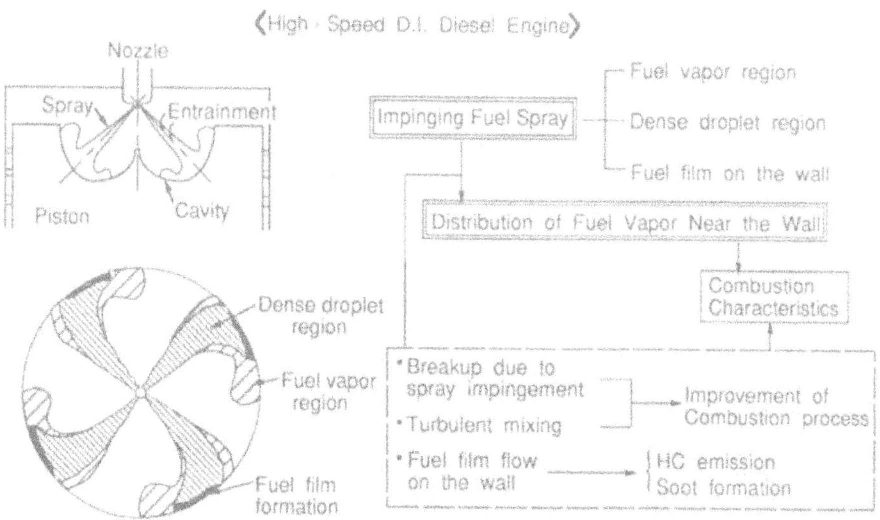

Fig.1 Background of this study

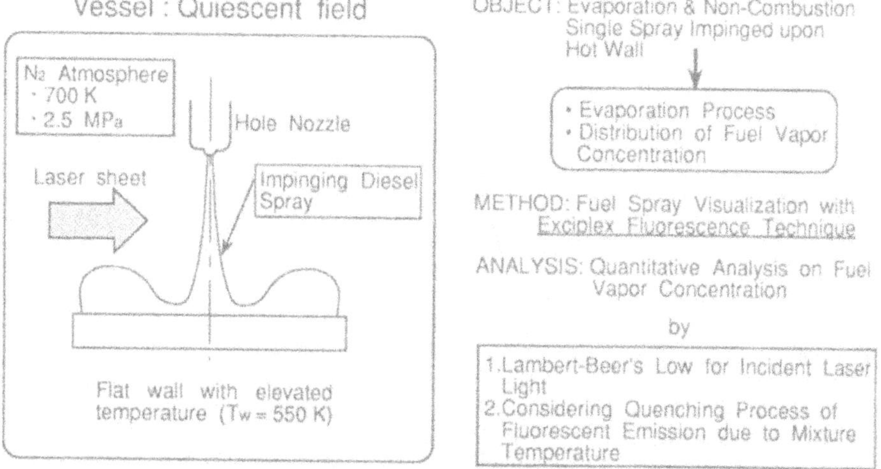

Fig.2 Outline of this study

2. Experimental Apparatus and Procedure

Figure 3 shows a schematic diagram of experimental apparatus. The experiments were conducted under the conditions of 700 K in ambient temperature and 2.5 MPa in back pressure inside a high pressure vessel with constant volume. This vessel had two quart glass windows which were installed perpendicular to each other. Thus , we could measure the fluorescence emissions from the spray irradiated with a laser light sheet. Nitrogen was utilized as the ambient gas to prevent quenching of liquid fluorescence by oxygen. This ambient gas was heated up using heaters inside the vessel.

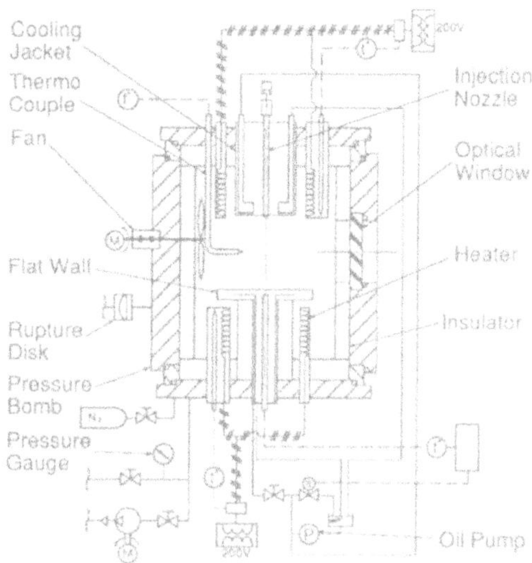

Fig.3 Schematic diagram of experimental apparatus

The naphthalene / TMPD exciplex system was used here. The fuel was 90% n-tridecane , 9% naphthalene and 1% TMPD by weight as indicated in Table 1. This fuel , which was pressurized by an injection pump , was dispensed into an injection nozzle in the center of the top of the vessel via an accumulator. The injection nozzle was a single hole nozzle with an 0.2 mm hole diameter. A single spray was injected vertically to a flat wall surface for a desired period by a solenoid valve system. The fuel temperature was kept at 373 K by a cooling system.

The flat wall surface made of an aluminum alloy was smoothed by buffing and it was kept at 550 K in surface temperature by cooling oil. The impingement distance between the nozzle to surface was 24 mm. Experiments were performed under the conditions shown in Table 2 simulating small sized D.I. diesel engines , varying the elapsed time t from the injection start.

Table 1 Fuel properties

		n-Tridecane ($C_{13}H_{28}$)	Naphthalene ($C_{10}H_8$)	TMPD ($C_{10}H_{16}N_2$)
Mass fraction	[%]	90.0	9.0	1.0
Boiling point	T_b [K]	509	491	533
Density (293K)	ρ [kg/m^3]	756	———	———
Viscosity (293K)	η [N·s/m^2]	1.88×10^{-3}	———	———
Surface tension (293K)	σ [N/m]	26.13×10^{-3}	———	———
Critical pressure	P_{cr} [MPa]	1.72	3.97	———
Critical temperature	T_{cr} [K]	671	478	———
Critical density	ρ_{cr} [kg/m^3]	240	314	———

Table 2 Experimental conditions

Injection nozzle	Type : Hole nozzle DLL-S	
	Diameter of hole d_n [mm]	0.2
	Length of hole L_n [mm]	1.1
Ambient gas		N_2 gas
Ambient temperature	T_a [K]	700
Ambient pressure	P_b [MPa]	2.5
Ambient density	ρ_a [kg/m^3]	12.3
Temperature of wall surface	T_w [K]	550
Injection duration	t_{inj} [ms]	1.9
Mean injection pressure	P_{inj} [MPa]	17.9
Quantity of injected fuel	Q_{inj} [mg]	7.4
Impingement distance	Z_w [mm]	24

3. Fuel Spray Visualization

Figure 4 shows a schematic diagram of the optical and photographic systems. The laser light of the third harmonic (355 nm) of the Nd:YAG (60 mJ / pulse in output , 8ns in pulse width) was focused into a sheet with about 50 mm in width and approximately 200 μm in thickness by way of three cylindrical lenses. The laser sheet was irradiated into a plane inclusive of the central axis in the spray to excite the exciplex-forming molecule. It was able to generate fluorescence emissions from each phase , in the two-dimensional section through the spray axis.

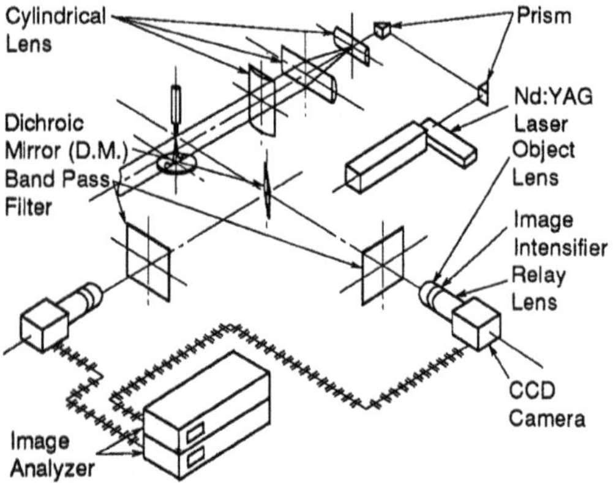

Fig.4 Optical and photographic systems

The emissions were photographed at right angles to the laser light plane. The fluorescence emissions from the liquid and vapor phases were spectrally separated by use of a dichroic mirror (D.M.; passing > 470 nm in wavelength) , as shown in Fig.5 (b). The exciplex emission dominates the condensed phase emission , while the TMPD monomer emission dominates the vapor phase emission. In this naphthalene / TMPD exciplex system , the liquid and vapor phase emissions peak at 480 nm and 410 nm in wavelength , respectively as shown in Fig.5 (a). Therefore , to isolate the exciplex signal due to the liquid phase , filter centered at 532 nm with a full width of half maximum of 2 nm was used . And to isolate the monomer signal due to the vapor phase a board band interference filter centered at 410 nm with a full width of half maximum of 19 nm was used . Thereafter , the emissions were amplified in intensity by an image-intensifier (I . I) . They were then photographed by the CCD camera system , with 540 × 480 pixel block and spatial resolution of 0.1 mm/pixel .

Then two-dimensional visualized images for both phases , corresponding to the semi-qualitative density information , were acquired respectively for the identical spray at the same time . Figure 6 shows the change in the fluorescence intensity for each phase with elapsed time t from the injection start.

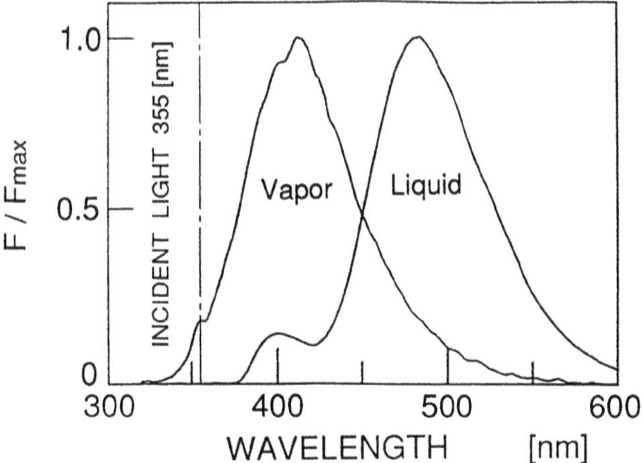

(a) Fluorescence spectra of the vapor and liquid phases

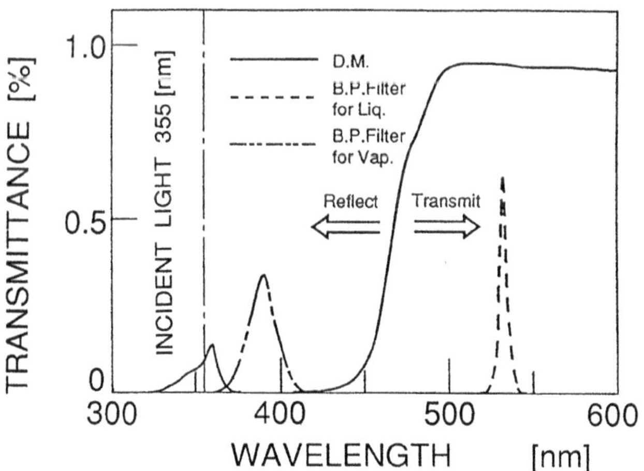

(b) Transmittance of the filters and mirror

Fig.5 Fluorescence spectra of the two phases and transmissivities of the filters and mirror

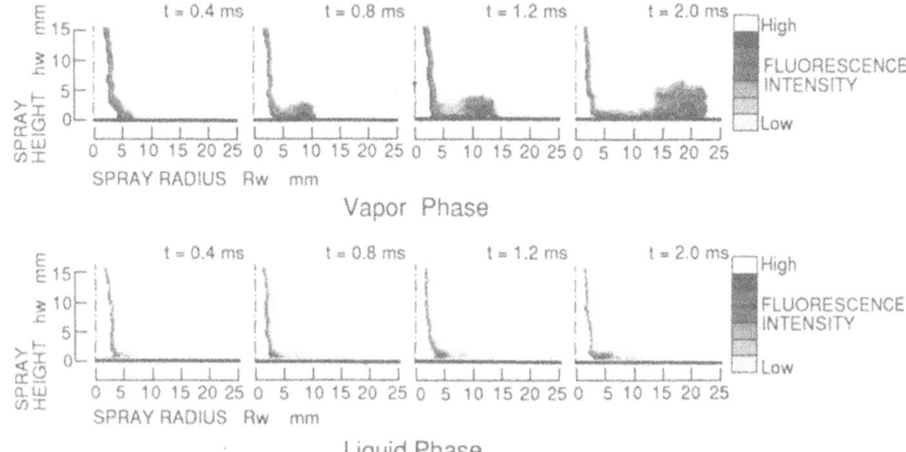

Fig.6 Change in fluorescence intensity with elapsed time from injectio (Zw = 24 mm)

4. Quantitative Analysis for Fuel Vapor Concentration

4.1. ASSESSMENT OF VAPOR CONCENTRATION BY LAMBERT-BEER'S LAW

In general , an emission intensity from a fluorescent substance is in proportion to the concentration of the substance. Figure 7 shows a model on vapor concentration analysis considering the absorption of an incident laser light. When a monochromatic incident laser light with intensity I_0 J/(m²• s) passes through the length of L m inside the substance with molar concentration C mol / m³ , the light with intensity I_t is just transmitted and the light with intensity I_a is absorbed by the substance. Then , the following equations are defined by the use of Lambert - Beer's law.

$$I_t = I_0 \cdot \exp(-\varepsilon \cdot C \cdot L) \quad \cdots (1)$$

$$I_a = I_0 - I_t = I_0 \{1 - \exp(-\varepsilon \cdot C \cdot L)\} \quad \cdots (2)$$

where, ε : molar absorption coefficient for the wave length of incident light (m³/(mol · m))
C : molar concentration of the substance (mol / m³)

Here , when the sectional area of the spray is divided into the mesh of number n with the same length L along incident light directions as shown in Fig.7(b) , the intensity I_i of incident laser light at the i - th mesh is given by the following equation.

$$I_i = I_0 \cdot \exp(-\varepsilon \cdot L \cdot \Sigma C_{i-1}) \quad \cdots (3)$$

Thus , it is able to assess the concentration C_i at the i - th mesh inside the fluorescent substance from the emission intensity F_i at the i - th mesh , by use of following equation.

$$F_i = A \cdot K \cdot I_0 \cdot \exp(-\varepsilon \cdot L \cdot \Sigma \cdot C_{i-1}) \times \{1 - \exp(-\varepsilon \cdot C_i \cdot L)\} \quad \cdots (4)$$

where , A : proportional constant concerning the optical systems

K : rate coefficient for the excited transition

C_i : molar concentration of the substance at the i - th mesh (mol / m³)

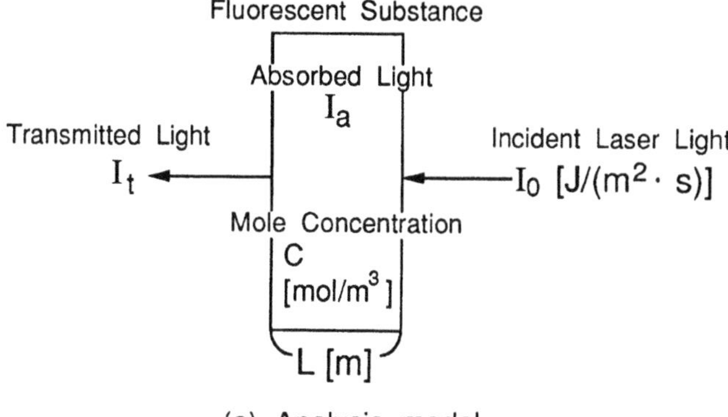

(a) Analysis model

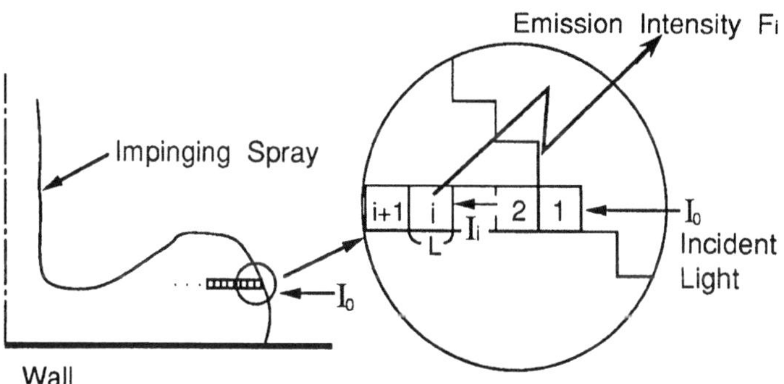

(b) Calculation mesh in impinging spray

Fig.7 Model on vapor concentration analysis with light absorption

Figure 8 shows the changes in fluorescence spectra with vapor concentration. These experiments were carried out under the high-temperature and high-pressure atmosphere of 700 K and 2.5 MPa , respectively varying the fuel vapor concentration in the state of homogeneous mixture obtained by fuel injection and well-stirred process inside the vessel. The fluorescent spectra were measured by a multi - channel analyzer (OMA : 1024ch. and 0.5 nm / ch.). The normalized fluorescent intensity F / Fmax , which is obtained with dividing the each emission intensity F by the maximum intensity Fmax in the case of Cmax = 9.8×10^{-2} mol / m³ (air excess ratio $\lambda = 0.5$) , is indicated in the ordinate. The maximum value of F / Fmax appears at 410 nm in wavelength for each vapor concentration , and the maximum emission at 410 nm allows an integration of the total fluorescence.

Figure 9 shows the calibration curve between the normalized fluorescent intensity F / Fmax at 410 nm and the vapor concentration ratio C / Cmax. In the figure , the plots are experimental data derived from Fig.8 , and the solid line represents the theoretical value calculated from Eq.(4). In this study , the length of each mesh equals to 2.0×10^{-4} m and it is assumed that the molar absorption coefficient ε is 65.0 $m^3/(mol \cdot m)$ at 355 nm in incident light wavelength referring to the result by Rotunno et al.[4]. The experimental results show very good agreement with the calculation values. Thus,the relation between the fluorescence intensity and the concentration of fluorescent substant could be provided by Eq.(4) in the gas phase. Accordingly,the fuel vapor concentration could be assessed by the monomer emission intensity from TMPD vapor for any measuring region.

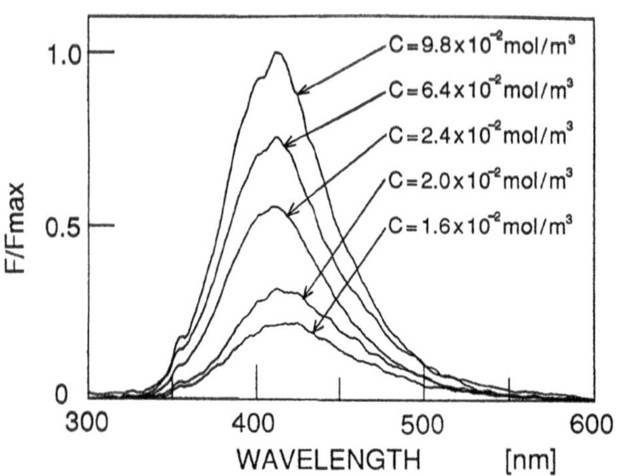

Fig.8 Change in fluorescence spectra with vapor concentration

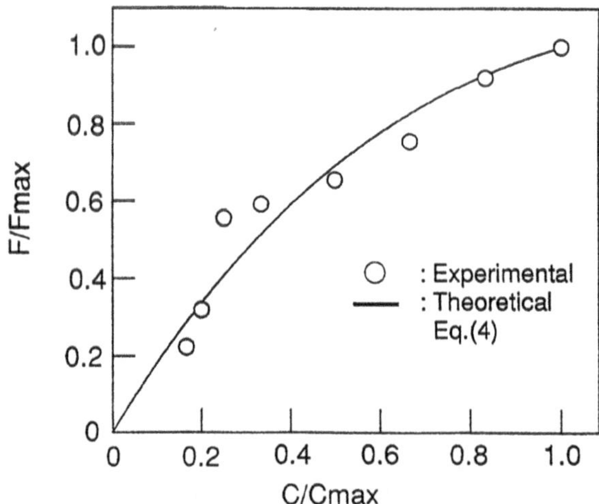

Fig.9 Calibration curve between normalized fluorescent intensity ratio and vapor concentration ratio

4.2. CORRECTION OF VAPOR CONCENTRATION FOR MIXTURE TEMPERATURE

In the case of mixture formation accompanied by evaporation of liquid fuel such as vaporizing diesel spray,the mixture temperature is lower than that of ambient gas. Also, the emission intensity from the fluorescent subject in vapor phase changes with the mixture temperature. In this study, mean temperature of the mixture Tr in the spray is estimated by the bulk mean temperature of liquid and ambient gas as follows.

$$Tr = \frac{-C_v\{c_{fl}(T_{sat}-T_{l0})+h_f\}+C_a \cdot c_a \cdot T_{ent}+C_v \cdot c_{fv} \cdot T_{sat}}{C_v c_{fv}+C_a c_a} \quad \cdots \quad (5)$$

where,

C_v: mass concentration of the fuel vapor (kg/m³) , C_a: mass concentration of the ambient gas (kg/m³)
c_{fv} : specific heat of the fuel vapor (J/(kg•K)) , c_{fl}: specific heat of the liquid fuel (J/(kg•K))
c_a: specific heat of the ambient gas (J/(kg•K)) , T_{l0}: initial temperature of the liquid fuel(K)
h_f: latent heat of fuel vaporization(J/kg) , T_{sat}: saturate temperature of the fuel (K)
T_{ent} : temperature of the ambient gas entrained into the spray.

Mixture temperature Tr and fuel vapor concentration Cv at each zone in the spray are calculated along incident light directions from the edge to the spray center so as to satisfy Eq.(4) and Eq.(5) simultaneously. Here , the temperature of the ambient gas entrained into each spray region is assessed by the averaged value of the mixture temperature at the adjacent outside zone and the ambient gas temperature.

It is well known that as the temperature of fuel vapor is increased , the emission intensity changes with a decrease owing to the decrease in the molar absorption coefficient. This relation is called as temperature quenching. Thus , the change in the emission intensity F(Tr) with the mixture temperature Tr is given as follows [5].

$$\frac{F_0-F(T_r)}{F(T_r)} = k_q \cdot \exp(-\frac{E}{R \cdot T_r}) \quad \cdots \quad (6)$$

where , F_0 : maximum emission intensity at a certain low temperature
k_q : temperature quenching coefficient
E : thermal energy level concerning vibration made (J / mol)
R : gas constant (J / (mol•K))

Let the emission intensity of the homogeneous mixture with temperature 700 K is replaced temporarily by F(700) , then Eq.(6) is transformed into the following equation.

$$\frac{F(T_r)}{F(700)} = \frac{1+k_q \cdot \exp\{-\frac{E}{R \cdot 700}\}}{1+k_q \cdot \exp\{-\frac{E}{R \cdot T_r}\}} \quad \cdots \quad (7)$$

As a result of an examination where the emission intensity from the mixture with a certain vapor concentration was detected by CCD camera system at each mixture temperature , we could obtain kq is 16 and E is 1.75×10^4 J / mol. Figure 10 shows the change in fluorescence intensity with the mixture temperature. The normalized fluorescent intensity F(Tr) / F(700) is indicated in the ordinate. This normalized intensity decreases linearly with increasing in the mixture temperature. Consequently , both of the mass concentration of the fuel vapor Cv and the mixture averaged temperature Tr for each measuring region are assessed quantitatively with an assumption that the saturated temperature and the evaporating rate for the base fuel of n-tridecane and the dopant of TMPD are the same value.

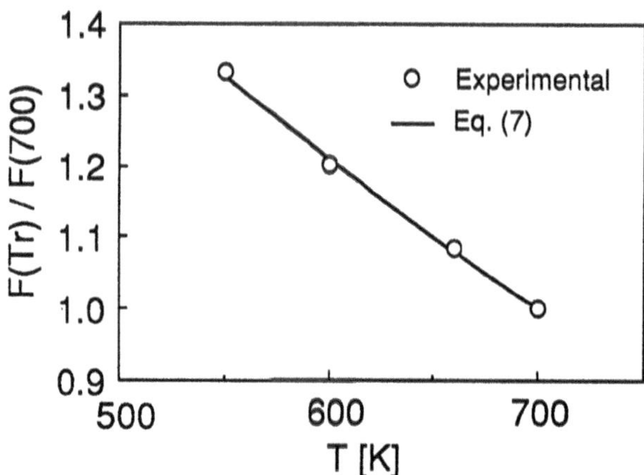

Fig.10 Change in fluorescence intensity with mixture temperature

5. Results and Discussions

Figure 11 shows the several images in the two - dimensional cross section of the impinging spray in the case of elapsed time t from the injection start is 2.0 ms. Measured fluorescence intensities of the vapor and the liquid phases are presented with seven gradations in Fig.11(a) and Fig.11(d) , respectively. Note that its density scale is not absolute , but only relative for each image. Figure 11(b) shows the vapor concentration result obtained by considering the absorption of incident laser light by the vapor phase using Eq.(4) from Lambert - Beer's law. And figure 11(c) represents the final result of the vapor concentration obtained by correcting the concentration with the mixture temperature by use of Eqs.(5) and (6). Here , in the central region of the impinging spray along the spray axis , a halation is occurred owing to the strong luminescence from dense droplets region. So , the quantitative results for the liquid density could not be discussed in this region. In Fig.11(b) , the higher concentration region is formed at the peripheral part of the wall jet region. This result seems to be attributed to the fact that the fluorescent emission is intensified due to the lower mixture temperature in this region. Thus , the relative homogeneous distribution of the vapor concentration is presented in Fig.11(c) , where the correction with the temperature is treated.

Figure 12 shows the temporal changes in vapor concentration and liquid fluorescence intensity. Then , an overlap in wavelength between the two fluorescence signals existed over a wavelength range from 380 nm to 510 nm as shown in Fig.5 (a). Therefore , the fluorescence emission near 390 nm from the liquid phase influences to a great extent the signal from the vapor phase. Thus , in the region where both fuel droplets and the vapor exist in the spray , the crosstalk in signals from the liquid phase to the vapor phase is unavoidable in measurements. However , it seems that a quantitative measurement of the vapor concentration is possible in the region where the droplets do not exist or a small number of droplets is present.

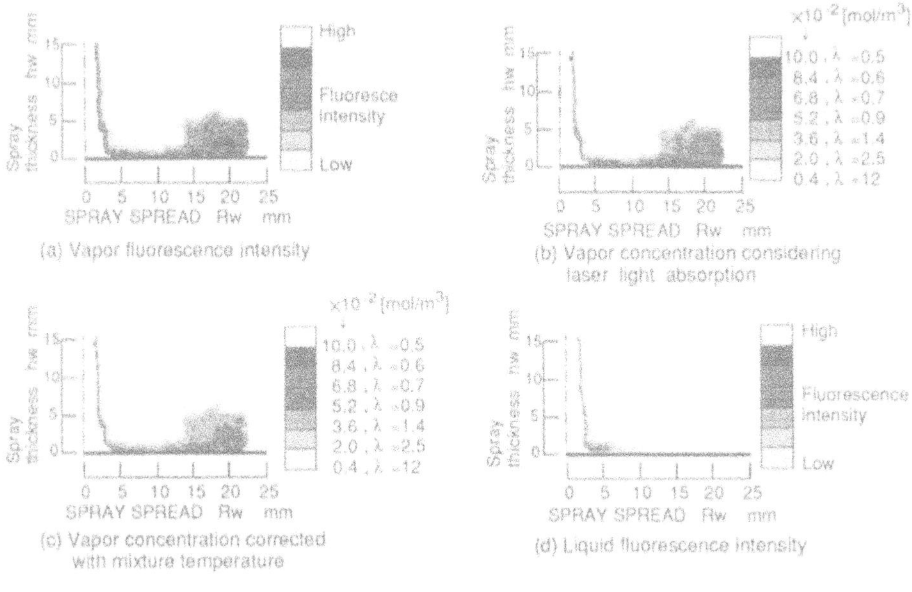

Fig.11 Images of impinging spray (Zw=24 mm, t=2.0 ms)

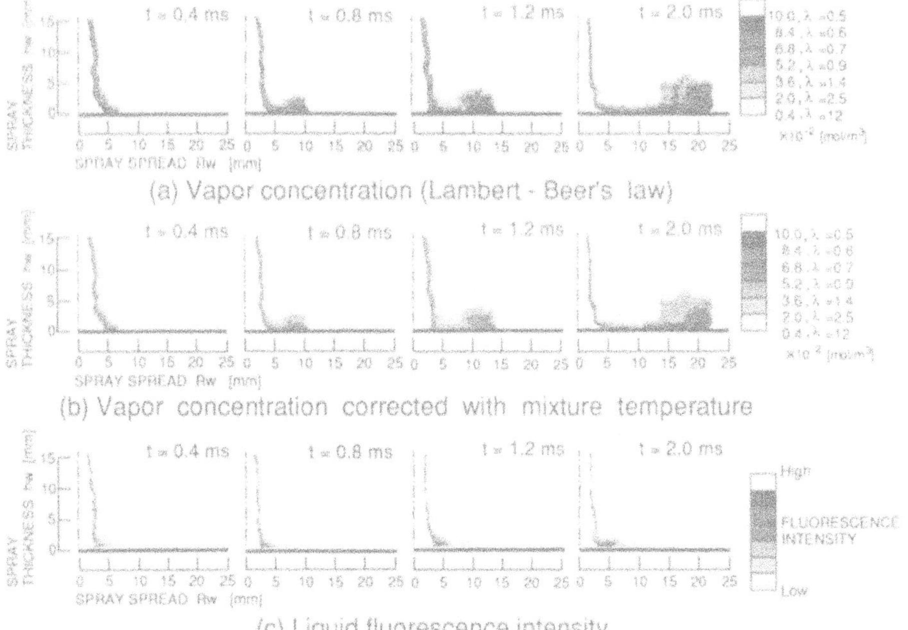

Fig.12 Temporal changes in impinging spray images(Zw=24 mm)

6. Conclusions

(1) The exciplex spray visualization technique has been applied to an impinging diesel spray upon a hot wall in a quiescent high temperature and pressurized atmosphere. Two - dimensional images of vapor and liquid phases were acquired simultaneously.

(2) Vapor concentration was assessed quantitatively by applying Lambert-Beer's law into measured fluorescence intensity in vapor phase and by considering the quenching process of the fluorescent emission due to the mixture temperature.

(3) Then, both of the mass concentration of fuel vapor and the mixture averaged temperature for each measuring region are assessed quantitatively with an assumption that the saturated temperature and the evaporating rate for the base fuel of n-tridecane and the dopant of TMPD are the same value.

Acknowledgements

The authors wish to express their appreciation to Mr. M Kobayashi for his cooperation in the experiments in this study.

References

[1] Katsura , N. , Saito , M. , Senda , J . and Fujimoto , H. (1989) , 'Characteristics of a diesel spray impinging on a flat wall ' SAE Trans. , sp-774 , 191-207.

[2] Melton , L. , A. (1983) , 'Spectrally separated fluorescence emissions for diesel fuel droplets and vapor ' , Applied Optics , vol. 21 , No. 14 , 2224-2226.

[3] Senda , J. , Fukami , Y. , Tanabe , Y. , and Fujimoto , H. (1992), 'Visualization of evaporative diesel spray impinging upon wall surface by exciplex fluorescence method ' , SAE Trans. , vol. , No.920578 , 1-10.

[4] Rotunno , A . , A. , Winter , M. , Dobbs , G. , M. , and Melton , L. , A. (1990) , 'Direct calibration procedures for exciplex-based vapor / liquid visualization of fuel sprays ' , Combust. Sci. and Tech. , vol. 71 , 247-261.

[5] Yagi , K. ,et al. (1958) ' Fluorescence (Theory-Measurement-Application)', Nankou dou 44 - 45

13. FLAME AND TURBULENCE INTERACTIONS IN A FOUR-STROKE SI ENGINE

J. H. Whitelaw and **H. M. Xu**
Imperial College of Science, technology and Medicine
London SW7 2BY
England

ABSTRACT. Flame and turbulence interactions have been examined in a four-stroke spark ignition engine, which was fuelled with propane with a nominal equivalence ratio of 0.9 and operated at 1000 rpm. The flame development was visualised with an intensified CCD camera and gas velocities in the combustion chamber were measured by laser Doppler velocimetry. An ionisation probe and piezoelectric transducer located in the cylinder head measured flame propagation and in-cylinder pressure respectively. The initial spherical shape of the flame kernel is shown to be distorted gradually by large scale turbulence from the expanding burned gas so that the front became increasingly corrugated before reaching the cylinder wall. Cyclic variations of the flame geometry increased correspondingly in the later stages of combustion and increased the ensemble averaged rms of velocity fluctuations in front of flame, especially at locations far from the spark plug; conditional averaging according to the flame arrival at the measuring location reduced this effect considerably.

1. Introduction

To compensate for the lower laminar burning velocity of lean-burn engines, high turbulence levels are required to increase the burn rate with limitations imposed by driveability, unburned hydrocarbon emissions, ignition and pumping losses associated with the greater dissipation of turbulence energy in the cylinder and passages [Bradley *et al*, 1988]. However, evaluation of turbulence intensity during the combustion process, especially in front of the flame, has been shown to be difficult due to the presence of the turbulent flame [Witze *et al*, 1984; Martin *et al*, 1985] and remains open to question [Heywood, 1988]. In our recent paper [Whitelaw and Xu, 1993], velocity measurements by laser Doppler velocimetry were combined with a flame ionisation probe to suggest that the rms of velocity fluctuations across the flame was enhanced to an extent which varied with velocity component, measurement location and burn rate, and the two dimensional velocity vectors suggested that the direction of flame propagation influenced the increased intensity caused by the flame front. The absence of visualisation of the flame

F. Culick et al., (eds.), Unsteady Combustion, 295–314.

propagation, however, prevented further interpretation of the turbulent motion of the mixture and its influence on measured turbulence intensity, which is also the case for many other velocity measurements, for example those of Rask, 1979; Liou *et al*, 1985; Witze *et al*, 1984; Martin *et al*, 1984; Obokata *et al*, 1992.

The importance of the flame structure has been recognised, for example by Abraham *et al*, 1985 and the combustion process in engines has been visualised for decades, Withrow *et al*, 1939; Nakanishi *et al*, 1975; Beretta *et al*, 1986, using high speed cinematography. Shadowgraph and Schlieren pictures have represented integrated light paths, as by Witze *et al*, 1980; Namazian *et al*, 1980; Baritaud, 1987 and two-dimensional visualisation techniques such as laser Mie scattering and laser induced fluorescence, have been developed to allow detailed studies of flame structures with high spatial resolution and some uncertainties, Baritaud and Green, 1986; zur Loye and Bracco, 1987; Zielgler *et al*, 1988; Mantzaras *et al*, 1988; Schipperijn *et al*, 1988. Also an intensified CCD camera makes it possible to record the combustion process images directly through flame radiation, Bates, 1989a, 1989b; Nakamura *et al*, 1989; Winklhofer *et al*, 1993.

In this paper, images of flame propagation have been obtained by an intensified CCD camera and related to the local velocity measurements of Whitelaw and Xu 1993. The implications of the images of the flame front for the interpretation of the velocity measurements are discussed. The experimental system and procedures are described in the next section, followed by presentation of the results and discussion, and the paper ends with a brief statement of conclusions.

2. Experimental Equipment and Procedures

The single cylinder four stroke spark ignition engine of Fig. 1 was developed by Fiat Research Centre specifically for combustion studies with extensive optical access. It features an extended, hollow piston incorporating a quartz window on its crown which allows a view of the disc-shaped combustion chamber through a fixed mirror on the engine block and two diametrically placed windows provide radial scan access to the combustion chamber. The cylinder head and inlet and exhaust ports have geometries typical of 1.6 litre automotive engine. Gaseous propane was injected into the port by a specially designed injector with an equivalence ratio of 0.9, defined in terms of the fuel and air charged during each cycle. The main characteristics of the engine are listed in table 1.

The engine was operated with skip firing which comprised five firing cycles followed by ten motored cycles in order to ensure the integrity of the engine optical components, and the water temperature was controlled at 323°K for the same reason. Whitelaw and Xu showed that the first 'firing' cycle tended to be a misfire and the second a fast-burning cycle caused by residual unburned fuel so that the results discussed here are those from cycles 3, 4 and 5, which are believed to be representative of the combustion process.

The pressure in the cylinder was monitored with an oscilloscope and measured by a

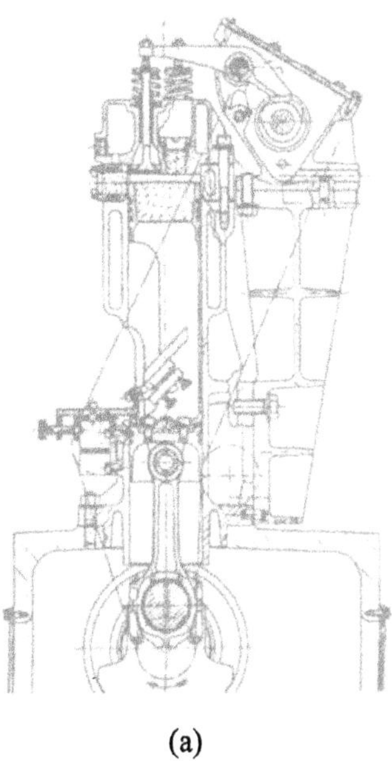

(a)

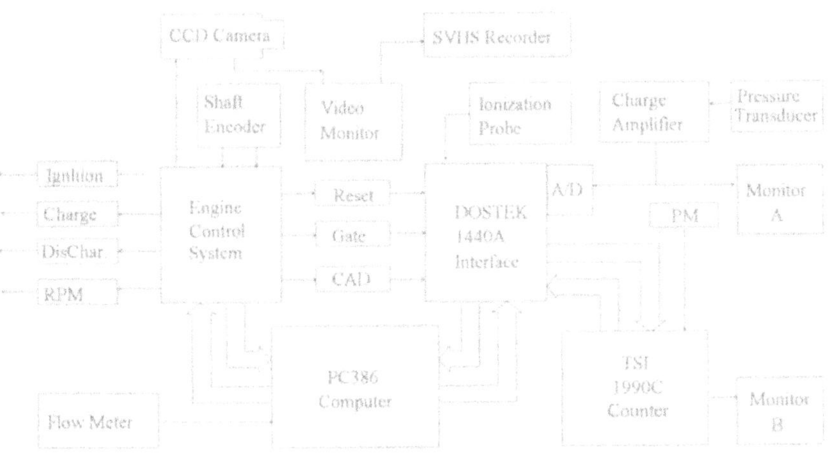

(b)

Fig. 1 Experimental arrangement.
(a) The single cylinder research engine.
(b) The instrumentation.

Table 1 Characteristics of the Fiat Research Engine

Bore	79.5	mm
Stroke	86	mm
Connecting rod length	145	mm
Clearance height	7.6	mm
Compression ratio	10	
Inlet valve opens	5°	BTDC
Inlet valve closes	225°	ATDC
Exhaust valve opens	225°	BTDC
Exhaust valve closes	5°	ATDC
Engine speed	1000	rpm
Mean piston speed	2.87	m/s
Ignition timing	320°	ATDC
Nominal equivalence ratio	0.90	

Kistler piezoelectric transducer and charge amplifier, and the A to D converted data were stored and post-processed. The arrival of the flame front was detected with an ionisation probe located in the cylinder head, Fig. 2, and connected to an amplifier circuit with output passed through a Schmidt trigger circuit to a data acquisition board.

Figure 2 provides a view of the combustion chamber through the piston crown window, which has a diameter of 56 mm, with the vectors indicating the locations and directions for which velocity was measured. The images of the flame were obtained withoptoelectronic shutter. The maximum frame rate was 25 Hz, which precluded

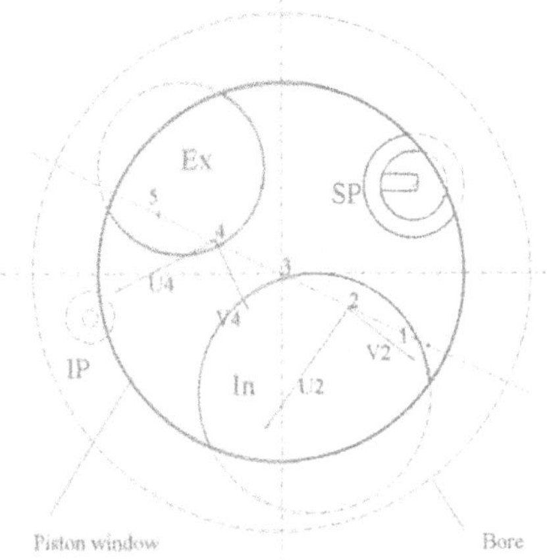

Fig. 2 A bottom view of the combustion chamber and sign conventions (number 1-5 indicate locations of LDV measurement).

continuous acquisition of the flame development in the same cycle, and the shutter was thus triggered externally by the signal from the engine management system for successive cycles. The triggering was varied from 5° CA after ignition to 372° CA at which the maximum pressure occurred, with exposure durations from 50 to 100 μs corresponding to 0.3 to 0.6 CA degrees. The video signal was monitored with a standard monitor and recorded by a video recorder (Panasonic Super-VHS AG-7350) and the signals on the tape were digitised later, image by image, with a frame grabber (MATROX IP-8) in 256-bit grey-scales. The data were then transferred to and stored in a computer (IBM 486 compatible) for further image processing.

The low luminosity of the flame light emissions from lean combustion of the engine was compensated by the intensifier. The resulting images were processed in two ways. They were filtered according to their grey scales at both ends of the spectrum until the best contrast was obtained with spots or noise removed and the images are presented in either positive or negative form. Also, images were inverted and the edges detected to produce images which resembled shadowgraphs with dark lines associated mainly with outlines of the flame curvature rather than with density variations.

The velocity measurements were obtained by a laser Doppler velocimeter, which has been discussed in detail by Whitelaw and Xu. The laser beam from an 1.5 W Argon-ion laser was delivered to the combustion chamber by a rotating diffraction grating-based optical unit through the mirror and quartz window in the piston. Light scattered from 0.56 μm diameter zirconium dioxide particles introduced into the inlet manifold was collected through one of the smaller side windows and passed to a photomultiplier with the resulting electrical signal processed by a counter (TSI Model 1990C), interfaced to a microcomputer (Dostek 1400A DMA board) with a resolution of 0.5 CA degrees. The data acquisition was triggered by the counter data-ready signal and enabled during a 128° CA window of each firing cycle, starting from 300° ATDC of induction. The measurements were made at five locations on the mid-plane of the TDC clearance volume and in the directions both along and orthogonal to the flame propagation (Fig. 2).

The instantaneous velocity data were conditionally ensemble averaged with averaged rms velocities sampled and calculated according to the maximum velocity for each cycle, which corresponded to the arrival of the flame front at the measuring volume with a resolution of 0.5 CA degrees. The results from these procedures are discussed with the flame images in the following section.

3. Results and discussion

The characteristics of combustion process in the engine are represented by real-time pressure traces of fifteen firing cycles in Fig. 3 and the flame development images taken at a sequence of crank angles, Fig. 4. The images at each of the crank angles were obtained from different individual cycles and are representative of hundreds of pictures; they tend to support the finding of Bates (1989b) that the flame shape is self-similar as the flame grows but with important differences of detail.

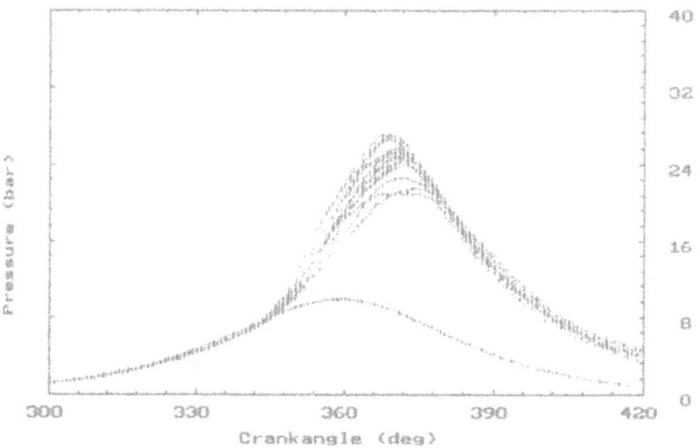

Fig. 3 Real time pressure traces of thirty firing cycles (#3-5) and one motored cycle.

With the spark discharged at 320° CA, the flame kernel initialisation is visible at 327.5° CA, Fig. 4a, and a spherical shaped flame kernel of diameter less than 10 mm in diameter appears at 330° CA, Fig. 4b. The flame kernel grows into a smoothly-curved flame "ball" and remains roughly spherical, Fig. 4c, d, over the next 10 CA degrees. The pressure traces do not show any change until 20 degrees after ignition when the flame has filled about 10% of the cylinder area. The images support the two-dimensional velocity measurement of Fig. 5, which shows that the flow field in this engine was nearly quiescent towards the end of compression. With the absence of strong bulk charge motion, the flame started as spherical and propagates radially, but it was gradually distorted by turbulence from both the preflame region and the expanding burned gas behind the flame front, as suggested by the velocity vectors in Fig. 5 and confirmed by the photographs taken at later stages of combustion. As the piston moved upwards and the flame volume grew, it was compressed against a glass window so that part of the image indicated a cooler region which changes in shape with increasing crank angle as can be seen from Fig. 4e-h. Thus, the edge of the dark region in the photographs may represent a better two-dimensional images of the flame front than does the edge of the bright region.

Twenty five CA degrees after ignition, the flame front is no longer smooth though the overall image remains generally spherical. It is possible to find a "best fit' circle, Beretta *et al*, 1986, for each of the later images with the centre close to the spark plug, but the scale of corrugation of the flame front clearly increases and cannot be ignored. Afterglow [Naknishi *et al*, 1975] is evident from the gases which burned earlier and, at later crank angles, is almost surrounded by the cooler region. At 365° CA, the flame encompasses almost the whole piston window but the combustion continues and the pressure reaches its maximum at 372° CA (Fig. 3) which corresponds to the time when the flame reaches the cylinder wall [Gatowski *et al*, 1984].

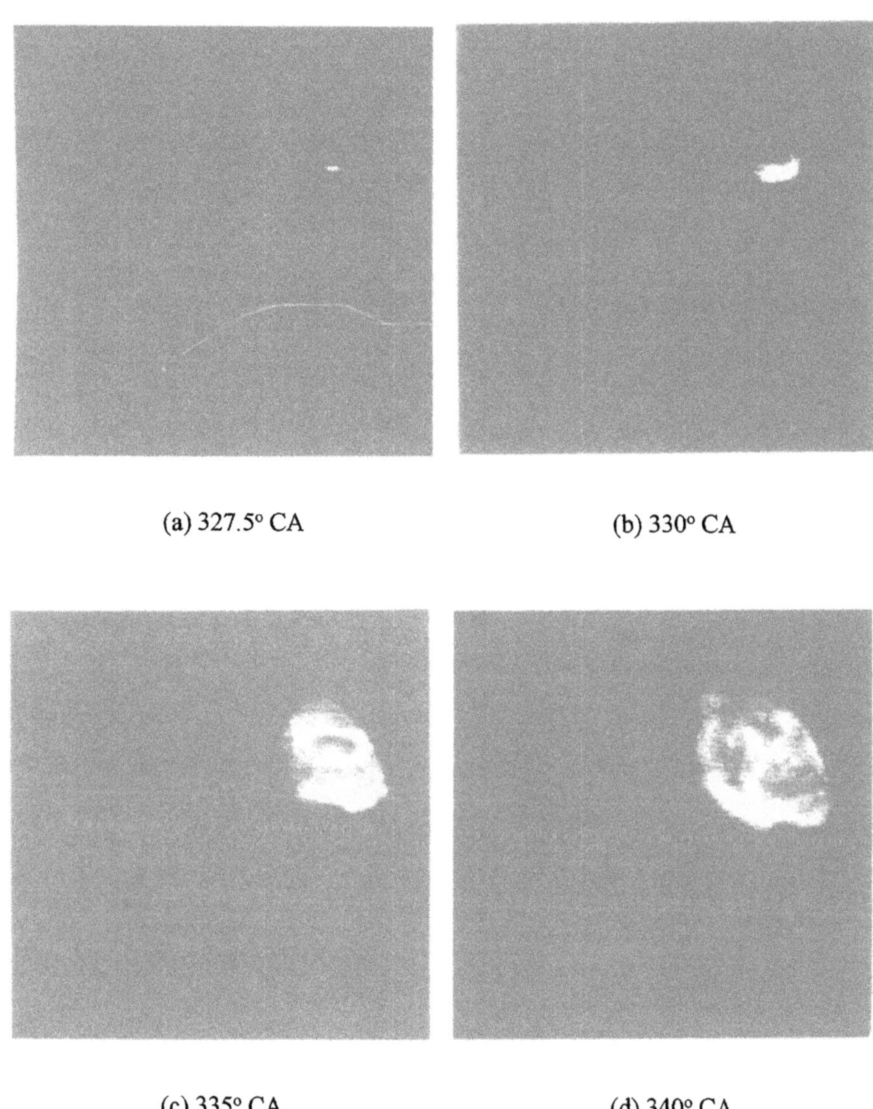

(a) 327.5° CA

(b) 330° CA

(c) 335° CA

(d) 340° CA

Fig. 4 Images of the flame obtained at a set of crank angles (Φ=0.9, θig=320° CA, 1000rpm). (scale bar: 20 mm)

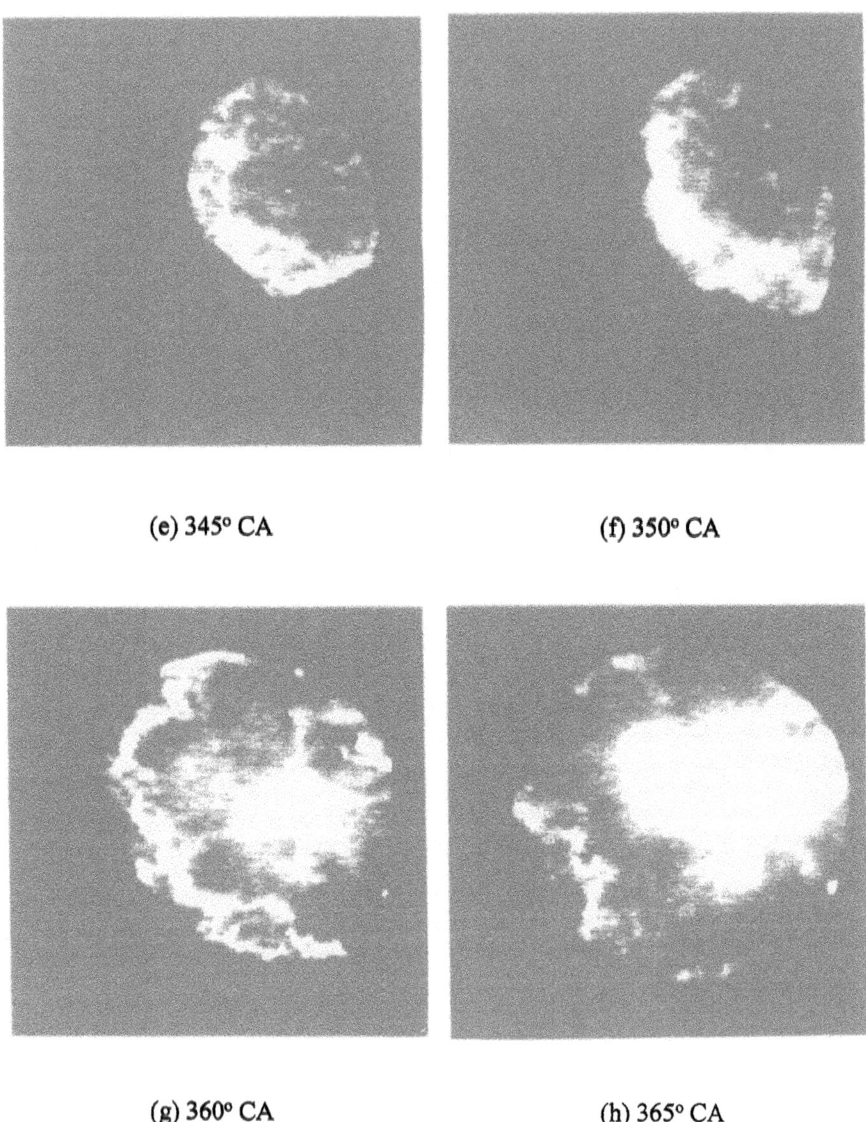

(e) 345° CA

(f) 350° CA

(g) 360° CA

(h) 365° CA

Fig. 4 Images of the flame obtained at a set of crank angles (Φ=0.9, θ_{ig}=320° CA, 1000 rpm). (continued, ├──┴──┤ 20 mm)

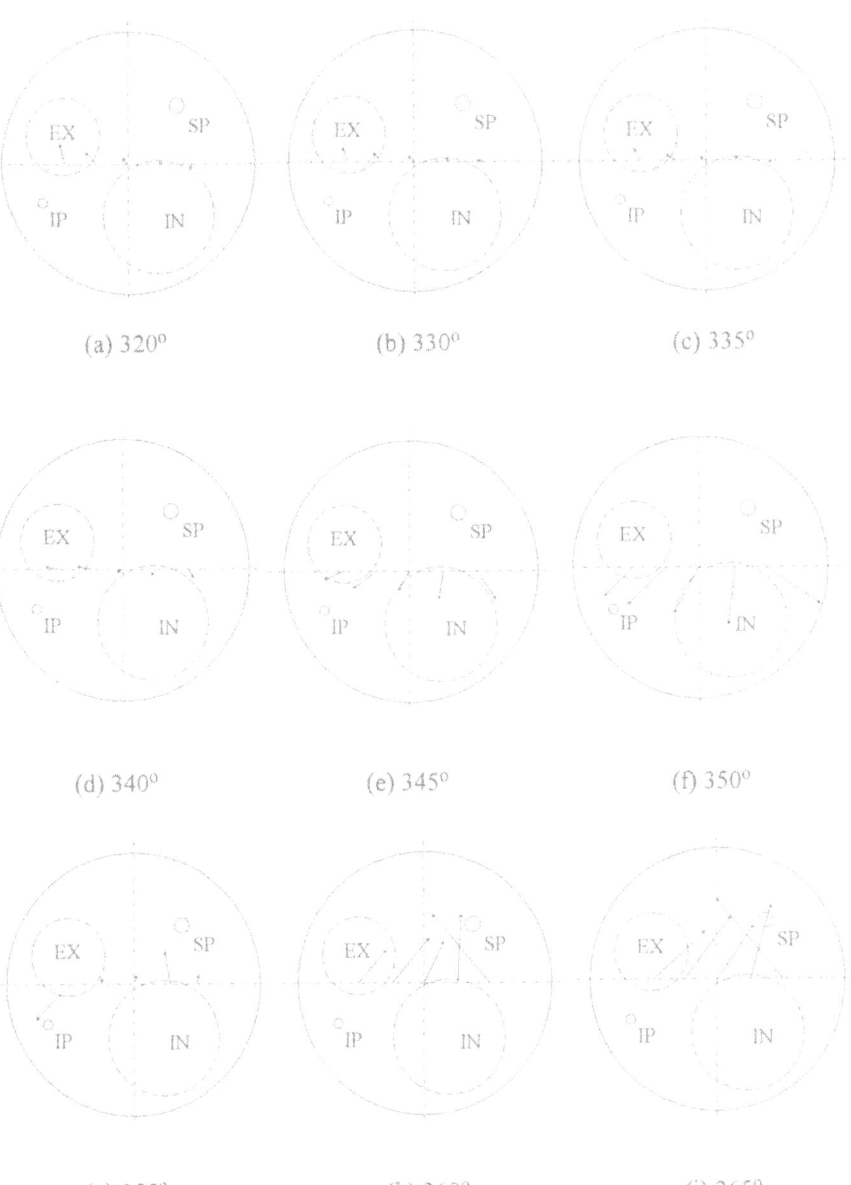

Fig. 5 Two-dimensional mean velocity vector plots (⊔ $4\bar{V}_p$) [Whitelaw and Xu, 1993].

It is evident that the position of the flame front frequently extended beyond the radial location of the two dimensional ensemble averaged velocity vectors of Fig. 5. This is to be expected since the flame is three dimensional and the front can be expected to be below the plane of the velocity measurements which is only 4 mm below the cylinder head. Also, the front has local gulfs of unburned gas so that the front at the LDV measuring plane is very likely to lag that of the visualisation. Thus correlation of velocity measurements with local flame front position is required in conditionally sampled measurements to avoid ambiguities.

The flame propagation influences turbulent velocity fluctuations in the combustion chamber [Witze *et al*, 1984; Martin *et al*, 1985] and the measurement of turbulence intensity requires knowledge of the location of the flame front [Whitelaw and Xu, 1993]. The expanding flame accelerates the unburned gases to a maximum velocity which occurs at the velocity measurement volume after which the gases decelerate rapidly and reverse, see Fig. 5. Because there are large velocity gradients and a nearly discontinuous velocity change across the flame front, the ensemble averaged turbulence intensity near a flame surface will be affected by cyclic variations in the combustion rate and more specifically, in the local flame structure. Ensemble averaged rms velocities, conditioned according to maximum pressure and the flame arrival at the ionisation probe, Fig. 6, show that the

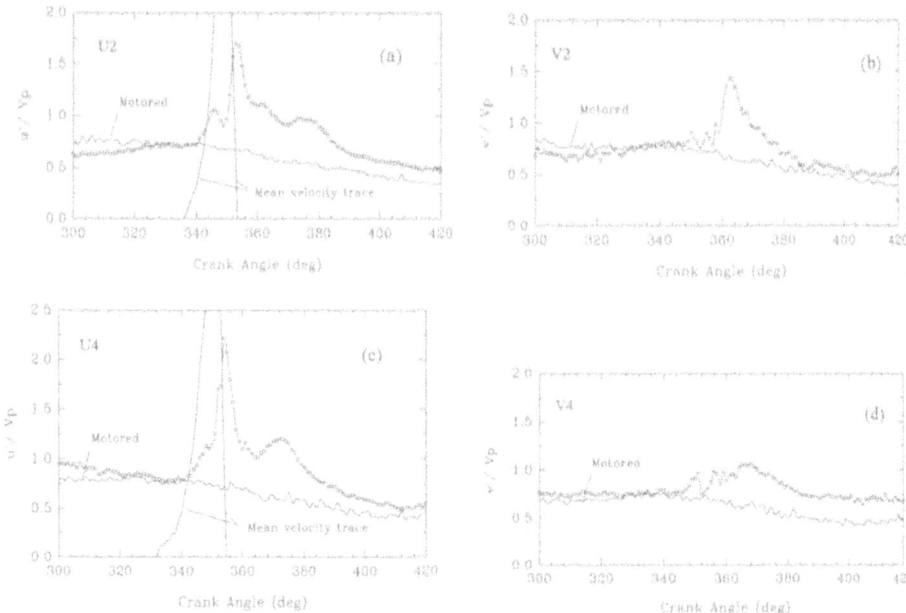

Fig. 6 Rms velocities in firing cycles corresponding to cycles #3-5, conditionally sampled by maximum pressure and the flame arrival at the ionisation probe [Whitelaw and Xu, 1993].

flame passage caused large velocity fluctuations at the LDV volume location, especially along the presumed direction of flame propagation (U component in Fig. 2), and the increase of velocity fluctuations becomes larger at location 4 which is further away from the spark plug. When the data are sampled according to the maximum velocities, which correspond to the flame arrival at the LDV probe volume, the rms velocity values across the flame are reduced although it seems that an increase remains (Fig. 7). This confirms that the shape of the flame has affected the measured turbulence intensity and that the flame structure varied as the flame propagated.

The direct emission photography assists the interpretation of the LDV measurements, but observation of the flame structure can best be made at the leading edge of the flame front where the superposition of reaction zones along the line of sight is at a minimum [Namazian *et al*, 1980]. The photographs of Fig. 8 to 10 are light-intensity inversed with best contrast visibility so that darkness corresponds to the light intensity and they show flame images taken at different stages of combustion. Cyclic variations of the flame geometry occur at each stage, including at 10 CA degrees after ignition, when the flame kernel is smaller than 10 mm, Fig. 8. The initial flame kernel is indeed spherical with the

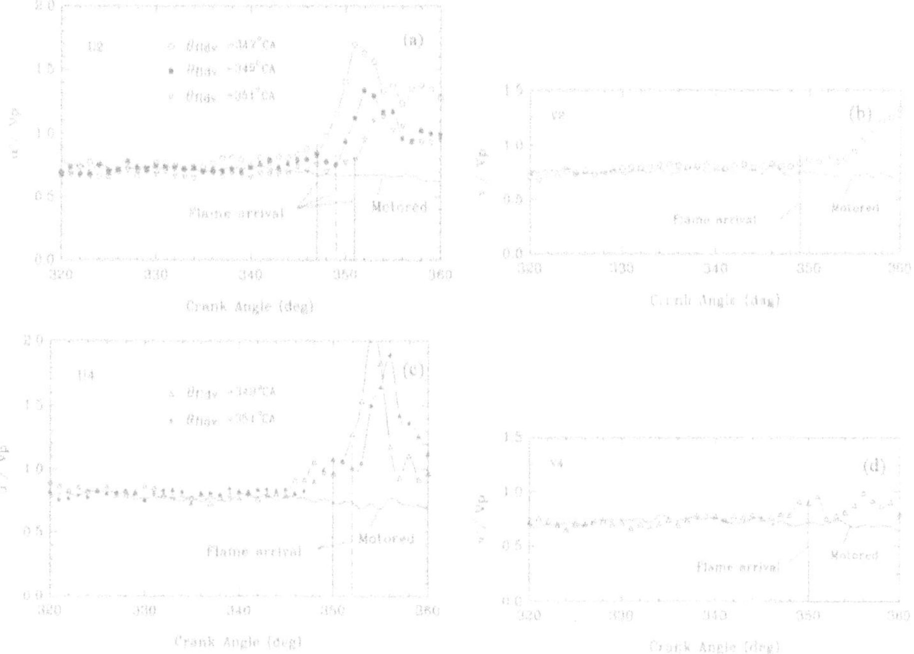

Fig. 7 Rms velocities with varying crank angles of flame arrival at the LDV measuring volume [Whitelaw and Xu, 1993].

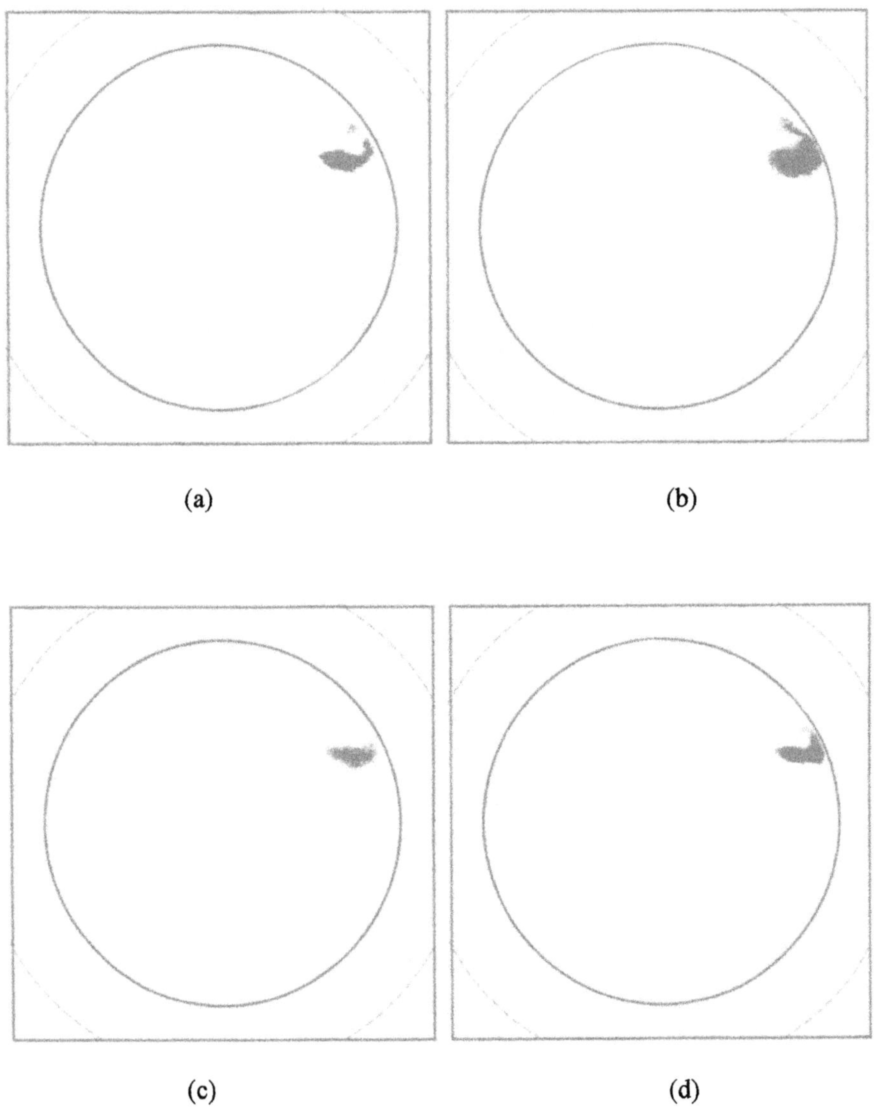

(a) (b)

(c) (d)

Fig. 8 Images of the flame kernel obtained at 330° CA (Φ=0.9, θig=320° CA, 1000 rpm).

(a-d): inversed. (⊔⊥⊔ 20 mm)

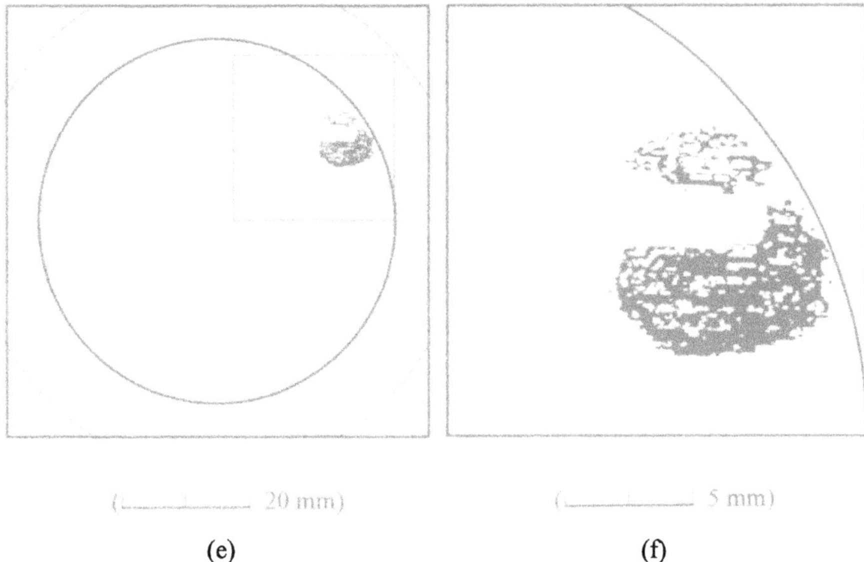

Fig. 8 Images of the flame kernel obtained at 330° CA (Φ=0.9, θ_{ig}=320° CA, 1000 rpm). (continued)
(e-f):edge detected.

absence of strong organised bulk charge motion but the size varies from one cycle to another, implying the variation of the initiation time of the kernel which is responsible for variations of the entire combustion process [Bates, 1989b]. The centre of the kernel is usually slightly away from the centre of the spark plug, in agreement with the weak swirl detected by LDV measurement around the ignition time and occasionally the flame kernel is non-spherical, Fig. 8d, possibly due to aspects of the electrodes, Pischinger and Heywood (1990).

The scale of the wrinkles of initial flame kernel is small in accord with the Schlieren photographs of Gatowski *et al* (1984) which show a matrix of dark lines on the spherical flame, which were believed to be the measures of the scale of the wrinkles of convolutions in the flame sheet, with thickness of the order 0.1-0.2 mm. The data-processing procedure called "edge detecting" was used to provide a special type of image, as in Fig. 8e and f, which shows that the integral shape and the wrinkling scale of the flame kernel are in accord with those of Gatowski *et al.* With a spherical flame kernel wrinkled to this small extent, the velocity measurements conducted near to the spark plug should be less influenced by the early flame geometry.

The images in Fig. 9 were taken at 345° CA, corresponding to the ensemble averaged time when the flame approached measurement location 2. The leading flame fronts remain close to spherical but with corrugations, and the average radius of the flame

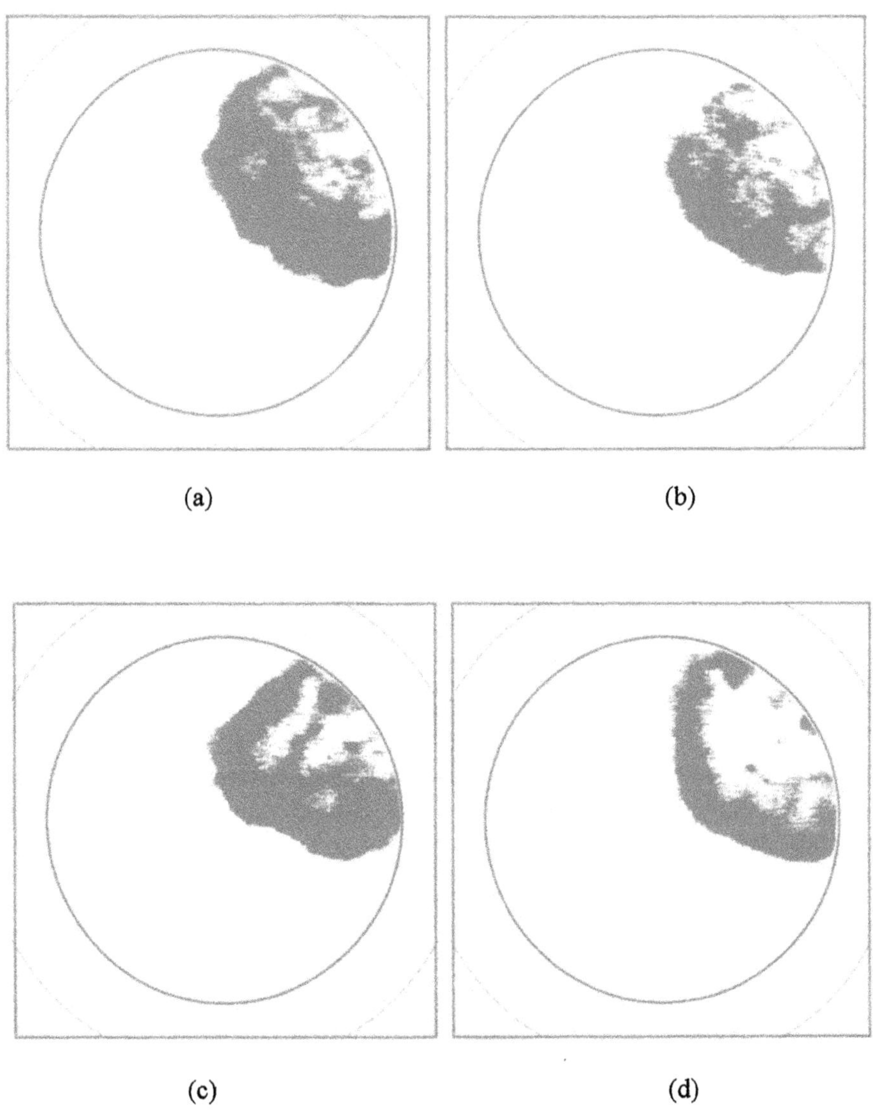

(a) (b)

(c) (d)

Fig. 9 Images of the flame obtained at 345° CA (Φ=0.9, θ_{ig}=320° CA, 1000 rpm).
(a-d): inversed (scale bar: 20 mm).

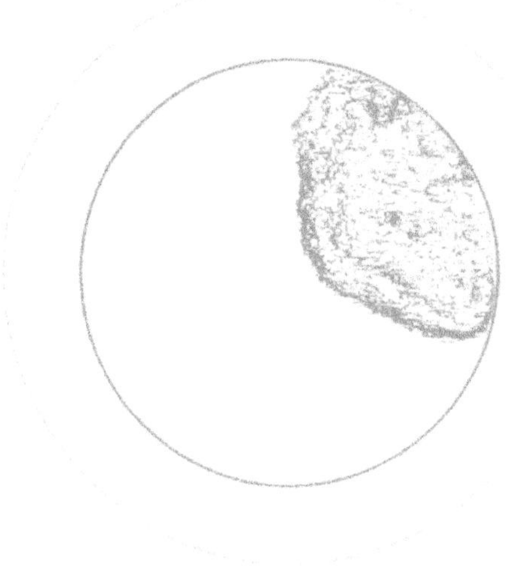

(e)

Fig. 9 Images of the flame obtained at 345° CA (Φ=0.9, θig=320° CA, 1000 rpm). (continued)
(e): edge detected. (⊢——⊥——⊣ 20 mm)

varies from cycle to cycle by up to 3 mm. For some cycles, for example those of Fig. 9c, the integral shape of the flame volume is more like a polygon, as observed by Bates (1989a), which is probably inherited from its early shape. Fig. 9e shows the structure of the flame at this stage more clearly and the intersection of the flame front with the piston crown window is responsible for the two-dimensional curvature with "fingers" of products, Mantzaras *et al* (1988). The scale of the wrinkles of the fully developed turbulent flame is around 1 mm, and the integral scale increases to nearly 5 mm as found by zur Loye and Bracco(1987).

The increase of the thickness of the flame brush is apparent on Fig. 10, where the images correspond to 360° CA and show that it becomes greater than the 10 mm suggested by Heywood 1988. This may be due to the greater effect of the turbulent flow on the present lean mixture and its dilution with residual gas. Undoubtedly, the increased convolution caused by the turbulence of the expanding burned gas in the preflame region increased the area of the flame front and this is beneficial to the combustion process.

Fig. 10e confirms that the flame is fully three dimensional with complicated structures. The engulfed flame volume is characterised by small and large scales of turbulent flow with a corrugated flame front 'brush'. The spatial difference between the leading edge of the flame and the gulf of unburned area implies large variations in the time of flame arrival detected by an ionisation probe, Fig. 2. In support

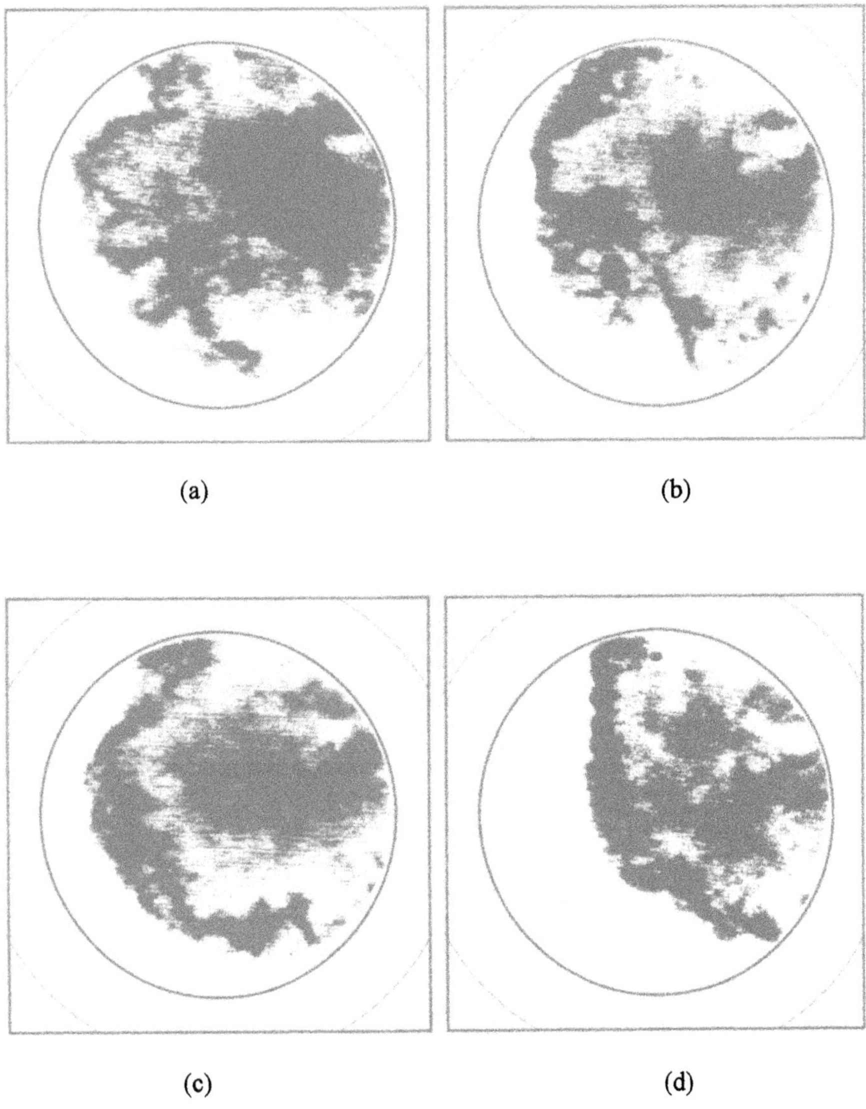

Fig. 10 Images of the flame obtained at 360° CA (Φ=0.9, θig=320° CA, 1000 rpm).
(a-d): inversed (20 mm).

(e)

Fig. 10 Images of the flame obtained at 360° CA (Φ=0.9, θ_{ig}=320° CA, 1000 rpm). (continued)
(e): edge detected. (scale bar 20 mm)

of this argument, Fig. 11a shows that the spread of the flame arrival time at the ionisation probe reaches 20 crank angle degrees with a close-to-normal probability distribution suggesting a random process. Fig. 11b shows that the cyclic variation of the time when the flame arrives at location 2, which is closer to the spark plug than the ionisation probe, is much smaller and this confirms that the flame becomes increasingly corrugated during thc propagation and explains why the increase of measured turbulence intensity in front of flame becomes larger at locations which are further from the spark plug. The observation supports the suggestion that velocity measurements should be synchronised with two dimensional visualisation since both the position and propagating direction of the local flame front affect the measuredturbulence level. Measurements of this type are likely to require considerable effort and data processing.

4. Conclusions

Investigations of the flame and turbulence interactions in a four-stroke spark ignition engine using imaging and laser Doppler velocimetry lead to the following conclusions:

(1) Flame and turbulence interactions result in cycle-by-cycle variations of flame geometry. At the initiation of the flame kernel, the flame tends to have the same shape

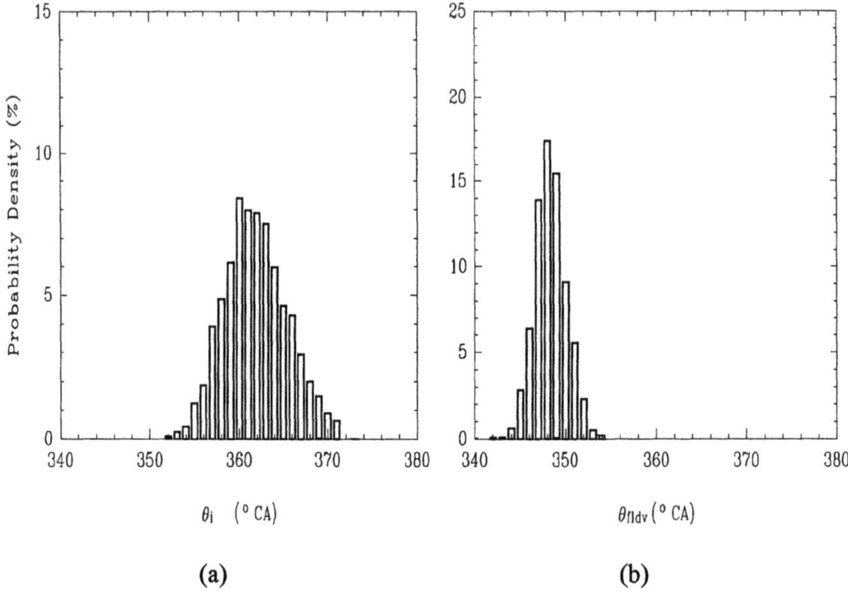

(a)

(b)

Fig. 11 PDFs of the crank angles at which the flame arrives at the ionisation probe and the LVD measurement location 2.

with very small scale wrinkles. In the absence of organised bulk flow, the flame kernel grows radially and the integral scales can be measured by direct emission visualisation with an image intensifier.

(2) The initial spherical shape of the flame kernel remains to about 20° CA after ignition as the flame volume grows and the smoothly-curved flame front is gradually distorted by large scale turbulence from both the expanding burned gas behind the reaction sheet and the preflame region. The scale of the flame corrugation increased with further propagation until it approached the cylinder wall. The large scales of the flame convolution exceed 10 mm in the present engine.

(3) The variations of flame geometry contribute to the measured increase of velocity fluctuations and to the apparent turbulence intensity across the flame. Conditional ensemble averages of the velocity data reduce the measured turbulence intensity in front of flame and the correctness of the results relies on the resolution of the flame arrival time and its direction of propagation.

(4) The flame is three dimensional and application of an ionisation probe in the cylinder head provides an incomplete identification of the local flame position at the velocity measuring locations. Synchronisation of two-dimensional visualisation in the plane of the velocity measurements is suggested to determine the true turbulence intensity and it seems likely that influence of cyclic variations of the flame geometry on the measured turbulence level will be small.

Acknowledgement

The authors are pleased to acknowledge the financial support from the Science and Engineering Research Council under Grant GR/H44943 and from the US Department of Energy under Grant DE-FG04-91/ER 75332. The engine was made available by the Fiat Research Centre and Dr. C Vafidis made considerable contributions to the arrangement of the engine and instrumentation. Thanks is also due to Dr. Hu, now with Brunel University, for his help in the photographic technique.

References

[1] Abraham, J, Williams, F. A. and Bracco, F. V. (1985), 'A Discussion of Turbulent Flame Structure in Premixed Charges, ' SAE paper 850345.
[2] Baritaud, T. A. and Green, R. M. (1986), 'A 2-D Flame Visualisation Technique Applied to the I.C. Engine, ' SAE paper 860025.
[3] Baritaud, T. A. (1987), 'High Speed Schlieren Visualization of Flame Initiation in a Lean Operating S.I. Engine, ' SAE paper 872152.
[4] Barr, P. K. and Witze, P. O. (1988), 'Some Limitations to the Spherical Flame Assumption Used in Phenomenological Engine Models, ' SAE paper 880129.
[5] Bartaud, T. A. and Green, R. M. (1986), 'A 2-D Flame Visualisation Technique Applied to the I.C Engine, ' SAE paper 860025.
[6] Bates, S.C. (1991), 'Further Insight Into SI Four-Stroke Combustion Using Flame Imaging, ' Combustion & Flame, 85, 331-352.
[7] Bates, S. C. (1989a), 'Flame Imaging Studies in a Spark-Ignition Four-Stroke Internal Combustion Optical Engine, ' SAE paper 890154.
[8] Bates, S. C. (1989b), 'Flame Imaging Studies of Cycle-by-Cycle Combustion Variations in a SI Four-Stroke Engine, ' SAE paper 892086.
[9] Beretta, G. P., Rashidi, M. and Keck, J.C. (1983), 'Turbulent Flame Propagation and Combustion in Spark Ignition Engines, ' Combust. Flame, Vol 52, pp. 217-245.
[10] Bradley, D. , Hynes, J., Lawes, M. and Sheppard, C. G . W. (1988), 'Limitations to Turbulence-Enhanced Burning Rates in Lean Burn Engines,' IMechE, C46/88.
[11] Gatowski, J. A., Heywood, J. B. and Delepace, C. (1984), 'Flame Photographs in a Spark-Ignition Engine,' Combustion and Flame, Vol. 56, p71-81.
[12] Hall, M. J., Dai, W. and Matthews, R. D. (1992), 'Fractal Analysis of Turbulent Premixed Flame Images from SI Engines, ' SAE paper 922242.
[13] Heywood, J. B. (1988), 'Internal Combustion Engine Fundamentals,' McGraw-Hill, ISBN 007-100499-8, 1988.
[14] Horie, K., Nishizawa, K, Ogawa, T, Akazaki S, and Miura, K (1992), 'The Development of a High Fuel Economy and High Performance Four-Valve Lean Burn Engine, ' SAE paper 920455.
[15] Lee, T.-W., North, G. L. and Sanivicca, D. A. (1993), 'Surface Properties of Premixed Propane/Air Flames at Various Lews Numbers, ' Combustion and Flame

93:445-456.

[16] Liou, T. -M, and Santavicca, D. A. (1983), 'Cycle Resolved Turbulence Measurements in a Ported Engine with and without Swirl,' SAE Paper 830419.

[17] Mantzaras, J., Felton, P. G. and Bracco, F. V. (1988), 'Three-Dimensional Visualization of Premixed-Charge Flames: Islands of Reactants and Products; Fractal Dimensions; and Homogeneity, ' SAE paper 881635.

[18] Martin, J. K., Witze, P. O. and Borgnakke, C. (1985), 'Combustion Effects on the Preflame Flow Field in a Research Engine,' SAE paper 850122.

[19] Martin, J. K., Witze, P. O. and Borgnakke, C. (1984), 'Multiparameter Conditionally Sampled Laser Velocimetry Measurements During Flame Propagation in a Spark Ignition Engine,' Sandia Report SAND84-8224, 1984.

[20] Nakamura, A., Ishii, K. and Sasaki, T. (1989), 'Application of Image Converter Camera to Measure Flame Propagation in S.I. Engine, ' SAE paper 890322, 1989.

[21] Nakanishi, K., Hirano, T., Inoue, T., and Ohigashi, S. (1975), 'The Effects of Charge Dilution on Combustion and Its Improvement - Flame Photograph Study, ' SAE paper 750054.

[22] Namazian, M., Hansen, S. Lyford-Pike, E., Sanchez-Barsse, J. Heywood, J and Rife, J. (1980), 'Schlieren Visualisation of the Flow and Density Fields in the Cylinder of a Spark-Ignition Engine, ' SAE paper 800044.

[23] Obokata, T., Hashimoto, T., Gojuki, S., Karasawa, T., Shiga, S. and Kurabayashi, T. (1992), 'LDA Characterisation of Gas Flow in a Combustion Chamber of a Four-Stroke S.I. Engine,' SAE Paper 920519.

[24] Rask, R. B. (1981), 'Comparison of Window, Smoothed-Ensemble and Cycle-to-Cycle Data Reduction Techniques for Laser Doppler Anemometer Measurements of In-Cylinder Velocity,', Proc. Symp. on Fluid Mechanics of Combustion Systems, Colorado.

[26] Schipperijn, F. W., Nagasaka, R., Sawyer, R. F. and Green, R. M. (1988), 'Imaging of Engine Flow and Combustion Processes, ' SAE paper 881631.

[27] Whitelaw, J. H. and Xu, H. M. (1993), 'An Experimental Study of Gas Velocity, Flame Propagation and Pressure in a Spark-Ignition Engine, ' SAE paper 932702.

[28] Winklhofer, E., Philipp, H., Fraidl, G. and Fuchs, H. (1993), 'Fuel and Flame Imaging in SI Engines, ' SAE paper 930871.

[29] Witze, P. O., Martin, J. K. and Borgnakke, C. (1984), 'Conditionally-Sampled Velocity and Turbulence Measurements in a Spark Ignition Engine,' Combustion Science and Technology, Vol. 36, pp, 301-317.

[30] Xu, H. M. (1994), 'Turbulence and Cyclic variations in a Spark Ignition Engine,' PhD Thesis, Imperial College, in Preparation.

[31] Ziegler, G.F.W., Zettlitz, A., Meinhardt, P., Herweg, R., Maly, R. and Pfister, W. (1988), 'Cycle-Resolved Two-Dimensional Flame Visualization in a Spark-Ignition Engine, ' SAE paper 8816634.

[32] zur Loye, A. O. and Bracco, F. V. (1987), 'Two-Dimensional Visualization of Premixed-Charge Flame Structure in an IC Engine, ' SAE paper 870454.

SPRAY TRANSPORT AND MIXING IN SPARK IGNITION ENGINES: SOME MODELLING AND EXPERIMENTAL TRENDS.

F.VANNOBEL
P.S.A. PEUGEOT CITROEN, DRAS, Route de Gizy, 78140 Velizy, FRANCE

ABSTRACT. This paper is dedicated to the study of sprays in the framework of manifold injection in an S.I. engine. Modelling of spray is considered and laser-based techniques were used to characterise droplets in an engine, and a simulated arrangement was built to maximise optical access for the phase-Doppler anemometer (PDA). The applications are, successively, a free air spray, a steady flow-rig and a cylinder-head without and with running camshaft. A strong correlation between cold-flow simulation obtained with an open-ended tube downstream the cylinder-head, and the motored engine is obtained in terms of droplets behaviour, concentration and trajectories. The quality of the air/fuel mixture is described by these means, in terms of the extent of non-homogeneity.

1. Introduction

In the field of mixture preparation, injection systems are needed to satisfy the exhaust emission standards, in association to a lambda probe and a catalytic converter. Engines which operate at an equivalence air/fuel ratio close to one, and with a three-way catalyst, require that the air/fuel ratio be determined to allow feedback from exhaust line to inlet manifold. Moreover, a compromise is required to decrease unburned Hydrocarbons (HC) emissions and improve combustion stability. In cold-start conditions, the catalytic converter is not effective, and 75% of HC emissions are produced in this period. After this limit in the exhaust gas temperature, 90% of HC emissions are removed by the catalyst. A better understanding of the structure and evolution of the spray in the context of real engine design is addressed in this paper. The liquid film, formed on the cylinder and manifold walls are known to play an important role on HC emissions. Heywood (1988) detailed the HC sources inside the manifold-injected SI engine by means of a fast response Flame Ionisation Detector and showed importance of crevices inside the combustion chamber. More recently Cheng et al (1993) have provided an overview of HC emissions, giving an estimation of the fraction of fuel escaping the primary combustion process, which is in the order of 9% of the fuel at part load. The present work contributes to the identification of HC sources in terms of droplets leading to incomplete combustion.

F. Culick et al., (eds.), Unsteady Combustion, 315–332.

2. Spray modelling in engines

In the field of mixture preparation inside Spark Ignition engines, research has usually been conducted from an experimental point of view, as can be explained by two statements. The first is historical: homogeneous mixtures were considered sufficient to implement a combustion model, as carburettors were commonly used, and spray modelling of port-injected gasoline engines was not required. The second is technical: the air-fuel systems have recently reached a technological step: single-point, then multi-point injection systems appeared; spray models then had to take into account many processes, specially in gasoline port-fuel-injected engines. Nevertheless, several attempts have been made, in the field of direct-injected engines -either in Diesel or gasoline- and will be discussed in the second part of this section. For port-injected engines, we can observe that phenomenological models, like that described by Lenz (1992) have been used in a first approach.
In the field of application inside reciprocating engines, the Diesel engine was the first considered and in the framework of direct-injection; the resulting spray models have been considered for the SI engine. The successive sub-models which are so far used are as follows: atomisation, collision and coalescence, break-up, evaporation and drag effects. The following sub-section gives an insight into the physical assumptions made in the framework of a spray model used for injectors encountered in reciprocating engines. The two engine cases basically consider the fuel injection pressure, as well as ambient air pressure: the Diesel case deals with several hundred bars inside the cylinder or a prechamber at, say, 60 bars, while the gasoline case is characterised by 3 bars injection pressure inside the manifold where the ambient pressure is less than one atmosphere, and depending on the throttle position. So, the two problems, although similar when considering physical mechanisms, deal with very different orders of magnitudes and must be dissociated.

2.1 DIRECT-INJECTION DIESEL ENGINE

Due to the high liquid injection pressure, cavitation acts from the beginning of the spray, as well as turbulence in the liquid phase inside the injector. Chaves et al (1991) have studied this phenomenon in a model direct-injection Diesel injector, and proposed a flow model which establishes the exit velocity of the liquid front, taking into account the cavitation within the injector. The spray angle is strongly affected by the design of injector, and by the gas-liquid density ratio. The atomisation process has been modelled from drop break-up experiments (see Reitz, 1987), and vaporisation using correlation from single droplet studies. In order to represent a real spray, an additional sub-model of diffusion of the vapour cloud away from the drop is required, according Reitz (1991). Also, multi-droplet interactions experiments are of interest but unfortunate and complex to be carried out. The actual models which are implemented in three-dimensional codes are generally tested by means of comparison of penetration-depth of both gaseous and liquid phases (Hodges et al, 1991). This method is not completely satisfactory, as uncertainties always exist in phases distinction as well as in the definition of liquid penetration in numerical result. The liquid/gaseous distinction could be solved by X-ray absorption photography, but this is not a trivial way. All other optical methods are known to be uncertain, as the very dense core of the spray absorbs light. On the other hand, the calculated radial expansion of the liquid phase is often under-estimated (Klingsporn and Renz, 1992). Several experiences of droplets impinging on a heated surface were also conducted. According wall temperature relative to the fuel boiling point, the surface is either wetted, accompanied by secondary droplets produced by splashing, or, at higher temperatures, surface boiling effects become predominant, preventing the wall from wetting, due to secondary atomisation mechanism. These findings were exploited by Gosman (1991) and currently examined with existing theories about droplet/wall interactions.

2.2 GASOLINE-SPRAY MODELLING

Spray modelling applicable to SI engines has essentially progressed outside the engine, but not yet inside the manifold, as wall wetting phenomena, evaporation, stripping of the liquid film by air, action of turbulence on film increase the complexity of the study.

2.2.1 Free air spray predictions. Considering interfacial instabilities of a liquid sheet delivered by a model-injector, Li and Tankin (1990) and Dumouchel and Ledoux (1991) have begun to describe the atomisation phenomenon as a disintegration process in terms of simple shapes of liquid sheets but allows several physical phenomena like for instance viscosity effects to be observed. Several images were compared to the theoretical predictions (Dementhon et al, 1992) and showed that the needle lift motion was interacting with the atomisation process.

Based on the same approach of an undulating liquid sheet, by a combination of integral conservation constraints and Maximum Entropy Formalism (MEF), the work of Sellens (1989) provides a physical model of break-up in a two-dimensional skim. Good agreement is found between experimental data and the prediction of drop-size and velocity fields, except in the region where drag forces strongly influence the distribution.

2.2.2 Atomisation processes. Reitz and Bracco (1982) have established a widely-used basis for the representation of free-air sprays. They classified several regimes of break-up, differentiated by the location, upstream or downstream of the nozzle, where first drops are emitted. This topic has also been well documented, for instance by Ruiz and Chigier (1985). Three phases are commonly identified in the atomisation process: primary and secondary break-up, followed by turbulence action.

The primary atomisation process consists of an exponential growth perturbation which yields droplets (Huh and Gosman, 1991). From the initial size or parent size -which numerically corresponds to the mean value of the population contained inside the given mesh cell- another break-up process is activated, resulting in a size distribution.

The secondary atomisation process occurs either at the scale of the drop size, called the bag effect, or through a circumferencial distortion called a strip effect, according Reitz and Diwakar (1987), and characterised by a Weber number. In the latter case, droplet formation results from the unstable growth of short wavelength surface waves on the jet surface which are caused by the relative motion of the jet and the ambient gas. In the mean time, collisions and coalescence between droplets interact. Another atomisation process involved in the engine context, due to the inlet valve is described in the paragraph 2.2.4.

2.2.3 Action of turbulence. The two break-up processes are followed by turbulence acting through two effects; turbulence generation and turbulent dispersion. The turbulence generation represents the distortion of a continuous-phase turbulence field by the wakes of drops passing through the flow. The turbulent flow provides enough energy to overcome surface tension and the viscous force on the droplet. Turbulent dispersion then transports drops according to their density. The particle dispersion is correlated to the ratio of the particle relaxation time to the time scale of turbulence.

The effect of particle residence time in the eddies of the gas phase is a complex problem and still requires validation of proposed numerical models, either Eulerian or Lagrangian, through detailed experiments. Gouesbet et al (1992) discussed these two approaches of modelling dispersed multiphase flows and encouraged that test-cases should be independently studied by several laboratories. Sommerfeld et al (1992) contributed to this procedure of comparing results obtained from numerical simulations and an experimental data-base.

2.2.4 Three-dimensional engine cycle simulation. Once droplets are correctly predicted in their evolution inside the manifold, they impinge onto the inlet valve or other surfaces. If the valve is closed, they form a liquid film which will be stripped by incoming air when the valve is opening: this "secondary atomisation process" has been studied by Posylkin et al (1993). The other phenomenon of importance is generated by air-flow effects, and has been described by Vannobel et al (1992).

Nevertheless, these various actions within spray models, even complementary, are not ready to be assembled, especially in the context of three-dimensional engine cycle calculations. First because they describe a partial aspect of the problem, without wall interaction nor transient conditions. Secondly, other physical processes involved by valve movement, exhaust back-flow causing evaporation, and specific aerodynamic field acting in a manifold, downstream of a throttle valve have to interact with results deduced from free-air spray study. Moreover, in order to address these questions, preliminary experimental tests have to be carried out in situ, for a better-understanding and handling of the complex situation.

3. Spray experiment in SI engines

An engine presents many complex conditions and the definition of flow initial boundary conditions raises difficulties and uncertainties, so that it is more efficient to address simple experiments. Then, if detailed measurements joined to a reliable theoretical modelling can be achieved, the result will be an improved understanding of spray behaviour and its representation.

This work has two objectives: first to begin an empirical analysis of spray behaviour inside manifold and cylinder of an SI engine with step-by-step evaluation of basic mechanisms relevant to air/fuel mixture preparation. Second to establish a preliminary test procedure with a transparent cylinder-head which allows more detailed investigations than a firing engine (Vannobel et al, 1994). Thermal and compression aspects are not reproduced here, but a relevant comparison between the two set-up studies is expected, since the steady concept has been fully exploited in the past for swirl characterisation of inlet pipes and cylinder-heads by several authors: Bicen et al (1984), Arcoumanis et al (1987) using Laser Doppler Velocimetry. In other cases, steady-flow has been used to examine gasoline droplet trajectories (see Kawazoe et al, 1990 and Dementhon and Vannobel, 1991). The obvious merit is the ability to address aerodynamics effects on spray behaviour inside a manifold without the complexity of an engine test-bench. Of course, the possible limitations have to be kept in mind, and our purpose is to establish a link between a running cam-shaft study on a simulation bench and an optically-accessible engine.

3.1 IMPLEMENTATION OF THE STEP-BY-STEP METHOD

The schedule adopted here is a compromise between a basic experiment, which is not exactly the purpose of this work, and a production firing engine. Measurements were obtained in a free air-spray, then inside a simulated-engine. Some aspects of the atomisation processes are observed in the first step by means of instantaneous video imaging which has been described in detail by Dementhon and Vannobel (1991), and with a Phase Doppler Anemometer, itself fully described by Bachalo et al (1988, 1990).

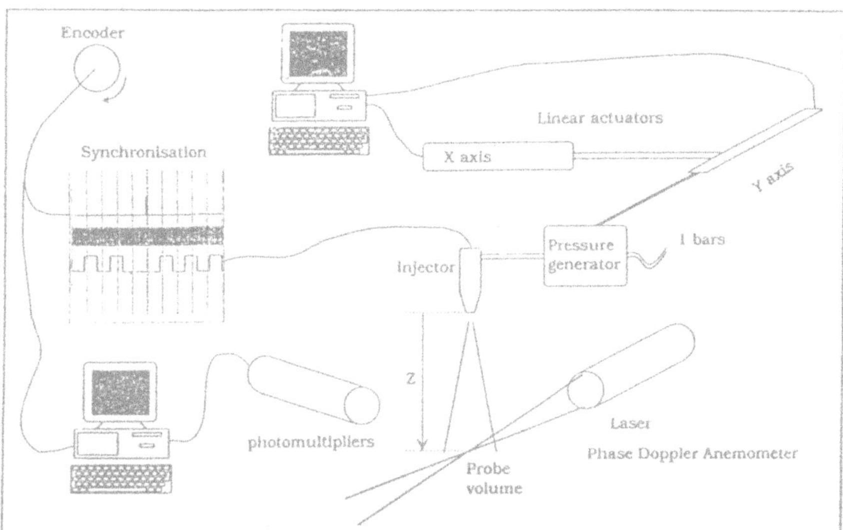

Figure 1: Device allowing simultaneous registration of spacing location X, timo, diameter and velocity of droplets of a free air spray .

A special device allowed registration of the X spatial location versus time as shown on **Figure 1**: time was registrated by an encoder run at a known constant speed, while spatial position of the injector was translated via a step-by-step linear actuator. The free air spray was scanned along a diameter at 100 mm downstream the nozzle. This distance corresponds to a typical value from the nozzle to the inlet valve in a port-injected engine. The PDA technique delivers a considerable information, as stated in **Figure 2**, of size, velocity, time, spatial location and size-velocity correlation which can be used to validate atomisation models. Both temporal and spatial evolution of droplets are inferred: when reading the Figure 2.a) from left to right, the reader can examin the figure 2.c) from bottom to the top, so that droplet velocity is deduced versus time, and its X location in the spray as well.

From measurements carried out at various distances from the nozzle (30, to 100 mm), the exit spray velocity can easily be deduced, as decaying along the axis from 30 to 10 m/s. Also a twin-measurements procedure, with and without size validation allows to evaluate spherical droplets contribution, which represents between 50 and 75 % of the whole population.

Nevertheless, the precise metering of instantaneous fuel quantity is desired, specially when data are dedicated to model validation. Simon et al (1992) use miniature pressure sensors installed in the valve-assembly dead-volume to obtain this specific information, and use it for optimisation of injection-valve.

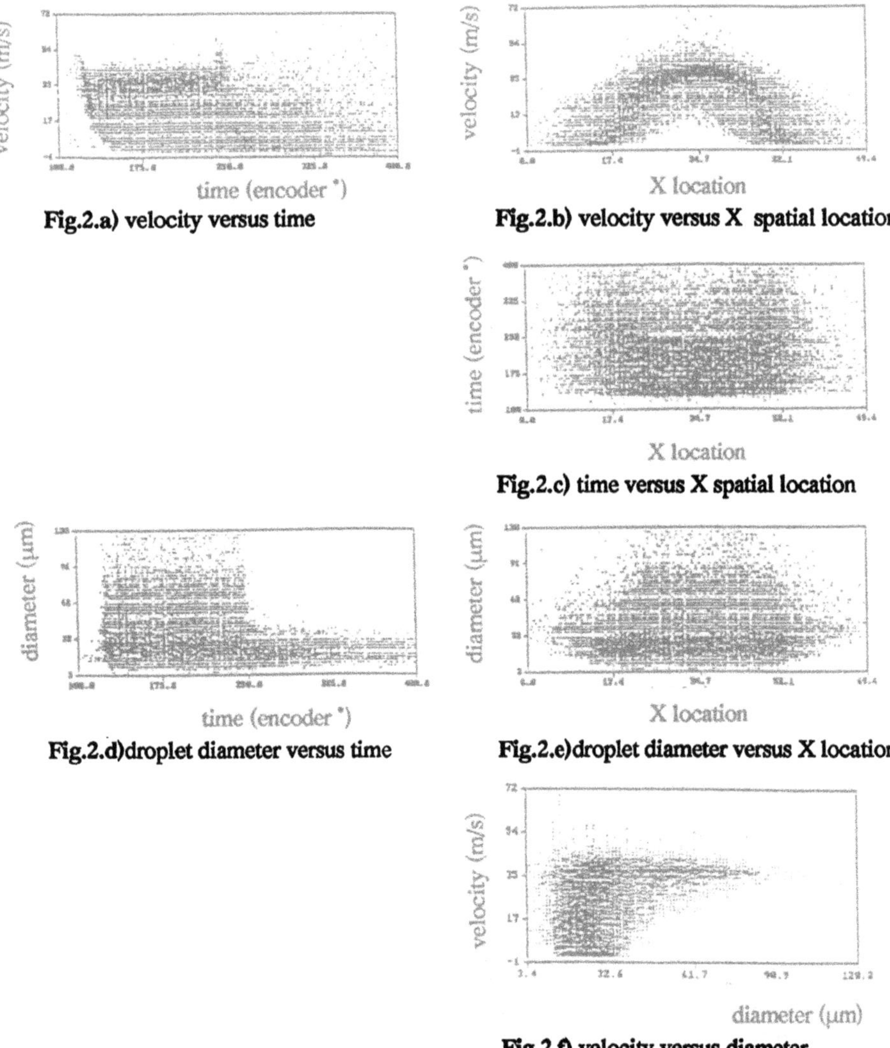

Fig.2.a) velocity versus time

Fig.2.b) velocity versus X spatial location

Fig.2.c) time versus X spatial location

Fig.2.d)droplet diameter versus time

Fig.2.e)droplet diameter versus X location

Fig.2.f) velocity versus diameter

Figure 2: Complete description of a spray in free atmosphere from Phase Doppler Anemometry measurements. 2.a): velocity versus time, 2.b): velocity versus X spatial location, 2.c): time versus X spatial location, 2.d): diameter versus time, 2.e): diameter versus X spatial location, 2.f): velocity versus diameter.

3.2 APPLICATION OF THE PHASE-DOPPLER TECHNIQUE INSIDE AN ENGINE

Taking into account the constraints and limitations of compression ratio and thermal effects on phase Doppler anemometry measurements performing inside a production, but optically-accessed engine, we considered necessary to perform theoretical work about the optical method. Based on the Generalised Lorenz Mie Theory developed by Gouesbet and Grehan (1982), a numerical code was assembled to simulate the Doppler signals detected by each photo multiplier, which provides more autonomy to the PDA user in terms of detection angles. Besides, the aim of this parallel task was to improve the PDA accuracy, above all with respect to the non uniformity of the intensity of the incident light. A first approach of this problem has been published in 1990 by Grehan et al, then improved by Dementhon (1992) and applied to experimental test by Vannobel et al (1992). In particular, the Gaussian beam effect has been recently addressed by Aizu et al (1993). The authors present a new apparatus allowing large particles to be measured with a modified PDA system without Gaussian beam errors. Pitcher et al (1990) have investigated the sensitivity of dropsize PDA measurements to refractive index changes in combusting fuel sprays and found a minimum at 70 degrees scattering angle for Diesel fuel.

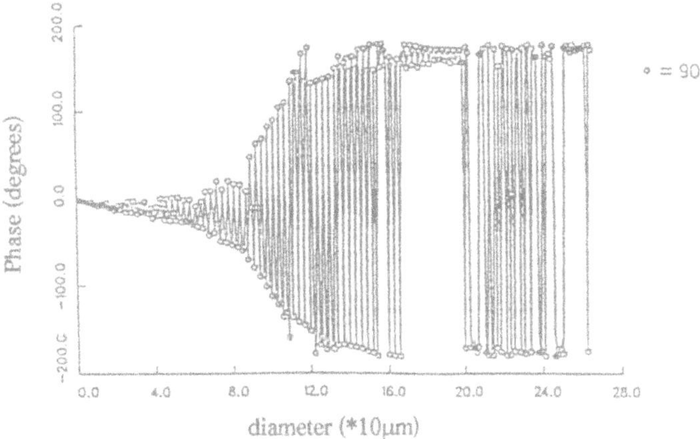

Figure 3 : Phase-diameter regression calculated by Lorenz Mie Theory for a 90 degrees scattering angle and refractive index of 1.42; from Dementhon (1992).

In the present work, a simulated gasoline was injected: the refractive index of this substitute fuel was determined equal to 1.42, and the scattering angle obtained from Lorentz Mie Theory calculations was 30 degrees. At this angle, a linear relationship between phase and diameter was foreseen. On the other hand, a scattering angle of 90 degrees was revealed discouraging, as indicated in **Figure 3**. The relationship between diameter and phase could not be considered linear. Although disappointing, this result will be issued with other scattering angles attempts. Of course, the PDA technique can be made a more flexible tool, by means of optical fibers and reduced detection cone angles, to allow it to be applied in a reciprocating engine.

However, we have seen in the above section that phase Doppler anemometry can be used in many ways: either only detection, where particles are simply counted, or only velocity, or both velocity and size. Also size only measurement allows to compare proportion of spherical drops compared to the whole population of particles.

4. Influence of aerodynamics on the spray behaviour

This part is now dedicated to realistic flow conditions in a Plexiglas cylinder-head: cold flow PDA measurements are carried out. Following its installation within the manifold, the interaction of spray with both air flow and moving inlet valve is examined. A motored camshaft reproduces the inlet valve movement, while the air-depression inside the cylinder -which value rises between 20 and 100 mbars- is controlled by operating condition of the settling chamber, and of throttle position adjustment. The results are outlined in two sections : steady and non-steady flow conditions, which respectively correspond to static or running camshaft.

4.1 STEADY FLOW CONDITIONS

4.1.1 Air-flow rate influence on spray characteristics inside the manifold. A clear dragging effect has been observed and described by Vannobel et al (1992), showing that while big droplets continue on their way, smaller ones deviate towards the upper region of the manifold. An imaginary lagrangian measurement procedure has been utilised in order to keep droplet detection on the spray axis at the maximum PDA validation rate while closing the throttle aperture from 0 to 60 degrees. This method ensures a data acquisition governed by experience and characterises the densest region within the spray.

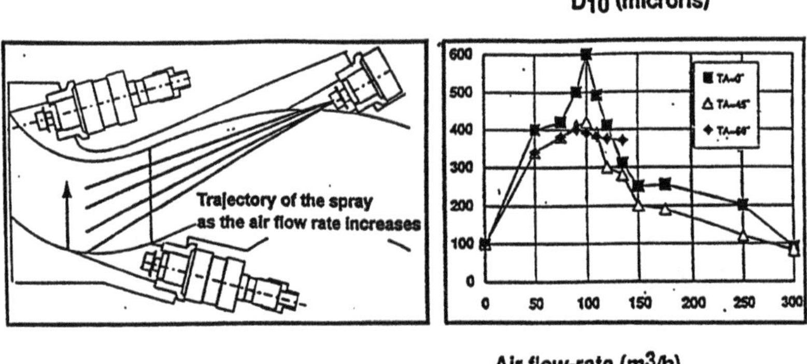

Figure 4. Air-flow rate influence on drop sizes detected on the spray axis for three throttle positions

The arithmetic sizes are given in **Figure 4** and show the emergence of two separate regions : the first one is characterised by an increasing D_{10} versus volume air flow rate, which means that droplet sizes are governed by injection pressure. Secondly, having reached a maximum value the droplets, giving rise to a secondary atomisation process which results in a D_{10} decreasing. This result confirms that relative liquid/air velocity is one of the relevant parameters for drop-size description. The drop size and velocity results obtained at this stage have been compared to the theoretical approach made by Lerat and Wu (1990) and Wu (1992).

4.1.2 Spray characteristics inside the cylinder. In order to evaluate the evolution of droplets emitted inside the manifold and in particular what happens at the valve gap crossing, PDA measurements have also been carried out inside the cylinder. The closest optical access to the cylinder-head plan was 21 mm (all the three detectors must be validated for the same drop), with a valve gap comprising between 6 and 10 mm.

Of course the air flow-rate effect has been studied, and the **Figure 5** indicates the results obtained. The various curves identified by the air cylinder under-pressure illustrate the trajectories followed by the drops. A size increase due to air-flow decreasing is seen and confirms some trends observed within the manifold : the second part of the curve presented in **Figure 4** is concerned, as air - liquid velocity is greater than 1. So, the number of droplets increases with air-flow rate. It must be noted that at 35 millibars a dramatic detection drop is evident. This can be interpreted as a consequence of liquid film deposit on the multiple walls (valve, cylinder) which subsequently limits the recognition degree of PDA detection. In fact, the effects on the wall are considerable. This clearly indicates that a liquid film thickness measurement technique has to be developed if a fuel budget is wanted in this study.

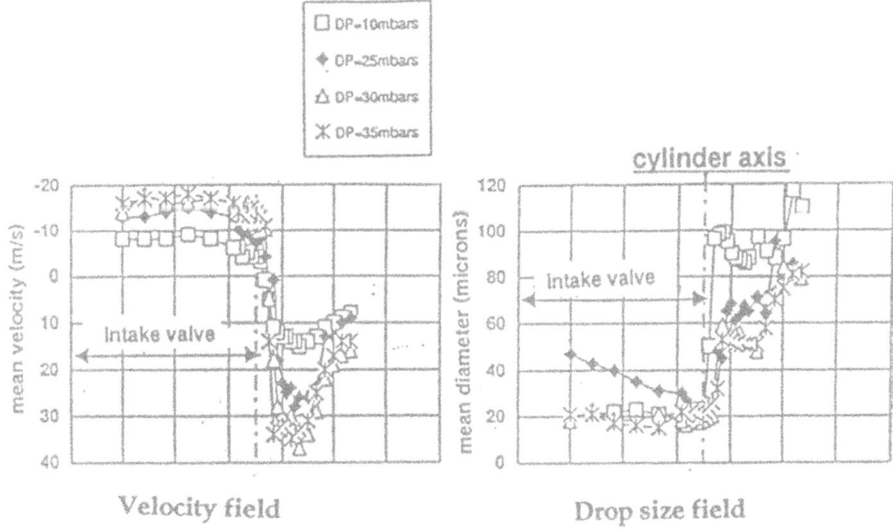

Figure 5. Air flow rate effect on drop sizes and velocity fields inside the cylinder

4.2 UNSTEADY FLOW CONDITIONS

In this context, the camshaft is motored at a constant speed up to 1500 revolutions per minute. For a constant engine speed, air velocities inside the manifold and inside the cylinder depend on the crank angle (shaft angle in this case). Thereby, the interaction between spray and air is hugely determined by the angle of injection. Nowadays, with respect to the production injection technology and the rising urge to control exhaust pollutants, injecting as the intake valve is closed (let's call this adjustment CVI : Closed Valve Injection) is more suitable, due to the pre vaporisation of the fuel in touch with the hot valve. Thus, any effort made to control the spray features upstream of the valve is largely inefficient. Moreover, one can forecast that such a method of injecting fuel will contribute to a liquid film build up in cold engine conditions. These are during the first few minutes following the starting of the engine. In order to increase the understanding of the phenomena related to the introduction of the liquid into the cylinder, our main interest variable focuses on the injection angle. The standard parameters featuring these tests are both engine speed which is remained at 500 rpm and engine load controlled by the throttle fully open in this case.

4.2.1 Spray characteristics inside the cylinder for non-steady air-flow conditions. First of all, some measurements have been taken scanning a diameter of the cylinder, 21 mm under the roof in order to spot the area of greatest liquid flux coming from the intake valve. The **figure** 6 displays the number of detected droplets versus probe volume location for two injection timing cases. These curves highlight clearly the main region of interest which is located at about Y=15 mm from the cylinder centre under the exhaust valve. It can be noticed that this location is close to the point detected in steady flow conditions: the spray cone is shifted by less than 5 mm.

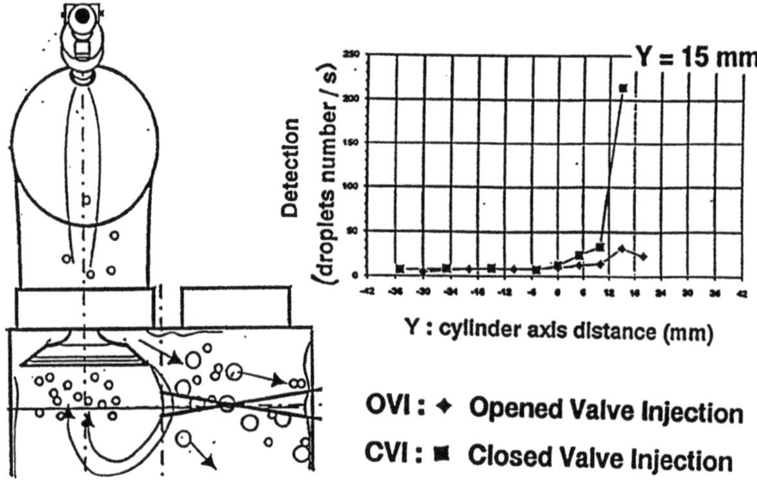

Figure 6. Detection droplet curves for two cases : Open Valve Injection and Closed Valve Injection.

4.2.2 Droplets detection at Y=15 mm versus injection timing. This location has been selected in order to investigate more intensively the effect of the injection timing. The **figure 7** shows the particle velocity versus shaft angle for several injection timing conditions.

It appears that for any injection angle, the droplets are detected with some delay. This could not be explained by their transit-time calculated from the measured velocity between the intake valve and the probe volume. In fact, it results from the valve acting as a buffer which delays the liquid introduction of the fuel into the cylinder because of a liquid film formation. This result which is understandable in the case where the injection occurs as the valve is closed (CVI) is more surprising during Open Valve Injection (OVI).

Indeed, one might have made the assumption that the dragging air effect would allow the droplets to pass around the valve. More than that, OVI condition is shown to induce a further delay than CVI : the fuel enters into the cylinder during the closure of the valve.

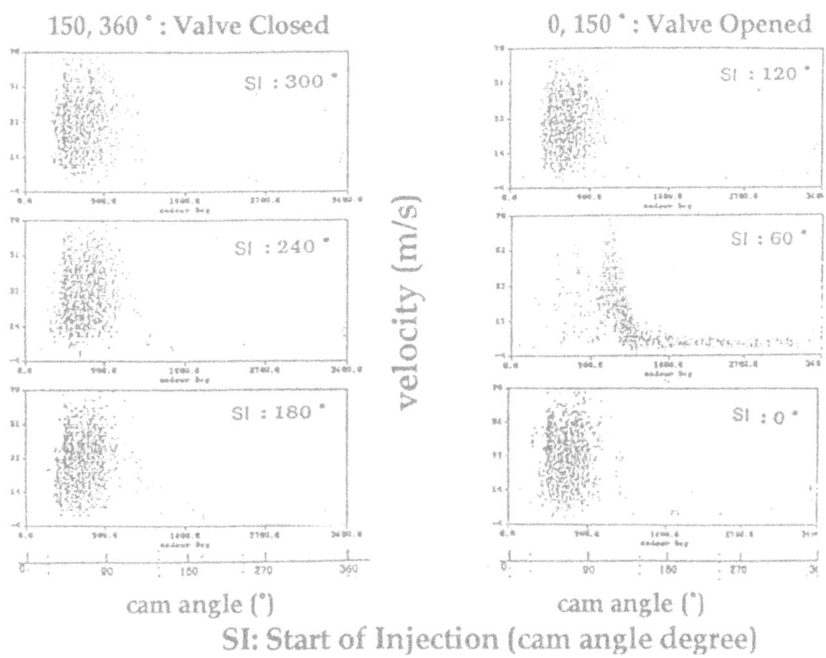

Figure 7. Effect of injection timing on velocity measurements versus shaft angle

4.2.3 Skip Injection effect. We have defined Skip Injection Parameter (SIP) as the injection frequency expressed as the number of cycles between 2 injections : SIP = 1 means every cycle, SIP = 2, every second cycle, etc.

For both OVI and CVI, the **figure 8** displays the measured velocities versus shaft angle. In order to assess the validation of our previous assumptions, let us see what happens for different injection timing values. In the CVI case, the delay between the valve opening (0° camshaft) and the first instant of droplet detection is independent of SIP. This means that the way the fuel is emitted inside the cylinder does not vary.

Impinging onto the closed valve, the liquid creates a liquid film that is secondarily atomised following the valve opening with a characteristic elapsed time featuring the film breaking. In the OVI case, on the other hand, a fairly different behaviour is seen. Indeed, one can observe that the detection delay after valve opening increases versus SIP.

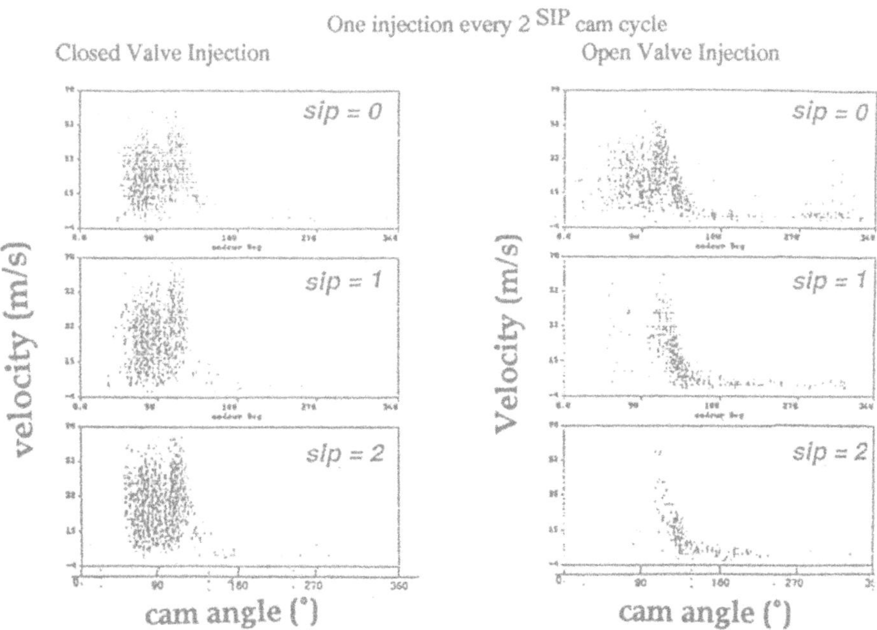

Figure 8. Effect of the Skip Injection Parameter on velocity measurements versus shaft angle

4.2.4 Visualisation of the spray within the manifold versus the injection timing. The photographs given in **figure** 9 display the spray behaviour for two conditions of injection timing. In the CVI case the liquid flies straight to the intake valve strained by the injection pressure (3 bars).

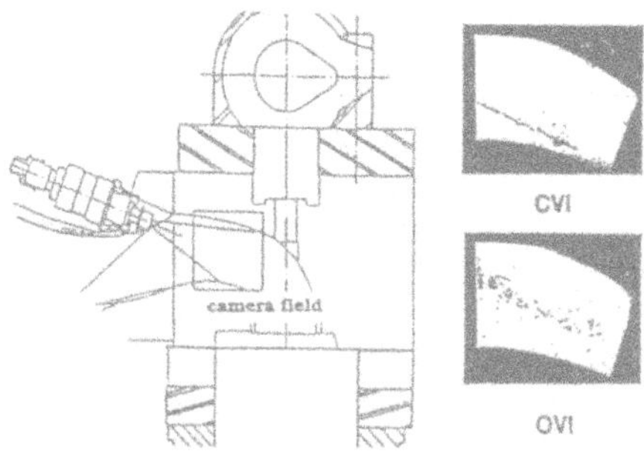

Figure 9. Spray trajectories within the manifold for two injection timing conditions: OVI and CVI .

Under OVI conditions, the droplets are diverted towards the upper region of the manifold, due to the dragging air effect. The impact of the liquid occurs more so on the valve stem rather than on the valve itself. Therefore the time required for the liquid to flow down through the cylinder is enhanced, which explains the supplementary delay noticed in this case.

5. Motored engine study

In the continuity of the above study with the transparent cylinder-head, a motored monocylinder optically-accessed engine has been built, equipped with the same cylinder-head geometry. This device allows us to compare droplets behaviour without and with compression effects due to the piston movement.

5.1 THE MOTORED ENGINE

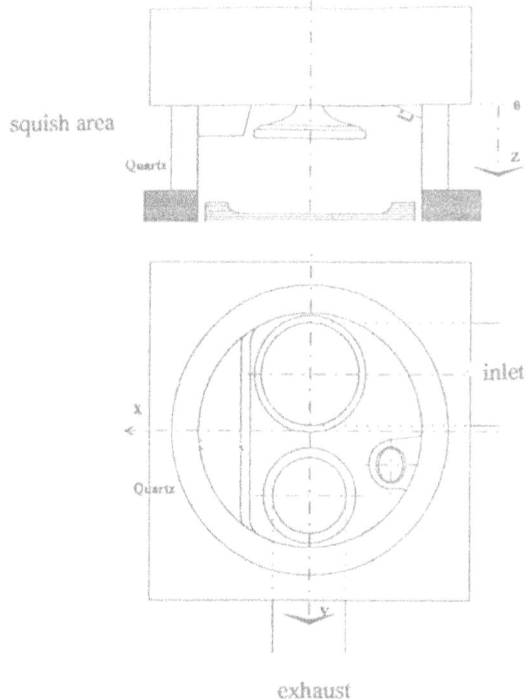

Figure 10: the motored monocylinder optically-accessed engine.

The **Figure 10** gives a schematic view of the engine, provided with a quartz liner allowing PDA measurements. The valves are parallel and vertical and the characteristics of the engine, operated at 1200 rpm, are as follows:

Bore	:	83 mm
Stroke	:	88 mm
Connecting rod length	:	143 mm
Compression ratio	:	7,3

The throttle is fully open, and a similar pintle injector is used.

The **Figure 11** describes a map of PDA measurements carried out close to the previously studied horizontal plane, i.e. 20 mm under the valve plane. Three main regions are distinguished, characterised by different droplets trajectories and size, as follows:

- the region, numbered 1: recirculation zone under the inlet valve; the AMD value is 15 μm. and SMD value is 60 μm. Droplets are directed upward at 12 m/s during the inlet stroke and then stagnant under the inlet valve during the compression stroke.
- the main region, numbered 2: main droplets flow, corresponding to the spray cone seen above; the size depends on the Start of Injection and is detailed in the paragraph 5.2.
- the region numbered 3: recirculation zone near the cylinder wall.

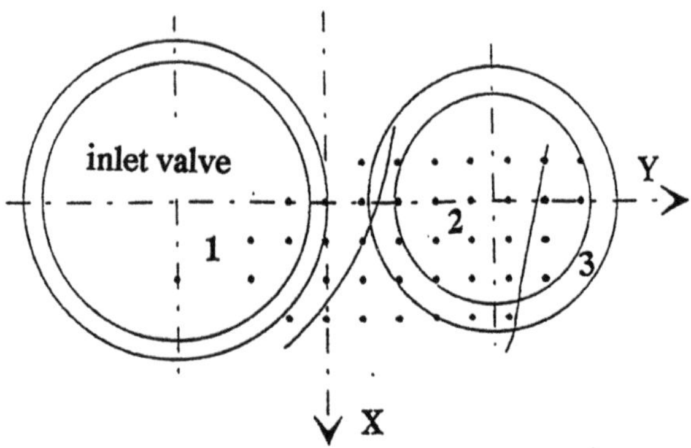

Figure 11: The three regions of interest in the cylinder.

5.2 THE MAIN REGION

The region 2 exhibits a data rate which is ten times higher than in the other regions. This maximum detection rate area corresponds to the spray cone previously observed on the flow-rig. On the droplets clouds -as they are instantaneous data cumulated on a half crank-revolution- given in **figure 12,** measurements are windowed during the intake and compression strokes: from 0 to 180 cam angle degrees. The instant 0 cam-angle corresponds here to the intake Top Dead center.

5.2.1. In the Closed Valve Injection case: it is seen from **figure 12a)** that droplets arrive early, as they are issued from the liquid film which was waiting on the valve stem during the previous cycle. When comparing the **figures** 7 and 12, the reader can observe that temporal sequences are very similar. Droplets come from a stripping process on the liquid film. The Sauter Mean Diameter is comprised between 40 and 50 μm, and the Arithmetic Mean Diameter between 80 and 120 μm. The time-lag between Start of Injection and first arrival of droplets is explained by the stripping process duration, added to the flight time from the liquid film on the inlet valve to the PDA probe. For this reason, the Injection Phasing does not influence the first detection. Also the SIP parameter is inactive on the temporal occurence of droplets cloud.

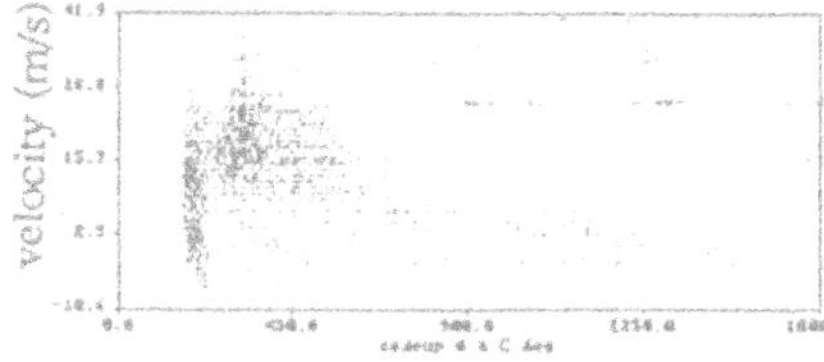

Figure 12.a) Closed Valve Injection

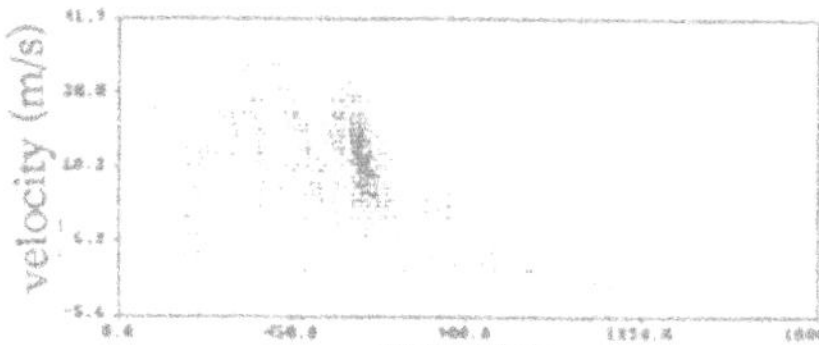

Figure 12.b) Open Valve Injection

Figure 12: Velocity droplets clouds in the main region, numbered 2 on Figure 11.

5.2.2. In the Open Valve Injection case: the Start of Injection is fixed at 36 cam angle degrees, and it is obvious from **Fig.12.b)** that droplets arrive during the closure of the valve, following the lift evolution when valve is coming back to its seat. The droplets cloud shown here is given for SIP value equal to zero. When varying the Start of Injection, the droplets cloud is shifted above the same amount, which corresponds to flight time between the injector and measurement probe location. The detection extent of the cloud is close to the injection duration, which in this case is 2.5 milliseconds. It can be inferred that droplets enter directly into the cylinder, without impinging a wall. This result is also confirmed by a "skip injection" procedure and cycle-to-cycle fuel behaviour study (Robart, 1994).
The corresponding AMD is 30 μm, and SMD 90 μm: the "direct spray" droplets are smaller than the droplets issued from liquid film, described in § 5.2.1.

6. Conclusion

This paper has described an investigation of the two-phase flow within manifold and cylinder, then extended to a motored engine. Yang et al (1993) have clearly relied droplet-sizes and Start of Injection to exhaust unburned Hydrocarbon emission, specially during open-valve injection, in the first half of the intake stroke. Their observations are consistent with the main features of the present study. The big droplets (300 μm) produce 50% higher HC emissions. Experiments carried out in engines by Lenz et al (1988), Kashiwaya et al (1990) and Koo and Martin (1991) can also provide useful data in this field. Martins and Finlay (1992) experimentally determined heat-transfer of impinging sprays on a heated instrumented valve.
As a main result of the present study, it has been shown that the location of spray cone -higher data rate region- in the motored engine is well approximated by the flow-rig experiment, as well as the temporal behaviour of droplets at their arrival into the cylinder. While complete results, extended to the comparison with a corresponding firing engine are fully described by Vannobel et al (1994), it can be argued now that a motored camshaft is a fairly convenient tool allowing understanding of liquid film and direct spray mechanisms. Cyclic repartition effects are also well reproduced with "skip injection" (SIP parameter) procedure. Fuel mixing preparation is proven to be a topic of interest when dealing with HC limitations at the exhaust of a port-injected SI engine.

7. References

Arcoumanis C., Hadjiapostolou A., Whitelaw J.H. (1987) Swirl center precession in engine flows. SAE'87 paper 870370

Aizu Y., Durst F., Grehan G., Onofri F., Xu T.H., 1993, PDA-system without Gaussian beam defects, Proceedings of Third International Congress on Optical Particle Sizing. August 23-26 1993, Yokohama, Japan.

Bachalo W. D., Sankar S. V., 1988: Analyser of the light scattering interferometry for spheres larger than the light wavelenght. Proceedings of 4th International Symposium on Applications of Laser Technics to Fluid Mechanics, Lisbon, Portugal.

Bachalo W.D., Brena de la Rosa A., Sankar S.V., 1990, Diagnostics for fuel spray characterization. Combustion Measurements, Hemisphere Publishing Corporation, N.Chigier editor, New York, pp 229-278

Bicen, A.F., Vafidis C., Whitelaw J., 1984, Steady and unsteady air flow through an intake valve of a reciprocating engine. Proc. ASME Winter Annual Meeting, New Orleans, December 1984.

Chaves H., Bode J.,Obermeier F., Schneider T., 1991, Influence of cavitation in a turbulent nozzle flow on atomization and spray formation of a liquid jet. Proceedings of Spray and Aerosol '91, ILASS - Europe, Guilford, U.K., pp.107-112.

Cheng W.K., Hamrin D., Heywood J.B., Hochgreb S., Min K., Norris M., 1993, An overview of hydrocarbons emissions mechanisms in spark-ignition engines, SAE'93, 932708

Dementhon J. B., 1992, L'injection essence dans un moteur à étincelle : mesures granulométriques par la méthode des phase oppler et phénomènes de pulvérisation, Thèse d'université, Rouen.

Dementhon J. B., Vannobel F., Dumouchel C., Ledoux M., 1992 A new approach of the pulverisation mechanisms involved in a gasoline spray: video imaging and modelling. Proceedings of Spray and Aerosol '92, ILASS - Europe, Amsterdam, The Netherlands.

Dementhon J. B., Vannobel F., 1991, Phase Doppler Anemometry in gazoline sprays : in atmosphere and in a steady flow rig, Spray and Aerosol '91, ILASS - Europe, Guilford, U.K., pp 48-53.

Dumouchel C., Ledoux M., 1991, Atomisation of flat and annular sheets: practical use of linear theories, ICLASS'91., Gaithersburg, U.S.A, paper 12.

Gosman A.D., 1991, Description of a subprogramme A-Fuel spray research. ATA Ingegneria Automotoristica, vol.44,n°3, pp120-123.

Gouesbet G., Gréhan G., 1982, Sur la généralisation de la théorie de Lorenz Mie, J. Opt., 13, 2, pp 97 - 103.

Gouesbet G., Berlemont A., Desjonqueres P., Grehan G., 1992, On modelling and measurements problems in the study of turbulent two-phases flows with particles. 6th workshop on Two-phase flow predictions, March 30-April 2 1992, Bilateral Seminars of the International Bureau, M.Sommerfeld editor, Erlangen, pp 365-387.

Gréhan G., Vannobel F., Dementhon J. B., Gouesbet G., 1990, Generalized Lorenz Mie theory : An application to phase Doppler technique, Proc. 5th Intl. Symp. on Appl. Laser Anemometry to Fluid Mech., Lisbon, Portugal.

Heywood J.B., 1988, Internal Combustion Engine Fundamentals, McGraw Hill editor, Chap 11., pp 567-625.

Hodges J.T., Baritaud T.A., Heinze, (1991) Planar liquid and gas fuel and droplet size visualisation in a DI diesel engine SAE910726. SAE Congress ,Detroit.

Huh K.Y., Gosman D., 1991, A phenomenological model of diesel spray atomisation. International Conference on multiphase flows, Sept 1991, Tsubuka, Japan.
Kashiwaya M., Kosuge T., Nakaggawa K., Okamoto Y., 1990, The effect of atomization of fuel injectors on engine performance, SAE'90, 909261.
Kawazoe H., Oshawa K., Kataoka M., 1990, LDA Measurement of Gasoline Droplet Velocities and Sizes at Intake-Valve Annular Passage in Steady Flow State, Bound Volume of Selected papers from Applications of Laser Techniques to Fluid Mechanics, pp 248-267, 9-12 July 1990, Springer-Verlag, Lisbon.
Klingsporn M., Renz U., 1992, Numerical simulation of a Diesel spray; 6th workshop on Two-phase flow predictions, March 30-April 2 1992, Bilateral Seminars of the International Bureau, pp 301-312, M.Sommerfeld editor, Erlangen.
Koo J. Y., Martin J. K., 1991, Comparisons of measured drop sizes and velocities in a transient fuel spray with stability criteria and computed PDF's, SAE'91, 910179.
Lenz H. P., Fraidl G. K., Friedl H., 1988, Fuel atomization with mixture preparation systems of S. I engines, SAE'88, 885015.
Lenz H.P., Mixture Formation in Spark-Ignition Engines , pp 66-111, SAE and Springer-Verlag Wien 1992.
Lerat A., Wu Z. N., Simulation numérique d'écoulements diphasiques air / essence avec modèles d'arrachement et d'impact de gouttelettes, 1989, Versailles.
Li X., Tankin S., 1987, Droplet size distribution : a derivation of a Nukiyama-Tanasawa type distribution function, Combustion Science and Technology, 56, 65.
Martins J.J.G., Finlay I.C., 1992, Fuel preparation in Port-Injected engines, SAE'92, 920518.
Nogi T., Ohyama Y., Yamauchi T., Kuroiwa H., 1988, Mixture Formation of Fuel Injection Systems in Gasoline Engines. SAE'88, 880558.
Nogi T., Ohyama Y., Yamauchi T., 1989, Effects of Mixture Formation of Fuel Injection Systems in Gasoline Engine. SAE'89, 891961.
Pitcher G., Wigley G., Saffman M., 1990, Bound Volume of Selected papers from Applications of Laser Techniques to Fluid Mechanics, 9-12 July 1990, Lisbon, Springer-Verlag, pp 227-247.
Posylkin M., Taylor A.M.K.P., Whitelaw J., 1993, Size and velocity characteristics of fuel droplets in the valve curtain area of a straight, steady flow inlet manifold, private communication, to be presented at the 7th International Symposium on the Application of Laser Techniques to Fluid Mechanics, Lisbon, 1994.
Reitz R.D., 1987, Modeling Atomization Processes in High-Pressure Vaporizing Sprays, Atomisation and Spray Technology 3, pp.309-337 .
Reitz R.D., 1991, Prospects and challenges for fuel spray research in the automotive industry. Atomisation and sprays 2000, National Science Foundation workshop, Gaithersburg, N.Chigier editor, pp 89-95.
Reitz R. D., Bracco F. V., 1982, Mechanisms of atomization of a liquid jet, Phys. fluids, 25, p.1730.
Reitz R. D., Diwakar R., 1987, Structure of high-pressure fuel sprays, SAE 870598
Robart D., 1994, Injection indirecte dans un moteur à étincelle: processus de pulvérisation et film liquide; granulométrie par phases Doppler. Thèse d'université, Rouen, in preparation.
Ruiz F., Chigier N., 1985, The mecanisms of high speed atomization, 1985, ICLASS'85.
Saikalis G., Byers R., Toshiharu N., 1993, Study on Air Assist Fuel injector Atomization and Effects on Exhaust Emission Reduction, SAE'93, 930323
Séllens R.W., 1989, Prediction of the Drop Size and Velocity Distribution in a Spray, Based on the Maximum Entropy Formalism Part.Part.Syst.Charact. 6, pp17-27.

Simon N., Arndt S., Ziegler E. 1992, Experimental Measurement Techniques to Optimize Design of Gasoline InjectionValves SAE'92, 920520

Sommerfeld M., Qiu H.H., 1992 Experimental studies on spray evaporation in a turbulent flow 6th workshop on Two-phase flow predictions, March 30-April 2 1992, Bilateral Seminars of the International Bureau. M.Sommerfeld editor, Erlangen.

Vannobel F., Dementhon J. B., Grehan G., 1992, Phases Doppler et ambiguité de trajectoire : Aspects théoriques et expérimentaux, 3eme congrès francophone de vélocimétrie laser, Toulouse, France.

Vannobel F., Dementhon J-B., Robart D. 1992, Bound Volume of Selected papers from Applications of Laser Techniques to Fluid Mechanics, 9-12 July 1992, Lisbon, Springer-Verlag.

Vannobel F., Robart D, Dementhon J-B., Taylor A.M.K.P., Whitelaw J., 1994, Fuel-droplets size-velocity fields inside the cylinder of a production firing engine. to be presented at COMODIA 94, 11-14 July, Japan.

Wu Z. N., 1992, Modélisation et calcul implicite multidomaine d'écoulements diphasiques gaz-gouttelettes Thèse d'Université, Paris VI.

Yang J., Kaiser E.W., Siegl W.O., Anderson R.W., 1993, Effects of Port-Injection Timing and Fuel Droplet Size on Total and Speciated Exhaust Hydrocarbon Emissions, SAE'93, paper 930711

15. IN-CYLINDER DIAGNOSTICS FOR PRODUCTION SPARK IGNITION ENGINES

P. O. Witze
Sandia National Laboratories,
Combustion Research Facility
Livermore, California 94551-0969
USA

ABSTRACT. A comprehensive review of diagnostic techniques for the study of in-cylinder phenomena in unmodified spark ignition engines is presented. The scope of the techniques considered is restricted to those that can be used without machining modifications to the engine. Thus, most of the diagnostics discussed are either integral with the spark plug or head gasket, or use the spark plug hole to study pre-ignition phenomena under motored operation. Among the more important instruments discussed are pressure sensors, ionization probes, luminosity detectors, gas sampling devices and gas velocity probes.

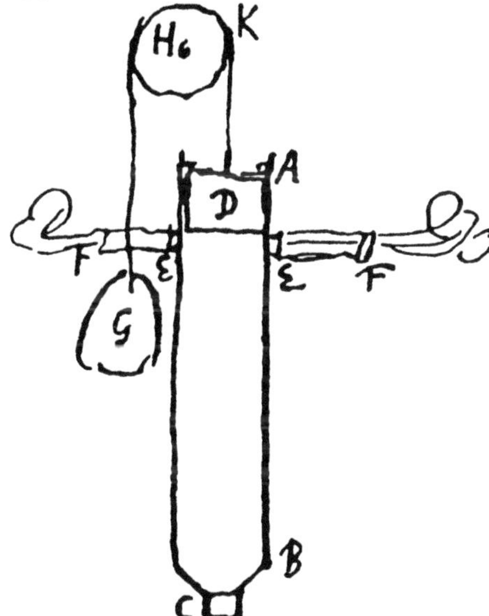

Figure 1. The gunpowder engine of Christiaan Huygens, from a freehand sketch in a letter to his brother in 1673, may have been the first internal combustion engine. The open-ended cylinder *AB* contained a sliding piston *D* that was driven to the top of the cylinder by ignition of a charge *C* of gunpowder and air. The rising piston uncovered exhaust ports *E*, allowing the heated gas to escape. The piston then descended, initially from its own weight and then from the cooling of the residual gas. Flexible leather hoses *F* served as check valves on the exhaust ports. Work was extracted by raising a weight *G* by a rope *K* looped over a rotating drum *H* (from Klemm, 1964).

F. Culick et al., (eds.), Unsteady Combustion, 333–368.

1. Introduction

In 1985 the Society of Automotive Engineers organized a collection of invited papers on new approaches to engine combustion analysis that included a tête-à-tête between two contrasting views on diagnostics for engine research. Amann (1985a) surveyed classical combustion diagnostics, purporting the premise that "there's no tool like an old tool". (This and other reviews by Amann (1983, 1985b) includes discussion of many fascinating experiments performed in the very early days of the internal combustion engine, and served as the source for many of the historical references used herein.) Representing the antithesis, Dyer (1985) surveyed new experimental techniques based on the use of lasers, purporting the premise that in situ measurements of temperature, species concentration, velocity and turbulence can be obtained without perturbing the process under study. Amann (1985a) calls this the "nonintrusion illusion", correctly pointing out that the installation of the windows necessary for optical access typically results in severe modification of the combustion chamber geometry. However, the classical diagnostics he discussed were all also intrusive to one degree or another and, in most instances, installation required machining of the head.

The scope of the current review is diagnostics that can be used in a production spark-ignition engine with no machining of the head, block or piston. It is recognized that this is a restrictive criteria, since some probes can be installed in relatively small holes, and for an instrument as important as a pressure transducer machining is surely worthwhile. However, if just a little bit of machining is allowed when establishing the ground rules, it becomes difficult to draw the line as to when enough is enough. At the same time, two liberal interpretations of the rules will be allowed: First, diagnostic tools that have not actually been demonstrated on *production* engines, but have been successfully applied to *research* engines, will be allowed. Second, if it is obvious that a technique could be used without engine modifications, even though it has not been explicitly demonstrated, it will be included. The breadth of diagnostic techniques that meet these requirements is surprisingly large, and I have chosen to organize them in the following categories: pressure, temperature, species, velocity, luminosity, ionization density, and visualization. In the introduction section to each of these categories, the earliest known work is briefly described for historical interest, similar to the inclusion of the early internal combustion engine presented in Fig. 1. This is then followed by brief descriptions of demonstrated applications of the particular sensor or technique for which engine modifications were not needed.

2. Pressure

From the advent of the Otto cycle engine, it was recognized that measurement of the cylinder pressure was essential to the understanding and optimization of the combustion process. In that era (e.g., Clerk, 1882) a mechanical indicator was used to record pressure, as shown in Fig. 2. Since that early period, pressure measurement technology has made great strides (see, e.g., Amann, 1985b), and today's devices are relative small, robust, and exhibit excellent frequency response. However, as will be shown, they still have their shortcomings.

McCullough (1953) appears to have been the first to report a pressure measurement obtained without engine modifications. His device was a specially-built combined spark plug and pressure sensor. Few details are given, other than that the sensor was a diaphragm-type strain gauge that is no longer in favor.

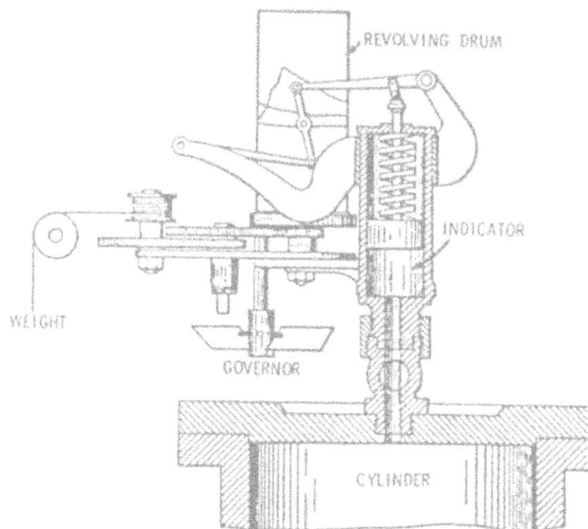

Figure 2. Nineteenth century mechanical pressure indicator. Displacement of a spring-loaded piston in the engine head is recorded on a revolving drum driven by a falling weight (from Midgley, 1920)

2.1 PIEZOELECTRIC SENSORS

Piezoelectric transducers are by far the most commonly used sensors for in-cylinder pressure measurements. They have a high frequency response, small size, light weight, and low power consumption (Randolf, 1990). The fundamental principle for their behavior is the piezoelectric effect, discovered by the Curie brothers in 1880, whereby electric charge is created on the surface of certain crystalline materials when mechanically loaded. While quartz crystals are most widely used, piezo-ceramic materials also exist. The main difference between the two is that the directional sensitivity of quartz sensors depends on the orientation of the cut, while for ceramics the piezoelectric effect is induced by the application of an electric field to align the dipoles (Morris, 1987).

For engine applications, the main problem with either piezoelectric material is the output sensitivity to sensor temperature (Brown, 1967, Alyea et al., 1969, Pischinger et al., 1985, Stein et al., 1987 and Randolf, 1990). This can occur as either long term drift during warmup of the engine, or intracycle drift caused by temperature transients during combustion. Schäfer et al. (1985) attribute the long term drift to sensitivity changes that in a quartz crystal are on the order of one-percent per 100°C temperature change. Short term drift is generally attributed to thermal strain in the thin diaphragm that protects the crystal from the combustion gas.

For installations where machining of the head is acceptable, the standard procedure is to use water-cooled flush-mounted pressure transducers. Water cooling solves the long term drift problem, and Brown (1967) has found that a thin layer of silicone rubber on the diaphragm eliminates short term drift. Brown did report a durability problem with the silicone rubber in a diesel engine application, but Lancaster et al. (1975) reported periods in excess of 50 hours of spark-ignition engine testing with no apparent problems or loss of effectiveness.

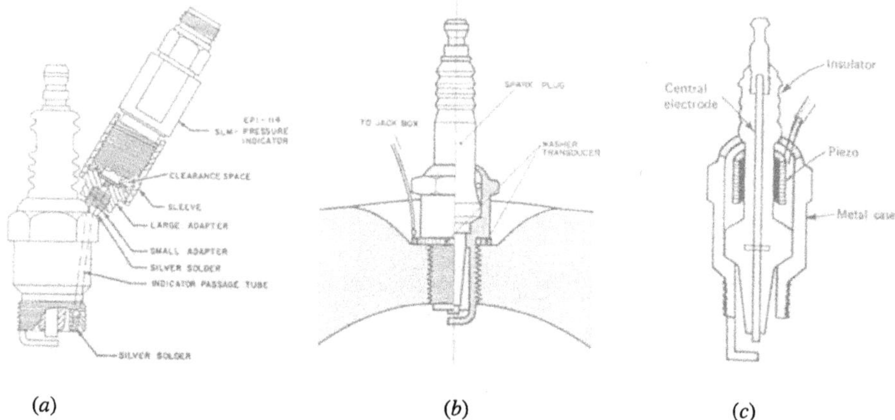

(*a*) (*b*) (*c*)

Figure 3. (*a*) Passage-type spark-plug sensor of Robison et al. (1958); (*b*) Ring-washer pressure sensor of Kondo et al. (1975); (*c*) Integrated spark-plug pressure sensor of Morris and coworkers (1985, 1987).

However, this paper is about instrumentation for unmodified engines, so the preferred transducer design and application procedures just described are not applicable. Instead, three different pressure-sensor installations that utilize the spark plug have been developed, as shown in Fig. 3. These designs and their application will be discussed in detail in the next three sections.

2.1.1. *Passage-Type Spark-Plug Sensor.* Robison et al. (1958) appear to have been the first to build and test a piezoelectric pressure sensor built in a commercial spark plug, as shown in Fig. 3*a*. Probes based on this design have been commercially available for many years. A major advantage of this approach is that the piezoelectric crystal is far removed from regions of temperature transients. The major disadvantage is that the passage distorts the response signal, and can also plug from deposits or fill with condensed water. Robison and coworkers investigated the effect of passage diameter and the size of the clearance space between the passage and the pressure sensor. They found they could eliminate pressure oscillations by reducing the diameter of the passage, but always had a small pressure "lag" because of the need to have a finite clearance volume to guarantee free movement of the pressure-sensitive element. In spite of this latter problem, however, they conclude that they could accurately measure both peak pressure and the maximum rate of pressure rise, whereas indicated mean effective pressure (IMEP) could not be measured with adequate accuracy because of the pressure lag. They used their probe to study engine knock and surface ignition.

Evers (1978) compared data obtained from a single-cylinder engine fitted with both a flush mounted pressure transducer and a spark plug pressure transducer. Three different commercially-available passage-type sensors were tested. For all three designs the pressure sensor was a quartz piezoelectric crystal, and the passage between the transducer and the combustion chamber included a slot cut through the spark plug threads. The major differences between the designs were the dimensions of the connecting passage. Three forms of pressure signal distortion were considered to be important, these being a lag time, frictional pressure drop along the connecting passage, and pressure resonance in the passage due to both acoustic and Helmholtz waves. Typically, the frequency of the acoustic waves is of the order of 10-100 kHz, and thus they present no real problems. On the other hand, the Helmholtz resonant frequencies are of the order of 1-10 kHz, which is similar to the characteristic frequencies of the transients in the pressure-time history and the pressure oscillations of engine knock. Evers also discussed the problems associated with water condensation in the passage, which can greatly change the performance characteristics. All three spark-plug sensors exhibited unacceptable pressure oscillations due to resonance in the passage; the amplitude of the oscillations varied greatly from cycle to cycle. The form of distortion he termed "lag time" was attributed to the acoustic transit time between the combustion chamber and the pressure sensor, and was shown to be insignificant. A frictional pressure loss was associated with the flow of gas in and out of the connection passage, and was shown to significantly delay the time of peak pressure as the engine speed was increased; however, at 2400 rpm it was shown to have little effect on the magnitude of peak pressure, and also only slightly delayed the fall in pressure during expansion.

Although their work was not done with a spark plug sensor, Brown (1967), Goto et al. (1986) and Randolf (1990) performed fundamental studies on the characteristics of connecting passages that have contributed to the understanding of this type of instrument. Brown investigated the errors associated with an idealized simulated passage having a diameter and clearance volume that was varied. From measurements of the pressure for a motoring diesel engine, he developed a theoretical equation that successfully predicted the error caused by passage restriction delays. Goto and co-workers doubled the natural frequency of the pressure oscillation in the passage by filling it with heat resistant oil. They also present a method for correcting pressure records for which it was not possible to eliminate the oscillation problem. Randolf was interested in the problem of piezoelectric pressure transducer output drift due to combustion-induced thermal shock. To study this problem, he compared the performance of a flush-mounted transducer with remote-mounted designs consisting of either a single passage or multiple slots, and found the latter to perform the best of the three.

2.1.2 *Ring-Washer Sensor.* Kondo et al. (1975) appears to have been the first to report a ring-washer pressure transducer, so named because it is installed like a washer between the spark plug and head (Fig. 3*b*). Although at first appearance this device would seem to be easy to install, in actuality the commercial versions (Rion Co. and NGK Spark Plug Co.) require a large clearance for the electrical cable, although it would seem that this problem could be reduced or eliminated by a more integrated design with the spark plug. The authors describe the principle of the device as being the reduction in the preload on the transducer from increased cylinder pressure on the spark plug. Because of the strong dependence of the response signal to the preload, they point out difficulties with both calibration and frequency response characteristics of the device.

Randall and Powell (1979) describe a piezoelectric ceramic ring-washer transducer similar to that of Kondo et al., but somewhat simpler in construction and more tolerant of high temperatures. Their paper is particularly useful for its in-depth presentation of the theoretical sensor characteristics, a dynamic model of the transducer response when installed under the spark plug, the temperature dependence of the piezoelectric constants, and the pyroelectric effect (a voltage generated by a rise in temperature). They compare their device to a laboratory grade, quartz piezoelectric pressure transducer, and conclude that the ring sensor is suitable to determine the crank angle of peak pressure and the high frequency pressure oscillations of knock.

Hata and Asano (1986) propose using a ring-washer transducer for electronic engine control. They evaluated both plastic and metal cased piezoelectric sensors, and concluded that the latter provided the necessary durability and performance characteristics. Individual cylinder timing control is proposed based on measurements of the crank angle of peak pressure and the occurrence of knock. An improvement of 1-5% in power output was achieved by controlling knock below a perceptible level.

Sawamoto et al. (1987) installed ring-washer transducers in two V-6 engines having different spark plug locations, one located on the cylinder axis and the other offset by an unspecified amount. Spectral analysis of the pressure signals during engine knock revealed that this type of sensor is capable of detecting the nodal surfaces of the lower-order modes of standing waves found in a circular cylinder. They also reported the frequency response characteristics of their transducer, and found it to be very constant out to 20 kHz. The experimental procedure that led to this result was not described.

Witze and Green (1993) did not observe similar frequency response behavior. They installed three NGK ring-washer transducers along a common diameter in the flat head of a disc-shaped combustion chamber, and compared their response to a water-cooled research grade pressure transducer located in the side wall of the combustion chamber. The ring-washer transducers were located on the cylinder axis and at 2/3 the bore radius, and were all torqued to the same preload. They found the crank angle of peak pressure to be slightly different for each ring-washer transducer, and that all three lagged the time of peak pressure as measured by the research grade sensor. Because the engine speed was only 600 rpm, this led them to conclude that the response frequency of ring-washer transducers is inadequate for practical engine applications.

Morris and Li-Chi (1985) evaluated the ring-washer transducer, and identified seven practical difficulties with its use: 1) Sensitivity to temperature; 2) Because it is thicker than the washer it replaces, the location of the spark gap is altered; this problem is compounded for spark plugs with tapered seats; 3) For most applications there is inadequate clearance to get the leads out; 4) Piezoelectric materials have specified pressure maxima, beyond which they lose their piezoelectric properties; normal spark plug torque specifications exceed this limit; 5) Sensitivity to mechanical noise; 6) High frequency noise on the order of 10 mv that presents problems if the signal needs to be differentiated; 7) Extreme sensitivity to ignition noise, due to the location of the sensor; they note that this problem can be resolved by placing the sensor under a head bolt, but head bolt torque specifications cannot be met. (Kistler Instrument Corp. (1993b) markets a thin ring-washer pressure transducer with integrated impedance converter for large-bore diesel engines that installs under the head bolt.)

2.1.3. *Integrated Spark-Plug Sensor.* Morris and Li-Chi (1985) proposed installing the piezoelectric sensor within the upper body of a spark plug, in the form of a ring around the ceramic insulator of the central electrode, as shown in Fig. 3*c*. This solved many of the inherent limitations of ring-washer transducers, but at the expense of a severe increase in the ignition noise problem. Low-pass filtering of the pressure signal was found to reduce this problem to a "manageable" level. In a later study, Morris (1987) proposed a variety of different designs for installing the sensor between the ceramic insulator and the metal body of the spark plug, but was never able to completely solve the ignition noise problem.

Kistler (1993a) has recently begun to market a spark plug with an integrated cylinder pressure sensor. The spark plug is asymmetric, permitting the sensor to be located near the electrode. The diaphragm that protects the piezoelectric sensor from the combustion gas is positioned to be flush with the combustion chamber. A high natural frequency of 120 kHz is maintained, allowing measurements without pipe oscillation even at speeds greater than 5000 rpm. Kuratle (1993) has reviewed the development history of Kistler spark plugs instrumented with pressure sensors, and presents a direct comparison between their most precise watercooled sensor and the new integrated sensor. Although he found that there was a small amount of short term thermal drift with the integrated spark plug sensor, it resulted in only a 0.2% error in the calculated IMEP.

2.2 FIBER-OPTIC SENSORS

Fiber-optic pressure sensors are a fairly new technology, and are just now beginning to be seriously considered for engine applications. Their great advantage is the potential for complete immunity from ignition noise. Because of the ignition noise problems Morris (1987) encountered with his integrated piezoelectric spark plug sensor, he investigated a fiber-optic sensing and signal transmission system. The principle he chose to explore was the micro-bending concept, where a loss in light transmitted through a fiber occurs when it is deformed by pressure. This he accomplished by wrapping a fine wire in a spiral fashion around the optical fiber; under compression, the relative movement of the two surfaces sandwiched about the fiber cause it to deform to a wavy shape. Several different installation procedures in a spark plug were tried, and the principal difficulty encountered was to achieve enough of a deformation (approximately 0.04 mm) to obtain an adequate loss in light transmission. Pressure signals obtained with this device are reported, but there is no comparison with a more conventional measurement technique.

He et al. (1993) have demonstrated a fiber-optic pressure sensor installed in a spark plug body with off-center electrode. The pressure sensor they use works on the principle of measuring the displacement of a diaphragm (Kyuma, 1989), as illustrated in Fig. 4. Light transmitted to the sensor is back reflected from the diaphragm and returned along the same fiber to a photodiode; the intensity of the returned light is directly related to the distance between the end of the fiber and the diaphragm, and thus is a measure of the force on the diaphragm. Because it is well known that external disturbances can cause significant variations in parasitic losses along the optical transmission line, a dual-wavelength system is used. The single fiber within the spark plug is terminated at the non-sensing end with a three-by-one coupler. Two of these fibers are used to launch light from two light-emitting diodes of different wavelengths λ to the sensor. A thin film optical coating at the end of the sensing fiber is transparent to one wavelength, and reflects the other. Thus, the light at both wavelengths share a common optical path up to the optical coating, but only the one wavelength is affected by displacement of the diaphragm; subsequent ratioing of the light detected at the

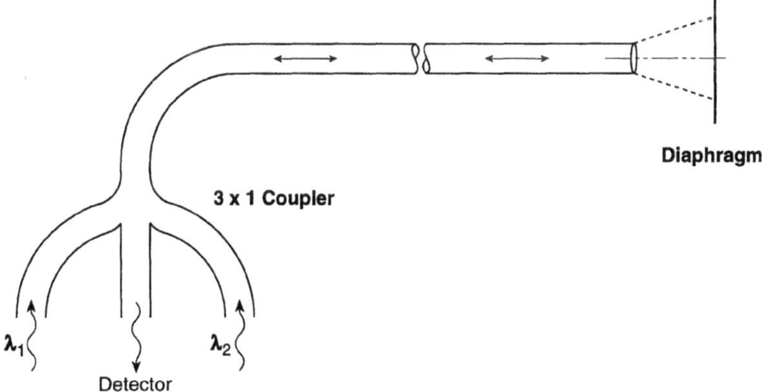

Figure 4. Fiber-optic displacement-type pressure sensor of He et al. (1993).

two wavelengths compensates for the errors from external disturbances. In practice, only a single photodiode is used to detect the light signals returned through the third fiber branch; the two LED sources are time-division multiplexed (chopped). They compare the performance of their probe with an instrumentation grade piezoelectric pressure transducer installed in the spark plug in a similar manner to the optical sensor. Both sensors were mounted in a retreated position along the guiding hole of the spark plug to keep temperatures below 200°C. They observed comparable signal-to-noise performance between the two, but the piezoelectric transducer output repeatedly drifted due to its inherent temperature sensitivity.

3. Temperature

The direct measurement of gas temperature in the cylinder is not as vital a need as cylinder pressure, but there are many instances where accurate temperatures are critical to the understanding of important combustion phenomena. These include heat transfer between the gas and the surrounding walls, the preflame chemical reactions that lead to autoignition and knock, and NO_X formation and crevice hydrocarbon burnup. Even something as seemingly simple as the gas temperature after valve closure during compression is important to know accurately, since in order to calculate the volume fraction burned from the measured pressure it is necessary to know the trapped mass.

According to Coker and Scoble (1913-1914), Clerk addressed the subject of combustion temperatures in 1882. He performed calculations of the temperatures reached when mixtures of air and town gas are exploded in closed vessels. Perhaps the earliest attempt to measure gas temperature in a firing engine was that of Burstall (1895). He used resistance thermometers with platinum wires 51 and 76 μm in diameter, with the engine running at 120 rpm under a very light load (approximately 80% misfires) to prevent failure of the sensor. His engine, which was of low compression ratio, led him to conclude that the highest combustion temperature was at least 1250°C. A few years later (Burstall, 1901) he made measurements as high as 1570°C, but recognized that these were undoubtedly too low because of thermal lag for the wire size used.

In an attempt to circumvent this problem of sensor durability at maximum combustion temperatures, Callendar and Dalby (1908) protected their resistance thermometer by installing it inside the hollow stem of a "jack-in-the-box" valve. During compression, the valve opened to allow the gas to pass through slots in the side of valve stem and contact the sensor. This protection system allowed the use of a fine 25 μm diameter platinum wire, which was tested for frequency response by motoring the engine at 100 rpm and comparing the measured temperature with that calculated from the measured pressure; at top-center, the measured temperature was low by about 20°C. With combustion, the highest temperatures they found with their procedure were between 2250 and 2500°C.

Coker and Scoble (1913-1914) may have been the first to measure gas temperature in an engine by means of a thermocouple. Platinum and rhodium or iridium thermocouples with diameters from 13 to 20 μm were used. The thermocouple junction was made by butt welding the wires, and the exposed wires were formed into a loop that hung in a vertical position in the combustion chamber, as shown in Fig. 5. They found that this loop design served remarkably well, although two failure modes were observed; in some cases the thermocouple was fused, and it others mechanical breakage occurred. The maximum gas temperature they measured successfully was 1705°C; they estimated the melt temperature of platinum to be 1750°C.

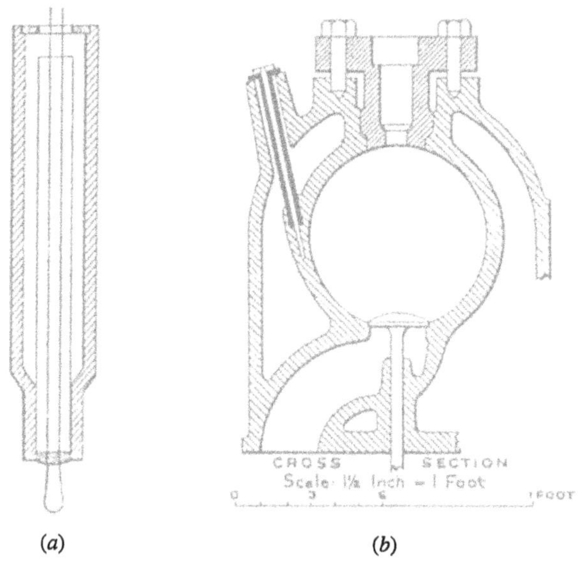

(*a*) (*b*)

Figure 5. (*a*) "Loop" thermocouple design of Coker and Scoble (1913-1914); (*b*) 1907 National Gas Engine Company engine, with 178 mm bore and 381 mm stroke. The thermocouple probe was installed in the port directly above the valve. Shown in the upper left is one of four mercury thermometers used to measure the wall temperature.

In 1936 Hershey reviewed the early attempts to measure in-cylinder gas temperatures, and the thermocouple and resistance thermometer of that era still appear to be the best techniques for motored engines. With combustion, the more recently developed fiber-optic thermometer appears to be the only available method for a production engine.

3.1 THERMOCOUPLE

Meyer and DeCarolis (1962) studied the effect of thermocouple wire size on frequency response using the four-thermocouple probe shown in Fig. 6. Wire diameters of 13, 20, 25 and 38 μm were used simultaneously in a single probe that installed in the fuel injector port of a diesel engine. (This work is included herein because this probe could have been installed in place of a spark plug.) They plotted the peak measured temperature as a function of thermocouple wire diameter, showing that the peak measured temperature asymptoted to a maximum value as the wire diameter was reduced. Although they only motored the engine at a speed of 100 rpm, they still found the frequency response of all but the 13 μm thermocouple to be inadequate.

Tsao et al. (1962) compared a 20 μm thermocouple with an infrared radiation pyrometer technique (Burrows et al., 1961) and found that the crank angle of peak gas temperature by the thermocouple increasingly lagged the pyrometer measurements as the engine speed was increased. Interestingly, when plotted in the time domain the difference between the two measurements decreased with increasing engine speed, apparently because of increased heat transfer to the thermocouple from increased turbulence. However, at even the lowest practical engine speed the frequency response of the thermocouple was shown to be inadequate.

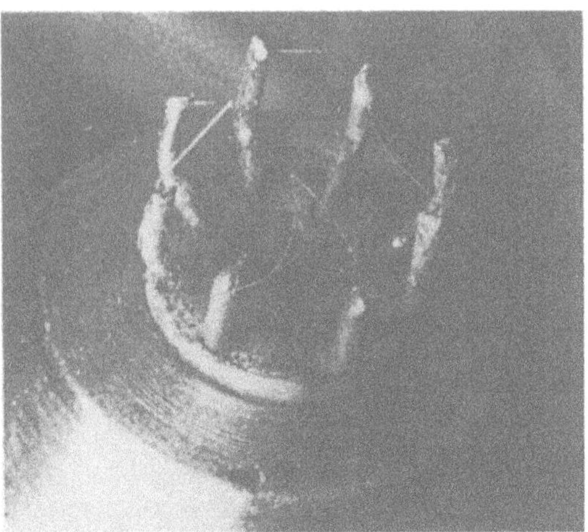

Figure 6. Four-thermocouple probe of Meyer and DeCarolis (1962) for insertion into a diesel engine combustion chamber in place of the injection nozzle.

Witze (1975) measured the gas temperature in a motored engine using a 7.6 μm diameter chromel/constantan thermocouple. The manufacturer (Hycal) rated the frequency response of the probe as better than 20 kHz. For an engine speed of 1775 rpm, peak temperature occurred at about 6 ms after top-center and, more importantly, only one successful data set was obtained before the four available probes were broken.

3.2 RESISTANCE THERMOMETER

The typical resistance thermometer illustrated in Fig. 7 consists of a fine sensing wire attached to two needle-like supports. A constant current is maintained through the wire, and a Wheatstone bridge is used to measure the voltage change across the wire as its resistance changes with temperature. It is necessary to calibrate the probe in an oven, and in a highly transient environment it has many of the frequency response limitations of a thermocouple. The sensing wire must be as thin as possible, which makes durability an issue. However, the supports for the wire cannot be small, which means that they will always lag the gas and sensor temperatures, resulting in an error due to heat transfer between the wire and the supports.

Despite the above mentioned efforts of Burstall (1895, 1901), the resistance thermometer is not suitable for measuring gas temperatures during combustion. However, it can be used to measure motored engine gas temperatures, which are critical to the application of hot wire anemometry discussed later in Section 5.1. Lancaster (1975) used a resistance thermometer installed in the

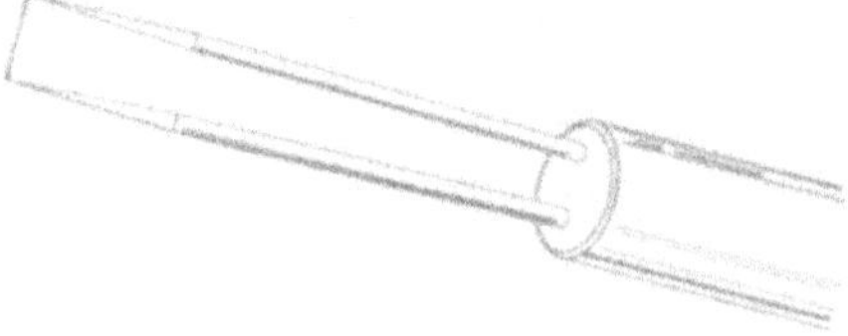

Figure 7. Typical resistance thermometer.

spark plug hole to measure the gas temperature during motored compression. He concluded that the air temperature calculated from cylinder pressure measurements through a polytropic relationship was sufficient, and thus used calculated temperatures to reduce his hot-wire anemometer measurements.

Witze (1980) compared resistance thermometer measurements made at two different locations in a motored engine with the adiabatic temperature computed from the measured cylinder pressure. He found spatial nonuniformities in the temperature that were significant enough to affect the velocity measurements obtained with a hot-wire anemometer.

3.3 FIBER-OPTIC THERMOMETER

Accufiber (1989a) has designed a probe made in a spark plug body that consists of an off-axis spark plug electrode and an fiber-optic thermometer. The thermometer consists a sapphire rod approximately 1 mm in diameter, the tip of which is coated with a noble-metal thin-film to form a small blackbody cavity (Fig. 8). When heated, the cavity emits light that is transmitted down the rod to a conventional low-temperature optical fiber terminated at the far end by a remote detector. The amount of light radiated from the cavity is a function of the cavity, or sapphire rod, temperature. Dils and Moore (1986) have developed procedures to solve the inverse problem of using the rod temperature to determine the heat transfer to the rod, which yields estimates of the gas temperature and velocity as well. Accufiber (1989b) has also shown that Fourier analysis of the heat flux signal can be used to detect the occurrence of knock.

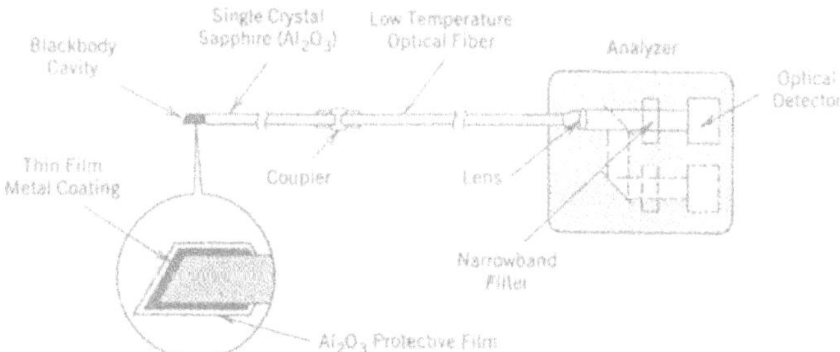

Figure 8. Accufiber sapphire blackbody sensor and optical analyzer (from Berthold, 1991).

4. Species

There are numerous instances when it is desirable to know the chemical composition of the in-cylinder gases. These include fuel/air/residual-gas mixing, sources of unburned hydrocarbon, and the preflame chemical kinetics leading to autoignition and knock. Unfortunately, this is perhaps the one area where modern technology has not produced any significant breakthroughs. In 1930, Withrow et al. used a mechanically actuated valve to withdraw gas samples from the cylinder during short crank angle windows in the engine cycle. Their purpose was to investigate the thickness of the reaction zone; they used oxygen as an indicator of the unburned mixture, and found that combustion was normally completed in a narrow zone, presumably a flame front. More than sixty years have passed, and today's technology still relies on sampling, only the sampling valves are now electronically controlled .

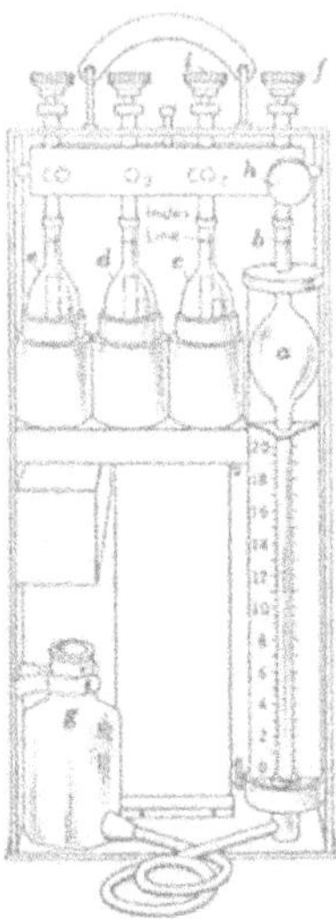

Figure 9. The Orsat gas analyzer served as the standard for many years for measuring the fuel-air equivalence ratio. A known volume of sample gas is trapped at atmospheric pressure in burette ***a***. The sample is then forced into pipette ***c*** which contains a solvent that absorbs CO_2, and then is transferred back to ***a*** to measure the volume change. The process is then repeated for burettes ***d*** and ***e*** to measure O_2 and CO, respectively (from Obert, 1973).

4.1 GAS SAMPLING

Matsui et al. (1979) developed a sampling probe and procedures to measure the local mixture strength in the spark gap on a cycle-resolved basis. Their gas sampling probe shown in Fig. 10 has an inwardly-opening needle valve in the center electrode of a spark plug. The valve is electronically controlled by a solenoid with a 3 ms minimum sampling time. A steady flow of helium carrier gas

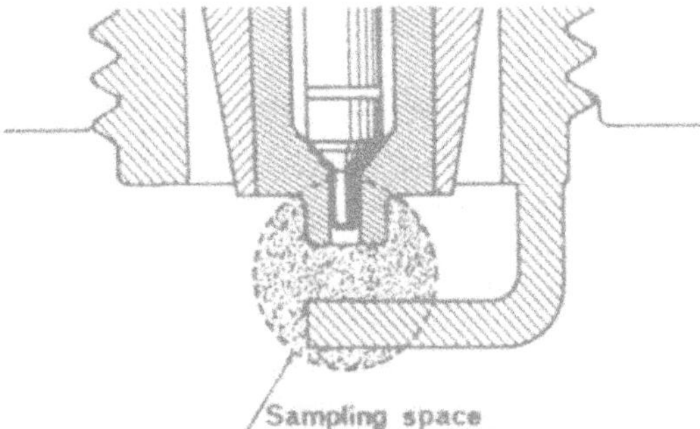

Figure 10. Electronically-controlled gas sampling valve embodied in the center electrode of a spark plug (from Matsui et al., 1979).

circulates through the probe to move the sampled mixture to the analyzing instruments. A flame ionization detector (FID) is used to measure the hydrocarbons, and a specially built infrared absorption device is used to measure carbon dioxide. For the tests that they performed the engine was operated on a lean mixture, such that it was unnecessary to measure carbon monoxide; however, they did confirm that an absorption technique would suffice for this measurement as well. To demonstrate the technique, they made measurements in both two- and four-stroke cycle engines, and observed a much larger variation in total hydrocarbon concentration in the former. This is the expected behavior, considering the greater difficulty in scavenging a two-stroke cycle engine and the shorter available time for mixing the fresh charge with the residual gases.

Collings (1988) has developed a similar technique, except that his procedure samples continuously. A 0.12 mm inside-diameter stainless steel capillary tube installed in the center electrode of a spark plug is used to transport the in-cylinder gases to an FID. While the probe is devoid of the mechanical complexities of the previous device, interpretation is far more difficult because the transit time for a sample to reach the FID is a function of the cylinder pressure driving the flow. Because the motoring pressure in the cylinder is very repeatable, it can be accounted for. Since the same is not true of the pressure rise due to combustion, the system was carefully designed to minimize the transit time (about 6 ms) to ensure that a sample of the mixture at the time of ignition reaches the FID before significant pressure rise occurs.

4.2 EMISSION SPECTROSCOPY

Ohyama et al. (1990) used a spark plug with a 1 mm diameter fiber cable in the center electrode to detect emissions from the CH and C_2 radicals, which they recorded simultaneously. They report results for the intensity of the emissions as a function of crank angle for different air-fuel ratios, and suggest that the intensity ratio CH/C_2 can be used to directly measure the air-fuel ratio. The strong dependence of CH and C_2 emissions on mixture strength is well known (Gaydon, 1957), and recently Chou (1993) has shown that the intensity ratio of the two eliminates the effects of pressure and temperature that occur in an engine.

Kalghatgi (1992) used a "Colortune" spark plug, a commercial product of Gunson Ltd., to gain optical access to a production engine to monitor the activity of a gasoline detergent additive package. This spark plug has a transparent annular ring between the center electrode and the metal plug body. An optical fiber was used to transmit emission light collected at the spark plug window to a bifurcated fiber bundle that split the signal for detection by two photodetectors with interference filters for CH and a marker specie, either potassium or sodium. A small amount of the marker was introduced in the inlet manifold at the start of a test, and after about one hour a quasi-steady state was reached for the emission of marker still trapped in deposits. Then the detergent-package fuel was turned on, and the resultant increase in marker emission was used as an indicator of detergent activity. The CH emission was used to monitor changes in the transparency of the window.

Itoh et al. (1995) used CH and C_2 emissions to study engine knock. Optical access was obtained by installing optical fibers in the head gasket. Seven 1 mm diameter fibers, with 200 μm core and 20 degree viewing angle, were positioned to observe the portion of the combustion chamber opposite the spark plug located in the sidewall. No special precautions were made to protect the fibers from the compression loads (Takagi, 1994), such as using metal clad fiber or insertion of the fibers in metal tubing. They found that the emission intensity of both radicals increased sharply when autoignition occurred. The emission intensity from C_2 also occasionally showed two peaks; the first was attributed to autoignition and the second to soot formation.

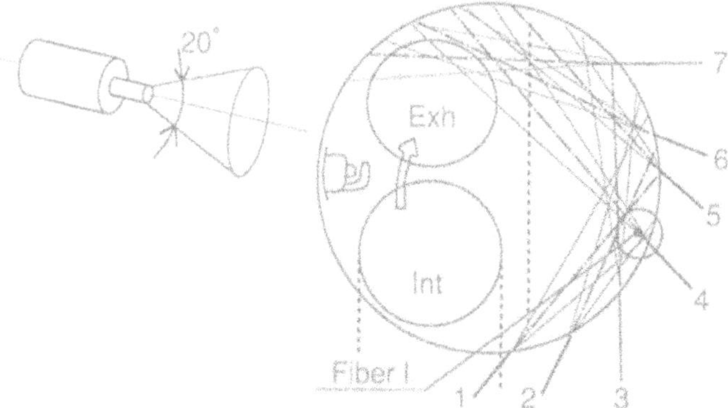

Figure 11. Field of view of multiple optical fibers installed in the head gasket to study autoignition of the end gas (from Itoh et al., 1995).

5. Velocity

Gas velocity and turbulence were recognized from the very beginning to be crucial to the successful operation of a piston engine. If the turbulent burning velocity did not increase with engine speed, engines could not be used over the wide range of speeds that we depend on. The time available to efficiently burn the charge in the cylinder decreases inversely with engine speed, and continually advancing the spark timing is not a practical solution. Fortunately, the turbulent fluctuations of the gas increase approximately linearly with engine speed, producing a corresponding increase in the burning rate. Clerk demonstrated the importance of turbulence to engine combustion in 1912, when he compared the cylinder pressure for ordinary engine operation with that obtained after several non-firing cycles during which the valves were disabled, as shown in Fig. 12.

Three velocity diagnostic techniques will be discussed in this section: hot wire anemometry, spark discharge anemometry, and laser Doppler velocimetry. In the two following sections, on luminosity and ionization density, some additional techniques for studying fluid motion will be discussed. Although these could be classified as velocity diagnostics, they have not been included in this section because they rely on additional diagnostic tools that have not yet been introduced.

5.1 HOT WIRE ANEMOMETRY

Hot wire anemometry is based on the change in electrical resistance that occurs in a fine, heated wire as it loses heat to its surroundings, which in this application is air moving with turbulent motion. In constant-temperature anemometry, a feedback control circuit adjusts the current through the sensor to maintain a constant resistance, or temperature. The anemometer output is the voltage necessary to drive this current, which is a nonlinear function of the temperature and pressure of the gas, and the component of gas velocity perpendicular to the wire. (Although the contribution from the velocity component parallel to the wire is assumed to be negligible, this is not always the case.) The exact direction of flow is not explicitly determined, although it can often be resolved by intui-

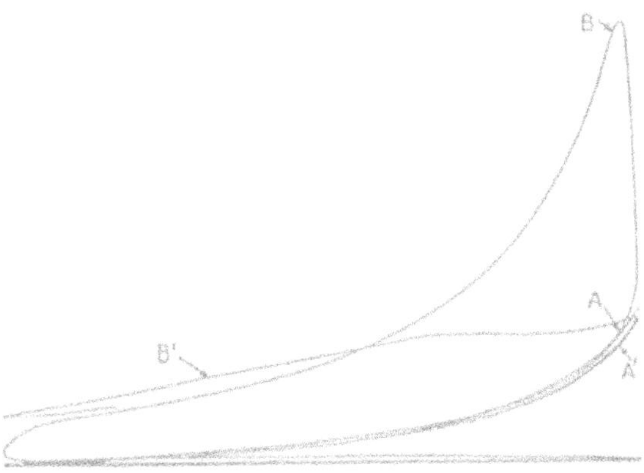

Figure 12. Indicator diagram (pressure vs. piston height from bottom center) of Clerk (1912), showing the influence of turbulence on the nature of the pressure development. The pressure rise from A to B is for ordinary running conditions, whereas A' to B' is for when turbulence had largely subsided.

tion and geometric considerations. The construction of a hot wire probe is identical to a resistance thermometer; the probe shown in Fig. 7 could be used as a single velocity-component probe, and the dual sensor "X" probe shown in Fig. 13 measures two velocity components simultaneously. Three sensor probes can also be built, and it is common practice in motored engine applications to use a resistance thermometer simultaneously with a one or two component velocity probe.

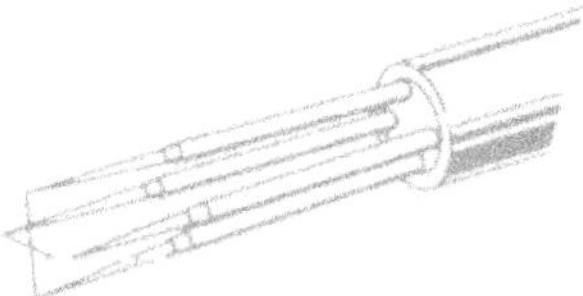

Figure 13. Dual sensor "X" probe for simultaneous hot wire anemometer measurements of two orthogonal velocity components.

Gas velocity is determined from the anemometer response through a calibration relationship that includes corrections for temperature and pressure variations that occur during compression and expansion. While it is technically possible to calibrate a probe over the complete range of velocities, pressures and temperatures expected during the compression and expansion strokes, this is not very practical. It is thus standard practice to use an empirically determined analytic model to correct indicated results for deviations from calibration conditions. The accuracy of hot wire anemometer measurements thus depends on the adequacy of an empirical heat transfer model, in addition to accurate knowledge of the gas and wire temperatures. Witze (1980) made a critical comparison between hot-wire anemometry and laser Doppler velocimetry in a swirling engine flow that minimized some of the inherent problems with hot wires, including a low compression ratio of 5.1, and still found unacceptable discrepancies in the hot wire data around top-center. More recently,

Catania and Mittica (1987) have reported hot wire anemometer studies in a diesel engine with a compression ratio of 16, operated at 3000 rpm. These are extremely difficult conditions for this technique because of the high gas temperatures (~800 K) and the very rapid transients in gas temperature. Their procedure (Catania, 1982) included measuring the gas temperature with the 10 μm diameter hot wire probe used as a resistance thermometer, and measuring the temperature of one of the prongs supporting the wire with a thermocouple. They estimate a maximum uncertainty of 10% in their computed velocities.

From the appearance of the probes shown in Figs. 7 and 13, it is evident that these probes are very delicate. Typical dimensions for the wire are 5-10 μm in diameter by 1.5-2.0 mm length. Because the wires cannot survive combustion temperatures, their use is restricted to motored engine studies, but this limitation permits their use in production engines with access through the spark plug hole. For some applications this is sufficient, since it is the flow field at the spark location at the time of ignition that is of greatest importance. At the same time, care must be taken in assuming that the fluid motion near the hot wire in a motored engine is similar to that near the spark plug in a fired engine. Witze et al. (1989) studied this problem in a four-stroke engine, and concluded that there are instances when these assumptions are permissible, but that there also are cases when significant differences exist between motored and fired velocities and turbulence. Miles et al. (1994) made a similar study in a ported two-stroke engine, and concluded that motored velocity measurements are not at all representative of a fired engine.

Semenov (1958a, 1958b) was first to use a hot wire anemometer in a motored engine, but he did machine an access port. Barton et al. (1971), Windsor and Patterson (1973), Dent and Salama (1975), and Lancaster (1976) all used the spark plug port for access, and pioneered the field of velocity measurements in spark ignition engines. In addition to mean velocity and turbulence intensity, time scales and frequency spectra have been measured, and the very important (and equally controversial) issue of the distinction between turbulence and cyclic variations was addressed for the first time.

5.2 SPARK DISCHARGE ANEMOMETRY

Lindvall (1934) developed the spark discharge anemometer to measure air velocity in a wind tunnel. In essence, the technique monitors the voltage required to sustain a spark that is being stretched by gas motion normal to the path between a pair of electrodes. The stronger the velocity, the longer the spark, and thus the greater the voltage rise. Maly et al. (1983) adapted the technique, which they call voltage rise anemometry (VRA), to measure gas velocity and turbulence in both motored and fired engines. For this application the nature of the spark is dependent on gas pressure, which is not too difficult to account for, and gas composition, which is a more difficult problem. They present measurements for the mean velocity and turbulence intensity during intake, as well as results showing the effect of flame propagation on flow velocity. Although this particular study was done with a modified engine, Maly and Meinel (1981) have demonstrated the technique using the standard spark plug in a production engine.

Ohigashi et al. (1968) have developed a spark discharge anemometer that is independent of gas properties. They create a spark with a pair of electrodes, and measure the time it takes for the ionized gas created by the spark to reach a separate pickup probe located at a known distance from the spark gap. Their initial measurements were performed in the inlet pipe of a two-stroke engine, but in 1969 they made in-cylinder measurements in a modified engine.

Ahmad et al. (1984) designed and tested a probe where the spark electrodes and pickup probe are located in a single device small enough to fit in the spark plug hole. Herrin (1984) used the version of this probe shown in Fig. 14 to study differences in fluid motion between two unmodified engine heads. Although the head geometries were symmetrically similar, his measurements revealed that this was not the case with the flow fields, and the differences he found helped explain the observed combustion performance.

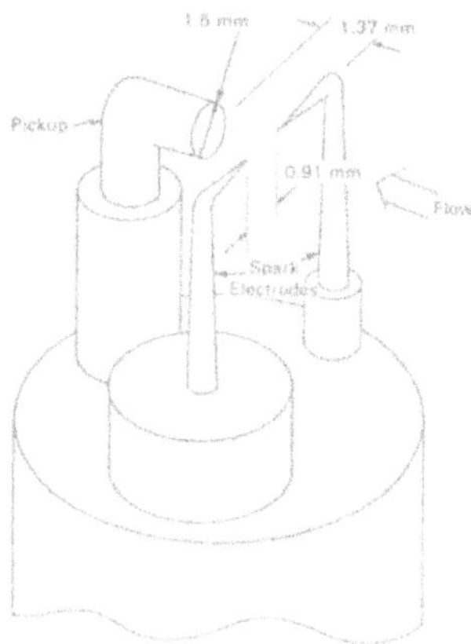

Figure 14 Spark discharge anemometer probe use by Herrin (1984).

5.3 LASER DOPPLER VELOCIMETRY

Although laser Doppler velocimetry (LDV) has never been demonstrated in an unmodified spark-ignition engine, it is technically possible, and because the technique is so powerful it has been included in this review. Vannobel (1992) has made LDV measurements in an unmodified production diesel engine by installing a fiber-optic probe in the fuel-injector access port. Commercial probes of this type are available in sizes as small as 14 mm in diameter, which is a standard spark plug size. Although it is doubtful that a small probe could be made with adequate thermal protection for a continually fired engine, it is conceivable that an LDV probe could be designed to withstand skip-fired operation.

The physical principle of the dual-beam LDV technique is illustrated in Fig. 15. Two coherent laser beams are crossed at the focal point of a common lens; the wave structure of the incident light creates interference fringes throughout the beam intersection region (*a*). Viewed in a plane perpendicular to the plane of the beams, the interference fringes appear as a pattern of straight light and dark lines (*b*). The spacing of these fringes ΔX is related to the wavelength of laser light λ and the half-angle of the two intersecting beams ϕ. When a small particle passes through this fringe pattern, it will scatter light in a manner indicated by Fig. 15*c*, where the frequency F of this "Doppler burst" is directly proportional to the particle velocity V_p. For a typical LDV setup, the fringe spacing ΔX is of order 2 μm, the probe volume length ℓ_m is of order 0.5 mm, and the probe volume diameter d_m is of order 0.1 mm.

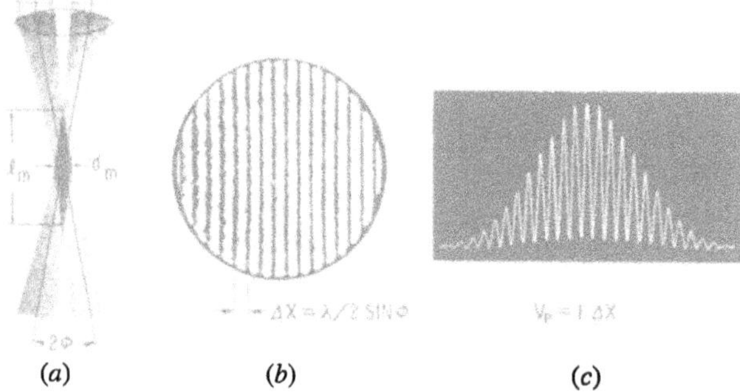

Figure 15. Dual beam laser Doppler velocimeter.

There are four very important features of LDV that make it the most powerful tool available for velocity measurements. First, it is independent of the thermodynamic state of the fluid under study; calibration is simply a matter of measuring the crossing angle of the two laser beams. Second, it is selective to only the velocity component normal to the fringes. Third, it is a straightforward matter to measure two orthogonal components of velocity simultaneously; this is usually done by creating two orthogonal and coincident fringe patterns using two different laser wavelengths (typically the blue and green lines of an Argon ion laser). Fourth, the velocity sense (i.e., direction of the particle crossing the fringes) can be determined by introducing a slight frequency difference between the two crossed laser beams; this makes the fringes move in space, creating a relative velocity of the fringes that allows the direction of particle motion to be determined.

To use LDV it is necessary to introduce artificial seed particles into the flow. For optimum performance they should be on the order of 1 μm in diameter; particles much larger than this may not adequately follow the dynamics of the flow. Typical seed materials include smoke, oil mist, and fine powders like titanium dioxide. Since the Doppler-shifted light scattered from the particles can be collected in direct backscatter with the same lens used to focus and cross the incident beams, only one access port is required.

Hutchinson et al. (1978) were the first to report crank-angle resolved LDV measurements in a reciprocating flow field; they used a plastic model engine driven at 200 rpm. Later in the same year, Rask (1979a) and Witze (1979) reported measurements obtained in modified production L-head engines of similar design. Then, one year later, Asanuma and Obokata (1979) and Rask (1979b) reported the first in-cylinder velocity measurements made in fired engines. Since that time, LDV has evolved to the point of being a routine tool widely used for engine research and development. Although, as mentioned above, it has not yet been applied to unmodified engines, it has been successfully demonstrated in production-like engines where optical access was gained through a window in the piston (e.g., Whitelaw and Xu, 1993).

6. Luminosity

From the very beginning, the pioneers of piston engine research longed to look in on the combustion process inside the cylinder, but it was not until 1936 that Withrow and Rassweiler succeeded in installing a window in an L-head engine that provided a full view of the combustion chamber. However, in 1924 Midgley and McCarty installed a single one-inch diameter quartz window in the wall of the combustion chamber to observe the flame color. A stroboscope was used to resolve the radiation as a function of crank angle. In 1927, Sir Harry Ricardo reported the installation of a row of six small windows in the head of an L-head engine, as shown in Fig. 16. Through these windows he could observe the progression of the flame as it traveled across the combustion chamber. With the exception that the optical fiber had not yet been invented, the concept of Ricardo's experiment is essentially identical to the recent work of Spicher and coworkers (1986, 1988), who installed numerous small windows in an engine head, cylinder liner and piston. They use optical fibers to transmit the visible emissions collected by the windows to individual photodetectors, enabling measurements of flame position versus time. Ricardo did not have such sophisticated electronics at his disposal, but in 1930 Glyde demonstrated a clever device that let him measure the time of flame passage past each window. A rotating disk with six holes that lined up with the windows in the head was geared to the ignition timing shaft. The phasing of the rotating disk, relative to ignition timing, was adjustable with a screw arrangement. For each window in the head, the angular orientation of the disk was adjusted until the flame was first visible when the corresponding hole in the disc was aligned with the window, resulting in a measure of flame position versus crank angle.

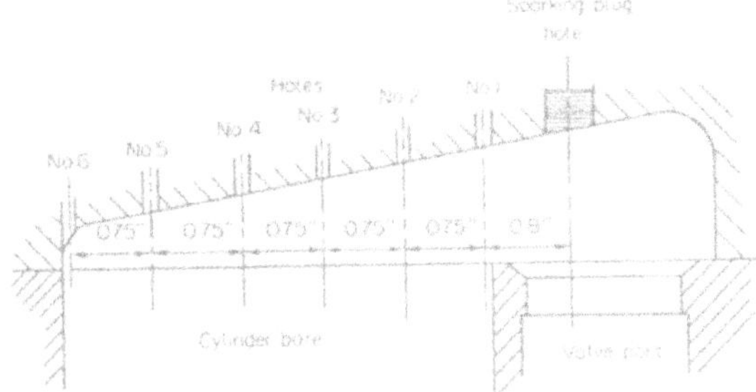

Figure 16. L-head engine with six small windows installed in a row in the head, in line with the spark plug (from Ricardo, 1927 and Glyde, 1930).

6.1 FLAME LOCATION

Access for optical measurements of flame position in unmodified engines is quite limited. One method that is beginning to see increased use employs multiple windows in the spark plug. Witze et al. (1988) installed eight optical fibers in the threaded base region of a standard 14 mm spark plug, as shown in Fig. 17. The fibers collect the light emitted from the flame as it crosses the field-of-view of the fibers, and transmit the light to individual photodetectors. The time from ignition until detection of the flame is used to compute the average flame velocity in the direction of each fiber relative to the spark location. In their prototype design the polished fiber ends were flush with the base of the spark plug, giving a 23° solid angle field-of-view characteristic of multimode fibers. Current commercially available designs protect the fibers with sapphire windows, and the resulting "tunnel vision" reduces the field-of-view to about 10°.

The fiber optic spark plug probe was developed specifically for studies of cyclic variation in flame kernel development. Of interest was the rate of growth and convective motion of the early flame kernel and the correlation of these quantities with pressure measurements. However, interpretation of measurements made with this probe is not straight-forward. What one would like is a measure of flame shape at a given time, from which the flame size and centroid location could be determined. Instead, the probe measures the arrival time of the flame at eight discrete locations

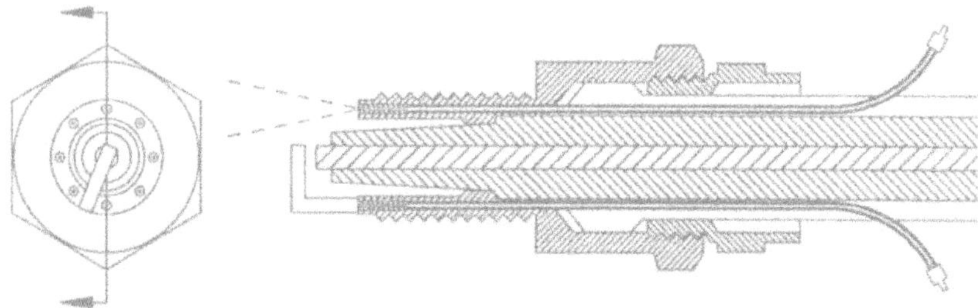

Figure 17. Spark plug probe instrumented with eight equally-spaced 0.2 mm diameter optical fibers. The field-of-view of each fiber is the 23° solid angle indicated by the dashed lines (from Witze et al., 1988).

relative to the spark gap. Kerstein and Witze (1990) developed an analytic model that assumes an elliptic flame shape, with constant but arbitrary growth rates for the major and minor axes. Most important, their model explicitly distinguishes between the growth rate and convective motion of the flame kernel. Witze et al. (1990) used the model to correlate cycle-resolved measurements of flame kernel growth rate and motion with pressure-derived measurements of combustion duration.

Hahn and Anderson (1991) used a fiber optic spark plug probe to study combustion performance in two and four valve-per-cylinder production engines. Using the elliptic-shaped flame kernel model, they observed a strong correlation between the flame growth rate and the 0-2% mass fraction burn duration, and less correlation with other combustion parameters derived from cylinder pressure. The two valve engine had a helical intake port to induce swirl in the cylinder, and the measured flame kernel convection velocity was approximately three times greater than that found in the four valve engine.

Bianco et al. (1991) used the spark plug probe to study combustion performance in a single cylinder engine configured with quiescent, swirl and tumble fluid motions. They analyzed their measurements with a model for flame kernel shape and centroid location determined from the average value of the eight flame arrival times. Because the time of arrival depends on both the growth rate and convective motion of the flame kernel, the resultant flame shape and centroid location is biased by convection's contribution to the average flame arrival time. The results of their study also found a strong correlation with the flame kernel growth rate and the 0-2% mass fraction burn duration, for all three flow field conditions. The flame kernel convection velocity was found to depend on the fluid motion in the expected manner. The 10-90% burn duration correlated weakly with the early flame kernel behavior.

Lord et al. (1993) applied this tool to study four different production engines designed to generate quiescent, swirl, and tumble charge motions. They present a comparison of the elliptic model with an average arrival time model, and chose the latter for analyzing their data. They found the expected behavior of largest flame kernel convection velocity in the swirl chambers, but also reported that the convection velocity direction varied considerably among cylinders in the same engine. Growth rate correlated well with 0-2% burn duration but showed negligible correlation with later burn times or the indicated mean effective pressure (IMEP).

Hinze and Cheng (1993) used an average time model to study flame kernel development in a four valve, four cylinder production engine run on a broad range of methanol/gasoline fuel mixtures. They found that the flame kernel growth rate was relatively unchanged for mixtures of 0-40% methanol; it increased by 30% from 40-60% mixtures; and then remained roughly constant from 60-100% methanol. From the statistics of the IMEP at lean idle conditions, they also found that large growth rates correlated with better combustion.

Meyer et al. (1993) used the spark plug probe in a four valve, four cylinder production engine. Their analysis did not attempt to separate flame kernel growth rate from convective motion. Like the previous authors, they found a strong correlation between the flame arrival time measurements and the 0-2% burn duration.

Witze and Bopp (1991) used a surface-gap version of the eight-fiber probe to measure the flame propagation speed in a single cylinder research engine. The instrumented spark plug was located in the center of the combustion chamber in the head, but ignition was at the perimeter with an electrode located in the gasket. As the flame passed beneath the ring of fibers, their discrete locations permitted calculation of the speed and direction of passage and, with lesser accuracy, the average curvature of the flame surface.

6.2 FLUID MOTION

Witze (1990) investigated the possibility of using the eight-fiber spark plug probe to measure gas velocity. Although the trial demonstration of the concept was performed in an optical research engine, the ultimate application was for production engines. The idea was to perform a spark sweep throughout the ignitable region of the engine cycle, and use the convection velocity of the flame kernel at each ignition angle to infer an average gas velocity in the vicinity of the spark plug. He attempted to calibrate the procedure with LDV measurements obtained within the spark gap, but concluded that the velocity gradients were too great to permit a correlation with this single point measurement. He then made LDV measurements at other locations within the field of view of the eight fibers, creating a "velocity range", or bandwidth, as a function of crank angle. To get the spark plug convection velocity measurements to fall within this range, it was necessary to apply a totally empirical correction based on flame growth rate data, which suggested inadequacies in the elliptic flame kernel model used to analyze the measurements.

6.3 COMBUSTION PERFORMANCE

In the two previous sections, multiple optical fiber arrays were used to measure the time of flame arrival at distributed locations. In this section, single fibers are used to collect visible emission from the combustion chamber for analysis of combustion performance.

Müller et al. (1983) have a patent proposing the installation of an optical fiber or rod in the center electrode of a spark plug, as well as in the head gasket. Their intent was to detect knock from the oscillations it induces in the visible emission from the burned gas. Their spark plug designs include the use of a solid glass rod or a fiber optic bundle, where the exposed end of the latter is spread out like a brush to give a wider field of view. No details were given of how fibers could be installed in the head gasket.

Remboski et al. (1989) made a comprehensive study of luminous emission at 927.7 nm. This near-infrared region was selected for several reasons: 1) Coincidence with the peak spectral response of an inexpensive silicon photodetector; 2) Weak influence by black-body emissions from the walls of the combustion chamber; 3) Correspondence to a strong water band in this wavelength region. The probe they used (Hartman et al., 1991) consisted of a small sapphire rod that protruded 4 mm into the combustion gases for the purpose of keeping it relatively free of deposits. Although the probe required a 10 mm hole for installation, it is evident that the light gathering element could be easily integrated with a spark plug. Their results show a good correlation between the measured luminosity and the location and magnitude of the maxima in pressure and heat release.

Ohyama et al. (1990) installed a 1.0 mm quartz glass optical fiber cable of 0.9 mm core in the center electrode of a J-gap spark plug. The tip of the fiber is machined into a convex shape to provide a 120 degree view of the inside of the combustion chamber. They used a monochrometer and photomultiplier to record emissions at 450 and 750 nm obtained from a four cylinder production engine. These measurements were compared with the measured cylinder pressure, looking for similarities in the shape and phasing of the signals. They conclude that the emission at 750 nm includes black body radiation from the walls, which cause the peak to occur significantly after pressure peak. The peak for 450 nm was adequately close to pressure peak until the mixture air/fuel ratio exceeded about 17, at which point the pressure peak increasingly lagged the emission peak.

Nutton and Pinnock (1990) installed a fused silica rod in the center electrode of a spark plug, as shown in Fig. 18. They investigated different designs for the end face of the rod in an effort to maximize durability and the field of view, and minimize the propensity to foul. The design that resulted has a polyhedron shape, since it was found that the corners between facets tended to break up deposits and help prevent the adhesion of contaminants to the end face. The emissions collected by the probe were detected by two photodiodes of different spectral response, centered at ~600 and ~900 nm. For ~900 nm they found a linear relationship between the crank angle for peak optical intensity and the crank angle for 90% mass-fraction-burned, as determined from the pressure. At the start of combustion they observed that the color of the emitted light changed much more rapidly than the intensity. This led them to use the intensity ratio of ~900/~600 as a measure of the start of combustion, and found it to be extremely sensitive and well correlated with the point of 2% mass fraction burned. However, they did observe that the start of combustion signal was dependent on the orientation of the probe within the combustion chamber. They also found that this intensity ratio correlated quite well with NO_X measurements, which they attribute to the strong temperature dependence of the NO_X formation mechanism. Finally, they studied engine knock by comparing

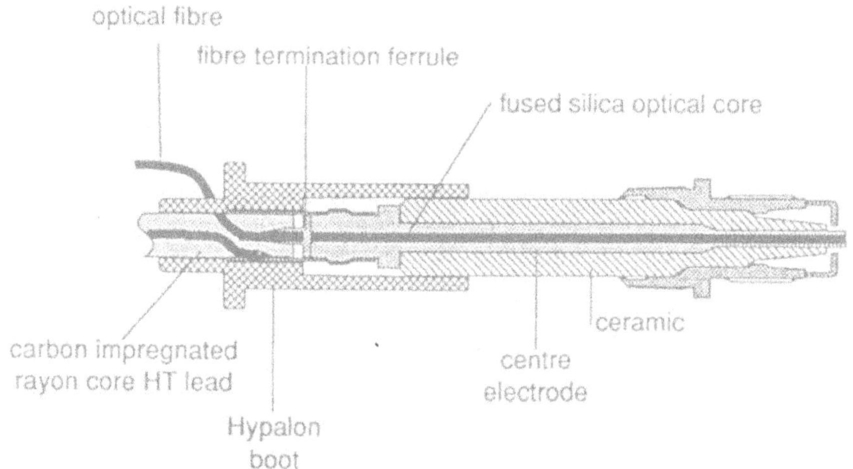

Figure 18. Optical spark plug sensor of Nutton and Pinnock (1990).

cylinder pressure measurements with the intensity of the emissions. For heavy engine knock they show that the signals are comparable, whereas for borderline knock they find the optical signal to be much more sensitive. As an example, for a case where the pressure record showed no indication of knock and the optical signal showed no evidence of oscillations, there still was a sudden change in the slope of the emission intensity that they suggest could be used in a feedback control scheme that could prevent knock from ever occurring.

7. Ionization Density

Before the technology was developed that allowed viewing of the flame through windows, the popular alternative was to 'touch' the flame with ionization probes. In 1920, MacKenzie and Honaman installed two ionization probes in a cylinder head, enabling the measurement of the average flame speed as it traveled from the spark gap past the two probes. Fourteen years later, Schnauffer (1934) installed twenty-four ionization probes throughout the head of an aviation engine. Because the electronics necessary to simultaneously record this many analog signals did not exist at this time, the ionization probes were wired to individual neon lamps arranged in a board in locations corresponding to the probe locations in the engine. As the propagating flame reached each probe, the lamps would light to reveal the evolution of the flame shape. The time of the lighting of the lamps could be determined one at a time using a stroboscope, or simultaneously by taking a high speed motion picture of the array of lamps.

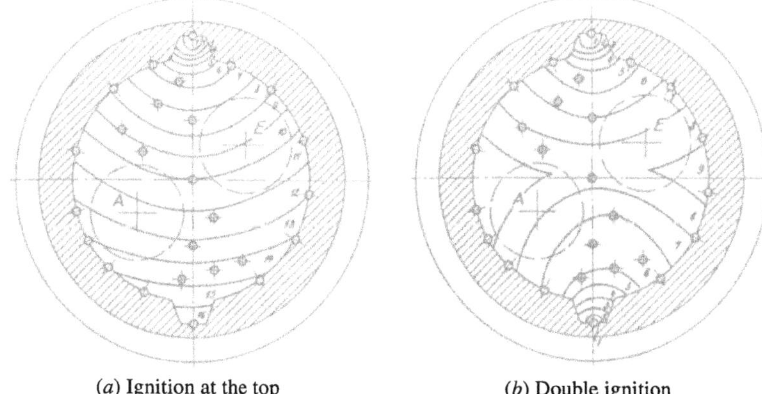

(*a*) Ignition at the top (*b*) Double ignition

Figure 19. Ionization probe measurements of Schnauffer (1934), comparing the evolution of the flame position for one and two point ignition. The locations of the 28 ionization probes are indicated by the circles.

In principle, an ionization probe is simply an insulated conductor at a voltage potential different from ground. When contacted by the free ions and electrons in the reaction zone of a flame, current flows through the conductor in the direction governed by the polarity of the bias voltage; because of the greater mobility of electrons, a positive voltage is commonly used. The strength of the current, and thus the detected signal, depends on the magnitude of the bias voltage, the probe surface area, and the ion density in the burned gas; signal strength is not a strong function of the gap size between the probe and ground.

7.1 FLAME LOCATION

Douaud et al. (1983) installed three ionization probes along the inner cylindrical wall of the threaded base of a spark plug. The probes were located 90° apart, with the middle probe directly opposite the ground strap. (Unfortunately, the authors neither mention the spark duration they used, nor discuss any problems they found with ignition noise interfering with the ionization probe signals. Our experience has shown that this can be a serious problem, particularly with the 1-2 ms spark duration used in today's engines.) The device was used to measure the time from ignition until detection of the flame by the ionization probes for parametric studies of spark advance, equivalence ratio, engine speed, and unshrouded and shrouded intake valve configurations.

McDougal (1990) has a patent proposing the use of ionization probes in the head gasket for the purpose of feedback control. The gasket design consists of a single sheet of conductive material sandwiched between two insulating layers. The conductive layer is also insulated from the combustion gas except for discrete tabs that protrude into the combustion chamber. A cluster of three tabs is located as close as possible to the spark plug, intended to serve as a measure of the early flame development period, and another cluster of five probes is located as far as possible from the spark plug, to measure the time of the end of combustion. Because the eight ionization probes are all part of a single conductive sheet, they operate in series. Problems in the operation of this series design may arise due to the relatively long lifetime of the ionization current after first contact of the flame with one of the tabs of the probe. Therefore, it may be necessary to use two separate circuits, in parallel.

Ma (1991) also has a patent describing a head gasket ionization sensor that is similar in purpose to McDougal's; however, the design is quite different. The sensor is a continuous annular ring around the cylinder, analogous to the fire ring of a standard head gasket with the exception that it is insulated from the head and block. It is proposed that a ceramic layer can be coated on selected regions of the face of the ring, so that only the uncoated parts of the surface of the gasket may act as sensing electrodes. The advantage of this design over McDougal's is that it provides the large surface area needed for strong signals without protruding into the combustion chamber, which is a potential source of preignition.

Witze (1989) installed eight ionization probes equally-spaced around the perimeter of a head gasket to measure the uniformity of the final stages of combustion. The gasket was made by sandwiching 0.35 mm diameter steel wire between two sheets of paper gasket material 0.75 mm thick. Silicone rubber sealant was used to bond the two sheets together and hold the electrodes in place. The electrode ends were positioned flush with the bore cutout in the assembled gasket, such that the surface area of the probes was quite small. A personal-computer-based data acquisition and analysis system was developed to permit real-time display of the eight flame arrival time measurements on a cycle-resolved basis. There is reason to believe that the survivability of this particular gasket design was dependent on the low 5:1 compression ratio of the research engine and the skip-fired mode of operation used.

Nicholson and Witze (1993) successfully demonstrated a head gasket instrumented with ionization probes made using printed-circuit-board technology in a single-cylinder development version of a production V-6 engine. Because the copper foil that served as the sensors was so thin (0.04 mm), the total surface area of the sensors was increased by using a multilayered board. As shown in Fig. 20, a pattern of 24 probes of three different widths was tested to determine the best compromise between signal strength and spatial resolution. The engine was operated at 1300 rpm light-load and 2000 rpm heavy-load conditions for more than forty hours without difficulty, although it was necessary to refurbish the sensor faces every 4-5 hours.

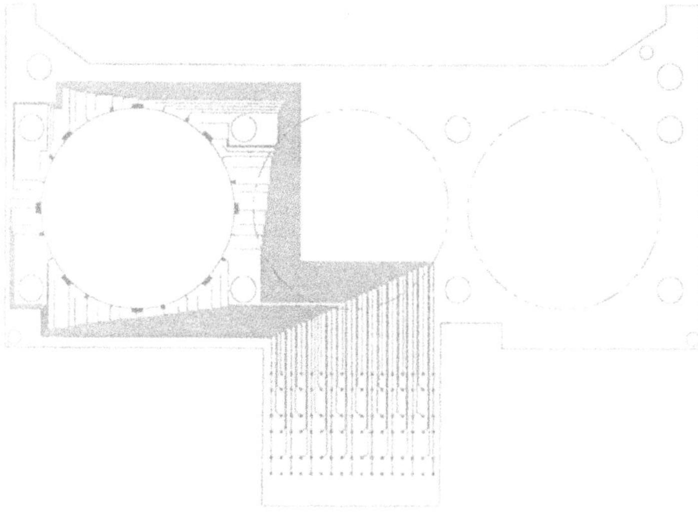

Figure 20. Ionization-probe printed-circuit-board head gasket tested by Nicholson and Witze (1993) in a production V-6 engine. The special development version of the engine had only one active piston.

Meyer et al. (1993) instrumented a production head gasket with ionization probes by bonding insulated wires into channels machined in one side of the gasket. Measurements were obtained in a production four cylinder engine operated at idle, intermediate-load 1500 rpm and road-load 2000 rpm conditions. In addition, the intake port was modified to produce different in-cylinder air motions of low swirl and tumble versus moderate swirl and high tumble. They simultaneously used a fiber-optic spark plug and looked for, but did not find, a correlation between the direction of flame kernel motion measured by the spark plug probe and the time of flame arrival at the ionization probes at the cylinder wall. What they did find was a strong correlation between the spark plug measurements and the 0-2% burn duration, and the ionization probe measurements and the 0-90% burn duration.

7.2 END-GAS LOCATION

Witze and Green (1993) used eight ionization probes built into the head gasket and uniformly distributed around the cylinder bore of a knocking, spark-ignition engine to locate the autoigniting end-gas region. They found that, as normal combustion evolved after spark ignition, the ionization probes individually responded to the arrival of the propagating flame; then, when autoignition occurred, the remaining probes located in the end-gas region responded in rapid succession. By utilizing pressure transducer measurements to determine when autoignition occurred, the ionization probe response becomes a means to locate the end-gas region (Fig. 21).

7.3 FLUID MOTION

Witze and Bopp (1991) showed that it was possible to use ionization probes in the head gasket to investigate in-cylinder fluid motion. Their technique was to use an ionization probe as the ignition

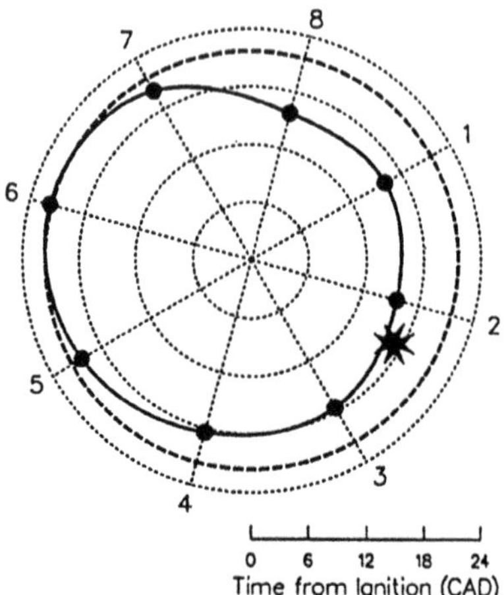

Figure 21. Ionization-probe flame arrival time measurements for a knocking engine. The dashed circle marks the onset of pressure oscillations, revealing that autoignition occurred in the vicinity of probe 6 (from Witze and Green, 1993).

electrode, creating a spark that arced to either the head or block. The flame that is created acts to "tag" the flow, and the remaining ionization probes detect the flame location as it evolves with time. For low levels of swirl, the degree and direction of swirl can be inferred by the asymmetry in the arrival time of the flame at the probes across the chamber from the ignition site. Also, by using the ignition and ionization probe arrangement shown in Fig. 22, they were able to measure the swirl velocity S with good accuracy, as determined by comparison with laser Doppler velocimetry (LDV) measurements of the gas velocity near the wall, corrected for the expansion velocity contribution from combustion. It is important to note, however, that while they were able to ignite the mixture in a 5:1 compression ratio research engine by using a printed-wiring trace as the electrode, they have not succeeded in doing this in a production engine. It is their belief that the greater breakdown voltage required at high compression ratios causes pyrolysis of the polyimide resin substrate material, leading to a cold surface spark that is unable to ignite the mixture.

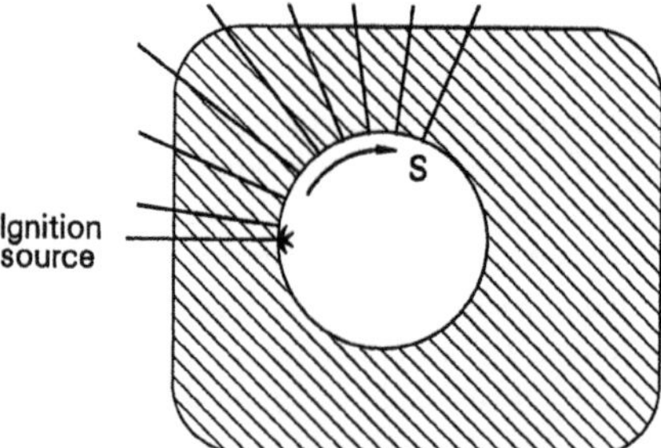

Figure 22. Ionization probe arrangement for measuring swirl velocity. Flame position versus time is measured in a small region along the wall downstream of the spark, before there has been time for considerable heat release (from Witze and Bopp, 1991).

Winch and Mayes (1953) appear to have been the first to use the spark plug in an engine to monitor the state of ionization of the gas in the cylinder. They applied 90 volts d.c. to the center electrode, in parallel with the standard ignition system, and monitored the current in this path with a load resistor. Normal ignition produced a distinct and repeatable wave form that, for normal combustion, was followed by a period of ionization current. If misfire occurred, this current was absent; if preignition occurred, ionization current was detected prior to ignition.

Rado and Johnson (1975) used essentially the same technique to measure combustion quality and its correlation with unburned hydrocarbons for various amounts of exhaust gas recirculation (EGR). They monitored the duration of the ionization current after ignition, and identified three burn types as good burns, slow burns, and misfires. Increased amounts of EGR were found to produce a transition from good burns to slow burns to misfires, with a corresponding increase in unburned hydrocarbons.

Anderson (1986) also used the spark plug as an ionization probe, in conjunction with a second, dedicated ionization probe and a passage-type spark-plug pressure sensor. Although the spark-plug ionization probe was not of primary interest, he looked for, but did not find, a correlation between the crank angle of maximum pressure and either a peak or inflection in the ionization current wave form.

Collings et al. (1986) used the spark plug as an ionization probe to detect knock. Previously, Curry (1963) had shown that an ionization probe will respond to knock with an oscillating signal similar to that of a pressure sensor. Collings et al. (1991) continued work in this area by showing that the ionization probe circuit could also be used to monitor the following: 1) spark voltage and duration; 2) spark plug leakage current due to plug fouling; 3) duration of the ionization current, which they use to characterize the combustion duration.

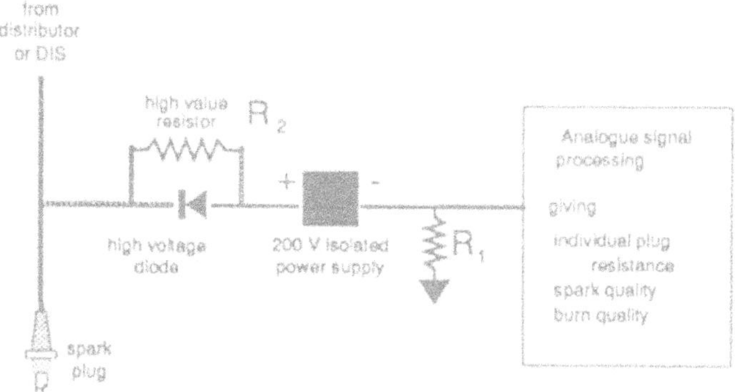

Figure 23. Electronic circuit used by Collings et al. (1991) to monitor the ionization current at the spark plug.

Shimasaki et al. (1993) and Miyata et al. (1993) also use the duration of the ionization current as a measure of combustion duration, and developed an analysis method that uses five different procedures to achieve reliable results that can be employed for engine control. The system they devised is capable of achieving optimal combustion even when the combustion fluctuates due to changes in the air/fuel ratio, EGR, ignition timing, etc., including for lean-burn engines.

8. Visualization

Direct visualization of in-cylinder phenomena has always been the dream of engine researchers. In 1872, Otto built a hand-cranked model engine with a transparent cylinder (Cummins, 1989). Lighted cigarettes were held by the intake port, and the action of the smoke entering the cylinder during the suction stroke was observed to study his ideas of mixture stratification. In 1885, Clerk (1921) motored a full-sized glass model of a spark ignition engine; smoke in the intake was again used as a marker of the gas motion in the cylinder. Woodbury et al. (1921) were the first to directly observe premixed-charge combustion, although the work was performed in a cylindrical combustion vessel. They filmed the combustion event using a rotating drum camera, obtaining the first continuous measurement of flame position versus time. In 1936, Withrow and Rassweiler obtained the first high speed movies of flame propagation in a reciprocating engine, as shown in Fig. 24. The L-head engine they used had a transparent head that gave complete visual access to the combustion chamber. In 1939, Lee made the first motion pictures of the air flow during intake for a motored engine. A glass cylinder was used to provide optical access to the intake process, and white goose down was used as the flow marker. In the same year, Rothrock and Spencer (1939) reported the first schlieren movies of flame propagation in a spark ignition engine. They used a mirror on the piston to double-pass the light field, which was generated by a battery of high-voltage condensers and a rotary distributor so arranged that the condensers were discharged consecutively, furnishing a series of spark discharges. Finally, a review of pioneering work in obtaining optical access to reciprocating engines would be remiss not to mention the invaluable contribution of Bowditch (1961), who first demonstrated the installation of a window in the crown of an elongated piston. While this obviously requires modification to the engine, a window installed in this manner minimizes perturbation to the combustion chamber geometry while still providing nearly complete optical access; even cup-in-piston configurations can be simulated. Bowditch pistons are now widely used for both engine research and development. Recently, Espey and Dec (1993) solved one of the limiting problems of this piston design, that being the time required to access the window for cleaning. Shown in Fig. 25 is their clever design for a sliding cylinder liner that permits quick access to all internal window surfaces.

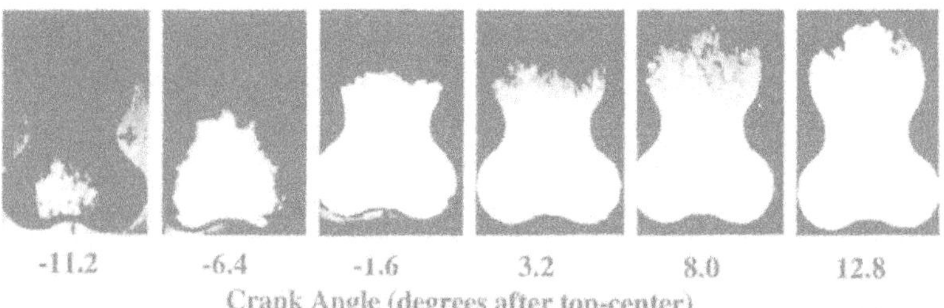

Figure 24. The first motion pictures of flame propagation in a spark ignition engine (Withrow and Rassweiler, 1936). Ignition was at 25 degrees before top-center.

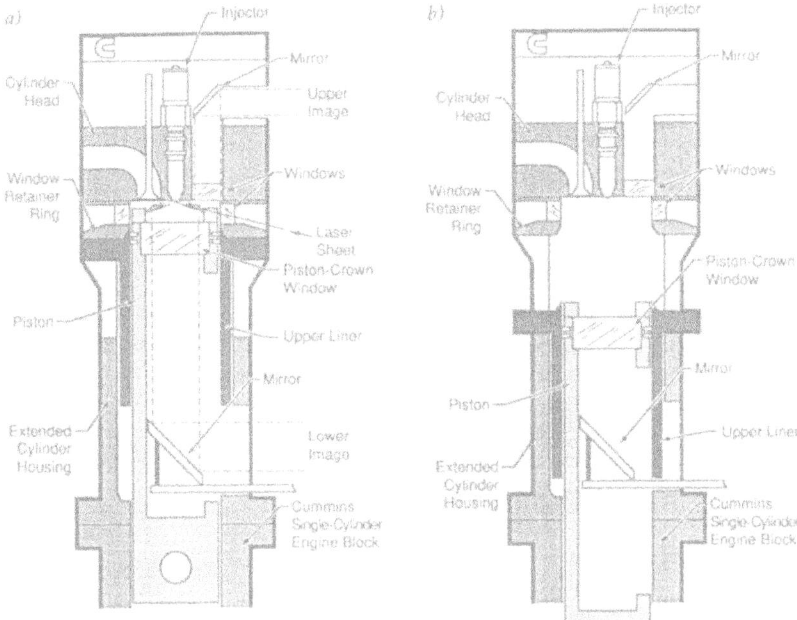

Figure 25. The Bowditch piston design of Espey and Dec (1993) for a direct-injection diesel engine: a) upper liner in the operating position; b) upper liner in the cleaning position, allowing access to interior optical surfaces through two large openings.

Although direct visualization apparently has never been performed in an unmodified engine, Clowater et al. (1992) have used a fiberscope installed in the spark plug hole to observe the motion of particles introduced in the intake charge of a motored engine. The cylinder of the engine was made of quartz, and a sheet of laser light normal to the cylinder axis was used to illuminate the particles. A fiberscope is a flexible device containing a coherent fiber bundle (Fig. 26*a*), that transmits an image from a receiving lens to an eyepiece or camera. Fiberscopes typically contain one or more additional incoherent fiber bundles (Fig. 26*b*) to transmit illuminating light to the object under observation. Fiberscopes as small as 6 mm in diameter are commercially available, such that a very feasible application would be to install a fiberscope in the spark plug hole to observe the spray pattern in a motored direct-injected engine. In addition, because these devices are so small, it would seem possible to build a functional spark plug containing a fiberscope protected by a window that could be used with combustion.

An optical fiber transmits light in the central portion of the fiber, called the core, which is surrounded by a layer of different refractive index, called the cladding, which does not transmit light. When packed into a bundle, typically 80-85% of the area of the bundle, called the packing fraction, is actually available for transmitting light. The lost area includes the cladding and the interstitial voids. Thus, the image quality obtained through a fiber-optic bundle suffers from the discretized nature of the core array and the lack of perfect coherence between the input and output ends of the fiber bundle (imperfections in the precise location of each fiber in the bundle). In addition, there are transmission losses on the order of 20% per meter that are not necessarily the same for each fiber in the bundle.

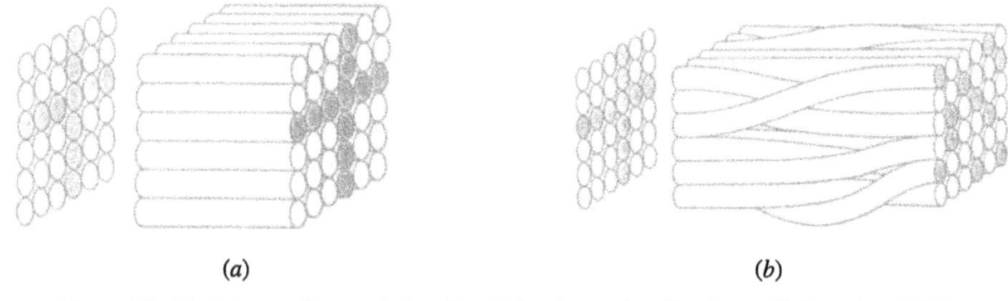

(*a*) (*b*)

Figure 26. (*a*) Coherent fiber-optic bundle; (*b*) Incoherent bundle; (from Chaimowicz, 1989).

These problems of image quality are all eliminated with technology like the Olympus Videoimagescope®. Rather than coupling the image to a camera through a fiber bundle, this device has the camera built into the tip of the flexible probe. The CCD (charge coupled device) image sensor is a 512 by 480 array of red, green and blue pixels that produces true-color video images in a probe as small as 6 mm diameter. Like the fiberscope, the Videoimagescope has the capability for external illumination of the object; unlike the fiberscope, it does not have guaranteed immunity from ignition-induced noise because it is a fully electrical device.

9. Summary

In reality, the ground rules established for the content of this review are too constraining to be applied to the decision process for instrumenting a production engine. For example, in-cylinder pressure measurements are far too valuable to permit any compromise in quality, so it is recommended that a hole be drilled for the installation of a research grade, water-cooled piezoelectric transducer. Similarly, if visualization of a direct-injection fuel spray might provide valuable insight, then install a fiberscope in the best possible location; the existing spark plug port should only be used if its field-of-view is satisfactory. On the other hand, the use of instruments like ionization probes in the head gasket, and any of the many instrumented spark plug probes reviewed, should be dramatically increased. It is one thing for researchers to demonstrate new tools, but it takes extensive use by practitioners before the tools ultimate value is reached. User-friendly analysis and interpretation procedures are essential before a diagnostic can be deemed as being truly useful. This review article has attempted to bring to the attention of the readers diagnostic tools that they may not have know of, or had forgotten about. The range of choices in this "for unmodified engines only" tool box is quite extensive, and it is apparent that only lack of imagination stands in the way of the tool box overflowing.

Acknowledgment

It is quite likely that this review article is not as complete as it could be, and I would like to offer my apologies to those authors whose contributions I have missed in my search for relevant work. I would also like to thank Paul Miles for all his help. This work was performed at the Combustion Research Facility of the Sandia National Laboratories, and was funded by the U.S. Department of Energy, Office of Transportation Technologies.

References

Accufiber (1989a) 'An Instrumented Spark Plug for In-Cylinder Optical Fiber Thermometer Measurement', Accufiber Application Note EAS100-A8, Luxtron Corporation, Beaverton, OR.

Accufiber (1989b) 'A New Knock Detection Technique', Accufiber Application Note EAS100-A9, Luxtron Corporation, Beaverton, OR.

Ahmad, T., Groff, E. G. and Myers, J. P. (1984) 'A Spark-Discharge Probe for Velocity Measurements in Engines', ASME Book No. G00279, Flows in Internal Combustion Engines II, p. 67.

Alyea, J. W., Uyehara, O. A. and Myers, P. S. (1969) 'The Development and Evaluation of an Electronic Indicated Horsepower Meter', *Trans. SAE* **78**, p. 784.

Amann, C. A. (1983) 'A Perspective of Reciprocating-Engine Diagnostics without Lasers', *Prog. Energy Combust. Sci.* **9**, p. 239.

Amann, C. A. (1985a) 'Classical Combustion Diagnostics for Engine Research', *Trans. SAE* **94**, Sec. 7, p. 168.

Amann, C. A. (1985b) 'Cylinder-Pressure Measurement and Its Use in Engine Research', SAE Paper No. 852067.

Anderson, R. L. (1986) 'In-Cylinder Measurement of Combustion Characteristics Using Ionization Sensors', *Trans. SAE* **95**, Sec. 3, p. 295.

Asanuma, T. and Obokata, T. (1979) 'Gas Velocity Measurements of a Motored and Firing Engine by Laser Anemometry', *Trans. SAE* **88**, p. 401.

Barton, R. K., Lestz, S. S. and Meyer, W. E. (1971) 'An Empirical Model for Correlating Cycle-by-Cycle Cylinder Gas Motion and Combustion Variations of a Spark Ignition Engine', *Trans. SAE* **80**, p. 695.

Berthold, J. W. III (1991) 'Industrial Applications of Fiber Optic Sensors', in E. Udd (ed.), Fiber Optic Sensors: An Introduction for Engineers and Scientists, John Wiley, New York, p. 415.

Bianco, Y., Cheng, W. K. and Heywood, J. B. (1991) 'The Effects of Initial Flame Kernel Conditions on Flame Development in SI Engine', *Trans. SAE* **100**, Sec. 3, p. 1852.

Brown, W. L (1967) 'Methods for Evaluating Requirements and Errors in Cylinder Pressure Measurement', *Trans. SAE* **76**, p. 50.

Burrows, M. C., Shimizu, S., Myers, P. S. and Uyehara, O. A. (1961) 'The Measurement of Unburned Gas Temperatures in an Engine by an Infrared Radiation Pyrometer', *Trans. SAE* **61**, p. 514.

Burstall, F. W. (1895) , *Phil. Mag.* **40**, p. 282.

Burstall, F. W. (1901) 'Second Report to the Gas-Engine Research Committee', *Proc. Inst. Mech. Engrs.*, parts 3-5, p. 1031.

Callendar, H. L. and Dalby, W. E. (1908) 'On the Measurement of Temperatures in the Cylinder of a Gas Engine', *Proc. Roy. Soc.*, Series A, **80**, p. 57.

Catania, A. E. (1982) '3-D Swirling Flows in an Open-Chamber Automotive Diesel Engine with Different Induction Systems', in T. Ukzan (ed.), Flows in I. C. Engines, ASME, New York, p. 53.

Catania, A. E. and Mittica, A. (1987) 'Induction System Effects on the Small Scale Turbulence in a High-Speed Diesel Engine', ASME Paper No. 87-ICE-38.

Chaimowicz, J. C. A. (1989) Lightwave Technology: an Introduction, Butterworths, London, p. 242.

Chou, T. (1993) 'Hydrocarbon Emission Sequence Related to Cylinder Mal-Distribution in a L-Head Engine', Ph.D. Dissertation, The University of Michigan, Ann Arbor, MI.

Clerk, D. (1882) 'The Theory of the Gas-Engine', *Proc. Instn. of Civil Engrs.* **69**, p. 220.

Clerk, D. (1912) Fifth Report Gaseous Explosions Committee, British Association Report; Reported in Bone, W. A. and Townend, D. T. (1927) Flame and Combustion in Gases, Longmans, Green and Co. Ltd., London, p. 243.

Clerk, D. (1921) 'Actions in Gas and Gasoline Engines', *SAE J.* **8**, p. 523.

Clowater, L., Hamady, F. and Schock, H. (1992) 'Fiber Optic Imaging System for Remote Location Flow Visualization Studies', SAE Paper No. 920305.

Coker, E. G. and Scoble, W. A. (1913-1914) 'Cyclical Changes of Temperature in a Gas-Engine Cylinder', *Proc. Inst. Civil Engrs.* **196**, part 2, p. 1.

Collings, N. (1988) 'A New Technique for Measuring HC Concentration in Real Time, in a Running Engine', SAE Paper No. 880517.

Collings, N., Dinsdale, S. and Eade, D. (1986) 'Knock Detection by Means of the Spark Plug', *Trans. SAE* **95**, Sec. 3, p. 969.

Collings, N., Dinsdale, S. and Hands, T. (1991) 'Plug Fouling Investigations on a Running Engine - An Application of a Novel Multi-Purpose Diagnostic System Based on the Spark Plug', *Trans. SAE* **100**, Sec. 4, p. 681.

Cummins, C. Lyle Jr., (1989) Internal Fire, Society of Automotive Engineers, Warrendale, p. 2.

Curry, S. (1963) 'A Three-Dimensional Study of Flame Propagation in a Spark Ignition Engine', *Trans. SAE* **71**, p. 628.

Dent, J. C. and Salama, N. S. 'The Measurement of Turbulence Characteristics in an Internal Combustion Engine', SAE Paper No. 750886.

Dils, R. R. (1991) 'Knock Detector Using Optical Fiber Thermometer', U. S. Patent No. 5,052,214.

Dils, R. R. and Moore, M. P. (1986) 'An Introduction to Optical Fiber Thermometer Measurements in Automotive Engines', Advances in Instrumentation, Vol. 41, Instrument Society of America, p. 1159.

Douaud, A., de Soete, G. and Henault, C. (1983) 'Experimental Analysis of the Initiation and Development of Part-Load Combustions in Spark-Ignition Engines', SAE Paper No. 830338.

Dyer, T. M. (1985) 'New Experimental Techniques for In-Cylinder Engine Studies', *Trans. SAE* **94**, Sec. 7, p. 196.

Espey, C. and Dec, J. E. (1993) 'Diesel Engine Combustion Studies in a Newly Designed Optical-Access Engine Using High-Speed Visualization and 2-D Laser Imaging', SAE Paper No. 930971.

Evers, L. W. (1978) 'Spark Plug Pressure Transducers for Measuring Indicated Work', SAE Paper No. 780148.

Gaydon, A. G. (1957) The Spectroscopy of Flames, John Wiley and Sons, NY, pp. 159-160.

Glyde, H. S. (1930) 'Experiments to Determine Velocities of Flame Propagation in a Side Valve Petrol Engine', *J. Inst. Pet. Technol.* **16**, p. 756.

Goto, S., Miyamoto, N. and Murayama, T. (1986) 'Improvement of the Dynamic Characteristics in the Connecting Passages for Measuring High Frequency Pressure Diagrams', *Trans. SAE* **95**, Sec. 5, p. 31.

Hahn, J. P. and Anderson, R. W. (1991) 'Correlation of Fiber Optic Spark Plug and Combustion Pressure Data on Two and Four Valve Per Cylinder Production Engines', Spring Central States Meeting of The Combustion Institute, April.

Hartman, P. G., Plee, S. L. and Bennethum, J. E. (1991) 'Diesel Smoke Measurement and Control Using an In-Cylinder Optical Sensor', *Trans. SAE* **100**, Sec. 3, p. 1259.

Hata, Y. and Asano, M. (1986) 'New Trends in Electronic Engine Control - To the Next Stage', *Trans. SAE* **95**, Sec. 3, p. 722.
He, G., Patania, A., Kluzner, M., Vokovich, D., Astrakhan, V., Wall, T. and Wlodarczyk, M. (1993) 'Low-Cost Spark Plug-Integrated Fiber Optic Sensor for Combustion Pressure Monitoring', SAE Paper No. 930853.
Herrin, R. J. (1984) 'An Application of Simple In-Cylinder Velocity and Temperature Measurement Techniques for Diagnosing Homogeneous-Charge-Combustion', General Motors Research Publication No. GMR-4911.
Hershey, A. E. (1936) 'Measurement of Gas Temperatures in an Internal Combustion Engine', *ASME Trans.* **58**, p. 195.
Hinze, P. and Cheng, W. K. (1993) 'Flame Kernel Development in a Methanol Fueled Engine', SAE Paper No. 932649.
Hutchinson, P., Morse, A. and Whitelaw, J. H. (1978) 'Velocity Measurements in Motored Engines - Experience and Prognosis', SAE Paper No. 780061.
Itoh, T., Nakada, T. and Takagi, Y. (1995) 'Emission Characteristics of OH and C_2 Radicals under Engine Knocking, *JSME International J.*, to be published.
Kalghatgi, G. T. (1992) 'A Study of Inlet System Detergency in a Gasoline Engine Using an Optical Method', SAE Paper No. 922256.
Kerstein, A. R. and Witze, P. O. (1990) 'Flame-Kernel Model for Analysis of Fiber-Optic Instrumented Spark Plug Data', SAE Paper No. 900022.
Kistler Instrument Corp. (1993a) 'Measuring Spark Plugs', *Automotive Engineering* **101**, No. 7, p. 92.
Kistler Instrument Corp. (1993b) 'Permanent Cylinder-Pressure Monitoring', *Automotive Engineering* **101**, No. 8, p. 60.
Klemm, A. (1964) A History of Technology.
Kondo, M., Niimi, A. and Nakamura, T. (1975) 'Indiscope--A New Combustion Pressure Indicator with Washer Transducers', SAE Paper No. 750883.
Kuratle, R. H. (1993) 'Measuring Spark Plugs with Integrated Cylinder Pressure Sensor', Experimental and Predictive Methods in Engine Research and Development, Institution of Mechanical Engineers Conference Publication C465/93, London, p. 27.
Kyuma, K. (1989) 'Physical and Chemical Sensors for Process Control', in B. Culshaw and J. Dakin (eds.), Optical Fiber Sensors: Systems and Applications, Artech House, Norwood, MA, p. 676.
Lancaster, D. R. (1975) 'Effects of Engine Variables on Turbulence in a Spark-Ignition Engine', *Trans. SAE* **85**, p. 671.
Lancaster, D. R., Krieger, R. B. and Lienesch, J. H. (1975) 'Measurement and Analysis of Engine Pressure Data', *Trans. SAE* **84**, Sec. 1, p. 155.
Lee, D. W. (1939) 'A Study of Air Flow in an Engine Cylinder', N.A.C.A. Report No. 653.
Lindvall, F. C. (1934) 'A Glow Discharge Anemometer', *Trans. Am. Inst. Elec. Eng.* **53**, p. 1063.
Lord, D. L., Anderson, R. W., Brehob, D. D. and Kim, Y. (1993) 'The Effects of Charge Motion on Early Flame Kernel Development', SAE Paper No. 930463.
Ma, T. (1991) 'Ionization Sensor', U. S. Patent No. 5,066,023.
MacKenzie, D. and Honaman, R. K. (1920) 'The Velocity of Flame Propagation in Engine Cylinders', *Trans. SAE* **15**, p. 299.

Maly, R., Meinel, H. (1981) 'Determination of Flow Velocity, Turbulence Intensity and Length and Time Scales from Gas Discharge Parameters', Proceedings of the 5th International Symposium on Plasma Chemistry, ISPC-5, Edinburgh, p. 552.

Maly, R., Meinel, H. and Wagner, E. (1983) 'Novel Method for Determining General Flow Parameters from Conventional Spark Discharges', Combustion in Engineering, Institution of Mechanical Engineers Conference Publication C67/83, London, p. 27.

Matsui, K., Tanaka, T. and Ohigashi, S. (1979) 'Measurement of Local Mixture Strength at Spark Gap of S. I. Engines', *Trans. SAE* **88**, Sec. 2, p. 1741.

McCullough, J. D. (1953) 'Engine Cylinder Pressure Measurements', *Trans. SAE* **61**, p. 557.

McDougal, J. A. (1990) 'Separation of Variables in an Ion Gap Controlled Engine', U. S. Patent No. 4,947,680.

Meyer, W. E. and DeCarolis, J. J. (1962) 'Compression Temperatures in Diesel Engines Under Starting Conditions', *Trans. SAE* **70**, p. 163.

Meyer, R. , Kubesh, J. T. and Shahed, S. M. (1993) 'Simultaneous Application of Optical Spark Plug Probe and Head Gasket Ionization Probe to a Production Engine', SAE Paper No. 930464.

Midgley, Jr., T. (1920) 'High-Speed Indicators', *Trans. SAE* **15**, p. 317.

Midgley, Jr., T. and McCarty, H. H. (1924) 'Internal-Combustion Engine Radiation Characteristics', *Trans. SAE* **19**, p. 18.

Miles, P. C., Green, R. M. and Witze, P. O. (1994) 'Two-Stroke Engine Blower and Crankcase Scavenging Flows Compared', *Combustion Research Facility News* **16**, No. 1, Sandia National Laboratories, Livermore, CA.

Miyata, S. Ito, Y. and Shimasaki, Y. (1993) 'Flame Ion Density Measurement Using Spark Plug Voltage Analysis', SAE Paper No. 930462.

Morris, J. E. (1987) ' Intra-Cylinder Combustion Pressure Sensing', *Trans. SAE* **96**, Sec. 5, p. 558.

Morris, J. E. and Li-Chi (1985) 'Improved Intra-Cylinder Combustion Pressure Sensor', *Trans. SAE* **94**, Sec. 3, p. 137.

Müller, K. Linder, E. and Maurer, H. (1983) 'Sensor Arrangement', U. S. Patent No. 4,393,687.

Nicholson, D. E. and Witze, P. O. (1993) 'Flame Location Measurements in a Production Engine Using Ionization Probes Embodied in a Printed-Circuit-Board Head Gasket', SAE Paper No. 930390.

Nutton, D. and Pinnock, R. A. (1990) 'Closed Loop Ignition and Fueling Control Using Optical Combustion Sensors', *Trans. SAE* **99**, Sec. 3, p. 1151.

Obert, E. F. (1973) Internal Combustion Engines and Air Pollution, Harper & Row, New York, p. 346.

Ohigashi, S., Hamamoto, Y. and Tanabe, S. (1968) 'Technique of Measuring Gas Stream Velocity by Means of Electric Discharges', *Bull. JSME* **11**, p. 1169.

Ohigashi, S., Hamamoto, Y. and Tanabe, S. (1969) 'New Digital Method for Measuring Gas Flow Velocity by Electric Discharge', SAE Paper No. 690180.

Ohyama, Y., Ohsuga, M. and Kuroiwa, H. (1990) 'Study on Mixture Formation and Ignition Process in Spark Ignition Engine Using Optical Combustion Sensor', *Trans. SAE* **99**, Sec. 3, p. 2002.

Pischinger, R., Kraßnig, G. and Glaser, J. (1985) 'Problems of Pressure Indication in Internal Combustion Engines', Proceedings of the International Symposium on Diagnostics and Modeling of Combustion in Internal Combustion Engines, COMODIA 90, September 3-5, Kyoto, Japan, p. 539.

Rado, W. G. and Johnson, W. J. (1975) 'Significance of Burn Types, as Measured by Using the Spark Plugs as Ionization Probes, with Respect to the Hydrocarbon Emission Levels in S. I. Engines', SAE Paper No. 750354.

Randall, K. W. and Powell, J. D. (1979) 'A Cylinder Pressure Sensor for Spark Advance Control and Knock Detection', *Trans. SAE* **88**, Sec. 1, p. 508.

Randolf, A. L. (1990) 'Cylinder-Pressure-Transducer Mounting Techniques to Maximize Data Accuracy', *Trans. SAE* **99**, Sec. 6, p. 201.

Rask, R. B. (1979a) 'Velocity Measurements inside the Cylinder of a Motored Internal Combustion Engine,' in H. D. Thompson and W. H. Stevenson (eds.), Laser Velocimetry and Particle Sizing, Hemisphere Press, Washington, p. 251.

Rask, R. B. (1979b) 'Laser Doppler Anemometer Measurements in an Internal Combustion Engine', *Trans. SAE* **88**, p. 371.

Remboski, D. J., Plee, S. L. and Martin, J. K. (1989) 'An Optical Sensor for Spark-Ignition Engine Combustion Analysis and Control', SAE Paper No. 890159.

Ricardo, H. R. (1927) 'Some Notes on Petrol Engine Development', *The Automobile Engineer* **17**, p. 149.

Robison, J. A., Behrens, M. D., Mosher, R. G. and Chandler, J. M. (1958) 'Investigating Combustion Phenomena in Unmodified Engines', *Trans. SAE* **66**, p. 549.

Rothrock, A. M. and Spencer, R. C. (1939) 'The Influence of Directed Air Flow on Combustion in a Spark-Ignition Engine', N.A.C.A. Report No. 657.

Sawamoto, K., Kawamura, Y., Kita, T. and Matsushita, K. (1987) Individual Cylinder Knock Control by Detecting Cylinder Pressure', *Trans. SAE* **96**, Sec. 5, p. 602.

Schäfer, H. J., Krull, O. and Maege, B. (1985) 'Thermal Stress on a Piezoelectric Pressure Transducer in the Combustion Chamber of an SI-Engine', *Trans. SAE* **94**, Sec. 3, p. 147.

Schnauffer, K. (1934) 'Engine-Cylinder Flame-Propagation Studied by New Method', *SAE J. (Trans.)* **34**, p. 17.

Semenov, E. S. (1958a) 'Device for Measuring the Turbulence in Piston Engines', *Instruments and Experimental Techniques* **1**, No. 1, p. 102.

Semenov, E. S. (1958b) 'Studies of Turbulent Gas Flow in Piston Engines', Otedelenie Technicheskikh Nauk, No. 8 (English Translation: NASA Technical Translation F97, 1963).

Shimasaki, Y., Kanehiro, M., Baba, S., Maruyama, S., Hisaki, T. and Miyata, S. (1993) 'Spark Plug Voltage Analysis for Monitoring Combustion in an Internal Combustion Engine', SAE Paper No. 930461.

Society of Automotive Engineers (1985) Engine Combustion Analysis: New Approaches, Warrendale.

Spicher, U. and Kollmeier, H.-P. (1986) 'Detection of Flame Propagation During Knocking Combustion by Optical Fiber Diagnostics', *Trans. SAE* **95**, Sec. 6, p. 552.

Spicher, U. Schmitz, G. and Kollmeier, H.-P. (1988) 'Application of a New Optical Fiber Technique for Flame Propagation Diagnostics in IC Engines', *Trans. SAE* **97**, Sec. 3, p. 803.

Stein, R. A., Mencik, D. Z. and Warren, C. C. (1987) 'Effect of Thermal Strain on Measurement of Cylinder Pressure', *Trans. SAE* **96**, Sec. 4, p. 442.

Takagi, Y. (1994) Nissan Motor Co., Ltd., Kanagawa, Japan, private communication.

Tsao, K. C., Myers, P. S. and Uyehara, O. A. (1962) 'Gas Temperatures During Compression in Motored and Fired Diesel Engines', *Trans. SAE* **70**, p. 136.

Vannobel, F. (1992) 'Backscatter Laser Doppler Velocimetry inside the Prechamber of a Diesel Engine', Proceedings of the Third International Conference on Vehicle Dynamics and Powertrain Engineering, June 11-13, Strasbourg, France.

Whitelaw, J. H. and Xu, H. M. (1993) 'An Experimental Study of Gas Velocity, Flame Propagation and Pressure in a Spark Ignition Engine', SAE Paper No. 932702.

Winch, R. F. and Mayes, F. M. (1953) 'A Method for Identifying Preignition', *Trans. SAE* **61**, p. 453.

Winsor, R. E. and Patterson, D. J. (1973) 'Mixture Turbulence - A Key to Cyclic Combustion Variation', *Trans. SAE* **82**, p. 368.

Withrow, L. , Lovell, W. G. and Boyd, T. A. (1930) 'Following Combustion by Chemical Means', *Ind. Eng. Chem.* **22**, p. 945.

Withrow, L. and Rassweiler, G. M. (1936) 'Slow Motion Shows Knocking and Non-Knocking Explosions', *Trans. SAE* **39**, p. 297.

Witze, P. O. (1975) 'Hot-Wire Turbulence Measurements in a Motored Internal Combustion Engine', Sandia Laboratories Report SAND75-8641.

Witze, P. O. (1979) 'Application of Laser Velocimetry to a Motored Internal Combustion Engine,' in H. D. Thompson and W. H. Stevenson (eds.), Laser Velocimetry and Particle Sizing, Hemisphere Press, Washington, p. 239.

Witze, P. O. (1980) 'A Critical Comparison of Hot-Wire Anemometry and Laser Doppler Velocimetry for I. C. Engine Applications', *Trans. SAE* **89**, p. 711.

Witze, P. O. (1989) 'Cycle-Resolved Multipoint Ionization Probe Measurements in a Spark Ignition Engine', *Trans. SAE* **98**, Sec. 3, p. 2074.

Witze, P. O. (1990) 'Calibration of Fiber-Optic Instrumented Spark Plug Velocity Measurements by Laser Velocimetry', Proceedings of the Fifth International Symposium on Application of Laser Techniques to Fluid Mechanics, July 9-12, Lisbon, paper 16.1.

Witze, P. O. and Bopp, S. C. (1991) 'Investigation of In-Cylinder Fluid Motion Using a Head Gasket Instrumented with Ionization Probes', *Trans. SAE* **100**, Sec. 3, p. 1227.

Witze, P. O. and Green, R. M. (1993) 'Determining the Location of End-Gas Autoignition Using Ionization Probes Installed in the Head Gasket', SAE Paper No. 932645.

Witze, P. O., Hall, M. J. and Bennett, M. J. (1989) 'Are Gas Velocities in a Motored Piston Engine Representative of the Pre-Ignition Fluid Motion in a Fired Engine?' *Experiments in Fluids* **8**, p. 103.

Witze, P. O., Hall, M. J. and Bennett, M. J. (1990) 'Cycle-Resolved Measurements of Flame Kernel Growth and Motion Correlated with Combustion Duration,' *Trans. SAE* **99**, Sec. 3, p. 74.

Witze, P. O., Hall, M. J. and Wallace, J. S. (1988) 'Fiber-Optic Instrumented Spark Plug for Measuring Early Flame Development in Spark Ignition Engines', *Trans. SAE* **97**, Sec. 3, p. 813.

Woodbury, C. A., Lewis, H. A. and Canby, A. T. (1921) 'The Nature of Flame Movement in a Closed Cylinder', *SAE Trans.* **15**, p. 299.

16. Cyclic Variation Due to Misfiring in a Small Two-Stroke Engine

T. Ohira*, Y. Ikeda and T. Nakagima+
*Suzuki Motor Corporation
Technical Engineering Division
P.O. Box 1, Hamamatsu 432-91
Japan
+Kobe University
Dept. of Mech. Engineering
Rokkodai, Nada, Kobe 657
Japan

ABSTRACT. The purpose of this study is to demonstrate cyclic variation of the combustion state including misfiring in a small two-stroke engine, focusing on the in-cylinder pressure variation and the gas constituent concentration at the exhaust port of the cylinder. It was found that the in-cylinder pressure at one point timing when its standard deviation became maximum value would be useful instead of indicated mean effective pressure (IMEP) in cycle-to-cycle varied combustion states. The cyclic variations of the CO and CO_2 emissions were demonstrated by the fast-response exhaust gas analyzer. The features of the pattern of cyclic variation of the CO and CO_2 emissions were similar to those of the IMEP. The difference level of CO and CO_2 emissions between the two successive cycles depended on the IMEP.

1. Introduction

Two-stroke engines for passenger vehicles and light electric power generators in space stations have been highlighted for their advantages such as high power output and its compactness. Developments of two-stroke engines depend on the achievement of hydrocarbon (HC) emission reduction and stable combustion (Duret, 1992,[1]). At present, conventional small two-stroke engines have a simple mechanism of piston-port valve, which make the control of flow in a two-stroke engine complicated in all operating conditions. Mostly, at low engine speed or partial load conditions, undesirable flow behavior might be observed during the scavenging process, as shown in Fig. 1. HC emission in the exhaust gas of the two-stroke engines is mainly caused by the short-circuited mixture gas through the cylinder to the exhaust pipe (Blair, 1990,[2], SAE publication, 1992[3], [4], Ohigashi, 1973,[5], Sher, et al., 1991,[6]) and the blow-out of unburnt-mixture gas after misfiring (Hamamoto, 1971,[7]) due to insufficient scavenging. The unstable combustion which generally occurred due to cyclic variation in a two-stroke engine can also easily induce noise (Gotoh, et al., 1993,[8]), vibration, knocking, auto-ignition problems (Patterson, 1966,[9], Aoyama, et al., 1977,[10], Tsuchiya, et al., 1983,[11]) or detonation which can damage the engine components. Therefore, cyclic variations of combustion including misfiring must be overcome for development of sophisticated engines.

Many investigations on cyclic variations of two-stroke engine combustion have been presented from the time-averaged mean value and standard deviation of operating conditions, namely IMEP, delivery ratio, air-fuel ratio, scavenging efficiency, and so on. Not withstanding these factors, the scavenging flow across the cylinder is very cardinal for understanding the phenomena of cyclic variation. Experimental and analytical investigations on the scavenging flow have been performed using various apparatus such as electrical

F. Culick et al., (eds.), Unsteady Combustion, 369–381.

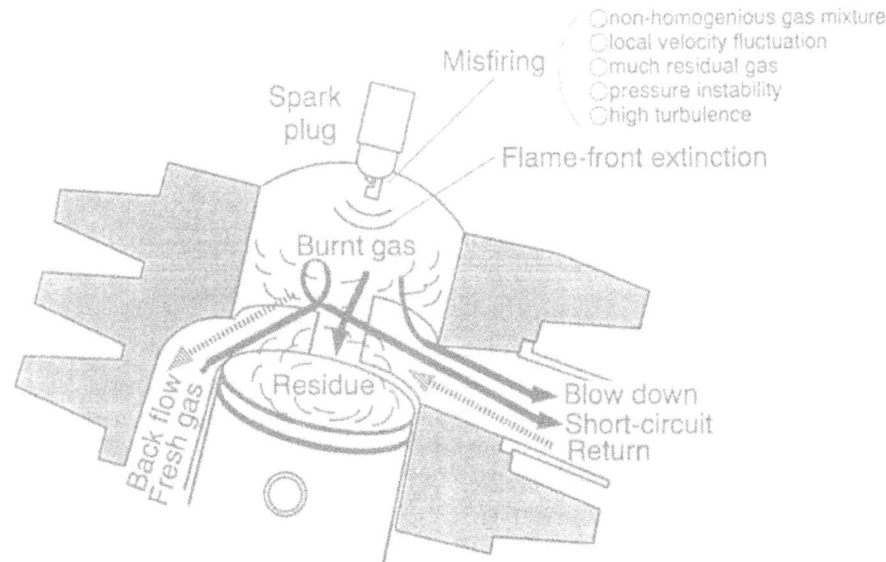

Fig.1 Illustration of flow in two-stroke engine

discharge velocimetry (Ohigashi, et al., 1969,[12]), LDV (Obokata, et al., 1987,[13], Fraser, et al., 1988,[14]) and computations (Sher, et al., 1990,[15], [16], Uzkan, 1988,[17], Abraham, et al., 1992,[18]) All of them discussed complicated flow behaviors in engines as averaged flow or steady flow, such that useful information to make clear causes and effect of cyclic variation could not be obtained from those reports.

To demonstrate cyclic variation in a two-stroke engine, the two-stage phenomena of cyclic variation should be understood in flow, pressure and concentration fields. One is the in-cylinder flow motion before ignition, that is, how the fresh unburnt-mixture gas flows into cylinder and the burnt gas blows down to the exhaust pipe or returns to be such as residual gas in the cylinder. This stage is the cause of cyclic variation. The other is the combustion intensity or heat release, that is, how the charged gas is ignited and the flame propagates. This stage can be regarded as the effect of cyclic variation.

Regarding the cyclic variation of flow behavior before ignition, the cycle-resolved analyses of the scavenging flow are needed at each cycle to observe the changes of in-cylinder gas motion. At present, however, the experimental measurements, e.g., LDV, are not powerful enough to satisfy both cycle-resolved and cyclic variation analysis in a practical fired engine. Although some papers reported high data rate of LDV measurements for a four-stroke engine (Bopp, et al., 1990,[19]) and I.C. engine (Fraser, et al., 1988,[14]), these papers did not show cyclic variation in the engine. In order to understand the cyclic variation and exactly support the developments of engines, we have the concept that the scavenging flow behavior with cyclic variation must experimentally be investigated by accurate temporal analysis in unison with variations of pressure and gas constituent concentration. Our research has focused on the scavenging flow in a practical two-stroke engine. A fiber-optic LDV system has been developed for the measurement of actual flow with wide velocity range in a fired engine (Ikeda, et al., 1991,[20],[21]). The system can also measure the pressure variations at each part in the engine, simultaneously with the velocity data. It has become clear that flows in the transfer port, at the exit of transfer port and in the exhaust pipe are greatly governed by the pressure propagation in the engine generated from combustion pressure (Ikeda, et al., 1991,[21], 1992,[22], Ohira, et al., 1993[23]). However, only these velocity data cannot yet demonstrate the scavenging flow behavior in the cylinder, and therefor are still insufficient to fully account for cyclic variation.

Concerning the cyclic variation of combustion state, temporary analyses of the heat release after ignition in each cycle must be done. Recently, some research groups tackled these in-cylinder phenomena through visualization analyses of the in-cylinder flow by the PIV technique (Green, et al., 1990,[24], Nino, et al., 1993,[25]), and the flame propagation (Witze, et al., 1988,[26], Winklhofer, et al., 1993,[27]). Nevertheless, these visualization techniques also have shortfalls, such as optical accessibility and limitation of measurement duration that are indispensable during the actual observation of the cyclic variation phenomena.

In our previous paper (Ikeda, et al., 1993,[28]), to understand the differences of the scavenging flow between firing and misfiring cycles, we have done conditional sampling on the flow velocity variations at the transfer port and the exhaust pipe under the partial load conditions, hence, succeeded in classifying the LDV data into misfiring and firing groups. However, the velocity variation of the scavenging flow across the cylinder is insufficient to completely help us to understand the flow behaviors, namely the short-circuit flow in a misfiring cycle. Gas constituent concentrations near the exhaust exit of the cylinder with a temporal and cycle-resolved analysis have to be undertaken in order to understand the period when the short-circuit flow and the burnt-gas return occur in relation to cycle-to-cycle combustion state. On the other hand, the detection method of combustion states or misfiring is very paramount to achieve the exact analysis of the flow data by conditional sampling. Pressures in the engine should be further used for the diagnoses of the flow variation and combustion fluctuation.

The purpose of this study is to demonstrate the cyclic variation of combustion states with respect to the in-cylinder pressure fluctuation and the relation between combustion state and emission level variation. Real-time measurements of in-cylinder pressure and CO and CO_2 concentrations near the exhaust exit of the cylinder were simultaneously carried out. A fast-response exhaust gas analyzer (Takeda, et al., 1991,[29]) with a non dispersive infrared (NDIR) type was used for observation of the cyclic variation of CO and CO_2 emission levels. The misfiring effect was investigated from the variations of CO and CO_2 concentrations.

2. Experimental Apparatus

The engine used in this study is a conventional crankcase-compressed loop-scavenged two-stroke motorcycle engine (Suzuki Motor Corporation : AX-100). The main specifications of the engine are shown in Table 1. A cross-sectional layout of the engine is illustrated in Fig. 2. The engine was operated under firing conditions at the engine speeds of 1,000 r/min to

Table 1 Engine specification

Type of Engine	Two-Stroke Spark Ignition Crankcase Compression with Reed Valve Schnürle Scavenging	
Displacement	98.2 (cm^3)	
Bore × Stroke	50.0 (mm) × 50.0 (mm)	
Exhaust Port Timing	Open	97.0 (C.A.)
	Close	261.7 (C.A.)
Scavenging Port Timing	Open	122.7 (C.A.)
	Close	240.0 (C.A.)
Ignition Timing	22.0 (C.A.) B.T.D.C. (4000r/min)	
Compression Ratio	6.59	

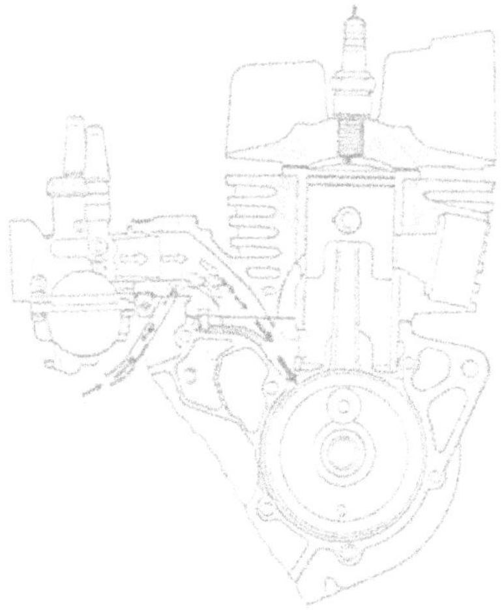

Fig.2 Layout of test engine

3,000 r/min. The throttle-opening ratio was 10 or 20 %. The operating conditions of the test engine are shown in Table 2. The fuel used in this experiment was a mixture of unleaded gasoline and motor oil (in the ratio of 40:1).

The measurement system in the present study is shown in Fig. 3. In-cylinder pressure measurements were carried out using a piezoelectric pressure transducer and a charge amplifier.

Gas near the exhaust exit of the cylinder was sampled and analyzed using a fast-response exhaust-gas analyzer (Horiba, Ltd. : MEXA-1300FRI). The main specification of this gas analyzer is shown in Table 3. The gas-sampling-probe orientation was set, as shown in Fig. 4, in order not to disturb the exhaust gas flow but to obtain large variation of the emission concentrations before the blow-out gas was mixed with the residual gas in the exhaust port, so that the measured emission level might represent the actual gas concentration. The gas was drawn into the sampling probe with 2-mm inner diameter and 6-mm outer diameter and chilled by the cold water flowing through the probe jacket to separate H_2O vapor from the gas. The sampling passage from the probe inlet to the concentration sensing cell is approximately 1 m long and 2-mm inner diameter except one point just behind the filter that is 1 mm diameter. This short and narrow passage design prevents the sampling gas from mixing and realizes the fast gas sampling. The sample volume was approximately 1.5 l/min which corresponded to 1.5 cm^3/cycle at the engine speed of 1,000 r/min. The drawing velocity of the sampling gas was 8.0 m/s at the probe inlet. Such small sample volume and velocity do not affect the flow behavior in the engine or the combustion states.

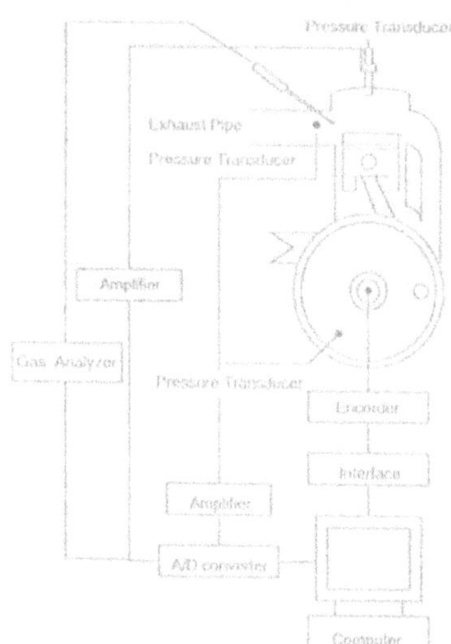

Fig.3 Mesurement system

Table 2 Engine Operating Condition

Engine Speed (r/min)	1000	1500	3000
Throttle Opening Ratio (%)	10	10	20
BMEP (MPa)	0.028	0.009	0.024
Delivery Ratio	0.16	0.13	0.23

Table 3 Gas analyzer specifications

Model	MEXA-1300FRI		
Measurement Method	NDIR		
Constituent	CO	CO_2	HC
Measurable Concentration	0-12%	0-16%	0-2%
Response Time	$T_D \leq 100ms$ $T_{90} \leq 30ms$ $T_{63} \leq 5ms$		

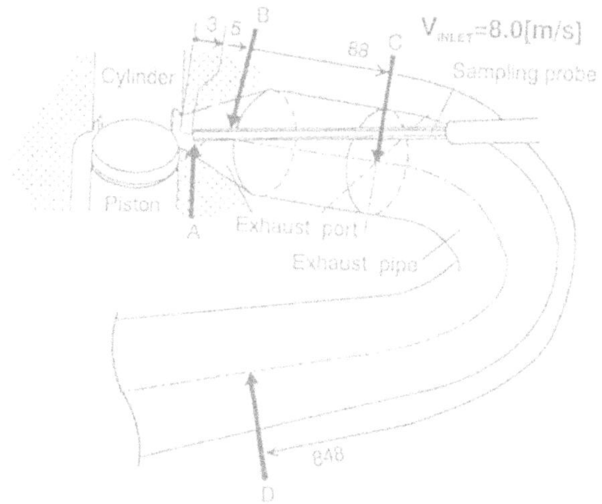

Fig.4 Gas sampling position

3. Results and Discussion

In order to demonstrate the cyclic variation of both the in-cylinder pressure and emission concentrations when misfiring or firing occurred, the in-cylinder pressure and the emission levels were measured simultaneously together with the crankangle degree. First, cyclic variation of the in-cylinder pressure will be discussed from its variation and IMEP. Next, CO and CO_2 concentration variations near the cylinder exit will be presented in comparison to the in-cylinder pressure variations or the cycle-to-cycle IMEP.

3.1 IN-CYLINDER PRESSURE

Misfiring in a two-stroke engine is caused by many factors in a cylinder, that is, much residual burnt gas due to insufficient scavenging, lack of fresh mixture with proper air-fuel ratio near the spark plug electrodes, quenching due to too much heat release to the spark plug and the cylinder wall, extinction due to high turbulence flow, and so on. Furthermore, these factors do not always lead to complete flame extinguish. These factors or formation of unburnt mixture and burnt gas in the cylinder simply control the size of the flame, ignition delay and flame propagation. Each cycle has different in-cylinder pressure variation under partial load condition, as shown in Fig. 5. It is very difficult to monitor combustion states in the cylinder and detect misfiring or incomplete firing from in-cylinder pressure measurements only.

However, misfiring detection is essential for analyzing the LDV measurement results whose data rate is not enough to satisfy the cycle-to-cycle analysis at the present, in order to achieve conditional sampling of the LDV data. It is well known that misfiring detection should be done using the IMEP (Shiomoto et al., 1978,[30]). The calculation of the IMEP requires the availability of the in-cylinder pressure data at every crank angle in each cycle at least. On the other hand, long duration of measured LDV-data evaluation imposes a limitation to memory capacity of the measuring system.

We found the misfiring detection method with the reduced number of the in-cylinder pressure data, considering memory capacity of the measurement system. The RMS. variation of the in-cylinder pressure was already shown in Fig. 5. The time at the maximum RMS variation was at ATDC 28 CA. The cycle-to-cycle variation of the in-cylinder pressure at

ATDC 28 CA (P_{CL28}) is shown together with the IMEP changes in Fig. 6. The changes in the P_{CL28} and the IMEP corresponded very well. The relationship between values of the P_{CL28} and the IMEP for each cycle are shown in Fig. 7. A linearity correlation between the P_{CL28} and the IMEP can be seen. This implies that the in-cylinder pressure at the time of the maximum RMS variation is a useful tool for the misfiring detection in experimental measurements.

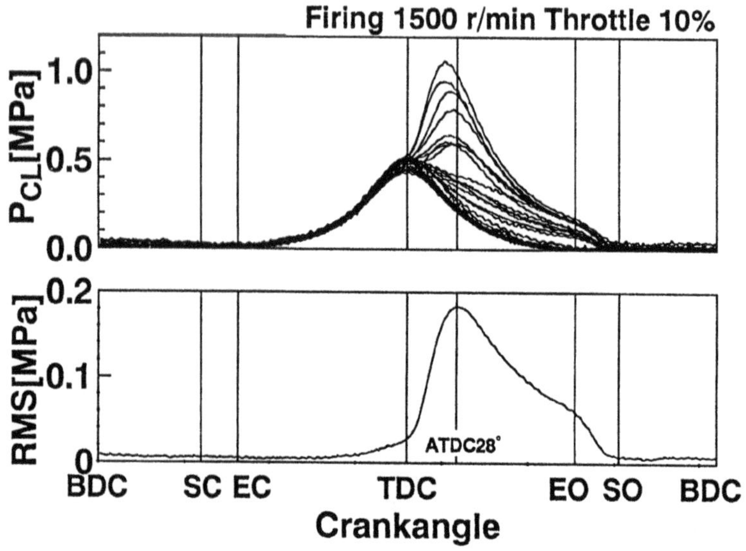

Fig.5 In-cylinder pressure variations and RMS

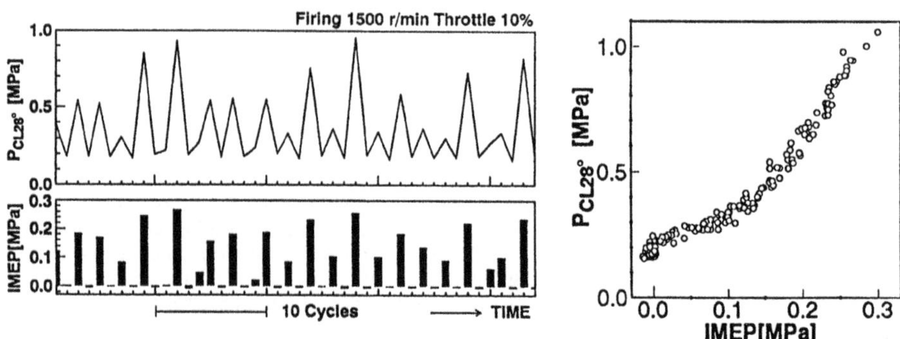

Fig.6 Cycle-to-cycle in-cylinder pressure at time of maximum RMS variation

Fig.7 Relation of in-cylinder pressure at time of maximum RMS to IMEP

3.2 TIME HISTORY OF EMISSION LEVEL

First of all, the fast-response exhaust gas analyzer had to be examined regarding its limitations of time response, optimum position of gas sampling, and time delay of gas sampling between the probe inlet and the sensing cell in the application to the practical fired engine. In this section, attention is paid to the emission variations in comparison to the in-cylinder pressure.

3.2.1 *Time Delay of Measurement Response*

The time-response of the gas analyzer in this study is displayed in Table 3. The used gas analyzer has a limitation of response time, as indicated in Table 3, namely, 90 %-response time is 30 msec and 63 %-response time is 5 msec. This response was examined using zero-span gas. In actual application to the practical fired engine, the response time should be confirmed. The emissions variation of CO and CO_2 at the engine speed of 1,500 r/min and 20 % throttle-opening ratio when the engine was successively fired, not ignited (misfired) and re-ignited (fired) are shown in Fig. 8.

At both conditions, when the engine was turned off and ignited again after being motored, the results of the time-delay showed the same time. The time delays of the measurement response showed 100 msec. The transitional variation of the emission level were different. After the ignition was turned off, the residual burnt gas in the exhaust pipe mixed with the unburnt gas flowing from the cylinder into the exhaust pipe, so that the CO and CO_2 concentration gradually decreased. On the contrary, after the engine was re-ignited, the unburned gas in the exhaust pipe mixed with the large volume of burnt gas blown out of the cylinder, therefore the CO and CO_2 level immediately increased.

3.2.2 *Variations at Different Positions*

The cyclic variation of the emissions at four positions were measured at 1,500 r/min and 10 % throttle-opening ratio, as shown in Fig. 9. The four positions of A, B, C and D were 3, 5, 88, and 848 mm from the exhaust exit of the cylinder. As shown in Fig. 9, the variation of CO and CO_2 emissions at A, B and C positions showed the same periodic time. This means that there was no periodical fluctuation of the CO and CO_2 variation. This is due to the absence of large recalculating flow in the exhaust pipe as presented in the LDV measurement results of our previous paper (Ohira , et al., 1993,[23]), the blown gas did not mix with residual gas in the exhaust pipe.

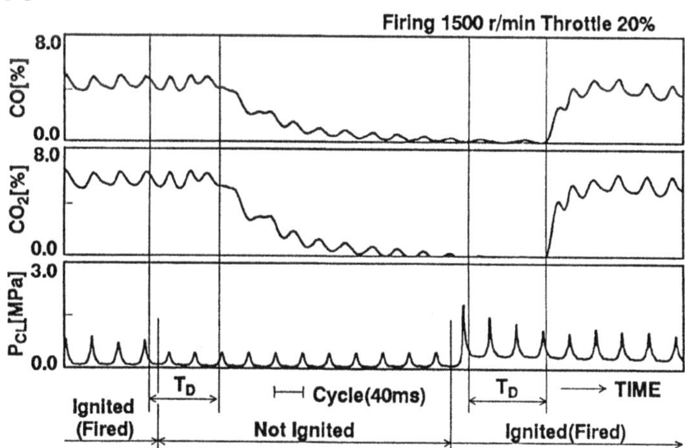

Fig.8 Time response of emission measurement

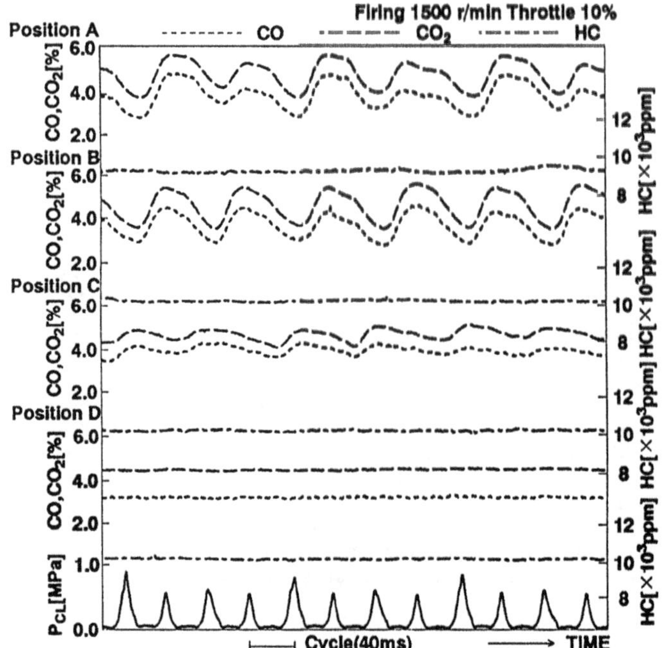

Fig.9 Emission variations at different positions

As shown in Fig. 9, position A, at which the emission level showed the largest variation, was most suitable for measuring cyclic variation of the emission level.

Even at position A, the HC emission level did not show any variation. Gas cooling by the NDIR method, as explained in the previous section, caused condensation of the high level concentration of unburnt-mixture gas in the sampling passage. A hydrogen flame ionization detector (FID) type that the sampling gas was drawn into a sensing cell after the gas was warmed up in a sampling passage will be useful for the measurement of the HC emission variation in a two-stroke engine. This study will focus on a decrease of CO and CO_2 emissions as an increase of unburnt HC emission, while discussing the measurement results of CO and CO_2 emission at point A.

3.2.3 *Emissions and in-cylinder pressure in different conditions*

The CO and CO_2 emission variations in the cases of (a) 1,000 r/min and 10 % throttle-opening ratio, (b) 1,500 r/min and 10 %, (c) 3,000 r/min and 20 % are shown together with the in-cylinder pressure (P_{CL}) variation in Fig. 10. One cycle takes 60 msec at 1,000 r/min and 20 msec at 3,000 r/min, so that it was not possible to achieve the cycle-resolved measurement, but the cycle-to-cycle variation can be seen in the figures under all conditions. The CO and CO_2 emission variation show similar curves. These emission variations and the P_{CL} variation have the same pattern of cycle-to-cycle variation at each condition: (a) shows a repeated pattern consisting of three cycles, (b) shows a repeated pattern consisting of four cycles and (c) shows a repeated pattern consisting of two cycles, respectively.

The mechanism of the emission variation between a misfired cycle and a fired cycle using the expanded drawing of the CO_2 variation in Fig. 10 (b) is shown in Fig. 11. The judged results are also shown in Fig. 11 as misfired and fired. In a fired cycle, the burnt gas blew out of the cylinder after exhaust-port opening (EO) and the CO_2 emission in the exhaust gas was generated by combustion and immediately increased as shown at point A. The CO_2 emission kept increasing after exhaust-port close (EC) till top dead center (TDC) due to the

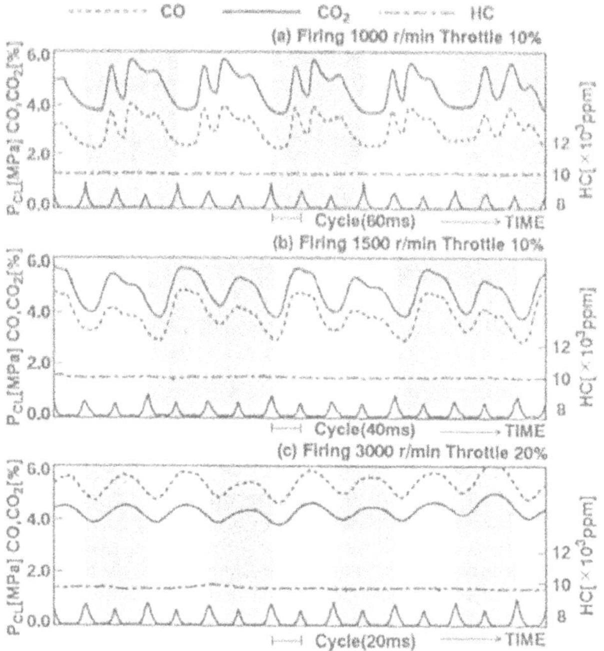

Fig.10 Time history of emissions and in-cylinder pressure in different conditions

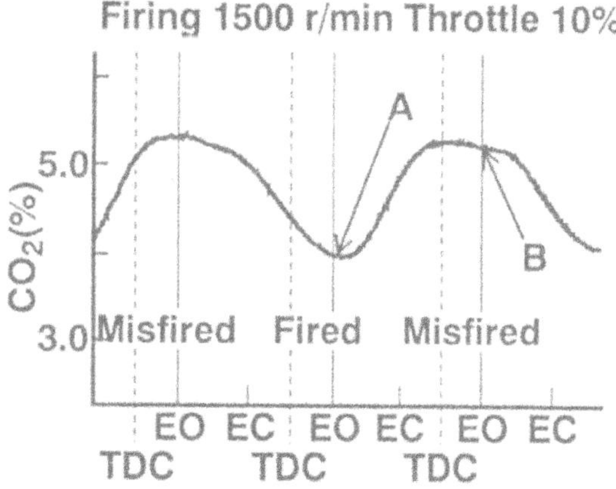

Fig.11 Emission variation between fired and misfired cycle

insufficient time response of the gas analyzer. In a misfired cycle, the CO_2 emission level decreased after EO as shown at point B, because the charged gas was considered to consist of some unburnt-mixture gas and the fresh short-circuit gas flowing into the exhaust port. The CO_2 emission, however, did not immediately decrease after EO due to two flow behaviors. One is the return flow of burnt gas from the exhaust pipe to the cylinder after bottom dead center (BDC) in the last firing cycle, and the other is the delay in starting of the blow-down flow from the cylinder to the exhaust port in misfired cycles. These flow behaviors have been reported in our previous paper (Ikeda , et al.,1993 ,[28]).

3.3 CORRELATION BETWEEN EMISSIONS AND COMBUSTION STATES

The difference levels of CO and CO_2 emissions between the two successive cycles induced firing or misfiring were demonstrated with respected to the intensity of the IMEP, and then the relation between CO / CO_2 as the combustion characteristics and the IMEP was investigated, in order to demonstrate the correlation between the emission and the combustion states.

First, the cycle-to-cycle differences of the CO and CO_2 emissions (ΔCO, ΔCO_2) at EO between the latest cycle and the subsequent cycle (i.e. A and B points in Fig. 11) are plotted in relation to the IMEP of the latest cycle, as shown in Fig. 12. The high correlation between ΔCO or ΔCO_2 and the IMEP can be apparently confirmed at 1,000 and 1,500 r/min. Because the IMEP is regarded as intensity of temperature, pressure and oxidization in combustion, both CO and CO_2 increased in proportion to the combustion intensity.

The reason why the correlation at 3000 r/min became lower than that at 1,500 r/min might be the limitation of the time-response and the time-delay adjustment of the sampling. The direct and sufficient time-resolved measurements of the exhaust gas analysis are required for the cycle-resolved investigations of high speed engines.

Finally, the time history of the ratio of CO to CO_2 was investigated with respect to the IMEP fluctuation, as shown in Fig. 13. The periodicity of the CO / CO_2 ratio equaled to the pattern duration of the cycle-to-cycle variation of IMEP at 1,000 and 1,500 r/min. The large increment of the CO / CO_2 ratio appeared in high IMEP cycles. This means that the intense combustion after the misfiring cycle led to high CO emission. The large increment of the CO / CO_2 may have been induced by richer air-fuel ratio gas mixture, lower temperature combustion or other factors which are hard to point out. These results show that there was no relation between the individual IMEP and the CO / CO_2 variation. The relation between IMEP and CO / CO_2 should be understood in the context of cyclic variation pattern of the combustion states.

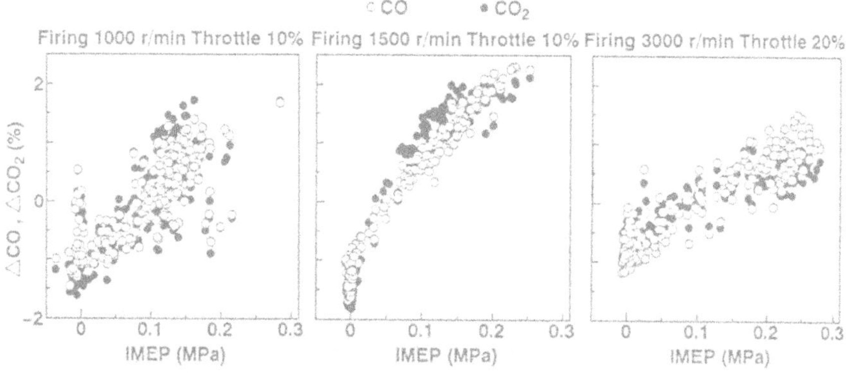

Fig.12 Correlation between emissions and IMEP

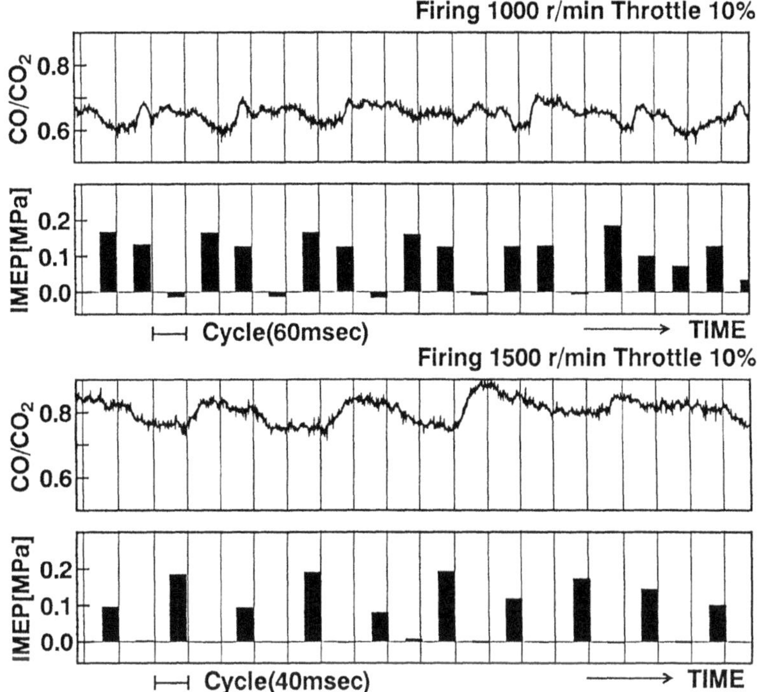

Fig.13 Variation of CO/CO_2 and IMEP

4. Conclusion

The cyclic variation of combustion states in a partially loaded two-stroke engine was investigated with respect to the in-cylinder pressure and CO or CO_2 emissions. The cyclic variations of the CO and CO_2 emission levels at the exhaust exit of the cylinder were measured by a fast-response exhaust gas analyzer. The knowledge obtained in the present experimental study is summarized as follows:

(1) The misfiring could not be detected by the in-cylinder pressure, but the in-cylinder pressure at the time of the maximum RMS variation could be an alternative criterion for the judgment of combustion states instead of the IMEP.

(2) The cyclic variations of CO and CO_2 emissions were demonstrated by the fast-response exhaust gas analyzer. The HC emission variation, however, could not be observed by the gas analyzer of NDIR type.

(3) The periodical pattern of the cyclic combustion variation can be estimated by the pattern of the cyclic variations of CO and CO_2 emissions and the CO / CO_2 ratio.

(4) The obvious correlation between the variation of the cycle-to-cycle level of the CO and CO_2 emissions and the IMEP demonstrated that firing increased the CO and CO_2 emission to the subsequent cycle in proportion to the IMEP intensity, and misfiring decreased the CO and CO_2 concentrations, that is, balk unburnt gas flew out to the exhaust port in the misfired cycles.

References

1. Duret, P., (1992) "The Key Points for the Development of an Automotive Spark Ignition Two-Stroke Engine," I Mech E, C389/278.

2. Blair, G. P., (1990) "The Basic Design of Two-Stroke Engines," SAE, Publication.

3. SAE, Publication, (1992) "Two-Stroke Engine Diagnostics and design," SAE, SP-901.

4. SAE, Publication, (1992) "Two-Stroke Small Engine and Emission Reduction," SAE, SP-883.

5. Ohigashi, S., (1973) "Irregularity Combustion in Engine Cylinder," (in Japanese) Trans. of JSAE, Vol.27, No. 27.

6. Sher, E., Hacohen, Y., Refael, S. and Harari, R., (1991) "Minimizing Short-Circuiting Losses in 2-S Engines by Throttling the Exhaust Pipe," SAE Paper No.901665.

7. Hamamoto, Y., (1971) "The measurement method of Combustion and Scavenging Pressure in a Two-Stroke Engine," (in Japanese) Trans. of Internal Combustion Engine, Vol.10, No.116.

8. Gotoh, T. and Maeda, O., (1993) "Reduction of Disagreeable Idle Sound in Two-Stroke Engines," SAE Paper No.930981.

9. Patterson, T. J.,(1966) "Pressure Variations, A Fundamental Combustion Problem," SAE Paper No.660129.

10. Aoyama, T.,Nakajima, M., Onishi, M. and Nagahiko, S., (1977) "Abnormal Combustion in Two-Stroke Motor Cycle Engines," SAE Paper No.770189, SAE Trans.

11. Tsuchiya, K., Nagai, Y. and Gotoh, T., (1983) "A Study of Irregular Combustion on Two-Stroke Cycle Gasoline Engines," SAE Paper No. 830091.

12. Ohigashi, S., Hamamoto, Y. and Tanabe, S., (1969) "New Digital Method for Measuring Gas Flow Velocity by Electrical Discharge," SAE Paper No. 690180.

13. Obokata, T., Hanada N. and Kurabayashi, T., (1987) "Velocity and Turbulence Measurements in a Combustion Chamber of S. I. Engine under Motored and Firing Operations by L.D.A. with Fiber-Optic Pick-Up," SAE Paper No.870166.

14. Fraser, R. A. and Bracco, F. V., (1988) "Cycle-Resolved LDV Integral Length Scale Measurements in an I.C. Engine," SAE Paper No.880381.

15. Sher, E., Hossain, I., Zhang, Q., and Winterbone, (1991) "Calculations and Measurements in the Cylinder of a Two-Stroke Uniflow-Scavenged Engine Under Steady Flow Conditions," Experimental Thermal and Fluid Science 1991, Vol.4,PP418-431.

16. Sher, E., (1991) "A Simple and Realistic Model for the Scavenged Two-Stroke Cycle Engine," IMechE, Vol.205 PP129-137.

17. Uzkan, T., (1988) "Analytically Predicted Improvements in the Scavenging and Trapping Efficiency of Two-Cycle Engine," SAE Paper No.880108.

18. Abraham, M. and Prakash, S. (1992) "Cyclic Variations in a Small Two-Stroke Cycle Spark-Ignited Engine - An Experimental Study," SAE Paper No.920427.

19 Bopp, S., Durst, F., and Tropea, Cam., "In-Cylinder Velocity Measurements with a Mobile Fiber Optic LDV System," SAE Paper No.900055.

20. Ikeda, Y., Hikosaka, M. and Nakajima, T., (1991) "Scavenging Flow Measurements in a Motored Two-Stroke Engine by Fiber LDV," SAE Transaction Journal of Engine, Vol.100, PP981-989, SAE Paper No.910669.

21. Ikeda, Y., Hikosaka, M., Nakajima, T. and Ohira, T., (1991) "Scavenging Flow Measurements in a Fired Two-Stroke Engine by Fiber LDV," SAE Transaction, Journal of Engine, Vol.100, PP990-998, SAE Paper No.910670.

22. Ikeda, Y., Ohira, T., Takahashi, T. and Nakajima, T., (1992) "Flow Vector Measurements at the Scavenging Ports in a Fired Two-Stroke Engine," SAE Transaction, Journal of Engine, Vol.101, PP635-645, SAE Paper No.920420.

23. Ohira, T., Ikeda, Y., Takahashi, T., Ito, T. and Nakajima, T., (1993) "Exhaust Gas Flow Behavior in a Two-Stroke Engine," SAE Transaction, Journal of Engine, Vol.102, SAE Paper No.930502.

24. Green, R. H. and Cousyn, B. J., (1990) "An Optical Research Engine for the Study of Two-Stroke Cycle In-Cylinder Phenomena," COMODIA 90, p.347-352.

25. Nino, E., Gajdeczko, B. F. and Felton, P. G., (1993) "Two-Color Particle Image Velocimetry in an Engine with Combustion," SAE Paper No.930872.

26. Witze, P. O., Hall, M. J. and Wallace, J. S., (1988) "Fiber-Optic instrumented Spark Plug for Measuring Early Flame Development in Spark Ignition Engines," SAE Paper No.881638.

27. Winklhofer, E., Philipp, H., Fraidl, G.and Fuchs, H., (1993) "Fuel and Flame Imaging in SI Engines," SAE Paper No.930871.

28. Ikeda, Y., Ohira, T., Takahashi, T. and Nakajima , T., (1993) "Misfiring Effects on Scavenging Flow at Scavenging Port and Exhaust Pipe in a Small Two-Stroke Engine," SAE Transaction, Journal of Engine, Vol.102, SAE Paper No.930498.

29 Takeda, K. and Aoki, J., (1991) "Exhaust Gas Measurement in Transient Mode by Fast Response Exhaust Gas Analyzer," (in Japanese) Trans. of Internal Combustion Engine, Vol.30, No.380.

30. Shiomoto, G .H., Sawyer, R. F. and Kelly, B. D., (1978) "Characterization of the Lean Misfire Limit," SAE Paper No.780235.

17. CYCLE-RESOLVED TWO-DIMENSIONAL LASER-INDUCED FLUORESCENCE MEASUREMENTS OF FUEL/AIR RATIO CORRELATED TO EARLY COMBUSTION IN A SPARK-IGNITION ENGINE

H.-M. Neij,* B. Johansson**, and M. Aldén*

*Dept. Comb. Physics
Lund Inst. of Tech.
PO Box 118, S-22100 Lund
Sweden

**Dept. of Heat and Power Engng (Combust. Engines)
Lund Inst. of Tech.
PO Box 118, S-22100 Lund
Sweden

*Dept. Comb. Physics
Lund Inst. of Tech.
PO Box 118, S-22100 Lund
Sweden

ABSTRACT. Planar Laser-Induced Fluorescence is used for two-dimensional visualization of the fuel concentration field near the spark gap prior to ignition in a spark-ignition engine. Special care has been taken to make the images, after post-processing and calibration, as quantitative as possible. Fuel concentration data are correlated to the early flame development through heat release data from pressure registrations. A strong correlation is found between the mean fuel concentration near the spark gap and the mean rate of combustion. No influences of small scale fuel inhomogeneity on combustion instability could be determined. At operating conditions resulting in homogeneous and reproducible fuel distributions, only very weak correlations can be found as expected. In these cases the flow velocity close to the spark plug becomes increasingly important.

1. Introduction

Motivated by the desire and need to understand combustion processes, several laser-based techniques have been developed and applied to combustion situations, see e.g. [1]. The application of laser techniques provides high temporal as well as spatial resolution. Also the non-intrusiveness and the possibility for species selectivity provide attractive features. During the last decades the Laser-Induced Fluorescence technique (LIF) has been developed, from a spectroscopic technique for point measurements, into a powerful tool for two-dimensional visualization of concentration fields of radicals and stable species, e.g. [2]. This technique generally uses sheet illumination from a pulsed laser source with right-angle detection of the fluorescence emitted from molecules in the beam path. Normally, a solid-state array detector is utilized for the detection of the scattered light.

The LIF technique has been widely used for combustion engine diagnostics, e.g. mapping of OH [3,4] and O_2 [5] concentration fields. It has also been demonstrated for visualization of fuel vapor or liquid/vapor phase fuel distributions, e.g. [5-7].

One of the major problems of combustion in spark-ignition engines is the lack of reproducibility of the event from cycle to cycle. Generally, there are considered to exist three major causes for this combustion instability [8]:

1. The total amounts of fuel, air and burned gas in the cylinder are not the same from cycle to cycle.

F. Culick et al., (eds.), Unsteady Combustion, 383–389.

2. The charge is inhomogeneous. This is especially important close to the spark plug.
3. The flow field in the engine varies from cycle to cycle.

All three do to some extent influence the very early flame development. This early flame development is especially sensitive to perturbations due to the smallness of the flame. The flame speed in the early stage of flame propagation is below the fully developed turbulent flame speed and can, to a first approximation, be considered as laminar [9].

The laminar flame speed is strongly dependent on the fuel/air ratio and the residual gas concentration [10]. The local fuel/air ratio close to the spark plug will consequently influence the flame development period. The relative importance of these three causes is not well understood but it is likely that the dominant cause changes depending on the operating parameters of the engine.

This paper is dealing with the influence of cyclic variations in the fuel concentration near the spark gap on the early flame development in spark-ignition engines. Two-dimensional visualization of the fuel concentration field is accomplished with LIF where special care has been taken to obtain as quantitative results as possible.

2. Experimental

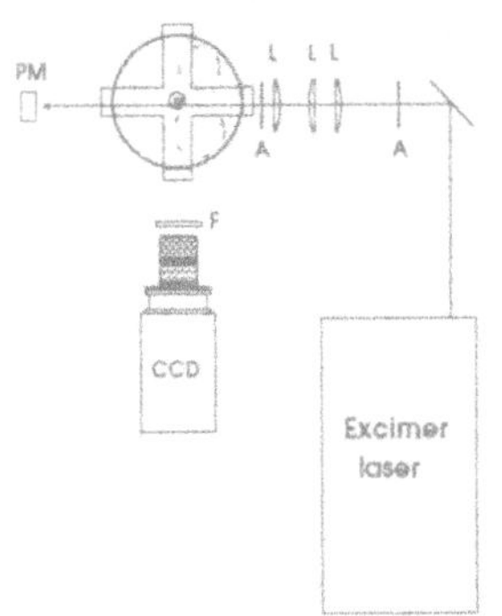

Figure 1. Experimental setup for the planar laser-induced fluorescence measurements. The positions of the spark plug and the inlet and exhaust valves are indicated. L: Lens, A: Aperture, F: Filter, PM: Power Meter.

The measurements were performed in a single-cylinder engine based on a six-cylinder VOLVO TD 102 diesel engine. Its main geometric properties are given in Table 1. Optical access was provided by four rectangular quartz windows in a spacer between the cylinder head and engine block. To maintain a reasonable compression ratio the original piston was extended. A disc-shaped combustion chamber geometry was used, where a swirl ratio of 2.8 was generated in the standard diesel intake port. The engine was fueled on iso-octane and run at a speed of 700 rpm. To reduce the influence of different amounts of residual gas from cycle to cycle, the engine was run in a skip-fire mode in which the engine was fired for 7 cycles and then motored for 3 cycles.

The pressure in the cylinder was measured with an AVL QC42D(X) piezo-electric transducer connected to a Kistler 5001 charge amplifier. The charge amplifier voltage output was connected to a 486/33 PC clone with a Data Translation DT2823 100 kHz 16-bit A/D-card.

The local fuel concentration was measured with Planar Laser-Induced Fluorescence (PLIF). The setup is given in Figure 1. Molecules of a species of interest are excited by the incoming laser light, which is formed into a vertical sheet, and the emitted fluorescence is collected at right angles. As excitation source a tunable excimer laser (Lambda Physik, EMG 150 MSC) was used yielding radiation around 248 nm with a maximum output energy of about 250 mJ/pulse. The laser sheet was placed approximately 1 mm in front of the spark plug. The fluorescence was imaged onto a two-dimensional detector, an image-intensified, fiber-coupled, 14-bit CCD camera (Princeton Instruments, ICCD-576S/RB-T) enabling acquisition of two-dimensional concentration distributions prior to ignition in the engine. The iso-octane was doped with approximately 2% 3-pentanone for simplified visualization. This compound is ideally suited for doping of iso-octane, concerning for example vapor pressure characteristics and molecular diffusion constants. Previous spectroscopic investigations performed on possible dopants indicate weak pressure and temperature dependencies [11]. The absorption at the excitation wavelength is very low for the concentrations and distances used in the engine experiments. Furthermore, the LIF signal intensity has shown a linear dependence on the laser intensity. For more details on the development of this technique, we refer to [12].

TABLE 1. Geometric properties of the engine.

Displaced volume	1600 cm^3
Bore	120.65 mm
Stroke	140 mm
Connection rod	160 mm
Compression ratio	7:1

3. Data processing

3.1. POST-PROCESSING OF 2-D FUEL DISTRIBUTIONS

Customized software was used for post-processing of the 2-D distributions. To eliminate influences from inhomogeneities in the spatial laser intensity distribution, the images were divided with a reference mode taken from accumulated images of uniform fuel distributions. Fluctuations in the total laser light energy were monitored on a shot-to-shot basis by a power meter and the images were afterwards normalized by the laser energy.

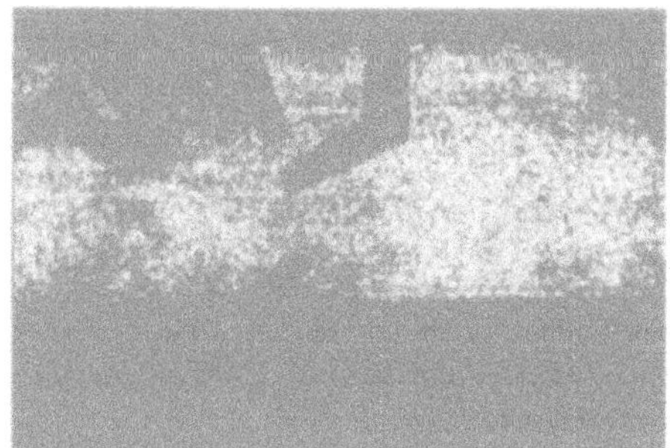

Figure 2a. Distribution of fuel, iso-octane + 2% 3-pentanone, 1 CAD before ignition. The area in front of the spark plug has been removed.

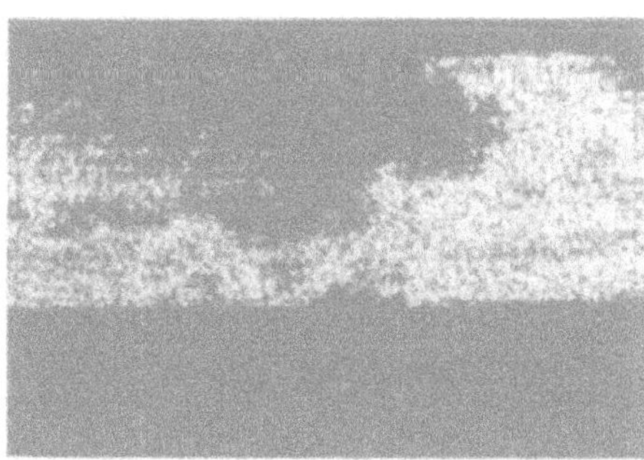

Figure 2b. Fuel distribution in another cycle within the same experiment.

The images from the LIF measurements were calibrated by means of using images taken with the engine running on a very homogeneous charge. The fuel/air ratio was simultaneously determined by exhaust gas analysis. Several images were accumulated and the mean intensity for a section of the image was used to convert fluorescence intensity to fuel/air equivalence ratio. During the experiments the LIF signal proved to depend linearly on the fuel concentration.

Examples of fuel distribution maps for two different cycles in the same experiment are shown in Figures 2a and 2b. Black corresponds to the lowest and white to the highest fuel concentration. The fuel distributions were captured 1 CAD before ignition. The ignition system was triggered at 14 CAD BTDC (Before Top Dead Center). To eliminate interferences from fluorescence scattered from the electrodes, the area in front of the electrodes was removed before evaluation. The total image area corresponds to 20×30 mm^2 inside the combustion chamber.

3.2. ONE-ZONE HEAT RELEASE MODEL

To extract information on the early flame development, a cycle-resolved heat-release calculation was performed. The parameters in the model are tuned in against the last motored cycle in the skip-fire sequence until the apparent heat released during motoring is below a maximum level [13]. As it is the very early flame, which is most sensitive to changes in the laminar flame velocity, it is desirable to detect the heat released as early as possible. The measurement noise does, however, set a minimum level which in this case was approximately 10-15 J. This corresponds to 0.5 % of the total energy released of the combustion. Figure 3a shows an example of the accumulated heat released for some 30 cycles when the engine is running lean. The ignition system was in this case triggered at 20 CAD BTDC and the 0.5% heat released was detected at approximately 10 CAD BTDC. Figure 3b shows the early phase of the combustion and the level of 0.5% heat released is indicated with a horizontal line.

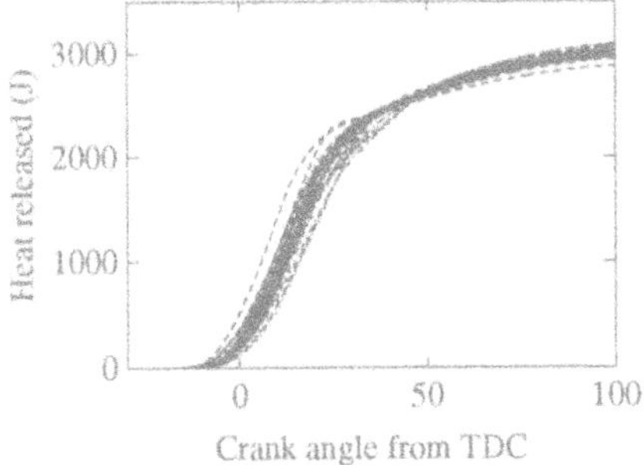

Figure 3a. Accumulated heat released for 30 cycles.

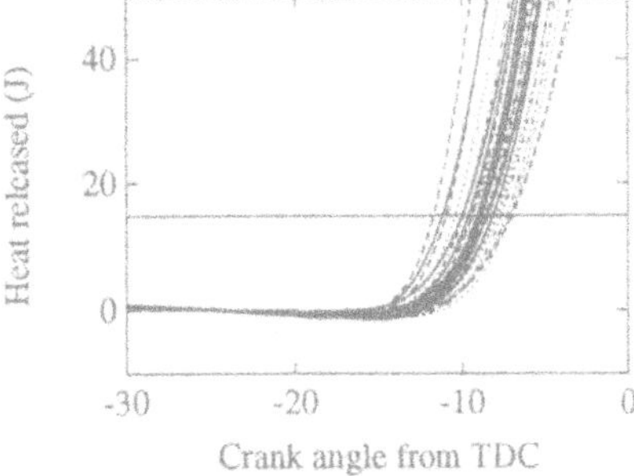

Figure 3b. Accumulated heat released in the early phase of the combustion. The horizontal line indicates the level of 0.5% of the total heat released.

4. Results

From the concentration maps obtained with the PLIF technique, the average fuel/air equivalence ratio was computed *within* a circle centered between the electrodes of the spark plug. The average fuel/air ratio within a circular area centered in the spark gap for increasing radial distance in the concentration map in Figure 2b is shown in Figure 4a.

For each radius the mean fuel/air ratio of all images (~50) within an experiment was correlated to the position of 0.5% heat released (CAD after ignition). A typical plot of the crank angle position of 0.5% heat released as a function of the derived mean equivalence ratio is found in Figure 4b. In Figure 4b is also indicated the second-order polynomial fitted to the experimental data. The corresponding correlation coefficient obtained as a function of radius is found in Figure 4c. The correlation coefficient depicts the square root of the ratio of the variance explained by the regression analysis to the total variance. For a more detailed description of the regression analysis, see e.g. [14].

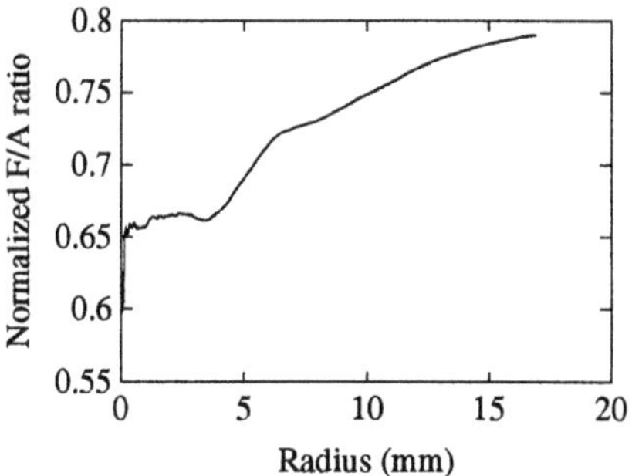

Figure 4a. The average fuel/air equivalence ratio within a circular area centered in the spark gap for increasing radius in the concentration distribution in Figure 2b.

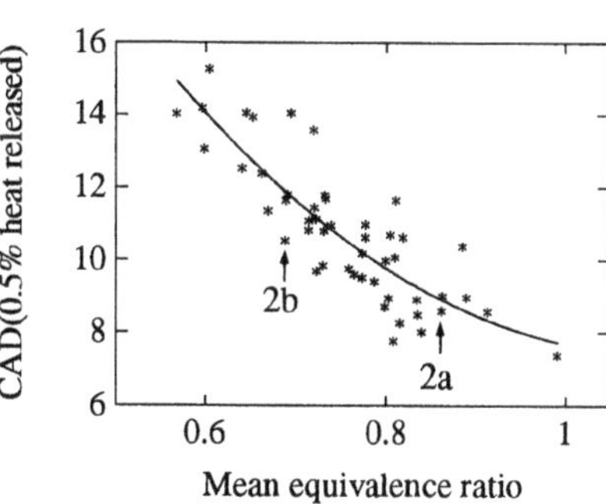

Figure 4b. Scatter plot of the crank angle position of 0.5% heat released as a function of derived mean fuel/air equivalence ratio. The line shows the second-order polynomial fit. The data points corresponding to the distributions in Figures 2a and 2b are also indicated.

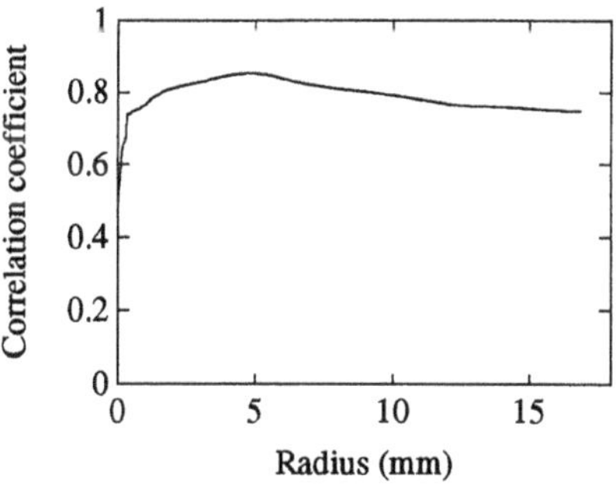

Figure 4c. The correlation coefficient as a function of radial distance from the spark gap.

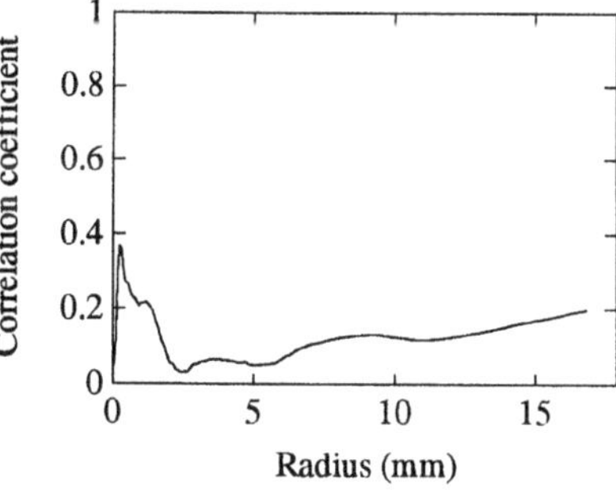

Figure 4d. The correlation coefficient as a function of radius in the case of a homogeneous and reproducible fuel/air mixture.

The reason for cyclic variations in the mean fuel concentration near the spark plug is attributed to variations in the large scale flow history from injection to ignition. There is also the possibility

that cyclic variations in the mean concentration can arise from cyclic variations in the total amount of injected fuel. However, the total amount of heat released, which could be used as an indicator of the total amount of injected fuel, exhibits cyclic variations of less than 1%.
An attempt was also made to investigate possible influences from small scale inhomogeneities in the fuel distribution on the cyclic combustion variations. However, no significant cycle-resolved correlations could be found between combustion parameters and the standard deviation or the normalized standard deviation in the fuel distribution maps ("the local inhomogeneity"). Thus, it seems, at least from this crude evaluation, that the influence from small scale inhomogeneity on combustion instability is too small to be detected at our engine conditions. This conclusion is furthermore supported by the fact that the increased combustion variations experienced when changing from a homogeneous to an inhomogeneous charge could be almost totally explained by the regression analysis, i.e. the variations in the mean equivalence ratio in the vicinity of the spark gap. The early flame development appears to be controlled by the mean fuel concentration near the spark gap; not the specific distribution of the fuel. However, more work on evaluation techniques will be needed before any definitive statements can be made.
In the case of a reproducible and homogeneous mixture, produced by fuel injection far from the engine intake port (>3 m), the correlation coefficient drops to a non-significant level as other parameters than the local fuel/air equivalence ratio become important in governing variations in the rate of combustion. The correlation coefficient as a function of radius around the spark-plug with a reproducible and homogeneous charge is shown in Figure 4d.

5. Conclusions

Laser-Induced Fluorescence can give useful two-dimensional images of the fuel concentration field in an operating engine when seeding iso-octane with a reliable tracer, 3-pentanone, in moderate quantities. Special care must also be taken to minimize artifacts due to inhomogeneities in the laser-mode cross section and shot-to-shot variations in laser light energy.
A surprisingly high correlation could be found between the mean concentration measured in a two-dimensional plane at 1 CAD before ignition and the early three-dimensional flame development which progress 8 to 15 CAD before being detected.
The high correlation indicates that one of the major reasons for the cyclic variations in the early flame development is variations in the local fuel/air equivalence ratio in the vicinity of the spark plug when the engine is operated on an inhomogeneous charge. The mean combustion rate seems to scale with the fuel/air ratio.
The large variations in the mean fuel concentration at the spark gap prior to ignition are most likely not due to variations in the amount of injected fuel but rather to cyclic variations in the flow history from injection to ignition.
The variations in the mean fuel concentration exert such a large influence on the combustion instability that the effect from small scale inhomogeneity, i.e. smaller than approximately 6 mm, is probably too small to be detected at our engine conditions.
At operating conditions resulting in homogeneous and reproducible fuel distributions, only very weak correlations could be found as expected. In these cases the flow velocity close to the spark plug becomes increasingly important as shown in previous LDV measurements in the engine [13].

6. Acknowledgments

This work was part of a CEC collaboration within the JRC-Homogeneous Combustion. We are indebted to this organization as well as to the Swedish Board for Industrial and Technical Developments, NUTEK, for financial support. We also thank Dr. I. Magnusson, VOLVO AB, for advice and helpful discussions.

7. References

[1] Eckbreth, A.C. (1987) Laser diagnostics for combustion and species, Abacus press, Cambridge.
[2] Hanson, R.K. (1986) 'Combustion diagnostics: Planar imaging techniques', Twenty-first Symposium (International) on Combustion/The Combustion Institute, Pittsburgh, pp. 1677-1691.
[3] Felton, P.G., Mantzaras, J., Bomse, D.S. and Woodin R.L. (1988) 'Initial two-dimensional laser induced fluorescence measurements of OH radicals in an internal combustion engine', SAE Paper 881633.
[4] Suntz, R., Becker, H., Monkhouse, P. and Wolfrum, J. (1988) 'Two-dimensional visualization of the flame front in an internal engine by laser-induced fluorescence of OH radicals', Appl. Phys. B47, 287-293.
[5] Andresen, P., Meijer, G., Schlüter, H., Voges, H., Koch, A., Hentschel, W., Oppermann, W. and Rothe, R. (1990) 'Fluorescence imaging inside an internal combustion engine using tunable excimer lasers', Appl. Opt. 29, 2392-2404.
[6] Baritaud, T.A. and Heinze, T.A. (1992) 'Gasoline distribution measurements with PLIF in a SI engine', SAE Paper 922355.
[7] Bardsley, M.E.A., Felton, P.G. and Bracco F.V. (1988) '2-D visualization of liquid and vapor fuel in an I.C. engine', SAE Paper 880521.
[8] Heywood, John B. (1989) Internal Combustion Engine Fundamentals, McGraw-Hill, Singapore.
[9] Beretta, G.P., Rashidi, M. and Keck, J.C. (1983) 'Turbulent flame propagation and combustion in spark ignition engines', Combustion and Flame 52, 217-245.
[10] Gülder, O.L. (1984) 'Correlation of laminar data for alternative SI engine fuels', SAE Paper 841000.
[11] Neij, H., Ossler, F., Agrup, S. and Aldén, M., in preparation.
[12] Neij, H. (1993) 'Development of laser-induced fluorescence for combustion engine diagnositics', Licentiate thesis, Department of Combustion Physics, Lund Institute of Technology, Sweden.
[13] Johansson, B. (1993) 'Correlation between velocity parameters measured with cycle-resolved 2-D LDV and early combustion in a spark-ignition engine', Licentiate thesis, Department of Heat and Power Engineering, Lund Institute of Technology, Sweden.
[14] Draper, N. and Smith, H. (1981) Applied regression analysis, second edition, John Wiley & Sons, New York.

Part III Advanced Tools: Experimental Techniques and Modelling Schemes

18. LASER DIAGNOSTICS FOR TEMPERATURE AND SPECIES IN UNSTEADY COMBUSTION

A. C. Eckbreth
United Technologies Research Center,
411 Silver Lane,
East Hartford, Connecticut 06108
USA

ABSTRACT. Nonintrusive laser probing of unsteady combustion processes provides the capability for the remote, in–situ, spatially and temporally precise measurement of, among other parameters, gas temperature and species concentrations. The predominant spatially–precise laser diagnostics for temperature and species include spontaneous Raman scattering (RS) and coherent anti–Stokes Raman spectroscopy (CARS) for measurements of major, i.e., $\geq\sim$ 1%, constituents, and laser–induced fluorescence spectroscopy (LIFS) and degenerate four wave mixing (DFWM) for minor, ppm level, species. Rayleigh scattering can provide total density measurements. These techniques are reviewed, including two–dimensional implementations, their state of maturity assessed, and major remaining research issues identified.

1. Introduction

Laser diagnostic techniques (Eckbreth, 1988; Taylor, 1993) are assuming an ever–increasing role in probing the hostile combustion processes characteristic of practical engines be they internal combustion, gas turbine or advanced propulsion, e.g., ramjets, scramjets, rockets. Physically intrusive probes can seriously perturb the combustion behavior being investigated, are often limited in spatial resolution and temporal response, and may not survive at high temperatures and pressures. Laser approaches are, by definition, non–intrusive and are capable of probing very high temperature/pressure environments. They permit remote, non–perturbing, in–situ examination of combustion phenomena and are capable of simultaneously high spatial (mm^3) and temporal resolution (10^{-8} to 10^{-6} sec), particularly important attributes for characterizing unsteady combustion processes. This paper will attempt to place the primary laser diagnostic approaches for spatially–precise temperature and species measurements in perspective, overview their capability and assess their state of maturity. The techniques to be described are spectroscopically–based and directly measure internal energy state populations, i.e., vibrational, rotational. From Boltzmann statistics, temperature derives from knowing the relative populations of two or preferably many of these internal energy states. Since the rotational modes equilibrate very rapidly, i.e., after a few collisions, measurement of the rotational temperature yields the translational temperature. Total concentration derives from knowledge of the absolute concentration in a single state and the temperature, or from integrating over all the individual state populations. In almost all practical combustion situations, Boltzmann statistics apply. Non–equilibrium distributions, should they exist, would be readily apparent in the spectral signatures and would be amenable to analysis as well.

Major species measurements and thermometry generally employ Raman–based approaches such as spontaneous Raman scattering (RS) and coherent anti–Stokes Raman spectroscopy (CARS). For low concentration, minority species measurements, diagnostic techniques need to

F. Culick et al., (eds.), Unsteady Combustion, 393–410.

exploit electronic resonances in the atoms/molecules to achieve the requisite sensitivities. The most prominent, spatially–resolved approaches for minority species are laser–induced fluorescence spectroscopy (LIFS) and degenerate four wave mixing (DFWM). RS and LIFS are incoherent processes, Fig. 1, whose signal properties derive from the nature of dipole radiation.

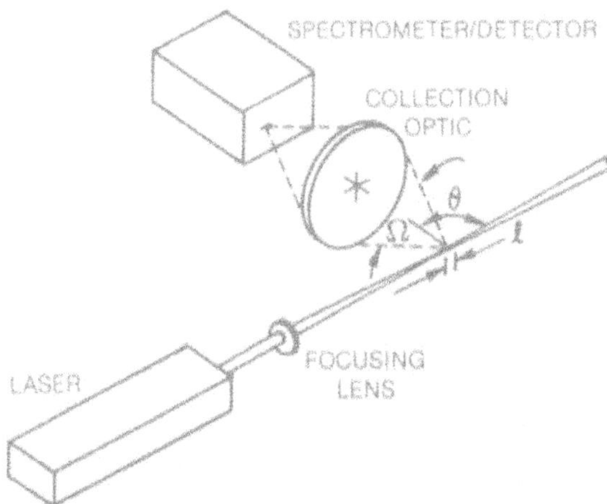

Figure 1. Typical geometric arrangement for incoherent laser light scattering diagnostic approaches.

Signal is scattered into essentially 4 π sr and the measurement volume is defined by the triangulation of the focused laser excitation beam and the optical signal collection system. Incoherent techniques lend themselves readily to two–dimensional field measurements, Fig. 2, via planar laser sheet illumination of the medium under study. Due to its weakness, 2D Raman imaging is seldom practical however.

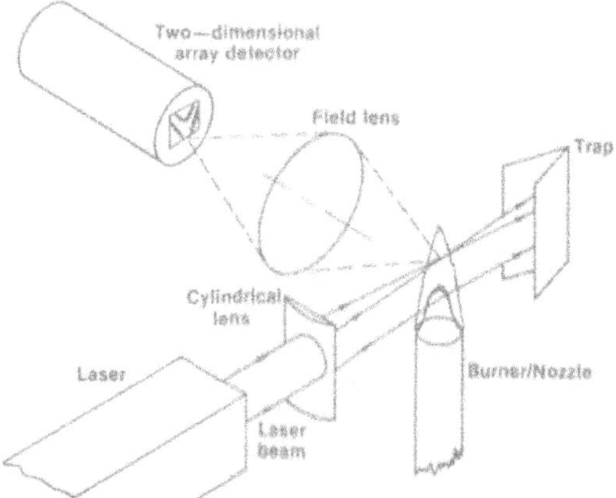

Figure 2. General approach to planar, two dimensional laser–illuminated imaging using incoherent light scattering.

CARS and DFWM are coherent processes originating from the third order nonlinear response of the probed species. Laserlike signal beams result, Fig. 3, and spatial resolution derives from the intersection region of the laser beams being "wave mixed." Coherent approaches possess the disadvantage of requiring opposed optical ports, are more difficult to calibrate due to line coherence and intensity effects, but are generally required for probing high interference environments, e.g., hydrocarbon–fueled diffusion flames, or for high pressure applications with restricted optical access (Eckbreth, 1988, 1989). In general, incoherent techniques are simpler to employ, calibrate and interpret. They are usually the diagnostic of first choice assuming the interferences, both naturally occurring and laser–induced, are tolerable. Incoherent methods are most often performed with illumination and signal collection at right angles (Fig. 1), although they are amenable to aligned illumination and collection axes, i.e., backscattering. All techniques are capable of two–dimensional field imaging; however, only electronic resonance methods, i.e., LIFS, DFWM, possess sufficient signal levels in moderate pressure combustion situations to be practically applicable. The four techniques are summarized in the diagnostic matrix shown in Fig. 4 which illustrates their general applicability. Capabilities can change with variants of the techniques however. For example, DFWM performed in the infrared can be used to probe major combustion constituents (Vander Wal, et al., 1992) while CARS can be electronically–resonantly enhanced to achieve minor constituent sensitivities (Druet, et al., 1978).

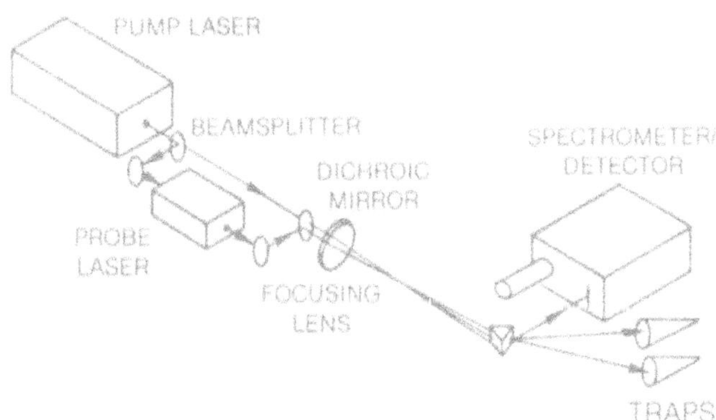

Figure 3. Generic arrangement for coherent wave mixing diagnostic approaches.

In the sections which follow, each of the foregoing techniques will be overviewed and their applicability to unsteady combustion devices and applications will be discussed. By design, the discussion will not go into detailed depth and the reader is encouraged to consult the cited references for in–depth treatments. Technique state of maturity will be assessed and major remaining research issues will be identified. Raman–based approaches for major species and temperature will be first considered followed by the electronic–resonance methods for minority species. In each area, the incoherent technique will be discussed first, hence the sequence will be RS, CARS, LIFS and DFWM. In conjunction with the Raman scattering discussion, the non–species–specific approach of Rayleigh scattering will also be included.

Level \ Approach	Incoherent	Coherent
Major species	Spontaneous Raman	CARS
Minor constituents	LIFS	DFWM

Figure 4. Typical diagnostic capabilities of incoherent and coherent laser diagnostic techniques. Code: CARS, coherent anti–Stokes Raman scattering; LIFS, laser–induced fluorescence spectroscopy; DFWM, degenerate four wave mixing.

2. Spontaneous Raman Scattering (RS)

2.1. BACKGROUND

Raman scattering is the inelastic scattering of light from molecules as illustrated in Fig. 5 and is termed rotational, vibrational or electronic depending on the nature of the energy exchange which occurs between the molecule and incident photons (Long, 1977). For combustion probing, ro–vibrational Raman processes are most widely employed. Light downshifted in frequency is termed Stokes scattering; light upshifted in frequency, which requires energy uptake from the molecule, is termed anti–Stokes and is observed only at high enough temperatures to produce sufficient excited state populations. Because so few molecules of the probed constituent actually participate in the scattering process, no significant perturbation of the medium occurs. The process is essentially instantaneous and arises due to the beat frequency of the oscillating dipole induced in the molecule by the incident light wave electric field and the molecular vibrations. Due to the quantization of the molecular energy states, the Raman spectrum resides at fixed frequency separations from the laser line characteristic of the molecule from which the scattering emanates. All molecules are Raman–active and simultaneously scatter at their distinctive Raman–shifted frequencies; hence, the technique is intrinsically capable of simultaneous, multiple species measurements. The Raman scattering is linearly proportional to the species number density and, with proper detection, many species can be monitored since spectral interferences between Raman bands in combustion gases are rare. The strength of the scattering scales as the fourth power of the Raman frequency mandating the use of energetic lasers in the visible or, preferably, in the ultraviolet. Unfortunately, Raman scattering is very weak resulting in Raman signal efficiencies, i.e., collected Raman energy to incident laser energy, of 10^{-15} to 10^{-16} for the major constituents in atmospheric pressure combustion with visible sources. This translates into signal levels from each constituent in the range of several hundred to a few thousand photons per laser pulse. Because of these low signal levels, broadband detection approaches, i.e., spectrally integrated, are employed for single pulse measurements. Band resolved measurements can be used generally only in laminar flames and involve averaging periods of one to several minutes.

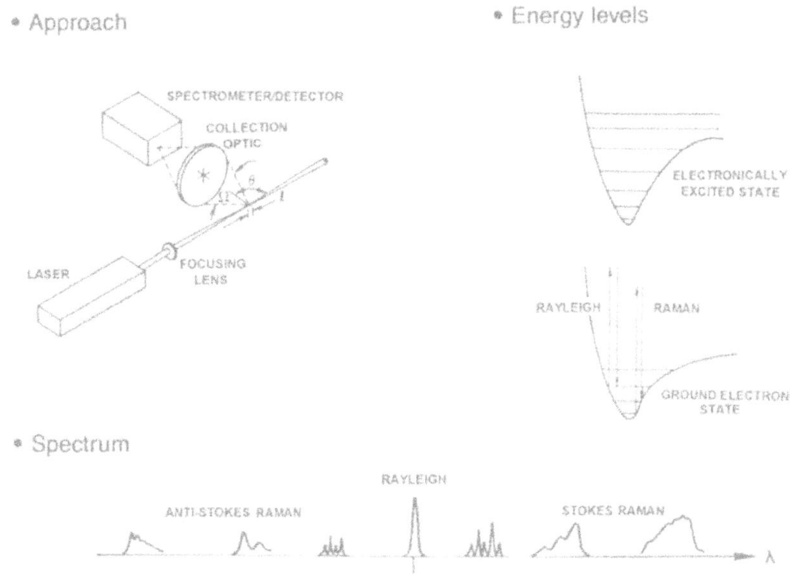

Figure 5. Spontaneous Raman and Rayleigh scattering.

2.2. UTILIZATION

Due to its inherent weakness, RS diagnostics are typically plagued by low signal to interference ratios in practical combustion situations or simulations, i.e., hydrocarbon fueled diffusion flames. Major interferences include background flame luminosity, laser modulated soot incandescence and unrejected particle scattering. To date, it has thus found major utilization in fundamental studies of turbulent, hydrogen fueled diffusion flames to perform single pulse measurements of temperature and all major constituents (Drake, et al., 1981; Dibble, et al., 1984; Wehrmeyer, et al., 1992). The studies utilize energetic flashlamp–pumped dye lasers in the visible or excimer lasers in the ultraviolet. Thousands of single pulse measurements are collected in a data base to statistically characterize the behavior of the flame.

Hydrocarbon species generally are strong Raman scatterers and Raman scattering has been used to measure fuel air ratios in internal combustion engines prior to ignition (Heinze, et al., 1992). If there is sufficient premixing of the fuel and air, luminosity and soot formation are constrained and Raman scattering can be employed subsequent to ignition as well.

Due to the weakness of Raman scattering, it is difficult to utilize it in a two–dimensional planar mode unless pressures are quite elevated. However, since the exciting laser is unattenuated in typical applications, linear imaging can be easily performed in all situations where point measurements are feasible (Reckers, et al., 1993). Linear imaging generally utilizes two dimensional array detectors fitted onto a spectrograph yielding spatial dimension along one axis and spectrally dispersed Raman signatures along the other. This permits temperature and species mapping along the linear dimension of the laser beam.

2.3. STATUS

RS is a very mature, well understood technique particularly with visible laser excitation. It is highly developed and widely used in "clean" flames and other low interference applications. In terms of diagnostic development, most of the recent efforts have focused on ultraviolet RS to exploit the high energy, high repetition rate and short wavelength of excimer lasers, particularly KrF at 248 nm (Wehrmeyer, et al., 1992; Shirley, 1990). Due to the scaling of the Raman cross section, RS at 248 nm is about twenty times stronger than it is in the green, 532 nm, region of the spectrum. However, in the uv, problems with fluorescence interferences, absorptions and perturbations can arise. These are now fairly well understood in H_2 fueled flames, particularly with 248 nm excitation, where O_2 and OH fluorescences can be excited and H_2O photodissociated via a two–photon process. Using narrowband excitation and proper tuning, these features can be suppressed or exploited (Wehrmeyer, et al., 1992; Shirley, 1990). More work is required in hydrocarbon fueled flames to characterize fluorescent interferences from hydrocarbon species, particularly polycyclic aromatics (PAHs), absorptions and the magnitude of laser induced/modulated soot incandescences (Shirley, 1992).

2.4. RAYLEIGH SCATTERING

Rayleigh scattering is the elastic, i.e., unshifted, scattering of laser light and accompanies Raman scattering. It is about 10^3 times stronger than vibrational Raman scattering and needs to be suppressed in the Raman detection channels for successful Raman measurements. Because the Rayleigh scattering is unshifted in frequency, it cannot be used to measure individual constituents but instead is a measure of the total gas density. In spatially–uniform pressure applications, e.g., low Mach number flows, Rayleigh scattering can be used to measure temperature via the density measurement and use of the perfect gas law (Dibble and Hollenbach, 1981). Although the total Rayleigh cross section is mixture sensitive, the dependence is weak in most airfed applications and with fuel tailoring, e.g., adding H_2 to CH_4, the cross section variation from reactants to products can be held to a few percent. Since particle scattering is also unshifted in frequency and significantly stronger than molecular scattering, Rayleigh scattering is restricted to particulate free applications. As long as soot formation is suppressed in hydrocarbon flame situations, e.g., lean, premixed, Rayleigh scattering can be used for two dimensional planar mapping of the total density and, hence, temperature field. Planar Rayleigh scattering is now commonly recorded with planar laser–induced fluorescence (PLIF) when energetic laser excitation is employed. Since the Rayleigh signal is always present, addition of a second camera permits its recording (Wolfrum, 1992; Schäfer, et al., 1993). It is finding use in turbulent flame studies as well as in internal combustion engines. In recent work in pulsed H_2–air detonations, planar Rayleigh scattering served as the primary diagnostic (Anderson and Dabora, 1992).

3. Coherent Anti–Stokes Raman Spectroscopy (CARS)

3.1. UTILIZATION

It is well established, through calculations and actual experience, that visible RS is incapable of probing the highly luminous, particle laden flames typical of hydrocarbon–fueled engines (Eckbreth, 1988). It is not yet known whether uv RS will provide this capability either. CARS, because of its strong signal levels and coherent nature, can successfully probe these harsh environments and has been employed for measurements in gas turbine combustors, afterburners, internal combustion engines — Otto cycle and diesel, burning solid propellants, ramjets and scramjets. In terms of applicability, it is the most widely utilized of the Raman based approaches. CARS signal levels are generally several orders of magnitude stronger than spontaneous RS and its beamlike character permits the entire signal to be captured affording large S/I (signal–to–interference ratio) increases over RS. Its coherent character also allows the use of relatively small diameter optical ports. Because of the manner in which CARS is implemented to achieve spatial and temporal precision, its detectivity capabilities are similar to RS and it is primarily restricted to measurements of major, i.e., ≥ 1%, constituents. At considerable experimental complexity, CARS can be electronically–resonantly enhanced permitting minor species sensitivity. This approach is considerably more complex than DFWM and resonance CARS to date has been restricted to laboratory studies.

3.2. CARS OVERVIEW

Detailed treatments of CARS are found in a number of reviews (Druet and Taran, 1981; Eckbreth, 1988, 1989; Greenhalgh, 1988; Taylor, 1993). Briefly, CARS is generated, as illustrated in Fig. 6, when laser beams at frequencies ω_1 and ω_2 are "mixed" via any one of several geometrical phase–matching schemes to generate the laser–like CARS signal at $\omega_3 = 2\omega_1 - \omega_2$. The laser interaction with the medium occurs through the third order nonlinear susceptibility giving rise to an oscillating polarization at ω_3 and, thus, the signal radiation. Phase matching ensures that the CARS signal generated at one location in the interaction volume will be phased with that generated in other locations so that constructive coherent signal growth occurs. ω_1 is termed the pump beam and ω_2, the Stokes, because it is downshifted in frequency by the vibrational/rotational Raman shift of the molecule being probed. Adjustment of the frequency difference ($\omega_1 - \omega_2$) to a particular spectral region permits different molecular constituents to be examined. Single pulse measurements are possible using a broadband dye laser, $\Delta\omega_2 \sim$ 5nm, to generate the entire CARS signature by accessing simultaneously all of the Raman resonances in a given molecular spectral band. Spectrally resolved multichannel detection then permits measurements with a time resolution of the laser pulse lengths employed, typically 10^{-8} sec. The spectral signatures are both constituent mole fraction and temperature dependent which provides the basis for their measurement. Typically employed frequency–doubled neodymium:YAG lasers, which provide the pump laser frequency, operate in the 10–50 Hz range permitting reasonable, but not real time, data rates.

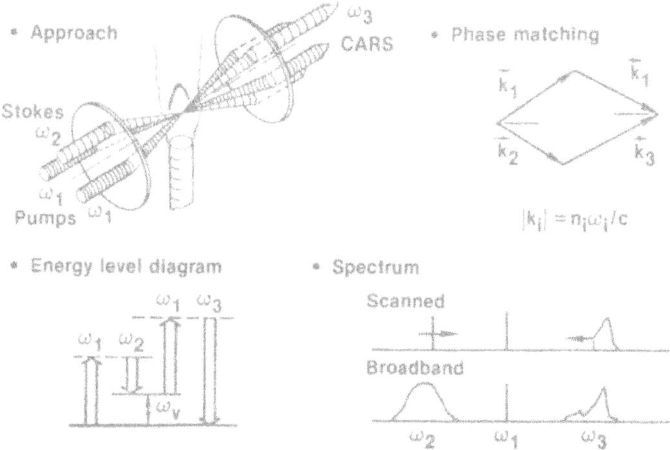

Figure 6. CARS—coherent anti–Stokes Raman spectroscopy.

Despite its practical applicability, there are several disadvantages relative to RS. Because CARS is a nonlinear, intensity dependent technique, its spectra are more complicated and its signals require intensity normalization in certain parameter ranges. CARS, due to the phase–matching requirement, is a double–ended technique requiring line–of–sight optical access with signal generation in the forward direction. Unlike RS, it cannot practically be implemented with a single optical port in backscattering. This line–of–sight requirement can be disadvantageous in some applications. There are other applications, e.g., high pressure, i.c. engines, where it is actually advantageous. Unlike RS, which is intrinsically multi–species, CARS, in its normal implementation described above, measures just a single constituent at a time. To overcome this limitation, multi–color CARS techniques have been developed which use laser beams at three different frequencies (Eckbreth and Anderson, 1987). These techniques are practically applicable but, nevertheless, add additional complexity. In fact, this is one of the motivations behind assessing uv RS for applicability to hydrocarbon–fueled engines. RS is simpler, intrinsically multi species, amenable to single port implementation, and, with excimer lasers, capable of one to two hundred Hz operation.

3.3. STATUS

CARS is quite a mature technique having been studied as a diagnostic for two decades and under intense development at a number of laboratories for more than a decade. The required hardware is highly developed and fairly reliable and there are even some commercial CARS instrument offerings. Most instruments are custom built, however, integrating commercial components in hardened, compact, mobile instruments which can survive the challenging environments of practical device test stands (Anderson, et al., 1986). CARS theory is reasonably well understood and spectroscopic software has been developed for most of the major diatomic and triatomic molecules of combustion interest. As documented earlier, it has been very widely applied to a variety of practical combustion devices.

Recent research emphasis has been oriented at simultaneous, multiple species measurements, electronic resonance CARS for minority species, single pulse spectral noise, and droplet/spray effects on CARS generation. There is also a lot of activity directed toward enhancing the spectroscopic data base, particularly Raman line broadening. There has also been work on spatially–resolved linear CARS imaging. That activity has been directed toward high pressure, solid propellant combustion and the linear spatial extents are quite modest, e.g., 200 to 400μ (Stufflebeam, 1991). In general, the imaging capabilities of CARS are very limited and, for the most part, CARS is restricted to pointwise measurements.

4. Laser–Induced Fluorescence Spectroscopy (LIFS)

4.1. OVERVIEW

Raman–based approaches, without electronic enhancement, are incapable of species detection below the percent or few tenths of a percent level. For detection of minor, trace constituents, e.g., pollutants, flame radicals, which occur on the ppm or tens of ppm level, molecular electronic resonances need to be exploited. The most widely used approach in this regard is the incoherent approach of laser–induced fluorescence spectroscopy (LIFS) schematically illustrated in Fig. 7 (Crosley and Smith, 1983; Kohse–Höinghaus, 1990). Fluorescence is the emission of light from an atom or molecule following promotion to an electronically–excited state by, our emphasis here, absorption of laser photons. Besides spontaneously emitting, i.e., fluorescing, the electronically–excited state can lose population by collisional quenching (transfer to the ground electronic state), stimulated emission, rotational and vibrational energy transfer, photoionization, and photodissociation, i.e., when excitation to a so–called predissociative state occurs. If upper state, rotational and vibrational energy transfer is important, it can be compensated for by monitoring the broadband fluorescence from the ensemble of collisionally excited states; to do this rigorously requires spectrally–resolved capture since the various transitions have linestrengths which are rotational level dependent. The fluorescence signal is proportional to the initial state population number density multiplied by a so–called fluorescence efficiency. For concentration measurements, initial states are selected which are reasonably temperature insensitive over a broad temperature range, obviating the requirement for a temperature measurement.

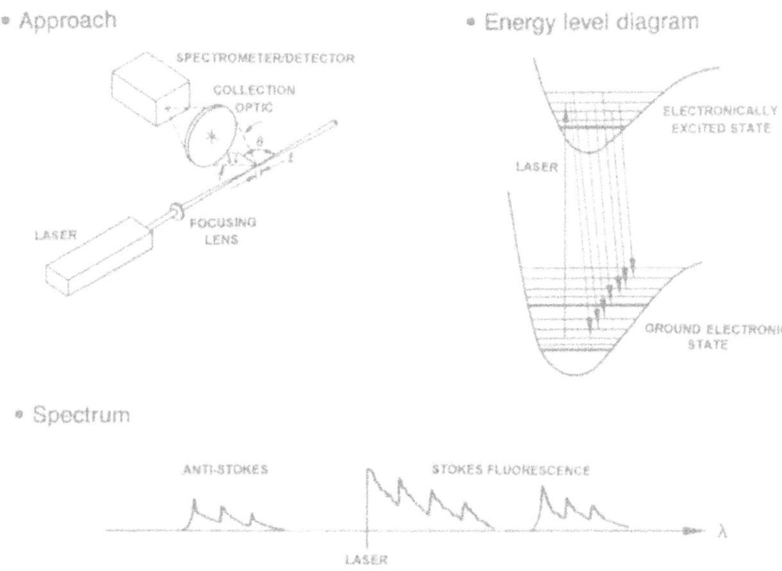

Figure 7. LIFS—laser–induced fluorescence spectroscopy.

The fluorescence efficiency is given by A/(A+Q+BI+P) where A is the spontaneous emission rate; Q, the quenching rate; BI, the stimulated emission rate where I is the laser intensity and B the Einstein coefficient for stimulated emission; and P is the predissociation rate. Assuming for the moment that the upper state is not predissociative, i.e., negligible P, and the laser intensity is low, the fluorescence efficiency is given by A/(A+Q), known as the Stern–Vollmer factor. Unfortunately, $A < Q$ typically, and Q cannot simply be ignored. Thus, in order to extract the species concentration from the fluorescence signal, one needs to know the quenching rate which is generally composition and temperature dependent. In laminar, steady combustion situations, there is the luxury of time to evaluate Q. This can be done by measuring the major constituents and temperature at each location and calculating the quenching correction. This assumes the quenching rate coefficients are known as a function of rotational state and temperature for each collider species. Relatively good data bases exist for several of the more important combustion intermediates, particularly OH, and quenching corrections can be applied to yield measurement inaccuracies in the range of $\pm$25–50%. This approach can, of course, also be applied in unsteady combustion situations. In this case, due to the temporal variation of the temperature and background gas composition at any location, one needs to measure the major collider species and temperature nearly simultaneously with the LIFS measurement (Barlow, et al., 1989). This is generally done with spontaneous Raman scattering which restricts its applicability to relatively clean flames. At low pressures, one can directly evaluate Q from the temporal decay of the LIFS signal. At atmospheric pressure, where typical decay rates are on the order of nanoseconds, one requires short pulse excitation and resolutions in the tens of picosecond range. This capability exists but is highly specialized and has not been routinely applied to LIFS measurements (Bergano, et al., 1983).

Ideally, one would prefer not to evaluate the quenching rate at all, particularly in unsteady combustion. There are two major avenues of approach in this regard. One can attempt to saturate the transition by increasing the laser intensity so that $BI >> A + Q$ which results in the LIFS signal being independent of intensity (there is an intensity term in the numerator which divides out) and Q, and linearly dependent only on the species concentration (Eckbreth, 1988; Carter, et al., 1987; Reisel, et al., 1993). This approach has been successfully applied in a number of steady and unsteady combustion situations. Inaccuracies enter since complete saturation is difficult to achieve in the spatial and temporal wings of the laser beams. The temporal wings can be avoided by using short detection time gates centered on the laser pulse. Spatial wing effects can be minimized by proper viewing geometry. With saturation approaches, one also has to be attentive to rotational relaxation effects handled by appropriate selection of detection bandwidth. A major advantage of operating in the saturation region, besides obviating the need for quenching information, is maximizing of the LIFS signal leading to optimum detection sensitivity. With the impressive advances still occurring in laser technology, particularly with solid state lasers and optical parametric oscillators, saturation laser intensities are becoming quite readily achievable.

The other approach to avoiding quenching is to excite a transition whose upper level is predissociative such that $P >> Q$. P depends only on the quantum states of the molecular species being probed and is invariant with background gas composition. This method has seen widespread utilization to OH detection, pumping the (3,0) band with KrF excimer lasers near 248 nm (Andresen, et al, 1988). The drawbacks to this approach are diminished fluorescence yield, i.e., A/P, and potentially low oscillator strength resulting in weaker laser absorption and fluorescence yield. This is indeed the case in OH but is compensated for by the large laser energies available at 248 nm from KrF, e.g., ~ 200 mJ. This approach, of course, requires the existence of a predissociative state with a good oscillator strength accessible by an energetic laser.

Temperature measurements can be made from the spectral distribution of the fluorescence signal although considerable care needs to be exercised to account for rotational level dependencies in order to get the proper temperature. Two line techniques can be also employed wherein two transitions, with the same upper state and, thus, the same quenching, rotational level effects, are sequentially excited and the resultant fluorescence signals ratioed (Cattolica, 1981). Saturated two–line approaches are also feasible (Lucht, et al., 1982).

4.2. UTILIZATION

Due to its incoherent nature and good signal strength, LIFS is readily implemented for two–dimensional field measurements, namely planar LIFS or PLIFS (Hanson, 1987; Hanson, et al., 1990). It is in this two–dimensional variant that LIFS has seen its greatest utilization in unsteady combustion. The ability to visualize planar slices of an unsteady combustion process yields tremendous insight into the fundamental phenomenology which is occurring. Pointwise measurements have a role, generally in concert with spontaneous Raman scattering to examine the statistics of turbulent combustion processes.

Most PLIFS investigations are performed in the linear or unsaturated regime due to an inability to saturate the transitions of interest. The measurements are generally semi–quantitative with regard to concentration since quenching corrections are not attempted. In these types of studies, this is not a serious drawback since spatial location information is of primary importance. Most investigations focus on OH because of its abundance, ease of excitation, i.e., spectral accessibility, and importance. These studies are performed in either the linear or saturated regime, the latter approach being enabled via the coincidence of powerful excimer lasers and OH spectral band absorptions, e.g., (3,0) at 248 nm, KrF laser; (0,0) at 308 nm, XeCl laser. Although OH is a marker of the flamefront, it is also a high temperature equilibrium product and care must be exercised in interpreting the images. For hydrocarbon fueled flames, CH is a better indicator of flamefront location but is more difficult to image because of its lower concentrations (relative to OH), weaker oscillator strengths (A coefficient) and spectral excitation region. NO imaging is also receiving considerable attention in an attempt to visualize NO formation regions. Since Rayleigh scattering accompanies the PLIFS, a second camera is often employed to permit its capture which, in low Mach number flows, leads to the temperature field assuming no soot is present.

4.3. STATUS

LIFS and PLIFS are quite mature and seeing widespread utilization in unsteady combustion studies. PLIFS is being experimentally enabled with the rapid advances being made in tunable, energetic laser sources throughout the blue and ultraviolet, the emergence of low noise, high sensitivity CCD (charge coupled device) cameras and the availability of relatively low cost, computational machines and visualization software. Work also continues on extending the fundamental data base, e.g., quenching rate coefficients as a function of temperature,rotational level, collider, etc. which will improve, over time, the quantitative accuracy of LIFS and PLIFS.

5. Degenerate Four Wave Mixing (DFWM)

As just seen, quenching is a major issue in LIFS and becomes more of a problem for elevated pressure applications typical of most practical unsteady combustion devices. In the linear or unsaturated region, LIFS produces a pressure independent signal for constant species mole fraction, i.e., the signal increase associated with increasing number density is offset by the concomitantly increasing quenching rate. In the saturation regime, saturation is harder to achieve at elevated pressures. At high pressures, LIFS can be more difficult to apply successfully due to signal scaling effects, particularly if optical access is limited, which is often the situation.

Although relatively early in its development as a combustion diagnostic, DFWM appears to be an attractive alternative to LIFS particularly for elevated pressures, ≥ 1 atm, and practical devices where high interference levels are likely. Furthermore, DFWM is relatively simple to implement for planar imaging. In work performed to date in the saturation regime, DFWM has sensitivities comparable to LIFS and corrections for collisional quenching were not necessary (Farrow and Rakestraw, 1992). Because it is a coherent effect, limited optical apertures are not problematical and high signal to interference ratios can be achieved by collecting the signal beam at some distance from the signal generation volume. CARS can also be extended to trace species detection if it is electronically–resonantly enhanced (Druet, et al., 1978). However, this requires at least two, preferably three, tunable lasers and is much more difficult to implement than DFWM in which only a single tunable laser is required.

5.1. OVERVIEW OF DFWM

Degenerate four wave mixing, Fig. 8, is a nonlinear optical effect (Shen, 1984; Abrams, et al., 1983) which, like CARS, originates through the third–order nonlinear polarization arising from the medium's radiation response. It may be viewed as a CARS process, $\omega_3 = \omega_1 + \omega_1' - \omega_2$, in which all the frequencies merge to a common value, i.e., ω_1', $\omega_2 \rightarrow \omega_1$, thus $\omega_3 = \omega_1 + \omega_1 - \omega_1$. Hence the DFWM signal is spectrally unshifted and occurs at the same frequency as the input lasers. One would think that this makes collection of the signal difficult, but good discrimination against spurious scattering and the incident laser beams is achievable through the appropriate phase matching geometry and the coherent nature of the signal. Resonant enhancement of the DFWM signal occurs as ω_1 coincides with allowed electronic transitions, in essence mapping out the absorption spectrum of the probed species. To date, most DFWM has employed scanned narrowband lasers for spectrum generation; such approaches for single shot measurements in unsteady combustion restrict one to monitoring only a single transition. For single pulse measurements, it is preferable to monitor a significant spectral region containing multiple transitions which requires the use of broadband laser techniques. Such approaches have been demonstrated in principle (Ewart and Snowdon, 1990; Yip, et al., 1992), but the spectral breadth has been restricted since it is difficult to frequency double broadband lasers with good efficiency. This situation will probably improve in time.

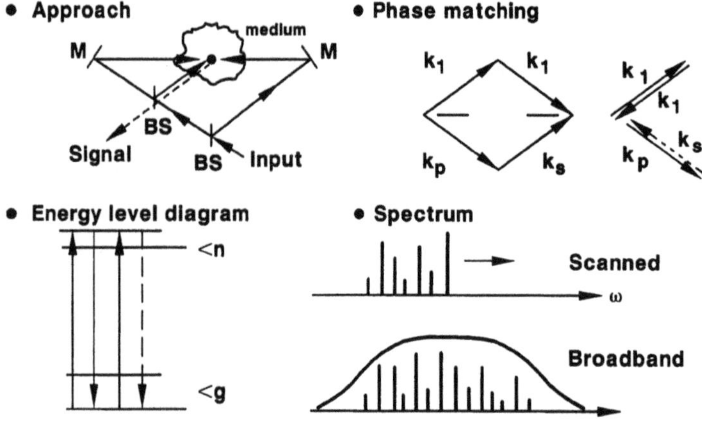

Figure 8. DFWM—degenerate four wave mixing.

As with CARS, DFWM must be phase matched which can be accomplished in a greater variety of geometries than CARS since all the frequencies in DFWM are equal permitting broader manipulation of the phase–matching diagram. Phase conjugate geometries are most common (right phase–matching geometry, Fig. 8) (Ewart and O'Leary, 1986; Drier and Rakestraw, 1990). In these approaches, two of the beams, called "pumps" counterpropagate and intersect with the input "probe" beam; the signal beam counterpropagates along the probe beam axis. Because of the phase conjugate nature, the signal beam emerges with no distortion from turbulence–induced refractive index variations. One does not need to necessarily arrange the counterpropagating pump beams as shown in Fig. 8. Using stimulated Brillouin scattering, a counterpropagating pump beam can be generated by focusing the forward going pump into a cell containing a fluid such as hexane (Winter and Radi, 1992; Feikema, et al., 1992). Conversion efficiencies of 40% are possible. Forward phase–matching geometries, similar to those used in CARS (Fig. 6), can also be employed and result in generation of the signal beam in the forward direction similar to CARS.

Another way of viewing DFWM is depicted in Fig. 9 in terms of an induced grating picture (Farrow and Rakestraw, 1992). The interaction between the probe laser and each pump laser forms an optical interference pattern as in laser Doppler velocimetry, resulting in a series of fringes where the light intensity undergoes a sinusoidal spatial variation in the beam overlap region. This forms a grating in the medium to coherently scatter the other pump laser into the direction from whence the probe laser originated in a process analogous to Bragg diffraction in a crystal. There are many types of induced gratings which can occur. Thermal gratings, caused by collisional quenching with attendant heat release, can generate signals which add to the DFWM signal and obscure interpretation. These effects become important at pressures above a few atmospheres but can be eliminated by using pumps with crossed polarizations.

Although insensitive to quenching, DFWM is sensitive to collisional effects which determine the transition linewidths (Farrow and Rakestraw, 1992). Unlike LIFS however, there is not a direct correlation between the uncertainty in the collisional quenching rate and measurement inaccuracy. Uncertainties in the linewidths cause measurement errors but desensitized from a one–to–one relationship. Based on studies to date, DFWM is best performed with pump intensities at the saturation level not only to maximize detection sensitivity but also to minimize measurement errors due to linewidth uncertainties.

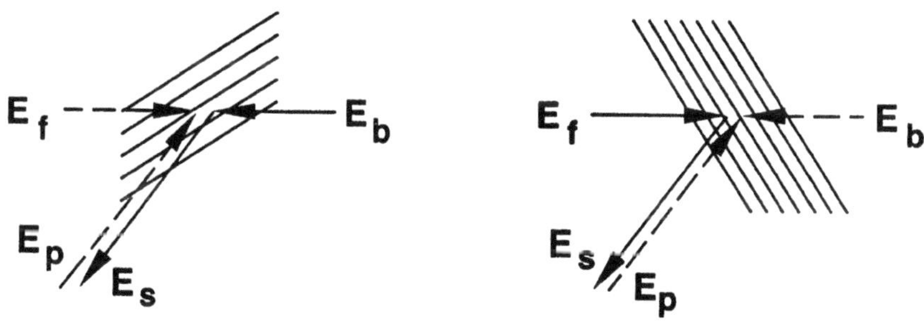

9. Induced grating picture of degenerate four wave mixing.

5.2. UTILIZATION

DFWM has been studied and utilized for quite some time to generate phase conjugate beams (Abrams, et al., 1983). Induced gratings have long been utilized by physical chemists to study collisional dynamics in a variety of media. However, it was only in the mid–eighties that DFWM was applied to flames, first from sodium in a flame (Pender and Hesselink, 1985) and shortly thereafter to OH detection (Ewart and O'Leary, 1986). Due to its relative infancy for combustion diagnostics, DFWM experiments have been performed in clean, laminar flames to study the physics of DFWM and examine diagnostic issues such as signal behavior in various intensity regimes and detection sensitivity compared to LIFS. Nevertheless, it has been shown capable of probing instrumentally hostile environments (Williams, et al., 1992) with measurement demonstrations in internal combustion engines (Rakestraw, 1992) and high temperature, diamond depositing plasmas.

By focusing the counterpropagating pump beams into thin sheets and using an unfocused probe beam, DFWM can be used for two–dimensional imaging. This has been applied to flame imaging of Na (Ewart, et al., 1989) and OH (Rakestraw, et al., 1990). As with pointwise DFWM, this approach is quite attractive for applications to high interference environments. Furthermore, in highly turbulent media, image distortion is removed due to the phase conjugate nature of the image signal beam. As with all imaging techniques, sheet illumination corrections need to be applied which is more complicated with DFWM due to the nonlinear intensity dependences involved in signal generation.

5.3 STATUS

Based upon the investigations conducted to date, DFWM appears very promising as a coherent approach to minor species measurements. If infrared DFWM emerges as a capable technique, it would have applicability to many major constituents as well (Vander Wal, et al., 1992). The technique is still immature diagnostically, but is being widely investigated and rapid resolution of a number of diagnostic issues can be expected. Laboratories possessing LIFS equipment can readily perform DFWM experiments, often in just a few days. Excellent experimental hardware is available. Theoretical understanding is fair to good but rapid progress there can be expected as well given the interest worldwide in this approach.

Diagnostically, there are several issues which require further study and resolution. These include the true viability of single pulse measurements and efficient, broadband source generation, particularly in the uv. Should measurements be made in the saturated or unsaturated regimes; recent experimental evidence seems to indicate that operation near the saturation intensity is preferable. What is the best approach to intensity referencing and calibration? Unlike CARS, there is not a strong nonresonant background signal that can be used for in–situ referencing. Is there an approach to in–situ referencing for concentration which would obviate the need for absolute intensity measurements? Is it more advantageous to work with isolated or clustered transitions from a signal strength standpoint? How does the signal scale at elevated pressures? Work performed to date indicates sharp decreases in signal with increasing pressure in the unsaturated regime, much less so in the saturated region. What is the preferred phase–matching approach? The phase conjugate geometries are Doppler free and do not excite, i.e., interact with, the entire constituent molecular bath if the transitions are inhomogeneously broadened, i.e., Doppler broadened. Forward phase–matching geometries may be preferable for diagnostics. How quantitative will DFWM imaging be? With its somewhat reduced, collisional sensitivity, will it be an attractive alternative to PLIFS in certain applications? The answers to these questions will be quite important in assessing the future utility of DFWM for combustion measurements.

6. Summary

The capabilities of the various approaches to spatially–precise measurements of temperature and species in unsteady combustion are summarized in Fig. 10. In general, one should use incoherent techniques as a matter of first choice in a given application due to their simplicity, ease of implementation and interpretation. Coherent techniques, which are more complex both experimentally and theoretically, are necessary in high interference environments and often are required for probing practical combustion devices.

	Constituents			
	Major	Minor	Simultaneous Multi–species	Multi–point
Raman Scattering	✓	—	intrinsic	1–D
CARS	✓	resonant CARS	multi–color CARS	unlikely
LIFS	some	✓	occasional	2–D
DFWM	some	✓	occasional	2–D

Figure 10. Summary capabilities of incoherent and coherent laser diagnostic approaches.

7. References

Abrams, R. L., Lam, J. F., Lind, R. C. and Steel, D. G. (1983), "Phase Conjugation and High Resolution Spectroscopy by Resonant Degenerate Four Wave Mixing," 211–284 in R. A. Fisher (Ed.), Optical Phase Conjugation, Academic Press, New York, NY.

Anderson, T. J. and Dabora, E. K. (1992), "Measurements of Normal Detonation Wave Structure Using Rayleigh Imaging," 1853–1860 in Twenty–Fourth Symposium (International) on Combustion, The Combustion Institute, Pittsburgh, PA.

Anderson, T. J., Dobbs, G. M. and Eckbreth, A. C. (1986), "Mobile CARS Instrument for Combustion and Plasma Diagnostics," Appl. Opt. 25, 4076–4085.

Andresen, P., Bath, A., Gröger, W., Lülf, H. W., Meijer, G. and ter Meulen, J. J. (1988), "Laser–induced Fluorescence with Tunable Excimer Lasers as a Possible Method for Instantaneous Temperature Field Measurements at High Pressures: Checks with an Atmospheric Flame," Appl. Opt. 27, 365–378.

Barlow, R. S., Dibble, R. W. and Lucht, R. P. (1989), "Simultaneous Measurement of Raman Scattering and Laser–induced OH Fluorescence in Nonpremixed Turbulent Jet Flames," Opt. Lett. 14, 263–265.

Bergano, N. S., Jaanimagi, P. A., Salour, M. M. and Bechtel, J. H. (1983), "Picosecond Laser Spectroscopy Measurement of Hydroxyl Fluorescence Lifetime in Flames," Opt. Lett. 8, 443–445.

Carter, C. D., Salmon, J. T., King, G. B. and Laurendeau, N. M. (1987), "Feasibility of Hydroxyl Concentration Measurements by Laser–Saturated Fluorescence in High–Pressure Flames," Appl. Opt. 26, 4551–4562.

Cattolica, R. (1981), "OH Rotational Temperature from Two–Line Laser–Excited Fluorescence," Appl. Opt. 20, 1156–1166.

Crosley, D. R. and Smith, G. P. (1983), "Laser–induced Fluorescence Spectroscopy for Combustion Diagnostics," Opt. Eng. 22, 545–553.

Dibble, R. W. and Hollenbach, R. E. (1981), "Laser Rayleigh Thermometry in Turbulent Flames," 1489–1499 in Eighteenth Symposium (International) on Combustion, The Combustion Institute, Pittsburgh, PA.

Dibble, R. W., Kollmann, W. and Schefer, R. W. (1984), "Conserved Scalar Fluxes Measured in Turbulent Nonpremixed Flames by Combined Laser Doppler Velocimetry and Laser Raman Scattering," Combust. Flame 55, 307–321.

Drake, M. C., Lapp, M., Penney, C. M., Warshaw, S. and Gerhold, B. W. (1981), "Measurement of Temperature and Concentration Fluctuations in Turbulent Diffusion Flames Using Pulsed Spontaneous Raman Spectroscopy," 1521–1531 in Eighteenth Symposium (International) on Combustion, The Combustion Institute, Pittsburgh, PA.

Dreier, T. and Rakestraw, D. J. (1990), "Degenerate Four–Wave Mixing Diagnostics on OH and NH Radicals in Flames," App. Phys. B 50, 479–485.

Druet, S., Attal, B., Gustafson, T. K. and Taran, J. P. E. (1978), "Electronic Resonance Enhancement of Coherent Anti–Stokes Raman Scattering," Phys. Rev. A18, 1529–1577.

Druet, S. A. J. and Taran, J. P. E. (1981), "CARS Spectroscopy," Prog. Quant. Elec. 7, 1–72.

Eckbreth, A. C. (1988), Laser Diagnostics for Combustion Temperature and Species, Abacus Press (now Gordon and Breach), Turnbridge Wells, Kent, UK.

Eckbreth, A. C. (1989), "Laser Diagnostics for Gas Turbine Thermometry and Species Measurements," 69–106 in D.F. G. Durao, J. H. Whitelaw and P. O. Witze (Eds.), Instrumentation for Combustion and Flow in Engines, Kluwer Academic Publishers, Dordrecht.

Eckbreth, A. C. and Anderson, T. J. (1987), "Multi–Color CARS for Simultaneous Measurements of Multiple Combustion Species," SPIE 742, 34–41.

Ewart, P. and O'Leary, S. V. (1986), "Detection of OH in a Flame by Degenerate Four–Wave Mixing," Opt. Lett. 11, 279–281.

Ewart, P. and Snowdon, P. (1990), "Multiplex Degenerate Four–Wave Mixing in a Flame," Opt. Lett. 15, 1403–1405.

Ewart, P., Snowdon, P. and Magnusson, I. (1989), "Two–Dimensional Phase–Conjugate Imaging of Atomic Distributions in Flames by Degenerate Four–Wave Mixing," Opt. Lett. 14, 563–565.

Farrow, R. L. and Rakestraw, D. J. (1992), "Detection of Trace Molecular Species Using Degenerate Four–Wave Mixing," Science 257, 1894–1900.

Feikema, D. A., Domingues, E. and Cottereau, M.–J. (1992), "OH Rotational Temperature and Number Density Measurements in High Pressure Laminar Flames Using Double Phase–Conjugate Four Wave Mixing," Appl. Phys. B 55, 424–429.

Greenhalgh, D. A. (1988), "Quantitative CARS Spectroscopy," 193–251 in R. J. H. Clarke and R. E. Hester (Eds.), Advances in Nonlinear Spectroscopy, J. Wiley and Sons, Chichester, UK.

Hanson, R. K. (1987), "Combustion Diagnostics: Planar Imaging Techniques," 1677–1691 in Twenty–first Symposium (International) on Combustion, The Combustion Institute, Pittsburgh, PA.

Hanson, R. K., Seitzman, J. M. and Paul, P. H. (1990), "Planar Laser Fluorescence Imaging of Combustion Gases," Appl. Phys. B 50, 441–454.

Heinze, T., Schmidt, T., Brüggemann, D. and Knoche, K.–F. (1992), "Mixture Formation in a Spray Observed by Spontaneous Raman Spectroscopy," 469–480 in Durao, D.F.G. et al. (Eds.), Combusting Flow Diagnostics, Kluwer Academic, Netherlands.

Kohse–Höinghaus, K. (1990), "Quantitative Laser–Induced Fluorescence: Some Recent Developments in Combustion Diagnostics," Appl. Phys. B 50, 455–461.

Long, D. A. (1977), Raman Spectroscopy, McGraw–Hill, New York, NY.

Lucht, R. P., Laurendeau, N. M. and Sweeney, D. W. (1982), "Temperature Measurements by Two–Line Laser–Saturated OH Fluorescence in Flames," Appl. Opt. 21, 3729–3735.

Pender, J. and Hesselink, L. (1985), "Phase Conjugation in a Flame," Opt. Lett. 10, 264–266.

Rakestraw, D. J. (1992). Private communication.

Rakestraw, D. J., Farrow, R. L. and Dreier, T. (1990), "Two–Dimensional Imaging of OH in Flames by Degenerate Four–Wave Mixing," Opt. Lett. 15, 709–711.

Reckers, W., Hüwel, L., Grünefeld, G. and Andresen, P. (1993), "Spatially Resolved Multi–species and Temperature Analysis in Hydrogen Flames," Appl. Opt. 32, 907–918.

Reisel, J. R., Carter, C. D., Laurendeau, N. M. and Drake, M. C. (1993), "Laser Saturated Fluorescence Measurements of Nitric Oxide in Laminar Flat, $C_2H_6/O_2/N_2$ Flames at Atmospheric Pressure," Combust. Sci. and Tech. 91, 271–295.

Schäfer, M., Dinkelacker, F., Buschmann, A. and Wolfrum, J. (1993), "Simultaneous Multi–dimensional Temperature and OH Concentration Field Measurements in Highly Turbulent Premixed Flames," Zeit. Phys. Chem., in press.

Shen, Y. R. (1984), The Principles of Nonlinear Optics, Wiley, New York, NY.

Shirley, J. A. (1990), "UV Raman Spectroscopy of H_2–Air Flames Excited with a Narrowband KrF Laser," Appl. Phys. B 51, 45–48.

Shirley, J. A. (1992), "UV Laser Spectroscopic Measurements in Jet Engine Combustion Exit Flows," AIAA Paper 92–0513.

Stufflebeam, J. H. (1991), "CARS Diagnostics of Reaction Pathways for Nitramine Combustion," AIAA Paper 91–2318.

Taylor, A. M. K. P. (1993), Instrumentation for Flows with Combustion, Academic Press, London, UK.

Vander Wal, R. L., Holmes, B. E., Jeffries, J. B., Danehy, P. M., Farrow, R. L. and Rakestraw, D. J. (1992), "Detection of HF Using Infrared Degenerate Four–Wave Mixing," Chem. Phys. Lett. 191, 251–258.

Wehrmeyer, J. A., Cheng, T.–S. and Pitz, R. W. (1992), "Raman Scattering Measurements in Flames Using a Tunable KrF Excimer Laser," Appl. Opt. 31, 1495–1504.

Williams, S., Green, D. S., Sethuraman, S. and Zare, R. N. (1992), "Detection of Trace Species in Hostile Environments Using Degenerate Four Wave Mixing: CH in an Atmospheric Pressure Flame," J. Am. Chem. Soc. 114, 9122–9130.

Winter, M. and Radi, P. P. (1992), "Nearly Degenerate Four–Wave Mixing Using Phase–Conjugate Pump Beams," Opt. Lett. 17, 30–322.

Wolfrum, J. (1992), Laser Diagnostics in Combustion: Elementary Processes and Complex Systems, World Scientific Publishing Co., Singapore.

Yip, B., Danehy, P. M. and Hanson, R. K. (1992), "Degenerate Four–Wave Mixing Temperature Measurements in a Flame," Opt. Lett. 17, 751–753.

19. MATHEMATICAL MODELING OF TURBULENT FLAMES

W. P. Jones and **M. Kakhi**
Department of Chemical Engng and Chemical Technology
Imperial College of Science, Technology and Medicine
London SW7 2BY
England

ABSTRACT.

The purpose of this article is to review methods for calculating the properties of combusting flows. A brief outline of the models needed to represent turbulent transport is provided and the density weighted forms of the k-ε model and a relatively simple second moment closure are discussed. The methods required to represent combustion in a turbulent environment are reviewed in detail and attention is focused on models for non-premixed and premixed flames. The joint probability density function (pdf) transport equation approach which, in principle, is applicable to all regimes of burning is also considered. For non-premixed flames, closures based on the conserved scalar formalism including the influence of flame stretch and laminar flamelets are described. In the case of premixed flames the theories covered encompass the Bray-Moss-Libby model and more recent techniques based on flame surface density. For joint pdf transport equation methods the closure of the 'molecular mixing' term represents a central difficulty and for this various proposals, including the linear mean square estimation closure, coalescence-dispersion models and mapping closures are appraised. Finally a comparison of calculated results (obtained with the joint pdf equation approach) with experimental measurements is provided. It is suggested that improved mixing models in which direct account is taken of the molecular transport properties of individual chemical species will be needed if the full potential of the pdf equation method is to be realised.

1. Introduction

The requirements imposed on mathematical models for predicting the performance of combustion systems nowadays are quite demanding. To predict with good accuracy the velocity and temperature fields is usually not sufficient. Limitations set out through legislation on the levels of combustion-generated pollutant emissions and energy consumption serve as a motivation for the development of calculation methods for turbulent reacting flows capable of reproducing finite-rate chemistry effects with good accuracy. An additional requirement is that the methods be sufficiently general so as to reproduce measurements in most flows of engineering interest. The achievement of such calculation methods, especially with regards to generality, is still some way off.

The categorization of combusting flows into premixed and non-premixed cases serves as an aid to analysis and constitutes an idealisation of reality. For example combustion in spark ignition engines occurs in the premixed mode, however the degree to which the fuel/air mixture has homogenized prior to entry into the combustion chamber may not be complete and so a totally premixed characterization would be a simplification. Combustion in equipment such as furnaces, Diesel engines and gas turbine combustors is classified as non-premixed. However, in such flames

F. Culick et al., (eds.), Unsteady Combustion, 411–491.

the presence of local extinction pockets can lead to fuel/air mixtures becoming partially premixed. Consequently a flame front would propagate into this partially premixed region.

Several reviews on various aspects of turbulent combustion have been performed in the past [64; 6; 7; 11; 12; 102; 103] and the present article attempts to augment these by considering new developments alongside the more traditional approaches so that the basic modelling problems can be highlighted. Recently work has appeared in the literature where direct numerical simulation (DNS) and large eddy simulation (LES) techniques have been applied to the problem of turbulent combustion. In the case of DNS, problems in idealized configurations and at low Reynolds numbers are considered, usually in homogeneous decaying turbulence (e.g. Vervisch et al. [126]). The use of such a method, with current computer capabilities, can only be directed at investigating the fundamental interactions occurring in flames. By carefully scrutinizing the turbulence-chemistry interactions, it may be possible to infer improved closure principles for the (model-dependent) techniques used to analyse the high Reynolds number reacting flows of engineering interest. With regards to LES, although the Reynolds number restrictions experienced by DNS do not pose a limitation, modelling is required. Because chemical reactions occur in the fine scales of turbulent motion, the source terms for chemical production appear in an unclosed form. Therefore at first sight it may appear that there is little to be gained from such a technique in combusting flows. More research is required to substantiate whether LES in combustion is beneficial, particularly in complex flows where it is claimed that the potential advantage of LES over more conventional (phenomenological) approaches in predicting the velocity field becomes apparent.

The aim of this review article is, primarily, to discuss some of the modelling approaches aimed towards the calculation of turbulent flames. Emphasis is placed on ideas behind closure principles. After a brief presentation of the governing equations and basic assumptions, combustion in non-premixed and premixed configurations is considered. This is then followed by a separate section on transported probability density function (pdf) equations for which predictions are presented and compared with experiment. This particular topic is considered separately since the formulation is not, in principle, restricted to non-premixed or premixed flame calculations.

2. The Governing Equations

The governing equations describing the flow of a fluid can be derived from the principles of conservation of mass, momentum and energy. Unless otherwise stated, they are written below using cartesian tensor notation :

Continuity

$$\frac{\partial \varrho}{\partial t} + \frac{\partial \varrho u_j}{\partial x_j} = 0 \tag{1}$$

where u_j is the instantaneous velocity vector, and ϱ is the density.

Navier - Stokes equation

$$\varrho \frac{\partial u_i}{\partial t} + \varrho u_j \frac{\partial u_i}{\partial x_j} = -\frac{\partial p}{\partial x_i} + \frac{\partial \tau_{ij}}{\partial x_j} + \varrho g_i \tag{2}$$

where τ_{ij} is the viscous stress tensor defined by :

$$\tau_{ij} = \mu \left(\frac{\partial u_i}{\partial x_j} + \frac{\partial u_j}{\partial x_i} - \tfrac{2}{3}\delta_{ij}\frac{\partial u_k}{\partial x_k} \right) \tag{3}$$

and where p is the pressure, μ is the mixture viscosity, δ_{ij} is the Kronecker delta and g_i is the gravitational acceleration vector.

Energy

$$\varrho\frac{\partial h}{\partial t} + \varrho u_j\frac{\partial h}{\partial x_j} = \frac{\partial p}{\partial t} + \frac{\partial}{\partial x_i}\left\{ \frac{\mu}{\sigma}\frac{\partial h}{\partial x_i} + \mu\left(\frac{1}{Sc} - \frac{1}{\sigma}\right)\sum_{\alpha=1}^{N} h_\alpha \frac{\partial Y_\alpha}{\partial x_i} \right\} \tag{4}$$

where h is the specific mixture enthalpy, Y_α is the mass fraction, Sc is the Schmidt number of the mixture and σ is the mixture Prandtl number. Low Mach number flows ($Ma \ll 1$) are considered and consequently the terms associated with the kinetic energy of the mixture and the viscous dissipation are neglected. Radiative heat losses are not considered. Fourier's and Fick's law for molecular fluxes are employed and work done against body forces is assumed to be negligible in comparison to thermal energy changes associated with chemical reaction. The term $\frac{\partial p}{\partial t}$ is assumed to be negligible in many practically occuring situations, e.g. jets, gas turbines and furnaces, but cannot be neglected in the flows occurring in internal combustion engines. The implication of $\frac{\partial p}{\partial t} \approx 0$ and the low Mach number assumption is that the density and the source term for reaction (in the species conservation equation, to be described later on) are only functions of the scalar field involving the specific enthalpy and the mass fraction of the species [99]. The effect of chemical reaction appears in the energy equation through the definition of h :

$$h = \sum_{\alpha=1}^{N} Y_\alpha h_\alpha \quad \text{where} \quad h_\alpha = c_{p,\alpha}T + \Delta_\alpha \tag{5}$$

The Δ_α's represent the heats of formation for the species involved in the reactions. The inclusion of this term accounts for the heat release (and absorption) effects due to the chemical reactions taking place. Finally, the Lewis number, defined as the ratio of the Schmidt number to the Prandtl number is close to unity in combusting flows, and it is a useful assumption to take it as exactly unity (since the third term on the rhs of equation (4) disappears). This simplifies the energy equation to :

$$\varrho\frac{\partial h}{\partial t} + \varrho u_j\frac{\partial h}{\partial x_j} = \frac{\partial}{\partial x_i}\left(\frac{\mu}{\sigma}\frac{\partial h}{\partial x_i}\right) \tag{6}$$

Species Conservation

$$\varrho\frac{\partial Y_\alpha}{\partial t} + \varrho u_j\frac{\partial Y_\alpha}{\partial x_j} = -\frac{\partial J_{i,\alpha}}{\partial x_i} + \dot{\omega}_\alpha \qquad \text{for} \qquad \alpha = 1,\ldots,N \tag{7}$$

where $\dot{\omega}_\alpha$ is the net formation rate of species α per unit volume. $J_{i,\alpha}$ is the diffusional flux of species α. Fick's law of diffusion, $J_{i,\alpha} = -\frac{\mu}{Sc}\frac{\partial \phi_\alpha}{\partial x_i}$, is strictly true for binary mixtures, and so the use of Fickian diffusion in the context of multi-component mixtures is only justified if the Schmidt number is the same for all species (in order to ensure that the mass fractions sum to unity). The assumption of uniform Sc (equal diffusivity) is known to be poor when species with widely differing molar masses are being considered. In the case of turbulent flows the discussion presented later demonstrates that equal (and constant) diffusivity is invariably invoked. To obtain a general expression for $\dot{\omega}_\alpha$, consider the system involving N chemical species and R reaction steps, expressed in the form [80] :

$$\sum_{\alpha=1}^{N} \nu'_{\alpha\beta} M_\alpha \rightleftharpoons \sum_{\alpha=1}^{N} \nu''_{\alpha\beta} M_\alpha \quad \beta = 1, \ldots, \mathrm{R} \tag{8}$$

where the rate constants for the forward and backward reactions are $k_{f\beta}$ and $k_{b\beta}$, respectively, and M_α denotes the chemical symbol of species α. If species α does not participate in a particular reaction then $\nu'_{\alpha\beta}$ and $\nu''_{\alpha\beta}$ are zero as appropriate. The rate constants can be expressed as :

$$k_{f\beta} = B_\beta T^{a_\beta} exp\left(-\frac{E_\beta}{RT}\right) \tag{9}$$

where B_β is the pre-exponential factor, a_β is a constant exponent, E_β is the activation energy and R is the universal gas constant. Clearly the rate of formation of species α is the sum of the formation rates of species α in all the individual reaction steps. In any particular reaction step, the rate of formation of species α is the net rate of formation of species α due to the forward and backward processes in reaction β. Mathematically this can be expressed as :

$$\dot{\omega}_\alpha = W_\alpha \sum_{\beta=1}^{\mathrm{R}} \left(\nu''_{\alpha\beta} - \nu'_{\alpha\beta}\right) \left\{ k_{f\beta} \varrho^{m_\beta} \prod_{\alpha=1}^{N} \left(\frac{Y_\alpha}{W_\alpha}\right)^{\nu'_{\alpha\beta}} - k_{b\beta} \varrho^{l_\beta} \prod_{\alpha=1}^{N} \left(\frac{Y_\alpha}{W_\alpha}\right)^{\nu''_{\alpha\beta}} \right\} \tag{10}$$

where $m_\beta \equiv \sum_{i=1}^{N} \nu'_{\alpha\beta}$ and $l_\beta \equiv \sum_{i=1}^{N} \nu''_{\alpha\beta}$ are the orders of the forward and reverse reactions respectively. In order to close the system of equations, an expression for the density is required. This is given by the equation of state for an ideal gas :

$$p = \varrho RT \sum_{\alpha=1}^{N} \frac{Y_\alpha}{W_\alpha} \tag{11}$$

The equations described so far constitute a closed set of equations and in principle they could be solved (numerically) with appropriate initial and boundary conditions. However, it can be shown that typically the cpu requirements for a direct numerical simulation (DNS) of inert mixing are $O(Re^3)$. It is quite clear that for the high Reynolds number flows of engineering interest, DNS is computationally prohibitive [105]. For statistically homogeneous 'low' Re flows, DNS is feasible and as discussed in section 5 it can be very useful. The traditional approach for treating high Re turbulent flows has been to average the conservation equations, with the consequence that all the fine details of the flow do not require spatial and temporal resolution. Different types of averaging can be applied, but they all share a common feature; the closed set of instantaneous equations are transformed by averaging into an 'open' set of equations. Terms appear which cannot be accounted for in terms of the dependent variables of the flow field. Modelling serves to approximate these terms, and the purpose of this article is to address the principles behind the representation of such terms. Two common types of averaging are the Reynolds (unweighted) and Favre (density-weighted) approach, defined as follows :

$$\bar{A}(\mathbf{x},t) \equiv \lim_{N_p\to\infty} \frac{1}{N_p}\sum_{i=1}^{N_p} A^{(i)}(\mathbf{x},t) \qquad \text{Reynolds averaging} \tag{12}$$

$$\tilde{A}(\mathbf{x},t) \equiv \frac{1}{\bar{\varrho}}\lim_{N_p\to\infty} \frac{1}{N_p}\sum_{i=1}^{N_p} \varrho^{(i)} A^{(i)}(\mathbf{x},t) \qquad \text{Favre averaging} \tag{13}$$

where $A^{(i)}(\mathbf{x},t)$ is the ith realisation. The standard practice is to decompose the instantaneous equations into mean and fluctuating parts and then to average them. With density-weighted averaging the velocity and scalars (but not the density and pressure) are decomposed according to $u_i = \tilde{u}_i + u_i''$, $\phi_\alpha = \tilde{\phi}_\alpha + \phi_\alpha''$. With the definition of Favre averaging in equation (13) it can easily be shown that $\overline{\varrho\phi_\alpha''} = \overline{\varrho u_i''} = 0$. For Reynolds averaging, $u_i = \bar{u}_i + u_i'$, $\phi_\alpha = \bar{\phi}_\alpha + \phi_\alpha'$ and where $\overline{u_i'} = \overline{\phi_\alpha'} = 0$. From the basic definitions, the following results can be established between weighted and unweighted statistics :

$$\bar{u}_i = \tilde{u}_i + \overline{u_i''}; \qquad \bar{\phi}_\alpha = \tilde{\phi}_\alpha + \overline{\phi_\alpha''} \tag{14}$$

$$\overline{u_i''} = -\frac{\overline{\varrho' u_i'}}{\bar{\varrho}}; \qquad \overline{\phi_\alpha''} = -\frac{\overline{\varrho'\phi_\alpha'}}{\bar{\varrho}} \tag{15}$$

$$\overline{\varrho' u_i''} = \overline{\varrho' u_i'}; \qquad \overline{\varrho'\phi_\alpha''} = \overline{\varrho'\phi_\alpha'} \tag{16}$$

The reason for employing Favre averaging is that the resulting conservation equations do not include terms involving correlations of density fluctuations. Consequently, the resulting equations are easier to interpret (and hence model) than their Reynolds averaged counterparts. Furthermore, Favre averaging leads to equations describing the mean of conserved quantities e.g., the resulting momentum equation is in terms of $\overline{\varrho u_i} \equiv \bar{\varrho}\tilde{u}_i$ rather than $\bar{\varrho}\bar{u}_i$ which is obtained from Reynolds averaging and is not conserved. Despite the advantages of adopting Favre averaging over Reynolds averaging, instruments may measure values close to density-weighted or unweighted averages [64]. Clearly the ability to calculate both unweighted and density-weighted quantities is desirable. But $\overline{\varrho u_i} \equiv \bar{\varrho}\tilde{u}_i = \bar{\varrho}\bar{u}_i + \overline{\varrho' u_i'}$, and $\overline{\varrho' u_i'}$ can be very difficult to calculate accurately; in terms of the pdf, unweighted and weighted statistics can be used interchangeably without approximation [6]. Applying Favre averaging to equations (1), (2), (6) and (7) yields :

$$\frac{\partial\bar{\varrho}}{\partial t} + \frac{\partial\bar{\varrho}\tilde{u}_j}{\partial x_j} = 0 \tag{17}$$

$$\frac{\partial}{\partial t}(\bar{\varrho}\tilde{u}_i) + \frac{\partial}{\partial x_j}(\bar{\varrho}\tilde{u}_j\tilde{u}_i) = -\frac{\partial\bar{p}}{\partial x_i} - \frac{\partial}{\partial x_j}\left(\bar{\varrho}\widetilde{u_j''u_i''}\right) + \bar{\varrho}g_i \tag{18}$$

$$\frac{\partial}{\partial t}\left(\bar{\varrho}\tilde{h}\right) + \frac{\partial}{\partial x_j}\bar{\varrho}\tilde{u}_j\tilde{h} = \frac{\partial}{\partial x_j}\left(\bar{\varrho}\widetilde{u_j''h''}\right) \tag{19}$$

$$\frac{\partial}{\partial t}\left(\bar{\varrho}\tilde{Y}_\alpha\right) + \frac{\partial}{\partial x_j}\bar{\varrho}\tilde{u}_j\tilde{Y}_\alpha = \frac{\partial}{\partial x_j}\left(\bar{\varrho}\widetilde{u_j''Y_\alpha''}\right) + \bar{\omega}_\alpha \qquad \text{for} \qquad \alpha = 1,\ldots,N \tag{20}$$

The above equations have been written for high Re, so that the averaged molecular fluxes can be treated as negligible in comparison with the turbulent fluxes $\bar{\varrho}\widetilde{u_i''u_j''}$ (Reynolds stress) and $\bar{\varrho}\widetilde{u_j''\phi_\alpha''}$ (scalar flux). Consequently the equations have a form identical to their constant density counterparts. The unknowns appearing in the equations are the Reynolds stress, scalar flux, mean density and the mean formation rate due to chemical reaction. This article is concerned mainly with the methods of determining $\bar{\varrho}$ and $\bar{\dot{\omega}}_\alpha$. However, a discussion of the turbulence models used to calculate $\bar{\varrho}\widetilde{u_i''u_j''}$ and $\bar{\varrho}\widetilde{u_j''\phi_\alpha''}$ is necessary since they form a crucial part of any calculation method. In what follows ϕ_α is used to denote the scalar field, i.e. species mass fractions and mixture enthalpy.

2.1. TURBULENCE MODELS

In turbulent combustion large fluctuations in fluid density arise and proper account must be taken of them. Most of the work to date concerning turbulence modelling is based on consideration of constant density flows - the problem of predicting inert turbulent flows is sufficiently demanding without the additional complications introduced by combustion. The most common approach to modelling variable density flows is to recast the Reynolds-averaged models in terms of density-weighted averaged quantities with the assumption that density fluctuations are accounted for by the averaging. Experimental verification of this assumption is a very difficult task owing to the level of detail required to determine the contribution made by density fluctuations in real flames and translating this into a modelling criterion. However, the results which have been obtained so far based on such an approach have been encouraging. An appropriate starting point is the density-weighted Reynolds stress equation (derivable from equation (2)) :

$$\begin{aligned}
\underbrace{\bar{\varrho}\frac{D}{Dt}\left(\widetilde{u_i''u_j''}\right)}_{\text{I}} + \underbrace{\bar{\varrho}\widetilde{u_i''u_k''}\frac{\partial \tilde{u}_j}{\partial x_k} + \bar{\varrho}\widetilde{u_j''u_k''}\frac{\partial \tilde{u}_i}{\partial x_k}}_{\text{II}} &= \underbrace{-\frac{\partial}{\partial x_k}\left\{\overline{\varrho u_i''u_j''u_k''} + \tfrac{2}{3}\delta_{ij}\overline{u_k''p'}\right\}}_{\text{III}} \\
&\quad \underbrace{-\left\{\overline{u_j''\frac{\partial \tau_{ik}}{\partial x_k}} + \overline{u_i''\frac{\partial \tau_{jk}}{\partial x_k}}\right\}}_{\text{IV}} \\
&\quad \underbrace{-\left\{\overline{u_i''\frac{\partial p'}{\partial x_j}} + \overline{u_j''\frac{\partial p'}{\partial x_i}} - \tfrac{2}{3}\delta_{ij}\overline{u_k''\frac{\partial p'}{\partial x_k}}\right\}}_{\text{V}} \\
&\quad \underbrace{-\left\{\overline{u_i''}\frac{\partial \bar{p}}{\partial x_j} + \overline{u_j''}\frac{\partial \bar{p}}{\partial x_i}\right\}}_{\text{VI}} + \underbrace{\tfrac{2}{3}\delta_{ij}\overline{p'\frac{\partial u_k''}{\partial x_k}}}_{\text{VII}}
\end{aligned} \tag{21}$$

In the above, $\frac{D}{Dt} \equiv \frac{\partial}{\partial t} + \tilde{u}_k\frac{\partial}{\partial x_k}$. The terms appearing in this equation have the following significance :

- I : Rate of change of $\widetilde{u_i''u_j''}$ including convection by the mean flow
- II : Production of $\widetilde{u_i''u_j''}$ by the interaction with the mean strain rate
- III : Diffusive transport of $\widetilde{u_i''u_j''}$ by velocity and pressure fluctuations

- IV : Viscous destruction of $\widetilde{u_i''u_j''}$
- V : Pressure redistribution of $\widetilde{u_i''u_j''}$ amongst its components
- VI : Production/destruction of $\widetilde{u_i''u_j''}$ by the mean pressure gradient
- VII : Production of $\widetilde{u_i''u_j''}$ by the fluctuating pressure and dilatation. This term is associated with noise and can be neglected for $Ma \ll 1$, [3]

Terms I and II appear in exact form whilst the rest of the terms require closure approximations. Terms VI and VII are identically zero in constant density flows. Density-weighted averaging eliminates the explicit appearance of g_i in the Reynolds stress equation, but the effect appears in the mean pressure gradient term. This can be shown by decomposing the mean pressure into a dynamic and hydrostatic contribution. The latter term can be expressed as, $\varrho_o g_i$, where ϱ_o is the density of the fluid at some position remote from the region of interest e.g. in the free stream. In equation (21), this contribution will appear as $-\varrho_o \overline{u_i''} g_i = \frac{\varrho_o}{\bar{\varrho}}(g_i \overline{\varrho' u_j'})$ using equation (15). It should be noted however that the effects of buoyancy also enter (indirectly) through the fluctuating pressure gradient terms as described by equations (37) and (38) in section 2.1.2. Term IV can be simplified for high Re turbulence; first the following manipulation can be applied :

$$\overline{u_j''\frac{\partial \tau_{ik}}{\partial x_k}} + \overline{u_i''\frac{\partial \tau_{jk}}{\partial x_k}} = \underbrace{\frac{\partial \overline{u_i''\tau_{jk}}}{\partial x_k} + \frac{\partial \overline{u_j''\tau_{ik}}}{\partial x_k}}_{\text{transport}} \underbrace{- \overline{(\bar{\tau}_{ik} + \tau_{ik}')\frac{\partial u_j''}{\partial x_k}} - \overline{(\bar{\tau}_{jk} + \tau_{jk}')\frac{\partial u_i''}{\partial x_k}}}_{\text{dissipation}} \tag{22}$$

Order of magnitude analysis demonstrates that $\overline{\tau_{ik}'(\partial u_j''/\partial x_k)} + \overline{\tau_{jk}'(\partial u_i''/\partial x_k)}$ is the term on the rhs of equation (22) which cannot be neglected as $Re \to \infty$ since it is directly affected by the fine scale motions of the turbulence. Assuming local isotropy of the fine scales implies that the off-diagonal components of these tensors are zero. Thus, the remaining dissipative terms can be written as :

$$\overline{\tau_{ik}'\frac{\partial u_j''}{\partial x_k}} + \overline{\tau_{jk}'\frac{\partial u_i''}{\partial x_k}} \equiv \bar{\varrho}\varepsilon_{ij} = \tfrac{2}{3}\delta_{ij}\bar{\varrho}\varepsilon \qquad \text{where} \qquad \varepsilon \equiv \frac{1}{\bar{\varrho}}\left(\overline{\tau_{lk}'\frac{\partial u_l''}{\partial x_k}}\right) \tag{23}$$

Following Lumley [84] the fluctuating velocity pressure gradient correlations are split up into a redistributive and a nonredistributive part :

$$\overline{u_j''\frac{\partial p'}{\partial x_i}} + \overline{u_i''\frac{\partial p'}{\partial x_j}} = \underbrace{(\overline{u_j''\frac{\partial p'}{\partial x_i}} + \overline{u_i''\frac{\partial p'}{\partial x_j}}) - \tfrac{2}{3}\delta_{ij}\overline{u_k''\frac{\partial p'}{\partial x_k}}}_{\text{redistributive}} + \underbrace{\tfrac{2}{3}\delta_{ij}\overline{u_k''\frac{\partial p'}{\partial x_k}}}_{\text{nonredistributive}} \tag{24}$$

The redistributive part has no role in the generation or destruction of turbulence as it is constructed to vanish on contraction of the indices (i.e. it disappears in the turbulence kinetic energy equation (27)). The non-redistributive term can be further manipulated to yield :

$$\tfrac{2}{3}\delta_{ij}\overline{u_k''\frac{\partial p'}{\partial x_k}} = \underbrace{\tfrac{2}{3}\delta_{ij}\frac{\partial \overline{u_k''p'}}{\partial x_k}}_{\text{transport}} - \underbrace{\tfrac{2}{3}\delta_{ij}\overline{p'\frac{\partial u_k''}{\partial x_k}}}_{\text{noise}} \tag{25}$$

For $Ma \ll 1$ the noise production term can be neglected. Using the results from equations (23) and (25), the equation for the turbulence kinetic energy, defined as,

$$k = \tfrac{1}{2}\left(\widetilde{u_1''^2} + \widetilde{u_2''^2} + \widetilde{u_3''^2}\right) = \frac{\widetilde{u_i''^2}}{2} \tag{26}$$

can be obtained by contracting the indices of equation (21) and dividing by two to yield :

$$\bar{\varrho}\frac{Dk}{Dt} + \bar{\varrho}\widetilde{u_i''u_j''}\frac{\partial \tilde{u}_j}{\partial x_i} = -\frac{\partial}{\partial x_j}\left\{\tfrac{1}{2}\overline{\varrho u_j''u_i''u_i''} + \overline{u_j''p'}\right\} - \bar{\varrho}\varepsilon - \overline{u_i''}\frac{\partial \bar{p}}{\partial x_i} \tag{27}$$

2.1.1. *Eddy-viscosity models.* The simplest turbulence closures are based on the eddy-viscosity approach. Essentially they attempt to represent turbulent transport using relations similar in form to the molecular flux laws. For a constant density thin shear layer situation the Reynolds stress $\overline{u'v'}$ is approximated by $-\nu_t\frac{\partial \bar{u}}{\partial y}$. Though similar in form to Newton's constitutive relation for momentum diffusion, $\nu_t \equiv \frac{\mu_t}{\varrho}$ is not a fluid property but depends on the local properties of the turbulence. Assuming the turbulence can be characterised in terms of a single characteristic length and time scale, l_c, u_c (respectively), $\nu_t \approx u_c l_c$ becomes a dimensional necessity, but at the same time forms the basis for constructing a modelled expression. Since the large scale motions are the energy containing motions, which are assumed to dictate the rate of spectral energy transfer, u_c can be identified with $\sqrt{k}$ and l_c with the integral length scale. The closure is complete provided k and l_c can be specified. In flows with appreciable density fluctuations it is assumed that the dimensional reasoning previously suggested still applies. Thus, the eddy-viscosity concept for the Reynolds stress terms can be expressed as :

$$\bar{\varrho}\widetilde{u_i''u_j''} = \underbrace{-\mu_t\left(\frac{\partial \tilde{u}_i}{\partial x_j} + \frac{\partial \tilde{u}_j}{\partial x_i}\right)}_{\text{I}} + \underbrace{\tfrac{2}{3}\bar{\varrho}k\delta_{ij} + \tfrac{2}{3}\mu_t\frac{\partial \tilde{u}_l}{\partial x_l}\delta_{ij}}_{\text{II}} \tag{28}$$

Term I ensures the symmetric property of the Reynolds stress tensor $\left(\widetilde{u_i''u_j''} = \widetilde{u_j''u_i''}\right)$ is preserved. Term II is included to ensure that the left and right hand sides of the equation are identical on contraction of the indices. The modelled scalar fluxes becomes :

$$\bar{\varrho}\widetilde{u_i''\phi_\alpha''} = -\frac{\mu_t}{\sigma_t}\frac{\partial \tilde{\phi}_\alpha}{\partial x_j} \tag{29}$$

where σ_t is a turbulent Prandlt/Schmidt number. To obtain the eddy-viscosity several levels of closure are possible, ranging from simple zero equation mixing length models to the widely used two equation models (Launder and Spalding [77]). Focusing attention on the latter (two-equation) specification, a popular choice of closure employs k and ε. Since $\varepsilon \approx \frac{k^{3/2}}{l_c}$, the eddy viscosity becomes, $\nu_t \approx k^{1/2}\frac{k^{3/2}}{l_c} \Rightarrow \nu_t \approx \frac{k^2}{\varepsilon}$, thus $\mu_t = C_\mu\bar{\varrho}\frac{k^2}{\varepsilon}$. This is referred to as the 'k-ε' model, Jones and Launder [59], where C_μ is a constant to be determined and where k and ε are obtained from the solution of modelled transport equations. In equation (27), the terms requiring closure appear on the rhs. The transport term and the mean pressure gradient term are approximated using gradient transport arguments :

$$-\left\{\frac{1}{2}\overline{\varrho u_j'' u_i'' u_i''} + \overline{u_j'' p'}\right\} \approx \frac{\mu_t}{\sigma_t}\frac{\partial k}{\partial x_j} \tag{30}$$

$$-\overline{u_i''}\frac{\partial \bar{p}}{\partial x_i} = \frac{\overline{\varrho' u_i'}}{\bar{\varrho}}\frac{\partial \bar{p}}{\partial x_i} = \frac{1}{\bar{\varrho}^2}\left(\bar{\varrho}\,\overline{\varrho' u_i'}\frac{\partial \bar{p}}{\partial x_i}\right) \approx -\frac{1}{\bar{\varrho}^2}\left(\frac{\mu_t}{\sigma_\varrho}\frac{\partial \bar{\varrho}}{\partial x_i}\right)\frac{\partial \bar{p}}{\partial x_i} \tag{31}$$

The approximation, equation (31), was introduced by Jones and McGuirk [61], but in many flows it makes only a small contribution and is often neglected. Substituting the above into equation (27) yields the modelled form of the k-equation :

$$\bar{\varrho}\frac{Dk}{Dt} + \bar{\varrho}\widetilde{u_i'' u_j''}\frac{\partial \tilde{u}_j}{\partial x_i} = \frac{\partial}{\partial x_j}\left(\frac{\mu_t}{\sigma_\varrho}\frac{\partial k}{\partial x_j}\right) - \frac{\mu_t}{\bar{\varrho}^2}\frac{\partial \bar{\varrho}}{\partial x_i}\frac{\partial \bar{p}}{\partial x_i} - \bar{\varrho}\varepsilon \tag{32}$$

The derivation of a transport equation for ε, as defined in equation (23), is complicated and exceedingly laborious. For constant density flows, where $\tau_{lk}' = \mu(\partial u_l'/\partial x_k + \partial u_k'/\partial x_l)$, the viscous dissipation rate can be expressed as $\varepsilon = \nu\overline{(\partial u_l'/\partial x_k)^2}$ by invoking local isotropy so that $\overline{(\partial u_l'/\partial x_k)(\partial u_k'/\partial x_l)} = 0$. A transport equation for this quantity can be obtained by straightforward manipulation of equation (2). For high Re, the equation for ε in constant density flows reduces to :

$$\frac{D\varepsilon}{Dt} = \underbrace{-\frac{\partial}{\partial x_k}\left\{\overline{\nu u_k'\left(\frac{\partial u_i'}{\partial x_l}\right)^2} + \frac{2\nu}{\varrho}\overline{\frac{\partial p'}{\partial x_l}\frac{\partial u_k'}{\partial x_l}}\right\}}_{\text{I}} \underbrace{-2\nu\overline{\frac{\partial u_i'}{\partial x_k}\frac{\partial u_i'}{\partial x_l}\frac{\partial u_k'}{\partial x_l}}}_{\text{II}} \underbrace{-2\left(\nu\overline{\frac{\partial^2 u_i'}{\partial x_l \partial x_k}}\right)^2}_{\text{III}} \tag{33}$$

The terms appearing on the rhs require modelling. An order of magnitude analysis shows that terms II and III individually vary as $Re^{1/2}$ compared with the other terms, and so in the limit, become infinite. But when taken together, they have to form a finite value in order for the equation to balance. Furthermore, term III is necessarily positive and consequently term II needs to be negative to balance the equation and in doing so acts as a production term. Terms II and III should be modelled together in order to avoid their respective singular behaviour as $Re \to \infty$. The final form of the modelled equation for density-weighted ε follows that of equation (32) very closely, partly because equation (33) is of very little aid for modelling. A widely used form of the ε-equation for variable density flows is :

$$\bar{\varrho}\frac{D\varepsilon}{Dt} + C_{\varepsilon_1}\frac{\varepsilon^2}{k}\bar{\varrho}\widetilde{u_i'' u_j''}\frac{\partial \tilde{u}_j}{\partial x_i} = \underbrace{\frac{\partial}{\partial x_j}\left(\frac{\mu_t}{\sigma_\varepsilon}\frac{\partial \varepsilon}{\partial x_j}\right)}_{\text{A}} - C_{\varepsilon_2}\bar{\varrho}\frac{\varepsilon^2}{k} + C_{\varepsilon_3}\frac{\varepsilon}{k}\frac{\mu_t}{\bar{\varrho}^2}\frac{\partial \bar{\varrho}}{\partial x_i}\frac{\partial \bar{p}}{\partial x_i} \tag{34}$$

The form of the A term in equation (34) is commonly used in k-ε closures. Hanjalić and Launder [65] proposed (in density-weighted form) : A $= \partial/\partial x_i(C_\varepsilon \bar{\varrho}\frac{k}{\varepsilon}\widetilde{u_i'' u_j''}\frac{\partial \varepsilon}{\partial x_j})$ with $\widetilde{u_i'' u_j''}$ being obtained from a second moment closure. In discussing the k-ε closure several constants were introduced. These constants are calibrated in simple (constant density) flows, such as decaying grid turbulence

($\Rightarrow C_{\varepsilon_2}$), near-wall flows (assuming production = dissipation and $\overline{u'v'}/k \approx$ constant $\Rightarrow C_{\varepsilon_1}, C_\mu$), and buoyancy dominated measurements ($\Rightarrow C_{\varepsilon_3}$). With computer optimization, the constants are assigned values so that the widest-range of flows possible can be predicted. This is in fact how constants are effectively determined in all phenomenological turbulence models. The following values are recommended : $C_\mu = 0.09$; $C_{\varepsilon_1} = 1.44$; $C_{\varepsilon_2} = 1.92$; $C_{\varepsilon_3} \approx 1.0$; $\sigma_k = 1.0$; $\sigma_\varepsilon = 1.3$. For free flows $\sigma_t = 0.7$ and near-wall flows $\sigma_t = 0.9$. For a round jet issuing into stagnant surrounds improved spreading rates in the self-similar region are obtained with $C_{\varepsilon_2} = 1.85$ and $C_\mu = 0.065$.

The k-ε model is robust and has been (and still is) widely used; it can lead to useful results in many circumstances. However it is clearly a very simple representation of what is undoubtedly an exceedingly complex phenomenon and consequently it does have its limitations. Specifically it is incapable of accurately reproducing the effects of mean swirling motions on mixing rates, the stabilizing/de-stabilizing influences of buoyancy forces and countergradient diffusion effects (as observed in premixed flames). In addition it predicts isotropy in statistically homogeneous flows (i.e. flows with no mean shear). A more detailed and accurate description of turbulent transport is needed.

2.1.2. *Second moment closures.* Higher order closures are employed to provide a more accurate description of turbulent transport, and second-moment (single-point) closures represent about the simplest level at which this can be achieved. The equations to be solved are the Reynolds stress equation (21) and the scalar flux equation :

$$\underbrace{\bar{\varrho}\frac{D}{Dt}\widetilde{u_i''\phi_\alpha''}}_{\text{I}} + \underbrace{\bar{\varrho}\widetilde{u_j''\phi_\alpha''}\frac{\partial\tilde{u}_i}{\partial x_j}\bar{\varrho}\widetilde{u_i''u_j''}\frac{\partial\tilde{\phi}_\alpha}{\partial x_j}}_{\text{II}} = \underbrace{-\frac{\partial}{\partial x_j}\left(\overline{\varrho u_i''u_j''\phi_\alpha''}\right)}_{\text{III}} - \underbrace{\overline{\phi_\alpha''\frac{\partial p'}{\partial x_i}}}_{\text{IV}} \\ \underbrace{-\overline{\phi_\alpha''}\frac{\partial\bar{p}}{\partial x_i}}_{\text{V}} + \underbrace{\overline{u_i''\dot{\omega}_\alpha}}_{\text{VI}} \tag{35}$$

As in equation (21) all terms on the rhs of equation (35) require closure. Term IV in equation (35) is referred to as pressure scrambling and term VI is a velocity-reaction source correlation term. As in the discussion of equation (21), order of magnitude and local isotropy arguments can be employed to neglect terms like $\overline{\phi_\alpha''(\partial\tau_{ik}/\partial x_k)}$ and $\overline{u_i''(\partial J_{k,\alpha}/\partial x_k)}$. The use equation (35) has so far been restricted to situations where $\overline{u_i''\dot{\omega}_\alpha} = 0$ (even in combusting flows) by making a suitable choice of ϕ_α, i.e. a strictly conserved scalar (section 3.2). For the time being a conserved scalar, ξ, can be viewed as a normalized element mass fraction which is, of course, neither consumed or produced. In order to complete the picture for combusting flows within the context of second-moment closures, the scalar variance equation is also required :

$$\bar{\varrho}\frac{D\widetilde{\phi_\alpha''^2}}{Dt} + 2\bar{\varrho}\widetilde{u_j''\phi_\alpha''^2}\frac{\partial\tilde{\phi}_\alpha}{\partial x_j} = \underbrace{-\frac{\partial}{\partial x_j}\left(\bar{\varrho}\widetilde{u_j''\phi_\alpha''^2}\right)}_{transport} - \underbrace{2\bar{\varrho}\tilde{\chi}_\alpha}_{dissipation} + \underbrace{\overline{\phi_\alpha''\dot{\omega}_\alpha}}_{\dot{\omega}_\alpha\ source} \tag{36}$$

Note that the scalar dissipation $\tilde{\chi}_\alpha$ in equation (36) is defined, for the case of Fickian diffusion, as $(\bar{\varrho})^{-1}\overline{(\mu/Sc)(\nabla\phi''_\alpha)^2}$. The closure problem is now set; all terms appearing on the rhs of equations (21), (35) and (36) require modelling.

Second moment closures (and to a certain extent eddy-viscosity closures) are based on the assumption of a dynamic equilibrium turbulence structure where the energy extracted from the mean motion at the large scales is balanced by transfer of energy through vortex stretching to the smaller scales. It is assumed that the structure of the flow can be uniquely determined by the characteristics of the large scale motions with the fine scale motions being unaffected by the mean strain. In rapidly evolving turbulent flows the concept of a dynamic equilibrium structure may lead to errors. In addition, characterization of the flow in terms of a single length and time scale is also subject to uncertainty for example in the viscous sub layer of a wall boundary layer. Only a brief discussion of second moment closures will be given here and a more detailed review can be found in the works of Jones [53; 55] and Launder [76]. Traditionally approximations have been constructed on a term-by-term basis and validated against measurements in simple configurations (decaying isotropic turbulence, nearly homogeneous turbulence subject to mean strain). There is clearly uncertainty in extending such approaches to variable density situations, particularly since very few measurements appear to exist in simple flows with large density variation. We now consider the closure approximations :

Viscous destruction of $\widetilde{u''_i u''_j}$

The treatment of term IV in equation (21) has already been discussed, leading to equation (23). Thus the modelled dissipation rate equation (34) is required to close this term.

Fluctuating Pressure terms

A principal difficulty in constant and variable density flows is the fluctuating pressure terms. These include the fluctuating velocity-pressure gradient correlation (term V in equation (21)) responsible for the transfer amongst the components of $\widetilde{u''_i u''_j}$, and the pressure scrambling (term IV in equation (35)). For constant density flows Chou [27] has shown that they can be related to volume integrals involving two-point correlations :

$$\overline{\frac{u_i}{\varrho}\frac{\partial p}{\partial x_j}} = \frac{1}{4\pi}\int \Bigg(\underbrace{\frac{\partial^3 \overline{u'_{l;\mathbf{x}'} u'_{m;\mathbf{x}'} u'_{i;\mathbf{x}}}}{\partial r_m \partial r_l \partial r_j}}_{A} + \underbrace{2\frac{\partial \bar{u}_{l;\mathbf{x}}}{\partial x_m}\frac{\partial^2 \overline{u'_{m;\mathbf{x}'} u'_{i;\mathbf{x}}}}{\partial r_l \partial r_j}}_{B} + \underbrace{\varrho_{\mathbf{x}'} g_l \frac{\partial^2 \overline{u'_{i;\mathbf{x}}}}{\partial r_l \partial r_j}}_{C} \Bigg) \frac{\mathrm{dVol}}{|\mathbf{x}'-\mathbf{x}|} + \text{Surface integral} \tag{37}$$

$$\overline{\frac{\xi}{\varrho}\frac{\partial p}{\partial x_j}} = \frac{1}{4\pi}\int \Bigg(\underbrace{\frac{\partial^3 \overline{u'_{l;\mathbf{x}'} u'_{m;\mathbf{x}'} \xi'_{\mathbf{x}}}}{\partial r_m \partial r_l \partial r_j}}_{A} + \underbrace{2\frac{\partial \bar{u}_{l;\mathbf{x}}}{\partial x_m}\frac{\partial^2 \overline{u'_{m;\mathbf{x}'} \xi'_{\mathbf{x}}}}{\partial r_l \partial r_j}}_{B} + \underbrace{\varrho_{\mathbf{x}'} g_l \frac{\partial^2 \overline{\xi'_{\mathbf{x}}}}{\partial r_l \partial r_j}}_{C} \Bigg) \frac{\mathrm{dVol}}{|\mathbf{x}'-\mathbf{x}|} + \text{Surface integral} \tag{38}$$

The terms on the rhs with subscript $\mathbf{x}'(\equiv \mathbf{x}+\mathbf{r})$ are evaluated at that position, with the integrations carried over $\mathbf{r}$ space. The surface integral is important only when the typical size of the energy-containing eddies is of the same order as the distance from a wall. In many combustion systems this contribution is ignored, partly because the ability to reproduce very accurately the wall friction and turbulence structure close to solid surfaces is often of secondary importance. Equations (37) and (38) contain three terms labelled as A, B and C. Term C is clearly a buoyancy contribution. Term B is affected by mean strain rates and is commonly referred to as the 'rapid' contribution - because it is the pressure term which appears in the linear problem of turbulence subjected to a rapid distortion. Term A involves turbulence quantities only. In the decay of homogeneous, strain-free and initially isotropic turbulence which has been distorted so that $u'^2 \neq v'^2 \neq w'^2$, the high Re (constant density) stress equations (21) with (23) become :

$$\frac{d\overline{u'^2}}{dt} = 2\overline{\frac{p}{\varrho}\frac{\partial u'}{\partial x}} - \tfrac{2}{3}\varepsilon; \qquad \frac{d\overline{v'^2}}{dt} = 2\overline{\frac{p}{\varrho}\frac{\partial v'}{\partial x}} - \tfrac{2}{3}\varepsilon; \qquad \frac{d\overline{w'^2}}{dt} = 2\overline{\frac{p}{\varrho}\frac{\partial w'}{\partial x}} - \tfrac{2}{3}\varepsilon \tag{39}$$

It is assumed that this type of turbulence 'returns to isotropy', an assumption which is supported by experiment [118]. Consequently term A is referred to as the return component. Denoting the three contributions A, B and C as $\mathcal{A}_{ij}^{(1)}$, $\mathcal{A}_{ij}^{(2)}$, $\mathcal{A}_{ij}^{(3)}$ respectively, closure approximations for these are obtained. The basis for modelling $\mathcal{A}_{ij}^{(1)}$ is the concept of return to isotropy. Rotta [107] assumed that the rate of return to isotropy, $\mathcal{A}_{ij}^{(1)}$, is proportional to the degree of anisotropy, $b_{ij} \equiv \left(\widetilde{u_i''u_j''}/k\right) - (2/3)\delta_{ij}$. In density weighted form, the model is :

$$\mathcal{A}_{ij}^{(1)} = -c_1 \varepsilon b_{ij} \tag{40}$$

where c_1 is the constant of proportionality obtained from measurements (e.g. [118]). It is found that the degree of scatter in the measurements leads to the following popular choices for c_1: 1.5, 1.8 and 3.0; this demonstrates the inadequacy of the linear theory, however its simplicity and the fact that it leads to reasonable results in practical flows justifies its use. More recently it has been argued that there is a 'tendency to isotropy' such that there is no linear return as the anisotropy goes to zero. This leads to refinements (with added complexity) in the modelled expression and are not discussed here.

For the contribution $\mathcal{A}_{ij}^{(2)}$, a simple approach is to assume that the mean strain has no direct influence on the return contribution; a widely-used form is that developed by Gibson and Launder [44] :

$$\mathcal{A}_{ij}^{(2)} = -c_2\left(P_{ij} - \tfrac{1}{3}P_{ll}\right) \qquad \text{where} \qquad P_{ij} = -\bar{\varrho}\left(\widetilde{u_i''u_l''}\frac{\partial \tilde{u}_j}{\partial x_l} + \widetilde{u_j''u_l''}\frac{\partial \tilde{u}_i}{\partial x_l}\right) \tag{41}$$

c_2 was originally assigned 0.6 and the return contribution, $c_1 = 1.6$; values chosen to conform with the exact results stemming from rapid distortion of initially isotropic turbulence. Gibson and Younis [88] later revised the constants by increasing the return constant at the expense of the rapid part ($c_1 = 3.0, c_2 = 0.33$), since conforming to rapid distortion theory was considered too restrictive as a modelling guideline and furthermore it was argued that the implications of rapid distortion theory were essentially inconsistent with the modelling concept behind second moment closures (i.e. the large scales do not affect directly the fine scales). The modification in the constant led to improvements in a wide range of complex flows [55].

Conventionally, the effects of buoyancy on turbulence structure are associated with $\mathcal{A}_{ij}^{(3)}$. The lack of data in simple flows with strong buoyancy effects implies that models have to be tested directly in inhomogeneous flows, making it difficult to isolate the contribution made by the different sub-models. A widely-used proposal is that originally proposed by Launder [75] :

$$\mathcal{A}_{ij}^{(3)} = c_3 \left(\bar{\varrho} g_i \overline{u_j''} + \bar{\varrho} g_j \overline{u_i''} - \tfrac{2}{3} \delta_{ij} \bar{\varrho} g_l \overline{u_l''} \right) \tag{42}$$

where $c_3 = 0.5$. Used in conjunction with (40) and (41) it yields good results over quite a wide range of conditions.

The modelling of the pressure scrambling term (IV in equation (35)) is difficult since its validation relies on an adequate representation of $\widetilde{u_i'' u_j''}$, and experimental data for scalar fields in simple situations is relatively scarce. Attempts at addressing the modelling through equation (38) and involving simple linear models are not fruitful and the following approximation has been widely used :

$$\overline{\phi_\alpha'' \frac{\partial p'}{\partial x_i}} = c_{\phi_1} \bar{\varrho} \frac{\varepsilon}{k} \widetilde{u_i'' \phi_\alpha''} - c_{\phi_2} \bar{\varrho} \widetilde{u_i'' \phi_\alpha''} \frac{\partial \tilde{u}_i}{\partial x_j} - c_{\phi_3} \bar{\varrho} g_i \overline{\phi_\alpha''} \tag{43}$$

with $c_{\phi_1} = 3$, $c_{\phi_2} = 0.5$ and $c_{\phi_3} = 0.5$. There is no correspondence between the three terms in equation (43) and terms A, B and C in equation (38).

Mean pressure gradient contribution

The modelling of term VI in equation (21) and term V in equation (35) is now considered. These terms do not appear in constant density flows, and are viewed as representing the preferential effect of a mean pressure gradient acting on light and heavy (or burnt and unburnt) 'packets' of gas. Its effect on the magnitude of the components of $\widetilde{u_i'' u_j''}$ and $\widetilde{u_i'' \phi_\alpha''}$ is an increase or decrease depending on the orientation of the pressure and density field. This term is considered as partly responsible for reproducing countergradient transport as observed experimentally [79]. It is possible to derive an exact equation for $\overline{u_i''}$ and then to introduce closure approximations as appropriate [53]. However there exists considerable uncertainty regarding the validity of the modelled equation and it may well be that the approximation,

$$\overline{u_i''} = \frac{\overline{\phi_\alpha''} \widetilde{u_i'' \phi_\alpha''}}{\widetilde{u_i''^2}} \tag{44}$$

will be sufficiently accurate in many circumstances. It is at least consistent with the exact result which may be derived for the case of mixing between two inert gases. $\overline{\phi_\alpha''}$ is closed through the combustion model (section 3). In terms of the density weighted pdf, the result is :

$$\overline{\phi_\alpha''} = \bar{\varrho} \int_{\phi_{\alpha,1}}^{\phi_{\alpha,2}} \left(\psi_\alpha - \tilde{\phi}_\alpha \right) \frac{\tilde{P}_\phi(\underline{\psi})}{\varrho(\underline{\psi})} \, d\psi_\alpha \tag{45}$$

where ψ_α is the sample space of the random variable $\phi_\alpha(\mathbf{x}, t)$.

Scalar variance dissipation rate

Following Spalding's [110] idea, the simplest method of approximating the scalar dissipation term ($\tilde{\chi}_\alpha$ in equation (36)) is by presuming that the ratio of the scalar to mechanical turbulence time scale is constant so that :

$$\tilde{\chi}_\alpha \approx \frac{C_D}{2} \frac{\varepsilon}{k} \widetilde{\phi_\alpha''^2} \tag{46}$$

where $(C_D/2)^{-1}$ is the time scale ratio with $C_D \approx 2.0$. In fact studies in decaying grid turbulence suggest that the time scale ratio can vary from 0.75 to 3.3. However, the approximation has been widely used and gives reasonable results in many thin shear layer flows. $\tilde{\chi}_\alpha$ can be interpreted as the rate at which fluctuations in $\phi_\alpha(\mathrm{x}, t)$ are reduced by molecular diffusion. A more general way of obtaining $\tilde{\chi}_\alpha$ is from a modelled transport equation [62].

Transport Terms

It is argued, mainly from experience in constant density thin shear layer flows, that the transport terms do not provide a major contribution to the second moment balance equations. The simplest level of approximation is a gradient transport hypothesis, originally proposed by Daly and Harlow [29] for the triple correlation $\overline{\varrho u_i'' u_j'' u_k''}$; in density weighted form the results can be expressed as :

$$\left(\overline{\varrho u_i'' u_j'' u_k''} + \tfrac{2}{3}\delta_{ij}\overline{u_k'' p'}\right) = -c_s \bar{\varrho} \frac{k}{\varepsilon} \widetilde{u_k'' u_m''} \frac{\partial \widetilde{u_i'' u_j''}}{\partial x_m} \tag{47}$$

$$\overline{\varrho u_i'' u_j'' \phi_\alpha''} = -c_s \bar{\varrho} \frac{k}{\varepsilon} \widetilde{u_j'' u_m''} \frac{\partial \widetilde{u_i'' \phi_\alpha''}}{\partial x_m} \tag{48}$$

$$\widetilde{u_i'' \phi_\alpha''^2} = -c_s \frac{k}{\varepsilon} \widetilde{u_i'' u_j''} \frac{\partial \widetilde{\phi_\alpha''^2}}{\partial x_j} \tag{49}$$

where $c_s \approx 0.22$.

The treatment of second moment closures is now complete. The model is considerably more elaborate than their eddy-viscosity counterparts and the resulting improvement in the ability to predict complex flow fields has been amply demonstrated, particularly in situations where k-ε is inadequate. Though the computational cost is larger than that required by the k-ε model, it is not prohibitive and the use of second moment closures is currently the most detailed approach feasible for complex and practical problems. The closure described is one of the simplest which can be constructed and more complex and general forms have been proposed and are being developed. However, to review these is beyond the scope of this article.

2.2. MEAN REACTION RATES

We conclude this section on the governing equations with a brief description of the other source of difficulty in the averaged conservation equations (17-20). This is associated with the description of the thermochemical field and manifests itself through the appearance of the terms $\bar{\dot{\omega}}_\alpha$ and $\bar{\varrho}$. In principle,

$$\bar{\dot{\omega}}_\alpha \equiv \lim_{N_p \to \infty} \frac{1}{N_p} \sum_{\mathrm{i}=1}^{N_p} \dot{\omega}_\alpha^{(\mathrm{i})}\left(T, \varrho, \phi_1, \ldots, \phi_N\right) \qquad \text{for} \qquad \alpha = 1, \ldots, N \tag{50}$$

The form of $\dot{\omega}_\alpha^{(\mathrm{i})}$ is given by equations (9) and (10). One possible approach is to decompose T, ϱ, ϕ_α $(\alpha = 1, \ldots, N)$ into mean and fluctuating parts, expand the exponential term (equation (9)) about the mean temperature using a Taylor series and then average the result. Typically the expansion is only valid for $|T'/T| < 1$ and this is not so severe in comparison to the requirement that for tractable closures the series must be truncated beyond terms involving quadratic powers; a truncation only justified for $E/RT \ll 1$, whilst for heat release reactions $E/RT \approx 10$. Consequently the method is unsuitable for combustion with significant heat release.

3. Non-premixed Combustion

3.1. INTRODUCTION

In this section on non-premixed combustion, attention will focus on the averaged conservation equations. The application of the averaging process to the species transport equation leads to the mean reaction rate term. It was pointed out in section 2.2 that direct attempts at closure through Taylor series expansion are unfeasible for the general case of highly exothermic combustion. In fact to date, no suitable deterministic approximation to evaluating this term has been obtained, and consequently it seems natural to resort to probabilistic approaches to achieve closure.

Based on the assumptions invoked in section 2, it is clear that the mean reaction rate term, $\bar{\dot{\omega}}_\alpha$, is in general a function of the scalar field $\underline{\phi} = \phi_\alpha$ $(\alpha = 1, \ldots, N)$ where ϕ_α includes the species mass fractions as well as the enthalpy. By introducing the pdf of the scalar field $P_{\underline{\phi}}(\underline{\psi})$, where $\underline{\psi}$ denotes the sample space of $\underline{\phi}$, closure for the mean reaction rate (and for the entire scalar field) can, in principle, be obtained from the following basic property of a pdf :

$$\bar{\dot{\omega}}_\alpha(\underline{\phi}) = \int_{\underline{\psi}} \dot{\omega}_\alpha(\underline{\psi}) P_{\underline{\phi}}(\underline{\psi}) d\underline{\psi} \quad ; \quad \bar{\phi}_\alpha(\mathbf{x}, t) = \int_{\underline{\psi}} \psi_\alpha P_{\underline{\phi}}(\underline{\psi}) d\underline{\psi} \tag{51}$$

Note that $P_{\underline{\phi}}(\underline{\psi})$ is formally defined as the single point, single time, Eulerian joint pdf of the scalar field. In writing the above, several features remain outstanding. Firstly, a thermochemical model is required so that $\dot{\omega}_\alpha$ is known, on an instantaneous basis, as a function of the scalar domain, $\underline{\psi}$. Secondly, the pdf $P(\underline{\psi})$ (the $\underline{\phi}$ subscript is dropped for clarity) needs to be specified. Assuming these quantities are known, the next stumbling block is the integration over the scalar space. The integrals written in equation (51) are multi-dimensional and the cpu requirements required for their computation is demanding. It is therefore desirable to reduce the dimensionality of problem, ideally to a single scalar which characterizes the thermochemistry of the reacting mixture.

3.2. THE CONSERVED SCALAR FORMALISM

The basic framework upon which most of the traditional theory of non-premixed combustion has been developed is the conserved scalar formalism. Under this formalism, a conserved scalar is a quantity whose instantaneous evolution equation satisfies the following form :

$$\frac{\partial}{\partial t}(\varrho z_\alpha) + \frac{\partial}{\partial x_k}(\varrho u_k z_\alpha) = \frac{\partial}{\partial x_k}\left(\frac{\mu}{Sc}\frac{\partial z_\alpha}{\partial x_k}\right), \quad \alpha = 1, \ldots, N \tag{52}$$

No source terms for chemical reaction appear in equation (52). This requirement is naturally satisfied by quantities such as element mass fractions, which are not created or destroyed by reaction, but are simply convected and diffused. Note, however, that chemical reaction influences the evolution of z through the density, ϱ. The next important feature of equation (52) is the presence of a single diffusion coefficient. Since z can represent elemental concentrations as well as enthalpy, the use of a single diffusion coefficient implies uniform Prandtl and Schmidt numbers as well as unity Lewis number and the effects of differential diffusion are thereby ignored. The reasons for such a fundamental assumption need to be elaborated.

The justification for the assumption of equal diffusivity may be attributed to Reynolds number similarity [6; 116]. The concept stems from the energy cascade arguments used to describe the interactions taking place in a turbulent flow (see for example Bradshaw [9]). It is well-accepted that Fourier components of small wave number (eddies of large size) that contribute most to the kinetic energy, contribute relatively little to the viscous dissipation. In addition an increase in the Reynolds number is accompanied by an extension of the large wave-number end of the spectral distributions (i.e. the fine scale motions re-adjust themselves), with little effect on the large-scale motions. An implication of this is that the interaction between the large and fine scale motions takes place via a spectrum of intermediate-sized eddies with the consequence that theses two extreme scales of motion are largely independent of each other. Based on such ideas, and in the limit of large Reynolds number, it is assumed that effects associated with the small scales of turbulence are unimportant with regards to the modelling of the averaged conservation equations. It is clear from order of magnitude estimates that at high Reynolds number, turbulent fluxes outweigh the averaged molecular fluxes. What is not so clear is whether highly diffusive scalar properties, such as H and H_2 (in hydrocarbon flames), calculated from a formalism where equal diffusivities are assumed is justifiable, particularly since chemical reaction, which is responsible for their formation, occurs at the small scales of turbulent motion. This issue will be re-discussed when considering transported pdf methods.

With the assumption of equal diffusivity, the element mass fraction transport equations all satisfy the same form of differential equation, differing only in the boundary condition specifications. The assumptions of low Mach number, constant thermodynamic pressure and negligible radiative heat losses to the surroundings (guaranteed by a no heat flux condition on all bounding surfaces) implies that the enthalpy also satisfies a conserved scalar transport equation. Following Bilger [2], for a two-stream mixing process, with each stream containing a uniform, constant and different composition, the following normalization can be applied :

$$\xi_\alpha(\mathbf{x}, t) = \frac{z_\alpha(\mathbf{x}, t) - z_{\alpha,2}(\mathbf{x}, t)}{z_{\alpha,1}(\mathbf{x}, t) - z_{\alpha,2}(\mathbf{x}, t)} \tag{53}$$

where z_α denotes element mass fraction or enthalpy and ξ is referred to as the mixture fraction. Through this normalization it is clear that the transport equations for the various ξ_α are of the same form and their boundary conditions are identical too. It can then be argued, Bilger [2], that after an initial transient ($t > t_o$) all ξ_α at any place and time are equal, even on an instantaneous basis. Thus,

$$\xi_{\alpha,1}(\mathbf{x}, t) = \xi_{\alpha,2}(\mathbf{x}, t) = \ldots = \xi(\mathbf{x}, t) \tag{54}$$

In summary, subject to certain assumptions, all conserved scalars are linearly related. Consequently the problem of calculating several conserved scalars, z_α, can be circumvented by solving for a single normalized conserved scalar, ξ, constructed from any one of the individual conserved z_α. Multiple feed problems can be considered in the same manner as the above provided the streams have the same composition and temperature as in streams '1' and '2'. Any relaxation of the assumptions presented above requires the specification of extra scalars. For example, if adiabatic flow is no longer assumed, enthalpy would have to be included as an extra scalar since it would no longer be conserved due to the presence of heat loss.

In combustion the characterization of the conserved scalar field in a flow domain is only of secondary importance and obtaining the concentration field of the reactive species is the main objective. Consequently thermochemical models are required which are compatible with the conserved scalar formalism, i.e. knowledge of ξ will specify instantaneously the value of any scalar, conserved or otherwise. These models all share the common assumption of 'fast chemistry'. In the case of the flame-sheet model (to be described in section 3.3.1), 'fast' chemistry should be interpreted as 'infinitely fast' chemistry and the flame approximates to a geometric surface 'flapping' through the mixture as a result of the influence of turbulent fluctuations. Note, however, that on a mean basis, the reaction zone is thick, and often described as a turbulent flame brush. For the more realistic proposal based on chemical equilibrium and laminar flamelets, 'thin' flame burning is inherently presumed with the transport (aerodynamic) time scales being much longer than the chemical time scales (i.e the Damköhler number , $Da \gg 1$, and the Karlovitz number $Ka \ll 1$).

3.3. THERMOCHEMICAL MODELS

3.3.1. *The Flame sheet approximation.*

This model stems from the initial work of Burke and Schumann [18]. One step irreversible global kinetics is employed in the following form :

$$1\,Kg\ \ Fuel \quad + \quad s\,Kg\ \ oxidant \quad \longrightarrow \quad (1+s)\,Kg\ \ Product$$

For the purpose of constructing thermochemical relations, chemical rate expressions are not required since it is assumed that the reaction proceeds at an infinite rate, implying that as soon as the reactants mix at the molecular scale, they burn instantly to form fully-burned products. As a result, fuel and oxidant cannot co-exist instantaneously anywhere except within an infinitesimally thin flame sheet. Note, however, that on a time averaged basis, fuel and air can be present in finite concentrations at a particular point; this is commonly referred to as 'unmixedness' [48]. In this situation, simple conserved scalars, known as Shvab-Zeldovich coupling functions can be used.

$$z_{fu,ox} = Y_{fu} - \frac{Y_{ox}}{s}, \quad z_{fu,pr} = Y_{fu} + \frac{Y_{pr}}{1+s}, \quad z_{ox,pr} = Y_{ox} - \frac{sY_{ox}}{1+s}$$

A suitable normalization of any of the above conserved scalars will lead to the definition of a mixture fraction. The representation of a flame zone as a geometric surface is clearly unrealistic, as substantiated by experimental studies in laminar and turbulent flames. Furthermore, in hydrocarbon flames this model precludes the determination of any intermediate species (e.g. CO and H_2) since the simplicity of the underlying global kinetics is restricted to the presence of fuel, oxidant and products as the participating species in the reactive mixture. However, the model is capable of reproducing gross flame characteristics (i.e mean temperature) reasonably well provided the combustion is unhindered by significant finite-rate effects. Its representation of the mean density also results in satisfactory flow field mixing patterns. However, the flame sheet methodology has now been superseded by more realistic thermochemical models.

3.3.2. *The Chemical Equilibrium Approach.* Studies in chemical kinetics, not only restricted to combustion, demonstrate that one-step irreversible chemistry is a gross simplification. Nearly all reactions are reversible to some degree. The concept of chemical equilibrium can be put to use in the conserved scalar formalism in the following manner. At equilibrium, if the elemental composition and two state variables for the mixture (e.g. the thermodynamic pressure and the enthalpy) are known, then the species concentrations can be determined [46]. Based on the discussion above relating to the conserved scalar concept, it is clear that the elemental composition and the enthalpy are known if the mixture fraction is specified. Because the thermodynamic pressure is presumed constant, knowledge of ξ is sufficient to characterize, instantaneously, the mixture composition at equilibrium. In principle, if the reactions are reversible, fuel and oxidant can co-exist; in this case it can no longer be assumed that the reactions are confined to an infinitely thin flame sheet discontinuity, but occur where necessary to keep the mixture in equilibrium. It must be stressed that in the chemical equilibrium approach, no input is required from a chemical mechanism. Identifying the participating elemental species, knowledge of ξ and the appropriate thermo-chemical data is sufficient.

The chemical equilibrium approach has been widely used, and like the flame sheet model, reasonable temperature and velocity field characteristics can be obtained. The major shortcoming of the approach is related to the prediction of CO levels in hydrocarbon-air flames. For hydrocarbon-air mixtures it is found (Jones [53]) that in the case of lean mixtures, negligible quantities of CO concentrations are predicted. The proportion of CO increases progressively with mixture strength and for equivalence ratios greater than about 1.2 the CO mole fraction is greater than that of CO_2; at equivalence ratios of around three, CO mole fractions of about 20% can be achieved. These levels are wholly unrealistic in laminar and turbulent hydrocarbon flames. The conclusion to draw is that although the presence of minor species may not affect the gross flame features significantly, their levels are very poorly reproduced if it is assumed that the fluid at a particular mixture strength is at its equilibrium composition. In reality mixtures too lean or too rich will simply not burn and the same applies if the temperature is too low. Based on these facts, certain ad-hoc modifications have been applied to the chemical equilibrium approach in order to yield more realistic CO levels.

Godoy [45] chose to clip the equilibrium concentrations so that for mixtures leaner than the rich flammability limit of the fuel the instantaneous chemical equilibrium composition was employed, and for richer mixtures the equilibrium composition at the rich flammability limit diluted with surplus fuel was assumed to be present. In the work of Eickhoff and Grethe [38] a model was employed whereby zones of chemical equilibrium were allowed to exist provided the temperature was above some pre-determined value ($\approx$ 1500K) calculated from an equilibrium solution employing an empirically specified mass fraction of fuel, $Y_{A,1}$. For mixtures where the fuel concentration $Y_A < Y_{A,1}$, the equilibrium composition was adopted. For richer mixtures the gas was assumed to consist of unreacted fuel and reaction products, characterised in terms of an averaged mole fraction for the species in the equilibrium zone. Godoy's proposal appears to be slightly more realistic than that of Eickhoff and Grethe since the latter model leads to discontinuous instantaneous profiles of temperature and species concentration as a function of Y_A. In addition, clipping the equilibrium concentration in terms of the rich flammability limit of the fuel (which is a physical characteristic) should be preferred relative to an empirically determined temperature. Nevertheless both models lead to CO levels which were in far better agreement with measurements than with the conventional equilibrium approach.

3.3.3. *The Laminar Flamelet Approach.* In this section the basic ideas behind the laminar flamelet technique are outlined. For a more detailed review of the theory, the papers by Williams [128] and Peters [94] should be consulted. At the end of section 3.2, it was pointed out that the laminar flamelet formulation was entirely consistent with (and restricted to) 'thin' flame burning. In fact the thickness of the burning zone must be less than or equal to the Kolmogorov length scale - this fundamental requirement for the existence of flamelets (i.e. laminar-like burning zones in a turbulent medium) is referred to as the Klimov-Williams criterion.

The flame stretch rate, γ, a quantity of central importance to laminar flamelet theory, can be defined as the fractional time rate of increase of the flame-zone area,

$$\gamma = \frac{1}{\Lambda}\frac{d\Lambda}{dt} \tag{55}$$

where Λ represents an area of flame surface. Effects of curvature on the flame front are usually ignored, i.e locally the flame front is presumed planar. Such an assumption is reasonable provided the ratio of a characteristic flame thickness to a characteristic radius of curvature is less than the product of the flame stretch rate and the residence time (Williams [128]). From the definition it is clear that flame stretch rate can be positive or negative, with the latter being referred to as flame compression. The effects of flame stretch and compression can be partly explained by the corresponding convective fluxes through the flame zone using a flame-attached coordinate system. The following results emerge : For stretch there must be a tendency for the flow to be towards the reaction plane. With nonpremixed combustion this implies that fuel and oxidiser, which are cooler than the flame zone, steepen the local temperature profile and make the region of heated gas thinner. The converse effect i.e. flame thickening of the heated region, occurs under compression although it is usually argued that stretch statistically predominates over compression in turbulence and it then emerges that γ plays the role of a strain rate or velocity gradient. The discussion above concentrates on non-premixed flames, but the concept of flame stretch is equally applicable to premixed flames.

A turbulent flame is thus viewed as an ensemble of strained laminar flamelets. The experimental observation that sections of a flame can break away from the bulk structure upstream, together with its highly contorted nature demonstrates that the flame zone can be viewed as a multiply-connected region. However, it is argued that even under these circumstances, provided $Da \gg 1$ and $Ka \ll 1$, the reaction zone is sufficiently thin so that it can instantaneously consume reactants in a 'laminar fashion'. It is clear that if the degree of flame stretch is too large, the flamelet will be quenched, leading to local extinction. Extinction studies of laminar diffusion flames show that the characteristic controlling parameter is Da. However, a unique quantitative definition of Da cannot be given and consequently Da is unsuitable for quantifying the flame stretch. The fact that gradients of scalars increase/decrease in magnitude depending on whether flame stretch/compression is operative suggests that the instantaneous scalar dissipation rate, χ, may be used as a quantitative parameter. If it is assumed that only the flame zone is critically affected by stretch, then, the external stretch imposed on the flamelet by the hydrodynamic field has the form of an instantaneous scalar dissipation rate evaluated at stoichiometric conditions $\chi_{st} = (\frac{\nu}{Sc})_{st}(\partial\xi/\partial x_j)^2_{st}$

Consequently if the local flame stretch increases, χ_{st} increases accordingly and the heat conduction from the reaction zone will increase since the local gradient across the flame zone increases.

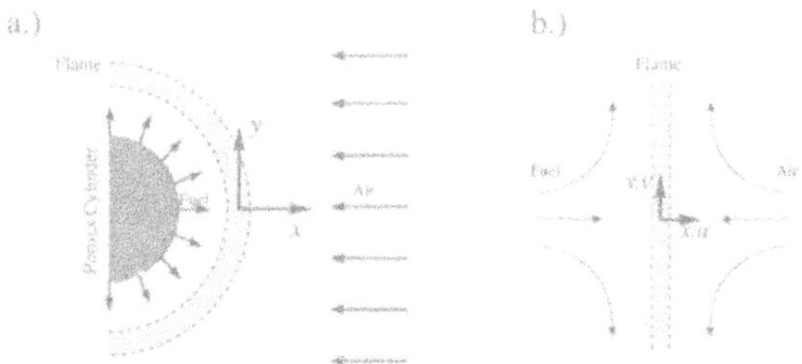

Fig. 1. a.) Schematic of a laminar counterflow diffusion flame geometry. The width of the flame is exaggerated for clarity b.) Locally the flame is treated as planar and the flow field is well represented by a potential flow solution.

If χ_{st} is increased beyond a critical value, χ_q , heat conduction to both sides of the flame can no longer be balanced by heat production due to chemical reaction and the flamelet is extinguished. In the literature analytical and experimental methods have been employed to determine χ_q and Da_q ([78; 117]). However, it should be borne in mind that χ_q is dependent on the thermodynamic and chemical-kinetic parameters, and as such it can be evaluated without consideration of turbulence. The basic geometry normally used to investigate such features is a laminar counterflow diffusion flame (fig 1) which is established in the forward stagnation region of a porous cylinder.

In fig 1, the flame thickness is sufficiently thin so that it can be treated as locally planar. Measurements suggest that where the fuel and air merge to form the reaction zone, the flow can be accurately viewed as an opposed jet whose velocity field is given by a potential flow solution, i.e. $u = -ax, v = ay$. The flame is increasingly stretched by raising the stagnation velocity gradient, a, which has dimensions of reciprocal time and is proportional to the inverse of a characteristic aerodynamic time - in other words, it has a similar interpretation to the scalar dissipation rate. However, 'a' represents a gross strain rate imposed on the flow, whereas χ_{st} is the local scalar dissipation rate at the flame front. The two quantities can be related (Peters and Williams [97]) :

$$\chi_{st} \approx 4a\xi_{st}^2 \left[\text{erfc}^{-1}(2\xi_{st})^2\right] \tag{56}$$

where erfc^{-1} denotes the inverse complementary error function. As 'a' is increased, the net rate of formation or consumption of the species and the heat release gradually rise. When the reaction rate reaches some critical value, the flame is blown off implying limitation due to the chemical kinetics.

In the above paragraphs, the basic ideas behind the laminar flamelet formulation have been expounded. In order to make use of such a formulation in turbulent flame calculations, a thermochemical model is required as discussed in sections 3.3.1 and 3.3.2. The following observation clarifies how this can be achieved. Bilger [4] examined data on laminar counterflow diffusion flames so as to determine the uniqueness of the $\phi_\alpha(\xi)$ relations. The implication for the various species profiles, re-plotted as a function of ξ and falling onto a single curve, was that the evolution of the laminar flame configurations satisfied a self-similarity solution, with ξ as the similarity variable. The definition of the mixture fraction against which one can replot the data raises an important point. For the data of Tsuji and Yamaoka [117], Bilger [4] reports that curves for ξ_C, ξ_H, ξ_O and ξ_N generally fell onto a single curve, with ξ_O showing the largest deviations. Given the similarity of the behaviour of these curves it may be (erroneously) concluded that a single mixture fraction can characterize the thermochemistry (and hence that differential diffusion effects are unimportant). In the case of a laminar flame this is strictly wrong, and in the case of high Re turbulence equal diffusivity is one of the fundamental assumptions. However, one non-radical specie which has a diffusivity significantly greater than the rest of the main participants in the reaction is H_2. But since H_2 appears in such low concentrations, it makes a very small contribution to ξ_H and consequently the fact the ξ's based on various elements show a similar behaviour does not provide definitive evidence that differential diffusivity is unimportant. In fact the main contributors to the various ξ_α are species which diffuse at much the same rate by virtue of their molecular masses being roughly equal.

The relations, $\phi_\alpha = \phi_\alpha(\xi)$, can therefore be used as a suitable thermochemical input for the modelling of turbulent diffusion flames. Its significant advantage over the previous models described in section 3.3 is that more realistic chemistry can be introduced into a turbulent flame calculation, and all the information (such as differential diffusion effects) are contained within $\phi_\alpha(\xi)$. Thus, the small-scale details of a turbulent flow (where chemical reaction occurs) can be accounted for in an indirect manner through $\phi_\alpha(\xi)$. Liew [81] employed experimental measurements in order to construct $\phi_\alpha(\xi)$. However, it is equally possible, and potentially more versatile, to obtain $\phi_\alpha(\xi)$ by solving the exact conservation equations for a laminar flame employing detailed kinetics and transport. The most favoured configurations (which have to conform to a similarity solution) are the steady two-dimensional counterflow diffusion flame geometry and the transient one-dimensional laminar flame. In this manner flamelet 'libraries' may be constructed on an once and for all basis and expressed in the instantaneous form, $\phi_\alpha(\xi, \chi_{st})$. It should be pointed out that one of the disadvantages of the flamelet technique is that for many fuels of engineering interest, the details of the reaction mechanisms are not adequately known, e.g kerosene in gas turbines and the variety of petrols used in internal combustion engines. Furthermore, many devices operate under lean conditions (e.g. gas turbine combustors) and in such situations the difference in the results from an equilibrium calculation and one based on flamelet thermochemistry are not significantly different, overall.

3.4. THE PRESUMED SHAPE PDF METHOD

In this section we are interested in the calculation of the averaged species concentrations, $\tilde{\phi}_\alpha(\mathbf{x}, t)$, using the basic idea behind equation (51). In the present case, $P(\underline{\psi})$ is interpreted as $P(\hat{\xi})$. For calculating density-weighted averages, the following relation can be employed :

$$\tilde{\phi}_\alpha(\mathbf{x}, t) = \int_0^1 \phi_\alpha(\hat{\xi})\tilde{P}(\hat{\xi})d\hat{\xi} \tag{57}$$

where the density-weighted pdf is defined as :

$$\tilde{P}(\hat{\xi}) = \frac{\varrho(\hat{\xi})P(\hat{\xi})}{\bar{\varrho}(\mathbf{x}, t)} \quad \text{and} \quad \bar{\varrho} = \left\{ \int_0^1 \frac{\tilde{P}(\hat{\xi})}{\varrho(\hat{\xi})} d\hat{\xi} \right\}^{-1} \tag{58}$$

The non-linearity of the instantaneous relations, $\phi_\alpha(\xi)$, implies that the use of $\tilde{\phi}_\alpha(\mathbf{x},t) = \phi_\alpha(\tilde{\xi})$ will lead to large errors - the Taylor series approximation referred to in section 2.2 also indicates this. In calculation methods the most popular approach is the presumed-shape pdf method. In this method a two parameter form of the pdf is presumed for the single random variable ξ in terms of its mean and variance, the values of which are obtained from the solution of their respective modelled transport equations. Since for combusting flows these equations are usually solved in density-weighted form, the resulting pdf that is constructed from these moments will be the density-weighted pdf. The transport equations for the mean and variance of ξ are very similar in form to equations (20) and (36) with $\dot{\omega}_\alpha = 0$. The modelled forms are written below for completeness.

$$\bar{\varrho}\frac{\partial\tilde{\xi}}{\partial t} + \bar{\varrho}\tilde{u}_j\frac{\partial\tilde{\xi}}{\partial x_j} = \frac{\partial}{\partial x_j}\left(\frac{\mu_t}{\sigma_t}\frac{\partial\tilde{\xi}}{\partial x_j}\right) \tag{59}$$

$$\bar{\varrho}\frac{\partial\widetilde{\xi''^2}}{\partial t} + \bar{\varrho}\tilde{u}_j\frac{\partial\widetilde{\xi''^2}}{\partial x_j} = \frac{\partial}{\partial x_j}\left(\frac{\mu_t}{\sigma_t}\frac{\partial\widetilde{\xi''^2}}{\partial x_j}\right) + 2\frac{\mu_t}{\sigma_t}\left(\frac{\partial\widetilde{\xi''^2}}{\partial x_j}\right)^2 - \bar{\varrho}C_D\frac{\varepsilon}{k}\widetilde{\xi''^2} \tag{60}$$

The turbulent fluxes appearing in the exact equations, $\widetilde{u_j''\xi''}$ and $\widetilde{u_j''\xi''^2}$, have been modelled using a gradient transport hypothesis and the scalar dissipation rate obtained from equation (46). In order to evaluate μ_t, the k-ε model is usually employed; alternatively a full second moment closure can be used in place of equations (59) and (60), as discussed in section 2.1.2.

Several different forms of the two-parameter pdf have been proposed. For example a form constituting two delta functions located at points ξ^+ and ξ^- has been applied by Khalil et al. [69] :

$$\tilde{P}(\hat{\xi}) = a\delta(\hat{\xi} - \xi^+) + (1-a)\delta(\hat{\xi} - \xi^-) \tag{61}$$

where a,ξ^+ and ξ^- are determined from $\tilde{\xi}$ and $\widetilde{\xi''^2}$. However, it is highly unlikely that the pdf in a reacting flow would resemble this form. On comparison with the measurements of Kent and Bilger [67] in H_2-air diffusion flames, Jones [53] concluded that such a pdf was unsatisfactory since it led to radial profiles of temperature in which two maxima appeared. Lockwood and Naguib [83] proposed a more realistic clipped Gaussian pdf. The clipping was required since the mixture fraction is a bounded random variable. An unweighted pdf was employed in their formulation, defined by :

$$P(\hat{\xi}) = \begin{cases} \frac{1}{\sigma\sqrt{2\pi}} \int\limits_{-\infty}^{0} \exp\left[-\frac{1}{2}\left(\frac{\hat{\xi}-\mu}{\sigma}\right)^2\right] d\hat{\xi} & \text{if } \hat{\xi} = 0 \\ \frac{1}{\sigma\sqrt{2\pi}} \exp\left[-\frac{1}{2}\left(\frac{\hat{\xi}-\mu}{\sigma}\right)^2\right] & \text{if } 0 < \hat{\xi} < 1 \\ \frac{1}{\sigma\sqrt{2\pi}} \int\limits_{1}^{\infty} \exp\left[-\frac{1}{2}\left(\frac{\hat{\xi}-\mu}{\sigma}\right)^2\right] d\hat{\xi} & \text{if } \hat{\xi} = 1 \end{cases} \tag{62}$$

where μ and σ cannot be obtained explicitly; an iterative procedure employing $\bar{\xi}$ and $\overline{\xi'^2}$ had to be used. A pdf which does not require arbitrary clipping in order to satisfy the bounds of the mixture fraction is the beta pdf :

$$\tilde{P}(\hat{\xi}) = \frac{\hat{\xi}^{a-1}(1-\hat{\xi})^{b-1}}{\int_0^1 x^{a-1}(1-x)^{b-1}dx} \tag{63}$$

$$\text{where} \quad a = \tilde{\xi}\left[\tilde{\xi}\frac{(1-\tilde{\xi})}{\widetilde{\xi''^2}} - 1\right] \quad \text{and} \quad b = (1-\tilde{\xi})\left[\tilde{\xi}\frac{(1-\tilde{\xi})}{\widetilde{\xi''^2}} - 1\right] \tag{64}$$

The integral in the denominator of equation (63) is referred to as the β function. This form of pdf has been widely used for engineering calculations of reacting turbulent flow. The implementation of equation (57) is particularly straightforward in the context of the β-pdf. The instantaneous relations, $\phi_\alpha(\xi)$, can usually be curve-fitted using a least-squares polynomial (of order ≈ 20) and the relations can then be expressed in the form,

$$\phi_\alpha = \sum_{j=0}^{N} \mathcal{A}_{j,\alpha}\xi^j \tag{65}$$

By substituting (65) into (57) and by employing the relations pertaining to the β- and Γ-function :

$$\beta(a,b) = \frac{\Gamma(a)\Gamma(b)}{\Gamma(a+b)} \quad \text{and} \quad \Gamma(a) = a\Gamma(a) \tag{66}$$

the following analytical expression for the mean of a dependent scalar variable is obtained :

$$\tilde{\phi}_\alpha = \sum_{j-0}^{N} \mathcal{A}_{j,\alpha}\frac{\prod_{i=0}^{j-1}(a+i)}{\prod_{i=0}^{j-1}(a+b+i)} \tag{67}$$

The numerator and denominator are defined for $j \geq 1$. According to comparisons made by Jones (1980) of the clipped Gaussian and the beta pdf, the measured temperature and species mass fractions were accurately predicted in diffusion flames. However the results were quite insensitive to the precise shape of the pdf once the mean and variance were determined.

The effect of intermittency, which manifests itself mathematically as a delta function at the bounds of a continuous pdf, has been examined in the formulation of Kent and Bilger [67]. The method involved splitting the pdf into its non-turbulent part where $\xi = 0$ (for a jet flow configuration) and a turbulent part characterized by $\tilde{\xi}_t$ and $\widetilde{\xi_t''^2}$. The Favre intermittency $\tilde{I}(\mathbf{x})$ is related to these quantities through the following relations :

$$\tilde{\xi} = \tilde{I}(\mathbf{x})\tilde{\xi}_t \tag{68}$$

$$\widetilde{\xi''^2} = \tilde{I}\left[\widetilde{\xi_t''^2} + \bar{\varrho}\left(\tilde{\xi}_t - \tilde{\xi}\right)^2\right] + \bar{\varrho}\left(1 - \tilde{I}\right)\tilde{\xi}^2 \tag{69}$$

Experimental data was used to obtain the intermittency function. From a plot of $\tilde{I}$ versus $(1 + \widetilde{\xi''^2}/\tilde{\xi}^2)^{-1}$ a linear relation was postulated :

$$\tilde{I} = \frac{1 + K}{\left[1 + \widetilde{\xi''^2}/\tilde{\xi}^2\right]} \tag{70}$$

where K can vary between 0 to 0.4. $\tilde{\xi}$ and $\widetilde{\xi''^2}$ were obtained from their respective transport equations and $\tilde{I}$ was then calculated from equation (70) and substituted into (68) and (69) to yield $\tilde{\xi}_t$ and $\widetilde{\xi_t''^2}$. These quantities were then used to construct the pdf in the turbulent region. The shape of the pdf was represented by ten delta functions (approximating a Gaussian pdf), each containing 10% of the probability. The value of K and the form of the pdf had little effect on the mean density, velocity and conserved scalar field. Good agreement with measurements was found in these cases.

3.5. TREATMENT OF TRACE SPECIES USING THE CONSERVED SCALAR FORMALISM

Attempts to calculate the concentration of trace species include the prediction of trace quantities such as nitrogen monoxide (NO). The formation of NO at high temperatures is accepted to be controlled by the extended Zeldovich mechanism :

$$\begin{aligned} O + N_2 &\rightleftharpoons NO + N \\ N + O_2 &\rightleftharpoons NO + O \\ N + OH &\rightleftharpoons NO + H \end{aligned}$$

The overall NO formation rate is much slower than those of the main heat release reactions and the concentrations of NO involved are such that neither the temperature nor the main species compositions are affected by its presence. In most combustion systems the NO concentration will be well below its equilibrium levels. In order to close the system of above equations, so that it is completely specified in terms of ξ, a steady state assumption for the N atom concentration and a partial equilibrium assumption for the O atom concentration is usually made. Consequently the mean rate of formation of NO per unit mass of mixture, $\overline{\dot{S}}_{NO}(\mathbf{x})$ is :

$$\overline{\dot{S}}_{NO}(\mathbf{x}) = \int_0^1 \dot{S}_{NO}(\hat{\xi})P(\hat{\xi};\mathbf{x})d\hat{\xi}$$

The results of Jones and Priddin [63] suggest that the above formulation is adequate at typical gas turbine operating pressures and inlet temperatures but the method does not produce accurate results at atmospheric conditions. Bilger and Beck [8] found for a series of measurements on vertical diffusion flames of hydrogen issuing into stagnant air that peak NO concentrations and NO formation rate are shifted towards the fuel-rich side of the flame. The phenomenon, referred to as the 'rich shift', is not detected using the above modelling. The main source of failure is thought to lie in the partial equilibrium assumption for the O atom concentration, because although the conversion of the major species may be 'fast' the recombination of free radicals is slow so that the free radical concentrations may be an order of magnitude above their equilibrium concentrations.

By constraining the thermochemical state of a mixture to be a function of the mixture fraction (as performed in the above modelling), it is not possible to account for finite rate effects rigorously. More independent reactive scalars have to be introduced in order to describe the thermochemical state. For example Janicka and Kollmann [52] applied a two-variable approach to the study of NO formation in H_2-air diffusion flames. The H_2-air kinetic scheme described by Dixon-Lewis et al. [32] was employed. Out of the seven reaction steps present in the mechanism, the four mole-conserving reactions are much faster than the three-body recombination reactions. Consequently it was assumed that the 'fast' reactions were in partial equilibrium. The 'slow' reactions had to treated so as to account directly for the highly non-linear formation rates. The participating species in the mechanism were H_2, O_2, H_2O, N_2, O, OH and H. Details of the thermochemical closure are provided in [52]. Briefly, the approach involved the specification of two variables; one for ξ and another for a normalized reaction progress variable, r. Knowledge of ξ and r together with three partial equilibrium expressions to calculate O, OH and H, balance equations for the total oxygen and hydrogen element concentration (related to ξ) and an equation of state characterised the thermochemistry. For nitric oxide formation, the extended Zeldovich scheme, shown above, was employed and the levels of O_2, OH, O and N_2 which appear in the scheme were given via the thermochemistry described above. Additional transport equations for NO and N were solved as a side calculation to the main H_2-air combustion. The average values of the dependent scalar field were obtained using the conventional presumed-shape pdf approach. However in this formalism, the joint pdf of ξ and r is required. The mean and the variances, $\bar{\xi}$, $\overline{\xi'^2}$, $\bar{r}$, $\overline{r'^2}$ are not sufficient to construct $P(\hat{\xi}, \hat{r}; \mathrm{x})$ and the values of the covariances of ξ and r are also required. This is where the presumed-shape pdf method demonstrates its main disadvantage. To perform calculations involving N independent scalars, the number of moment equations (which contain unclosed terms) required to construct the pdf requires the solution of $N(3+N)/2$ equations [64]. This makes the method very expensive and potentially inaccurate unless the number of independent scalars are kept to a minimum. With regards to $P(\hat{\xi}, \hat{r}; \mathrm{x})$, Janicka and Kollmann [52] simplified the problem by assuming that ξ and r are statistically independent, so that $P(\hat{\xi}, \hat{r}) = P(\hat{\xi})P(\hat{r})$. A β-function was employed for $P(\hat{\xi})$ and three δ-functions arbitrarily situated at r=0,1,$\bar{r}$ to represent $P(\hat{r})$ and in order to construct the pdf, modelled transport equations for $\bar{\xi}$, $\overline{\xi'^2}$, $\bar{r}$, $\overline{r'^2}$ had to be solved. Predictions for temperature, major species and H_2 agreed very well with measurements in jet flame configurations. NO formation was well predicted in the 'medium' range of the flames, but was somewhat underpredicted in the initial region and overpredicted far downstream. It was also concluded that the NO-formation, within the framework of the thermochemistry employed, was very sensitive to $\bar{\xi}$ and $\bar{r}$, but the shape of the pdf chosen for $P(\hat{\xi})$ and $P(\hat{r})$ did not appear to have an important effect.

Bilger [5] proposed a perturbation approach where he defined the species composition as : $\phi_\alpha = \phi^e_\alpha(\xi) + \vartheta_\alpha$. Here, $\phi^e_\alpha(\xi)$ is the equilibrium value of ϕ_α (obtained from a fast chemistry/equilibrium assumption) and ϑ_α represents the departure of species 'α' from its equilibrium value. Substituting this expression into the species transport equation and averaging leads to :

$$\bar{\varrho}\frac{\partial \vartheta_\alpha}{\partial t} + \bar{\varrho}\tilde{u}_k\frac{\partial \vartheta_\alpha}{\partial x_k} = \frac{\partial}{\partial x_k}(\overline{\varrho u_k'' \vartheta_\alpha}) + \overline{\varrho D \frac{\partial \xi}{\partial x_k}\frac{\partial \xi}{\partial x_k}\frac{d^2\phi_\alpha^e}{d\xi^2}} + \bar{\dot{r}}_\alpha \quad \text{for} \quad \alpha = 1, \ldots, N \tag{71}$$

The second term on the rhs of equation (71) is referred to as the microscale mixing source term, and represents the rate at which out-of-equilibrium material was produced by fine scale turbulent mixing. Closure for this term was achieved by assuming that the correlation between the scalar dissipation, χ, and the mixture fraction, ξ, was small. This left the term $\bar{\dot{r}}_\alpha$ for which the proposed closure took the form $\bar{\dot{r}}_\alpha \approx -\bar{\varrho} f_1(\xi)\vartheta$. This form of reaction rate was found to occur for the photochemical smog system of dilute NO_2, NO and O_3 in air under sunlight. Such a form was also found for H_2-air combustion through the specification of partial equilibrium assumptions and a progress variable in the manner of Janicka et al. [52] described earlier.

3.6. APPLICATIONS OF THE FLAMELET TECHNIQUE WITH FLAME STRETCH

In section 3.3.3 the laminar flamelet technique was briefly discussed and its implementation in turbulent flame calculations using the presumed shape pdf method (section 3.4) focussed attention on flamelet thermochemistry at a fixed strain rate. Examples of such calculations can be found for example in Fairweather et al. [40] who employ the conventional conserved scalar approach with flamelet thermochemistry at a fixed value of strain rate. In Fig(2) predictions are shown for a turbulent propane-air jet diffusion flame [54]. Clearly the results demonstrate that the laminar flamelet technique reproduces the CO levels relatively well.

Various implementations of flamelet thermochemistry have been employed for the calculation of turbulent flames. For example, Liew [81] employs unstrained ($\chi_{st} = 0$) as well as strained flamelets, though in his work, χ_{max} was used in place of χ_{st}, with the former being calculated where the temperature was at a maximum. This modification was introduced in order to account for the translation of the reaction zone to higher values of $\xi (> \xi_{st})$ as the degree of flame stretch increased. For $\phi_\alpha(\xi, \chi_{max})$ a bivariate pdf is required for closure, $\tilde{P}(\hat{\xi}, \hat{\chi}_{max})$. The conventional approach for obtaining mean properties is to assume statistical independence of the random variables, ξ and χ_{max}. Without recourse to measurements it is difficult to estimate the accuracy of this assumption and how strongly it influences the results. It was pointed out in section 3.4 that the statistics of the flow are often not strongly affected by the choice of the pdf, $\tilde{P}(\hat{\xi})$. However, it is not clear that this follows for $\tilde{P}(\hat{\xi}, \hat{\chi}_{max})$, since ξ is effectively a variable influenced primarily by gross mixing patterns, whereas χ_{st} (or χ_{max}) is strongly influenced by the non-equilibrium effects that may be present. Consequently, in flows with appreciable finite-rate effects, the choice of the shape for $\tilde{P}(\hat{\xi}, \hat{\chi}_{max})$ may well be an important consideration. Statistical independence implies that :

$$\tilde{\phi}_\alpha = \int_0^\infty \int_0^1 \phi_\alpha(\hat{\xi}, \hat{\chi}_{max}) \tilde{P}(\hat{\xi}) \tilde{P}(\hat{\chi}_{max}) d\hat{\xi} d\hat{\chi}_{max} \tag{72}$$

where $\phi_\alpha(\xi, \chi_{max})$ can be obtained from a library of stretched laminar flamelets. In order to avoid computing the double integral in equation (72), Liew modelled the burning process by

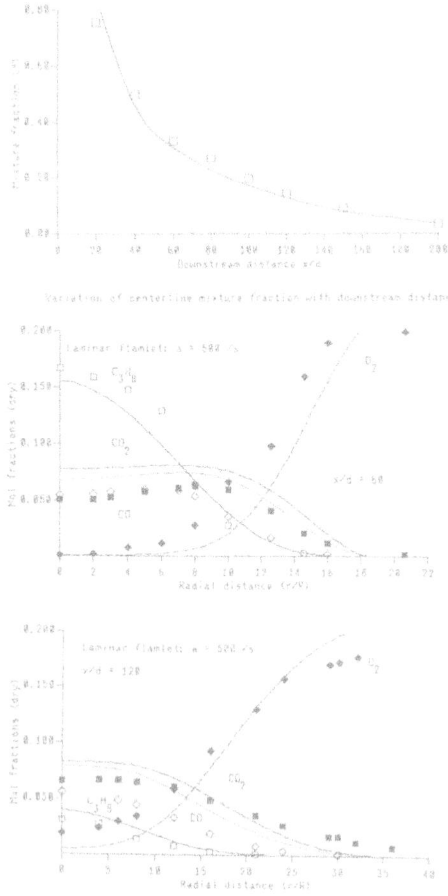

Fig. 2. Laminar flamelet calculations for a turbulent Propane-air Jet diffusion flame. Measurements from Godoy [45]

introducing two phases represented by undisturbed flamelet burning, $\phi_{\alpha,f}(\xi)$ and a post-extinction phase $\phi_{\alpha,I}(\xi)$. Consequently,

$$\tilde{\phi}_{\alpha,f} = \int_0^{\chi_q} \tilde{P}(\hat{\chi}_{max}) d\hat{\chi}_{max} \int_0^1 \phi_{\alpha,f}(\hat{\xi}) \tilde{P}(\hat{\xi}) d\hat{\xi} \tag{73}$$

The upper limit of χ_{max} in (72) has been replaced by (the physically correct value of) χ_q in (73). The quantity $\int_0^{\chi_q} \tilde{P}(\hat{\chi}_{max}) d\hat{\chi}_{max} \equiv \wp_c$ was regarded as the probability of undisturbed flamelet burning. Although the integral in (73) involves the undisturbed flamelet parameter, $\phi_{\alpha,f}(\xi)$, the effect of stretch appears through the term $\wp_c$; as stretch increases, $\wp_c$ decreases. The probability of finding the alternative phase of non-burning is $(1 - \wp_c)$. Thus for the post-extinction phase :

$$\tilde{\phi}_{\alpha,I} = (1 - \wp_c) \int_0^1 \phi_{\alpha,I}(\hat{\xi}) \tilde{P}(\hat{\xi}) d\hat{\xi} \tag{74}$$

The relation $\phi_{\alpha,I}(\xi)$ was modelled as pure mixing process. The quantity of interest, $\tilde{\phi}_\alpha$, was obtained from :

$$\tilde{\phi}_\alpha(\xi, \chi_{max}) = \tilde{\phi}_{\alpha,f}(\xi) + \tilde{\phi}_{\alpha,I}(\xi) \tag{75}$$

While $\tilde{P}(\hat{\xi})$ was assumed to be a β-pdf, knowledge of $\wp_c$ was required for closure; this in turn required the specification of $\tilde{P}(\hat{\chi}_{max})$. By invoking Kolmogorov's third hypothesis (originally proposed for the viscous dissipation rate), $\tilde{P}(\hat{\chi})$ was assumed to be log-normal with the additional assumption that $\chi \approx \chi_{max}$. $\tilde{\chi}$ was modelled using equation (46) and the variance of the pdf was given by the third hypothesis directly and relied considerably on empiricism with the appearance of geometry dependent constants. However Liew claimed that the quantity which was of interest, $\wp_c$, demonstrated little sensitivity to significant variations of the model constants.

In the work of Rogg et al. [106], the following specification was employed : $\phi_\alpha = \phi_\alpha(\xi_p, \chi_{st}, \zeta)$, with $\zeta = \xi_l/\xi_{st}$, $\xi_p = (\xi - \xi_l)/(\xi_r - \xi_l)$ and where ξ_l, ξ_r were defined as the mixture fractions on the lean and rich side (respectively) of the flamelet. ζ was a quantity referred to as the 'premixedness parameter', and quantified the degree to which the reactants entering the flamelet zone were premixed. When $\xi_r = 1, \xi_l = 0$ then $\xi_p = \xi$ and $\zeta = 0$ as expected. The motivation for introducing such a parameter is related to the steepening of the scalar gradients in the flame zone as a result of increased stretch. This may suggest that at very large strain rates (particularly near extinction) the diffusion flamelet burns in a partially premixed fashion. In this case, $\tilde{P}(\hat{\xi}_p, \hat{\chi}_{st}, \hat{\zeta})$ needs to be constructed and by assuming statistical independence [106], and relating ξ_p to ξ and ζ, the following quantities remained to be determined : $\tilde{\xi}$, $\widetilde{\xi''^2}$, $\widetilde{\xi''\zeta''}$, $\tilde{\zeta}$ and $\widetilde{\zeta''^2}$. The first two quantities were obtained from modelled transport equations, $\widetilde{\xi''\zeta''}$ was tentatively neglected, $\widetilde{\zeta''^2}$ was prescribed through a double δ-function pdf and $\tilde{\zeta}$ assigned the value of $\Upsilon\tilde{\xi}(1 - \tilde{\xi})x/L$, where x was the streamwise distance from the burner (jet diffusion flame configuration), L was the flame height and Υ was empirically assigned various values (0, 5 and 10). The method by which averaged quantities were obtained was essentially the same as that described by Liew [81]. The instantaneous relations were strongly affected by the premixing and for certain values of ζ, the composition structure demonstrated double reaction zones. The turbulent flame predictions were also very sensitive to the premixedness. The results suggest that partial premixing of laminar diffusion flamelets is significant in controlling the evolution of the scalar field in a turbulent flame. But the degree of empiricism makes it difficult to draw any concrete conclusions.

4. Premixed Combustion

4.1. INTRODUCTION

Premixing implies that the composition of the fuel is essentially uniform - the fuel and oxidant have mixed down to the molecular scale prior to combustion. As a consequence a strictly conserved scalar

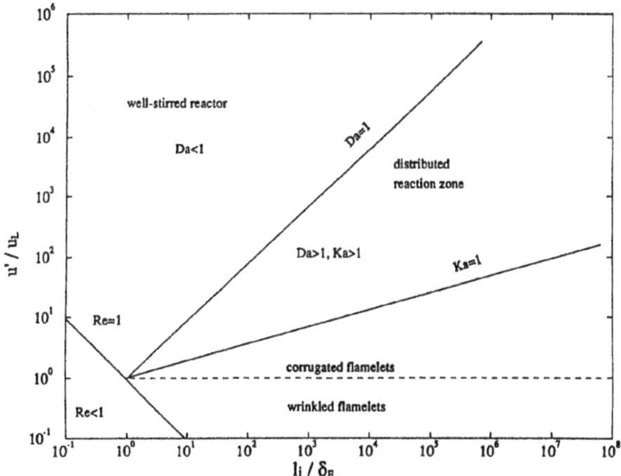

Fig. 3. Phase diagram showing the different regimes of combustion in premixed flames. u' - characteristic velocity of energy-containing motions, u_L - laminar flame speed (strained or unstrained), l_i - integral length scale, δ_f - flame thickness

(such as the mixture fraction, ξ) will take a uniform value set by the initial fuel/air ratio. Therefore the closure problem associated with mean formation rates of non-conserved scalars cannot be avoided. The representation of a premixed flame propagating into the cold unburned fuel/air mixture is another difficulty for modelling and is still the subject of research. The structure of premixed flames has received considerable attention and phase diagrams have been constructed [94], as shown in figure (3), to illustrate the different regimes of combustion as a function of dimensionless quantities such as the Damköhler (Da), Karlovitz (Ka) and Reynolds (Re) numbers.

For large Da and small Ka ($u' < u_L$), the effect of turbulence is to wrinkle the flame surface, increasing its area and hence the effective flame speed (Damköhler [30]). Increases in Ka ($Ka < 1$), such that $u' > u_L > v_k$ (where v_k denotes the Kolmogorov velocity scale, characteristic of the smallest dissipative motions in the flow), implies that the large eddies (that scale with u') will cause substantial convolution of the flame surface. However, the smaller eddies cannot corrugate the flame front by virtue of their smaller velocity; this regime is referred to as corrugated flamelets. To characterise this regime, Peters [94] introduced the Gibson length scale, $L_G \equiv u_L^3/\epsilon$ such that an eddy of 'size' L_G has a velocity of the order of the local flame velocity. L_G is a measure of the size of burnt gas pockets that move into the unburnt mixture, try to grow there due to the advancement of the flame front normal to itself, but are reduced in size again by newly arriving eddies of size L_G. It is argued that flame stretch effects are most effective at this particular scale. As the time scale of the turbulence becomes comparable to that of the chemistry, ($Da \approx 1$, $Ka > 1$), the distributed reaction regime prevails. Under these circumstances the turbulent flame can no longer be envisaged as an ensemble of strained laminar flamelets, but instead, the instantaneous region of combustion is 'distributed' throughout the time-averaged combustion zone (Summerfield et al. [114]). Further reduction in the turbulent time scale ($Da < 1$) moves the combustion into the well-stirred reactor regime, where kinetic control significantly influences the evolution of the flow.

4.2. SPALDING'S EDDY BREAK-UP (EBU) MODEL

Early concepts concerning premixed flames centred around the prediction of the turbulent flame speed and its relation to dimensionless parameters (e.g. the Reynolds number and the ratio of a characteristic turbulence velocity to the laminar flame speed u'/S_L). Such work has been reviewed by Bray [11]. For the purpose of predicting turbulent premixed flames in realistic geometries, knowledge of the flame speed is not sufficient. An appropriate starting point in the mathematical modelling of turbulent premixed flames is the work of Spalding [111], who employed the averaged form of the conservation equations to predict a bluff body stabilized flame spreading obliquely across a confined duct into the unburned stream until the whole duct cross-section was covered by the flame. A major motivation for the study was to obtain an accurate prediction of the angle of spread where experimental findings showed that at any downstream position, the distance from the symmetry plane to the flame edge was approximately one tenth of the distance from the baffle, regardless of the mixture strength, the approach velocity and the initial temperature of the mixture.

Prandtl's mixing length hypothesis was used for turbulent transport and two combustion models were employed. The first assumed kinetic control, with a bimolecular reaction rate (one step irreversible chemistry) employing the time averaged concentrations and temperatures. Comparison of concentration profiles with experimental data showed that the predicted concentration curves, albeit qualitatively correct given the simplicity of the turbulence/chemistry model, were too steep, implying that the chemical reaction was too densely concentrated in the high temperature part of the flame. Comparison with flame spread data showed that the predictions were quite poor since an increase in flame spread with increasing approach temperature was predicted while a slight decrease was measured. Spalding recognised that the reaction rate had to be formulated in a way which increased the influence of the turbulence field, and diminished that of the chemical kinetic constants. As a consequence the eddy break-up (EBU) model was formulated using ideas analogous to the energy cascade arguments of turbulence; namely that the rate controlling step is the breaking up of large eddies of burned and unburned gas into successively smaller ones. At sufficiently small scales, the flame zone (which can be identified as the interface between burned and unburned fragments of gas) can consume reactants at an effective rate. In the EBU model it is presumed that the rate of species consumption/production is proportional to the rate of turbulent mixing of the unburnt and burnt gases,

$$\bar{\omega}_{pr} \propto \frac{(\overline{Y_{pr}'^2})^{\frac{1}{2}}}{\tau} \tag{76}$$

Due to the assumption of one-step chemistry (involving fuel, oxidant and products) and adiabatic flow, knowledge of the product mass fraction, Y_{pr}, characterizes the thermochemistry. The time scale τ is associated with an integral time scale representative of the energy-containing motions and is consistent with the idea of turbulent mixing being the rate controlling parameter. Cast in terms of k and ε, $\tau \approx k/\varepsilon$, so that,

$$\bar{\omega}_{pr} \propto \frac{\varepsilon}{k}(\overline{Y_{pr}'^2})^{\frac{1}{2}} \tag{77}$$

The EBU model, based on conventional modelling practice (as for equations (59) and (60) in section 3.4) can be summarized as :

$$\bar{\varrho}\frac{\partial \bar{Y}_{pr}}{\partial t} + \bar{\varrho}\bar{u}_j\frac{\partial \bar{Y}_{pr}}{\partial x_j} = \frac{\partial}{\partial x_j}\left\{\frac{\mu_t}{\sigma_t}\frac{\partial \bar{Y}_{pr}}{\partial x_j}\right\} + \bar{\omega}_{pr} \tag{78}$$

$$\bar{\varrho}\frac{\partial \overline{Y'^{\,2}_{pr}}}{\partial t} + \bar{\varrho}\bar{u}_j\frac{\partial \overline{Y'^{\,2}_{pr}}}{\partial x_j} = \frac{\partial}{\partial x_j}\left\{\frac{\mu_t}{\sigma_t}\frac{\partial \overline{Y'^{\,2}_{pr}}}{\partial x_j}\right\} + 2\frac{\mu_t}{\sigma_t}\left(\frac{\partial \bar{Y}_{pr}}{\partial x_j}\right)^2 - C_D\bar{\varrho}\frac{\varepsilon}{k}\overline{Y'^{\,2}_{pr}} \tag{79}$$

where $\bar{\omega}_{pr} = \min[\dot{\mathcal{R}}, C_{EBU}\bar{\varrho}(\varepsilon/k)(\overline{Y'^{\,2}_{pr}})^{1/2}]$, $\dot{\mathcal{R}} = a\bar{\varrho}\bar{Y}_{fu}\bar{Y}_{ox}\exp[-E/(R\bar{T})]$. and the model 'constant', $C_{EBU} \approx 0.35$. The source term for reaction incorporates some kinetic control. For example in the cold regions of the flow $\dot{\mathcal{R}} \approx 0$, but the EBU rate term is generally finite, so that $\bar{\omega}_{pr} = 0$. However, this constitutes an ad-hoc addition to the modelling in order to account for kinetic control. Furthermore, a potentially large term representing the correlation between the instantaneous reaction rate and the scalar fluctuation, $\overline{Y'_{pr}\dot{\omega}_{pr}}$, has been omitted from equation (79) (c.f. equation (36)). Application of the model to the previously described test problem yielded concentration profiles in much better qualitative agreement with experiment than those resulting from the kinetically controlled combustion model. The predictions of the flame spread were also appreciably better and in the case of approach temperature the slight drop in flame spread was correctly predicted. The ideas behind the EBU concept, although plausible, are essentially intuitive - no formal derivations were performed. Because of its reliance on the assumption of combustion being effectively mixing-controlled, the model is consistent with the concept of thin flame burning.

4.3. THE PREMIXED FLAME MODEL OF BRAY AND MOSS

A further, important, development in the theory of premixed combustion was the work of Bray and Moss [17] (also reviewed in terms of Favre averaging in Bray [11]). A key assumption in the work was that combustion occurred through a global, one-step irreversible chemical reaction. It was argued that since the chemical kinetics of combustion rarely conforms to such a global description, the analysis would be most successful when the combustion was rate-limited by turbulent mixing rather than chemical kinetics. In what follows only the main features of the model are presented. Two reactive chemical species are considered, namely 'reactant' (fuel and oxidant) which form 'product'; the mixture may also include diluents . A progress variable, c , for the global combustion reaction is defined as the mass fraction of product normalized so that $c = 1$ when combustion is complete (i.e. either the fuel and/or oxidant is exhausted), while $c = 0$ in the unburned mixture which has no product. So $Y_{pr} = Y_3 = cY_{3,\infty}$ where $Y_{3,\infty}$ is the mass fraction of product in the fully burned gas whose temperature,T_∞, is the adiabatic flame temperature. From the species transport equation (7) for Y_3, a transport equation for c can be easily derived :

$$\frac{\partial}{\partial t}(\varrho c) + \frac{\partial}{\partial x_k}(\varrho u_k c) = \frac{\partial}{\partial x_k}\left(\varrho D\frac{\partial c}{\partial x_k}\right) + \frac{\dot{r}_3}{Y_{3,\infty}} \tag{80}$$

From the definition of element mass fraction, $Z_\alpha \equiv \sum_{\beta=1}^{N}\mu_{\alpha\beta}Y_\beta$, where $\mu_{\alpha\beta}$ denotes the number of Kg of the αth element in 1Kg of the βth species,

$$Y_1 = \frac{1}{\mu_{11}}\left(Z_1 - \mu_{13}Y_{3,\infty}c\right) \quad \text{and} \quad Y_2 = \frac{1}{\mu_{22}}\left(Z_2 - \mu_{23}Y_{3,\infty}c\right) \tag{81}$$

and Y_4 is clearly independent of c. In the present notation, α=1,2,3 and 4 represents fuel, oxidant, products and diluents, respectively. Denoting T_o and ϱ_o as the temperature and density of the reactants, it can be shown (Bray [11]) that :

$$T = T_o\,(1+\tau c) \quad \text{and} \quad \varrho = \frac{\varrho_o}{(1+\tau c)} \quad \text{where} \quad \tau = \frac{T_\infty}{T_o} - 1 \tag{82}$$

and τ can vary from 4 to 9. In general, $\dot{r}_3 = \dot{r}_3(\varrho, T, Y_1, Y_2)$, but because ϱ, T, Y_1 and Y_2 are all functions of c, this can be replaced by, $\dot{r}_3 = \dot{r}_3(c)$. The statistics of the flame are calculated by way of a pdf (similar in principle to the presumed-shape pdf method described earlier). It is argued that the pdf $P(c;\mathbf{x})$, must represent gas which is unburnt and gas which has burned completely, as well as gas mixture which is in the process of burning. It is convenient then to separate these three contributions such that :

$$P(c;\mathbf{x}) = \alpha(\mathbf{x})\delta(c) + \beta(\mathbf{x})\delta(1-c) + \left[H(c) - H(c-1)\right]\gamma(\mathbf{x})f(c;\mathbf{x}) \tag{83}$$

where $\delta(c)$ and $H(c)$ are Dirac delta and Heaviside step functions, respectively. The coefficients α, β and γ are non-negative functions of position and $f(c)$ satisfies the normalization condition, $\int_0^1 f(c)dc = 1$. α, β and γ may be regarded as the probabilities that unburned mixture, fully burned product, and burning mode, respectively, exist at the point in question - it can be shown by integrating $P(c)$ with respect to c that $\alpha + \beta + \gamma = 1$. In this *a priori* pdf method, $f(c)$ is to be represented empirically and the functional form for $P(c)$ is assumed to have at most two parameters, $\tilde{c}$ and $\widetilde{c''^2}$.

An approximation which drastically simplifies the model is that $\gamma \ll 1$. This represents, in the limiting case, a flow which fluctuates between unburnt and fully-burnt mixtures. Bray [11] expressed the transport equations for $\tilde{c}$ and $\widetilde{c''^2}$ in non-dimensional form and in the expression for the source terms identified the grouping γDa. It was argued that γDa was of order unity and thus the approximation $\gamma \ll 1$ was accompanied by $Da \gg 1$; this implies that the characteristic chemical time is much smaller than the characteristic turbulence time and so the rate of combustion is controlled mainly by turbulent mixing. In fact it can be shown that with the condition $\gamma \ll 1$, an expression for $\dot{r}_3/Y_{3,\infty}$ which is very similar to the form of Spalding's EBU model is obtained. The condition $\gamma \ll 1$ leads to a pdf $P(c)$ consisting mainly of two delta functions with $\alpha+\beta \approx 1$. Using the relation for the density, equation (82) in averaged form and the equation of state expressed as $\varrho_\infty T_\infty = \varrho_o T_o$ there results :

$$\beta = \frac{(1+\tau)\tilde{c}}{1+\tau\tilde{c}} \quad \text{and so} \quad \alpha = 1 - \beta = \frac{1-\tilde{c}}{1+\tau\tilde{c}} \tag{84}$$

The only term remaining in the specification of the pdf $P(c)$ is the burning mode pdf. Consistent with the $\gamma \ll 1$ approximation is the idea that the burning mode pdf makes little contribution to the pdf $P(c)$; under these circumstances $\widetilde{c''^2} \approx \tilde{c}(1-\tilde{c})$ and as shown in equation (83), to construct $P(c)$ only α and β would be required, which are functions of $\tilde{c}$ only. It can be shown [11], however, that $f(c)$ does influence $\bar{\dot{\omega}}$ and $\overline{c''\dot{\omega}}$. Furthermore, the source terms are related to $f(c)$ through the ratio of its moments, and in this respect it is argued that the results are insensitive to the details of the specification of $f(c)$.

Many of the applications of the Bray-Moss model have been confined to an infinite plane one-dimensional reaction zone which is swept relative to an incoming stream of reactants. Further details of the application and modelling can be found in Bray and Libby [13; 79], and Bray et al. [15]. Turbulent transport was modelled initially through an eddy-viscosity concept, and subsequently through the joint pdf of velocity and reaction progress variable. The model has been used to study the effects on the turbulence kinetic energy, k, as the flow passed through the reaction zone. In such a situation the effects to be accounted for include dilatation associated with the heat release (which tends to bring about a reduction in k), and interaction between the mean motion and the Reynolds stresses which generates k; this latter effect was neglected (due to an order of magnitude argument based on angular deflection of a typical streamline passing through a planar flame) in comparison to the production of k from buoyancy effects associated with the mean pressure gradient. The use of the joint pdf implies that closure of the turbulent transport terms can be achieved without invoking an eddy viscosity assumption and at the same time it allows the validity of such a hypothesis to be investigated in turbulent combustion. The second moment closure of Bray et al. [15] focused attention on the solution of the equations governing the evolution of $\overline{\varrho u'' u''}$ and $\overline{\varrho u'' c''}$. The numerical treatment was facilitated by replacing the x-coordinate as the independent variable with the mean product concentration $\tilde{c}$; this manipulation also had the advantage of eliminating $\bar{\dot{\omega}}_{pr}$ from the formulation and thereby obviating the need for its specification (unless the spatial structure in the form of $x(\tilde{c})$ was required). The results demonstrated that as τ (a measure of the heat release) was increased, extensive regions within the flame showed signs of counter-gradient diffusion - i.e diffusion of opposite sign to that required by gradient transport. The physical process responsible for this effect was due to a 'buoyancy' mechanism in which packets of low-density burned gas were preferentially accelerated relative to packets of reactants by the mean pressure gradient acting across the flame. The resulting relative motion between burned and unburned packets took place in the direction opposite to that predicted by gradient diffusion. The predictions of the model based on the joint pdf reproduced with good accuracy the changes in the intensity of the velocity and scalar fluctuations across the flame. The mean pressure gradient provided the dominant term which augmented k throughout the interior of the flame, but its effect diminished at the boundaries of the flame where the effects of dilatation and viscous dissipation (both sinks of k) became relatively more dominant. Despite the success of the above treatment, the generalization of this approach to two and three dimensional flows is likely to require further effort, firstly because of the increased dimensionality of the joint pdf (more velocity components have to be accounted for) and secondly the Reynolds stress terms require more detailed modelling.

The original Bray-Moss model has been extended to situations which involve more realistic chemistry. Within the framework of multi-step chemistry, a redefinition of the reaction progress variable is required. Clearly the normalized mass fraction of product is unsuitable since with complex chemistry there is no single product. In addition some species may have been depleted where others just begin to appear and therefore the state of any particular reactive species is not (necessarily) indicative of the state of the reaction. c is therefore redefined in terms of temperature :

$$c = \left(\frac{T - T_o}{T_\infty - T_o}\right) \tag{85}$$

where T_o and T_∞ are the temperatures upstream and downstream of the reaction zone, respectively. One of the most notable applications of complex chemistry is the work of Champion et al. [22]

in which the chemistry is assumed to proceed sequentially. The semi-global chemical mechanism of Edelman and Fortune (described in [22]) was employed to investigate a reacting mixture of propane and air in a ducted, stationary and one-dimensional flow. It was assumed that the complete chemical scheme could be partitioned into two distinct zones (not necessarily spatially distinct). Firstly a delay zone which was represented by the global reaction for the propane break down. This reaction led to small heat release and had to be the fastest of the reactions for sequential chemistry to be postulated. In the next combustion zone, the chemical description included all the remaining reactions in the semi-global scheme (involving slower radical recombinations). Thermochemical closure was achieved by expressing conservation of species in terms of c, invoking partial equilibrium assumptions and by taking the concentration of oxygen to hydrogen radicals as a constant.

The Bray-Moss model [17] has drawn upon ideas from laminar flamelet theory; this is related to the specification of $f(c)$. For example, if the distribution of product or temperature through a laminar flame can be represented by $c = c(x/\delta_f)$ where $c(x/\delta_f)$ is a monotonically increasing function from $c = 0$ to $c = 1$, then it can be shown that :

$$f(c) = \left\{ \frac{dc}{d(x/\delta_f)} \right\}^{-1} \tag{86}$$

Therefore, given a representative laminar flame profile, it is possible to determine $f(c)$. The original theory makes no further use of the laminar flamelet idea thereafter. However, more recently, Bray [12] has extended the flamelet pdf model to include effects of flame stretch. The motivation for this line of approach is that the basic EBU type expression involves the time scale k/ε. This implies that $\bar{\dot{\omega}}$ will increase without limit if the turbulence time scale is reduced, since no allowance is made for flamelet quenching. It is argued that in a turbulent flame burning may be represented by a number of strained laminar flames with Karlovitz numbers $Ka_0, Ka_1, \ldots, Ka_n$ which are assumed to occur with probabilities $\wp_0(\mathbf{x}), \wp_1(\mathbf{x}), \ldots, \wp_n(\mathbf{x})$ such that $\sum_{s=0}^{n} \wp_s(\mathbf{x}) = 1$. Under these circumstances, the mean reaction rate can be written as :

$$\bar{\dot{\omega}} = \gamma \sum_{s=0}^{n} \left\{ \wp_s \int_0^1 \dot{\omega}_s(\hat{c}) f_s(\hat{c}) d\hat{c} \right\} \tag{87}$$

where $\dot{\omega}_s$ is the source term for the flamelet with Karlovitz number Ka_s and $f_s(c)$ is given by equation (86) with the conditioning on the Ka_s, which affects the flame structure and thickness. The mean species concentration then becomes :

$$\bar{\phi}_i(c) = \alpha\phi_i(0) + \beta\phi_i(1) + \gamma \sum_{s=0}^{n} \left\{ \wp_s \int_0^1 \phi_{i,s}(\hat{c}) f_s(\hat{c}) d\hat{c} \right\} \tag{88}$$

where $\phi_{i,s}(c)$ is determined from a flamelet library - in order to use such a thermochemical specification, the same ideas invoked in the work of Champion et al. [22] are required, i.e. a redefinition of c in terms of temperature and sequential chemistry. Note that even if $\gamma \ll 1$, the last term on the rhs of equation (88) must be retained, particularly if reaction intermediates are to be calculated, since these species are predominantly created in the reaction zone. Bray [12] assumed that γ was independent of flamelet quenching and controlled by the scalar dissipation mechanism, for which the modelling principles leading to equation (46) were employed. In this case equation (46) is recast in terms of c rather than ξ. The effects of stretch are necessarily coupled to the position of the flamelet and thus to the progress variable c. It was assumed that Ka was uncorrelated with c, which is similar in principle to an assumption of statistical independence. The implication of such an assumption is that the integral $\int_0^1 \dot{\omega}_s(\hat{c}) f_s(\hat{c}) d\hat{c}$ can be taken out of the summation in equation (87), and in doing so, the subscript 's', characterising the degree of stretch can be replaced by the unstrained flamelet subscript 'o'. By invoking a log-normal distribution for the viscous dissipation rate, relating this to the Karlovitz number under quench conditions and assuming that the quenched flamelets make no contribution to $\bar{\dot{\omega}}$, $\sum_{s=0}^{n} \wp_s$ can be replaced by the probability of unquenched flamelets, $\wp_{uq}$. The final result is :

$$\bar{\dot{\omega}} = \bar{\varrho} \left(\frac{\varepsilon}{k} \right) \frac{\tilde{c}(1-\tilde{c})}{(2c_m - 1)} \wp_{uq} \quad \text{where} \quad c_m = \frac{\overline{c\dot{\omega}_o}}{\bar{\dot{\omega}}_o} \tag{89}$$

This is a basic EBU-type expression multiplied by the proportion of flamelets that are unquenched. How well such a modification to the basic Bray-Moss theory improves predictions has yet to be fully determined.

4.4. MORE RECENT DEVELOPMENTS IN PREMIXED FLAME THEORY

At large Damköhler numbers the source term for chemical reaction has been shown to be proportional to the scalar dissipation rate in premixed [10] and non-premixed [2] configurations; in the premixed situation, $\bar{\dot{\omega}} \propto \tilde{\chi} \equiv (\bar{\varrho})^{-1} \overline{\frac{\mu}{Sc} (\nabla c'')^2}$. The main difficulty in exploiting this relation for the purpose of calculating the mean reaction rate is in the representation of $\tilde{\chi}$. Conventional modelling based on equation (46) leads to a rate expression similar to that given by the Bray-Moss model. In section 3.4 it was noted that a similar modelling expression in terms of ξ yields good results for calculating $\widetilde{\xi''^2}$. However, in this section we are concerned with a reactive scalar, c, and the appearance of the time scale (k/ε) as the controlling parameter is questionable since it ignores the distribution of scalar time scales in the reaction zone which can be very different to (k/ε).

Recently EBU-type combustion models, as embodied in the original Bray-Moss theory, have come under careful scrutiny, particularly with regards to their prediction of flame propagation. For example Weller et al. [127] pointed out that the use of such a combustion model led to problems of high flame speed predictions near a wall whereas observations show the flame to slow down in such regions. Catlin and Lindstedt [21] studied flame propagation through a uniform turbulence field in a one-dimensional planar open-ended tube geometry where the turbulent reaction rate was quenched near the cold front of the flame. The computations demonstrated that the burning velocity and the reaction zone thickness increase rapidly as the quench point approaches the cold front and, although the burning velocity remained bounded, the reaction zone thickness became indefinitely large. The conclusion was that cold front quenching was necessary for the prediction of experimentally observed flame behaviour in tubes using mixing-controlled reaction models.

It has been noted (Bray et al. [16]) that in the Bray-Moss theory length scale information, particularly with reference to the local scalar field, is not present. An attempt to remedy this situation is described in [16] and [14]. The formalism centres around the representation of the mean reaction rates as $\bar{\dot{\omega}}(\mathbf{x}) = \dot{\omega}_f(\mathbf{x})\nu(\mathbf{x})$, where ν represents the mean frequency at which flamelets cross the spatial location under consideration and where $\dot{\omega}_f$ is the mean reaction rate per crossing. The main task is the determination of ν which is achieved through the solution of a modelled transport equation at the one-point, two-time level for the autocorrelation of the progress variable, c, with time advance τ i.e. $\overline{c(\mathbf{x},t)c(\mathbf{x},t+\tau)} \equiv \mathcal{P}_{11}(\mathbf{x},\tau)$. It was assumed that $c(\mathbf{x},t)$ observes a zero-one function or square-wave time series behaviour, by virtue of the $\gamma \ll 1$ approximation. In the modelled equation for $\mathcal{P}_{11}(\mathbf{x},\tau)$, terms involving molecular diffusion were neglected (high Reynolds number assumption) and correlations such as $\overline{c\dot{\omega}}$ were treated by analysing the covariance between a delta function (simulating $\dot{\omega}$ as a pulse train) and a square-wave time series (for c). For the remaining terms, the limiting behaviour for $\tau = 0$ and $\tau \to \infty$ was either known or postulated, and consequently closure was achieved by assuming that an exponential function could be employed to interpolate between these limits. By assuming a pdf for the passage times for the reactant and product modes at a point, an expression for ν was obtained. In a one-dimensional planar flame geometry it was found that the predictions of mean passage times for reactants and products as well as the normalized power spectral density of c were in very good agreement with the measurements of Shepherd and Moss [108]. It must be stressed, however, that for the validation of the model, data from the measurements were employed to evaluate some of the constants associated with the model. Although this is one of the advantages of the theory (i.e. the quantities appearing in the model are directly measurable), it is not clear how easily the experimental input can be supplemented by theory. In order to calculate $\bar{\dot{\omega}}$, some means of prescribing $\dot{\omega}_f$ is necessary and this can be achieved through the use of a flamelet library [12]. In this approach a term, $\bar{V}_{n,s}$, representing the mean speed normal to a flamelet with a characteristic Karlovitz number, Ka_s, is required. This was modelled by assuming that this mean speed is independent of stretch effects, and then related to the mean flow velocity (one-dimensional situation) with : $\bar{V}_n \approx \bar{V}/\mathcal{K}$, where $\mathcal{K}$ is an empirical constant. No predictions of $\bar{\dot{\omega}}$ have yet been obtained to demonstrate whether the flamelet theory incorporated in the model behaves correctly. Furthermore, the extension of this model to two-dimensional and three-dimensional situations requires further effort.

Recent work on premixed combustion has attempted to characterize the burning rate through the evaluation of a flame area and burning velocity. One such proposal is based on the theory of fractals. Fractals are 'objects' that display self-similarity over a range of scales (referred to as the 'inner' and 'outer' cutoffs). (For a brief introduction to the concept of fractals, the work of Sreenivasan and Meneveau [113] may be consulted.) It is felt that because flamelet surfaces appear 'rough', with multiple scales of wrinkling, the theory of fractals could be employed in characterizing flamelet surface area. Gouldin [47] considered homogeneous, isotropic turbulence where the influence of combustion on turbulence was neglected. The flame front was treated as a passively convected interface for length scales lying between an outer cutoff, ϵ_o, based on the integral length scale of turbulence (L) and an inner cutoff, ϵ_i, relating to the Kolmogorov scale (η). The use of a passive interface implied that the effect of flame propagation and curvature on the geometrical structure of the flame front was negligible (i.e. $\delta_f/l \to 0$ and $u'/u_l \to \infty$). By making use of a power-law expression characteristic of fractal behaviour, Gouldin proposed :

$$\frac{A_i}{A_o} = \frac{u_t}{u_{l,o}} = \left(\frac{L}{\eta}\right)^{D-2} \tag{90}$$

where D is the fractal dimension, u_t is the turbulent flame speed, $u_{l,o}$ is the unstrained laminar flame speed, A_i is 'true' flame area measured at length scales down to the inner cutoff and A_o is the (smaller) area taking account only of the wrinkling at the outer cutoff. Simple correction functions were then incorporated to account for finite values of u'/u_l and flame thicknesses. A fractal dimension of 2.37 was chosen, based on an average value taken from measurements and analysis in non-combusting flows. The fractal model (with and without correction) was validated against the measurements of turbulent flame speed of Abdel-Gayed, Bradley and coworkers, where combustion bomb data were reported for six fuels for a broad range of turbulence conditions. With the exception of the C_8H_8 (where doubts were raised concerning the validity of the measurements) the model predictions for $u_t/u_{l,o}$ versus $u'/u_{l,o}$ were within 30% of the data. Generally, correct trends were predicted but the inclusion of the correction terms did not necessarily improve the results. Furthermore, taking C_3H_8 as an example, it was found that extremely good predictions, both in magnitude and trends, were obtained for lean conditions but not for rich mixtures. It was also found that small changes in D had a significant effect on the predictions. This final point is of particular interest, since North and Santavicca [90] demonstrate that fractal dimension increases with increasing turbulence intensity and decreasing laminar flame speed. North and Santavicca point out that Gouldin's modelling of the effect of u'/u_l implies that at the lowest Reynolds numbers experimentally investigated, the predictions would lead to $\epsilon_i > L$ suggesting that fractals should no longer be observed, whereas in reality, fractal character was noticed. Kerstein [68] suggests that the assumption that the flame front behaves like a passive interface is physically implausible, and proposes a macroscopic picture of the propagating flame, with the flame structure being viewed as a diffusion-reaction zone. His analysis leads to a value for the fractal dimension equal to $7/3$ which is very close to the value used by Gouldin.

The work of Darabiha et al. [31] serves as an example of an approach to premixed flame modelling based on the solution of an equation for the flame area per unit volume (also referred to as the flame surface density). The principle of the flame area technique is to obtain closure for the mean reaction rate in terms of the mean flame area per unit volume, $\tilde{\Sigma}$. Instantaneously, this quantity is defined as : $\Sigma = \lim_{\delta V \to 0}(\partial S/\partial V)$, where S denotes a material surface element with area δS. A balance equation for Σ can be derived :

$$\frac{\partial}{\partial t}(\varrho\Sigma) + \frac{\partial}{\partial x_k}(\varrho u_k \Sigma) = -\frac{1}{2}\varrho n_i n_j \left(\frac{\partial u_i}{\partial x_j} + \frac{\partial u_j}{\partial x_i}\right) \tag{91}$$

where n_i is the unit vector normal to the surface. The Favre-averaged form of this equation requires closure. The turbulent flux of Σ, $-\bar{\varrho}(\widetilde{u_k''\Sigma''})$ was closed using as a conventional gradient transport model including a turbulent Schmidt number for the surface density. The Favre-averaged form of the rhs of equation (91) was modelled as a function of the strain rate (assumed $\propto \varepsilon/k$), $\bar{\varrho}$ and $\tilde{\Sigma}$. In order to complete the closure, further modelled terms had to be added to account for the reduction in flame area brought about by the interaction of two flamelets separated by fresh reactant and removal of flame area by quenching when the flamelets are situated in a highly-strained field. Intuitive arguments were employed to represent these complicated processes. The mean consumption rates were calculated from : $\tilde{\omega}_i = -\bar{\varrho} V_{r,i} \tilde{\Sigma}$, where $V_{r,i}$ is the volume rate of consumption per unit flame area of the ith species (presumably based on mean concentrations). For the purpose of calculating a turbulent flame, a flamelet library was constructed based on the calculation of a laminar premixed counterflow flame of fresh reactants and burnt products, using a detailed propane-air mechanism. The flamelet model was characterised by the strain rate, temperature of the unburnt gas stream and the equivalence ratio of the incoming reactant stream. A baffle-stabilized ducted turbulent premixed flame was investigated supported by measurements of the local heat release rate, obtained by light scattering techniques. For the purpose of prediction, the empirical constants emerging from the modelling were assigned the value of unity (without further optimization/calibration). The reacting flow calculations were initiated by introducing a temperature zone with a Gaussian shape and a localized distribution of flame surface density in a region of recirculation of the cold flow. How sensitive the predictions were to variations in the specification of the reaction zone is not clear. Qualitatively, the predictions followed the correct trends; for example a reduction in the equivalence ratio brought about the observed reduction in the length of the reaction zone. However, the model predicted heat release in regions not observed in the measurements. Despite the extensive modelling employed to close the Σ equation, the shortcomings of the prediction were attributed to the use of the gradient transport hypothesis.

Cant et al. [20] also investigated the surface-to-volume ratio equation in premixed combustion. They considered the flame surface to be defined by all points $\underline{X}$ such that $c(\underline{X}, t) = c^*$, where c^* was a specified and fixed value of the progress variable c. In this respect the flame surface was a constant-property surface. An exact equation for the mean surface area per unit volume was proposed, although comparison with the equation appearing in the work of Darabiha et al [31] was rendered difficult due to the use of surface averaging in the former case. Cant et al. identified realizability conditions which had to be satisfied and modelled the unclosed terms appearing in their formulation of the Σ-equation. In particular, the strain rate in the tangent plane of the surface was modelled to be consistent with the limiting behaviour of a material surface and a fixed surface in isotropic turbulence. The mean curvature, influenced by small scale turbulence effects and instabilities in laminar flame propagation, was represented as a linear function of Σ and the local mean surface laminar flame speed. A one dimensional test case with open boundaries was considered. At time $t = 0$ the flame was assumed to exist in a field of statistically homogeneous isotropic turbulence with zero mean flow and no density variations. The results yielded a flame which quickly settled down to a steady propagation speed and a self-preserving structure. In addition the effect on the turbulent flame speed as a function of the laminar flame speed and the Reynolds number (based on the Taylor microscale), showed the correct trends. Cant [19] has recently extended the formulation to the variable-density case. The basic modelling follows on from [20]. Certain results from the Bray-Moss theory were employed together with a flamelet library for CH_4-air combustion in order to investigate the same simplified test case as described above. The results once again yielded a plausible evolution for the flame profiles.

Peters [95] investigated premixed turbulent combustion in the flamelet regime on the basis of an equation which described the instantaneous flame contour as an iso-scalar surface of the scalar field $G(\mathbf{x}, t)$. The condition for the flame front $G(\mathbf{x}, t) = G_o$ (an arbitrary constant) divides the flow field into two regions where $G > G_o$ is the region of burnt gas and $G < G_o$ that of the unburnt mixture. The scalar difference $G - G_o$ was interpreted as the distance from the flame surface. The Markstein length, $\mathcal{L}$, which is a characteristic length scale for flame response and proportional to the flame thickness, was introduced into the G-equation by expressing the laminar burning velocity (subjected to stretch) in terms of the unstretched burning velocity and terms associated with flame curvature and flow divergence. The result was :

$$\frac{DG}{Dt} \equiv \frac{\partial G}{\partial t} + \underline{v} \cdot \nabla G = u_{l,o}|\nabla G| + \mathcal{D}_{\mathcal{L}} \nabla^2 G - \mathcal{L}\left(\frac{D|\nabla G|}{Dt}\right)_{\mathcal{L}\to 0} \tag{92}$$

where $\mathcal{D}_{\mathcal{L}} = u_{l,o}\mathcal{L}$, with $\mathcal{D}_{\mathcal{L}}$ being referred to as the Markstein diffusivity. $u_{l,o}$ and $\mathcal{L}$ were defined with respect to the unburnt mixture, so that $\underline{v}$ was the velocity conditioned ahead of the flame. Peters then constructed (through Reynolds decomposition and ensemble-averaging) transport equations for $\bar{G}$ and $\overline{G'^2}$ which contained several terms requiring closure. In particular, unclosed terms were identified in the $\overline{G'^2}$ equation which were associated with the smoothing of the flame front and were argued to be most effective at different scales of turbulence. Guidance for the modelling of the $\overline{G'^2}$ equation was obtained from consideration of constant-density homogeneous isotropic turbulence. The existence of an equilibrium range for the scalar field under consideration was hypothesized, and consequently between wavenumbers (κ) representing the reciprocal of the integral and Gibson scale, the scalar spectrum function $\Gamma(\kappa, t)$ assumed a universal form. $\Gamma(\kappa, t) = 4\pi\kappa^2 \widehat{g^2}(\kappa, t)$, where $\widehat{g^2}(\kappa, t)$ is the Fourier transform of $g^2(r, t) \equiv \overline{G'(r, t)G'(\mathbf{x} + r, t)}$. Introduction of a gradient transport hypothesis permitted closure of the $\Gamma(\kappa, t)$ equation, thereby constraining the modelled terms to be a function of the local wavenumber. The resulting linear differential equation was solved exactly by the method of characteristics and integration of the differential equation over wavenumber space and comparison with the constant-density homogeneous analog of the $\overline{G'^2}$ equation completed the closure. Modelling of the $\bar{G}$ equation remained. A term representing the product of the flame surface area per unit volume and the unstretched laminar flame speed was modelled using intuitive arguments and dimensional analysis, with the result that the closed term involved k/ε as a characteristic time scale. The turbulent transport and Markstein diffusion terms were combined and modelled as a curvature term incorporating a turbulent diffusivity. Calculations were performed for plane and oblique flames for which the curvature term disappeared and while no direct comparison with measurements was made, the predicted flame brush thickness and position agreed qualitatively with experimental observations.

5. The Transported Pdf Approach

5.1. INTRODUCTION

In the techniques which employ the concept of a pdf, described up to now, the closure problem associated with the mean reaction rate has been avoided by assuming that the instantaneous thermochemical state of the flow is a function of one (or more) scalar(s) and averages are obtained when the instantaneous relations are weighted by a presumed-shape pdf. In this section a method is described which involves the solution of a transport equation for the pdf. Attention will focus on the scalar pdf, $P(\underline{\psi}; \mathbf{x}, t)$, referred to in section 3.1 and results will be presented in jet diffusion flame configurations demonstrating the capabilities (and shortcomings) of the formalism.

Pdf transport equation modelling has been investigated for well over two decades. However it is relatively recently that the methodology has emerged from its rather deep-rooted theoretical background to become a potential tool for engineering calculations. This is mainly due to the development of Monte Carlo algorithms used to solve the pdf equation. Reviews of the topic can be found in the works of O'Brien [91], Pope [102] and Kollmann [71; 72].From the statistical point of view, the specification of $P(\underline{\psi}; \mathbf{x}, t)$ makes available all single point, single time moments of the scalar field. In contrast, moment closures yield only those moments for which modelled transport equations are solved. Where pdf methods appear most advantageous is in the description of turbulent reacting flows where the thermochemistry is invariably a non-linear function of several independent scalars. In the conserved scalar approach (section 3.2) the pdf of a single, strictly conserved scalar is employed; however, in order to reproduce finite-rate effects (e.g. emissions and extinction effects) accurately, more independent scalars need to be incorporated in the characterization of the pdf. Furthermore the extension of presumed shape pdf methods to multi-scalar problems is difficult; firstly the number of modelled transport equations required to construct the pdf rapidly rises with the number of independent scalars and secondly, presuming a shape for a multi-dimensional pdf which satisfies the appropriate bounds becomes increasingly difficult and uncertain as the number of independent scalars rises. Such difficulties are averted through the use of pdf transport methods.

5.2. DERIVATION OF THE SCALAR PDF EQUATION

The transport equation for $P(\underline{\psi})$ (the dependence on $\mathbf{x}$ and t is omitted for clarity) is an exact, unclosed equation and its derivation relies on the use of the conservation equations (presented in section 2). Consequently the equation has a physical foundation. The most widely-used method of derivation is based on the concept of a fine-grained density function $\hat{p}(\underline{\psi}, \underline{\phi}; \mathbf{x}; t)$, Lundgren [85], and is defined such that $\hat{p}(\underline{\psi}, \underline{\phi})d\underline{\psi}$ is the probability that at position $\mathbf{x}$ and time t, $\psi_\alpha \leq \phi_\alpha(\mathbf{x}, t) \leq \psi_\alpha + d\psi_\alpha$ for all $\alpha = 1, \ldots, N$. Note that ψ_α denotes an independent variable (like $\mathbf{x}$ and t) and represents the sample space of the dependent random variable $\phi_\alpha(\mathbf{x}, t)$; $\underline{\psi}$ denotes the set of ψ_α for $\alpha = 1, \ldots, N$. Clearly $\hat{p}(\underline{\psi}, \underline{\phi})$ is a multi-dimensional quantity and it represents the pdf for a single realization of a turbulent flow, so that if $\psi_\alpha \leq \phi_\alpha(\mathbf{x}, t) \leq \psi_\alpha + d\psi_\alpha$ is not satisfied for all α, then $\hat{p}(\underline{\psi}, \underline{\phi}) = 0$, otherwise $\hat{p}(\underline{\psi}, \underline{\phi}) = 1$. As a consequence the fine-grained density can be expressed in the following manner :

$$\hat{p}(\underline{\psi}, \underline{\phi}) \equiv \prod_{\alpha=1}^{N} \delta\left\{\phi_\alpha(\mathbf{x}, t) - \psi_\alpha\right\} \tag{93}$$

In contrast, the conventional pdf $P(\underline{\psi})$ is representative of an ensemble of realizations, such that its integral over a finite region of its sample space represents the likelihood of occurrence for an event. It is a straightforward matter to show [102] that :

$$P(\underline{\psi}) = \langle p(\underline{\psi}, \underline{\phi}) \rangle \tag{94}$$

The details of the derivation of the transport equation for $P(\underline{\psi})$ can be found in O'Brien [91] and Kollmann [72]. (For a derivation which does not involve the concept of the fine-grained density, the work of Pope [102] should be consulted). Kollmann [70; 71] derives the pdf equation in terms of the characteristic function, which is the Fourier transform of the pdf. Whatever the method of derivation the transport equation for the density weighted scalar pdf (see equation (58) for a definition of the density-weighted pdf) is given by :

$$\underbrace{\bar{\varrho}\frac{\partial\tilde{P}(\underline{\psi})}{\partial t}}_{\text{I}}+\underbrace{\bar{\varrho}\tilde{u}_k\frac{\partial\tilde{P}(\underline{\psi})}{\partial x_k}}_{\text{II}}+\underbrace{\bar{\varrho}\sum_{\alpha=1}^{N}\frac{\partial}{\partial\psi_\alpha}\left\{\dot{\omega}_\alpha(\underline{\psi})\tilde{P}(\underline{\psi})\right\}}_{\text{III}}=\underbrace{-\frac{\partial}{\partial x_k}\left\{\bar{\varrho}\langle u_k''|\underline{\phi}=\underline{\psi}\rangle\tilde{P}(\underline{\psi})\right\}}_{\text{IV}}$$
$$+\underbrace{\sum_{\alpha=1}^{N}\frac{\partial}{\partial\psi_\alpha}\left\{\left\langle\frac{\partial J_{i,\alpha}}{\partial x_i}\middle|\underline{\phi}=\underline{\psi}\right\rangle\tilde{P}(\underline{\psi})\right\}}_{\text{V}}\tag{95}$$

The terms have the following significance :

- Term I : Rate of change in physical space
- Term II : Mean convection in physical space
- Term III : Chemical source production in compositional space
- Term IV : Turbulent transport in physical space
- Term V : 'Molecular mixing' in compositional space

Assuming Fickian diffusion the 'molecular mixing' term can be expressed as :

$$-\sum_{\alpha=1}^{N}\sum_{\beta=1}^{N}\frac{\partial^2}{\partial\psi_\alpha\partial\psi_\beta}\left\{\left\langle\frac{\mu}{Sc}\frac{\partial\phi_\alpha}{\partial x_k}\frac{\partial\phi_\beta}{\partial x_k}\middle|\underline{\phi}=\underline{\psi}\right\rangle\tilde{P}(\underline{\psi})\right\}\tag{96}$$

With the assumption of constant μ it can also be shown [72] that the term can be expressed in terms of a two-point pdf :

$$-\frac{\mu}{Sc}\sum_{\beta=1}^{N}\frac{\partial}{\partial\psi_\beta}\lim_{\mathbf{x}'\to\mathbf{x}}\frac{\partial^2}{\partial x_k'\partial x_k'}\left\{\int\psi_\beta' P(\underline{\psi},\underline{\psi}';\mathbf{x},\mathbf{x}',t)d\underline{\psi}'\right\}\tag{97}$$

where the primes denote terms based on conditions at $\mathbf{x}'$.

5.3. TERMS APPEARING IN THE PDF EQUATION

Inspection of equation (95) reveals that the chemical source production term appears in exact and closed form. In other words it is known in terms of $\underline{\psi}$ and $\tilde{P}(\underline{\psi})$. This feature of equation (95) has always constituted the main reason for using the pdf approach in reacting flows. The implication is that realistic chemistry involving finite-rate kinetics can be incorporated explicitly in turbulent flow calculations without an assumed flame structure to characterize the thermochemistry. In comparison to other forms of thermochemical closure, this is clearly a novel, advantageous and unique feature. However, the molecular mixing term appears in unclosed form due to the appearance of the conditional expectation. To evaluate this term, the joint pdf of the scalars and their gradient (equation ((96)) at the single point level, or the two point joint pdf of the scalars (equation ((97)) is required. Needless to say, transport equations for such higher-order pdf's contain further unknown terms and increase the dimensionality and/or computational cost of the problem dramatically. The molecular mixing term involves diffusion coefficients and composition gradients which stem directly from the molecular diffusion term in equation (7) and an order of magnitude argument demonstrates that it cannot be neglected, even at high Re. Since diffusion is important predominantly amongst the small scales of turbulent motion, an accurate representation of these scales is clearly essential to the determination of the evolution of the pdf and hence of the averaged scalar field. In hydrocarbon combustion, this is particularly true for species such H_2 and H which are formed and consumed by reaction (in the fine scales of turbulent motion) and which possess much larger diffusivities than other scalars. As a consequence it can be argued that the classical closure problem associated with turbulent combustion, which manifests itself as a mean reaction rate in moment closures is simply re-expressed in terms of the molecular mixing term in pdf closures. In fact the appearance of the chemical source term in exact form in the pdf formalism is merely a consequence of the mathematical manipulations employed to derive such an equation. As pointed out by Kollmann [72], in pdf methods nonlinear terms are transformed into linear terms with variable coefficients. This is achieved by converting the associated dependent variables (which make the nonlinear terms unclosed upon averaging) into independent variables of the pdf at the cost of increasing the dimensionality of the formalism. The results of such a manipulation do not overcome the basic physical problems inherent in any statistical approach to turbulence, whether it be Reynolds/Favre averaging, spatial averaging or transported pdf modelling. The pdf $\tilde{P}(\underline{\psi})$ is only a function of the scalar field and includes no velocity information; consequently the turbulent diffusion of the pdf (term V) appears in unclosed form. The joint velocity-composition pdf allows this term to be expressed in exact form, but at the expense of introducing further unknowns and extra independent variables to the equation. Finally it is to be noted that at the level of closure corresponding to equation (95), $\tilde{u}_k$ is unknown. This quantity is conventionally supplied through the use of a turbulence model (section 2.1).

5.4. GENERAL SOLUTION FEATURES OF PDF TRANSPORT EQUATIONS

At this stage it may appear somewhat premature to discuss the solution algorithm for pdf transport equations, particularly since the modelling of the unclosed terms has not yet been considered. However, the reality is that the method of numerical solution and the modelling under this formalism are strongly interconnected. For problems involving a single scalar dimension and in simple geometries, solutions of pdf equations using conventional finite-difference techniques have been obtained (Janicka et al. [50; 51]). However, in order to exploit the full potential of pdf equations, it is necessary to solve for a multi-dimensional pdf involving several non-conserved scalars. Pope [100] estimated that the number of computer operations at each node (in compositional space) and at each time step, if finite-difference equations (fde) were to be employed, was of the order of $exp(6N)$, where N is the number of scalar variables. In the same paper he described a solution algorithm for performing a Monte Carlo (particle-method) simulation of the scalar pdf equation. In this approach the computational cost rises only linearly with N and the pdf is simulated via an ensemble of stochastic particles. That a continuous pdf $\tilde{P}(\underline{\psi})$ has a discrete counterpart is a widely used principle; for example equations (12) and (13) demonstrate how an average can be constructed from an ensemble of samples. The histogram constructed from $A^{(i)}(\mathbf{x}, t)$ is an approximate discrete representation of the pdf. In the limit of an infinite number of realizations (with the sampling 'bin' sizes tending to zero), the histogram tends to the pdf - this result is formalized by equation (94).

With regards to the closures for the mixing term in equations (95) and (96) to be discussed below, the implication is that given suitable approximation (written in pdf form), an equivalent particle model representation needs to be constructed. Thus in the section dealing with the mixing models, both interpretations will be given for the case of a single scalar. For certain mixing models the extension to multi-scalar problems is simple; in other models it is still not clear how this can be achieved in a rigorous and/or tractable manner. However, the existence of a link between the particle method and the pdf approach is essential if the advantages of the former method are to be exploited. Such a link and the necessary criteria for its use have been investigated by Pope [98; 100; 102]; it is argued that the fundamental basis for 'equivalence' of these two methodologies is that the resulting statistical moments (mean, variance, etc...) arising from the particle-method approach should evolve in an identical manner to its deterministic counterpart in the limit of an infinite number of notional particles.

5.5. BASIC MODELLING REQUIREMENTS OF THE MOLECULAR MIXING TERM

Proposals for closing the molecular mixing term are invariably tested in the simplest of flows e.g. inert, homogeneous and isotropic turbulence. Under these circumstances the scalar pdf is no longer a function of position and simply evolves, in time, with the result that the scalar variance decays to zero. It is argued that if pdf methods cannot reproduce the statistics of the decay process satisfactorily in this simple flow, then the claim that such a formalism provides a more complete statistical representation of fluid flows would appear somewhat paradoxical. In the situation described above, the most basic requirement of the mixing model is that it leaves the mean of the scalar unchanged (i.e. $\Delta\langle\phi(\mathbf{x},t)\rangle \equiv \langle\phi(\mathbf{x},t+\delta t)-\phi(\mathbf{x},t)\rangle = 0$) and predicts a variance decay rate in accordance with experimental observations :

$$\frac{d}{dt}\langle\phi'^2\rangle = -C_D\frac{\varepsilon}{k}\langle\phi'^2\rangle \tag{98}$$

where C_D has a value around 2.0. All models are designed to satisfy the above (minimum) requirements. In discussing the statistical behaviour of the various models, reference will be made to the following parameters : skewness, flatness and superskewness. These are the normalized moments of the pdf, defined as follows :

$$\Omega_n = \frac{\int_\psi(\psi-\langle\phi\rangle)^n P(\psi)d\psi}{\left\{\int_\psi(\psi-\langle\phi\rangle)^2 P(\psi)d\psi\right\}^{\frac{n}{2}}} \qquad \begin{cases} n=3 \Rightarrow \text{skewness} \\ n=4 \Rightarrow \text{flatness} \\ n=6 \Rightarrow \text{superskewness} \end{cases} \tag{99}$$

The real test of any mixing model lies in its ability to predict correctly the evolution of the scalar pdf. The test case utilised by most workers is based on initial conditions corresponding to segregated fluid parcels (in composition space), idealized as a double δ-function distribution. Here it is to be expected that the process of mixing/homogenization will result in the initial pdf relaxing to a smooth, bell-shaped distribution and in the limit of $t \to \infty$, converging to a δ function at the mean composition. Though there is no formal proof that it should be so, it is conventional to assume that the bell-shaped distribution is Gaussian. Tavoularis and Corrsin [115] investigated (experimentally) the evolution of a scalar (temperature) in a turbulent flow with uniform mean velocity and temperature gradients, and nearly homogeneous fluctuating velocity and temperature fields. The overheat of the fluid by heating rods was considered small enough to have negligible effect on the turbulence, and hence the temperature can be viewed as a passive scalar. It was found that typical values of the skewness and flatness factors of the temperature measured on the centre-line of the tunnel, far downstream, were indistinguishable from the values for a normally distributed random variable (which are zero and three,respectively); it was found that a Gaussian with the same mean and variance fitted the pdf's nearly perfectly. The velocity-scalar joint pdf also closely resembled a jointly-normal distribution. Eswaran and Pope [39] performed a DNS of decaying passive scalar fields in inert, homogeneous and isotropic turbulence. A statistically stationary velocity field was maintained by 'forcing' the simulations i.e. energy was added to the velocity field at low wave numbers, allowing the simulations to reach a quasi-equilibrium state where the rate at which energy was dissipated at the small scales was equal to the rate at which energy was added at the large scales. The results demonstrated that as time progressed, $P(\psi; t)$ evolved from a double δ function distribution to an inverted parabola and subsequently to a bell-shaped distribution. The evolution of the pdf shapes appeared to be independent of initial conditions (such as the ratio of the mechanical and scalar turbulence length scales, which had a significant effect on the scalar variance decay rate). It was argued that this observation should considerably simplify the task of constructing a realistic mixing model and it was found that after about eight eddy turnover times (a relatively long period of time), the skewness, flatness and superskewness of ϕ tended very closely to the Gaussian values of zero, three and fifteen respectively. The tendency to Guassianity was accompanied by an ever increasing lack of correlation between χ (defined in equation (36)) and $\phi(\mathbf{x}, t)$. This is an important result since the molecular mixing term (equation (96)) is effectively χ conditioned on $\phi(\mathbf{x}, t)$. As a modelling recommendation it is argued that the assumption of χ being independent of $\phi(\mathbf{x}, t)$, although valid at long times, is not valid in the short-term range of mixing and this raises questions concerning the validity of the statistical independence assumption invoked in sections 3.5 and 3.6.

Despite the above comments concerning the evolution of a decaying scalar field to Gaussianity, it must be borne in mind that the scalars considered here are bounded quantities and so their pdf's are bounded; therefore a scalar pdf cannot be exactly Gaussian. The implication is that the pdf approaches a Gaussian distribution in the limit of zero variance where most of the interacting fluid 'parcels' are very close to the mean composition. However the distribution near the bounds of the scalar are unlikely to resemble Gaussian behaviour, but this is immaterial given the degree of homogenization at that stage of the mixing process. A number of proposals have been made for closing the molecular mixing term and these are discussed below.

5.6. LINEAR MEAN SQUARE ESTIMATION (LMSE) CLOSURE

Using equation (97) and the definition of conditional probability, $P(\underline{\psi}, \underline{\psi}') = P(\underline{\psi}' | \underline{\phi}(\mathbf{x}, t) = \underline{\psi}) P(\underline{\psi})$, the pdf equation in homogeneous turbulence can be written as :

$$\varrho \frac{\partial P(\underline{\psi}; t)}{\partial t} = -\frac{\mu}{Sc} \sum_{\beta=1}^{N} \frac{\partial}{\partial \psi_\beta} \lim_{\mathbf{x}' \to \mathbf{x}} \frac{\partial^2}{\partial x'_k \partial x'_k} \left\{ \int \psi'_\beta P(\psi'_\beta | \underline{\phi} = \underline{\psi}) d\psi'_\beta \, P(\underline{\psi}) \right\}$$

$$= -\frac{\mu}{Sc}\sum_{\beta=1}^{N}\frac{\partial}{\partial\psi_\beta}\lim_{\mathbf{x}'\to\mathbf{x}}\frac{\partial^2}{\partial x'_k\partial x'_k}\left[\mathrm{E}(\phi'_\beta|\underline{\phi}=\underline{\psi})P(\underline{\psi})\right] \tag{100}$$

Dopazo [33; 35] and Dopazo and O'Brien [36] consider the case of the single scalar, where $\mathrm{E}(\phi'|\phi=\psi)$ is the expectation of the random variable $\phi(\mathbf{x}',t)$ at a point $\mathbf{x}'$, conditioned on the event that $\phi(\mathbf{x},t)=\psi$ at a point $\mathbf{x}$. If it is assumed that $P(\psi'|\phi=\psi)$ is conditionally Gaussian, the theory of stochastic processes implies that :

$$\mathrm{E}(\phi'|\phi=\psi) = \langle\phi\rangle + r(\mathbf{x},\mathbf{x}',t)\left(\psi-\langle\phi\rangle\right) \tag{101}$$

where $\langle\phi\rangle$ is the expectation (mean) of the random variable $\phi(\mathbf{x},t)$ and $r(\mathbf{x},\mathbf{x}',t)$ is the correlation coefficient of $\phi(\mathbf{x},t)$ and $\phi(\mathbf{x}',t)$,

$$r(\mathbf{x},\mathbf{x}',t) = \frac{\langle[\phi(\mathbf{x},t)-\langle\phi(\mathbf{x},t)\rangle]\,[\phi(\mathbf{x}',t)-\langle\phi(\mathbf{x}',t)\rangle]\rangle}{\sigma_\phi(\mathbf{x},t)\sigma_\phi(\mathbf{x}',t)} \tag{102}$$

where σ_ϕ denotes the square root of the variance of ϕ. Dopazo and O'Brien [37] point out that the above assumption does not imply that $\phi(\mathbf{x},t)$ and $\phi(\mathbf{x}',t)$ are normally distributed; it is claimed that it constrains the expected value of $\phi(\mathbf{x}',t)$, given $\phi(\mathbf{x},t)$, to a certain prescribed behaviour. More recently O'Brien [91] reformulated the problem in terms of linear mean square estimation, which involves less restrictive assumptions but gives rise to the same result. The requirement of homogeneity and isotropy together with equations (101) and (102) leads to the following result for the rhs of equation (100) :

$$-3\left(\frac{\partial^2 r}{\partial\ell^2}\right)_{\ell=0}[\psi-\langle\phi(\mathbf{x},t)\rangle] \equiv \frac{6}{\lambda_\phi^2}[\psi-\langle\phi(\mathbf{x},t)\rangle] \tag{103}$$

λ_ϕ^2 can be interpreted as a (Taylor) microscale for the scalar field given by : $-2\left(\partial^2 r/\partial\ell^2\right)_{\ell=0}^{-1}$ where $\ell=|\mathbf{x}'-\mathbf{x}|$. Equating the result of equation (103) to the lhs of equation (100), the closure can be expressed as :

$$\frac{\partial P(\psi;t)}{\partial t} = A\frac{\partial}{\partial\psi}\left[\left(\psi-\langle\phi\rangle\right)P(\psi;t)\right] \quad \text{where} \quad A=\frac{6\nu}{Sc\,\lambda_\phi^2} \tag{104}$$

Multiplying equation (104) by $\left(\psi-\langle\phi\rangle\right)^2$ and integrating over ψ space leads to :

$$\frac{d\langle\phi'^2\rangle}{dt} = -2A\langle\phi'^2\rangle \tag{105}$$

Comparing the result in equation (105) with that of equation (36) (in the case of constant density, homogeneous, isotropic and inert turbulence) implies the following identity :

$$\chi_\phi \equiv A\langle\phi'^2\rangle \tag{106}$$

Referring to equation (104) as the LMSE model, the particle method simulation corresponding to this closure becomes :

$$\frac{d\phi^{(p)}}{dt} = -\frac{C_d}{2}\frac{\varepsilon}{k}\left(\phi^{(p)}-\langle\phi\rangle\right) \tag{107}$$

where A has been replaced with $C_d\frac{\varepsilon}{k}$, and by selecting $C_d \approx 2$ a variance decay rate in conformity with measurements can be predicted as suggested by equation (98). The model clearly constitutes a linear, deterministic simulation of the mixing phenomenon, continuous in time. Analytical solutions for equation (104) are discussed in [37; 91; 102]; the results of such analysis show that if the initial condition is a normal distribution, the model preserves the Gaussianity which is considered as a satisfactory result. However, the drawback is that the initial functional form of the pdf is never relaxed and consequently, in the case of an initial double δ function specification for the pdf, the subsequent time evolution remains as two δ functions which approach each other in composition space. Pope [102] points out that such behaviour is to be expected since the model contains no information about the shape of the distribution - only the mean appears in equation (107). Despite this drawback, Dopazo [35] argues that the modelling is applicable provided the turbulence is "given a chance" to induce randomness and smear out the initial discontinuous state. In other words it is applicable if the pdf is continuous. Predictions based on LMSE were compared against measurements in [35] for grid-generated homogeneous turbulence with an asymmetric mean temperature profile. In the far downstream region good agreement was found for the conditional expected value of temperature versus temperature.

5.7. COALESCENCE-DISPERSION MIXING MODELS

5.7.1. *Curl's Mixing Model.* The work of Curl [28] was concerned with the analysis of the mixing of clouds of droplets in a two-liquid phase chemical reactor. It was assumed that all the droplets in the system were of the same size and that coalescence took place between two droplets having a solute concentration of c_1 and c_2. Redispersion occurred immediately to produce two equal droplets of concentration $\frac{1}{2}(c_1 + c_2)$. The total concentration range was divided into L intervals of concentration width δc and $P(l)\delta c$ was defined as the fraction of the droplets with concentrations within the $l^{th}(1 \leq l \leq L)$ interval. The basic balance equation constructed for $P(c)$ stated that the rate of change of the number of droplets in particular interval is influenced by :

- the rate of creation of new droplets in the interval $c, c+\delta c$ due to collisions of droplets between symmetrically and asymmetrically located concentration intervals
- the rate of loss of droplets by collision of droplets from $c, c+\delta c$ with those from all other concentration intervals

and this led to :

$$\frac{\partial P(c)}{\partial t} = 4Q\left\{\int_0^c P(c+\alpha)P(c-\alpha)d\alpha - P(c)\right\} \tag{108}$$

Curl defined Q as twice the collision rate divided by the total number of droplets in the system and solved equation (108) for the case of inert and reacting flows using conventional finite-difference techniques. Subsequently Spielman and Levenspiel [112] proposed a particle (Monte Carlo) method for the solution of such an equation and used the methodology to investigate the flow characteristics of an ideal stirred tank reactor involving a dispersed phase system. Curl's mixing closure has also been used in the context of continuous systems, to describe the turbulent mixing of gases. The work of Flagan and Appleton [41] and the references cited therein may be consulted as examples of such applications.

Dopazo [35] pointed out that since Curl's model was originally intended for interactions of droplets, the direct extrapolation of this model to the mixing of a single-phase gaseous scalar field was questionable. An implication would be that two fluid 'parcels' at different temperatures, say, would mix instantaneously to form fluid parcels with a temperature equal to the average of the two original temperatures. This feature of the model is clearly unrealistic and has also been highlighted by Pope [101] on performing the stochastic simulations; considering the test situation discussed in section 5.5, an initial condition corresponding to a double δ-function was specified and the mixture allowed to evolve. Of the N_p particles in the system, $N_p/2$ were assigned $\phi^{(p)} = 1$ and the remainder as $\phi^{(p)} = -1$. In any particular time step, pairs of particles were selected, and the number chosen was the nearest integer to :

$$N_c = \mathcal{B}\delta t \frac{\varepsilon}{k} N_p \tag{109}$$

This expression was chosen to ensure a variance decay rate consistent with equation (98) and for Curl's model $\mathcal{B} \approx 2$. The particles were then allowed to interact along the straight line connecting them in composition space, such that their new concentrations 'jump' from their initial values to their joint mean value. The values of particles not selected remain unaltered. The results demonstrated that after the first time step ϕ took the values -1, 0 and 1; after the second step they were -1, -1/2, 0, 1/2 and 1; after the jth step they were all integer multiples of 2^{1-j}. Therefore a continuous distribution is never achieved since certain values of the scalar sample space (those not corresponding to multiples of 2^{1-j}) cannot be reached. In spite of this, after a sufficient period of time, a bell-shaped (but non-Gaussian and discrete) distribution is obtained. In contrast to the LMSE closure however, the pdf does relax from its initial shape.

5.7.2. *Modified Curl.* Dopazo [35] and Janicka et al. [51] independently proposed modifications to Curl's model (section 5.7.1) in order to avoid some of its shortcomings. Although the starting point motivating their modelling was different, the end results were very similar indeed largely because their approach followed the ideas behind the derivation of Curl's model. Essentially the difference between Curl's original model and the modified forms lies in the treatment of the dispersion of the material point properties. In discussing these differences it is useful to introduce the concept of the transition pdf. The transition pdf is defined such that $T_r(\varphi, \Psi|\psi)d\psi$ is the probability that the interaction of a material point $\phi = \varphi$ with a point $\phi = \Psi$ produces $\psi \leq \phi \leq \psi + d\psi$ and $(\varphi + \Psi - \psi) \leq \phi \leq (\varphi + \Psi - \psi) + d\psi$. It is clear from the definition that the transition pdf has to satisfy the following properties :

- In coalescence-dispersion models the mixing process is effectively represented by the pairwise interaction of material points. These material points are only allowed to mix along the line joining them in composition space and this is consistent with the notion of physical mixing. Therefore, if ψ is outside the interval (φ, Ψ), it follows that $T_r(\varphi, \Psi|\psi)d\psi$=0. This implies that the interval (φ, Ψ) denotes a local convex domain for ψ.
- $T_r(\varphi, \Psi|\psi) = T_r(\varphi, \Psi|\varphi + \Psi - \psi)$. This effectively states that after pairwise interaction, the new scalar properties ψ and $\varphi + \Psi - \psi$ are formed with the same likelihood. This condition, together with the previous requirement ensure that the mean of the dependent scalar, ϕ, remains unchanged by the molecular mixing process.

- $T_r(\varphi, \Psi|\psi)$ is necessarily a pdf with respect to ψ. Therefore the normalization condition follows; when the integration is performed over the domain of ψ consistent with the comments made above : $\int_\psi T_r(\varphi, \Psi|\psi) d\psi = 1$

In terms of the transition pdf, coalescence-dispersion models can be expressed in the general form (using a density-weighted pdf) [58] :

$$\bar{\varrho}\frac{\partial \tilde{P}(\psi)}{\partial t} \approx \bar{\varrho} C_D \frac{\varepsilon}{k} \left\{ \int_\varphi d\varphi \int_\Psi d\Psi \tilde{P}(\varphi)\tilde{P}(\Psi) T_r(\varphi, \Psi|\psi) - \tilde{P}(\psi) \right\} \tag{110}$$

where C_D is assigned the value of 6 so that the observed variance decay rate is predicted (equation (98)).

In terms of Curl's original specification, where material points/particles coalesce and disperse with properties involving their mean value, the transition pdf takes the following form :

$$T_r(\varphi, \Psi|\psi) = \delta \left[\psi - \tfrac{1}{2}(\varphi + \Psi) \right] \tag{111}$$

In modified forms of Curl's model the material points that interact are not constrained to mix to their common mean values. In fact the degree of mixing can vary between no mixing and complete mixing (up to the mean composition) and it can be assumed that all composition states can be attained with equal likelihood between these limits. Thus, the transition pdf will then be a uniform distribution [51] over the range of the local convex domain (φ, Ψ) :

$$T_r(\varphi, \Psi|\psi) = \begin{cases} \frac{1}{|\varphi - \Psi|} & \psi \in [\varphi, \Psi] \\ 0 & \psi \notin [\varphi, \Psi] \end{cases} \tag{112}$$

In this case discontinuous pdf's cannot arise (in the sense that Curl's original model produces) since all possible composition states can be reached. It must be noted however that mixing is still represented as a jump process and such closures are essentially discontinuous representations of the mixing phenomenon.

The proposal of Dopazo [35] and Janicka et al. [51] has been examined further by Pope [101]. In the particle method analogy of the generalised coalescence-dispersion model, a random selection of a pair of particles $\phi^{*(p)}(t)$ and $\phi^{*(q)}(t)$ is followed by 'mixing' such that the new concentrations are given by :

$$\begin{aligned} \phi^{*(p)}(t + \delta t) &= \phi^{*(p)}(t) + \tfrac{1}{2}x \left(\phi^{*(q)}(t) - \phi^{*(q)}(t) \right) \\ \phi^{*(q)}(t + \delta t) &= \phi^{*(q)}(t) - \tfrac{1}{2}x \left(\phi^{*(q)}(t) - \phi^{*(q)}(t) \right) \end{aligned} \tag{113}$$

x controls the extent of mixing ($x \in [0, 1]$); with $x = 0$ no mixing occurs, whilst Curl's model is recovered with $x = 1$. x has to be a random variable with a continuous distribution, $\mathcal{A}(x)$, otherwise the problems associated with Curl's original formulation (given by $\mathcal{A}(x) = \delta(x - 1)$) arise. The simplest choice is to make $\mathcal{A}(x)$ a uniform distribution, i.e. $\mathcal{A}(x) = 1$. With regards to equation (109), $\mathcal{B}$ changes according to the choice of $\mathcal{A}(x)$. Pope [101] shows that :

$$\mathcal{B} = \left(a_1 - \tfrac{1}{2}a_2\right)^{-1} \quad \text{where} \quad a_m = \int_0^1 x^m \mathcal{A}(x)dx \tag{114}$$

Thus for $\mathcal{A}(x) = 1$, $\mathcal{B} \approx 3$. Calculations of scalar mixing in inert, homogeneous and isotropic turbulence fields demonstrate that starting from a double δ function distribution, continuous pdf's can be obtained with the generalised coalescence-dispersion model. However, the flatness (equation (99)) and higher *even* moments of the resulting pdf rise exponentially with time, whilst for a Gaussian pdf, these values are constant. Curl [28] demonstrated this analytically for his particular specification and Kosaly [73] has generalized the result regardless of the form of $\mathcal{A}(x)$.

Kosaly [73] argued that the 'proper' choice of $\mathcal{A}(x)$ was a potential method of incorporating more physics into coalescence-dispersion mixing models. In the initial stages of mixing in homogeneous turbulence, the pdf resembles an inverted parabola, and so a rational choice for $\mathcal{A}(x)$ should peak at $x = 0$ and monotonically decrease for increasing values of x. This ensures that the degree of mixing is controlled so that concentrations at the mean value do not appear in the early stages. Most proposals for $\mathcal{A}(x)$ do not satisfy this requirement in the early stages of mixing (e.g. $\mathcal{A}(x) = 1$) and furthermore, the question of how fast $\mathcal{A}(x)$ falls with x is important; Kosaly demonstrates analytically that under the condition of $\frac{a_n}{a_1} \ll 1, n > 1$ (following equation (114)), coalescence-dispersion modelling yields an identical result to LMSE (section 5.6) plus terms $O(\mathcal{B}a_2)$. It should not infered from this that LMSE is simply the limiting behaviour of coalescence-dispersion models. Both these class of models are fundamentally different; LMSE is a linear, deterministic and continuous representation of mixing, whilst coalescence-dispersion models are nonlinear, jump processes whose derivations are somewhat phenomenological. The idea of improving the modelling of the coalescence-dispersion closure through more elaborate specifications of $\mathcal{A}(x)$ has not received widespread attention, since it is not regarded as the primary factor affecting the performance of the model. Kosaly and Givi [74] investigated the sensitivity of predictions to the choice of $\mathcal{A}(x)$ and considered LMSE and Curl's model to represent opposite extremes in mixing behaviour. The former was represented through the following specification : $\mathcal{A}(x) = \delta(x - \epsilon), \epsilon \to +0$ though in fact, $\epsilon \approx 0.1$ was employed. Calculations of a two-dimensional, stationary, incompressible, turbulent round jet discharging into stagnant surrounds demonstrated that the pdf's arising out of the use of $\mathcal{A}(x) = \delta(x - \epsilon)$, $\mathcal{A}(x) = 1$ and $\mathcal{A}(x) = \delta(x - 1)$ were quite similar at about 10 diameters downstream of the nozzle.

Due to the shortcomings of the coalescence-dispersion models so far described with regard to the predictions of the normalized moments, much effort has been devoted to towards improving them so that finite values of flatness and superskewness, for example, can be obtained.

5.7.3. *Age Biased coalescence-dispersion model.* Pope [101] formulated an age biased model such that continuous pdf's with finite flatness and superskewness could be obtained. In this model the selection of elements for mixing was biased towards the age of the element, whereas in the conventional coalescence-dispersion modelling described previously, the likelihood of sampling (without replacement) a particular element is equal for all particles. The age of an element was defined as the normalized time since it mixed with another element. Denoting $t^{(n)}$ as the last time that the element n was selected for mixing, the age of the nth element was defined as : $\tau^{(n)} = \frac{\varepsilon}{k}(t - t^{(n)})$ where t is the current time. An age distribution $\mathcal{T}(s)$ was assigned as the pdf of τ, where s is the sample space of τ. The age distribution can be changed by biasing the sampling according to the age of the elements. This was achieved through the sample bias, $\mathcal{Z}(s)$, normalized so that $\int_0^\infty \mathcal{T}(s)\mathcal{Z}(s)ds = 1$. $\mathcal{T}(s)$ was obtained from the following evolution equation :

$$\mathcal{T}(s + \frac{\varepsilon}{k}\delta t; t + \delta t) \approx \mathcal{T}(s;t) + 2\mathcal{B}\frac{\varepsilon}{k}\delta t \left[\delta(s) - \mathcal{T}(s)\mathcal{Z}(s)\right] \tag{115}$$

The argument $s + \frac{\varepsilon}{k}\delta t$ in equation (115) reflects the fact that an element (not selected for mixing) ages an amount $\frac{\varepsilon}{k}\delta t$ in a time interval δt. The fraction of elements selected is $2\mathcal{B}\frac{\varepsilon}{k}\delta t$ and the effect is to remove an amount $\mathcal{T}(s)\mathcal{Z}(s)$ from $\mathcal{T}(s)$ and to add it to $\mathcal{T}(0)$; in other words the age of the mixed elements is reset to zero. $\mathcal{Z}(s)$ is interpreted as the relative probability of sampling the element n, of age $\tau = s$; for modified Curl $\mathcal{Z}(s) = 1$, so that $\mathcal{T}(s) = 2\mathcal{B}exp(-2\mathcal{B}s)$. The proposed mixing model was defined by two pdf's, $\mathcal{A}(x)$ and $\mathcal{Z}(s)$ which could be expressed as an evolution equation for the joint pdf of ϕ and τ. Pope demonstrated that the equation admitted self-similar solutions for the test problem in homogeneous turbulence. These solutions were examined in terms of the moments of the pdf to yield the flatness, superskewness and the constant $\mathcal{B}$ (which was no longer given by equation (114)). A stochastic model algorithm, simulating the behaviour of the resulting pdf equation was proposed and solved for the binary mixing problem. Six different $\mathcal{A}(x)$ specifications were employed with six different $\mathcal{T}(s)$. The results demonstrated that the 'best' combination of $\mathcal{A}(x)$, and $\mathcal{T}(s)$, in terms of yielding normalized moments closest to the Gaussian value involved, paradoxically, $\mathcal{A}(x) = \delta(x-1)$ (i.e. Curl's original form) with an age distribution $\mathcal{T}(s) = 2\mathcal{B}$. It was found that a quartic function yielded the best compromise between good overall behaviour (in terms of pdf evolution) and acceptable values of the normalized moments.

Chen and Kollmann [25] investigated the application of Pope's age-biased model and chose a power law function for the biasing function and calculated joint pdf's of mixture fraction and temperature for reacting and non-reacting turbulent jets. Very simplified kinetics was employed so that temperature was the only indicator of the progress of the reaction. The results of their investigation suggested that age-biasing had little influence on the statistics of the mixture fraction up to fourth moment for the non-reacting jet flows. Examination of the joint pdf's of temperature and mixture fraction (in the form of scatter diagrams) also demonstrated little sensitivity to the choice of mixing model. An implication of this result is that convection and turbulent transport may moderate any improvements which may otherwise arise from an improved mixing model in simple (e.g. homogeneous and isotropic) flows.

5.7.4. *Reaction Zone Conditioning.* Coalescence-dispersion closures represent jump processes i.e. they are discontinuous in time. This feature of the model is problematic partly because mixing is a continuous process and 'interacting' eddies do not suddenly jump to new concentrations. Furthermore in the flamelet regime of combustion, where chemical reaction is confined to a thin region of both physical and compositional space, a jump process allows for the possibility of two material points (or particles) to be mixed through the reaction zone *without* any reaction (figure 4). In this sense, the process of combustion is never 'felt' by the interacting particles.

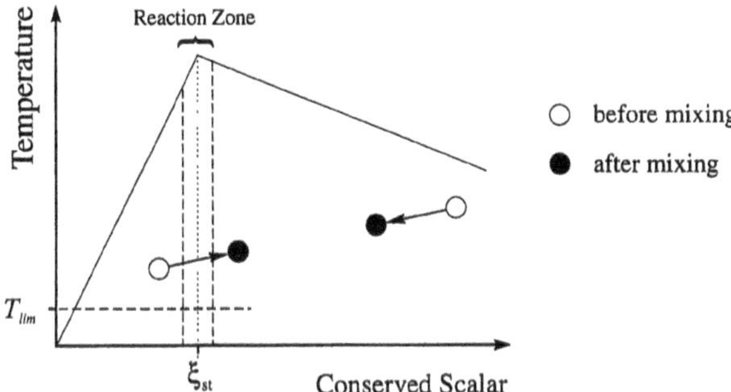

Fig. 4. Schematic of the mixing process based on coalescence-dispersion closures showing how a particle mixes and thereby crosses the reaction zone without reacting. T_{lim} is the temperature below which the reaction rates are negligibly low.

The problem highlighted by figure 4 is of concern if the particles have a sufficiently high temperature for reaction rates to be effective. Thus, with finite-rate kinetics 'cold' particles ($T < T_{lim}$) can, in principle, mix across the stoichiometric ξ-plane without appreciable reaction, but this is inconsistent with the flame sheet regime since the 'mixed equals burned' criterion applies. Chen and Kollmann [25] have proposed a form of reaction zone conditioning which can account for the combined effects of chemical reaction and mixing, and requires no substantial modifications to current coalescence-dispersion proposals. With reference to figure 4, the conditioning involves performing mixing and reaction sequentially, in sub-steps, such that a particle is forced to pass through the flame zone, react and then resume mixing.

As discussed in section 5.7.3 with reference to [25], joint pdf's of temperature (T) and mixture fraction (ξ) were examined for turbulent jet diffusion flames for the configuration found in [86]. The mixing models used, which are of relevance here, involved the modified form of Curl's model with and without reaction zone conditioning. Two extreme reaction mechanisms were used; a flame sheet type reaction zone where $\frac{dT}{dt} = 10^{10}$ K (ms)$^{-1}$ inside the 'thin' reaction zone until T reaches its maximum value at a fixed ξ, and a slow chemistry case where $\frac{dT}{dt} = 10^3$ K (ms)$^{-1}$. Outside the reaction zone the rate was set to zero. For the slow chemistry case (with a relatively broad reaction zone) conditioning the mixing process on the reaction zone did not alter the joint pdf (T, ξ) noticeably relative to the pdf predicted by the unconditioned model. In the fast chemistry test case, reaction zone conditioning showed a very large effect, such that all particle properties lay on the equilibrium $T = T(\xi)$ line. In contrast, without conditioning, a significant number of points fell below the $T = T(\xi)$ line, demonstrating the deficient behaviour of the original model.

Norris and Pope [89] also addressed the problem of reaction zone conditioning but followed a different approach. They identified the criteria under which the effect is important : when particles are selected with values of ξ that lie on either side of ξ_{st}, and whose difference in ξ-property $\Delta\xi$ was large compared to the width of the reaction zone. It was therefore proposed to re-order the initial sub-ensemble of particles selected (without replacement) in ascending values of ξ and to pair off adjacent particles - in this manner, $\Delta\xi$ is minimized between any particle pair prior to mixing. It was claimed that this approach has physical justification since the instantaneous conservation equation for ξ, $\frac{D\xi}{Dt} = \Gamma\nabla^2\xi$, indicates that $\xi(\mathbf{x})$ is affected by the ξ field within an infinitessimal neighbourhood of $\mathbf{x}$, where ξ differs infinitessimally from $\xi(\mathbf{x})$. The application of adjacent pairing of particles in ascending order of ξ in homogeneous turbulence with flame-sheet thermochemistry led to far fewer 'jumps' across ξ_{st}. However, in the non-reacting homogeneous test case it was found that such re-ordering led to pdf's with bi-modal shapes, and consequently bore no qualitative resemblance to a Gaussian distribution. Consequently a compromise pairing scheme, which effectively involved trying out different combinations of ordered-pairing sequences, was performed. At the expense of introducing extra 'jumps' across the flame zone, improved pdf shapes were obtained. However, it does not appear possible to extend the ordered-pairing approach to multi-scalar pdf equations without ambiguity - against which scalar property does one choose to re-order the randomly selected set of particles ? The choice of a particular scalar will lead to different sequences of particles.

How much improvement can be gained in practice through the use of reaction zone conditioning has yet to be demonstrated, and since its inception little development appears to have been undertaken. Such modifications to coalescence-dispersion closures are important, however, if only because they highlight that mixing models developed in non-reacting flows may not be applicable when chemical reaction is present.

5.8. LANGEVIN MODELS

Non-linear integral models (such as coalescence-dispersion models) can be interpreted stochastically as Poisson processes which form the basis of particle interaction (jump) models, Pope [102]. Random diffusion processes can also be used to simulate the behaviour of molecular mixing, and these form the basis of Langevin models. Pope [102] proposed a model for the mixing term using a Langevin equation :

$$\delta\phi(t) = \phi(t+\delta t) - \phi(t) = G\left[\phi(t) - \langle\phi(t)\rangle\right]\frac{\delta t}{\tau} + \left(B\frac{\langle\phi'^2\rangle}{\tau}\right)^{\frac{1}{2}}\delta W_t \qquad (116)$$

The first term on the rhs is a linear deterministic term followed by a random diffusive term. G and B are constants to be determined, and B is a non-negative diffusion coefficient for this process. τ is a characteristic turbulent time scale and δW_t is the increment of a Weiner process which is defined by the properties : $\langle\delta W_t\rangle = 0$ and $\langle\delta W_t^2\rangle = \delta t$ where $\delta W_t = W_{t+\delta t} - W_t$. A function which satisfies these properties is : $W_t \equiv (\delta t)^{1/2}\sum_{i=1}^{N_s} v_i$, where v_i $(i = 1, \ldots, N_s)$ represents a set of independent standardized (i.e. zero mean, unity variance) normal random variables. W_t is a continuous and non-differentiable function of time. The mean is preserved in this formulation, and the requirement for the correct variance decay rate leads to a relation between G and B leaving one constant to be specified. Pope points out that from any initial condition and in the absence of any disturbing influences, the pdf relaxes to Gaussian with this model. The rate at which the pdf relaxes depends on the value of B; for $B = 0$, LMSE is recovered for which there is no relaxation. A drawback of the model is that the term involving B can lead to the violation of scalar bounds and for this reason, the expression shown in equation (116) is not considered as useful for representing scalar mixing. However, for the velocity field which is not bounded, such a Langevin model can be (and has been) employed.

5.9. BINOMIAL MODELS

Valiño and Dopazo [119] proposed a sequential LMSE and random jump process model capable of producing a Gaussian relaxation for bounded scalars. During a time step δt, two additive subprocesses take place involving $\phi^{(n)}(t)$: first all N_p particles modify their values according to a LMSE variation and then $N_p\mathcal{B}'\delta t/\tau$ of the particles are randomly selected out of the N_p stochastic particles (where τ is a characteristic time scale.). $\mathcal{B}'$ [1] is a positive dimensionless constant whose value strongly affects the relaxation rate of the pdf and its value requires calibration against measurements. The $N_p(1 - \mathcal{B}'\delta t/\tau)$ remaining particles do not alter their values, whilst those randomly sampled modify their values according to :

$$\phi^{(p)}(t+\delta t) - \langle\phi(t)\rangle = \varsigma^{(p)}\langle(\phi - \langle\phi\rangle)^2\rangle^{\frac{1}{2}} \tag{117}$$

where $\varsigma^{(p)}$ is either a standardized Gaussian random variable if ϕ is an unbounded scalar, or a standardized binomially distributed random variable if ϕ is bounded. Since the latter case is relevant to reactive scalars, we will concentrate on its representation :

$$\varsigma^{(p)} = \frac{\eta - M\mathcal{P}}{\sqrt{M\mathcal{P}(1-\mathcal{P})}} \tag{118}$$

η is a binomially distributed *integer* random variable with mean $M\mathcal{P}$ and variance $M\mathcal{P}(1-\mathcal{P})$. $\mathcal{P}$ is a real number in $[0, 1]$ and η can take integer values between zero and M. To ensure boundedness, the minimum and maximum values of ϕ correspond to η equal to zero and M, respectively. With this requirement it can be shown :

$$M = \frac{(\langle\phi\rangle - \phi_{min})(\phi_{max} - \langle\phi\rangle)}{\langle\phi'^2\rangle} \tag{119}$$

$$\mathcal{P} = \frac{(\langle\phi\rangle - \phi_{min})}{(\phi_{max} - \phi_{min})} \tag{120}$$

In a homogeneous turbulence $\langle\phi'^2\rangle$ decreases with time and thus, for fixed bounds, equation (119) implies that M increases. Furthermore as $M \to \infty$, the de Moivre-Laplace theorem shows that $\varsigma^{(p)}$ tends to a normal continuous random variable; thus a distribution close to Gaussian at long times is implied. For the test case corresponding to an initial double δ function with $\phi_{min} = 0$ and $\phi_{max} = 1$, difficulties arise in the early stages of mixing with regards to equation (119). Valiño and Dopazo show that if the result of equation (119) is truncated to the nearest integer (M is an integer), no relaxation of the pdf is observed until $M \geq 2$. This difficulty was overcome by allowing for random samplings of stochastic particles around each of the initial δ functions. As $\langle\phi'^2\rangle$ decreases, these two samplings converge towards that originally proposed with a single M and $\mathcal{P}$ specification (given by equations (119) and (120)). Monte Carlo simulations of homogeneous turbulence demonstrated that the moments of the pdf evolved correctly; i.e. to within statistical

[1] Note that $\mathcal{B}'$ is not the same as the $\mathcal{B}$ appearing in equations (109) and (114), since the second operation that is applied in the binomial model affects neither the mean nor the variance (which are only affected by the application of LMSE in the first step). In equations (109) and (114), $\mathcal{B}$ is chosen so as to yield the correct variance decay rate.

error (based on finite sample size) the mean and skewness were invariant, the variance decayed exponentially and the flatness tended asymptotically to 3 - the Gaussian value. However, the rate at which the moments evolved were strongly affected by $\mathcal{B}'$; for $\mathcal{B}' = 0.5$, the pdf's evolved towards a bell-shaped distribution, but with the appearance of two spikes corresponding to the initial δ functions. With $\mathcal{B}' = 2$ qualitatively better pdf evolution was obtained.

Subsequently Valiño and Dopazo [120] combined a LMSE and a binomial diffusion process to obtain a modified Langevin model :

$$\delta\phi(t) = \underbrace{\left\{ K \left[1 - \left(\frac{\phi'(t)}{\phi'_*} \right)^2 \right] \frac{\delta t}{\frac{1}{2}\tau} \langle \phi'^2 \rangle \right\}^{\frac{1}{2}} \varsigma^{(p)}}_{\text{Random diffusion term}} - \underbrace{\left[1 + K \left(1 - \frac{\langle \phi'^2 \rangle}{\phi'^2_*} \right) \right] \frac{\delta t}{\tau} \left(\phi(t) - \langle \phi \rangle \right)}_{\text{Linear deterministic term}} \tag{121}$$

where ϕ'^2_* is equal to ϕ^2_M whenever $\phi^{(p)}(t) > 0$ and equal to ϕ^2_m for $\phi^{(p)}(t) < 0$. ϕ^2_M and ϕ^2_m are the maximum (positive) and minimum (negative) fluctuations allowed, respectively. K is a constant to be adjusted to yield the correct evolution of the pdf shape and to be consistent with the scalar variance decay rate, equation (98). It is worth noting that the binomial model described in [119] is sequential but is neither continuous nor differentiable; the present (non-sequential) model is continuous but non-differentiable, and is therefore a more physically realistic representation of mixing. Equation (121) is similar to the proposal of Pope [102], equation (116), except that the Weiner process (which may generate unbounded scalar values) is now replaced by a standardized binomially distributed random variable, $\varsigma^{(p)}$, and defined so as to ensure boundedness as in the previous binomial model. As time increases $\langle \phi'^2 \rangle$ decreases, so that $\langle \phi'^2 \rangle / \phi'^2_* \to 0$, $\phi'(t)/\phi'_* \to 0$ and for large values of t, equation (121) asymptotically yields relaxation to a Gaussian pdf. Comparison of the pdf forms (using $K = 2.1$) with the DNS data of Eswaran and Pope [39] demonstrated very good agreement. However, the sensitivity of the results to the value of K was not discussed.

Binomial models have been demonstrated to work well for pdf's of single scalars with fixed bounds. However, the ability of the model to handle multi-scalar pdf's is not clear. The bounds of a reactive scalar are not fixed; each scalar property (e.g. the specific mole number of H_2, CO, etc...) has its own bounds fixed by conservation principles and the allowable scalar domain is a complicated hypervolume in multi-dimensional composition space. Now, in Monte Carlo simulations, equations (119) and (120) would be used to select η from a random number generator. For a given stochastic particle, a binomial distribution characteristic of each scalar property (in terms of M and P) is required from which η is to be selected. Furthermore the bounds for each property can differ from one particle to another, and these bounds evolve as a result of mixing and reaction. Thus, in principle it is possible for different values of M and P to be needed for each property and for each particle over any particular time step. This has the implication that only one sample would be selected from any given binomial distribution characterised by M and P. The consequences of this are at present unclear, but it appears inconsistent with the idea of sampling random numbers with a fixed distribution.

5.10. MAPPING CLOSURES

In the literature dealing with mixing models, the quest for predicting Gaussian pdf's in homogeneous turbulence appears to have culminated in the proposal for mapping closures. This closure has received a considerable amount of attention over a relatively short period of time, partly because of its ability to predict (in analytical form in certain cases) the correct evolution and asymptotic behaviour of passive scalar mixing in homogeneous fields. Chen et al. [23] addressed the problem of closure for the mixing term by constructing $\phi(\mathbf{x},t)$ as a time dependent mapping of a Gaussian reference field; the evolution of the mapping characterizes the evolution of the scalar field for which the pdf is sought. The mapping technique is capable of handling strongly non-Gaussian $\phi(\mathbf{x},t)$. The pdf equation for homogeneous turbulent scalar fields is considered in the following form :

$$\frac{\partial P(\psi;t)}{\partial t} + \frac{\partial}{\partial \psi}\left[\mathrm{E}\left\{D\nabla^2\phi|\psi\right\} P(\psi;t)\right] = 0 \tag{122}$$

where D is the diffusivity and $\mathrm{E}\left\{D\nabla^2\phi|\psi\right\}$ is the expectation of $D\nabla^2\phi$ conditioned on $\phi(\mathbf{x},t)=\psi$. The cumulative distribution function (cdf) of $P(\psi;t)$ is defined as :

$$F(\psi;t) \equiv \int_{-\infty}^{\psi} P(\hat{\psi};t)d\hat{\psi} \tag{123}$$

Equation (123) taken together with (122) leads to the transport equation for the cdf :

$$\frac{\partial F(\psi;t)}{\partial t} + \mathrm{E}\left\{D\nabla^2\phi|\psi\right\}\frac{\partial F(\psi;t)}{\partial \psi} = 0 \tag{124}$$

The closure establishes the following mapping relation [104; 72] :

$$\phi_s(\mathbf{x},t) \equiv X[\theta(\mathbf{z}),t] \tag{125}$$

$\phi(\mathbf{x},t)$ is the dependent scalar field and $\theta(\mathbf{z})$ is a time-independent standardized Gaussian reference field residing at a spatial location $\mathbf{z}$. $\phi_s(\mathbf{x},t)$ is referred to as the surrogate field [104] and is effectively defined through equation (125). The mapping X does not determine the relation between the locations of $\phi(\mathbf{x},t)$ and $\phi(\mathbf{z})$. However, if the relation between the locations in physical and reference space is restricted to a stretching transformation uniform in physical space, the following relation exists :

$$\frac{\partial z_\beta}{\partial x_\alpha} = \delta_{\alpha\beta} m(t) \tag{126}$$

where $m(t)$ accounts for the average rescaling produced by the turbulent stretching and molecular diffusion [43]. The mapping, X, is constructed as follows : the value of the cdf of a turbulent field at $X(\eta,\mathbf{x},t)$ (η is the sample space of the Gaussian reference field) is equal to the value of the standardized Gaussian cdf at $\theta(\mathbf{z})=\eta$:

$$F(X(\eta,\mathbf{x},t)) = F_G(\eta) \tag{127}$$

implying the relation $\psi = X(\eta, \mathbf{x}, t)$. A direct consequence of the monotonicity of the cdf's is that the mapping increases monotonically too, i.e. $X(\eta_1) < X(\eta_2)$ for $\eta_1 < \eta_2$. Differentiating equation (127) with respect to η :

$$\frac{\partial F}{\partial \psi}\frac{\partial X}{\partial \eta} = \frac{\partial F_G}{\partial \eta} \quad \Rightarrow \quad P(\psi;t) = P_G\left(\frac{\partial X}{\partial \eta}\right)^{-1} \tag{128}$$

Performing chain differentiation of equation (127) with respect to time yields :

$$\frac{\partial F}{\partial t} + \frac{\partial F}{\partial \psi}\frac{\partial X}{\partial t} = 0 \tag{129}$$

This result employs the concept of time independence of the reference cdf F_G. Valiño et al. [123] consider the case of time dependent statistics i.e. $\phi_s(\mathbf{x}, t) \equiv X[\theta(\mathbf{z}, t), t]$ (c.f. equation (125)). In this case the equation for $P_G(\eta)$ satisfies :

$$\frac{\partial P_G(\eta, t)}{\partial t} = -\frac{\partial}{\partial \eta}\left[\langle\frac{\partial \theta}{\partial t}\bigg|\theta = \eta\rangle P_G(\eta, t)\right] \tag{130}$$

A closure hypothesis for $\langle\frac{\partial \theta}{\partial t}\big|\theta = \eta\rangle$ was proposed which preserved the Gaussianity for the reference field and involved a free parameter β adjusted to produce the correct variance decay rate (obtained from DNS results). For $\beta = 0$ the results of Chen et al. [23] are recovered. However, it was also demonstrated that the shape of $P(\psi;t)$ was largely independent of the value of β. Substitution of equation (124) into (129) coupled with the fundamental assumption [104; 43] that the unknown statistics of the turbulent field are the same as the known statistics of the surrogate field, leads to :

$$\frac{\partial}{\partial t}X(\eta, \mathbf{x}, t) = \mathrm{E}\left\{D\nabla^2\phi_s|\psi_s = X(\eta, \mathbf{x}, t)\right\} \tag{131}$$

It is a straightforward matter to show that the Laplacian of the surrogate field is :

$$\nabla^2\phi_s(\mathbf{x}, t) = \nabla^2 X(\theta(\mathbf{z}), t) = m^2\left(X''\frac{\partial \theta}{\partial z_i}\frac{\partial \theta}{\partial z_i} + X'\frac{\partial^2 \theta}{\partial z_i \partial z_i}\right) \tag{132}$$

where $X' \equiv \frac{\partial X}{\partial \theta}$. The conditional expectation of equation (132) can be established provided the derivatives of the mapping are completely specified by the condition $\phi_s(\mathbf{x}, t) = \psi$. This implies [72] that the mapping must be local, i.e. the variation of the reference field at locations $\mathbf{z}' \neq \mathbf{z}$ has no influence on $X(\theta(\mathbf{z}), t)$. Thus, the condition $\phi(\mathbf{x}, t) = \psi$ is the image of the condition $\theta(\mathbf{z}) = \eta$. This, taken with the results for Gaussian statistics of the reference field [104] leads to the evolution equation for the mapping :

$$\frac{\partial X}{\partial t} = m^2 D\left\langle\frac{\partial \theta}{\partial z_i}\frac{\partial \theta}{\partial z_i}\right\rangle\left(X'' - \frac{\eta}{\langle\theta^2\rangle}X'\right) \tag{133}$$

where $\langle\theta^2\rangle = 1$. D is treated as spatially uniform. Since equation (122) makes use of the mixing term employing Fickian diffusion, this strictly implies that D cannot vary from one specie to another (refer to equation (7))- thus it appears that this level of closure is constrained to an equal diffusivity assumption. The mapping closure described above relates to single point statistics. As in all other one-point, one-time closures, length scale information is not available; $\langle \frac{\partial \theta}{\partial z_i} \frac{\partial \theta}{\partial z_i} \rangle$ in equation (133) requires external information and is similar in form to a scalar dissipation. Valiño and Gao [122] suggest the use of a characteristic time k/ε, which is what is used in other mixing models. Another possibility is the modelling of the joint scalar and its gradient [121]. In this case $\tau_\phi \equiv \langle \phi'^2 \rangle / \langle D \nabla \phi' \nabla \phi' \rangle$ can be obtained directly from the joint pdf. How easily the latter option can be implemented is still not clear.

Gao [42] presented a closed form solution to equation (133). Two important features of the solution were that it preserved the boundedness of the scalar field and also predicted relaxation to a Gaussian distribution. Pope [104] and Valiño and Dopazo [120] obtained analytical solutions to equation (133) for the case of $P(\psi; 0) = \frac{1}{2}[\delta(\psi + 1) + \delta(\psi - 1)]$. Excellent agreement was found with the relaxation of pdf's arising from DNS. Monte Carlo simulations of the mapping closure for this test case [123; 122] led to equally good agreement. The basic assumption used to obtain equation (133) can be expressed as :

$$\underbrace{\mathrm{E}\left\{\nabla^2 \phi_s | \phi_s = \psi\right\}}_{\mathrm{E}_s} = \underbrace{\mathrm{E}\left\{\nabla^2 \phi | \phi = \psi\right\}}_{\mathrm{E}} \text{ or } \mathrm{E}\left\{(\nabla \phi_s)^2 | \phi_s = \psi\right\} = \mathrm{E}\left\{(\nabla \phi)^2 | \phi = \psi\right\} \qquad (134)$$

Gao [43] tested this assumption by comparing predictions of $\mathrm{E}_s(\phi_s)/\mathrm{E}_s(\frac{1}{2})$, for which an analytical expression existed (using the initial double δ function specification), against DNS results. Excellent agreement was obtained suggesting that the underlying assumption behind equation (134) is sound. In section 5.5 it was pointed out that the DNS results of Eswaran and Pope [39] predicted Gaussianity after a relatively long period of time, implying the existence of a substantial interim period of non-Gaussianity. Gao [43] demonstrated that the evolution of E_s was closely related to its initial value, which can be affected by different initial length scales, consistent with DNS results. The observed persistence of non-Gaussianity in the flatness and superskewness is clearly related to the behaviour of the tails of the pdf. While near the mean the behaviour is strongly Gaussian, near the bounds it is not (but after substantial mixing very little probability, i.e area under the pdf, exists near the tails). Miller et al. [87] used mapping closures together with other mapping relations (referred to as Johnson-Edgeworth translations) and DNS results to show that the level of agreement of the normalized conditional dissipation of the scalar at the bounds was rather poor. The deficiency was attributed to the inability of the closures to allow for variations in the scalar bounds as the inert mixing process evolved. The DNS results they employed for validation demonstrated that the scalar maxima and minima migrated towards the mean with time; these were used as an empirical input for calculations where the bounds were allowed to relax, and the subsequent improvement in the predictions at the scalar extremities were noted.

How useful mapping closures will be for prediction methods in practical flows is still an open question and the problems associated with incorporating extra scalars (inert and reactive) in a mathematically and computationally tractable manner need to be resolved. Furthermore, although such a technique yields excellent results in homogeneous turbulence, it is not clear whether it will necessarily yield better results in inhomogeneous flows where the turbulent time scales can vary and are not known. The use of k/ε as the only characteristic time scale may limit its usefulness relative to simpler and cruder models previously discussed.

5.10.1. *Closing Remarks concerning mixing models.* Several techniques for closing the molecular mixing term have been discussed, each with varying degrees of success in simple configurations. For practical applications to inhomogeneous reactive flows involving several scalars and incorporating realistic finite-rate chemistry, LMSE and coalescence-dispersion models appear to be the most widely used. This is essentially because they can be extended without any significant additional effort to more complicated situations. In this regard, many of the models previously discussed are still in the early development stages and their capabilities have yet to be demonstrated.

5.11. MODELS FOR THE TURBULENT TRANSPORT OF THE PDF

Traditional modelling practice suggests that in many flows turbulent transport can be adequately represented through a gradient transport hypothesis. For many of the applications of the pdf method to thin shear flows it appears to yield acceptable results for quantities relating to gross mixing patterns (e.g. mixture fraction). Hence we can write :

$$\underbrace{-\frac{\partial}{\partial x_k}\left\{\bar{\varrho}\langle u_k''|\underline{\phi}=\underline{\psi}\rangle\tilde{P}(\underline{\psi})\right\}}_{\text{IV}} \approx \frac{\partial}{\partial x_k}\left[\frac{\mu_t}{\sigma_t}\frac{\partial\tilde{P}(\psi)}{\partial x_k}\right] \tag{135}$$

where $\sigma_t = 0.7$ and μ_t is given by the k-ε model (section 2.1.1). With a second moment closure being used for the velocity field, Chen and Kollmann [24] used the following approximation :

$$\underbrace{-\frac{\partial}{\partial x_k}\left\{\bar{\varrho}\langle u_k''|\underline{\phi}=\underline{\psi}\rangle\tilde{P}(\underline{\psi})\right\}}_{\text{IV}} \approx \frac{\partial}{\partial x_k}\left[c_s\bar{\varrho}\frac{k}{\varepsilon}\widetilde{u_k''u_j''}\frac{\partial\tilde{P}(\underline{\psi})}{\partial x_j}\right] \tag{136}$$

with $c_s = 0.3$.

In the past other proposals for closing the turbulent diffusion term have been made. Dopazo [34] tentatively proposed an LMSE type approximation and such an expression can be expected to work well in nearly Gaussian cases. Janicka et al. used a modelled transport equation for $\langle u_k'|\phi = \psi\rangle$ (i.e Reynolds averaging and for a single conserved scalar) - though the modelling involved a formidable task. Certain terms involving the pressure were either neglected through ad-hoc assumptions or through order of magnitude analysis and conditional Reynolds stresses were modelled using gradient transport hypothesis. However, these methods have now been largely replaced by the techniques referred to in the previous paragraph.

5.12. THE MONTE CARLO METHOD FOR SOLVING THE SCALAR PDF EQUATION

Having discussed in some detail the modelling of the scalar pdf equation, the problem of solving the equation has to be addressed. In doing so, no assumptions on the form of the mixing model to be employed are made, although it is assumed that a particle method can be constructed to represent it. However, it is assumed that the turbulent diffusion of pdf is represented by gradient modelling as discussed in section 5.11.

The solution formalism for the scalar pdf transport equation is based on the method of fractional steps. First equation (95) is expressed in the following form :

$$\frac{\partial \tilde{P}}{\partial t} + \tilde{u}_k \frac{\partial \tilde{P}}{\partial x_k} + \frac{\partial \tilde{P}\dot{\omega}_\alpha}{\partial \psi_\alpha} = \frac{1}{\bar{\varrho}} \frac{\partial}{\partial k} \left\{ \frac{\mu_t}{\sigma_t} \frac{\partial \tilde{P}}{\partial x_k} \right\} + \Xi(\underline{\psi}; \mathbf{x}, t) \tag{137}$$

where summation is implied over repeated indices (including greek subscripts). Equation (137) is then discretised using an explicit forward time finite-difference approximation which may be written in the following form :

$$\tilde{P}(\underline{\psi}; \mathbf{x}, t + \Delta t) = \{\mathbf{I} + \Delta t(\mathbf{T} + \mathbf{\Xi} + \mathbf{\Omega})\}\, P(\underline{\psi}; \mathbf{x}, t) \tag{138}$$

where $\mathbf{I}$ is the unit matrix and where $\mathbf{T}$, $\mathbf{\Xi}$ and $\mathbf{\Omega}$ are difference coefficient matrices associated with transport (convection and diffusion), molecular mixing and chemical reaction. For numerical stability the convective contribution to $\mathbf{T}$ must either correspond to upwind differencing or alternatively hybrid (upwind-central) differencing can be used for the complete term. Performing an approximate factorization on equation (138) yields :

$$\tilde{P}(\underline{\psi}; \mathbf{x}, t + \Delta t) = \underbrace{(\mathbf{I} + \mathbf{\Omega}\Delta t)}_{\text{Source}} \underbrace{(\mathbf{I} + \mathbf{\Xi}\Delta t)}_{\text{Mixing}} \underbrace{(\mathbf{I} + \mathbf{T}\Delta t)}_{\text{Transport}} \tilde{P}(\underline{\psi}; \mathbf{x}, t) \tag{139}$$

Equation (139) can now be solved via a sequence of fractional steps. First the effect of transport is calculated from : $\tilde{P}(\underline{\psi}; \mathbf{x}, t + \Delta t_T) = (\mathbf{I} + \mathbf{T}\Delta t)\, \tilde{P}(\underline{\psi}; \mathbf{x}, t)$ where Δt_T is a notional time step indicating that transport has been applied. A similar procedure is then applied, first to represent mixing and then to account for chemical reaction. On completion of these operations an approximation to $\tilde{P}(\underline{\psi}; \mathbf{x}, t + \Delta t)$ is obtained. In the work of Pope [100] a slightly different factorisation was applied and the convection and diffusion operators are represented separately. There appears to be no advantage in adopting such a practice since convection would then have to be represented using (first-order accurate) upwind differencing. In the approach outlined presently, second-order accurate central differencing can be used for cell Peclet numbers (Pe) less than two, with upwinding being applied for larger Pe, giving a somewhat more accurate representation of convection and diffusion.

In order to minimise the computational effort, a particle method (or Monte Carlo simulation) is employed. For mixing this has already been discussed and the simulation of transport will now be described. It is important to appreciate that for simple problems Monte Carlo methods are inefficient compared with standard numerical techniques, but for multi-dimensional problems they are practicable relative to the other methods which become computationally prohibitive. In the present context the Monte Carlo method is not a statistical simulation of the pdf transport equation, but of the finite-difference analogue of the pdf equation. The simulation is performed on a finite-difference grid and rather than considering $\tilde{P}(\underline{\psi}; \mathbf{x}, t)$ explicitly, an ensemble of N_p 'representative' particles are located at each grid node. Each particle carries with it N properties corresponding to the random variables of the pdf. The ensemble $\Phi(\mathbf{X_i}, t)$ at a point can be compactly represented in terms of its individual particles $\phi_\alpha^{(i)}$ as follows :

$$\Phi(\mathbf{X_i}, t) = \begin{pmatrix} \underline{\phi}^{(1)} \\ \vdots \\ \underline{\phi}^{(N_p)} \end{pmatrix} = \begin{pmatrix} \phi_1^{(1)}, & \ldots, & \phi_N^{(1)} \\ \vdots & \vdots & \vdots \\ \phi_1^{(N_p)}, & \ldots, & \phi_N^{(N_p)} \end{pmatrix} \tag{140}$$

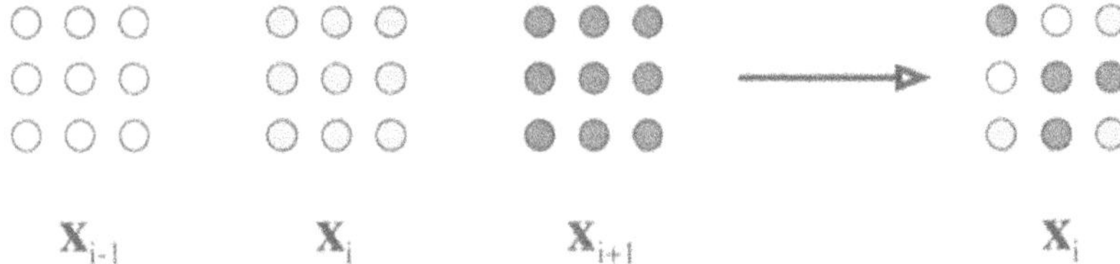

Fig. 5. Schematic of representative particles. Random sampling without replacement from $\mathbf{X_{i-1}}$, $\mathbf{X_i}$ and $\mathbf{X_{i+1}}$ leads to the new ensemble at $\mathbf{X_i}$. The shading of the particles only identifies where the particles originate from.

where $\mathbf{X_i}$ refers to the coordinates pertaining to the finite-difference grid. From such a representation, average values of any function of $Q[\underline{\phi}(\mathbf{X_i}, t)]$ are given by :

$$\langle Q[\underline{\phi}(\mathbf{X_i}, t)]\rangle = \frac{1}{N_p}\sum_{i=1}^{N_p} Q\left[\underline{\phi}^{(i)}\right] \tag{141}$$

A very important question is whether the simulation is actually a correct representation of the pdf equation. Pope [100] defined "equivalence" as follows : If for all functions Q, the ensemble average $\langle Q\rangle$ converges in probability to the expected value $\tilde{Q}$, then $\Phi(\mathbf{X_i}, t)$ and $\tilde{P}(\mathbf{X_i}, t)$ are equivalent. Convergence in probability means [92] that for a number $\epsilon > 0$, the probability of the event $\left\{\left|\langle Q\rangle - \tilde{Q}\right| > \epsilon\right\}$ tends to zero as $N_p \to \infty$. The Monte Carlo method simulates the process of transport (convection plus diffusion) in physical space by shifting element properties (chosen at random) from node to node; this means that node $\mathbf{X_i}$ would interact with its nearest neighbours and figure 5 shows schematically the nature of such interactions.

The groups of particles on the lhs of the arrow are representative of the situation at time t, whilst the application of the transport operator leads to the new ensemble at node $\mathbf{X_i}$ at time $t + \Delta t_T$ consistent with equation (139). The number of representative elements chosen from $\mathbf{X_{i-1}}$, $\mathbf{X_{i+1}}$ and referred to as $\mathcal{N}_{\mathbf{X_{i-1}}}$, $\mathcal{N}_{\mathbf{X_{i+1}}}$ (respectively) is very important if equivalence is to be achieved between the Monte Carlo simulation and the finite-difference solution. The analysis of Pope [100] demonstrated that $\mathcal{N}_{\mathbf{X_{i\pm1}}} = N_p\times$finite-difference coefficient for the transport term (contributing to the effect on $\mathbf{X_i}$ from $\mathbf{X_{i\pm1}}$). Clearly the two methods (f.d.e and Monte Carlo) are complementary i.e. even though f.d.e's are not solved, the coefficients have to be computed prior to the application of the diffusion and convection simulation using the Monte Carlo technique. Thus the new ensemble at node $\mathbf{X_i}$ will comprise $\mathcal{N}_{\mathbf{X_{i-1}}}$ particles from node $\mathbf{X_{i-1}}$, $\mathcal{N}_{\mathbf{X_{i+1}}}$ from node $\mathbf{X_{i+1}}$ and $(N_p - \mathcal{N}_{\mathbf{X_{i-1}}} - \mathcal{N}_{\mathbf{X_{i+1}}})$ from $\mathbf{X_i}$. In order to form the new ensemble at $\mathbf{X_{i+1}}$, the neighbouring nodes $\mathbf{X_i}$ and $\mathbf{X_{i+2}}$ (not shown in figure 5) are employed. In the case of $\mathbf{X_i}$, the newly formed ensemble is *not* used for the purpose of randomly selecting particles, but the original ensemble prior to modification (as illustrated in figure 5 on the lhs of the arrow) is employed. This avoids the possibility of returning the same particles that were copied over from $\mathbf{X_{i+1}}$ when forming the new ensemble at $\mathbf{X_i}$ in the previous step. Note that in accordance with hybrid differencing, if the cell Peclet number is greater than two, central differencing switches to upwind differencing; under these circumstances, $\mathbf{X_i}$ would interact with $\mathbf{X_{i+1}}$ or $\mathbf{X_{i-1}}$ only (depending on the direction of the local velocity field) in order to form the new ensemble at $\mathbf{X_i}$.

Since the source terms for chemical reaction appear in exact form in the pdf equation, this process is simulated deterministically. In principle, each particle reacts independently for an interval of time Δt according to a system of ordinary differential equations (ODE's),

$$\frac{d}{dt}\left(\phi_\alpha^{(i)}\right) = \frac{\dot{\omega}_\alpha(\underline{\phi}^{(i)})}{\varrho(\underline{\phi}^{(i)})} \quad \text{where} \quad \alpha = 1, \ldots, N \quad \text{and} \quad 1 \le i \le N_p \tag{142}$$

For combusting flows, these ODE's correspond to a stiff set of equations and their solution is non-trivial [49]. Limitations of computer speed and memory preclude the implementation of detailed kinetics, thereby necessitating the use of reduced reaction schemes [96]. Even so, Chen et al. [26] have estimated that 1000 cpu hours on a Cray XMP-48 computer would be required for the simulation of jet flames with five scalars and $N_p \approx 50000$. One strategy for significantly reducing the very large computational demand is the calculation of the integrals of the reaction rates for all possible chemical states and then storing the results in a 'look-up' table. During Monte Carlo simulations, the changes of the scalar properties as a result of reaction are obtained from interpolations based on the previously generated look-up table, thereby replacing the repeated time integration of the stiff ODEs. This strategy is computationally very efficient, but, the overriding constraint becomes the size of the 'look-up' table. As the complexity of the reduced mechanism increases, the number of independent scalars required to specify the thermochemistry increases. For interpolation tables including more than six scalars the demands on memory can become quite unmanageable if a sufficiently fine resolution of the chemistry is required; coarse grids may lead to solutions casting doubts on the accuracy of the results [109]. The scalar bounds for the interpolation tables are constructed using mass conservation principles and the assumption of constant carbon-to-hydrogen (C/H) and oxygen-to-nitrogen (O/N) atom ratios [26; 109]; this latter assumption is consistent with the idea of equal diffusivity which is implied both by the turbulence models and the mixing models. Based on these requirements it is possible to construct a table where a substantial part of the scalar domain would be redundant i.e. reaction would never occur in those regions, because the temperature is too low. For this reason, Sion and Chen [109] clipped some of the scalar bounds (e.g. for H-radicals) based on an arbitrary factor (typically three to four) of the maximum concentration of the scalar which would be found in a typical laminar flame. In very recent work [1] the concept of the interpolation table has been replaced by direct solution of the ODE's during Monte Carlo simulations and the computational effort is made manageable through the use of parallelization techniques. Nevertheless, the calculations reported in the next section involve the use of an interpolation table.

5.13. APPLICATION OF THE PDF FORMALISM TO JET DIFFUSION FLAMES

Rather than providing a detailed review of the applications of pdf methods to realistic flows (which can be found in [91; 102; 71; 72], results from an application of the scalar pdf equation to jet flames using reduced reaction schemes are presented. The results are representative (in certain respects) of those obtained by other workers [58; 24; 26; 109; 124; 125]. It will also highlight some interesting deficiencies of the pdf formalism at the current level of closure. A detailed description of the methodology behind the calculations and the complete set of results can be found in Kakhi [66].

The solution algorithm employed for the calculations involved a Monte Carlo simulation to obtain the scalar field, coupled with a finite-volume method for the velocity field. The latter field affects the former through k and ε and the density-weighted velocity. In return, the scalar field provides a mean density for use in the momentum equations. Two-dimensional axisymmetric jet flow configurations were analysed using a code which allowed for grid evolution as the jet spread downstream (of the burner), into stagnant or coflowing surrounds [93]. The constants appearing in the turbulence models ensured an accurate prediction of the spreading rate of the jets. The solutions of the velocity, k and ε equations were performed using a marching procedure based on a semi-implicit scheme. k was calculated by way of a modelled transport equation as described in section 2.1.1, equation (32) and through a Reynolds stress closure [62]. For the purpose of investigating jet diffusion flames, the prediction of the velocity field showed little sensitivity to the choice of the turbulence model, since both closures are known to work well for this configuration in constant and variable density flows. Backward Euler differencing was employed for the axial convection and hybrid differencing for the cross-stream transport; for the conditions in the flow field calculation (e.g. fifty cross-stream grid nodes and relatively small angular spread), $Pe < 2$ was predominant so that central differencing was mainly operative for approximating the cross-stream transport. The Monte Carlo simulation was effectively an explicit scheme and hence restrictions on the forward step were imposed by the algorithm. At any particular node the pdf was simulated by typically 1000 stochastic particles which carried the scalar properties. In order to calculate combusting flows, the evolving jet had to be ignited at some arbitrarily located region downstream of the burner. At typically four diameters downstream, those regions falling within the flammability limits of the fuel were assigned the composition of fully-burnt products consistent with the local mixture fraction. With a grid of 50 cross-stream nodes and a forward step size of 1% of the half jet width, the simulations required between two to seven hours on Silicon Graphics (Iris Indigo) workstations (with the interpolation table).

Tests were performed using two different reduced reaction schemes. One was based on a global alkane hydrocarbon combustion mechanism involving no radical species, Jones and Lindstedt [60], and employing rate constants derived from semi-empirical arguments. The thermochemistry was characterized in terms of four independent scalars (implying a four scalar pdf), chosen to be the mixture fraction, C_nH_{2n+2}, CO and H_2. The other scheme investigated was based on a systematic reduction of a detailed mechanism for methane-air combustion to a six-scalar description of the chemistry, Chelliah et al. in [96]. In addition to the four scalars used in the global scheme, H and CH_3 were also specified for the latter. This mechanism represents a more realistic representation of hydrocarbon burning, although both have been validated by other workers (cited above) in laminar flame calculations employing detailed molecular transport. The results to be described below relate only to the use of the global scheme, because difficulties were encountered in sustaining the flames subsequent to ignition with the six-scalar mechanism. This latter scheme involves only one (reversible) fuel consumption step, via combination with H-radicals (the concentration of which has to be arbitrarily specified in the ignition plane) while in contrast the global scheme incorporates two irreversible fuel consumption steps (involving O_2 and H_2O). It was apparent from the outset that the six scalar mechanism would be far more sensitive to the initial conditions and subsequently it was found very difficult to sustain combustion with the mechanism unless the mixture was ignited at about 15 diameters downstream of the burner. At this location the mixture had undergone a sufficient degree of fuel/air interdiffusion to prevent subsequent quenching of the reaction zone. Ignition at 4 diameters resulted in the flame going out by about 10 diameters. The cause of this behaviour is probably not the scheme itself, rather, in the present context, the synergistic effect of the sensitivity of the reaction scheme and the mixing term closure approximation creates the problem. Nevertheless, it does raise doubts with regards to the use of such 'improved' reaction schemes in conjunction with transported pdf methods.

Results are presented for two turbulent jet diffusion flames corresponding to the experimental configurations of Godoy [45] and Masri et al. [86]. The former is a propane-fueled jet issuing into stagnant surrounds and the latter corresponds to a methane jet surrounded by a thick shroud of hot ($\approx$ 2600 K) pilot gas evolving in a coflowing stream of air, also investigated in Jones and Kakhi [56; 57]. Because of the presence of a pilot flame, this particular test case does not require ignition of the fuel/air mixture as discussed earlier. For the Masri flame, two sets of results are shown corresponding to flame L and B in their notation. The B flame experiences local extinction at around 20 diameters downstream followed by subsequent relighting. This flame is subjected to strong finite rate effects and acts as a very severe test of the modelling. Only the mean concentrations are shown (the rms properties are detailed in [66]). From figures 6 to 8 the following observations can be made concerning the propane-air flame. The gross features of the flame, such as mixing patterns (mixture fraction, N_2), C_3H_8, CO_2 and O_2 are well reproduced. However, figure 7 demonstrates that the carbon monoxide field is clearly overpredicted. Jones and Kollmann [58] had also investigated the same flame but using a different and less realistic chemical scheme. A reduced scheme for the oxidation of propane was employed; the thermochemical state (including composition, density and temperature) was obtained via a partial equilibrium assumption in which the free energy was minimized subject to the constraints that the mixture fraction, total mole number and mass fraction of C_3H_8 (three-scalar chemistry) take predetermined values. Such a constrained equilibrium approach led to underpredictions of CO and the agreement of the CO_2 predictions was poorer than displayed in figure 7. More recently Chen and Kollmann [24] extended the work by incorporating a four scalar description of the chemistry. The major difference between the three and four scalar description lies in the assumption of a partial equilibrium for the CO oxidation process. In [58] the CO/CO_2 ratio was determined entirely by the H/OH ratio and the equilibrium constant, whilst in [24] CO was treated as an independent scalar. The effect was a significant improvement in CO and CO_2 predictions (although they were still under and overpredicted, respectively) with little effect on mixture fraction profiles. Clearly, the treatment of the chemistry has little effect on the conserved scalars, but a greater effect on the reactive scalars. However, in the present treatment, no assumptions relating to constrained equilibrium theory are made. A more striking observation in the present predictions is that the mixing models LMSE and modified Curl yield results in very close agreement to each other.

Figures 9 to 12 correspond to the predictions of the Masri (L) flame. Once again the general features of the flame are adequately reproduced with little difference between the predictions obtained with LMSE and the modified Curl models. However, figure 11 demonstrates a significant degree of overprediction in the H_2 field, accompanied by a similar behaviour in the CO field (figure 10). Given that CO levels are rarely above 6% in hydrocarbon-air laminar flames, a separate calculation was performed for a confined laminar jet diffusion flame (without the pdf equation, of course), *and assuming equal diffusivity of the scalars*. Once again it was noticed that unrealistically high concentrations of CO and H_2 were obtained; of the same order as that found in the turbulent flame calculations. Sensitivity of the results to the incorporation and relaxation of differential diffusion effects in laminar flames has also been observed with regard to the H_2 field when using detailed kinetic mechanisms, Lindstedt [82]. In this case a detailed scheme was employed and so the effect on the CO field was less pronounced, presumably due to the additional paths available for CO consumption. While these results suggest that reduced schemes may show a greater sensitivity to the assumption of equal diffusivity, a realistic description of molecular transport at the fine scales of turbulence may well be necessary if accurate predictions of CO levels are to result. It has been argued that at high Reynolds number the large scale motions are unaffected by the details of the fine scale fluctuations; an idea which stems from Reynolds number similarity and has motivated the suggestion that equal diffusivity is a plausible assumption in turbulent flow calculations. This argument is of questionable validity, however, for highly diffusive reactive scalars (such as H_2 and H) which are produced (and largely consumed) entirely in the reaction zone where scalar gradients and diffusion processes are important. The results for the flame affected by local extinction and subsequent downstream relighting, Masri (B) flame, are shown in figures 13 to 15. Figure 13 shows that very good mixture fraction and UHC predictions were obtained. For the temperature, and minor species field (CO and H_2), however, the predictions are variable in the sense that LMSE and modified Curl do not yield the same result. LMSE predicts a continual drop in temperature (figure 14), whereas modified Curl overpredicts the temperature throughout. As a consequence, LMSE predicts acceptably low values of CO, but the flame is effectively going out, whilst modified Curl predicts unrealistically high CO (and especially H_2) levels, but with reasonable temperature profiles. In this sense, neither model can be claimed to be superior. In the light of the results for the flames unaffected by extinction (figures 6 to 12), it would be too bold to expect reasonable (or even consistent) predictions in the latter flames experiencing strong finite-rate effects.

Vervisch [124; 125] also reports substantial overpredictions in the Masri (B) flame using the scheme of Jones and Lindstedt [60]. But the LMSE closure was used whereas the results of figure 13 to 15 suggest that with LMSE the flame is extinguishing - the reasons for this discrepancy are not clear. It is difficult to say why such a large difference in behaviour is observed in the Masri (B) flame whilst in the (L) flame the mixing models evolve with similar characteristics. Whether the mixing model is the main factor or the synergistic effect of the mixing/chemistry representations can only be determined by implementing improved schemes for both. However, the difficulties associated with such an endeavour have yet to be overcome.

6. Concluding Remarks

In this review mathematical modelling approaches for predicting single phase, gaseous, low Mach number, high Reynolds number turbulent combustion has been discussed in the context of non-premixed and premixed systems. Developments in scalar pdf modelling have also been addressed.

For non-premixed flames the conserved scalar approach provides a reasonably accurate and computationally economic method of predicting the properties of non-premixed turbulent flames. In this respect classical equilibrium and laminar flamelet approaches probably provide equally good representation of the mean density, and hence, the velocity field. The laminar flamelet approach has the advantage of providing much more realistic CO levels in hydrocarbon-air jet diffusion flames partly because the maximum levels of CO found in laminar flames are typical of those found in such turbulent flames. The flamelet libraries, calculated from detailed computations of stretched laminar flames, incorporate the details of the complex chemistry and differential diffusion effects (in terms of the mixture fraction and a measure of the local stretch rate) which other turbulent combustion models invariably ignore. In this manner, the flamelet approach allows such features to be indirectly incorporated into predictive methods for combusting flows without addressing the difficulties associated with the unclosed terms (e.g. mean reaction rates). The penalty for this is the range of applicability of the formalism. This is partly demonstrated by the fact that for many fuels of engineering importance, e.g. kerosene and the various grades of petrol, the details of the kinetics are unknown and extremely difficult to determine due to their complexity. For such configurations it is not clear how well the laminar flamelet formulation is suited to describe the thermochemistry. Furthermore, in engineering devices such as gas turbine combustors, the mixture strength is overall lean, and it is found (experimentally) that the emission levels of CO and UHC in such devices can be significant; however, equilibrium and flamelet thermochemistry suggests negligibly small emission levels of CO and UHC. The emissions of UHC and CO are associated with finite-rate chemistry effects and for these, the conserved scalar - flamelet approach is clearly not appropriate. The generality of the flamelet approach to the modelling of extinction and other flame-stretch effects in terms of the mixture fraction and the local scalar dissipation rate also needs to be determined. In particular the validity of the assumption of statistical independence of the mixture fraction and the scalar dissipation rate requires greater scrutiny. If the assumption proves to be unsatisfactory, it may well be that the transported pdf approach is the only suitable alternative.

While the conserved scalar approach dominates non-premixed combustion theory, in contrast, there are several theories of premixed combustion. The problem of adequately representing the flame propagation in premixed mixtures complicates matters considerably. Until recently the theory of premixed turbulent combustion was dominated by the Bray-Moss-Libby (BML) formalism, which can be considered to place Spalding's original eddy break-up theory on a rigorous foundation. The usefulness of this theory is limited to 'thin' flame burning, and the applicability of its more recent extensions to include stretch effects need to fully determined, especially in practical configurations. In this respect, the lack of measurements in premixed flames introduces substantial uncertainties for the development of model improvements and the detailed validation of existing theories. In the flamelet regime of combustion, recent approaches based on flame area models seem to demonstrate a potential to yield good results and it will be of interest to see how well these methods cope in more realistic configurations.

From the combustion literature it is clear that transported pdf methods are becoming more widely used. In this review scalar pdf modelling has been considered at the single point, single time level. This constitutes the simplest level of closure in terms of the pdf formulation applied to combusting flows although more elaborate proposals have been suggested by other workers, e.g. velocity-composition and velocity-composition-viscous dissipation rate pdf equations. It is important to recognize that although such proposals alleviate some of the problems faced by scalar pdf modelling (e.g. the convection appears in exact form and in the latter case time scale information can be included) additional unknown terms are introduced. In turbulence, one does not get something for nothing; treating convection exactly is advantageous, but modelling of the conditional expectations of the fluctuating pressure gradient and viscous stress terms in velocity pdf closures is no simpler than modelling the conventional Reynolds stresses in moment closures. In fact the modelling of the former seems more daunting since second-moment closures have received considerable attention for quite some time, and have demonstrated their ability in predicting velocity field characteristics in complex turbulent fields. It is for this reason that at the moment, a combination of second-moment turbulence modelling coupled with a scalar pdf evolution equation is considered as the most practical level of closure since advantage can be taken of the merits of both approaches in calculating their respective fields (i.e. velocity and scalar fields).

In scalar pdf modelling, the closure of the molecular mixing term has posed a serious problem for quite some time. The molecular mixing term is central to the turbulence-chemistry interactions primarily because it involves molecular diffusivities and scalar gradients, which scale with the dissipative motions in a turbulent flow, and this is precisely where combustion occurs. The most widely used class of mixing models are those based on binary interaction of fluid parcels (coalescence-dispersion closures), and linear mean square estimation (LMSE). In addition to their poor representation of pdf evolution in homogeneous, isotropic and inert turbulence, these mixing models have a serious drawback, ultimately related to the fact that they employ k/ε as the (only) characteristic turbulent time scale; a time scale representative of the large scale energy-containing motions. Furthermore the diffusivities of the reacting scalars are taken to be effectively uniform, whilst it is well-known that species such as H_2 and H diffuse considerably faster than others by virtue of their significantly lower molar masses. Consequently these models do not address the problem of mixing in the presence of a flame front where information concerning local scalar gradients and molecular diffusivities can be important. Recently several studies on 'improved' mixing models (e.g binomial sampling, mapping closures) have appeared which ensure that an arbitrary initial passive scalar pdf relaxes to a Gaussian pdf in a homogeneous turbulence field. However, the results presented here suggest that relaxation to a Gaussian pdf, although desirable, may not be the primary factor in accurately predicting turbulent flame properties where the reactive scalar pdf's can be far from Gaussian. If this is the case then these newer mixing models may not yield improved CO and H_2 predictions unless a more detailed description of the fine scales is incorporated.

Suggestions are increasingly made that the use of the joint pdf of the scalar and its gradient represents a promising approach and alleviates the problem of evaluating turbulent scalar time scales. However, it must borne in mind that even when the hurdle of approximating the non-closed terms in this equation is crossed, the dimensionality becomes another consideration. For the four-scalar chemistry and two-dimensional axisymmetric flows, for which predictions were presented, $(4+2\times 4 =)$ 12 independent scalars would be required. Tabulation of the chemistry would be out of the question, and parallelization would be essential. However, if substantially improved predictions can be obtained then the intensive computational requirements can be justified.

This review has presented some of the current shortcomings and problem areas in pdf methods. In spite of the difficulties which remain, the formulation represents an excellent tool to investigate modelling concepts without the theoretical drawbacks associated with conventional techniques and it appears to be only approach which is equally applicable to premixed and non-premixed combustion. In addition the study of the evolution of pdf's in turbulent flows is of interest in its own right and may lead to improved understanding of turbulent transport and combustion phenomenon. Finally, in contrast to many approaches pdf transport methods (at least in principle) are not restricted to any particular mode of burning, e.g. the flamelet regime.

References

[1] R. Armstrong, M.I. Koszykowski, F. Dai, J.-Y. Chen, N.J. Brown, and J. Macfarlane. Full chemistry in turbulent combustion dynamics. In *Abstracts from Fifth International Conference on Numerical Combustion*, page 11, Garmisch-Partenkirchen, Germany, September 29 - October 1 1993.

[2] R.W. Bilger. The structure of diffusion flames. *Combustion Science and Technology*, 13:155–170, 1976.

[3] R.W. Bilger. Turbulent jet diffusion flames. *Progress in Energy and Combustion Science*, 1:87–109, 1976.

[4] R.W. Bilger. Reaction rates in diffusion flames. *Combustion and Flame*, 30:277–284, 1977.

[5] R.W. Bilger. Perturbation analysis of turbulent non-premixed combustion. *Combustion and Flame*, 30:277–284, 1980.

[6] R.W. Bilger. Turbulent flows with non-premixed reactants. In F.A Libby and F.A. Williams, editors, *Turbulent Reacting Flows*, pages 65–113. Springer-Verlag, 1980. Appeared under 'Topics in Applied Physics', vol.44.

[7] R.W. Bilger. Turbulent diffusion flames. *Annual Review of Fluid Mechanics*, 21:101–135, 1989.

[8] R.W. Bilger and R.E. Beck. Further experiments in turbulent jet diffusion flames. In *15th Symposium (International) on Combustion / The Combustion Institute*, page 541, 1975.

[9] P. Bradshaw. *An Introduction to Turbulence and its Measurement*. Pergamon, 1971.

[10] K.N.C. Bray. The interaction between turbulence and combustion. In *17th Symposium (International) on Combustion / The Combustion Institute*, pages 223–233, 1978.

[11] K.N.C. Bray. Turbulent flows with premixed reactants. In F.A Libby and F.A. Williams, editors, *Turbulent Reacting Flows*, pages 115–183. Springer-Verlag, 1980. Appeared under 'Topics in Applied Physics', vol.44.

[12] K.N.C. Bray. Methods of including realistic chemical reaction mechanisms in turbulent combustion models. In *Second workshop on modelling of chemical reaction systems*, pages 356–375, 1986. Meeting held in Heidelberg, August 1986.

[13] K.N.C. Bray and P.A. Libby. Interaction effects in turbulent premixed flames. *Physics of Fluids*, 19(11):1687–1701, 1976.

[14] K.N.C. Bray and P.A. Libby. Passage times and flamelet crossing frequencies in premixed turbulent combustion. *Combustion Science and Technology*, 47:253–274, 1986.

[15] K.N.C. Bray, P.A. Libby, G. Masuya, and J.B. Moss. Turbulence production in premixed turbulent flames. *Combustion Science and Technology*, 25:127–140, 1981.

[16] K.N.C. Bray, P.A. Libby, and J.B. Moss. Flamelet crossing frequencies and mean reaction rates in premixed turbulent combustion. *Combustion Science and Technology*, 41:143–172, 1984.

[17] K.N.C. Bray and J.B. Moss. A unified statistical model of the premixed turbulent flame. AASU Report 335, University of Southampton, 1974.

[18] S.P. Burke and T.E.W. Schumann. Diffusion flames. *Industrial and Engineering Chemistry*, 20(10):998–1004, 1928.

[19] R.S. Cant. Turbulent reaction rate modelling using flamelet surface area. In *Proceedings of the Anglo-German Combustion Symposium*, page 48. The Combustion Institute, British Section of the Combustion Institute, April 1993. meeting held held at Queen's College, Cambridge.

[20] R.S. Cant, S.B. Pope, and K.N.C. Bray. Modelling of flamelet surface-to-volume ratio in turbulent premixed combustion. In *23rd Symposium (International) on Combustion / The Combustion Institute*, pages 809–815, 1990.

[21] C.A. Catlin and R.P. Lindstedt. Premixed turbulent burning velocities derived from mixing controlled reaction models with cold front quenching. *Combustion and Flame*, 85:427–439, 1991.

[22] M. Champion, K.N.C. Bray, and J.B. Moss. The turbulent combustion of a propane-air mixture. *Acta Astronautica*, 5:1063–1077, 1978. Previously presented at the 6th Colloquium on Gas dynamics of Explosions and Reacting Systems, Stockholm 1977.

[23] H. Chen, S. Chen, and R.H. Kraichnan. Probability distribution of a stochastically advected scalar field. *Physical Review Letters*, 63(24):2657–2660, December 1989.

[24] J.-Y. Chen and W. Kollmann. Pdf modelling of chemical non-equilibrium effects in turbulent non-premixed hydrocarbon flames. In *22nd Symposium (International) on Combustion / The Combustion Institute*, pages 645–653, 1988.

[25] J.-Y. Chen and W. Kollmann. Mixing models for turbulent flows with exothermic reactions. In *Turbulent Shear Flows 7*, pages 277–292. Springer-Verlag, 1991. Selected papers from the seventh international symposium on turbulent shear flows.

[26] J.-Y. Chen, W. Kollmann, and R.W. Dibble. Pdf modelling of turbulent non-premixed methane jet flames. *Combustion Science and Technology*, 64:315–346, 1989.

[27] P.Y. Chou. On velocity correlation and the solution of the equation of turbulent fluctuation. *Quaterly of Applied*

Mathematics, 3:38, 1945.

[28] R.L. Curl. Dispersed phase mixing : 1. Theory and effects in simple reactors. *A.I.Ch.E. Journal*, 9(2):175–181, 1963.

[29] B.J. Daly and F.H. Harlow. Transport equations in turbulence. *Physics of Fluids*, 13:2634, 1970.

[30] G. Damköhler. *Z. Elecktrochem.*, 46:601–626, 1940. English translation : NACA Technical Memorandum, 1112, 1947.

[31] N. Darabiha, V. Giovangigli, A. Trouvé, S.M. Candel, and E. Esposito. Coherent flame description of turbulent premixed ducted flames. In R. Borghi and S.N.B. Murthy, editors, *Turbulent Reactive Flows*, pages 591–637. Spinger-Verlag, 1989.

[32] G. Dixon-Lewis, F.A. Goldsworthy, and J.B. Greenberg. Flame structure and flame reaction kinetics. *Proceedings of the Royal Society of London Series A*, 346:261–278, 1975.

[33] C. Dopazo. *Non-isothermal turbulent reactive flows : stochastic approaches*. PhD thesis, State University of New York at Stony Brook, 1973.

[34] C. Dopazo. Probability density function approach for a turbulent axisymmetric heated jet. Centreline evolution. *Physics of Fluids*, 18(4):397–404, 1975.

[35] C. Dopazo. Relaxation of initial probability density functions in the turbulent convection of scalar fields. *Physics of Fluids*, 22(1):20–30, 1979.

[36] C. Dopazo and E.E. O'Brien. An approach to the autoignition of a turbulent mixture. *Acta Astronautica*, 1:1239–1266, 1974.

[37] C. Dopazo and E.E. O'Brien. Statistical treatment of non-isothermal chemical reactions in turbulence. *Combustion Science and Technology*, 13:99–122, 1976.

[38] H.E. Eickhoff and K. Grethe. A flame-zone model for turbulent hydrocarbon diffusion flames. *Combustion and Flame*, 35:267–275, 1979.

[39] V. Eswaran and S.B. Pope. Direct numerical simulations of the turbulent mixing of a passive scalar. *Physics of Fluids A*, 31(3):506–520, 1988.

[40] M. Fairweather, W.P. Jones, R.P. Lindstedt, and A.J. Marquis. Predictions of a turbulent reacting jet in a cross-flow. *Combustion and Flame*, 84:361–375, 1991.

[41] R.C. Flagan and J.P. Appleton. A stochastic model of turbulent mixing with chemical reaction : nitric oxide formation in a plug-flow burner. *Combustion and Flame*, 23:249–267, 1974.

[42] F. Gao. An analytical solution for the scalar probability density function in homogeneous turbulence. *Physics of Fluids A*, 3(4):511–513, April 1991.

[43] F. Gao. Mapping closure and non-Gaussianity of the scalar probability density function in isotropic turbulence. *Physics of Fluids A*, 3(10):2438–2444, October 1991.

[44] M.M. Gibson and B.E. Launder. Ground effects on pressure fluctuations in the atmospheric boundary layer. *Journal of Fluid Mechanics*, 86(3):491, 1978.

[45] S. Godoy. *Turbulent Diffusion Flames*. PhD thesis, University of London, 1982.

[46] S. Gordon and B.J. McBride. Computer program for calculation of complex equilibrium composition, rocket performance, incident and reflected shocks and chapman-jouget detonations. SP - 273, NASA, 1971.

[47] F.C. Gouldin. An application of fractals to modelling premixed turbulent flames. *Combustion and Flame*, 68:249–266, 1987.

[48] W.R. Hawthorne, D.S. Weddell, and H.C. Hottel. Mixing and combustion in turbulent gas jets. *Third Symposium on Combustion, Flame and Explosion Phenomena*, pages 266–288, 1949.

[49] A.C. Hindmarsh. LSODE and LSODI, Two new initial value ordinary differential equation solvers. *ACM SIGNUM Newsletter*, 15(4), 1980.

[50] J. Janicka, W. Kolbe, and W. Kollman. The solution of a pdf transport equation for turbulent diffusion flames. In *Proceedings of the Heat Transfer and Fluid Mechanics Institute*, pages 296–312. Stanford University Press, 1978.

[51] J. Janicka, W. Kolbe, and W. Kollman. Closure of the equations for the probability density function of turbulent scalar fields. *Journal of Non-Equilibrium Thermodynamics*, 4:47–66, 1979.

[52] J. Janicka and W. Kollmann. A two-variable formalism for the treatment of chemical reactions in turbulent H_2-air diffusion flames. In *17th Symposium (International) on Combustion / The Combustion Institute*, page 421, 1979.

[53] W.P. Jones. Models for turbulent flows with variable density and combustion. In W. Kollmann, editor, *Prediction methods for Turbulent Flows*, pages 380–421. Hemisphere Publ. Corp., 1980. A collection of the notes for a lecture series on prediction methods for turbulent flows held at the VKI in January 1979.

[54] W.P. Jones, 1989. unpublished work.

[55] W.P. Jones. Turbulence modelling and numerical solution methods for variable density and combusting flows.

In P.A. Libby and F.A. Williams, editors, *Turbulent Reactive Flows*, pages 309–374. Academic Press, 1994. to appear.

[56] W.P. Jones and M. Kakhi. Pdf modelling for the prediction of turbulent diffusion flames. In *Proceedings of the Anglo-German Combustion Symposium*, page 92. The Combustion Institute, British Section of the Combustion Institute, April 1993. meeting held held at Queen's College, Cambridge.

[57] W.P. Jones and M. Kakhi. Scalar pdf modelling of finite-rate effects in turbulent non-premixed methane-air flames. In *Abstracts from Fifth International Conference on Numerical Combustion*, page 81, Garmisch-Partenkirchen, Germany, September 29 - October 1 1993.

[58] W.P. Jones and W. Kollmann. Multi-scalar pdf transport equations for turbulent diffusion flames. In Durst et al., editor, *Turbulent Shear Flows 5*, pages 296–309. Springer-Verlag, 1987. Selected papers from the fifth international symposium on turbulent shear flows.

[59] W.P. Jones and B.E. Launder. The prediction of laminarisation with a two-equation model of turbulence. *International Journal of Heat and Mass transfer*, 15:301–314, 1972. also AIAA selected reprint series XIV, p.119-132, 1973.

[60] W.P. Jones and R.P. Lindstedt. Global reaction schemes for hydrocarbon combustion. *Combustion and Flame*, 73:233–249, 1988.

[61] W.P. Jones and J.J. McGuirk. Mathematical modelling of gas turbine combustion chambers. *AGARD Proc.*, No. 275:4.1–4.11, 1979.

[62] W.P. Jones and P. Musonge. Closure of the reynolds stress and scalar flux equations. *Physics of Fluids*, 31(12):3589–3603, 1988.

[63] W.P. Jones and C.H. Priddin. Predictions of the flow field and local gas composition in gas turbine combustors. In *17th Symposium (International) on Combustion / The Combustion Institute*, pages 399–409, 1978.

[64] W.P. Jones and J.H. Whitelaw. Calculation methods for reacting turbulent flows. *Combustion and Flame*, 48:1–26, 1982.

[65] Hanjalić K. and Launder B.E. A reynolds stress model of turbulence and its application to thin shear flows. *Journal of Fluid Mechanics*, 52(4):609–638, 1972.

[66] M. Kakhi. *The transported probability density function approach for predicting turbulent combusting flows*. PhD thesis, University of London, 1994. in preparation.

[67] J.H. Kent and R.W. Bilger. The prediction of turbulent diffusion flame fields and nitric oxide formation. In *16th Symposium (International) on Combustion / The Combustion Institute*, page 1643, 1977.

[68] A.R. Kerstein. Fractal dimension of turbulent premixed flames. *Combustion Science and Technology*, 60:441–445, 1988.

[69] E.E. Khalil, D.B. Spalding, and J.H. Whitelaw. The calculation of local flow properties in two-dimensional furnaces. *International Journal of Heat and Mass transfer*, 18:775, 1975.

[70] W. Kollmann. Pdf transport equations for chemically reacting flows. In R. Borghi and S.N.D. Murthy, editors, *Turbulent Reactive Flows*, pages 715–730. Spinger-Verlag, 1989.

[71] W. Kollmann. The pdf approach to turbulent flow. *Theoretical and Computational Fluid Dynamics*, 1:249–285, 1990.

[72] W. Kollmann. Pdf transport modelling. Modelling of combustion and turbulence Von Karman Institute for fluid dynamics, lecture series, 1992.

[73] G. Kosaly. Theoretical remarks on a phenomenological model of turbulent mixing. *Combustion Science and Technology*, 49:227–234, 1986.

[74] G. Kosaly and P. Givi. Modelling of turbulent molecular mixing. *Combustion and Flame*, 70:101–118, 1987.

[75] B.E. Launder. On the effects of a gravitational field on the turbulent transport of heat and momentum. *Journal of Fluid Mechanics*, 67(3):569, 1975.

[76] B.E. Launder. Phenomenological modelling : Present and future ? In J.L. Lumley, editor, *Whither Turbulence ? Turbulence at the Crossroads*, pages 439–485. Spinger-Verlag, 1990. Proceedings of a workshop held at Cornell University, Ithaca, March 1989. Published by Spinger-Verlag as Lecture Notes in Physics 357, 1990.

[77] B.E. Launder and D.B. Spalding. *Mathematical Models of Turbulence*. Academic Press, 1972.

[78] A. Liñán. The asymptotic structure of counterflow diffusion flames for large activation energies. *Acta Astronautica*, 1:1007–1039, 1974.

[79] P.A Libby and K.N.C. Bray. Countergradient diffusion in premixed turbulent flames. *AIAA Journal*, 19(2):205–213, 1981. Originally presented as Paper 80-0013 at the AIAA meeting Jan 1980, Pasadena, California.

[80] P.A. Libby and F.A. Williams. Fundamental aspects. In F.A Libby and F.A. Williams, editors, *Turbulent Reacting Flows*, pages 1–43. Springer-Verlag, 1980. Appeared under 'Topics in Applied Physics', vol.44.

[81] S.K. Liew. *Flamelet models of turbulent non-premixed combustion.* PhD thesis, University of Southampton, 1983.

[82] R.P. Lindstedt, 1994. Private Communication.

[83] F.C. Lockwood and A.S. Naguib. The prediction of the fluctuations in the properties of free, round-jet, turbulent diffusion flames. *Combustion and Flame*, 24:109–124, 1975.

[84] J.L. Lumley. Prediction methods for turbulent flows. Introduction. Von Karman Institute lecture series no. 76., 1975.

[85] T.S. Lundgren. Distribution functions in the statistical theory of turbulence. *Physics of Fluids*, 10(5):969–975, 1967.

[86] A.R. Masri, R.W. Bilger, and R.W. Dibble. Turbulent non-premixed flames of methane near extinction : Mean structure from Raman measurements. *Combustion and Flame*, 71:245–266, 1988.

[87] R.S. Miller, S.H. Frankel, C.K. Madnia, and P. Givi. Johnson-Edgeworth translation for probability modelling of binary scalar mixing in turbulent flows. *Combustion Science and Technology*, 91:21–52, 1993.

[88] Gibson M.M. and B.A. Younis. Calculation of swirling jets with a reynolds stress closure. *Physics of Fluids*, 29(1):38, 1986.

[89] A.T. Norris and S.B. Pope. Turbulent mixing model based on ordered pairing. *Combustion and Flame*, 83:27–42, 1991.

[90] G.L. North and D.A. Santavicca. The fractal nature of premixed turbulent flames. *Combustion Science and Technology*, 72:215–232, 1990.

[91] E.E. O'Brien. The probability density function (pdf) approach to reacting turbulent flows. In F.A Libby and F.A. Williams, editors, *Turbulent Reacting Flows*, pages 185–218. Springer-Verlag, 1980. Appeared under 'Topics in Applied Physics', vol.44.

[92] A. Papoulis. *Probability, random variables and stochastic processes.* Mc Graw-Hill, 1965.

[93] S.V. Patankar and D.B. Spalding. *Heat and mass transfer in boundary layers : A general calculation procedure.* Intertext books, 2nd edition, 1970.

[94] N. Peters. Laminar flamelet concepts in turbulent combustion. In *21st Symposium (International) on Combustion / The Combustion Institute*, pages 1231–1250, 1986.

[95] N. Peters. A spectral closure for premixed turbulent combustion in the flamelet regime. *Journal of Fluid Mechanics*, 242:611–629, 1992.

[96] N. Peters and B. Rogg, editors. *Reduced kinetic mechanisms for applications in combustion systems.* Lecture notes in Physics m15. Springer-Verlag, 1993.

[97] N. Peters and F.A. Williams. Liftoff characteristics of turbulent jet diffusion flames. *AIAA Journal*, 21(3):423–429, 1983.

[98] S.B. Pope. The relationship between the probability approach and particle models for reaction in homogeneous turbulence. *Combustion and Flame*, 35:41–45, 1979.

[99] S.B. Pope. The statistical theory of turbulent flames. *Philosophical Transactions of the Royal Society of London Series A*, 291:529–568, 1979.

[100] S.B. Pope. A Monte Carlo method for the pdf equations of turbulent reactive flows. *Combustion Science and Technology*, 25:159–174, 1981.

[101] S.B. Pope. An improved mixing model. *Combustion Science and Technology*, 28:131–135, 1982.

[102] S.B. Pope. Pdf methods for turbulent reactive flows. *Progress in Energy and Combustion Science*, 11:119–192, 1985.

[103] S.B. Pope. Turbulent premixed flames. *Annual Review of Fluid Mechanics*, pages 237–270, 1987.

[104] S.B. Pope. Mapping closures for turbulent mixing and reaction. *Theoretical and Computational Fluid Dynamics*, 2:255–270, 1991.

[105] W.C. Reynolds. The potential and limitations of direct and large eddy simulation. In J.L. Lumley, editor, *Whither Turbulence ? Turbulence at the Crossroads*, pages 313–343. Spinger-Verlag, 1990. Proceedings of a workshop held at Cornell University, Ithaca, March 1989. Published by Spinger-Verlag as Lecture Notes in Physics 357, 1990.

[106] B. Rogg, F. Behrendt, and J. Warnatz. Turbulent non-premixed combustion in partially premixed flamelets with detailed chemistry. In *21st Symposium (International) on Combustion / The Combustion Institute*, pages 1533–1541, 1986.

[107] J.C. Rotta. Statistische theorie nichthomogener turbulenz i und ii. *Zeitschrift für Physik*, 129:547–573, 1951. Also in vol. 131 pp. 51-77.

[108] I.G. Shepherd and J.B. Moss. Characteristic scales for density fluctuations in turbulent premixed flames. *Combustion Science and Technology*, 33:231, 1983.

[109] M. Sion and J.-Y. Chen. Scalar pdf modelling of turbulent non-premixed methanol-air flames. *Combustion Science and Technology*, 88:89–114, 1992.

[110] D.B. Spalding. Concentration fluctuations in a round turbulent free jet. *Chemical Engineering Science*, 26:95–107, 1971.

[111] D.B. Spalding. Mixing and chemical reaction in steady confined turbulent flames. In *13th Symposium (International) on Combustion / The Combustion Institute*, pages 649–657, 1971.

[112] L.A. Spielman and O. Levenspiel. A Monte Carlo treatment for reacting and coalescing dispersed phase systems. *Chemical Engineering Science*, 20:247–254, 1965.

[113] K.R. Sreenivasan and C. Meneveau. The fractal facets of turbulence. *Journal of Fluid Mechanics*, 173:357–386, 1986.

[114] M. Summerfield, S.H. Reiter, V. Kebely, and R.W. Mascolo. The structure and propogation mechanism of turbulent flames in high speed flow. *Jet Propulsion*, 25(8):377–384, 1955.

[115] S. Tavoularis and S. Corrsin. Experiments in nearly homogeneous turbulent shear flow with a uniform mean temperature gradient. Part 1. *Journal of Fluid Mechanics*, 104:311–347, 1981.

[116] A.A. Townsend. *The Structure of Turbulent Shear Flows*. Cambridge University Press, 2nd edition, 1976.

[117] H. Tsuji and I. Yamaoka. Structure analysis of counterflow diffusion flames in the forward stagnation region of a porous cylinder. In *13th Symposium (International) on Combustion / The Combustion Institute*, pages 723–781, 1971.

[118] M.S. Uberoi. Equipartition of energy and local isotropy in turbulent flows. *Journal of Applied Physics*, 28:1165–1170, 1957.

[119] L. Valiño and C. Dopazo. A binomial sampling model for scalar turbulent mixing. *Physics of Fluids A*, 2(7):1204–1212, 1990.

[120] L. Valiño and C. Dopazo. A binomial Langevin model for turbulent mixing. *Physics of Fluids A*, 3(12):3034–3037, 1991.

[121] L. Valiño and C. Dopazo. Joint statistics of scalars and their gradients in nearly homogeneous turbulence. In A.V. Johansson and P.H. Alfredsson, editors, *Advances in Turbulence 3*, pages 312–323. Springer-Verlag, 1991.

[122] L. Valiño and F. Gao. Monte Carlo implementation of a single-scalar mapping closure for diffusion in the presence of chemical reaction. *Physics of Fluids A*, 4(9):2062–2069, September 1992.

[123] L. Valiño, J. Ros, and C. Dopazo. Monte Carlo implementation and analytic solution of an inert-scalar turbulent mixing test problem using a mapping closure. *Physics of Fluids A*, 3(9):2195–2198, September 1991.

[124] L. Vervisch. *Prise en compte d'effets de cinétique chimique dans les flammes de diffusion turbulentes par l'approche fonction densité de probabilité*. PhD thesis, University of Rouen, 1991.

[125] L. Vervisch. Applications of pdf turbulent combustion models to real non-premixed flame calculations. Modelling of combustion and turbulence Von Karman Institute lecture series 1992, 1992.

[126] L. Vervisch, J.H. Chen, S. Mahalingam, and I.K. Puri. Numerical study of finite-rate chemistry effects and unequal schmidt numbers in turbulent non-premixed combustion. In *Ninth Symposium on Turbulent Shear Flows*, Kyoto, Japan, August 16-18 1993.

[127] H.G. Weller, C.J. Marooney, and A.D. Gosman. A new spectral method for calculation of the time-varying area of a laminar flame in homogeneous turbulence. In *23rd Symposium (International) on Combustion / The Combustion Institute*, pages 629–636, 1990.

[128] F.A. Williams. Recent advances in the theoretical descriptions of turbulent diffusion flames. In S.N.B. Murthy, editor, *Turbulent Mixing in Non-reactive and Reactive Flows*. Plenum Press, 1975.

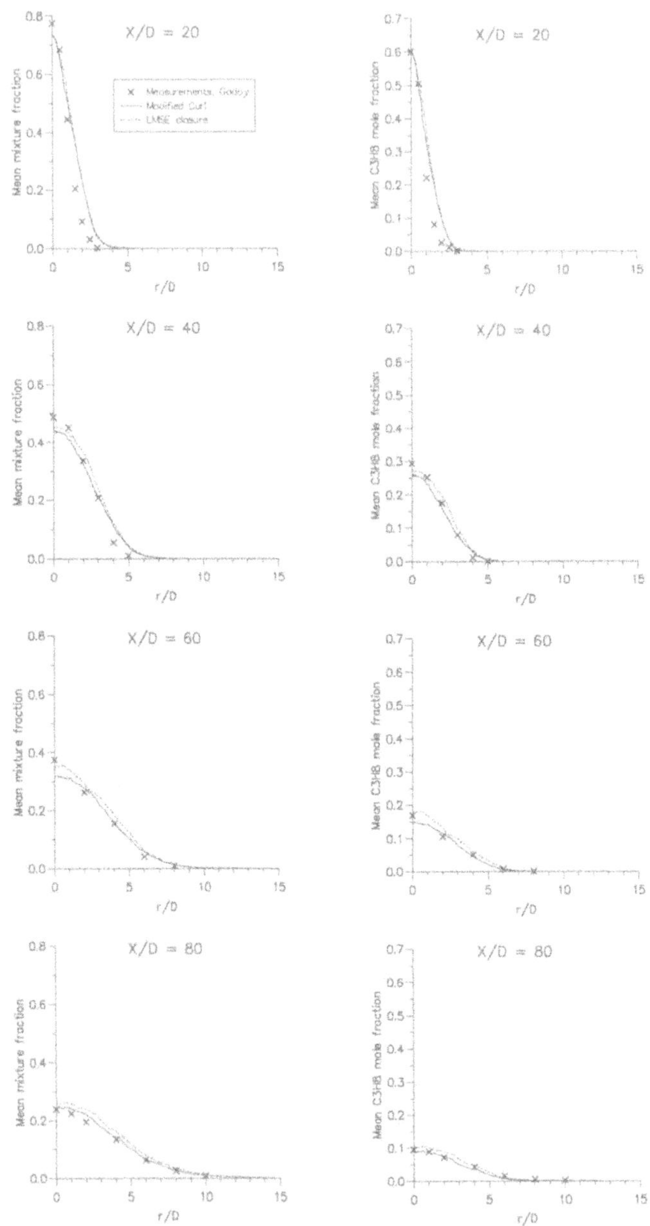

Fig. 6. Radial profiles of the measurements (Godoy [45]) and predictions for ξ and C_3H_8 in a propane-air jet diffusion flame issuing into stagnant surrounds. The k-ε model and the scheme of Jones and Lindstedt [60] are employed. Predictions using the LMSE and coalescence-dispersion ($\mathcal{A}(x) = 1$) closures are shown. $Re \approx 24000$

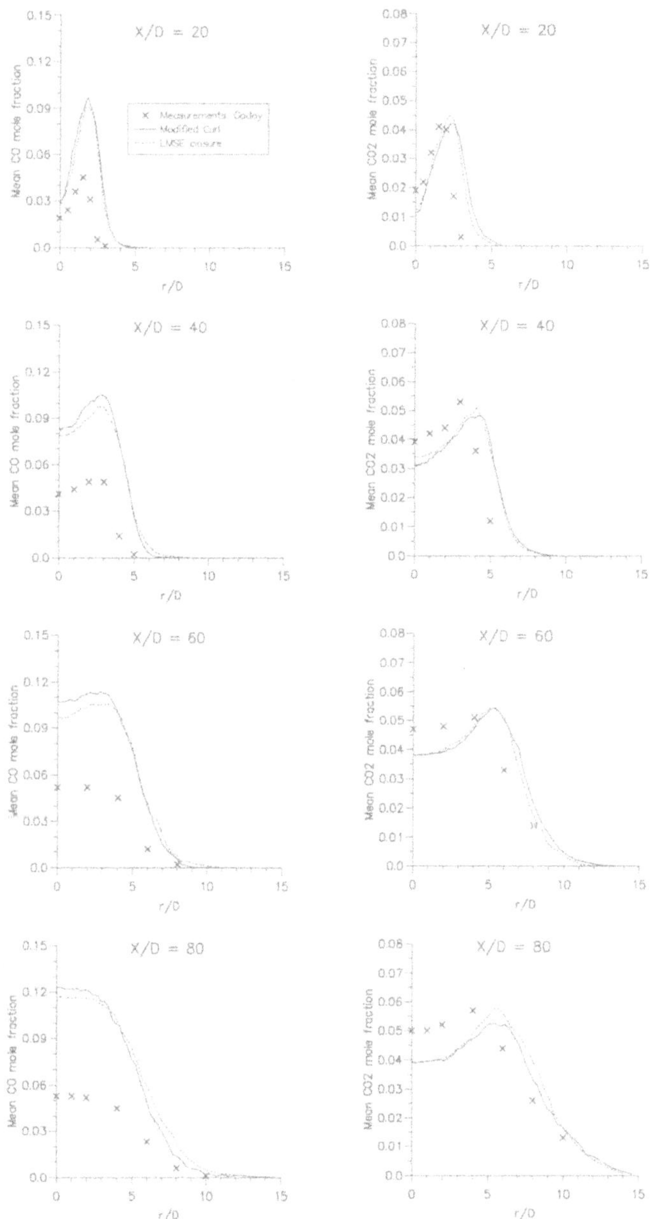

Fig. 7. Radial profiles of the measurements (Godoy [45]) and predictions for CO and CO_2 in a propane-air jet diffusion flame issuing into stagnant surrounds. The k-ε model and the scheme of Jones and Lindstedt [60] are employed. Predictions using the LMSE and coalescence-dispersion ($\mathcal{A}(x) = 1$) closures are shown. $Re \approx 24000$

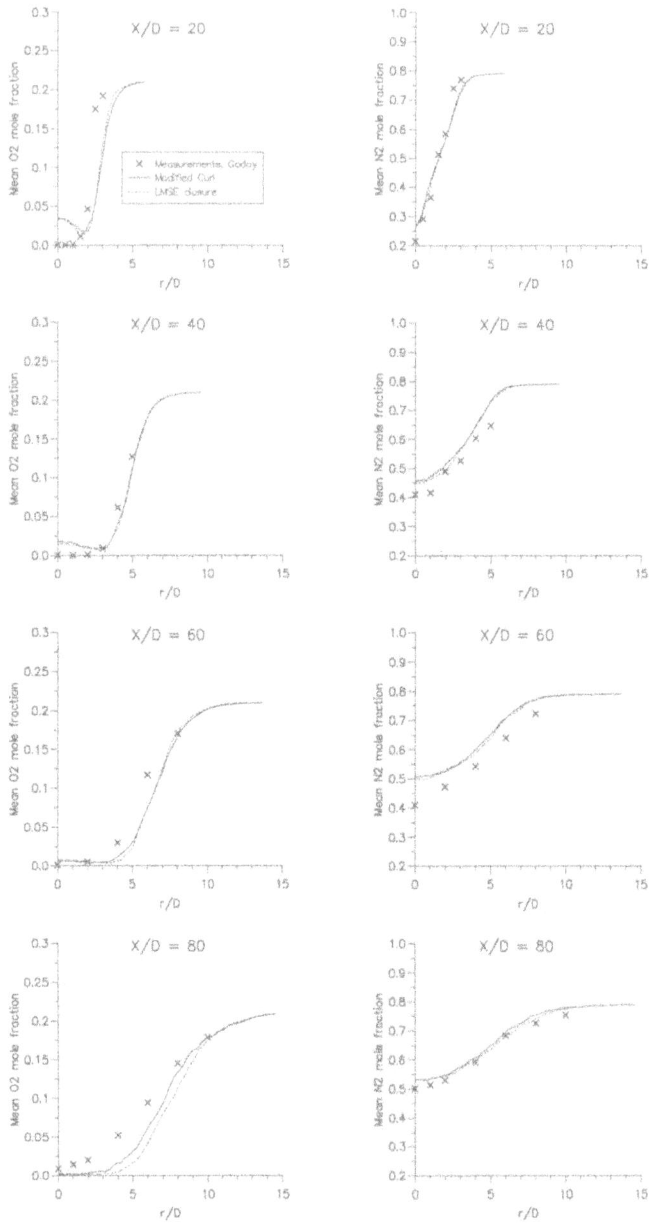

Fig. 8. Radial profiles of the measurements (Godoy [45]) and predictions for O_2 and N_2 in a propane-air jet diffusion flame issuing into stagnant surrounds. The k-ε model and the scheme of Jones and Lindstedt [60] are employed. Predictions using the LMSE and coalescence-dispersion ($\mathcal{A}(x) = 1$) closures are shown. $Re \approx 24000$

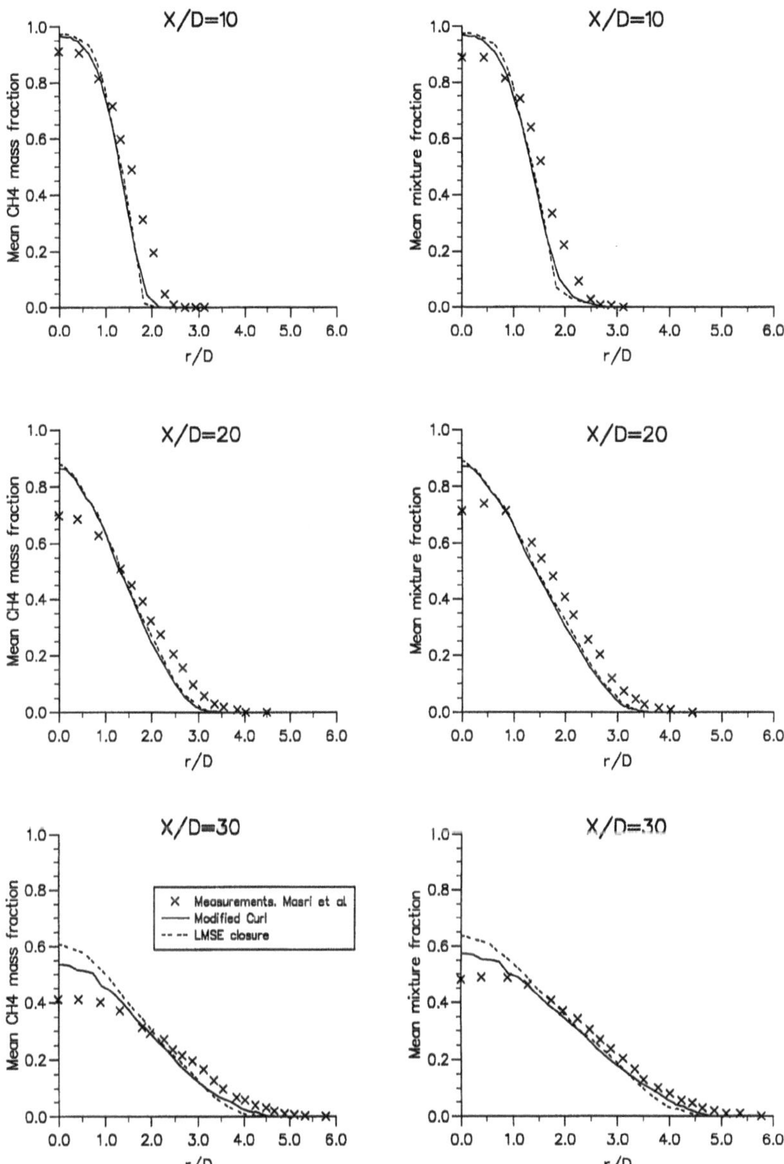

Fig. 9. Radial profiles of the measurements (Masri et al. [86], 'L-flame') and predictions for CH_4 and ξ in a (piloted) methane-air jet diffusion flame issuing into coflowing surrounds. The k-ε model and the scheme of Jones and Lindstedt [60] are employed. Predictions using the LMSE and coalescence-dispersion ($\mathcal{A}(x) = 1$) closures are shown.

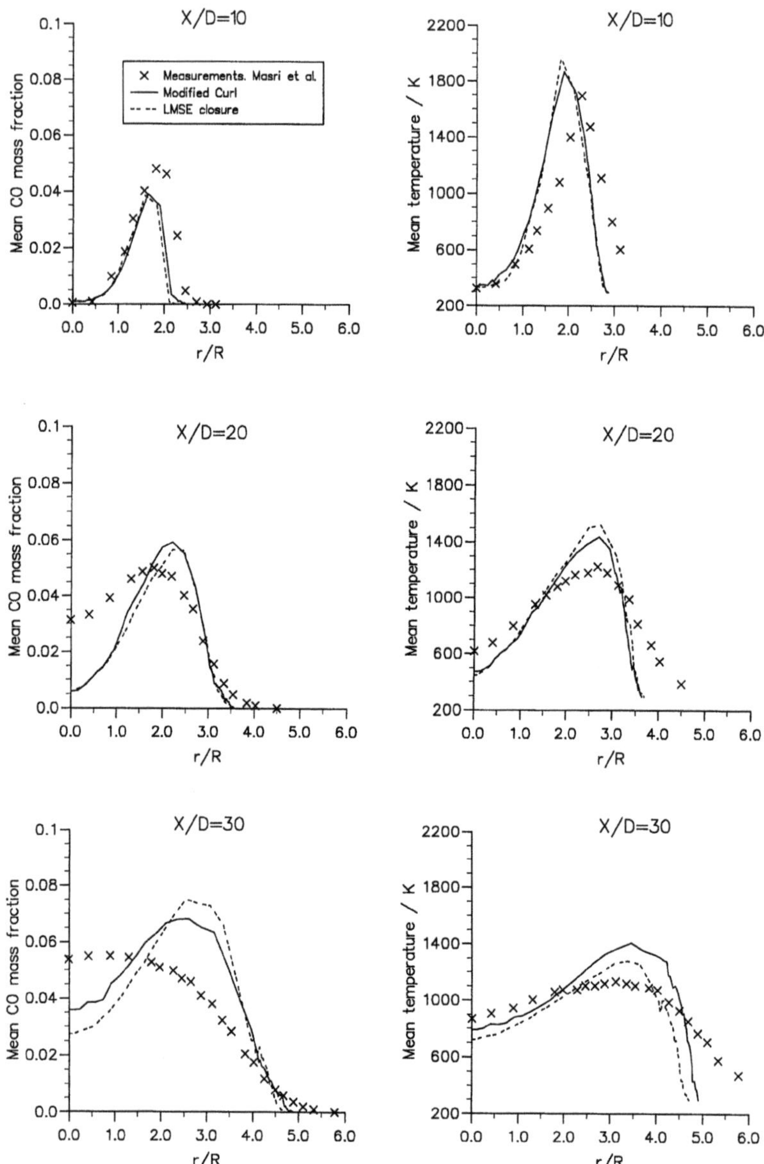

Fig. 10. Radial profiles of the measurements (Masri et al. [86], 'L-flame') and predictions for CO and temperature in a (piloted) methane-air jet diffusion flame issuing into coflowing surrounds. The k-ε model and the scheme of Jones and Lindstedt [60] are employed. Predictions using the LMSE and coalescence-dispersion ($\mathcal{A}(x) = 1$) closures are shown.

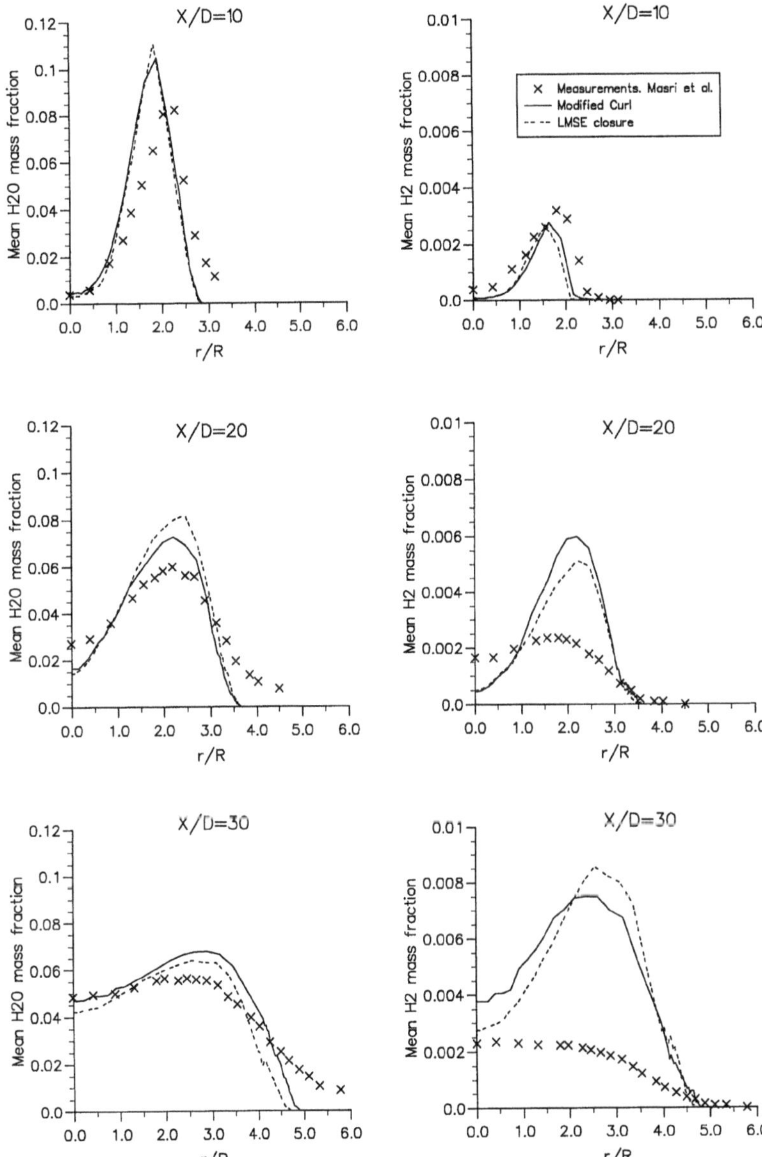

Fig. 11. Radial profiles of the measurements (Masri et al. [86], 'L-flame') and predictions for H_2O and H_2 in a (piloted) methane-air jet diffusion flame issuing into coflowing surrounds.The k-ε model and the scheme of Jones and Lindstedt [60] are employed. Predictions using the LMSE and coalescence-dispersion ($\mathcal{A}(x) = 1$) closures are shown.

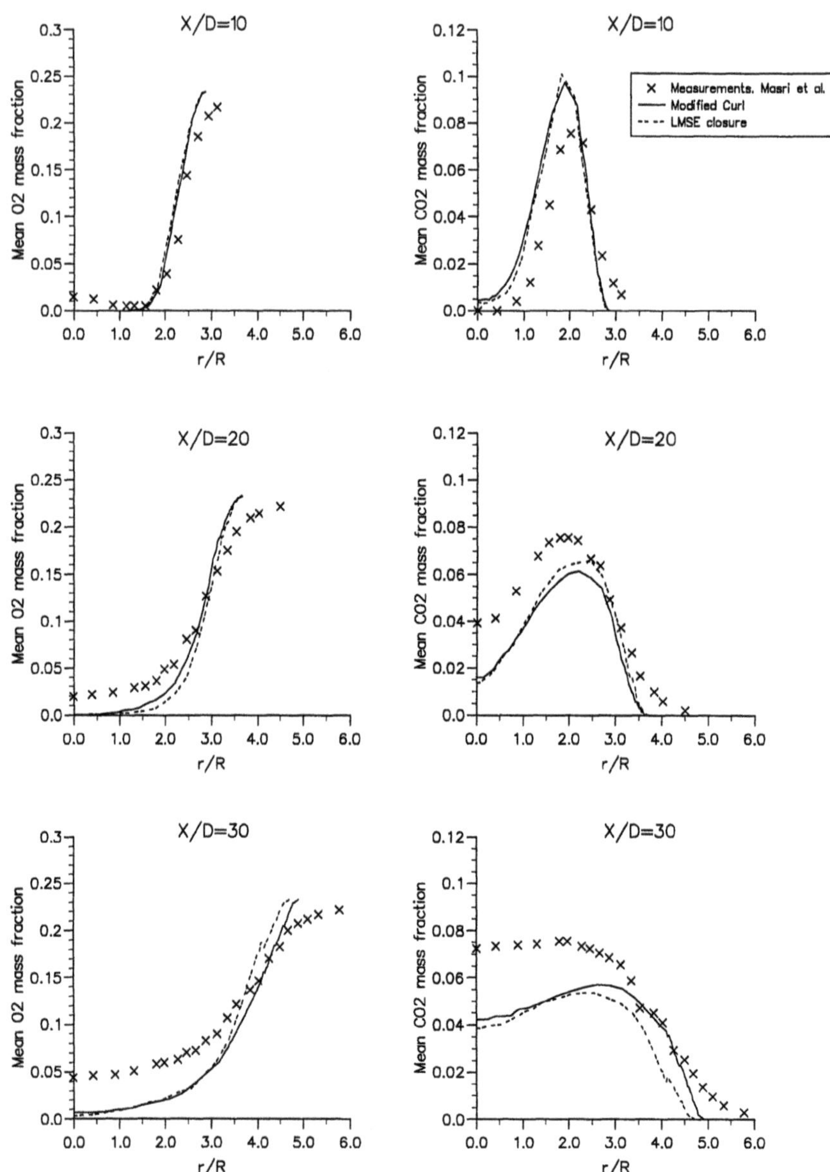

Fig. 12. Radial profiles of the measurements (Masri et al. [86], 'L-flame') and predictions for O_2 and CO_2 in a (piloted) methane-air jet diffusion flame issuing into coflowing surrounds.The k-ε model and the scheme of Jones and Lindstedt [60] are employed. Predictions using the LMSE and coalescence-dispersion ($\mathcal{A}(x) = 1$) closures are shown.

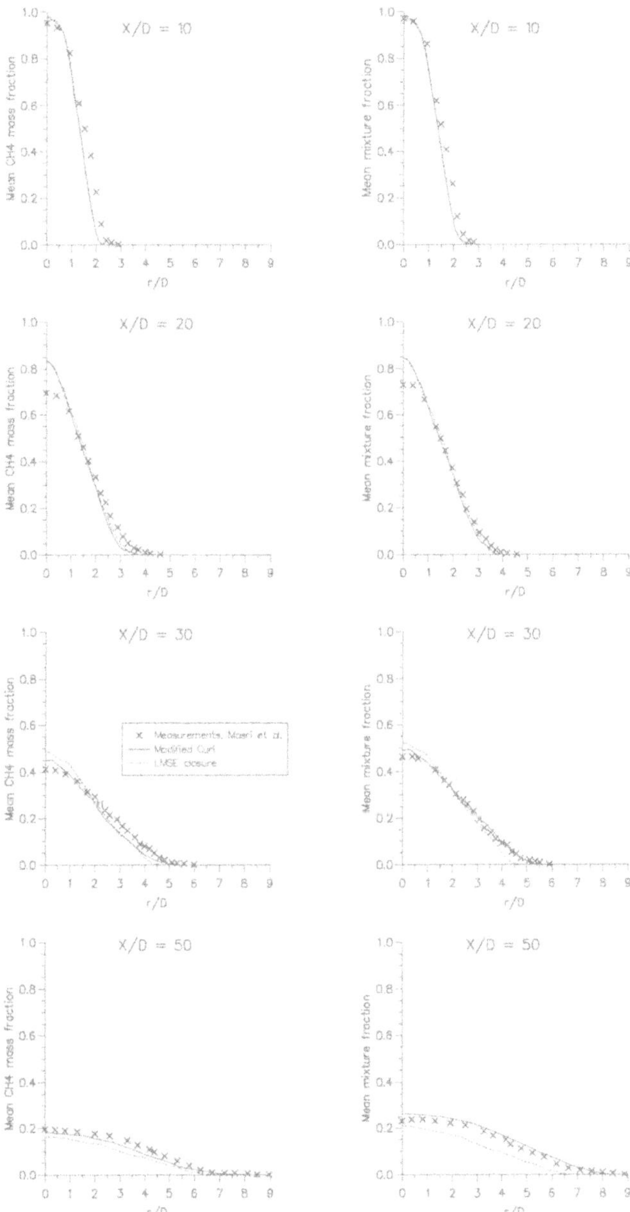

Fig. 13. Radial profiles of the measurements (Masri et al. [86], 'B-flame') and predictions for CH_4 and ξ in a (piloted) methane-air jet diffusion flame issuing into coflowing surrounds.The k-ε model and the scheme of Jones and Lindstedt [60] are employed. Predictions using the LMSE and coalescence-dispersion ($\mathcal{A}(x) = 1$) closures are shown.

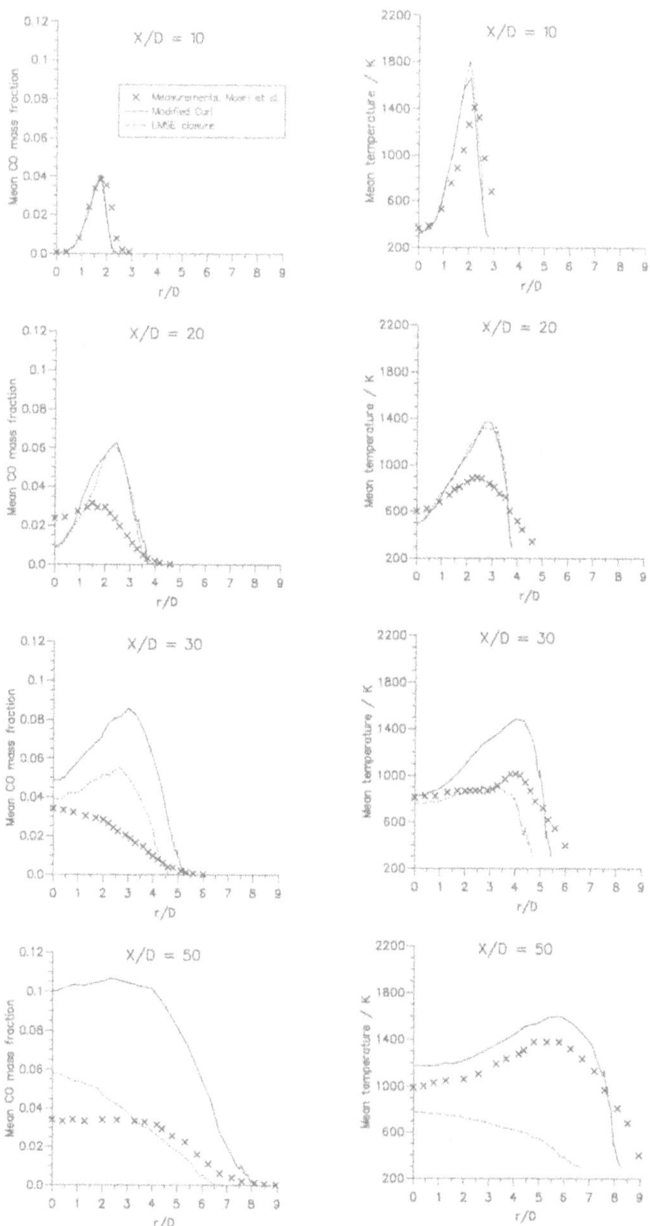

Fig. 14. Radial profiles of the measurements (Masri et al. [86], 'B-flame') and predictions for CO and temperature in a (piloted) methane-air jet diffusion flame issuing into coflowing surrounds. The k-ε model and the scheme of Jones and Lindstedt [60] are employed. Predictions using the LMSE and coalescence-dispersion ($\mathcal{A}(x) = 1$) closures are shown.

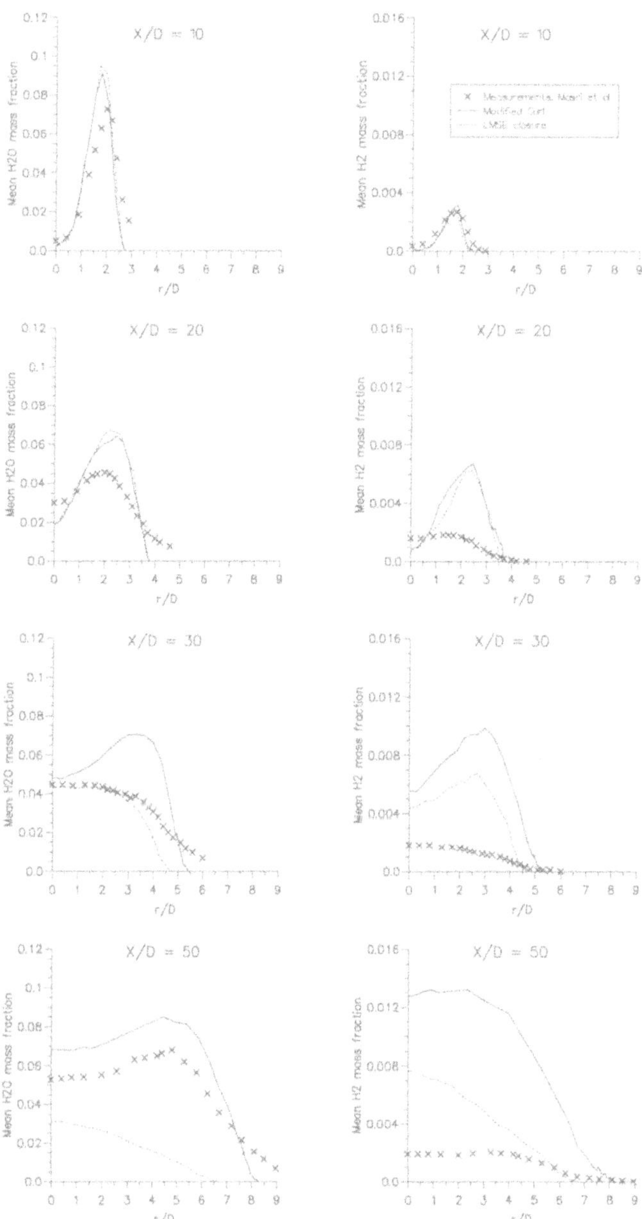

Fig. 15. Radial profiles of the measurements (Masri et al. [86], 'B-flame') and predictions for H_2O and H_2 in a (piloted) methane-air jet diffusion flame issuing into coflowing surrounds. The k-ε model and the scheme of Jones and Lindstedt [60] are employed. Predictions using the LMSE and coalescence-dispersion ($\mathcal{A}(x) = 1$) closures are shown.

20. THE MODELING OF COMBUSTION AND POLLUTANT FORMATION IN ENGINE FLOWS

N. Peters
Institut für Technische Mechanik
RWTH Aachen
52056 Aachen
Germany

The state-of-the-art of combustion models for implementation into 3D codes is presented. Flamelet models for premixed, non-premixed and partially premixed combustion are described and their implementation into 3D-codes is discussed. Flamelet modeling has the advantage of separating the numerical effort associated with the resolution of fast chemical time scales from the 3D-computation of the engine combustion cycle. A maximum of four scalar field equations have to be solved in the engine code in order to determine the flamelet position and its statistical distribution.

A new aspect proposed here is to use so-called "representative interactive flamelets" that are solved on-line with the 3D code. The parameters and boundary conditions that govern the unsteady evolution of these flamelets are extracted from the 3-D engine calculation by statistical averaging over a representative domain of interest.

1. Introduction

Due to the importance of automotive transportation, large efforts have been made to enhance the efficiency and exhaust gas quality of SI- and Diesel-engines. Reciprocating engines already reached a very high technological standard, but future work has to be done in order to increase this standard even further.

To meet this aim a major key lies in a deeper understanding of the combustion process itself. Important work has been done in the field of modeling turbulent combustion processes mostly by considering characteristic flame configurations, which has led to significant improvements of the predictive power of combustion models. These models are essential for a successful computer simulation of the 3-D combustion process within the cylinder of an engine. Based on those improved combustion models, the numerical simulation will become a powerful working tool in engine development.

In addition, numerical predictions should always be validated against experimental data. Here optical, non-intrusive measurement techniques like Particle Image Velocimetry (PIV) and Laser-Induced Fluorescence (LIF) have given a deeper insight in the microstructure of the turbulent combustion processes and could supply validations of the theoretical modeling attempts.

F. Culick et al., (eds.), Unsteady Combustion, 493–512.

2. Regimes in Turbulent Combustion

An important factor in the progress of understanding turbulent combustion is the identification of different regimes defined by different time and length scale ratios. From a technical point of view the interaction between turbulence and chemistry may be classified by two criteria: premixed or non-premixed combustion, slow or fast chemistry. Combustion in Diesel-engines is essentially non-premixed while combustion in spark ignition engines occurs in the premixed regime. Slow chemistry is of little practical interest in engine combustion. An important exception is pollution formation, such as the formation of NO or the oxidation of soot in the hot combustion gases or the incomplete combustion of hydrocarbons desorbed from the oil layer at the cylinder walls. The main combustion process, however, occurs nearly always in the fast chemistry regime. The reason is simple: for combustion to be stable and complete it must be rapid and therefore the chemical time scales must be sufficiently short under all circumstances. Therefore the engine is designed such that only at limit conditions, i.e. at very high engine speeds, turbulent time scales may become as short as the chemical time scales of the main combustion process.

A given turbulent flow field may locally be characterized by the root-mean-square velocity fluctuation v' and the turbulent macroscale ℓ_t, yielding a turbulent time scale $t_t = \ell_t/v'$. Specifically, if Favre-averaged quantities (denoted by a tilde) are used, one may relate v' and ℓ_t to $\widetilde{k}$ and $\widetilde{\varepsilon}$, where $\widetilde{k}$ is the turbulent kinetic energy and $\widetilde{\varepsilon}$ its dissipation by

$$v' = \sqrt{\frac{2\widetilde{k}}{3}}, \quad \ell_t = c_d v'^3/\widetilde{\varepsilon}, \quad t_t = \frac{2}{3} c_d \frac{\widetilde{k}}{\widetilde{\varepsilon}}. \tag{2.1}$$

Based on these integral scales the turbulence Reynolds number is defined

$$\mathrm{Re}_t = \frac{v'\ell_t}{\nu} \tag{2.2}$$

where ν is the kinematic viscosity. In terms of ν and $\widetilde{\varepsilon}$ the Kolmogorov length and time scales are

$$\eta = \left(\frac{\nu^3}{\widetilde{\varepsilon}}\right)^{1/4}, \quad t_\eta = \left(\frac{\nu}{\widetilde{\varepsilon}}\right)^{1/2}. \tag{2.3}$$

Furthermore, for non-premixed combustion, the non-homogeneous mixture field must be considered. Fluctuations of the mixture fraction, to be defined below, are characterized by

$$Z' = \sqrt{\widetilde{Z''^2}} \tag{2.4}$$

where $\widetilde{Z''^2}$ is the mixture fraction variance.

Diagrams defining combustion regimes in terms of length and velocity scale ratios have been developed for premixed and non-premixed flames and are presented elsewhere [1]. In the flamelet regime for premixed combustion the flame thickness

ℓ_F must be smaller than the Kolmogorov scale, enabling a certain range of turbulent eddies to corrugate the flame front. The Gibson scale

$$\ell_G = \frac{{s_L}^3}{\widetilde{\varepsilon}} \tag{2.5}$$

can be defined as the order of magnitude of the smallest turbulent structures in a flame front [2]. Here s_L is the laminar burning velocity. It can be shown that ℓ_G must be larger than ℓ_F in the flamelet regime. In Fig. 1 taken from [3], different turbulent length scales measured in an SI-engine at speeds varying from 500 to 2500 rpm are presented. The largest one is the flame brush thickness $\ell_{F,t}$. The next one is the crossing length scale λ_G equivalent to the Taylor-scale of the flame front corrugations. The smallest turbulent scale of the front is the Gibson length scale ℓ_G. In these engine experiments it is always larger than the Markstein length $\mathcal{L}$, estimated as four times the laminar flame thickness ℓ_F which was calculated from approximations in [3]. Since the Gibson scale is typically one order of magnitude larger than the laminar flame thickness, it may be concluded that combustion in spark ignition engines occurs typically in the flamelet regime.

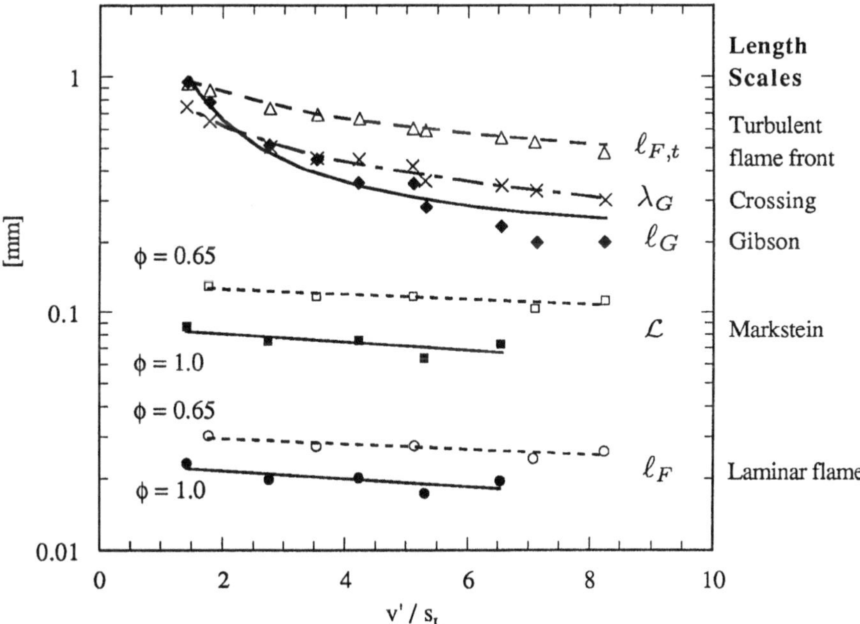

Fig. 1: Characteristic length scales of turbulent premixed combustion in SI-engines

3. Numerical Simulation of Turbulent Combustion

Present numerical simulations of the combustion process in engines is in most cases limited to the cycle-resolved calculation of the thermodynamic process, assuming two-zone-models for the description of flame propagation. The flame is viewed as a spherical surface which is traveling from the spark plug through the combustion chamber, dividing the volume into burnt and unburnt regions. Similar models are also used for Diesel engine combustion. These models nearly always require some empirical input from measurements, such as the pressure over crank angle, and therefore provide no self-contained prediction.

Since more powerful computers with a still exponentially growing capacity and computational speed continue to become available, there is a strong move towards 3D simulations of the engine combustion process. A number of 3D codes have been developed [5], [6], but the spatially resolved numerical simulation of the intake process, compression and combustion in an engine combustion chamber is still an outstanding task that needs an enormous amount of CPU-time. Most of the currently available 3D simulations are therefore restricted to the prediction of the cold flow in the intake ports and the combustion chamber. A significant part of the necessary CPU-time is needed for generation of appropriate grids if real geometries have to be modeled in detail.

Turbulence modeling usually relies on the k-ε-model which for engine simulations has to be modified for compressibility. The various attempts for this modification are mainly resulting in different model constants. Another approach, in the spirit of large eddy or direct simulation, has been performed in [7] in order to resolve the small structures of the flow. Here, the Navier-Stokes-equations have been solved without any turbulence modeling in a highly resolved 3D-grid. Complex phenomena like separated flows and small vortices caused by the curvature of the intake port and its inclination could be identified in the intake ports and cylinder under realistic conditions of moving valves and piston. Near the end of the compression process, the large vortices were observed to break up into small ones and the turbulence intensity increases due to compression.

In [8], the more complicated intake flow in a Diesel engine with helical port and reentrant bowl in the piston has been simulated. The results showed the development of two characteristic swirl motions in the cylinder and piston bowl and the buildup and stabilization of the swirl during compression. A strong interaction was found between the swirling motion and the shape and position of the piston bowl.

A computer code which is in use for engine combustion simulation in several groups is the KIVA code originally from Los Alamos National Laboratory [6]. In [9] it was used for simulations in conjunction with experimental investigations using a subgrid scale model for turbulence modeling and an eddy-break-up model for treatment of combustion, assuming combustion chemistry as infinitely fast and therefore the turbulent time scale as the important time scale of the problem. Simulations at the present stage are capable to show tendencies, but quantitative results need a further improvement in turbulence and combustion modeling to be able to assist engine development.

4. A Flamelet Formulation for Premixed Combustion

For the simulation of combustion in SI-engines, the regime of premixed combustion has to be considered. For this regime, recent developments led to a flamelet concept based on a field equation for a scalar G that represents the flame surface location. This formulation is derived from first principles and is likely to replace previous formulations based on a balance equation for a reaction progress variable, which uses the eddy break-up model. The introduction of the scalar G leads to similar coordinate transformation as the mixture fraction Z in non-premixed combustion, presented below, and therefore allows to split the problem into a one-dimensional calculation of the premixed flame structure and a 3D-calculation of the flow and scalar field to obtain the flame location. The field equation for this scalar G can be deduced from kinematic arguments [10]–[12]

$$\frac{\partial G}{\partial t} + \boldsymbol{v}^c \cdot \nabla G = s_L |\nabla G| \,. \tag{4.1}$$

Here, $\boldsymbol{v}^c$ is the velocity immediately ahead of the flame front. The laminar burning velocity s_L depends on local flame stretch by flow divergence and local flame front curvature. The flame front can be viewed as an iso-surface $G(x,t) = G_0$. The scalar difference $G - G_0$ is proportional to the distance from the flame surface. In the turbulent case, the field equation can be treated like any equation for a scalar: It can be split into an equation for the mean $\overline{G}$ and an equation for the fluctuation G'. The arising closure problem was addressed by transforming the equation into wavenumber space, where a suitable modeling is possible [12].

The resulting equation for $\overline{G}$ is

$$\frac{\partial \overline{G}}{\partial t} + \overline{\boldsymbol{v}^c} \cdot \nabla \overline{G} = s_T |\nabla \overline{G}| \,. \tag{4.2}$$

Here, s_T is the turbulent burning velocity and $\overline{\boldsymbol{v}^c}$ a conditioned mean flow velocity immediately ahead of the flame front. A simply modeling approach for this velocity will be presented in section 6. Numerous investigations of the influence of turbulence on the propagation speed of a turbulent premixed flame front have shown that the linear dependence between turbulence intensity and turbulent flame speed, which was already suggested in 1940 by Damköhler, is not sufficient to recover experimental results namely from engine experiments. It is evident, that additional effects of the flow field on the flame are important and must be taken into account. The turbulent burning velocity in eq.(4.2) was approximated in [13] as

$$s_T = \left(s_L + b_2 (s_L v')^{1/2} + b_1 v' \right) \left(1 - b_3 \frac{\mathcal{L}}{\ell_t} \frac{v'}{s_L} \right) . \tag{4.3}$$

The second term in brackets of eq. (4.3) accounts for the influences of flame stretch and leads to the bending effect in the relation between turbulence intensity and turbulent burning velocity.

5. A Flamelet Formulation for Non-Premixed Combustion

Combustion in a Diesel engine essentially occurs under non-premixed conditions. In using the mixture fraction as independent coordinate, it is possible to transform the balance equations for temperature and species into mixture fraction space and to resolve the inner structure of the reaction zone in a one-dimensional calculation of these equations in the mixture fraction coordinate [14]. Combustion essentially takes place in the vicinity of the surface of stoichiometric mixture. Let us consider locally a coordinate system where the coordinates x_2 and x_3 are within that surface and the coordinate x_1 is normal to it. We now replace x_1 by Z and retain the previous independent coordinates x_2, x_3 and t. This leads to the temperature equation in the form [1], [14]

$$\begin{aligned} \rho\left(\frac{\partial T}{\partial t}+\sum_{k=2}^{3} v_k\frac{\partial T}{\partial x_k}\right) &= \rho\frac{\chi}{2}\frac{\partial^2 T}{\partial Z^2}+\sum_{k=2}^{3}\frac{\partial}{\partial x_k}\left(\rho D\frac{\partial T}{\partial x_k}\right) \\ &- \frac{1}{c_p}\sum_{i=1}^{n} h_i\dot{m}_i + 2\rho D\sum_{k=2}^{3}\frac{\partial Z}{\partial x_k}\frac{\partial^2 T}{\partial Z\partial x_k}+\frac{1}{c_p}\frac{\partial p}{\partial t}-\frac{1}{c_p}\frac{\partial q_{Rk}}{\partial x_k}\,. \end{aligned} \tag{5.1}$$

In this equation the scalar dissipation rate χ is defined as

$$\chi = 2D\left(\frac{\partial Z}{\partial x_k}\right)^2 . \tag{5.2}$$

If the flamelet is thin in the Z-direction, an order of magnitude analysis similar to that for a boundary layer shows that the second derivative with respect to Z is the dominating term on the left hand side of eq.(5.1). To leading order in an asymptotic analysis this term must balance the reaction term on the right hand side. The two terms containing the time derivative are important if very rapid changes occur. If the time derivative terms and the last term, related to radiative transport, are retained, the flamelet structure is to leading order described by the one-dimensional time-dependent temperature equation

$$\rho\frac{\partial T}{\partial t} = \rho\frac{\chi}{2}\frac{\partial^2 T}{\partial Z^2} - \frac{1}{c_p}\sum_{i=1}^{n} h_i\dot{m}_i + \frac{1}{c_p}\frac{\partial p}{\partial t} - \frac{1}{c_p}\frac{\partial q_{Rk}}{\partial x_k}\,. \tag{5.3}$$

Similar equations may be written down for the chemical species. The validity of the flamelet approach for non-premixed combustion has recently been discussed in [15].

When the flamelet structure has been resolved by a one-dimensional, time dependent calculation as a function of prescribed parameters, all scalars are known.

In turbulent non-premixed combustion it has become common practice to use Favre (density weighted) averaged equations. In addition to the continuity and momentum equation, and equations describing turbulence quantities like $\widetilde{k}$ and $\widetilde{\varepsilon}$ and thereby the turbulent length and time scales, we need the balance equations for the mixture fraction

$$\overline{\rho}\frac{\partial \widetilde{Z}}{\partial t} + \overline{\rho}\,\widetilde{v}_k\frac{\partial \widetilde{Z}}{\partial x_k} = \frac{\partial}{\partial x_k}\left(\overline{\rho}D_T\frac{\partial \widetilde{Z}}{\partial x_k}\right) \tag{5.4}$$

and the mixture fraction variance

$$\overline{\rho}\frac{\partial \widetilde{Z''^2}}{\partial t} + \overline{\rho}\,\widetilde{v}_k\frac{\partial \widetilde{Z''^2}}{\partial x_k} = \frac{\partial}{\partial x_k}\left(\overline{\rho}D_T\frac{\partial \widetilde{Z''^2}}{\partial x_k}\right) - 2\overline{\rho}\,\widetilde{v_k''Z''}\frac{\partial \widetilde{Z}}{\partial x_k} - \overline{\rho\chi}\,. \tag{5.5}$$

In previous presentations the enthalpy has been related to the mixture fraction. For a general formulation, however, it is preferable to include the mean enthalpy as an additional variable

$$\overline{\rho}\frac{\partial \widetilde{h}}{\partial t} + \overline{\rho}\,\widetilde{v}_k\frac{\partial \widetilde{h}}{\partial x_k} = \frac{\partial}{\partial x_k}\left(\overline{\rho}D_T\frac{\partial \widetilde{h}}{\partial x_k}\right) + \frac{\partial \overline{p}}{\partial t} + \widetilde{u}\frac{\partial \overline{p}}{\partial x_\alpha} - \frac{\partial \overline{q}_{Rk}}{\partial x_\alpha}\,. \tag{5.6}$$

The term describing temporal mean pressure changes $\partial\overline{p}/\partial t$ is important in internal combustion engines operating under non-premixed conditions, such as the Diesel engine. The last term describes the radiative exchange in the combustion chamber.

6. A Flamelet Formulation for Partially Premixed Combustion

While the combustion phase at later stages in Diesel engine combustion occurs under non-premixed conditions, there is an early phase after injection where gaseous fuel and air inter-diffuse to generate a mixture with relatively small mixture fraction gradients. When this mixture ignites, a flame propagates through those regions, that are within the flammability limits. These regions are larger for early injection times and therefore rapid ignition and flame propagation with the consequence of large NO_x formation takes place predominantly under these conditions. Later on and for late injection times, ignition and flame propagation are strongly influenced by mixture fraction gradients or, in view of eq. (5.2), by the local scalar dissipation rate. We will outline the general features of a theory of partially premixed flamelets by assuming that the local burning velocity s_L depends only on the local mixture fraction and the mixture fraction gradient, expressed in terms of the scalar dissipation rate. It is expected to decrease with increasing χ and to vanish at a value χ_q, which by dimensional arguments, should be of the same order of magnitude as the dissipation rate at quenching for a non-premixed diffusion flamelet [14]. Assuming a linear decrease, one may express the dependence of the burning velocity on Z and χ as

$$s_L(Z,\chi) = s_L(Z)\left(1 - \alpha\frac{\chi}{\chi_q}\right)\,, \tag{6.1}$$

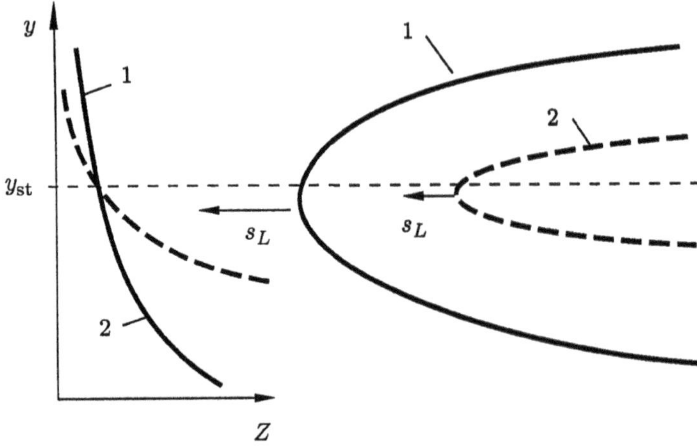

Fig. 2: A schematical illustration of triple flamelets for two different mixture fraction gradients

where α is assumed to be constant. For two different presumed mixture fraction gradients in y-direction the corresponding triple flamelet structure is schematically shown in Fig. 2. If, as in case 1, the mixture fraction gradient in the vicinity of Z_{st} is small (and therefore χ is small), the triple flamelet is broad in physical space and its propagation speed relative to the flow is mainly determined by the premixed burning velocity $s_L(Z_{st})$. If, on the other hand, as in case 2, the mixture fraction gradient and χ are large, the triple flamelet structure is narrow. Its leading edge will propagate with a lower burning velocity since it looses heat not only by preheating the unburnt mixture upstream but also that on the both sides in y-direction. As χ approaches χ_q the heat loss mechanism towards the sides becomes dominant and the triple flamelet structure is quenched in a similar way as an one-dimensional diffusion flamelet.

We now want to consider eq. (4.1) for the mean scalar field $\overline{G}$ to model flame propagation through a partially premixed region. For simplicity we will model the conditioned velocity in that equation by $\overline{v^c} = \bar{\rho}\tilde{v}/\rho_u$, where $\tilde{v}$ is the Favre average velocity and $\rho_u = \rho_u(\widetilde{Z})$ is the density that would correspond to the unburnt mixture at the local mean mixture fraction. The remaining problem is the modeling of the r.h.s. of eq. (4.1). Denoting the scalar gradient $|\nabla G|$ by σ, the ensemble average of the source term in eq. (4.1) is

$$\overline{s_L(Z,\chi)|\nabla G|} = \int_0^\infty \int_0^\infty \int_0^1 s_L(Z,\chi)\sigma P(Z,\chi,\sigma)\, dZ\, d\chi\, d\sigma\,. \tag{6.2}$$

Here $P(Z, \chi, \sigma)$ is a joint probability density of Z, χ and σ. In general, one may not assume these to be statistically independent. In fact, scalar gradients of G will tend to align with gradients of Z in very lean and rich parts of the mixture, while the gradient of G is nearly normal to that of Z around $Z = Z_{\mathrm{st}}$.

We will, for simplicity, however assume statistical independence and furthermore neglect fluctuations of χ. The joint pdf is then written

$$P(Z, \chi, \sigma) = P(Z)P(\sigma)\delta(\chi - \overline{\chi}). \tag{6.3}$$

Then the integral in eq. (6.2) takes the form

$$\overline{s_L(Z, \chi)|\nabla G|} = \overline{|\nabla G|} \left(1 - \alpha \frac{\overline{\chi}}{\chi_q}\right) \int_0^1 s_L(Z)P(Z)\, dZ. \tag{6.4}$$

Here one may exploit the fact, that the burning velocity $s_L(Z)$ has a strong peak in the vicinity of Z_{st} and decreases rapidly for $Z < Z_{\mathrm{st}}$ towards the lean and for $Z > Z_{\mathrm{st}}$ towards the rich side. A convenient approximation for the integral in eq. (6.4) is therefore

$$\int_0^1 s_L(Z)P(Z)\, dZ = s_L(Z_{\mathrm{st}})P(Z_{\mathrm{st}})(\Delta Z)_{s_L}. \tag{6.5}$$

which defines the width $(\Delta Z)_{s_L}$. This is essentially the range in mixture fraction space where the burning velocity is a significant fraction of the maximum burning velocity. It will depend on the shape of $s_L(Z)$ and via $P(Z)$ on $\widetilde{Z}$ and $\widetilde{Z''^2}$. However, using a beta-function pdf and assuming the mixture fraction variance to be proportional to the square of the mean mixture fraction, one obtains for methane flames a nearly constant value for $(\Delta Z)_{s_L}$, which in the vicinity of stoichiometric mixture is close to 0.06.

If eq. (6.5) is inserted into eq. (6.4) the product $s_L(Z_{\mathrm{st}})\overline{|\nabla G|}$ appears. For premixed turbulent combustion it has been argued [13] that

$$s_L\overline{|\nabla G|} = s_T|\nabla \overline{G}|, \tag{6.6}$$

where s_T is the turbulent burning velocity and $|\nabla \overline{G}|$ the absolute gradient of the mean G-field. This is equivalent to Damköhler's suggestion that the ratio of the turbulent to the laminar burning velocity should be proportional to the ratio of the flame surface area to the cross sectional area of the mean flame.

For the turbulent burning velocity an ansatz of the form

$$s_T = s_L + b_2(s_L v')^{1/2} + b_1 v' \tag{6.7}$$

has been used [13] when stretch effects are neglected. Inserting eqs. (6.5)–(6.7) into eq. (6.4) this leads with $s_L = s_L(Z_{\mathrm{st}})$ to

$$\overline{s_L(Z, \chi)|\nabla G|} = s_T|\nabla \overline{G}| \tag{6.8}$$

where

$$s_T = \underbrace{[s_L(Z_{st}) + b_2(s_L(Z_{st})v')^{1/2} + b_1 v']}_{\text{premixed flame propagation}} \underbrace{P(Z_{st})(\Delta Z)_{s_L}}_{\substack{\text{partial}\\\text{premixing}}} \underbrace{\left(1 - \alpha\frac{\overline{\chi}}{\chi_q}\right)}_{\substack{\text{flamelet}\\\text{quenching}}} . \tag{6.9}$$

This formulation contains three contributions:

1. A term accounting for premixed flame propagation with the maximum laminar burning velocity $s_L(Z_{st})$ which is enhanced by turbulent velocity fluctuations.
2. A term due to partial premixing which restricts flame propagation to regions where the probability for stoichiometric conditions is large. Therefore the first term in eq. (15) is multiplied by $P(Z_{st})$. This conditioning is less severe if $(\Delta Z)_{s_L}$ is larger.
3. A term accounting for flamelet quenching. The turbulent burning velocity decreases if the mean scalar dissipation rate increases, indicating that laminar triple flamelets are less able to propagate due to increasing local mixture fraction gradients.

7. Numerical Calculation of Partially Premixed Flame Propagation

The model described in paragraph 6 has been used to calculate the unsteady flame propagation and the stabilization in turbulent jet flames [6]. For this purpose the KIVA II code [24] was modified to include eqs. (4.1) and (5.4)–(5.6) together with eqs. (6.8) and (6.9). The $\overline{G}$ field was initialized with $\overline{G} = 0$ for unburnt conditions. Combustion was initiated by setting $\overline{G} = 0.1$ in one cell. The evolution of the $\overline{G}$ field then led to values $\overline{G} > G_0$, where G_0 was chosen as $G_0 = 1.0$. If $\overline{G}$ increased beyond $\overline{G} = 2.0$ it was set equal to that value.

The combustion model of KIVA was removed, but equations for the mean mixture fraction $\widetilde{Z}$, its variance $\widetilde{Z''^2}$ and the mean enthalpy $\widetilde{h}$ were included.

The mass fraction of the chemical species were determined by using a flamelet library for laminar counterflow diffusion flames for different velocity counterflow gradients. There are two possible states for a diffusion flamelet which are conditioned by the value of $\overline{G}$: If in a computational cell $\overline{G} \gg G_0$, it is considered to be completely burnt and the mass fractions are determined by

$$Y_{i,b}(\widetilde{Z}, \widetilde{Z''^2}, \overline{a}) = \int_0^1 Y_i(Z, a)\widetilde{P}(Z)\, dZ\,, \tag{7.1}$$

where $Y_i(Z, a)$ is taken from the library of burning flamelets setting the velocity gradient a of the flamelet equal to the local strain rate

$$\overline{a} = \frac{\widetilde{\varepsilon}}{\widetilde{k}} \tag{7.2}$$

of the turbulent flow. In eq. (7.1) a beta function pdf for $\widetilde{P}(Z)$ was used. The integration was performed in advance and values for $\widetilde{Y}_{i,b}$ were tabulated as functions of $\widetilde{Z}$, $\widetilde{Z''^2}$ and $\overline{a}$.

If $\overline{G} \ll G_0$ in a computational cell, the mass fractions are those of fuel and air in an unburnt mixture at the local value on $\widetilde{Z}$

$$\widetilde{Y}_{i,u} = Y_{i,u}(\widetilde{Z}) . \tag{7.3}$$

If the computation cell is located within the flame brush thickness, the weighted sum

$$\widetilde{Y}_i = f\widetilde{Y}_{i,b} + (1-f)\widetilde{Y}_{i,u} \tag{7.4}$$

is used. The fraction of burnt flamelets in each cell calculated by assuming that G fluctuations are Gaussian distributed [17]. Then

$$f = \int_{G=G_O}^{\infty} \frac{1}{\sqrt{2\pi\overline{G'^2}}} \exp\left\{-\frac{(G-\overline{G})^2}{2\overline{G'^2}}\right\} dG , \tag{7.5}$$

where the variance $\overline{G'^2}$ is assumed proportional to the square of the integral length scale. The proportionality factor was varied between 0.5 and 1.0 depending on the numerical grid size in order to obtain stable solutions. The temperature $\widetilde{T}$ in the cell was calculated from the mean enthalpy by

$$\sum_{i=1}^{n} \widetilde{Y}_i h_i(\widetilde{T}) = \widetilde{h}(\boldsymbol{x}, t) , \tag{7.6}$$

where the specific enthalpies are taken from NASA polynomials. The numerical constants used in eq. (6.9) were $b_2 = 0.8$ and $b_1 = 1.6$, respectively, and $\alpha = 1$ with $s_L(Z_{st}) = 39$ cm/sec for methane and $\chi_q = 17$/sec [24]. The scalar dissipation rate $\overline{\chi}$ in eq. (6.9) does not take mixture fraction variations into account [17] but was defined as

$$\overline{\chi} = 2\frac{\widetilde{\varepsilon}}{\widetilde{k}}(\Delta Z)_F \quad , \tag{7.7}$$

where $(\Delta Z)_F$ is the flame thickness in mixture fraction space, assumed here as $(\Delta Z)_F = 2Z_{st}$. The conditioned velocity fluctuation in eq. (6.9) was approximated by

$$v' = \left(\frac{2\overline{\rho}\widetilde{k}}{3\rho_u}\right)^{1/2} . \tag{7.8}$$

The temperature profile and the velocity change due to thermal expansion at the flame front $\overline{G} = G_0$ is continuous over a few grid points due to the weighting used in eq. (7.4).

In order to simulate a propagating flame front through a turbulent methane jet into still air, one computational cell of the steady cold flow field was initiated on the centerline downstream of the position of mean stoichiometric mixture. The flame ignition and propagation is the driven by the gradient of $\overline{G}$ towards the neighboring cells, which were originally set to $\overline{G} = 0$. An example calculation is shown in Fig. 3 where the axial distance between the flame base and the nozzle is plotted versus time. The solid line denotes the calculation and the symbols represent the measured values from different experimental runs.

It is seen that the slope of the calculated curve and therefore the propagation velocity agrees well with that of the experimental data taken from high speed cinematography. A closer inspection of the calculations shows that flame propagation follows the surface of mean stoichiometric mixture and depends on the velocity fluctuations there. The experiments, on the other side, seem to show a transport of the flame by the large structures at the edge of the jet. Therefore there remains a substantial difference between the very crude turbulence modeling based on the k-ε-model and the large scale dynamics in a jet.

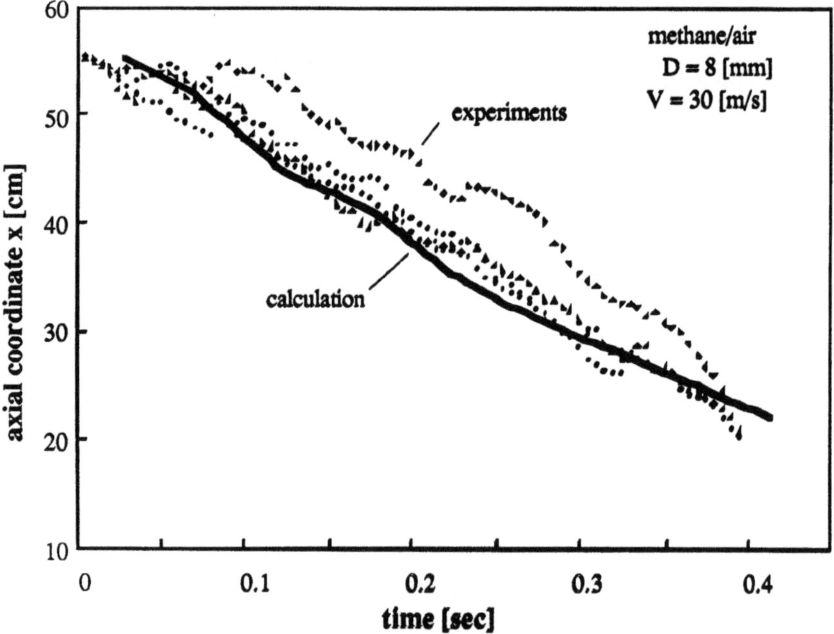

Fig. 3: Flame base position during unsteady upstream propagation of a partially premixed flame in a methane jet

8. Pollutant Formation

The most urgent problems in engine combustion are the mechanisms of pollutant formation, specifically soot- and NO_x-formation and HC-emission. Soot formation in Diesel engines turns out to be the most complicated task for reaction kinetics and work on the evaluation of appropriate reaction mechanisms is still in progress [18]. Few detailed data exist for soot formation and oxidation at high pressures as in Diesel engines. For ethylene flames the soot volume fraction was found in [19] to increase with p^2 for pressures up to 10 bar but approximately linearly at higher pressures up to 70 bar. An important step in modeling was done by combining an elementary kinetic mechanism for lower melocular species with models for fast polymerisation [20] and models for coagulation and surface growth. This kinetically based soot model is able to predict soot formation of a Diesel fuel within the correct order of magnitude. Furthermore a definite sensitivity was found on the aromatic content of the fuel.

The other urgent emission problem is NO_x. There are two major contributions to NO_x-formation: Thermal NO-production through the extended Zeldovich-mechanism and the "prompt-NO" mechanism wherein hydrocarbon fragments attack bimolecular nitrogen, producing atomic nitrogen, cyanides and amines, which are subsequently oxidized to nitric oxides. The flame temperature will determine which one of these mechanisms is dominant. The prompt-NO mechanism is relatively insensitive to temperature and will contribute about 10 to 30 ppm to the total NO-production. Thermal NO-production can contribute up to several hundred ppm and therefore, NO-reduction in engines will have to be focused on thermal NO. Since this mechanism is very sensitive to temperature, it is essential to incorporate radiative heat losses into the calculation.

9. The Calculation of Representative Interactive Flamelets

In premixed combustion it is often sufficient to assume not only the flame thickness, but the entire non-equilibrium structure of the premixed flame as thin compared to the turbulent scales. Then the premixed flamelet structure is made of step function profiles jumping from the unburnt to the burnt state as G crosses G_0. In the case of non-premixed combustion, however, the flamelet profiles are a function of the mixture fraction and the scalar dissipation rate and are obtained by solving eq.(5.3) and similar equations of the chemical species with the appropriate boundary conditions. Typically for engine calculations, the boundary conditions, for instance the air temperature T_2 at $Z = 0$ and the fuel temperature at T_1 at $Z = 1$, change with time due to bulk gas compression or expansion, or due to droplet evaporation.

Furthermore, the scalar dissipation rate χ, characterized by its mean value $\overline{\chi}$, changes with time. In order to account for these imposed transient parameter changes we propose to perform on-line flamelet calculations together with the 3D-code. The concept of representative interactive flamelets differs from the previous concept of a flamelet library [17] that is useful for flamelets with time-independent parameters and boundary conditions. For instance, in Diesel engine combustion, droplets from the fuel spray reduce the local enthalpy during the droplet heating phase. Therefore the enthalpy that is available for the ignition process is small at early stages of the evaporation process and in regions where there is a large amount of vaporized fuel. This is shown in Fig. 4, where the difference between the mean enthalpy $\widetilde{h}$ of the gas phase and the enthalpy on the air side h_2 in a fuel spray is shown in a scatter plot over the mean mixture fraction $\widetilde{Z}$. Both are taken from locations in the vicinity of the spray. At an early time of 0.1 ms, where only 10% of the fuel is vaporized, there are entries with very low enthalpies, indicating that cold droplets in the computational cell reduce the enthalpy. Later on, when more of the fuel is vaporized, the enthalpy increases. If all the droplets had reached the evaporation temperature the exchange of mass and heat by evaporation balance and one obtains a straight line

$$h = h_2 + Z(h_2 - h') \tag{9.1}$$

where h' is the enthalpy of the liquid phase at the evaporation temperature. This linear relation, which is equally valid between $\widetilde{h}$ and $\widetilde{Z}$, is also plotted in Fig. 4 as a solid line. The temperature T_u of the unburnt mixture resulting from eq. (9.1) is

$$T_u = T_2 + Z(T_2 - T_1) \tag{9.2}$$

where T_1 is the extrapolated temperature at $Z = 1$

$$T_1 = T_L - (h'' - h')/c_p\,. \tag{9.3}$$

Here T_L is the evaporation temperature of the liquid, h'' is the enthalpy of the gas phase and c_p is the specific heat of the gas phase at that temperature. If one assumes a linear relationship between h and Z or between T_u and Z also during the heating phase, one may approximate enthalpy data taken from the 3D code as a function of time and thereby impose the available enthalpy of an unsteady flamelet evolution. Similarly, one may approximate the evolution of the scalar dissipation rate from the representative regions in the 3D-calculation and feed it into the flamelet calculation. The resulting species profiles $Y_i(Z,a,t)$ from the flamelet calculation can be used in a similar way as in eq. (7.1) to calculate the evolution of the mean mass fractions $\widetilde{Y}_i(x,t)$ in the 3-D calculation. Using the local values of $\widetilde{Z}(x,t)$, $\widetilde{Z''^2}(x,t)$ to calculate the mixture fraction pdf $\widetilde{P}(Z)$ and the flamelet strain rate a equal to the mean value $\overline{a}(x,t)$ according to eq. (7.2) one obtains

$$\widetilde{Y}_i(x,t) = \int_0^1 Y_i(Z,a,t)\widetilde{P}(Z)dZ\,. \tag{9.4}$$

The temperature in the 3D-code may then be calculated iteratively from eq. (7.6).

Flamelet calculations, though one-dimensional, may involve a large number of chemical species, if ignition of higher hydrocarbons and pollutant formation is to

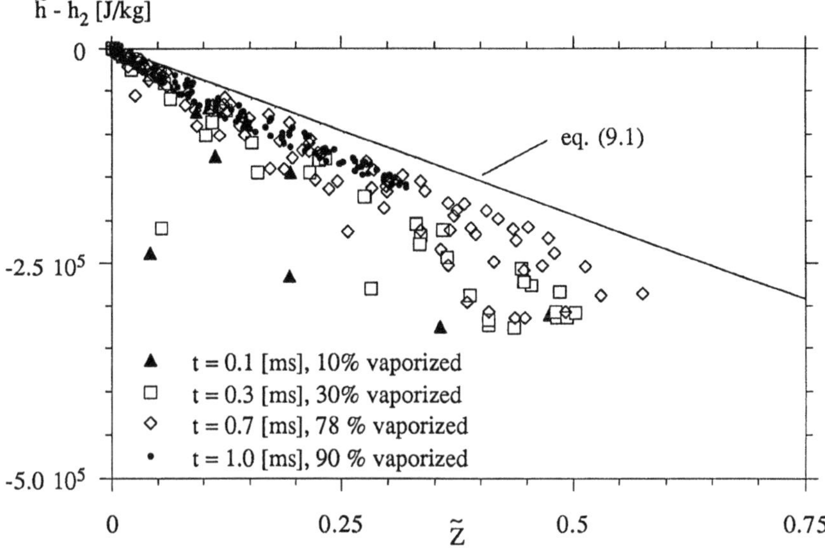

Fig. 4: Scatter plot of the available mean enthalpy as a function of the mean mixture fraction in spray of a n-heptane

be included. It is then necessary to reduce the chemical mechanism such hat only the important steps are included. A systematic procedure for reducing chemical mechanisms has been proposed in [21] and has been applied to flame calculations in [22]. Recently, Müller [23] has developed a 16-step reduced mechanism for the ignition of n-heptane. It is believed that n-heptane serves as an one-component model fuel for Diesel fuel. This mechanism was derived by

1. formulating a base mechanism of 83 elementary reactions which reproduces the ignition delay times of a large mechanism of 1011 reactions with reasonable accuracy.
2. introducing steady state assumptions for 21 of the 41 chemical species of the base mechanism by computing their production and consumption rates. If the consumption rate is much faster than the production rate, the species is assumed in steady state.

The resulting global mechanism is shown in Table 1.

One-dimensional flamelet calculations with this mechanisms have been performed by imposing a time dependent enthalpy profile which is linear in Z and a time dependent scalar dissipation rate. The scalar dissipation rate depends on Z in the same way as in counterflow flames [1]. The temperature T_1 on the fuel side and the

1*a*	$n\text{-}C_7H_{16}$	$\rightarrow$	$3\,C_2H_4 + CH_3 + H$
1b_1	$n\text{-}C_7H_{16} + H$	$\rightarrow$	$3\,C_2H_4 + CH_3 + H_2$
1b_2	$n\text{-}C_7H_{16}$	$\rightarrow$	$C_3H_6 + 2\,C_2H_4 + H_2$
1b_3	$n\text{-}C_7H_{16} + H + O_2$	$\rightarrow$	$1'C_7H_{15}O_2 + H_2$
1b_4	$n\text{-}C_7H_{16} + H + O_2$	$\rightarrow$	$2'C_7H_{15}O_2 + H_2$
2	$1'C_7H_{15}O_2 + O_2$	$\rightarrow$	$O_2C_7H_{14}O_2H$
3	$2'C_7H_{15}O_2 + O_2$	$\rightarrow$	$O_2C_7H_{14}O_2H$
4	$O_2C_7H_{14}O_2H + H_2$	$\rightarrow$	$OC_7H_{13}O_2H + H_2O + H$
5	$OC_7H_{13}O_2H + H_2$	$\rightarrow$	$2\,C_2H_4 + CH_3 + CH_2O +$ $CO + H_2O + H$
6	$C_3H_6 + H_2O$	$\rightarrow$	$C_2H_4 + CO + 2\,H_2$
7*a*	$C_2H_4 + 2\,H$	$\rightarrow$	$2\,CH_3$
7*b*	$C_2H_4 + O_2$	$\rightarrow$	$2\,CO + 2\,H_2$
8	$C_2H_2 + O_2$	$\rightarrow$	$2\,CO + H_2$
9	$CH_3 + O_2 + 2\,H_2$	$\rightarrow$	$CH_2O + H_2O + 3\,H$
10	CH_2O	$\rightarrow$	$CO + H_2$
11	$CO + H_2O$	$\rightarrow$	$CO_2 + H_2$
12	$H_2O_2 + 2\,H_2$	$\rightarrow$	$2\,H_2O + 2\,H$
13	$H + O_2$	$\rightarrow$	HO_2
14	$2\,H_2 + O$	$\rightarrow$	$H_2O + 2\,H$
15	$3\,H_2 + O_2$	$\rightarrow$	$2\,H_2O + 2\,H$
16	$2\,H$	$\rightarrow$	H_2

Table 1: 16-Step reduced mechanism for *n*-heptane ignition and combustion

scalar dissipation rate χ_{st} at stoichiometric mixture are shown as a function of time in Fig. 5. These functions were approximated from the data plotted in Fig. 4 and the corresponding evolution of χ_{st} in that spray. The temperature T_2 on air side was constant since the pressure was constant in that calculation. The temperature T_1 increases linearly up to the evaporation temperature and remains constant thereafter. The resulting temperature, fuel and oxygen profiles, plotted in Fig. 6, 7 and 8, respectively, illustrate the ignition of the flamelet. At first, there is a small temperature rise and a corresponding consumption of fuel and oxygen in the lean mixture $Z < Z_{st}$. Ignition occurs in that region first, because of the high temperature there and the rapid decrease of temperature with the mixture fraction. Then at 0.24 msec, the maximum temperature shifts towards the slightly rich mixture $Z > Z_{st}$, associated with a burn-out of all the fuel on the lean side. Afterwards there is a rapid consumption of fuel and oxygen and a corresponding temperature rise at stoichiometric mixture which reaches a maximum value at 0.32 msec. For larger times the oxygen on the fuel side is gradually consumed and the adiabatic flame temperature will be approached in the entire flamelet.

One may perform such representative interactive flamelet calculations for different regions in the vicinity of the fuel spray. In practice, less than 10 flamelet calculations should be sufficient to represent the regions of interest. Experience will be needed in defining these regions appropriately within the 3D-calculations and in generating smooth transitions of the profiles at the boundaries between them.

10. Conclusions

The modeling of combustion and pollutant formation in engine flows is still in the stage of development. Flamelet formulations, when put into practice, may reduce the numerical effort considerably by separating the 3D-flow calculation from the calculation of the chemistry. The success of the combustion model will then depend on its validation by comparison with experimental data.

Acknowledgements: This work would not have been possible without the many discussions with U.C. Müller, B. Rogg, K.N.C. Bray, K.P. Schindler, I. Magnusson and J.A.J. Karlsson during the IDEA programme, which was sponsored by the Commission of the European Communities in the frame of the Joule II programme, the Swedish National Board for Industrial and Technical Development and the Joint Research Committee of European automobile manufacturers (JRC), namely Fiat, Peugeot SA, Renault, Volkswagen and Volvo.

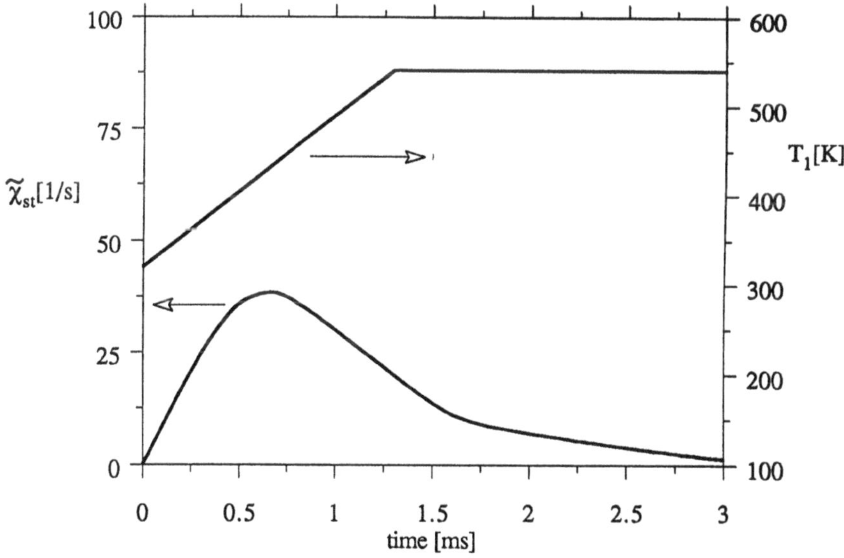

Fig. 5: Time dependent values of the temperature T_1 on the fuel side and the stoichiometric scalar dissipation rate in a transient flamelet calculation

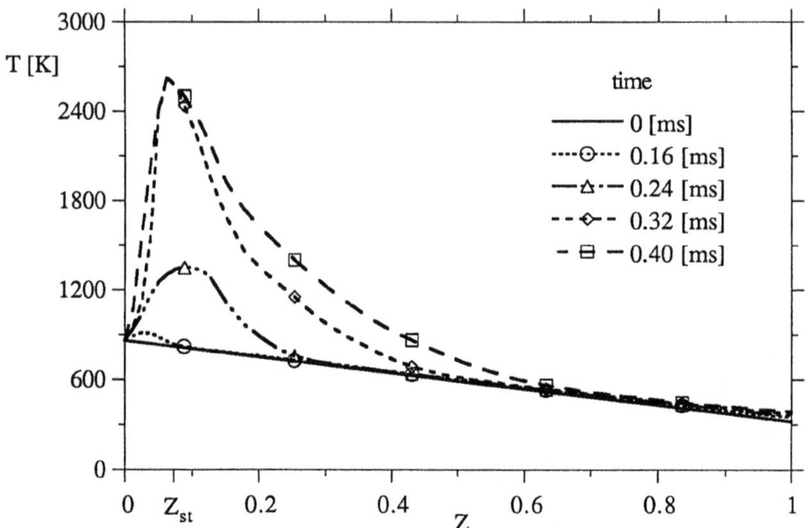

Fig. 6: Temperature profiles in an igniting flamelet with imposed transient boundary conditions and scalar dissipation rate

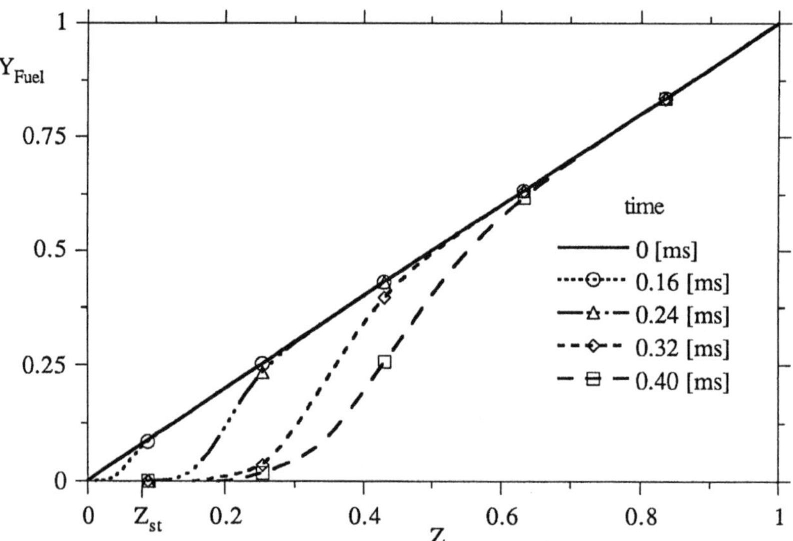

Fig. 7: n-Heptane profiles in the igniting flamelet corresponding to Fig. 6

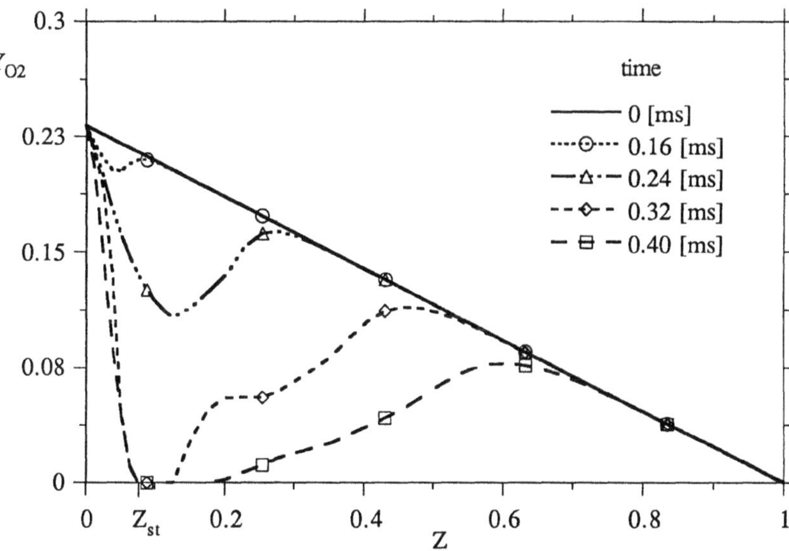

Fig. 8: Oxygen profiles in the igniting flamelet cooresponding to Fig. 6

11. References

[1] Peters, N.: Length scales in Laminar and Turbulent Flames. In: Numerical Approaches to Combustion Modeling (E.S. Oran, Jay P. Boris, Eds.), Progr. Astronautics and Aeronautics **135** (1991), 155–182.

[2] Peters, N.: Laminar Flamelet Concepts in Turbulent Combustion. 21st Symposium on Combustion, The Combustion Institute, Pittsburgh, 1231–1250 (1987).

[3] Wirth, M., Keller, P., Peters, N.: A Flamelet Model for Premixed Turbulent Combustion in SI-Engines, SAE paper 932646, 1993.

[4] Göttgens, J., Mauss, F., Peters, N.: Analytic approximations of burning velocities and flame thickness of lean hydrogen, methane, ethylene, ethane, acetylene and propane flames, Twenty-Fourth Symposium (International) on Combustion. The Combustion Institute (1992), pp. 129–135.

[5] Gosman, A.D.: Aspects of the simulation of combustion in reciprocating engines, in Numerical Simulation of Combustion Phenomena, Lecture Notes in Physics **241** Springer Verlag, Berlin (1985), p. 46–73.

[6] Amsden, A.A., O'Rourke, PJ., Butler, T.D.: Kiva II: A Computer Program for Chemically Reactive Flows with Sprays, Los Alamos National Laboratory report LA-11560-MS, 1989.

[7] Naitoh, K., Fujii, H., Urushihara, T., Takagi, Y., Kuwahara, K.: Numerical

Simulation of the Detailed Flow in Engine Ports and Cylinders, SAE 900256.

[8] Aita, S., Tabbai, A., Munck, G., Montmayeur, N., Takenaka, Y., Aoyagi, Y., Obana, S.: Numerical Simulation of Swirling Port-Valve-Cylinder Flow in Diesel Engines, SAE 910263.

[9] Henriot, S., Le Coz, J.F., Pinchon, P.: Three-Dimensional Modeling of Flow and Turbulence in a Four-Valve Spark Ignition Engine—Comparison with LDV-Measurements, SAE 890843.

[10] Markstein, G.H.: Nonsteady flame propagation, p. 8, Pergamon Press, Oxford 1964.

[11] Williams, F.A. in: The Mathematics of Combustion (J. Buckmaster, Ed.), SIAM, Philadelphia (1985), pp. 97.

[12] Peters, N.: A spectral closure for premixed turbulent combustion in the flamelet regime, J. Fluid Mech. **242** (1992), 611–629.

[13] Wirth, M. and Peters, N.: Twenty-Fourth Symposium (International) on Combustion, The Combustion Institute, Pittsburgh (1992), pp. 493–501.

[14] Peters, N.: Laminar diffusion flamelet models in non-premixed turbulent combustion, Prog. Energy Combust. Sci. **10**: 319–339 (1984).

[15] Buriko, Yu., Ya., Kuznetsov, V.R., Volkov, D.V., Zuitsev, S.A., Uryvsky, A.F.: Comb. Flame **96** (1994), 104–120.

[16] Müller, C.M., Breitbach, H., Peters, N.: Partially premixed turbulent flame propagation in jet flames, submitted to the 25th Symposium (International) on Combustion (1994).

[17] Bray, K.N.C. and Peters, N. in: Turbulent Reactive Flows (P.A. Libby, F.A. Williams, Eds.), Academic Press, to appear 1994.

[18] Frenklach, M., Wang, H.: Twenty-Third Symposium (International) on Combustion, The Combustion Institute, Pittsburgh (1990), p. 1559–1566.

[19] Böhm, H., Feldermann, Chr., Heidermann, Th., Jander, H., Luers, B., Wagner, H. Gg.: Twenty-Fourth Symposium (International) on Combustion, The Combustion Institute, Pittsburgh (1992), p. 991–998.

[20] Mauss, F., Trilken, B., Peters, N. in: Mechanism and Models of Soot Formation (H. Bockhorn, Ed.), Springer Verlag, in press.

[21] Peters,N.: Numerical and Asymptotic Analysis of Systematically Reduced Reaction Schemes for Hydrocarbon Flames, in Numerical Simulation of Combustion Phenomena, R. Glowinski et al., Eds., Lecture Notes in Physics, Springer Verlag, p.90, 1985

[22] Peters, N. in: Reduced Kinetic Mechanisms for Applications in Combustion Systems (N. Peters, B. Rogg, Eds.), Lecture Notes in Physics, Springer Verlag (1993).

[23] Müller, U.C.: Reduzierte Reaktionsmechanismen für die Zündung von n-Heptan und iso-Oktan unter motorrelevanten Bedingungen, Doktorarbeit RWTH Aachen, 1993.

21. MODELING OF TURBULENT COMBUSTION FOR I.C. ENGINES: CLASSICAL MODELS AND RECENT DEVELOPMENTS

R. Borghi, L. Delamare and T. Mantel
CORIA
Faculté des Sciences de Rouen
France

1. Introduction

1.1. A SHORT PICTURE OF THE STATE OF THE KNOWLEDGE

Turbulent combustion in a premixed medium has been recognized from 1941 (Damköhler) as a very interesting and very useful research domain. One of its field of application is the combustion in spark ignited internal combustion engines, but there are others...
The first studies on turbulent premixed combustion have been focused on the velocity of propagation of the combustion zone, by looking at the similarities and differences with the classical laminar normal flame speed. Does this quantity exist for turbulent flames as an intrinsical quantity that does not depend on the ignition devices ? What are the direct links of this quantity with the laminar flame speed and the turbulence characteristics ? These questions are not yet answered in a convenient way, fifty years after, although a large amount of knowledge has been accumulated.

One of the reasons why these (apparently) simple questions have not been solved is that the flame displays several type sof structure, depending on the laminar flame characteristics compared with the turbulence characteristics. This fact has been very early recognized and discussed ; apparently such discussions have been very active both in US and in USSR, during the fifties (see ref [1], Summerfield et al., and ref [2], Zimont). An other reason is that the flame itself, during its propagation, is able to modify the turbulence characteristics ; this is obviously due to the heat release and volume expansion, but also to the fact that the increase of temperature leads also to an increase of viscosity ; the first effect has been discussed very early,

F. Culick et al., (eds.), Unsteady Combustion, 513–542.

by Karlovitz et al. [3]. A related effect that has been recognized also is that the definition and use of a turbulent diffusion coefficient is much less valid for turbulent flames than for turbulent jets ; this has been the essence of the so-called BML-model, in 1980 (see ref [4]).

In order to be able to understand all these effects, it has been recognized from 1975 approximately that turbulent premixed combustion needed to be studied by means of balance equations that include turbulent diffusion and mean reaction rate as well as convection. Such an approach is not limited to the finding of the velocity of propagation of the turbulent flame, but is able also to calculate its thickness and its "structure", in the sense of statistical averages, of course. The pionnering papers for this approach have been probably the ones of D.B. Spalding [5] K. Bray and B. Moss [6], and R. Borghi and P. Moreau [7]. At this time the problem of defining and finding a model for the mean reaction rate has been mainly discussed, and the so-called Eddy Break Up model has been proposed by D.B. Spalding, for the case where the chemical reactions are very fast with respect to any turbulent process. This model has been given under two or three slightly different forms, and has been recovered by Bray and Moss in their first paper [6]. Together with one turbulence model (and any turbulence model could be used, even the simplest Prandtl mixing length model), this model is able, in principle, to calculate the thickness and the velocity of propagation of any turbulent premixed flame, in any geometry, including an effect of expansion on the turbulence field ; to take into account viscous effects and non gradient diffusion would need, however, relatively deep improvements. A modification of the EBU model for taking into account non infinitely fast chemistry has also been proposed : it is the so-called "presumed pdf" model of [8] ; even this very simple modelling can lead to some success in the prediction of turbulent combustion in I.C. Engines : the flow-field and temperature field that are predicted are realistic, and it is possible to find pressure traces in fair agreement with experiments. But the calculations cannot be claimed as predictive, because the modelling constant needs some adjustment when the equivalence ratio of the engine is changed, for instance in addition the ignition processes have also to be represented empirically.

A further step in the development of the researches in turbulent combustion modelling has occured from 1980. First, the approaches involving a balance equation for the probability density function (PDF) of reactive species concentrations, and eventually velocities, have been studied by Anand and Pope [9], following the pionnering paper of Dopazo and O'Brien [10] in 1976. The potential of such an approach is to replace the "presumed pdf" model, for the case of non infinitely fast chemistry, as well as to allow non gradient diffusion (if the velocity is included in the PDF). Secondly, the so-called "flamelets approaches" have been proposed, in order to include more physics in the EBU model and to attempt to enlarge its validity domain. Several approaches, which all invoke the presence of flamelets, have been proposed, but they are all different, both on the point of view of principles and of practice.

In parallel to these studies concerning turbulent combustion models, other researches have been devoted to the turbulence models themselves, and specially to their ability to take into account the flow field modifications that are due to the combustion : large density changes and density fluctuations, expansion or compression of parts of the flows. These studies are of particular interest for the calculation of turbulent combustion in I.C. engines ; indeed any turbulent combustion model needs at least one length scale and one time scale of the turbulence, and they have to be provided through a turbulence model ; if the turbulence model is not able to give right values, including the true influence of the physical flow phenomena induced by combustion, we can expect a failure of the calculations whatever can be the turbulent combustion model. The first studies of this type has been the one of Gosman et al.[11].

One of the most important problem for turbulent combustion modelling is the difficulty to obtain simple and relevant experimental data. There are several well known experimental configurations for studying turbulent flames : the so-called "V-flame", the "Bunsen burner", the "counterflow flame". All of them involve a relatively complex turbulence-combustion interaction in a non homogeneous geometry. The turbulent bomb with fans could be the simplest one, but the spatial homogeneity and isotropy is achieved only in the central part of the device. In addition, the relevance of such basic experiments for application to I.C. engines is to be questioned : the flow and turbulence fields very different from the ones that we can find in an engine, and the unsteady character of the flame propagation in engine cylinder is a very important peculiarity ; it may be suspected that during a large fraction of the total duration of the combustion period, the flame has not yet reached its quasi-steady regime. This unsteady character is the cause, on the other hand, of the need of statistical averages that cannot be replaced by time averages ; it is not easy to define adequately what type of average is needed, and probably the simple ensemble average is not enough.

1.2 THE SCOPE OF THIS PAPER

This paper will be divided in three parts. In the second section we will show the results that can be obtained with classical models of the type of Eddy Break Up model. We will not describe here models based only on the turbulent flame speed. The results will be discussed and the shortcomings of the model will be emphasized, based on theoretical or practical questions. The first question concerns the most convenient type of average ; the second deals with the mean reaction rate model, and the third with the non gradient diffusion. Finally, the fourth is related to the turbulence model itself.

The third section shall enter deeper in the problem of turbulent combustion model, showing how a new model based on a balance equation for the dissipation rate for the fluctuations of the "progress variable" can be derived. This model is suitable only for "wrinkled" flames, that is for fast chemistry, for the moment. But it takes into account possible unsteady properties of the flame, and allows for non gradient diffusion.

The fourth part will deal with revisiting the k-ϵ turbulence model. It is shown that this model can give unrealistic results (namely $\widetilde{v'^2} < 0$!) due to its inability to handle compressed turbulence; Based on results of direct numerical simulation and rapid distortion theory, a new model where algebraic stress modelling is added to k-ϵ equations is proposed.

2. Classical models and their problems

2.1 THE PROBLEM OF AVERAGING

Any theoretical model for any turbulent flow, and consequently for turbulent combustion, can predict only one "averaged flow", and not one unsteady individual experiment. This follows because the Navier-Stokes equations are unstable to any small scale perturbations in the boundary or initial conditions. For turbulent flows that are steady in the average, the theory can deal with time-averaged quantities (and the average can be density-weighted or not) ; for flows that are spatially homogeneous in the average, spatial averaged quantities can be considered. For turbulent flows neither steady nor homogeneous (in the averaged sense), ensemble averaged quantities can be used. Ensemble averaged quantities are classically considered in theoretical models for I.C. engines, and the ensemble of event that is considered is the ensemble of cycles of four strokes.

But considering this type of average is not necessary, and is probably not the most convenient way of doing. Indeed,any type of average can be used in theoretical calculations provided that the averaged quantities do have a mathematical existence. Some of the possible averaged quantities would be without any interest in practice, and have consequently to be avoided : for instance, the time averaged quantity would give only a mean value of any quantity within one cycle, and would be unable to bring some information on the combustion period itself... Similarly, one can often question on the relevance of ensemble averaged quantities. On the other hand, any averaging procedure leads to unclosed equations (because the Navier-Stokes equations are non linear), and must to be complemented with some model that closes the set of equations. It is clear that the type of average that will lead to the simplest and the most approximate models will be the best one. With this in mind, one conceives that any averaging procedure that includes fluctuations caused by two or three different physical phenomena, has to be combined with complex closure assumptions, because they have to take into account all these different phenomena. This can be the case for ensemble averages, having to include cyclic variations provoked by turbulence, but also by some remaining gases from the previous cycle, and by the spark variability ; these three causes are not at all similar in nature, and then the building of one simple model including all of them is not an easy task.

Let us assume that we can consider an engine with a very fine reproductibity of the spark, and that we are sure that the residual gases are mixed to a very small scale with the fresh mixture. Would we be perfectly happy with ensemble averages ? Our answer is no ! Indeed, the turbulent spectrum is very broad, involving some very large length scales : only the scales smaller than a certain value are sufficiently intrinsical and free of the influence of the actual geometry of the engine. Only an averaging procedure that contains only these small scale fluctuations can lead to sufficiently universal models (i.e. with modelling constants that are really constant !). This point is relevant for any turbulent flow, and not only for flows in I.C. engines ; but in the case of relatively simple turbulent flows (like planar jets, round jets, boundary layers...), the close resemblance of large scales within each of these classes of flows allows to find relatively universal models, even if the model includes the largest scales. But the different types of I.C. engines display sufficiently different geometries to prevent the building of a sufficiently universal model. The unsteady character of the flow increases the difficulty, because the "universal" properties are attained only after some transient behavior., and the time is lacking, in

I.C. engines to be sure that a quasi-steady régime isobtained.

The consequence is that a satisfactory model of turbulent combustion has probably to lie in the category of "subgrid scale models", which have to be used in some kind of "large eddy simulation". Only a part of the turbulent fluctuations, the fluctuations whose length scales are below a given size, have to be modelled while the large scale fluctuations are calculated numerically. Such an approach is now used for classical turbulent flows (see ref [12]); Very often, the cut-off lengthscale has been chosen very small, and very simple models for the eddy-diffusivity have been used. But this is not the only choice for L.E.S. simulations, and because we need to keep as large as possible the grid size for computations in I.C. engines, more elaborate subgrid scale models are needed. It can even be considered that a convenient SGS model can be similar with a classical turbulent combustion and turbulence model, with the same closure formulae and eventually the same additional partial differential equations ; but the difference will lie in the value of the modelling constants, as in ref [13].

The use of such a "large eddy simulation" for I.C. engines will allow, hopefully, to apply the same model with the same constants to very different engine geometries. In addition, they will be able to reproduce the main characteristics of cylic variations, when the calculation are submitted to adequate perturbations of initial conditions (and ignition characteristics). However, the comparison of their results with ensemble averaged experiments will demand also ensemble averaged calculations with randomly perturbed initial conditions, and will be consequently more time consuming.

At present time, almost all numerical calculations of turbulence and turbulent combustion that have been performed for I.C. engines are not such L.E.S. calculations. Some of them have been compared with experiments, showing a relatively good agreement. This does not contradict the previous discussion. A classical model, with some "adjusted constants", can indeed represent quite well a relatively small class of experimental results. But such a model has to be considered only as preliminary ; it has to be improved by simultaneously reducing the grid size and changing the modelling constants (reducing the "eddy viscosity", in particular), and this is necessary in order to enlarge its validity domain for different geometries.

2.2. RESULTS WITH CLASSICAL MODELS OF "EDDY BREAK UP" TYPE

The "Eddy Break Up" model leads essentially to calculate the mean reaction rate for one "mean progress variable" $\tilde{c}$ (average weighted by the density) as :

$$\widetilde{\dot{W}_C} = + \frac{C_{EBU}}{\tau_t} \tilde{c}\,(1 - \tilde{c}) \tag{1}$$

$\tilde{c}$ is zero in the unburned mixture and one in the fully burned gases. τ_t is a characteristic turbulence time scale, to be provided from any turbulence model. C_{EBU} is a modelling constant.

The balance equation for $\tilde{c}$ is classical :

$$\frac{\partial \bar{\rho}\, \tilde{c}}{\partial t} + \frac{\partial}{\partial x_\alpha} (\bar{\rho}\, \tilde{u}_\alpha\, \tilde{c}) = \frac{\partial}{\partial x_\alpha} \left(-\bar{\rho}\, \widetilde{u'_\alpha c'}\right) + \bar{\rho}\, \widetilde{\dot{\omega}_c} \tag{2}$$

and $\widetilde{u'_\alpha c'}$ has to be modelled also, for instance with a diffusion coefficient :

$$\widetilde{u'_\alpha c'} = - \frac{C_\mu}{S_c} \frac{k^2}{\epsilon} \frac{\partial \tilde{c}}{\partial x_\alpha} \tag{3}$$

C_μ and S_c being the well known constants.

Equations (1), (2), (3), together with the k-ϵ model and the continuity, Reynolds and energy equations, allow to compute the flow field and temperature field.

The first version of the model, due to D.B. Spalding [5] used the square root of $\tilde{c}\,(1 - \tilde{c})$ (equal to $\widetilde{c'^2}$ when the instantaneous c is either 0 or 1), but the power unity has been revealed necessary from the paper of Bray and Moss [6]. In recent papers of Bray et al. (for instance [14]), the constant and the turbulence time scale have been found to display an additional dependance on $\tilde{c}$, which would disappears in the (academic) case of no heat release.

The timescale τ_t is not well precisely defined in (1). All we know about it is that it has to be proportional to k/ϵ, if we use a k-ϵ turbulence model, or the another time sacle of the turbulence, if we use an other model. D.B. Spalding, in [5], used directly the modulus of the transverse velocity gradient as $1/\tau_t$, in agreement with the Prandtl mixing length model.
Nevertheless, the use of (1), with C_{EBU} adjusted, is enough to calculate realistic flow field and flame shape in I.C. engines. Fig. 1 shows, as an example, the flame shape computed with such a model in the "cubic piston engine" of MIT. The details of calculations are in ref.[15]. It has to be

emphasized, however that this model alone cannot always reproduce well the ignition period, when the flame size is less than 0.5 or 1 cm ; some additional adjustment has to be done, for instance by imposing an ignition crank angle that is not the real one ...

Anyway, if we concentrate only on the flame propagation period, the formula (1) is not as powerful as fig. 1 seems to show because it cannot reproduce satisfactorily the influence of the equivalence ratio of the initial mixture, or the influence of changing the turbulence kinetic energy at the time of the spark. This can be put into evidence particularly in studying the flame propagation in a constant volume combusiton chamber, in which a given turbulence has been established before the ignition. Such an experiment has been built and studied in ref [16] ; a schematic drawing of the chamber is given fig. 2. In this device, it is easy to change the equivalence ratio and/or the turbulence kinetic energy at the ignition, and so the influence different ratios ($k^{1/2}/U_L$), where U_L is the laminar flame velocity, can be tested ; in addition, due to the compression of unburned gases, U_L increases during the flame propagation and $k^{1/2}/U_L$ decreases. The use of formula (1), with the same constant EBU, together with the k-ϵ model and numerical integration of the full set of averaged balance equations, allows to calculate the mean flame propagation in this device. The details of such a calculation are given in ref [17]. The "trajectories" of the flame are shown fig. 3, for three different cases. If C_{EBU} is adjusted for one case, the calculation does not follow well the experiments in other cases.

Similar discrepancies have been found also when the same model has been applied to combustion in I.C. engines. But in such a complex case, because we do not know really what is the turbulence kinetic energy and length scale (or time scale) at the time of the spark, it is more difficult to point out the deficiency of the model.

2.3. DISCUSSIONS ON THE SHORTCOMING OF THE "EDDY BREAK UP" MODELS

The models assuming that the chemistry is very fast are often considered as perfectly insensitive to any chemical charecteristic, and consequently the lack of sensitivity to equivalence ratio can be thought as normal. But this is true only if k^{s}/U_L is very large, and for internal combustion engines, although the turbulence kinetic energy may be large (of the order of 1 to 4 m/sec for $k^{1/2}$), due to the high temperature of the unburned compressed gases U_L is also of the same magnitude. In this case, the propagation of the flamelets within the mean flame brush plays an important role (see ref [18] on the structure of premixed flames, and ref [19] for a numerical simulation) ; one can then conceive that C_{EBU} depends on $k^{1/2}/U_L$. This effect may explain the discrepancy of fig. 3, and may allow to remove it. Indeed, in fig. 3 are plotted new curves calculated with using in the C_{EBU} of eq. (1) a dependance on $k^{1/2}/U_L$ found from the simulation of ref [19] ; the agreement is now very satisfactory.

This influence of $k^{1/2}/U_L$ is neither due to any finite rate chemistry effect, nor to the stretching of flamelet that can induce local extinctions ; here the value of $k^{1/2}/U_L$ is of order one, and the Karlovitz number (τ_C/τ_K) is sufficiently close to zero in order to prevent any extinction or finite chemistry effect (see ref.[18]). The influence of $k^{1/2}/U_L$ is just due to the propagation of flamelets (with respect to the unburned gases) that influences the flamelet surface area per unit of volume within the flame brush. The mean reaction rate per unit of volume, when local extinctions are negligible, is nothing but the mean flamelet surface area per unit of volume (say Σ) multiplied by U_L. Σ is clearly proportional to $1/l_t$, where l_t is the integral length scale of the turbulence, but is also influenced by $k^{1/2}/U_L$; when $k^{1/2}/U_L$ decreases, the surface area decreases because it is less wrinkled (see ref [19]) . So the mean reaction rate is expected to be U_L/l_t multiplied by a function of $k^{1/2}/U_L$, that is $1/\tau_t$ times a function of $k^{1/2}/U_L$.

This discussion is only a rough qualitative discussion ; more quantitative justification is needed, and can be attempted considering the "flamelet surface area per unit of volume" in more details. The new model that will be discussed in the third section will deal with this question.

Another important shortcoming of the classical models is their inability to represent satisfactorily the ignition period. There are three different aspects here : first, the spark release some additional energy in the flow, and the gaseous medium, during the spark operation, is no more adiabatic. The duration of the spark, in practice, is not very short : durations of the order of 1 m/sec are normal, corresponding to more than 10 degrees of crankangle ! Secondly, the chemistry is not always fast in this period, specially for low equivalence ratio where the spark is not able to ignite the flame ! Thirdly, the flame brush does not experience suddently all the turbulence fluctuations, and in addition its response time to these fluctuations is not very small ; it is probably of the order of the turbulence time scale, that is again 1 or 2 milliseconds, and during this delay time the flame brush displats a transient evolution. It is possible to consider the transient evolutions of the flame brush in the framework of infinitely fast chemistry ; this will be done in the third section of this paper. But the other aspects described here do need to release the fast chemistry assumption ; they can be studied in the framework of PDF models with finite rate chemistry. We will not consider these problems here, although they are very important in practice.

The presence of counter-gradient diffusion is also an important question that can contradict classical models using a diffusion coefficient as in (3). This counter-gradient diffusion has been the core of the BML model ; it has been experimentally shown by Sheperd and Moss [20] in a Bunsen burner. The presence of this effect in propagating quasi-planar flames has not been proved, but we can expect that it will exist when considering the balance equations for the turbulent mass flux itself, at least in the second part of the flame brush, on the products side. This will probably affect quite significantly the structure of the flame brush. Anyway, even without counter-gradient diffusion, it appears from the balance equation for the flux that the reaction will possibly change the diffusion coefficient when it exists. The model described in section three will take into account, to some extend, this fact.

Classical Eddy Break Up like models are also deficient close to walls. There are two reasons for that. First, their insensitivity to temperature prevent them to take into account the heat losses and the quenching close to walls. Second, even with adiabatic walls, one cannot claim, in principle, that chemistry is very fast close to the wall : indeed, in an usual boundary layer τ_t is decreasing to zero close to the wall, and this will lead necessarily to the point where the chemical characteristic time will be lower than τ_t, and consequently the structure of the flame will change and the Eddy Break Up model will have to be released. The remedy to this situation is to build a model valid outside the domain of wrinkled flames ; this is not easy, because it has to take into account flamelet stretching, flamelet interactions, and also heat losses.

The presumed PDF model of [8] is one candidate for that, but very crude. This problem of finite rate chemistry will occur also if we would like to calculate the formation of pollutants in I.C. engines. Generally speaking, the PDF models are well suited when finite rate chemistry is invoked ; but there is a lot of work to do with them.

Finally, the last problem to be discussed here concerns the turbulence model. The k-ϵ model is well known to be very robust, and often realistic. But the turbulent flows within the cylinders of I.C. engines are very peculiar, specially within the combustion period. Very often, the turbulence models have been tested during the intake stroke,with the valve open (see for instance [21]) : the flow is then very similar to a circular jet surrounding a recirculation zone. The swirling character of the flow, with a "swirl" or a "tumble" have also been studied. But the turbulence in the combustion period is only a small part of the turbulence in the intake stroke. The turbulence is further dissipated, compressed by the piston, and then compressed and expanded at different location during the flame propagation period. Very few is known on the behavior of the real turbulence, as well as the modelled turbulence, during the propagation of the flame brush in an enclosure. We will discuss this point and show some results in the fourth section of this paper.

3. A new model for turbulent wrinkled flame propagation

3.1. PRESENTATION OF THE NEW MODEL

We consider here only the case of wrinkled flames, when the turbulent flame brush is made with flamelets that are wrinkled and displaced by the turbulent fluctuations (see Borghi [18] for more details on this régime). In this case the modelling of the physical mechanisms involved needs to consider, on one side the extent of the flamelets, i.e. the surface area of flamelets by unit of volume, and on the order side the internal structure of flamelets ; if the turbulence is not too strong in such a way that a characteristic chemical time is much smaller than τ_K, the Kolmogorov time, the strain experienced by the flamelet is too weak to perturb their velocity, and their surface area per unit of volume, say Σ, is the controlling quantity. It can be shown that Σ is directly proportional to the dissipation rate of the fluctuations of the progress variable ; this is discussed in ref [22]. It can be shown also that the mean reaction rate $\widetilde{\omega}_c$ is nothing but $U_L \overline{\Sigma}$, where $\overline{\Sigma}$ is the mean value of Σ. This allows to build a new model of turbulent combustion based on a balance equation for the mean dissipation rate $\widetilde{\epsilon_c}$. Such a model has been proposed recently, see ref [23], in the (academic) limit of constant density flame ; it is presently being extended to the general case. An improvement of this model avoiding any gradient assumption for the turbulent flux of mass has been also proposed in [24].

3.2. THE MODEL

The set of equations that concerns the mean progress variable $\tilde{c}$ is the following :

$$\frac{\partial}{\partial t}(\bar{\rho}\,\tilde{c}) + \frac{\partial}{\partial x_\alpha}(\bar{\rho}\,\widetilde{u_\alpha}\,\tilde{c}) = \frac{\partial}{\partial x_\alpha}\left(-\bar{\rho}\,\widetilde{u'_\alpha c'}\right) + \bar{\rho}\,\frac{\tilde{\epsilon}_c}{\frac{1}{2} - b_0} \tag{4}$$

$$\frac{\partial}{\partial t}\left(\bar{\rho}\,\widetilde{\epsilon_c}\right) + \frac{\partial}{\partial x_\alpha}\left(\bar{\rho}\,\widetilde{u_\alpha} - \widetilde{u_{l\alpha}\epsilon_c}\right) = \frac{\partial}{\partial x_\alpha}\left(\frac{\nu_t}{\sigma_{\epsilon c}}\frac{\partial \tilde{\epsilon}_c}{\partial x_\alpha}\right) + C_{pu}\frac{\epsilon_c}{k}\frac{\nu_t}{\sigma_t}\frac{\partial \widetilde{u_\alpha}}{\partial x_i}\frac{\partial \widetilde{u_\alpha}}{\partial x_i}$$

$$+ C_{pc}\frac{\epsilon}{k}\frac{\nu_t}{\sigma_c}\frac{\partial \tilde{c}}{\partial x_\alpha}\frac{\partial \tilde{c}}{\partial x_\alpha} + \alpha_0\,Re_T^{1/2}\,\bar{\rho}\,\tilde{\epsilon}_c\frac{\epsilon}{k} - \beta_0\,Re_T^{1/2}\,\bar{\rho}\,\frac{\tilde{\epsilon}_c^{\,2}}{\tilde{c}\,(1-\tilde{c})}\left(1 + C_u\frac{U_L}{k^{1/2}}\right)^{-2\gamma} \tag{5}$$

and

$$\frac{\partial \bar{\rho}\,\widetilde{u'_\alpha c'}}{\partial t} + \frac{\partial}{\partial x_\beta}\left(\widetilde{u'_\alpha c'}\,\widetilde{u_\beta}\right) = \frac{\partial}{\partial x_\beta}\left(\bar{\rho}\,\frac{\nu_t}{\sigma_c}\frac{\partial \widetilde{u'_\alpha c'}}{\partial x_\alpha}\right) - \bar{\rho}\,\widetilde{u'_\alpha u'_\beta}\frac{\partial \tilde{c}}{\partial x_\beta} - \bar{\rho}\,\widetilde{u'_\beta c'}\frac{\partial \widetilde{u_\alpha}}{\partial x_\beta}$$

$$+ \overline{c'}\,\frac{\partial \bar{p}}{\partial x_\alpha} \quad - C_D\frac{\bar{\rho}\,\widetilde{u'_\alpha c'}\ \tilde{\epsilon}_c}{\tilde{c}\,(1-\tilde{c})} + \frac{\bar{\rho}\,\widetilde{u'_\alpha c'}\,b_0}{\left(\frac{1}{2} - b_0\right)\tilde{c}\,(1-\tilde{c})} \tag{6}$$

b_0, C_{pu} , C_{pc} , α_0, β_0, C_u, σ_c, $\sigma_{\epsilon c}$, C_D are modelling constants, whose tentative values are given in [24]. $\widetilde{u_{l\alpha}}$ is the mean value of the projection of the instantaneous velocity of propagation of the flamelets in the direction of α axis ; a tentative formula for it is given in [24]. These equations are just transposed from constant density equations ; the influence of density fluctuation shas to be assessed. In the case of wrinkled flames (where $c = 0$ or $c = 1$) $\overline{c'}$ is just $\bar{\rho}\,\tilde{c}\,(1-\tilde{c})\left(\frac{1}{\rho_b} - \frac{1}{\rho_0}\right)$. Of course, these equations have to be complemented with a turbulence model and the continuity and Reynolds equations.

One recognize that (4) replaces (2), with the $\widetilde{w}_c$ related to $\widetilde{\epsilon}_c$. The equation for $\widetilde{\epsilon}_c$ is almost directly an equation for Σ ; it takes into account the stretching effect of turbulence, and the influence of the propagation of flamelets is embedded in the last term, the one involving $U_L/k^{1/2}$. More details on the justifications of eq (5) (which is a model !) can be found in [23].

It is interesting to notice that the equation (5) can be reduced to an algebraïc equation for $\widetilde{\epsilon}_c$ when some of the productions and dissipation terms of the right hand side are predominant. For instance, when the two terms with the Reynolds number dominate, we obtain simply :

$$\widetilde{\epsilon_c} = \frac{\alpha_0}{\beta_0}\frac{\epsilon}{k}\left(1 + C_u \frac{U_L}{k^{1/2}}\right)^{2\gamma} \tilde{c}\,(1 - \tilde{c}) \tag{7}$$

Keeping in mind that $\widetilde{\epsilon_c} / \left(\frac{1}{2} - b_0\right)$ is nothing but $\tilde{\dot{w}}_c$, we recover here some kind of generalization of the Eddy Break Up formula (1), but with a dependance on $U_L/k^{1/2}$.

On the other hand, a similar limit can be taken for the turbulent flux equation. For instance, if the $\frac{\partial \tilde{c}}{\partial x_\beta}$ term is the predominant production term, we have :

$$\widetilde{u'_\alpha c'} = - \frac{\widetilde{u'_\alpha u'_\beta}\, \tilde{c}\,(1 - \tilde{c})}{\left[C_D - \frac{b_0}{\frac{1}{2} - b_0}\right] \widetilde{\epsilon_c}} \cdot \frac{\partial \tilde{c}}{\partial x_\beta} \tag{8}$$

The formula (8) allows to define diffusion coefficients, but they clearly depend on the chemical reactions, through b_0, and U_L, through $\widetilde{\epsilon_c} / \tilde{c}\,(1 - \tilde{c})$, see (7).
In addition, if the term $\overline{c'} \frac{\partial \overline{p}}{\partial x_\alpha}$ is not negligible, it can provoke a non gradient diffusion, according to the discussion of Bray et al. [4].

3.3. SOME EXAMPLE OF APPLICATIONS OF THE NEW MODEL

The $\widetilde{\epsilon_c}$. model is only in a first phase of testing, and the values of the constants as well as the precise expressions of some terms cannot considered as definitely chosen. However, first tests in simple cases are very encouraging.
The following results are related to the case of the propagation of a turbulent flame brush after ignition from a point. The academic assumption ρ = cste has been chosen, and the turbulence has been considered given and constant.
Fig. 4 shows the profiles of $\tilde{c}$ during the propagation of the flame brush. The radius is given in mm, but the physical lengthscale is l_t, the integral lengthscale of the turbulence, chosen here as 1 mm ; the physical timescale is $l_t/k^{1/2}$, here chosen as 1 msec ; $k^{1/2}/U_L$ has been, here, equal to 3.37. Fig. 5 shows the profiles of $\overline{\epsilon_c}$, and fig. 6 the profiles of $\overline{u'c'}$. The curves show that, after a transient period, a quasi-steady propagation occurs ; the variations of $\overline{\epsilon_c}$ and mainly $\overline{u'c'}$ in the transient period are quite significant.

More details can be found in ref [24]. It is shown also that the velocity of propagation of the turbulent flame is in agreement with experimental knowledge (due to the right choice of some of the modelling constants).

4. Test of turbulence models in a constant volume combustion chamber

4.1. SOME RESULTS WITH THE CLASSICAL K-ϵ MODEL

The k-ϵ model is widely used for computations in I.C. engines, but it has been mainly tested during the intake stroke, for the calculation of the flow without any flame. Indeed, we don't know if this model is realistic in a highly strained flow field as a flow field created by a turbulent flame propagation in a constant volume chamber is.

The turbulent flame propagation in the experimental device of fig. 2 constitutes a good test case for turbulence models, even with a classical Eddy Break Up model (with the constant adequately adjusted). Such a calculation has been done and the results are shown fig. 7 to fig. 9. Fig. 7 shows, at given times during the propagation, the flame brush (indicated with iso $\tilde{c}$ lines) superimposed with the mean velocity field. We remark that the flame brush propagates, but it is very thick (approximately 3 cm) and its shape is not regular. Fig. 8 shows the field of turbulent kinetic energy ; essentially there is an increase of k in the burned gases close to the end wall, and also in the fresh mixture ahead of the flame in the "squish " region, where the wall and the flame are not perpendicular. Fig. 9 shows the $\widetilde{v'^2}$ component of k only ; it appears clearly that in one location of high k, $\widetilde{v'^2}$ is negative ! This is clearly unrealistic, and due to the Boussinesq relations used in k-ϵ model.

One could argue that we are only interested in k and in ϵ, and not in $\widetilde{v'^2}$. But $\widetilde{u'^2}$, $\widetilde{v'^2}$, $\widetilde{w'^2}$ do play an important role in the production terms of the k equation. So if $\widetilde{v'^2}$ is evidently wrong, these terms and consequently k are wrongly calculated !

If we consider a simple one dimensional flow compressed in such a way that only $\frac{d\tilde{u}}{dx}$ exist, the Boussinesq relations give :

$$-\widetilde{u'^2} = \frac{4}{3} \nu_t \frac{d\tilde{u}}{dx} - \frac{2}{3} k, \text{ with } \nu_t = C\mu \frac{k^2}{\epsilon}$$

As both $\widetilde{u'^2}$ and $\widetilde{v'^2}$ are to be positive, that implies that

$$-\frac{1}{C_\mu} \leq \frac{k}{\in}\frac{d\tilde{u}}{dx} \leq \frac{1}{2C_\mu}$$

This emphasizes that the classical k-∈ can work only if the flow distorson timescale $\left|\frac{d\tilde{u}}{dx}\right|^{-1}$ is not too large with respect to the turbulence timescale .

In the constant volume combustion chamber, the flame produces strong compressions or expansions of the flows in different parts of the chamber ; if these distorsions of the flow are too strong, that leads to a failure of the classical k-∈ model.

4.2. A NEW ASM MODIFIED K-∈ MODEL

In order to try to correct this shortcoming, it is necessary to use more exact informations concerning the components of the Reynolds stress. Algebraïc stress models are the simplest way to get such informations, without needing additional partial differential equations. If we follows the classical procedure, [25] and [26], one can obtain

$$\widetilde{u_i' u_j'} = k\left(\frac{2}{3}\delta_{ij} + \frac{(1 - C_2) P_{ij} - \frac{2}{3}\delta_{ij} P_k}{C_1 \in + P_k - \in - C_6 k \frac{\partial \tilde{u}_K}{\partial x_K}}\right) \tag{9}$$

with $P_{ij} = -\widetilde{u_i' u_k'}\frac{\partial \tilde{u}_j}{\partial x_k} - \widetilde{u_j' u_k'}\frac{\partial \tilde{u}_i}{\partial x_k}$ and $P_K = -\widetilde{u_i' u_j'}\left(\frac{\partial \tilde{u}_i}{\partial x_j}\right)$

C_1, C_2 and C_6 are coefficients usually taken as constants. C_1 and C_2 are 2.5 and 0.6 respectively, from comparisons with experiments in shear layers with $\frac{\partial \tilde{u}_K}{\partial x_K} = 0$. C_6 is directly related to the effect of comparison or expansion.

The rapid distortion theory can give some informations for choosing C_6 [27]. When we identify its results with the Reynolds stress equations, neglecting dissipation, it is possible to obtain

$C_6 = 0.6(\widetilde{u_i'^2}\frac{\partial \tilde{u}_i}{\partial x_i})/\left(k\frac{\partial \tilde{u}_k}{\partial x_k}\right)$ for uni-axial compression. Other expressions are found for expansion, or for spherical distorsion. We will consider here that uni-axial compression is well representative of the flow distortions induced in the cylinder during the propagation of the flame ; the release of this approximation would imply to use a full second order model.

The final equation for $\tilde{k}$ is then :

$$\frac{\partial}{\partial t}\bar{\rho}\,\tilde{k} + \frac{\partial}{\partial x_j}(\bar{\rho}\,\tilde{u}_j\,k) = \frac{\partial}{\partial x_j}\left(\bar{\rho}\frac{\nu_t}{\sigma_k}\frac{\partial k}{\partial x_j}\right) - \bar{\rho}\widetilde{u_i u_j}\frac{\partial \tilde{u}_i}{\partial x_j} - \bar{\rho}\epsilon \tag{10}$$

with equation (9) giving the Reynolds stress components.

If the turbulence is "distorted" strongly with high $\frac{k}{\epsilon}\frac{\partial \tilde{u}_K}{\partial x_K}$, the turbulence is not in an usual equilibrium state, so the usual ϵ equation cannot be kept as valid, and has also to be modified. This is the reason why a new "compressibility term" has to be added ; from the first work, it has been written under the form $C_3\bar{\rho}\frac{\partial \tilde{u}_k}{\partial x_k}\epsilon$. Usually, C_3 is estimated also from the rapid distorsion theory (see for instance [28]), but also some influence of the variation of the molecular viscosity seems also to be taken into account [29]. The values of C_3 that have been recommanded so far range from 0.04 to -1.333, dependig on the assumptions used . Here we will consider the results of Direct Numerical simulations of Wu et al. [30], and their three equation model that agree well with DNS. Their model uses an additoinal time scale, say τ, together with k and ϵ , as characterising quantities for the turbulence. If we integate the three equation model in the case of compression of an homogeneous turbulent field, it is possible to find an explicit relation between τ and k and ϵ (depending also of the flow distortion, of course), and this relation can be used to get the additional "compressibility term" ; we obtain finally

$$C_3 = 0.293 + 0.56\,P_c\;/\,k\frac{\partial \tilde{u}_K}{\partial x_K} \quad \text{with} \quad P_c = \frac{4}{3}C_\mu\frac{k^2}{\epsilon}\left(\frac{\partial \tilde{u}_k}{\partial x_k}\right)^2 - \frac{2}{3}k\frac{\partial \tilde{u}_k}{\partial x_k}$$

For spherical compression that gives C_3 = - 0.666 and the value of rapid distorsion theory with variable viscosity gives - 0.633. More details concerning these calculations can be found in [31].

The ability of this new ASM/k-ϵ model to reproduce the evolution of compressed homogeneous turbulence is shown on figures 10. This new model is compared to DNS, and to the full Reynolds stress equations that include the new coefficients C_3 and C_6.

4.3. CALCULATIONS WITH THE NEW MODEL OF THE PROPAGATING FLAME

This new model has been implemented into the computer code in order to calculate the experiment of fig. 2. The new flow field and flame shape is shown fig. 11. One see the modification of the flame shape, if we compare with fig. 7 obtained with the standart k-ϵ model : the flame looks thiner in the middle of chamber ; their is a thickening close to lateral well, which shows that the model is probably not perfect. Fig. 12 shows the turbulent kinetic energy ; there are large differences with fig. 8 ; one see at the first look that there is no more turbulence created at the closed end of the chamber...

It cannot be claimed now that this new model is fully satisfactory ; we need for that more detailed comparisons with measurements of k in burned gases behind the flame brush.

5. Concluding remarks

Although very simple models based on the so-called "Eddy Beak Up" model of D.B. Spalding are able to give very realistic shapes of flames and flows, the quantitative calculation of turbulent premixed flame propagation within the cylinder of I.C. engines demands several further studies.

We have discussed here in detail the different aspects of the needed modelling : ignition, wrinkled flame propagation, flame wall interaction, non-gradient diffusion, turbulence, finite rate chemistry effects.

We have presentetd the results of the recent work that has been done in our laboratory concerning some of these aspects namely flame propagation including possible non gradient diffusion and turbulence. They clearly show that the problems are not minor ones, but they appear as very encouraging.

Acknowledgments : This paper is based on studies supported by the European community, within the programme Joule I, and also by the CNRS and the ADEME, in France.

6. References

[1] SUMMERFIELD, M. et al
The Structure and Propagation Mechanism of Turbulent Flames in High Speed Flows.
Jet Propulsion, **25**, p. 377 (1955)

[2] ZIMONT, V.L.
Theory of Turbulent Combustion of a Homogeneous Fuel Mixture at High Reynolds Numbers.
Fizika Goreniya i Vzryva, Vol. 15 n° 3, pp. 23-32, (1979).

[3] KARLOVITZ, B., DENNISTON, J.R., KNAPSCHAEFFER, D.H., WELLS, F.E.
4th Symp. (Int) on combustion,
The Combustion Institute, Pittsburgh, p. 613, (1953).

[4] BRAY, K.N.C., MOSS, J.B., LIBBY, P.A.
Turbulent Transport in Premixed Flames.in : Convective Transport and Instability Phenomena
Ed. J. Zierep, H. Oertel
Braun Verlag, Karlsruhe, (1982)

[5] SPALDING, D.B.
Mixing and Chemical Reaction in Steady Confined Turbulent Flames.
13th Symp. on Combustion.
The Combustion Institute, Pittsburgh, p. 649, (1971).

[6] BRAY, K.N.C., MOSS, J.B.
A Unified Statistical Model of Premixed Turbulent Flame.
Acta Astronautica, Vol. 4 n° 3-4, pp. 291-319, (1977).

[7] BORGHI, R., MOREAU, P.
Turbulent Combustion in a Premixed Flow.
Acta Astronautica, Vol. 4, pp. 321-341, (1977).

[8] BORGHI, R., DUTOYA, D.
On the Scales of the Fluctuations in Turbulent Combustion.
17th Symp. (Int) on Combustion,
The Combustion Institute, Pittsburgh, pp. 235-244, (1978).

[9] POPE, S.B.
The Statistical Theory of Turbulent Flames.

Philos. Trans. R. Soc.London,
Series A, Vol. 291, pp. 529-568, (1979).

[10] DOPAZO, C., O'BRIEN, E.E.
A Probabilistic Approach to the Autoignition of Reactive Turbulent Mixtures.
Acta Astronautica, Vol. 1, p. 1239, (1974).

[11] GOSMAN, A.D., JOHNS, R.J.R., WATKINS, A.P.
Development of Prediction Methods for in Cylinder Processes in Reciprocating Engines.
"Combustion Modelling in Reciprocating Engines"
Eds. J.N. Mattavi and C.A. Amann, Plenum, N.Y., pp. 69-129, (1980).

[12] DEARDORFF, J.W.
A Numerical Study of the Three-Dimensional Turbulent Channel Flow at Large Reynolds Number.
J. F. M. Vol. 41, part 2, p. 453, (1970).

[13] HA MINH, H. KOURTA, A.
Semi Deterministic Turbulence Modelling for Flows Dominated by Strong Organized Structures.
9th Turbulent Shear Flows, August 16-18, Kyoto, Japan, (1993).

[14] BRAY, K.N.C.
Studies of the Turbulent Burning Velocity.
Proc. R. Soc. London,
Series A, Vol. 431, pp. 315-335, (1990).

[15] BORGHI, R., ARGUEYROLLES, B., GAUFFIE, S., SOUHAITE, P.
Application of a "presumed p.d.f." Model of Turbulent Combustion to Reciprocating Engines.
21th Symp. (Int) on Combustion, pp. 1591-1599, Munich, (1986).

[16] FLOCH, A., TRINITE, M., FISSON, F., KAGEYAMA, T., WOON, C.H., POCHEAU, A.
Flame Propagation Behaviour in a Variable hydrodynamic Constant Volume Combustion Chamber. Presented at 9th ICDERS, Ann Artbor, (August 1989).

[17] DELAMARE, L., BORGHI, R., GONZALEZ, M.
The Modelling and Calculation of a Turbulent Premixed Flame Propagation in a Closed Vessel : Comparisons of Three Models with Experiments.
SAE n° 910265, SAE International Congress, Feb. 25-March 1, Detroit, (1991).

[18] BORGHI, R.
On the Structure and Morphology of Turbulent Premixed Flames in : Recent Advances in the Aersopace Sciences .
C. Casci (Ed.) , Plenum Pub. Corp., pp. 117-138, (1985).

[19] SAID, R., BORGHI, R.
A Simulation with a "Cellular Automaton" for Turbulent Combustion Modelling.
22nd Symposium (International) In Combustion,
The Combustion Institute, pp. 569-577, (1988).

[20] SHEPERD, I.G., MOSS, J.B.
Measurements of Conditioned Velocities in a Turbulent Premixed Flame.
19th *AIAA Aerospace Science Meeting*, St-Louis, Missouri, (1981).

[21] EL TAHRY, S.H., HAWORTH, D.C.
Directions in turbulence Modeling for in Cylinder Flows in Reciprocating Engines.
Journal of Propulsion and Power, vol.8, 5, pp.1040-1048, (1992)

[22] BORGHI, R.
Turbulent Premixed Combustion : Further Discussions on the Scale of Fluctuations.
Combustion and Flame, Vol. 80, pp. 304-312, (1990).

[23] MANTEL, T., BORGHI, R.
A New Model of Premixed Wrinkled Flame Propagation Based on a Scalar Dissipation Equation.
ICDERS, Nagoya, (August 1991).

[24] MANTEL, T., BORGHI, R., PICART, A.
Turbulent Premixed Flame Propagation Revisited : Results with a New Model.

9th Turbulent Shear Flows, Kyoto, August 16-18, (1993).

[25] JONES, W.P.
Models for Turbulent Flows with Variable Density and Combustion in : Prediction Methods for Turbulent Flows, Ed. W. Kollmann,
A Von Karman Institute Book, Hemisphere Pub. Corp. (1990).

[26] RODI, W.
ZAMM n° 56, p. 219, (1975).

[27] BORGNAKKE, C., XIA, Y.
SAE Paper 910260, (1991).

[28] COLEMAN, G.N., MANSOUR, N.N.
Phys. Fluids **A3**, (9), (1991).

[29] WU, C.T., FERZIGER, J.H., CHAPMAN, D.R.
Simulation and Modelling of Homogeneous Compressed Turbulent.
5th Symposium on Turbulent Shear Flows, Cornell University, August (1985).

[30] DELAMARE,L.
Modélisation numérique de la propagation d'une flamme turbulente en milieu confiné.
Thèse, Faculté des Sciences de ROUEN, (1992)

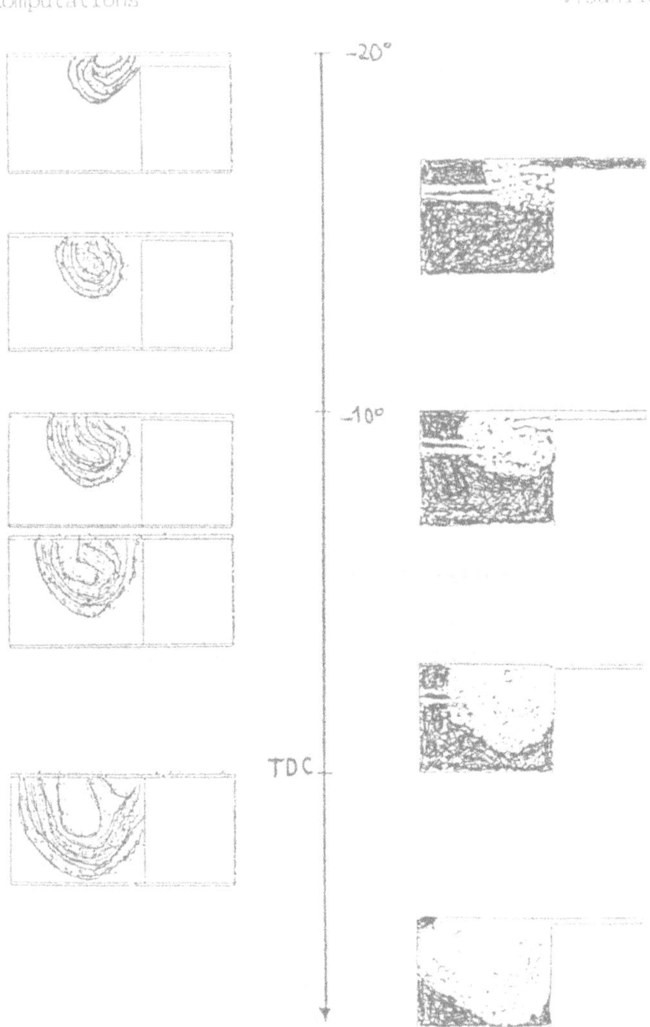

Fig. 1 : Calculations with the Eddy Break Up model in the square piston engine of M.I.T. : mean isothermal lines are compared with schlieren pictures.

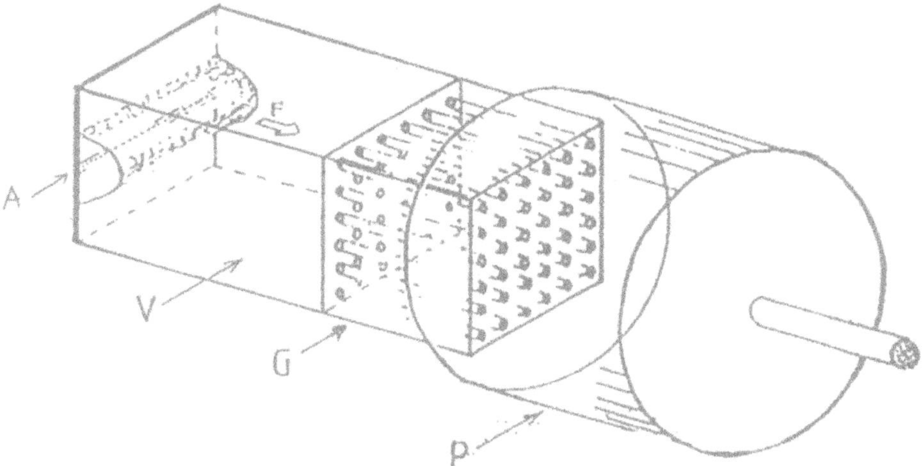

Fig. 2 : Sketch of the constant volume turbulent combustion chamber. A : multispark ignition ; V : large windows for visualization ; G : grid for turbulence generation ; P : Piston for chamber filling.

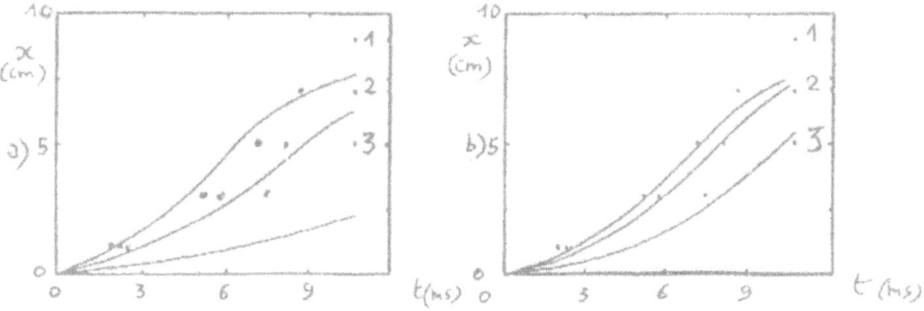

Fig. 3 : Mean position of the flame at different times in the experiment of Fig. 2. The position of the isotherm 600 K is measured on the symmetry plane of the chamber. 1 : $k^{1/2} / U_L)_0 = 1.95$; 2 : $k^{1/2} / U_L)_0 = 1.34$; 3 : $k^{1/2} / U_L)_0 = 0.8$ a) Calculation with the constant C_{EBU} ; b) Calculation with C_{EBU} function of $k^{1/2} / U_L$.

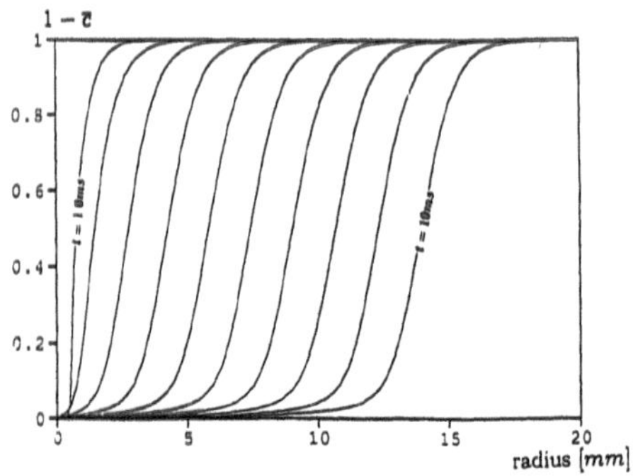

Fig. 4 : Flame brush propagation calculated from the ignition. 1 msec between each profile of $\overline{c}$. $k^{1/2} = 1$ m/s ; $l_t = 1$ mm.

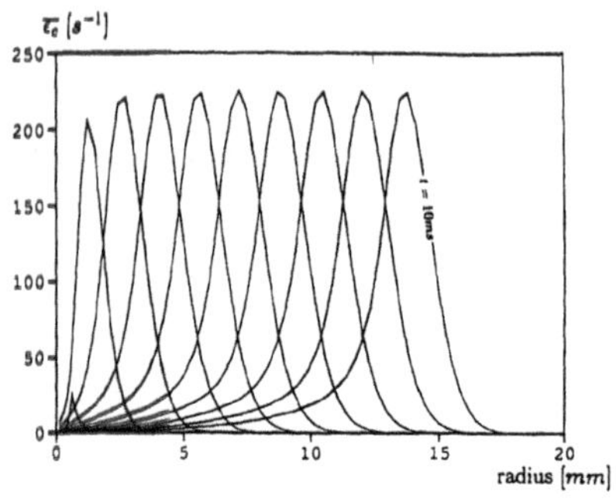

Fig. 5 : Profiles of $\overline{\epsilon}_c$ during the flame propagation. 1 msec between each profile.

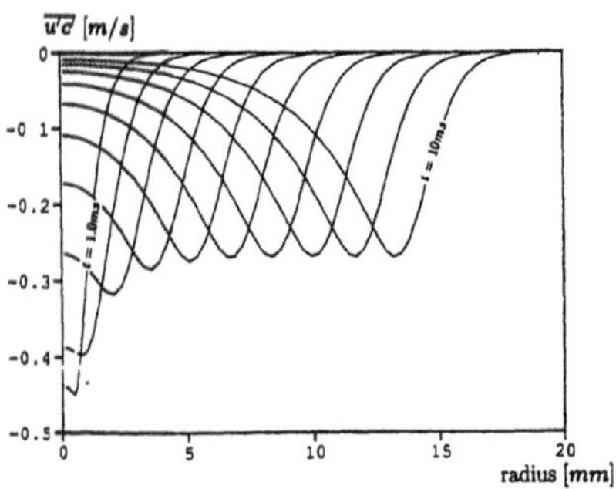

Fig. 6 : Profiles of $\overline{u'c'}$ during the flame propagation. 1 msec between each profile.

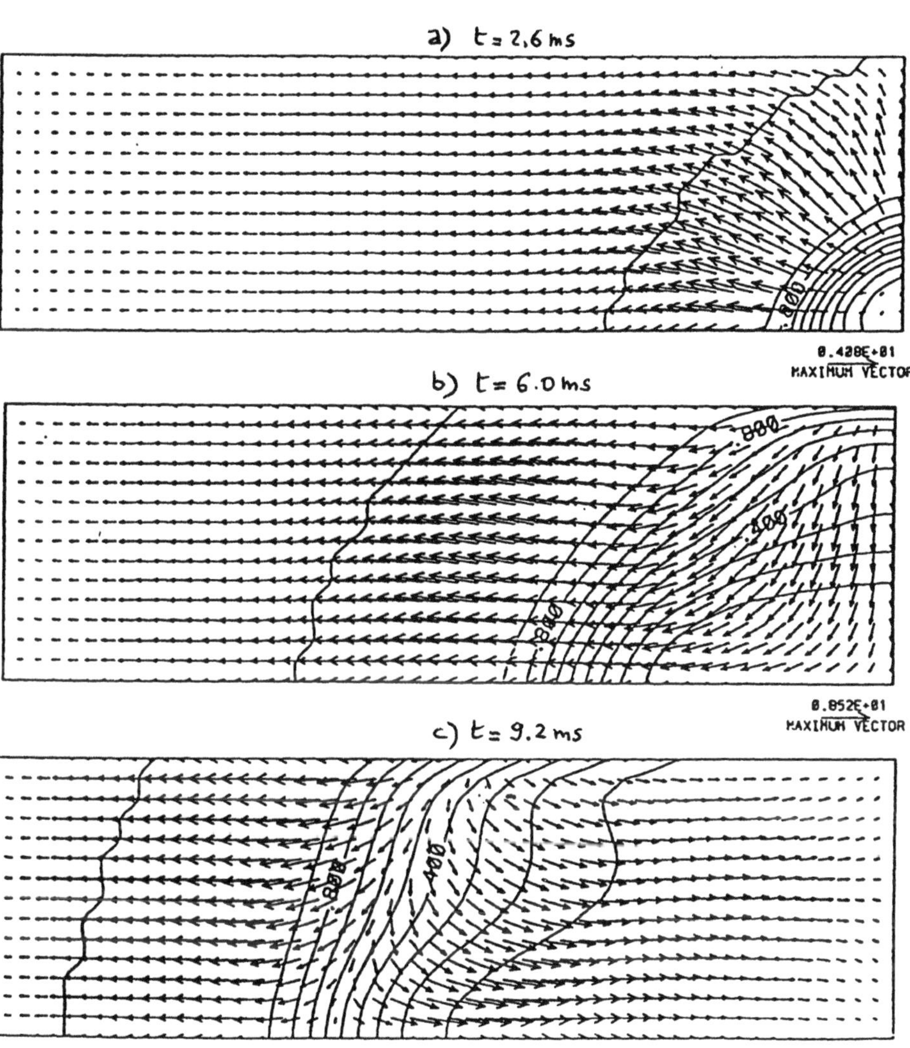

Fig. 7 : Calculated mean velocity and iso-$\tilde{c}$ lines in the chamber of Fig. 2. Only the top half is calculated ; the ignition is on the right. The classical k-ϵ model is used.

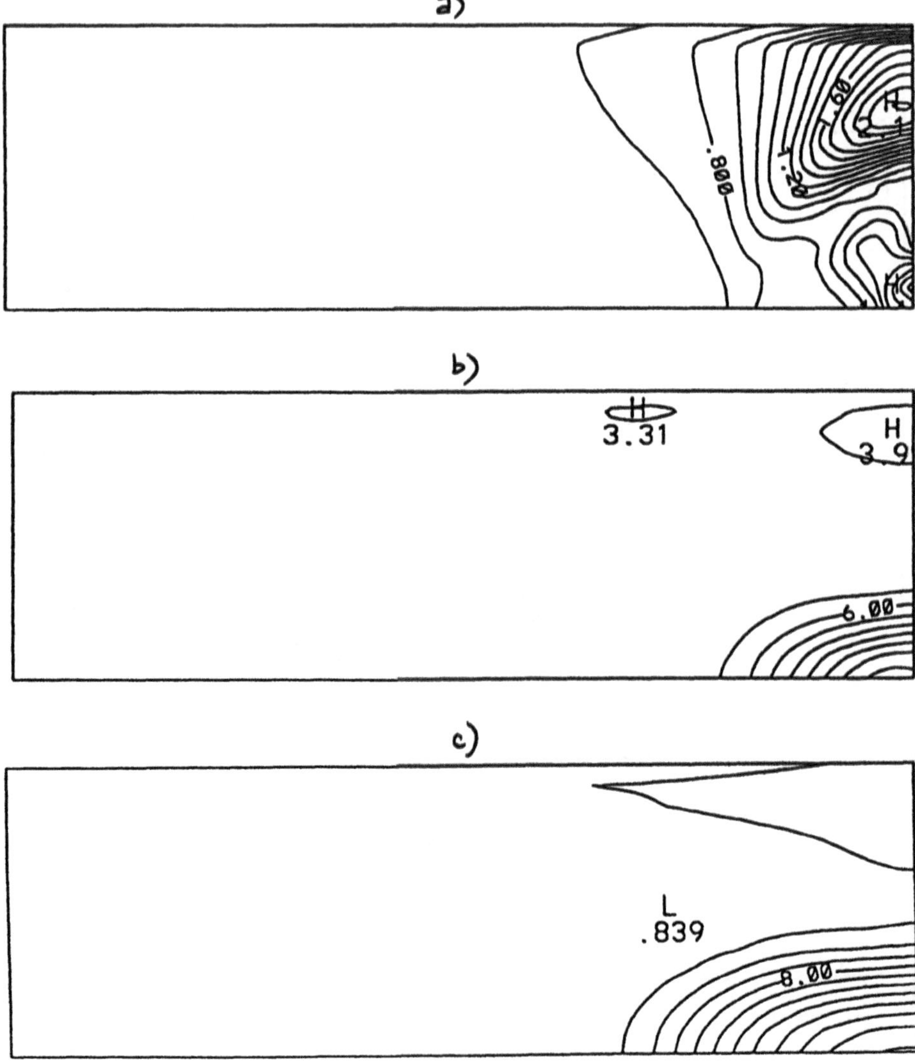

Fig. 8 : Calculated turbulent kinetic energy in the chamber, at the same instants as Fig. 7. The classical k-∊ model is used.

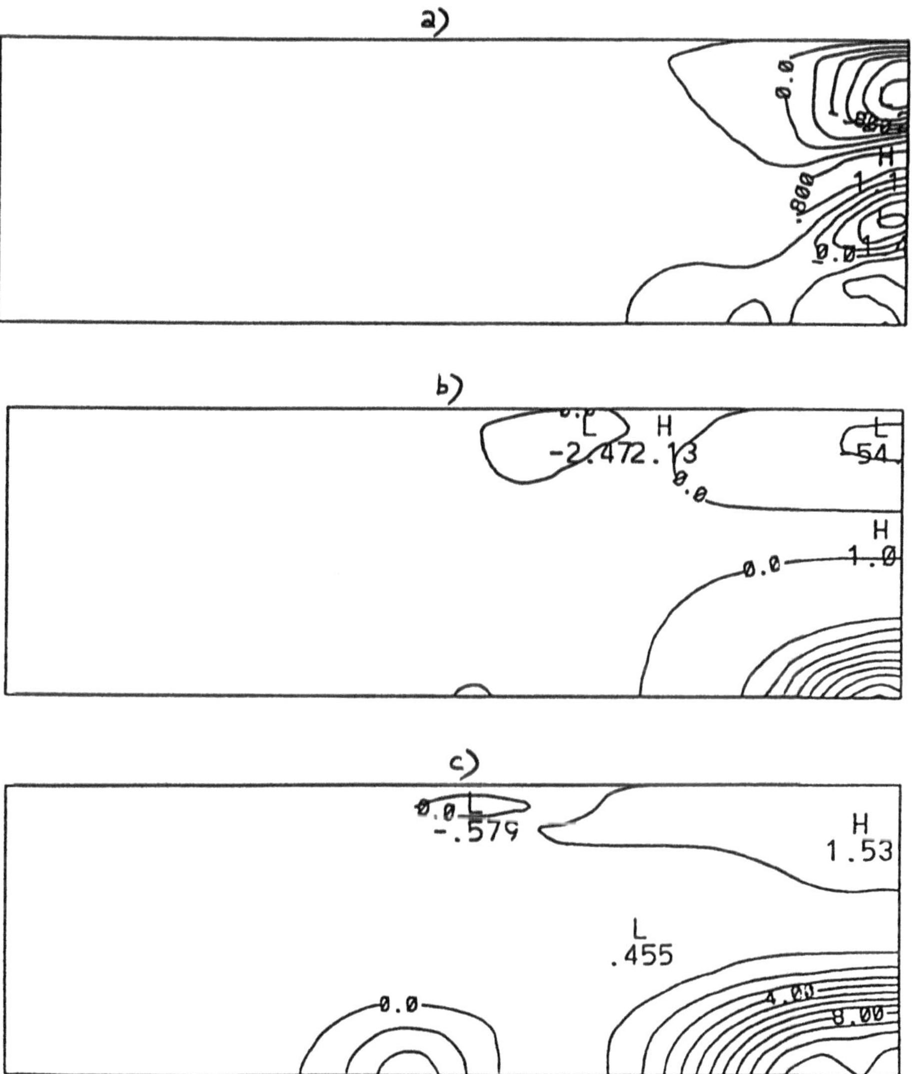

Fig. 9 : Calculated $\widetilde{V'^2} = R_{22}$ in the chamber, at the same instants as Fig. 7. The classical k-ϵ model gives negative values !

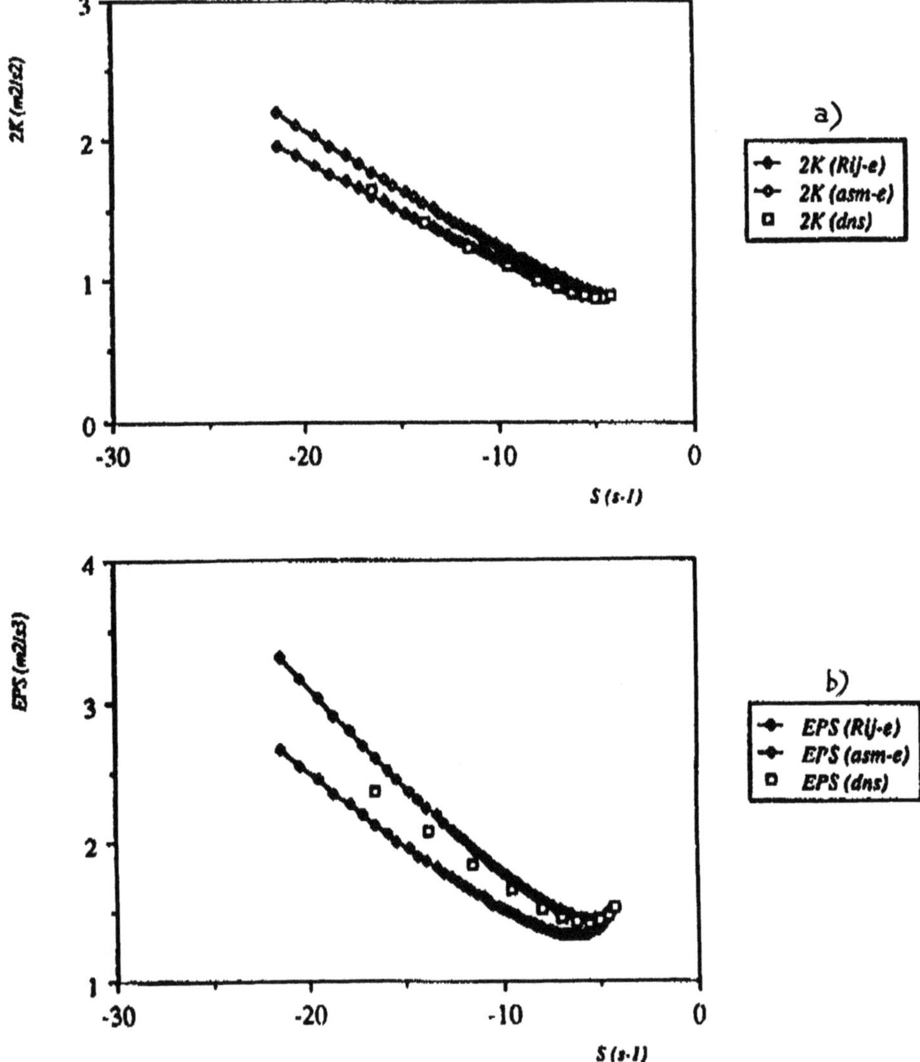

Fig. 10 : Tests of the new model for compressed (uniaxially) homogeneous turbulence. k and ϵ evolutions.

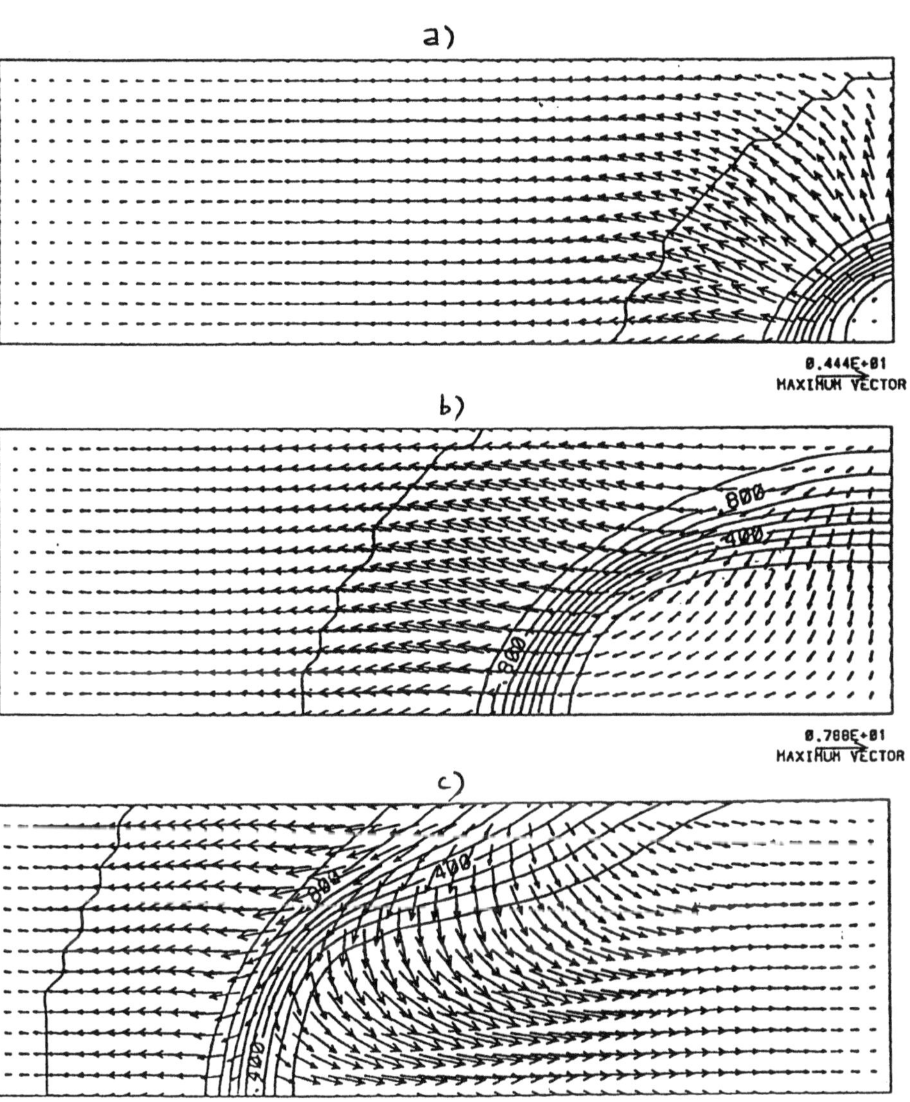

Fig. 11 : Calculated mean velocity and iso-$\tilde{c}$ lines in the chamber of Fig. 2. The new ASM-k-ϵ model is used. a), b), c) correspond to the same instants as Fig. 7.

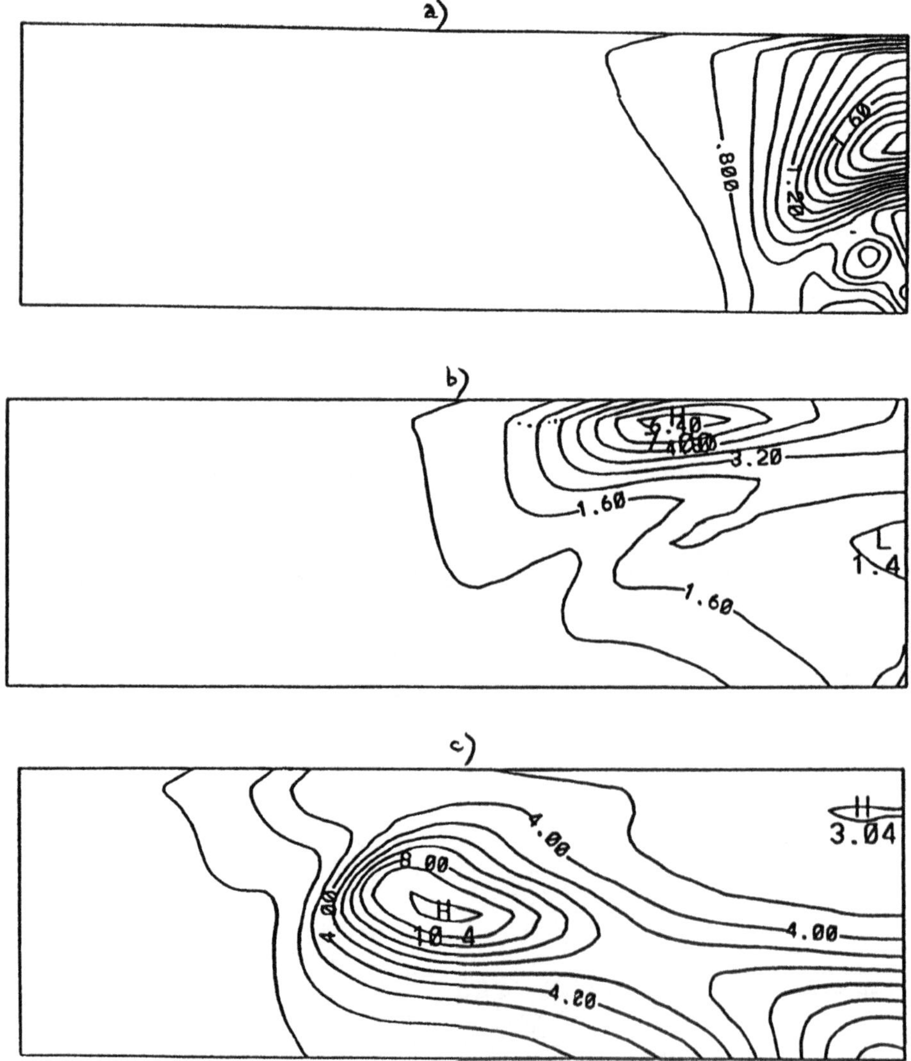

Fig. 12 : Calculated turbulent kinetic energy in the chamber of Fig. 2. The new ASM-k-∊ model is used. a), b), c) correspond to the same instants as Fig.11.

22. SOME EFFECTS OF DENSITY VARIATIONS IN TURBULENT FLOWS WITH PRESSURE FLUCTUATIONS

A.A. Burluka and V.A. Frost
Institute for Problems in Mechanics, Russian Academy of Sciences
117526, 101, Av. Vernadskogo, Moscow, Russia

V. Meytlis
University of Hartford, College of Engineering,
200 Bloomfield Avenue, West Hartford, CT 06117

ABSTRACT. To analyse the possible effects of density variations, which are induced by combustion and heat exchange processes, on turbulent flows with varying pressure field it is proposed to employ the balance equation for conditional probability density function (PDF) of velocity at concentration value fixed. The latter is derived from the governing set of Langevin's type equations which replaces the commonly used Navier-Stocks and molecular diffusion equations set. By using it the balance equations for conditionally averaged velocity and fluctuation intensity are obtained and these values have been shown to depend upon density and turbulent Schmidt number for uniform homogeneous turbulence, even when there is no pressure wave. And, naturally, lighter gas volumes have greater velocity fluctuation intensity.

Also in the acoustic wave an amplitude of the velocity fluctuations depends on density and Schmidt number, that leads to the additional velocity shear between the lighter and heavier gas moles in the wave. The latter due to the turbulent friction, causes the acoustic energy dissipation to increase. It was found that conditionally averaged velocity fluctuations intensity is function on frequency in acoustic wave, if the latter is greater than some threshold value. It is worth noting that similar behaviour of Nusselt number has been found in experiments of Keller (1993). The turbulence time scale should be greater than half a period of the acoustic oscillations for acoustics to affect a turbulence. That fact may be employed to explain the existence of the lowest frequency threshold after attaining of which the acoustic oscillations commence to affect the heat and mass transfer processes. Besides, it was supposed that velocity shear in acoustic wave decreases the characteristic non-uniformity length scale, and qualitative picture of turbulent scales evolution is presented.

1. INTRODUCTION

Approach based on PDF transport equation using has been originally proposed to study the combustion phenomena in turbulent gases. However, since then, the field of its applicabilty was extended and now it covers also the studies of turbulence in liquids with rather slow chemistry, stratified flows and variable density effects and not only for combustion tasks.

A main feature of a combustion process with pressure oscillations in combustion chamber, which will be of interest here, is an interaction between density variations, e.g. induced by mixing and chemical reactions and varying pressure field. The different moles of gases may have densities differing greatly, so unburnt mixture or initial components density may be a several times greater than of the products at high temperature. Also a gas cooling due to a contact with heat exchanger surface may lead to sufficient density changes. When the flow is turbulent non-uniformity of the density field and the large scale turbulent fluctuations make the density to be stochastic value. At

F. Culick et al., (eds.), Unsteady Combustion, 543–552.

the same time non-zero pressure gradients (whether average or in acoustic waves) lead the moles of different densities to acquire the different accelerations. Besides any bulk forces like the buoyancy or centrifugal reveal the same effect. And such the effects should be regarded in analysis of the combustion processes, e.g. in the pulse combustors.

Pulse combustion heating systems have many advantages over conventional burners, such as thermal efficiencies of 95 % or more, low pollutant emissions, self aspiration and high rates of convective heat transfer in the combustion chamber. However, they have not seen widespread use, primarily because of a lack of understanding of the heat transfer and other fundamental processes has resulted in the desigh proceeding largely by trial and errors Putnam et al.(1986). It connects with rather evolved interaction between the pulsating pressure field and variable density as well as with complex mixing phenomena in pulse combustor chamber in which there exists unsteady permanent macro-scale mixing between the fresh mixture and burnt products.

An experience of the usage of pdf equation for concentration for modelling of the various cases of turbulent combustion [Chung (1969), Frost (1975, 1977)], allow to create the model to reckon these phenomena. In particular, model proposed may be applied for estimations and forecasts in describing the pulse combustors work (see e.g. Arpaci et al. (1991)).

We present a qulitative picture of the phenomena and make some assumptions we consider to be plausible. The estimates of the main relevant physical quatities are then obtained based on these assumptions and used to establish some regularities of conditional velocity field behaviour. The effects found may not play the principal role in heat- and mass-tansfer in pulse combustors, but one needs to understand them to complete the description. For example, as far as it concerns the turbulence decay, the variable density effects should not affect it in linear approximation. This is because the turbulent kinetic energy spectra are the same for the moles of different densities. However they may change the characteristic time and length scales and by virtue of this affect the decay in a non-straightforward manner.

2. GOVERNING EQUATIONS OF THE MODEL

Let us use the model developed in works of Chung (1969) and Frost (1975), which based on Langevin's type balance equations (moment and diffusion) with the net averaged bulk forces actions on the particle in hand retained. This set is distinguished with the presence of the instant unaveraged density in terms including the averaged pressure gradient. In lieu of commonly used (see e.g. Kuznetsov(1976), Dopazo (1993), Pope and Anand (1987)) Navier-Stockes and molecular diffusion equations the following governing set is used :

$$d\mathbf{X}/dt = \mathbf{u}(\mathbf{x}, t) \tag{1}$$

$$d\mathbf{u}/dt = \alpha \cdot (\mathbf{u}(\mathbf{x}, t) - \langle \mathbf{u}(\mathbf{X}, t)\rangle) + \nu\langle \Delta \mathbf{u}\rangle_{u,c} + \varepsilon \cdot \frac{\langle \rho \rangle}{\rho} \cdot \mathbf{f} - grad\langle p\rangle/\rho - \mathbf{k}g \tag{2}$$

$$dc/dt = -\beta \cdot (c(\mathbf{x}, t) - \langle c(\mathbf{X}, t)\rangle) + W(c) + d\langle \Delta c\rangle_{u,c} \tag{3}$$

$$\rho = \rho(c) \tag{4}$$

Here angle brackets denote the conditional averaging operation made with **u** and **c** fixed; and **X**, **x**, **u** denote the current and initial particle position and its velocity; p is a pressure, c is the concentration, which defines the density ρ, ν and d -are kinematic viscosity and molecular diffusivity, α and β are the semi- empirical coefficients which define the intensity of the turbulent moment exchange and concentration non-uniformity decay, $W(c)$ - is the chemical reactions rate. **f** is spontaneous disturbance of the pressure and friction forces. It is assumed to be δ-correlated with time and ε stands for its' intensity.

Terms depending on density are shown in (2) with an assumption that the net forces effects are the same for the particles of different densities but the accelerations of those particles are inversely proportional to their densities. In (2) the ε coefficient has a dimension of turbulent dissipation but arises from pressure fluctuations and is nothing but a partial "return" of turbulent kinetic energy, lost by particle stagnating in media, due to the energy exchange between the turbulent motion and pressure fluctuation field. As a matter of fact only the part of the kinetic energy of particle stagnating in turbulent media goes over into the internal energy of media. The sketch of the phenomena is as follow. Fluctuating pressure field accelerates differently the moles of different density. There are friction forces between the neighbour media volumes moving with different velocities (e.g. as it is for shear layer). When the slower moles slow down the faster ones they acquire the additional impetus.

That is while $1/\alpha$ is the time scale for the decay of velocity fluctuations, the value of ε characterisethe turbulent kinetic energy flux from pressure oscillations field. This value has been previously used in work of Chung(1969) to construct the stationary turbulence model.

The governing set (1)-(4) describes the moments and mass exchange between the moving particles and main stream as well as friction and pressure fluctuations actions on those particles, i.e. in other words (1)-(4) is a Lagrangian flow description.

This set is used because it is complete and close statistical description of some system, including all the principal phenomena which have to be explored. Instead of making the numerous closure hypothesis needed for joint pdf obtained from molecular diffusion and Navier-Stockes governing set [Kollmann(1990)] we need to determine three parameters α, β and ε which have obvious physical meaning (see Chung(1969), Frost(1975)).

As a result of established in Chung(1969), Frost(1975) transformations an equation for the joint velocity and concentration pdf $P = P(\mathbf{u}, c|\mathbf{X}, t)$ can be derived from (1)-(4) in an Eulerian frame:

$$\begin{aligned} \frac{\partial P}{\partial t} + u_\chi \frac{\partial P}{\partial X_\chi} = \varepsilon \cdot \frac{\langle \rho^2 \rangle}{\rho^2} \cdot \frac{\partial^2 P}{\partial u_\chi \partial u_\chi} + \frac{\partial}{\partial c}[(\beta(c - \langle c \rangle) - W(c) - d\langle \Delta c \rangle_{\mathbf{u},c})P] \\ \frac{\partial}{\partial u_\chi}[(-\nu \cdot \Delta \langle u_\chi \rangle_{\mathbf{u},c} + \frac{1}{\rho} \cdot \frac{\partial \langle p \rangle}{\partial X_\chi} + (g \cdot \mathbf{j}) + \alpha \cdot (u_\chi - \langle u_\chi \rangle))P] \end{aligned} \tag{5}$$

Hereinafter the summation is assumed over the repeated greek indices.

The dependence on density of term including ε assumes that the random accelerations are inversely proportional to the density while the force action does not depend on density. Herebeneath spatial and temporal arguments are omitted in the pdf denoting for the sake of brevity.

The analogous equation obtained from a governing set of Navier-Stockes and molecular diffusion equations has been derived in Dopazo(1976), Kuznetsov(1976).

By using Bayes theorem, one may write

$$P(\mathbf{u}, c) = P_u(\mathbf{u}|c) \cdot P_c(c) \tag{6}$$

where $P_u(\mathbf{u}|c)$ is the conditional PDF for the velocity with the c value fixed while $P_c(c)$ is the concentration PDF. By integrating (5) with respect of $\mathbf{u}$ PDF equation for $P_c = P_c(c)$ is obtained:

$$\frac{\partial P_c}{\partial t} + \frac{\partial \langle u_\chi \rangle_c \cdot P_c}{\partial X_\chi} = \frac{\partial}{\partial c}(\beta \cdot (c - \langle c \rangle) - W(c) - d\langle \Delta c \rangle_c)P_c) \tag{7}$$

Then by substituting (6) into (5) and using (7), the balance equation for conditional velocity pdf $P(\mathbf{u}|c)$ is obtained. This equation is written below for the spatially uniform case of zero buyoancy force and when there is no chemical reactions:

$$\frac{\partial P_u}{\partial t} = \frac{\partial}{\partial u_\chi}[-\nu \cdot \Delta \langle u_\chi \rangle_{\mathbf{u},c} + \alpha \cdot (u_\chi - \langle u_\chi \rangle)) \cdot P_u] + \varepsilon \cdot \frac{\langle \rho^2 \rangle}{\rho^2} \cdot \frac{\partial^2 P_u}{\partial u_\chi \partial u_\chi} + \frac{\partial}{\partial c}[\beta \cdot (c - \langle c \rangle) \cdot P_u] \tag{8}$$

If using Navier-Stocks coupled with molecular diffusion of a reacting scalar equations as a governing set instead of (1)-(4) the balance equation for $P_u(\mathbf{u}|c)$ under the same conditions is:

$$\frac{\partial P_u}{\partial t} = \frac{\partial}{\partial u_\chi}[\langle -\nu\cdot\Delta u_\chi + \frac{1}{\rho}\frac{\partial \rho}{\partial c}(\nabla c\cdot\nabla u_\chi)\rangle_{u,c}\cdot P_u] + \frac{\partial}{\partial c}[\langle -d\cdot\Delta c + \frac{1}{\rho}\frac{\partial \rho}{\partial c}(\nabla c)^2\rangle_{u,c}\cdot P_u] \tag{9}$$

When comparing this equation with (8) term by term one can realise that $1/\alpha$ and $1/\beta$ are some characteristic temporal scales for turbulent kinetic energy dissipation and for the decay of concentration non-uniformities, correspondingly. At the same time we could not take into consideration the fluctuating pressure effects explicitily, without making an additional assumption, if using the equation (9) for the spatially uniform case.

The main advantage of balance equation for $P = P(\mathbf{u}, c)$ form, adopted here is that terms, which are responsible for turbulent transfer and mixing are closed in a rather simple way. In that connection it is worth to note that up to now there are no satisfying enough and commonly adopted closure hypotheses for those terms in (5) (see e.g. Dopazo(1993)). Perhaps it may be assumed because of rather complicated nature of the turbulent exchange processes.

When density is constant, equation (8) with the constant values of α and β has a stationary solution for the velocity PDF, presented in Chung(1969) as the normal distribution. However, if α and ε are considered to be the functions of the turbulent kinetic energy E (in Frost(1992) the following relations have been proposed: $\alpha = 0.45\cdot E^{1/2}/L$ and $\varepsilon = 0.2\cdot E^{3/2}/L$, where L is turbulent integral scale) it appears that a balance equation for E derived from (5) (see Frost(1992)), coincides with that used in semi-empirical models like the $k-\varepsilon$ model. For the uniform case it has a solution with E decreasing proportionally t^{-1} when an turbulent integral scale L depends on time as $t^{1/2}$. If using the experimental dependence for L $\sim t^{0.4}$ obtained in experiments of Sreenivasan et al. (1980), the resulting dependence $E \sim t^{-1.2}$ turns out to be close to the trends observed in that work.

3. CONDITIONAL VELOCITY FLUCTUATIONS

The dependence of velocity fluctuations intensity upon density can be found in the statistically uniform (spatially and temporally) case, when no mean movement is assumed, by computing $\langle \mathbf{u}'^2\rangle_c$. For that purpose by multiplying (8) with $\mathbf{u}'^2$ and by integrating with respect of $\mathbf{u}$, one can obtain:

$$\beta c\cdot\frac{\partial\langle \mathbf{u}'^2\rangle_c}{\partial c} - 2\cdot\alpha\langle \mathbf{u}'^2\rangle_c = -2\cdot\varepsilon\frac{\langle\rho^2\rangle}{\rho^2} \tag{10}$$

Herebeneath c stands for deviation from mean $(c - \langle c\rangle)$ for the sake of brevity. A solution of which (for positive c):

$$\langle \mathbf{u}'^2\rangle_c = \frac{2\cdot\varepsilon}{\beta}\cdot c^{-2\alpha/\beta}\int_0^c c^{-2\alpha/\beta-1}\cdot\frac{\langle\rho^2\rangle}{\rho^2}\cdot dc \tag{11}$$

has an apparent transition to the $\langle \mathbf{u}'^2\rangle_c = \varepsilon/\alpha = \langle \mathbf{u}'^2\rangle$ for the case of constant density. Further the following dependence for density is used:

$$\rho(c) = (a + b\cdot c)^{-1}; \qquad \langle\rho^2\rangle \simeq a^{-2} \tag{12}$$

which is true for iso-thermal mixing of the different density components and can be treated as a good approximation for other cases. The value of b is a density sensitivity to concentration variations. Then by substituting (12) into (11) we obtain

$$\langle \mathbf{u}'^2\rangle_c = \langle \mathbf{u}'^2\rangle\cdot[1 + \frac{4\cdot b/a}{1-\beta/2\alpha}\cdot c + \frac{b^2/a^2}{1-\beta/\alpha}\cdot c^2] \tag{13}$$

It was found in Chung(1969) and Frost(1975) that turbulent Schmidt number Sc is defined with α and β as Sc $= (\alpha + \beta)/(2 \cdot \alpha)$. The value of β changes from zero (when a micro scale mixing is absent) to α (when the intensities of turbulent friction and micro scale mixing are equal) and thus the Schmidt number is varied from 0.5 to 1. Thus the velocity fluctuations distribution depends upon density and Schmidt number Sc in a following way:

$$\langle \mathbf{u}'^2 \rangle_c = \langle \mathbf{u}'^2 \rangle \cdot [1 + \frac{8 \cdot b/a}{3 - 2 \cdot Sc} \cdot c + \frac{b^2/a^2}{2 \cdot (1 - Sc)} \cdot c^2] \qquad (14)$$

The dependecies given by this formula are plotted on Fig. 1 for different Schmidt numbers.

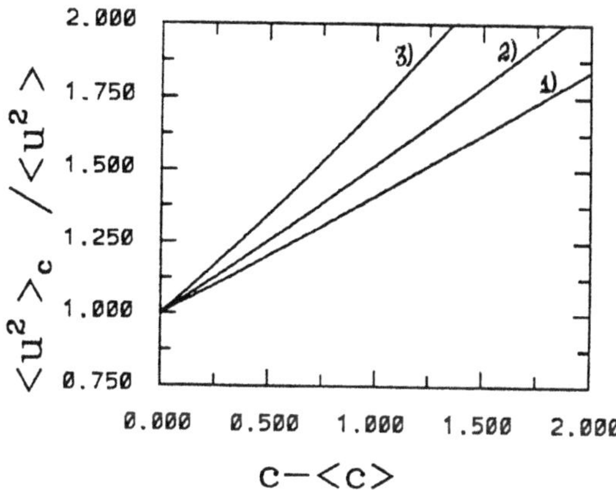

Fig.1. The conditional velocity fluctuations *vs.* concentration deviation from mean value for 1) $Sc = 0.5$ 2) $Sc = 0.7$ 3) $Sc = 0.9$

The lowest curve for $\langle \mathbf{u}'^2 \rangle_c$ corresponds to $Sc = 0.5$, i.e. it is the case where there is no scalar dissipation. It can be easily seen that the greater conditional velocity fluctuations correspond the lighter volume of media. This result seems to be reliable and obvious. However, on increasing the Schmidt number or, in other words, on increasing the scalar dissipation (let us remind that β is a coefficient determining its rate) the corresponding $\langle \mathbf{u}'^2 \rangle_c$ curves recline above. That is the greater Schmidt number, the greater $\langle \mathbf{u}'^2 \rangle_c$ for the media volume of the same density. This result seems to be suprising, but it can be explained with following consideration.

Let us suppose the media volume is characterised with certain rates of scalar and velocity fluctuations dissipations. The ratio of those is defined with Schmidt number, i.e. with ratio β/α. The increase of Sc means that concentration fluctuations decay more rapidly while dissipation of velocity fluctuations, the rate of which is determined with α parameter, remains the same. Thus on increasing the Schmidt number the greater velocity fluctuations intensity correspond the same intensity of concentration fluctuations.

In other words, the greater values of β correspond the greater rate of disribution $P_c(c)$ width decreasing. This width is characterised with $\langle c'^2 \rangle$ value. At the same time moles of media characterised with this $P_c(c)$ distribution keep on the intensity of velocity fluctuations. Thus on decreasing of $\langle c'^2 \rangle_c$ or shrinking of $P_c(c)$ distribution the value of $\langle \mathbf{u}'^2 \rangle_c$ remains the same, and we obtain the dependency on Sc, presented on Fig. 1. To a certain degree one can say that on decreasing the $P_c(c)$ distribution width the flow moles transport their properties (e.g. the $\langle \mathbf{u}'^2 \rangle$ value) in the concentration space.

4. CONDITIONAL VELOCITY IN ACOUSTIC WAVE

Let us consider an acoustic wave propagation through turbulent media consisting of the different density volumes. The velocity amplitude vs. density is calculated with the using the conditionally averaged velocity $\langle \mathbf{u} \rangle_c = \int_0^\infty \mathbf{u} \cdot P(\mathbf{u}|c)\ d^3\mathbf{u}$ at the concentration level fixed. The balance equation for $\langle u \rangle_c$ is derived from (8) into which the term with acoustic pressure gradient $1/\rho \cdot (\text{grad } p)_a$ is added, by multiplying on $\mathbf{u}$ and integrating with respect of $\mathbf{u}$:

$$\frac{\partial \langle \mathbf{u} \rangle_c}{\partial t} = \beta c \cdot \frac{\partial \langle \mathbf{u} \rangle_c}{\partial c} - \alpha \cdot (\langle \mathbf{u} \rangle_c - \langle \mathbf{u} \rangle) - \frac{1}{\rho} \cdot (\text{grad } p)_a \tag{15}$$

The gradient of acoustic field pressure is the only source for an oscillating motion. If the concentration does not fluctuate then (15) goes over into the usual acoustic relation:

$$\frac{\partial \langle \mathbf{u} \rangle}{\partial t} = -\frac{1}{\langle \rho \rangle} \cdot (\text{grad } p)_a \tag{16}$$

If suppose that

$$(p, \mathbf{u}) = (\hat{p}_0, \hat{\mathbf{u}}_0) \cdot exp(2\pi \cdot \mathbf{i} \cdot (\omega \cdot t + \mathbf{X} \cdot \omega / \mathbf{c_a}))$$

where $\mathbf{c_a}$ is a sonic speed and

$$\hat{\mathbf{u}}_0 = -\frac{\hat{p}}{\mathbf{c_a} \cdot \langle \rho \rangle}$$

is the velocity in the acoustic wave for the case of constant density, then (15) has a solution for in which the effects of Sc can be easily traced:

$$\frac{\langle \hat{\mathbf{u}} \rangle_c - \hat{\mathbf{u}}_0}{\hat{\mathbf{u}}_0} = \left(\frac{\langle \rho \rangle}{\rho} - 1 \right) \cdot \frac{1 + 2\mathbf{i} \cdot [\frac{\alpha}{2\pi \cdot \omega}] \cdot (1 - Sc)}{1 + 4 \cdot [\frac{\alpha}{2\pi \cdot \omega}]^2 \cdot (1 - Sc)^2} \tag{17}$$

The polar graphic of this dependence is depicted on Fig.2 for ω varying from zero to $2\pi\omega/\alpha = 10$ for different values of ratio $\langle \rho \rangle / \rho$.

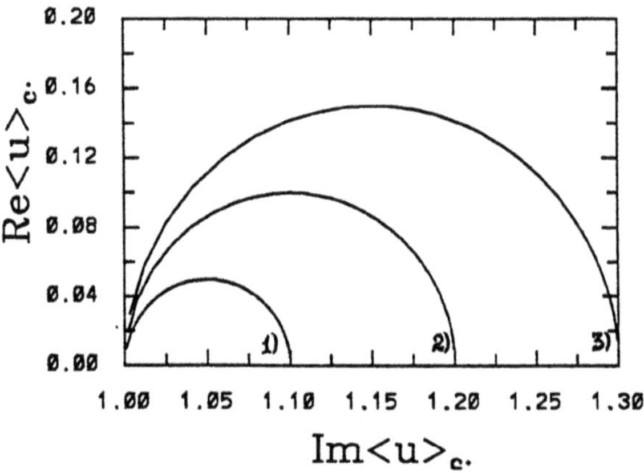

Fig.2. The dependence of conditional velocity in the acoustic wave for 1) $\langle \rho \rangle / \rho = 1.1$ 2) $\langle \rho \rangle / \rho = 1.2$ 3) $\langle \rho \rangle / \rho = 1.3$

The inverse of the turbulent time scale α has an order of magnitude $10 \div 10^3 sec^{-1}$. and, consequently, the value of $2\pi\omega/\alpha$ range chosen corresponds to the acoustic frequencies from zero to a several hundreds Hz.

The results obtained say that sufficient relative movement of the heavier and lighter gas particles exists in the pressure wave propagating in the turbulent media. That is additional shear stresses arise induced by this movement that, in turn, increases the acoustic energy dissipation. The value of latter may be calculated as function of the frequency and density dispersion, if some density distribution is known, in a following way.

If assume the equilibrium between the turbulent and acoustic motion to be attained one can draw the conclusion that acoustic wave energy supplies some 'quasi-stationary' level of turbulent fluctuations. In other words, the dissipations of acoustic energy and kinetic energy of turbulence are equal. For $P(\mathbf{u}, c)$ being known turbulent kinetic energy E is given by:

$$E = \frac{1}{2} \cdot \int \mathrm{d}c \int (u_X - \langle u_X \rangle) \cdot (u_X - \langle u_X \rangle) \cdot P(\mathbf{u}, c) \mathrm{d}^3\mathbf{u} \tag{18}$$

or

$$E = \frac{1}{2} \cdot \int (\langle u \rangle_c - \langle u_X \rangle)^2 \cdot P_c(c) \mathrm{d}c \tag{19}$$

and its dissipation is defined (see Emelianov et al.(1992)) as:

$$\partial E / \partial t = -\alpha \cdot E + \varepsilon \tag{20}$$

To find averaged dissipation of the acoustic energy one needs to integrate the last equation over the period of acoustic oscillations. Here it should be noted that as (17) is a linear function on c the intensity of velocity fluctuations in the acoustic wave may be given as:

$$\langle \mathbf{u}'^2 \rangle = \langle c'^2 \rangle \cdot \frac{(b \cdot \hat{\mathbf{u}}_0 \rho_0)^2}{2} \cdot \frac{1}{1 + 4 \cdot [\frac{\alpha}{2\pi \cdot \omega}]^2 \cdot (1 - Sc)^2} \tag{21}$$

The dependencies on frequency given by (21) are presented on Fig. 3 for different Schmidt numbers.

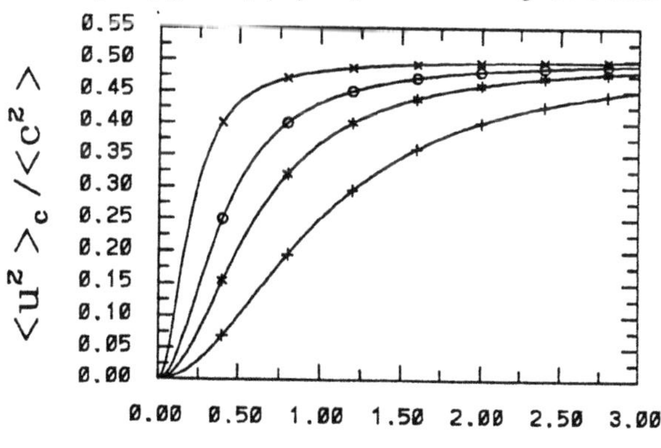

Fig.3. The dependence of conditional velocity fluctuations upon frequency for 1) $Sc = 0.5$ 2) $Sc = 0.7$ 3) $Sc = 0.8$ 4) $Sc = 0.9$

If flow is characterised with some "natural" turbulence intensity, the acustics begins to affect the heat- and mass-transfer in that flow when the "acoustic-induced" turbulent fluctuations level exceeds that "natural" one. It is interesting to note that the dependence of Nusselt number on frequency found in experiments of Keller (1993) (see Fig. 4, which is taken from that work), reveals the similar trends, though it is hard to establish the precise relation between the $\frac{\alpha}{2\pi\cdot\omega}$ and experimental frequency dependent values for which Fig. 4 is plotted.

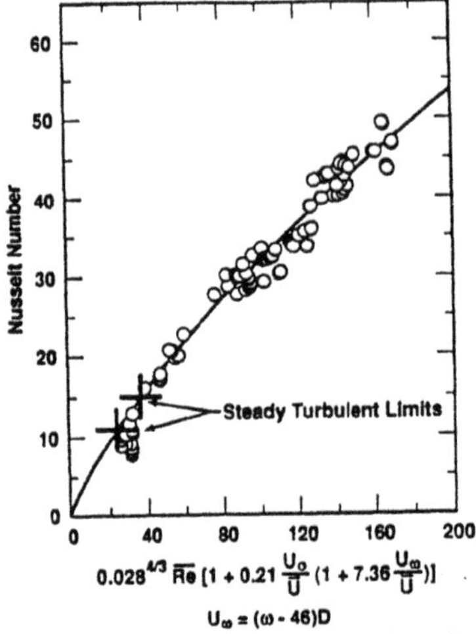

Fig. 4 (taken from Keller (1993).
The dependence of Nusselt number upon frequency.

Further the conditional intensity of velocity fluctuations increases first with the frequency raise, until it attains a plateau, when frequency becomes greater than near doubled the inverse turbulent time scale $1/\alpha$. Also the greater Schmidt number the lower value of this frequency threshold exists. However one can see that the frequency range in above-mentioned experiments is far from the values at which plateau, at which acoustics cancels affect the turbulence, may exist. One can predict that there exists the frequency threshold after which further increasing of frequence does not lead the the raise of Nusselt number. However, in the frequency range of practical importance (Keller (1993), Putnam et al. (1986)) there exists the strong interaction between acoustics and turbulence, which may be explored with the methods like presented here.

5. EVOLUTION OF CHARACTERISTIC SCALES IN ACOUSTIC WAVE

In addition to their action on the acoustic energy dissipation differences in the velocities of different density moles lead to the changes of the characteristic turbulent scales for concentration and density fields and, over that, to the changes of the exchange processes intensity. Such a mechanism is clear enough. Non-uniformity of $\langle \mathbf{u} \rangle_c$ induces a field of velocity gradients, which stretch the initial moles of different densities into the lines and layers strained along the acoustic wave direction. When a backwards motion takes place, no reverse shrinkage is in effect. Instead, these layers and lines of density non-uniformities are rotated and stretched again. After some relaxation time passing an equilibrium is attained between the mixing processes, which are responsible for the increase of scale, and stretch, which reduces the characteristic scales. Thus the velocity and acceleration shifts between heavier and lighter gas volumes leads to the broadening of the acoustic front. It is a formidable task to simulate such process directly. However one can realize, that sufficient influences of acoustic wave on characteristic scales occur when half a period of acoustic vibration is less than turbulence time scale. The existence of the lowest frequency threshold in mass and heat transfer in pulse combustion is possibly explained with the mechanism presented above.

6. SUMMARY

A mathematical model which is based upon the conditional pdf for velocity at concentration fixed has been constructed. A considerations of some special cases allow to obtain analytical expressions for:
i) the distribution of the conditionally averaged velocity fluctuations intensity in homogeneous case;
ii) the density effects on a acoustic velocity distribution inside the pressure wave. Also the influence of turbulent Schmidt number Sc on the acoustic effects can be traced;
iii) the level of the acoustic energy losses, connected with the relative movement of the different density particles, if some density distribution is presumed.
iv) the "acoustic-induced" level of turbulence which is also a function of density and Schmidt number.
An assumption was made that variance of conditionally averaged velocity at value of concentration fixed due to the density variation may result to a decrease of turbulent time and length scales after some frequency threshold attaining.

7. ACKNOWLEDGEMENTS.

The authors are indebtfully grateful to Prof. R. Borghi for support and for many valuable discussions.

Laboratory 230 of C.N.R.S.-CORIA and Russian fond of fundumental research are acknowledged for financial support.

8. REFERENCES

Arpaci V.S., Dec J.E., and Keller J.O. (1991) International Symp. on Pulsating Combust. August 5-8, 1991, Monterey, California, USA.
Chung P.M. (1969), AIAA J. 7, 1982.
Dopazo C. (1976), 'A Probabilistic Approach to Turbulent Flame Theory', Acta Astronaut., **3**, 853.
Dopazo C. (1993), 'Recent Developments in PDF Methods', in Libby P.A. and Williams F.A. (eds.), 'Turbulent reactive flows', to be published.
Emelianov V.M., Frost,V.A., and Nedoroob S.A. (1992), in Proc. of 24th Symp.(Int.) on Combust., 5-10 July 1992, The Combust. Inst., Pittsburgh, PA, 435-442.
Frost V.A. (1975), Fluid Mech.,Sov. Res., 4, 124.
Frost,V.A. (1977), in Proc. of 4th USSR Symp. on Combust. and Explos., 23-27 September 1974, Nauka, Moscow, 361-365.
Keller J.O. (1993), in F. Culick, M.V. Heitor and J.H. Whitelaw (eds.), "Unsteady Combustion", Kluwer Acad. Publ., Dordrecht (to be published).
Kollmann W. (1990), "The pdf approach to turbulent flow", Theoret. Comput. Fluid Dyn.,**1**, p.249.
Kuznetsov V.R. (1976), 'Probability Distribution of Velocity Difference in the Inertial Interval of a Turbulence Spectrum', Izv. Akad. Nauk SSSR, Mekh. Zhid. i Gaza, **1**, 32.
Pope S., Anand M.S. (1987), 'Calculations of Premixed Turbulent Flames by PDF Methods', Comb. and Flame, **67**, 67.
Putnam A.A., Belles F.E., and Kentfield J.A.C. (1986) 'Pulse Combustion', Prog. Energy Combust. Sci., **12**, 43-79.
Sreenivasan K.R., Tavoularis S., and Corrsin S. (1980), J. Fluid Mech., **100**, 597-621.

AUTHOR INDEX

SUBJECT INDEX

GPSR Compliance
The European Union's (EU) General Product Safety Regulation (GPSR) is a set of rules that requires consumer products to be safe and our obligations to ensure this.

If you have any concerns about our products, you can contact us on

ProductSafety@springernature.com

In case Publisher is established outside the EU, the EU authorized representative is:

Springer Nature Customer Service Center GmbH
Europaplatz 3
69115 Heidelberg, Germany

www.ingramcontent.com/pod-product-compliance
Ingram Content Group UK Ltd.
Pitfield, Milton Keynes, MK11 3LW, UK
UKHW021901190726
13853UKWH00003B/1375

* 9 7 8 9 4 0 0 9 1 6 2 1 0 *